Investigating College Algebra with Technology

KATHY BURGIS
Aquinas College

JEFF MORFORD
Henry Ford Community College

Key College Publishing
Innovators in Higher Education

Kathy Burgis
Aquinas College
Department of Mathematics
Grand Rapids, MI 49506

Jeff Morford
Henry Ford Community College
Mathematics Division
Dearborn, MI 48128

Key College Publishing was founded in 1999 as a division of Key Curriculum Press® in cooperation with Springer Science and Business Media. We publish innovative texts and courseware for the undergraduate curriculum in mathematics and statistics as well as mathematics and statistics education. For more information, visit us at www.keycollege.com.

Key College Publishing
1150 65th Street
Emeryville, CA 94608
(510) 595-7000
info@keycollege.com
www.keycollege.com

Published by Key College Publishing, an imprint of Key Curriculum Press.

Student CD-ROM
Key College Publishing guarantees that the CD-ROM that accompanies this book is free of defects in materials and workmanship. A defective disk will be replaced free of charge if returned within 90 days of the purchase date. After 90 days, there is a $10.00 replacement fee.

Library of Congress Cataloging-in-Publication Data on file

Printed in the United States of America
10 9 8 7 6 5 4 3 2 1   09 08 07 06 05

ISBN 1-931914-37-0

Development Editor
Erika Shaffer

Production Director
McKinley Williams

Editorial Production Project Manager
Laura Ryan

Copyeditor
April Wells-Hayes

Text Designer
Marilyn Perry

Compositor
Thompson Type

Senior Art Editor
Jason Luz

Cover Designer
Jensen Barnes

Cover Artwork
Ramble, 2003 Benjamin Edwards

Printer
RR Donnelly

Editorial Director
Richard J. Bonacci

General Manager
Mike Simpson

Publisher
Steven Rasmussen

About the Cover Art: Benjamin Edwards (b. 1970, Iowa City, Iowa) reconfigures strip malls, gas stations, and super stores, into composites that assume the theatricality and beauty of classical architecture. His paintings are composites of numerous digital photographs, which Edwards programs into a computer, simplifies, and color-codes. He then superimposes the images upon one another and deletes and reorders details. This process creates a template for the painting, which is then constructed over hundreds of hours using acrylics and textured materials. Edwards currently lives and works in Washington, D.C.

To Rich, Steph, Ben, and David

and

To Cassie, Alex, and Sasha, who sometimes had to wait an extra night to find out what happened next in Middle Earth, Narnia, Shannara, and elsewhere.

Contents

Annotated Contents vii
Preface xix
To the Student xxvii
To the Instructor xxix
Acknowledgments xxxi

CHAPTER 1 **Problem Solving** 1
1.1 Pictures, Graphs, and Diagrams 2
1.2 Symbolic Representation 9
1.3 Organizing Information 16
1.4 Measures of Central Tendency and Box Plots 24
1.5 Measures of Spread 31
Chapter 1 Review 45

CHAPTER 2 **Patterns and Recursion** 53
2.1 Recursively Defined Sequences 54
2.2 Modeling Growth and Decay 64
2.3 A First Look at Limits 71
2.4 Graphing and Sequences 77
2.5 Loans and Investments 86
Chapter 2 Review 99

CHAPTER 3 **Linear Models and Systems** 103
3.1 Linear Equations 104
3.2 Revisiting Slope 111
3.3 Fitting a Line to Data 119
3.4 Linear Systems 127
3.5 Substitution and Elimination 133
Chapter 3 Review 142

CHAPTER 4 **Functions, Relations, and Transformations** 148
4.1 Interpreting Graphs 149
4.2 Function Notation 155
4.3 Lines in Motion 163
4.4 Translations and the Quadratic Family 170
4.5 Reflections and the Square Root Family 177
4.6 Stretches and Shrinks and the Absolute Value Family 184
4.7 Transformations and the Circle Family 192
4.8 Compositions of Functions 200
Chapter 4 Review 211

CHAPTER 5 **Exponential, Power, and Logarithmic Functions** **215**

5.1 The Exponential Function 216
5.2 Properties of Exponents 223
5.3 Fractional Exponents and Roots 229
5.4 Applications of Power Equations 239
5.5 Building Inverses of Functions 243
5.6 The Logarithmic Function 251
5.7 Properties of Logarithms 257
5.8 Applications of Logarithms 264
Chapter 5 Review 273

CHAPTER 6 **Quadratic and Other Polynomial Functions** **277**

6.1 Polynomial Degree and Finite Differences 278
6.2 Equivalent Quadratic Forms 286
6.3 Completing the Square 294
6.4 The Quadratic Formula 301
6.5 Complex Numbers 307
6.6 Factoring Polynomials 314
6.7 Higher-Degree Polynomials 320
6.8 More about Finding Solutions 327
Chapter 6 Review 337

CHAPTER 7 **Matrices and Linear Systems** **341**

7.1 Matrix Representations 342
7.2 Matrix Operations 349
7.3 The Row Reduction Method 360
7.4 Solving Systems with Inverse Matrices 368
7.5 Systems of Linear Inequalities 378
7.6 Linear Programming 385
Chapter 7 Review 395

CHAPTER 8 **Parametric Equations and Trigonometry** **401**

8.1 Graphing Parametric Equations 402
8.2 Converting Parametric to Nonparametric Equations 411
8.3 Right Triangle Trigonometry 418
8.4 Using Trigonometry to Set a Course 431
8.5 Projectile Motion 438
8.6 The Law of Sines 445
8.7 The Law of Cosines 453
Chapter 8 Review 462

CHAPTER 9 **Conic Sections and Rational Functions** 466

9.1 Using the Distance Formula 467
9.2 Circles and Ellipses 474
9.3 Parabolas 486
9.4 The Hyperbola 494
9.5 Nonlinear Systems of Equations 503
9.6 Introduction to Rational Functions 508
9.7 Graphs of Rational Functions 516
9.8 Operations with Rational Expressions 524
Chapter 9 Review 533

CHAPTER 10 **Series** 539

10.1 Arithmetic Series 540
10.2 Infinite Geometric Series 547
10.3 Partial Sums of Geometric Series 554
Chapter 10 Review 562

CHAPTER 11 **Probability** 565

11.1 Randomness and Probability 566
11.2 Counting Outcomes and Tree Diagrams 577
11.3 Mutually Exclusive Events and Venn Diagrams 586
11.4 Random Variables and Expected Value 593
11.5 Permutations and Probability 600
11.6 Combinations and Probability 607
11.7 The Binomial Theorem and Pascal's Triangle 613
Chapter 11 Review 623

Selected Answers 627
Glossary 675
Photo Credits 687
Index 688

Resources on the Student CD

Prerequisite Review: Numbers and Figures • Operations on Numbers • The Ideas that Motivate Algebra • Exponents • Radicals • Polynomials • Factoring Polynomials • Rational Expressions

Calculator Notes

Microsoft® Excel Notes

Annotated Contents

Preface xix
To the Student xxvii
To the Instructor xxix
Acknowledgments xxxi

CHAPTER 1 **Problem Solving** 1

This introductory chapter sets a solid foundation for the data-driven, cooperative, technological, problem-solving pedagogy students will experience throughout the book. First students will see that they must solve problems throughout the course and that they will frequently work with one another to solve these problems. Later in the chapter, the work on real-world problem solving is extended with the introduction of statistical tools for dealing with larger amounts of information. In addition to statistics, students will use algebra to make meaningful models of real-world data. The graphing calculator and spreadsheets are introduced in this first chapter so students become familiar with these powerful aids to problem solving. By the end of this chapter, students will know math is relevant and that they can do math.

1.1 Pictures, Graphs, and Diagrams 2

This lesson introduces problem-solving skills such as working together, working backwards, and drawing a diagram or picture. The Cartesian Coordinate System is used to solve a number of problems.

Investigation: Camel Crossing the Desert *(Problem Solving)* 4

1.2 Symbolic Representation 9

This lesson introduces the power of symbolic algebra for solving problems. Examples demonstrate the use of a standard four-step process for using algebra as a problem-solving tool.

Investigation: Problems, Problems, Problems *(Problem Solving Using Algebra)* 12

1.3 Organizing Information 16

Students develop strategies for organizing information including making a table and making a tree diagram. Students will use these strategies to solve problems in this section and throughout the textbook—notably in Chapter 10 when they investigate probability.

Investigation: Who Owns the Zebra? *(Problem Solving Using a Tree Diagram or other Organization Method)* 19

1.4 Measures of Central Tendency and Box Plots 24

This lesson introduces the statistical concepts of mean, median and mode. It also introduces box plots so that students can begin to visualize data distributions.

Investigation: Pulse Rates *(Box Plots)* 27

1.5 Measures of Spread 31

This lesson introduces the concepts of standard deviation and variance. Students will see that data with the same mean and median can have very different distributions. Students will also explore the difference between the standard deviation of a large population and the standard deviation of a small sample of the population.

Investigation: A Good Design *(Variance and Standard Deviation)* 31

Investigation: Standard Deviation Experiment *(Standard Deviation of a Population and Standard Deviation of a Sample)* 35

Chapter 1 Review 45

CHAPTER 2 **Patterns and Recursion** **53**

Chapter 2 focuses on discrete functions. It includes a basic introduction to arithmetic and geometric sequences. It introduces the concept of limits by examining sequences that model phenomena such as drug dosing. Finally it demonstrates that sequences and series can model loans and annuities.

2.1 Recursively Defined Sequences 54

This lesson introduces recursion, arithmetic, and geometric sequences. At first sequences are defined recursively. In Chapters 3 and 5, explicit formulas are introduced.

Investigation: Monitoring Inventory *(Arithmetic Sequences)* 58

2.2 Modeling Growth and Decay 64

This lesson opens a discussion of how geometric sequences can represent decay or growth. It includes practical situations in which a growth model or decay model is appropriate.

Investigation: Looking for the Rebound *(Geometric Sequences)* 65

2.3 A First Look at Limits 71

This lesson defines a shifted geometric sequence and explains how a shifted geometric sequence can often approach a limiting value.

Investigation: Doses of Medicine *(Limiting Values)* 72

2.4 Graphing and Sequences 77

In this lesson students discover how to identify arithmetic, geometric, and shifted geometric sequences from their graphs. Students discover that arithmetic sequences grow at a constant rate, and shifted geometric sequences approach a horizontal line representing the limiting value.

Investigation: Match Them Up *(Multiple Representations of Sequences)* 77

2.5 Loans and Investments 86

This lesson explores how shifted geometric sequences can model loans and investments. Students will amortize loans using shifted geometric sequences.

Investigation: Life's Big Expenditures *(Loans)* 86

Chapter 2 Review 99

CHAPTER 3 **Linear Models and Systems** **103**

Chapter 3 moves from the recursive formulas to explicit formulas and from the discrete to the continuous while focusing on linear models. Best fit lines are found to model data of various sorts. Students learn how to solve two-variable linear systems.

3.1 Linear Equations 104

Students will learn about explicit formulas for sequences and match sequences to their graphs. The emphasis is on linear equations and arithmetic sequences.

Investigation: Match Point *(Matching Recursive and Explicit Formulas)* 106

3.2 Revisiting Slope 111

This lesson reintroduces students to slope. The intercept form of a line's equation is discussed.

Investigation: Balloon Blastoff *(Slope)* 112

3.3 Fitting a Line to Data 119

The lesson discusses general strategies for finding a line to model data. Calculator regression is introduced as one method to find these lines.

Investigation: The Wave *(Modeling Nearly Linear Data)* 122

3.4 Linear Systems 127

Students see uses for solving linear systems in this lesson. Then they learn to solve systems of equations with the dependent variable isolated.

Investigation: Popular Trends *(Linear Systems)* 128

3.5 Substitution and Elimination 133

This lesson introduces the substitution and elimination methods for solving linear equations. A few systems with no solution, or infinitely many solutions are included.

Investigation: It All Adds Up *(Elimination)* 135

Chapter 3 Review 142

CHAPTER 4 **Functions, Relations, and Transformations** **148**

Chapter 4 shows how transformations to graphs appear in relations equations. Horizontal shifts, vertical shifts, stretches, and reflections are included. This chapter also formalizes the concept of function and classifies some of the more familiar functions by function family—linear, quadratic, square root, or absolute value. The chapter closes with a lesson on composition of functions.

4.1 Interpreting Graphs 149

Students create qualitative graphs in this section. They also invent stories that could describe a qualitative graph.

Investigation: Graph a Story *(Qualitative Graphs)* 150

4.2 Function Notation 155

This lesson introduces the concept of function and function notation. Step functions are explored through a project.

Investigation: To Be or Not to Be *(Multiple Representations)* 157

4.3 Lines in Motion 163

Students see how translations affect the equations of the line. The exercise set includes translations of nonlinear functions represented as graphs by applying the same rules.

Investigation: Movin' Around *(Translations)* 163

4.4 Translations and the Quadratic Family 170

Students discover that the translations from Lesson 4.3 apply to more than linear functions. Lesson 4.4 applies those same translations to quadratic functions.

Investigation: Make My Graph *(Translations)* 171

4.5 Reflections and the Square Root Family 177

This lesson introduces reflections through the study of the square root function. Both horizontal and vertical reflections are included.

Investigation: Take a Moment to Reflect *(Reflections)* 177

4.6 Stretches and Shrinks and the Absolute Value Family 184

Students explore the non-rigid transformations of stretching and shrinking. The emphasis is on the absolute value function.

Investigation: The Pendulum *(Stretching and Shrinking)* 187

4.7 Transformations and the Circle Family 192

This lesson summarizes the various transformations and shows that they apply to ellipses. In fact, students find equations of ellipses by viewing them as transformations of circles.

Investigation: Modeling an Ellipse *(Transformations)* 193

4.8 Compositions of Functions 200

This lesson expands what students know about function notation. It introduces function composition.

Investigation: Looking Up *(Composition of Functions)* 202

Chapter 4 Review 211

CHAPTER 5

Exponential, Power, and Logarithmic Functions 215

Just as Chapter 3 moved from arithmetic sequences to continuous linear functions, Chapter 5 moves from geometric sequences to exponential functions. Inverse functions are introduced to define the logarithm. The last two lessons describe the properties of logarithms and their use in solving exponential equations.

5.1 The Exponential Function 216

Just as section 3.1 gave the general form for a linear equation, this lesson gives the general formula for an exponential function. It includes examples from finance.

Investigation: Radioactive Decay *(Exponential Functions)* 216

5.2 Properties of Exponents 223

Students review the properties of exponents. These include the Product Property, the Quotient Property, the Power of a Power Property, and the Power of a Quotient Property. Zero and negative exponents are included.

Investigation: Exponent Rules! *(Properties of Exponents)* 223

5.3 Fractional Exponents and Roots 229

This lesson introduces rational exponents. It goes on to show how they can be used to solve equations involving power functions.

Investigation: Getting to the Root *(Rational Exponents)* 229

5.4 Applications of Power Equations 239

Students see that power functions can model real-world situations. The focus is on applications from finance and physics.

5.5 Building Inverses of Functions 243

Students explore how the graphs of inverse functions are related. It continues with a discussion of one-to-one functions and recognizes functions with inverses from their graphs. Finally, students learn how to find inverse functions algebraically.

Investigation: The Inverse *(Inverse Functions)* 243

5.6 The Logarithmic Function 251

Students learn that the common logarithm is the inverse function of the base 10 exponential function. Other logarithms are introduced, along with the change of base formula.

Investigation: Exponents and Logarithms *(Logarithms are Inverses of Exponential Functions)* 251

5.7 Properties of Logarithms 257

This lesson focuses on the properties of logarithms. These logarithmic properties are compared with the corresponding properties of exponents.

Investigation: Logarithm Rules! *(Properties of Logarithms)* 258

5.8 **Applications of Logarithms** 264

This lesson revisits some exponential applications from earlier in the chapter. Students find exponential models for data. An exploration introduces the number *e* and the natural logarithm.

Investigation: Cooling *(Exponential Modeling)* 265

Chapter 5 Review 273

CHAPTER 6 **Quadratic and Other Polynomial Functions** **277**

Chapter 6 begins by showing that degree 2 polynomials have a common second difference just as degree 1 polynomials (lines) have a common first difference (slope). Students learn that different forms for expressing quadratic equations show different information more clearly. Students also learn to solve quadratic equations using completing the square and the quadratic formula. This includes equations with complex solutions. Finally students apply similar techniques along with polynomial long division to solve higher-degree polynomials.

6.1 **Polynomial Degree and Finite Differences** 278

This lesson introduces polynomials and shows how to use finite differences to determine a polynomial's degree. This lesson also introduces using the calculator to find quadratic regressions.

Investigation: Free Fall *(Quadratic Modeling)* 281

6.2 **Equivalent Quadratic Forms** 286

Students learn that it is frequently useful to express quadratic equations in vertex, or factored form.

Investigation: Rolling Along *(Quadratic Modeling)* 289

6.3 **Completing the Square** 294

Completing the square is described geometrically in the first instance. The lesson shows that completing the square is useful not only for solving equations, but also for rewriting equations in vertex form.

Investigation: Square Me: The Details *(Completing the Square)* 296

6.4 **The Quadratic Formula** 301

First this lesson derives the quadratic formula. Students then apply the quadratic formula to solve quadratic equations.

Investigation: How High Can You Go? *(Projectile Motion)* 303

6.5 **Complex Numbers** 307

This lesson discusses complex numbers. First it shows how they can appear as solutions to quadratic equations. Then it discusses the history of complex numbers.

Investigation: Complex Arithmetic *(Arithmetic with Complex Numbers)* 309

6.6 Factoring Polynomials 314

This lesson generalizes what students learned about factored form in Lesson 6.2. Students explore the connection between zeros of a polynomial function and factors of a corresponding polynomial expression.

Investigation: The Box Factory *(Writing Equations for Polynomials)* 315

6.7 Higher Degree Polynomials 320

Students explore how a polynomial's equation affects its graph. Special emphasis is placed on finding extremes based on the graph and knowing how degree and end-behavior are related.

Investigation: The Largest Triangle *(Modeling with Polynomials)* 321

6.8 More about Finding Solutions 327

This lesson describes polynomial long division, the rational root theorem, and the factor theorem—the tools most commonly used to factor higher-degree polynomials in college algebra. Students use these tools to find the zeros of polynomial functions.

Chapter 6 Review 337

CHAPTER 7 **Matrices and Linear Systems** **341**

Chapter 7 centers mostly on matrices and their applications to processes involving transitions and to solving linear systems of equations. Problems use linear systems of equations and inequalities. The last lesson discusses using systems of linear inequalities to solve linear programming problems.

7.1 Matrix Representations 342

This lesson introduces matrices. Students discover how a transition matrix can help model the changes in a system over time.

Investigation: Chilly Choices *(Transition Matrices)* 342

7.2 Matrix Operations 349

Students investigate addition, scalar multiplication, and multiplication of matrices. In addition to transition matrices, examples include matrices that represent ordered pairs of vertices of polygons.

Investigation: Find Your Place *(Multiplying Matrices)* 353

7.3 The Row Reduction Method 360

Students see how the row-reduction method provides a short-hand notation for solving systems by elimination. Students solve systems using this technique.

Investigation: A Martian Experiment *(Solving Systems, Modeling)* 362

7.4 Solving Systems with Inverse Matrices 368

Students learn about identity matrices and inverse matrices. Then students solve systems of equations using inverse matrices.

Investigation: The Inverse Matrix *(Finding Inverses of 2 by 2 Matrices)* 369

7.5 Systems of Linear Inequalities 378

This lesson shows how to graph systems of linear inequalities that model real situations. Students will graph more linear inequalities in Lesson 7.6 as part of solving linear programming problems.

Investigation: Paying for College *(Systems of Inequalities)* 378

7.6 Linear Programming 385

This lesson includes examples of linear programming. It has a special focus on problems related to business and nutrition science.

Investigation: Maximizing Profit *(Linear Programming)* 385

Chapter 7 Review 395

CHAPTER 8 **Parametric Equations and Trigonometry** **401**

Chapter 8 begins with a discussion of parametric equations. Then it introduces right triangle trigonometry and its applications to course setting and projectile motion. Finally the Law of Sines and Law of Cosines are used to solve arbitrary triangles.

8.1 Graphing Parametric Equations 402

Parametric equations are introduced in this lesson. Many of the examples and exercises focus on concrete problems about modeling motion.

Investigation: Simulating Motion *(Parametric Equations)* 404

8.2 Converting Parametric to Nonparametric Equations 411

This lesson compares parametric equations to a single equation for modeling motion. Students convert parametric equations to single equations.

Investigation: Parametric Walk *(Converting Parametric Equations to a Single Equation)* 411

8.3 Right Triangle Trigonometry 418

This lesson introduces the functions sine, cosine, and tangent using right triangles. Many applications involve the bearing of a moving object.

Investigation: Two Ships *(Right Triangle Trigonometry)* 420

8.4 Using Trigonometry to Set a Course 431

Students realize the usefulness of trigonometry in setting ship and airplane courses. Examples involving current or wind are included.

Investigation: Motion in a Current *(Setting a Course)* 431

8.5 Projectile Motion 438

This lesson gives another example of the uses of parametric equations. It breaks projectile motion into its horizontal and vertical components using trigonometry.

Investigation: Basketball Free Throw *(Projectile Motion)* 441

8.6 The Law of Sines 445

The lesson begins with a proof of the Law of Sines. Then the lesson gives several examples of the use of the Law of Sines for students to read and solve.

Investigation: Oblique Triangles *(Law of Sines)* 445

8.7 The Law of Cosines 453

This lesson develops the Law of Cosines. Students proceed to solve triangles using the Law of Sines and the Law of Cosines.

Investigation: Around the Corner *(Law of Cosines)* 455

Chapter 8 Review 462

CHAPTER 9 **Conic Sections and Rational Functions** **466**

In Chapter 9's first few lessons, students study conic sections. Each conic section is introduced using the locus of points definition. Then formulas for each conic section are derived when their foci and vertices lie on lines parallel to the axes. Students will use these formulas when they learn to solve nonlinear systems. Knowing the formulas allows students to sketch many of the systems to confirm that they have found the correct number of solutions. Finally the hyperbola $1/x$ provides a connection to the study of rational functions. A section on rational expression arithmetic is included should students need to review the topic.

9.1 Using the Distance Formula 467

The distance formula helps students model situations comparable to the classic problem of finding the least expensive way to lay cable partially underwater and partially above ground.

Investigation: Bucket Race *(The Distance Formula)* 467

9.2 Circles and Ellipses 474

Students sketch circles and ellipses and learn to recognize and find their equations. Relations between foci and vertices are explored.

Investigation: A Slice of Light *(Finding an Ellipses Equation)* 480

9.3 Parabolas 486

Students sketch parabolas and learn to recognize and find their equations. Connections between the focus, vertex, and directrix are explored.

Investigation: Fold a Parabola *(Definition of a Parabola)* 490

9.4 The Hyperbola 494

Students sketch hyperbolas and learn to recognize and find their equations. Connections between the foci and vertices are explored.

Investigation: Passing By *(Hyperbolas)* 498

9.5 Nonlinear Systems of Equations 503

The lesson gives examples of solving nonlinear systems using substitution and elimination. Students also learn to examine the graph of a system to find the number of solutions to the system.

Investigation: Systems of Conic Equations *(Numbers of Solutions for Systems)* 503

9.6 Introduction to Rational Functions 508

Students explore end behavior of rational functions and behavior near asymptotes using tables of values. Mixture problems are solved using rational equations to give another example of the use of rational functions.

Investigation: The Breaking Point *(Rational Functions)* 508

9.7 Graphs of Rational Functions 516

Students discover when the graph of a rational expression has a hole and when it has an asymptote. Two examples of graphing general rational functions are given.

Investigation: Predicting Asymptotes and Holes 517

9.8 Operations with Rational Expressions 524

This lesson shows how to add, subtract, multiply, and divide rational expressions. Students also simplify complex rational expressions.

Chapter 9 Review 533

CHAPTER 10 **Series** **539**

Sequences are explored in Chapters 2, 3, and 5. Chapter 10 is about series, especially arithmetic and geometric series. A formula is developed for summing infinite geometric series. This chapter also develops formulas for partial sums of geometric and arithmetic series.

10.1 Arithmetic Series 540

This lesson introduces series. It focuses primarily on arithmetic series. It develops the formula for the partial sums of an arithmetic series.

Investigation: Arithmetic Series Formula *(Partial Sums of Geometric Series)* 542

10.2 Infinite Geometric Series 547

Students investigate infinite geometric series. The lesson concludes by presenting the formula for summing any infinite geometric series with a common ratio whose absolute value is less than one.

Investigation: The Infinite Geometric Series Formula
(Infinite Geometric Series) 548

10.3 Partial Sums of Geometric Series 554

The lesson expands on the series sum formula from Lesson 10.2. It generalizes student investigation results to a general formula for partial sums of geometric series.

Investigation: The Geometric Series Formula
(Partial Sums of Geometric Series) 554

Chapter 10 Review 562

CHAPTER 11 **Probability** **565**

Chapter 11 introduces the basics of theoretical and experimental probability. After introducing counting principles, it shows how permutations and combinations aid in probability calculations. It is shown that Pascal's triangle gives combination numbers. And because of the connections between the Binomial Theorem, combinations, and Pascal's triangle, this is the natural place to include the binomial theorem.

11.1 Randomness and Probability 566

The lesson gives examples of experimental and theoretical probability and describes how to calculate each.

Investigation: Coin Flip *(Outcomes and Events)* 567

11.2 Counting Outcomes and Tree Diagrams 577

This lesson reintroduces students to tree diagrams from Lesson 1.3. Students use tree diagrams to calculate probabilities. The lesson also defines independent events and conditional probability.

Investigation: The Multiplication Rule *(Multiplication Rule for Independent Events)* 578

11.3 Mutually Exclusive Events and Venn Diagrams 586

This lesson introduces Venn diagrams. Students learn that Venn diagrams are useful tools for solving probability problems about mutually exclusive events.

Investigation: The Addition Rule *(Addition Rule for the Probability of Events)* 587

11.4 Random Variables and Expected Value 593

This lesson introduces the concepts of random variables. Students calculate the expected value of a random variable.

Investigation: "Dieing" for a Four *(Expected Value)* 593

11.5 Permutations and Probability 600

The lesson centers on the counting principle; therefore it also includes an introduction to factorials.

Investigation: Order and Arrange *(Permutations)* 600

11.6 **Combinations and Probability** 607

Students explore the differences between combinations and permutations. They then calculate a number of combinations.

Investigation: Winning the Lottery *(Combinations)* 610

11.7 **The Binomial Theorem and Pascal's Triangle** 613

The lesson introduces Pascal's triangle. The rows of Pascal's triangle are also compared to the coefficients of binomial expansions. The lesson closes by describing how the binomial theorem can be used to calculate the coefficients of a binomial expansion.

Investigation: Pascal's Triangle and Combination Numbers *(Combinations)* 613

Chapter 11 Review 623

Selected Answers 627

Glossary 675

Photo Credits 687

Index 688

Resources on the Student CD

Prerequisite Review: Numbers and Figures • Operations on Numbers • The Ideas that Motivate Algebra • Exponents • Radicals • Polynomials • Factoring Polynomials • Rational Expressions

Calculator Notes

Microsoft® Excel Notes

Preface

Success in today's world demands that students learn mathematics in a meaningful way. Concepts learned in the classroom must be applied to their real-life situations using the technology available to them. *Investigating College Algebra with Technology* combines a sequence of topics, hands-on activities, and use of technology to create a program that gives students a solid, lasting foundation in algebra. In the approach of this textbook, the integration of technology takes the focus off the tools themselves and places it on decision making, reflection, reasoning, and problem solving.

Investigating College Algebra with Technology enriches the traditional algebra curriculum with data analysis, functions, and probability. Students work with data-rich applications, making the connection to the relevancy in their lives.

Key Features

Investigations are at the heart of this book.

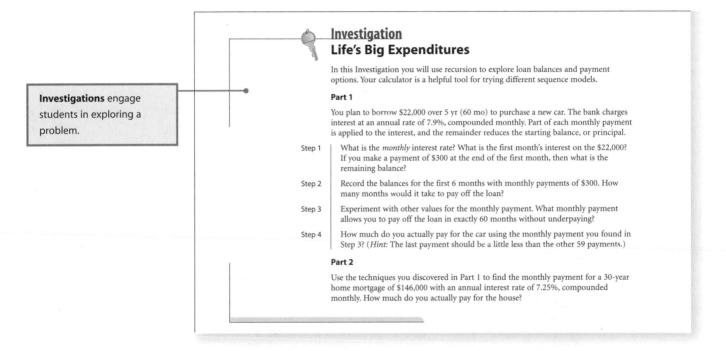

Investigations engage students in exploring a problem.

Investigation
Life's Big Expenditures

In this Investigation you will use recursion to explore loan balances and payment options. Your calculator is a helpful tool for trying different sequence models.

Part 1

You plan to borrow $22,000 over 5 yr (60 mo) to purchase a new car. The bank charges interest at an annual rate of 7.9%, compounded monthly. Part of each monthly payment is applied to the interest, and the remainder reduces the starting balance, or principal.

Step 1 What is the *monthly* interest rate? What is the first month's interest on the $22,000? If you make a payment of $300 at the end of the first month, then what is the remaining balance?

Step 2 Record the balances for the first 6 months with monthly payments of $300. How many months would it take to pay off the loan?

Step 3 Experiment with other values for the monthly payment. What monthly payment allows you to pay off the loan in exactly 60 months without underpaying?

Step 4 How much do you actually pay for the car using the monthly payment you found in Step 3? (*Hint:* The last payment should be a little less than the other 59 payments.)

Part 2

Use the techniques you discovered in Part 1 to find the monthly payment for a 30-year home mortgage of $146,000 with an annual interest rate of 7.25%, compounded monthly. How much do you actually pay for the house?

Whenever possible, the lessons engage students in exploring a problem before telling them how to solve the problem. As students ask questions, see patterns, make conjectures, and ask more questions, they gain the confidence to approach new problems and apply new concepts to solve them. Not all investigations in the book ask students to discover a new concept. In some investigations, students use the mathematics they already know to analyze data—either given data or data they gather themselves. As students look at data and try to make sense of the relationships they see, they learn to use mathematics to model phenomena in the real world.

Examples come before or after investigations as needed. Some are presented within the context of applications, others in a purely mathematical form. All examples are fully worked out and include reasons that justify the process.

Examples are often presented in the context of applications.

EXAMPLE

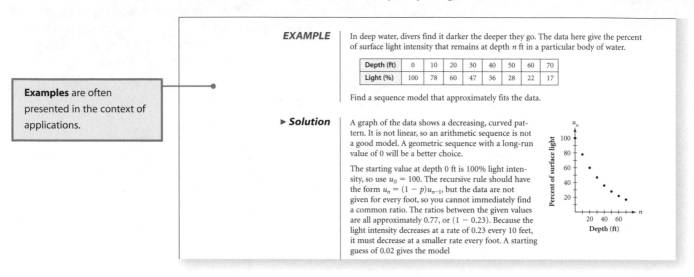

In deep water, divers find it darker the deeper they go. The data here give the percent of surface light intensity that remains at depth n ft in a particular body of water.

Depth (ft)	0	10	20	30	40	50	60	70
Light (%)	100	78	60	47	36	28	22	17

Find a sequence model that approximately fits the data.

▶ ***Solution***

A graph of the data shows a decreasing, curved pattern. It is not linear, so an arithmetic sequence is not a good model. A geometric sequence with a long-run value of 0 will be a better choice.

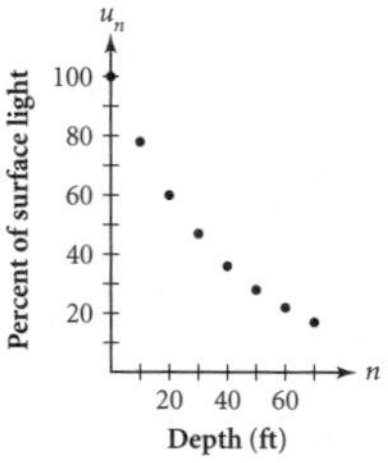

The starting value at depth 0 ft is 100% light intensity, so use $u_0 = 100$. The recursive rule should have the form $u_n = (1 - p)u_{n-1}$, but the data are not given for every foot, so you cannot immediately find a common ratio. The ratios between the given values are all approximately 0.77, or $(1 - 0.23)$. Because the light intensity decreases at a rate of 0.23 every 10 feet, it must decrease at a smaller rate every foot. A starting guess of 0.02 gives the model

Exercise sets are developmental. **Practice Your Skills** exercises come first and involve direct application of what students have learned in the lesson. **Reason and Apply** problems are more challenging, developing reasoning and the transfer of knowledge. Finally, **Review** problems keep previous learning current.

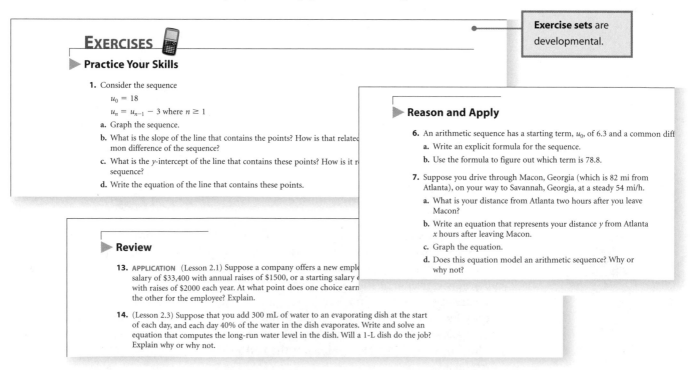

Exercise sets are developmental.

EXERCISES

Practice Your Skills

1. Consider the sequence

$u_0 = 18$

$u_n = u_{n-1} - 3$ where $n \geq 1$

a. Graph the sequence.

b. What is the slope of the line that contains the points? How is that relatec mon difference of the sequence?

c. What is the y-intercept of the line that contains these points? How is it r sequence?

d. Write the equation of the line that contains these points.

Reason and Apply

6. An arithmetic sequence has a starting term, u_0, of 6.3 and a common diff

a. Write an explicit formula for the sequence.

b. Use the formula to figure out which term is 78.8.

7. Suppose you drive through Macon, Georgia (which is 82 mi from Atlanta), on your way to Savannah, Georgia, at a steady 54 mi/h.

a. What is your distance from Atlanta two hours after you leave Macon?

b. Write an equation that represents your distance y from Atlanta x hours after leaving Macon.

c. Graph the equation.

d. Does this equation model an arithmetic sequence? Why or why not?

Review

13. APPLICATION (Lesson 2.1) Suppose a company offers a new emplo salary of $33,400 with annual raises of $1500, or a starting salary o with raises of $2000 each year. At what point does one choice earn the other for the employee? Explain.

14. (Lesson 2.3) Suppose that you add 300 mL of water to an evaporating dish at the start of each day, and each day 40% of the water in the dish evaporates. Write and solve an equation that computes the long-run water level in the dish. Will a 1-L dish do the job? Explain why or why not.

Objectives at the beginning of each chapter outline the math to be learned, while **More Background** sections preceding some lessons list prerequisite high school algebra skills for those lessons.

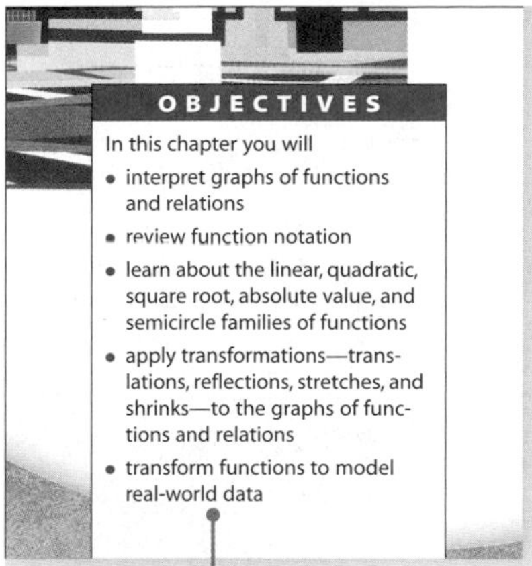

OBJECTIVES

In this chapter you will

- interpret graphs of functions and relations
- review function notation
- learn about the linear, quadratic, square root, absolute value, and semicircle families of functions
- apply transformations—translations, reflections, stretches, and shrinks—to the graphs of functions and relations
- transform functions to model real-world data

Objectives outline the math to be learned.

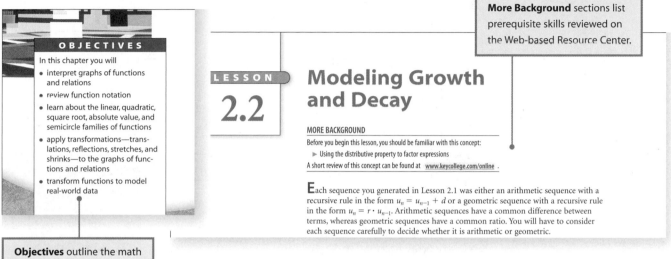

More Background sections list prerequisite skills reviewed on the Web-based Resource Center.

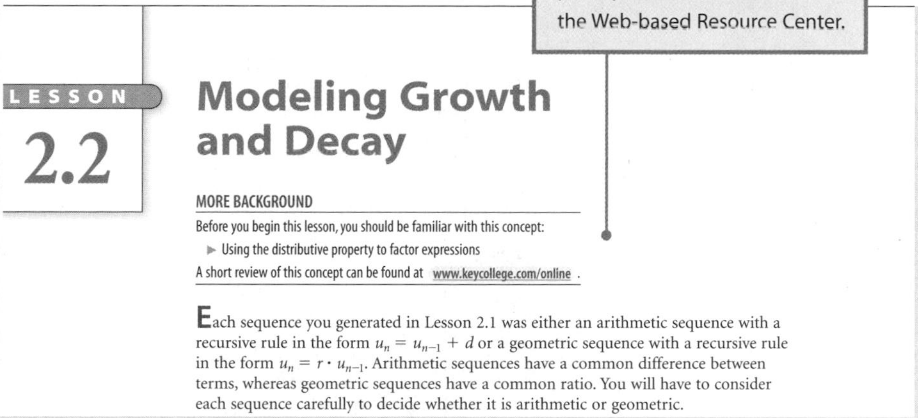

LESSON 2.2

Modeling Growth and Decay

MORE BACKGROUND

Before you begin this lesson, you should be familiar with this concept:

▶ Using the distributive property to factor expressions

A short review of this concept can be found at www.keycollege.com/online .

Each sequence you generated in Lesson 2.1 was either an arithmetic sequence with a recursive rule in the form $u_n = u_{n-1} + d$ or a geometric sequence with a recursive rule in the form $u_n = r \cdot u_{n-1}$. Arithmetic sequences have a common difference between terms, whereas geometric sequences have a common ratio. You will have to consider each sequence carefully to decide whether it is arithmetic or geometric.

Reviews of each basic algebra concept can be found at the Web-based **Resource Center** for *Investigating College Algebra with Technology.* In addition, a complete **Prerequisite Review** is available on the Student CD.

Explorations give students a chance to apply what they know to new topics, extending their understanding of the underlying concepts.

Explorations extend understanding to new topics.

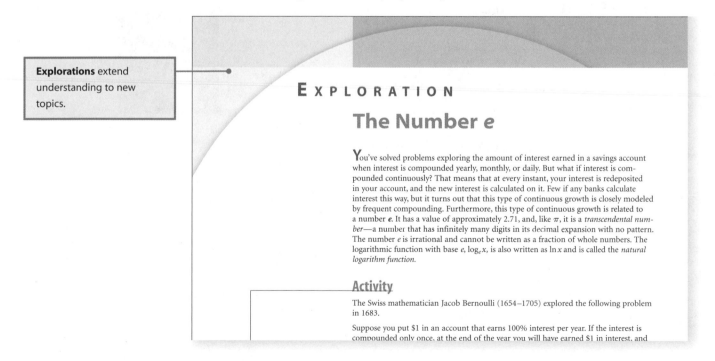

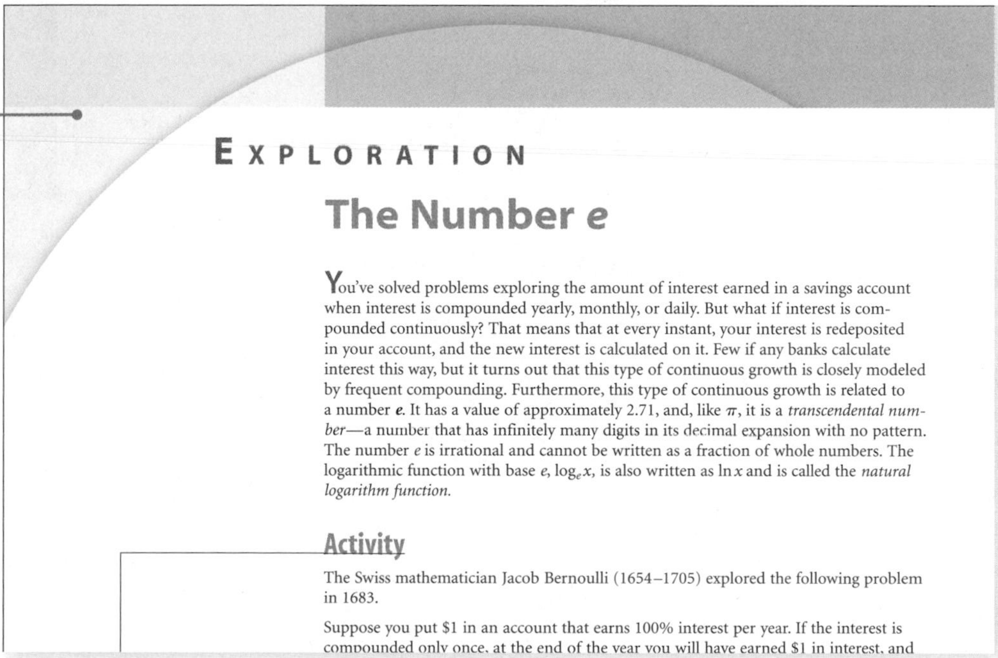

EXPLORATION

The Number *e*

You've solved problems exploring the amount of interest earned in a savings account when interest is compounded yearly, monthly, or daily. But what if interest is compounded continuously? That means that at every instant, your interest is redeposited in your account, and the new interest is calculated on it. Few if any banks calculate interest this way, but it turns out that this type of continuous growth is closely modeled by frequent compounding. Furthermore, this type of continuous growth is related to a number ***e***. It has a value of approximately 2.71, and, like π, it is a *transcendental number*—a number that has infinitely many digits in its decimal expansion with no pattern. The number e is irrational and cannot be written as a fraction of whole numbers. The logarithmic function with base e, $\log_e x$, is also written as $\ln x$ and is called the *natural logarithm function.*

Activity

The Swiss mathematician Jacob Bernoulli (1654–1705) explored the following problem in 1683.

Suppose you put $1 in an account that earns 100% interest per year. If the interest is compounded only once, at the end of the year you will have earned $1 in interest, and

Projects engage students in practical research, while **Improving Your Reasoning Skills** sections offer puzzles that approach critical thinking from different points of view.

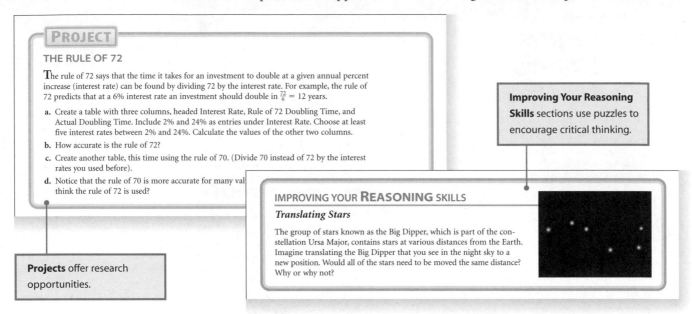

PROJECT

THE RULE OF 72

The rule of 72 says that the time it takes for an investment to double at a given annual percent increase (interest rate) can be found by dividing 72 by the interest rate. For example, the rule of 72 predicts that at a 6% interest rate an investment should double in $\frac{72}{6} = 12$ years.

a. Create a table with three columns, headed Interest Rate, Rule of 72 Doubling Time, and Actual Doubling Time. Include 2% and 24% as entries under Interest Rate. Choose at least five interest rates between 2% and 24%. Calculate the values of the other two columns.

b. How accurate is the rule of 72?

c. Create another table, this time using the rule of 70. (Divide 70 instead of 72 by the interest rates you used before).

d. Notice that the rule of 70 is more accurate for many val
think the rule of 72 is used?

Projects offer research opportunities.

Improving Your Reasoning Skills sections use puzzles to encourage critical thinking.

IMPROVING YOUR **REASONING** SKILLS

Translating Stars

The group of stars known as the Big Dipper, which is part of the constellation Ursa Major, contains stars at various distances from the Earth. Imagine translating the Big Dipper that you see in the night sky to a new position. Would all of the stars need to be moved the same distance? Why or why not?

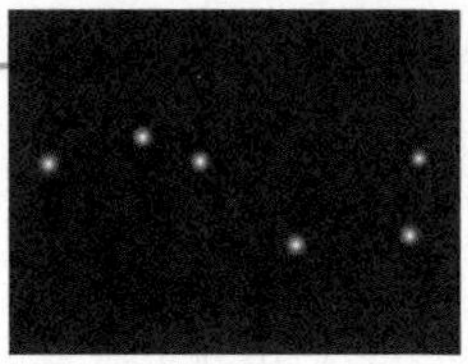

Connections encourage students to make the links between the mathematics they're learning and the world around them.

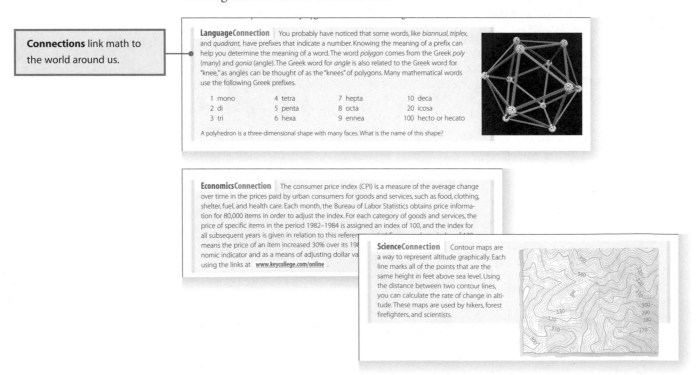

Connections link math to the world around us.

LanguageConnection You probably have noticed that some words, like *biannual, triplex,* and *quadrant,* have prefixes that indicate a number. Knowing the meaning of a prefix can help you determine the meaning of a word. The word *polygon* comes from the Greek *poly* (many) and *gonia* (angle). The Greek word for *angle* is also related to the Greek word for "knee," as angles can be thought of as the "knees" of polygons. Many mathematical words use the following Greek prefixes.

1 mono	4 tetra	7 hepta	10 deca
2 di	5 penta	8 octa	20 icosa
3 tri	6 hexa	9 ennea	100 hecto or hecato

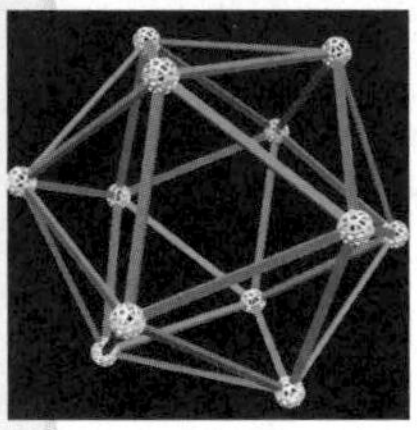

A polyhedron is a three-dimensional shape with many faces. What is the name of this shape?

EconomicsConnection The consumer price index (CPI) is a measure of the average change over time in the prices paid by urban consumers for goods and services, such as food, clothing, shelter, fuel, and health care. Each month, the Bureau of Labor Statistics obtains price information for 80,000 items in order to adjust the index. For each category of goods and services, the price of specific items in the period 1982–1984 is assigned an index of 100, and the index for all subsequent years is given in relation to this referen… means the price of an item increased 30% over its 198… nomic indicator and as a means of adjusting dollar va… using the links at www.keycollege.com/online .

ScienceConnection Contour maps are a way to represent altitude graphically. Each line marks all of the points that are the same height in feet above sea level. Using the distance between two contour lines, you can calculate the rate of change in altitude. These maps are used by hikers, forest firefighters, and scientists.

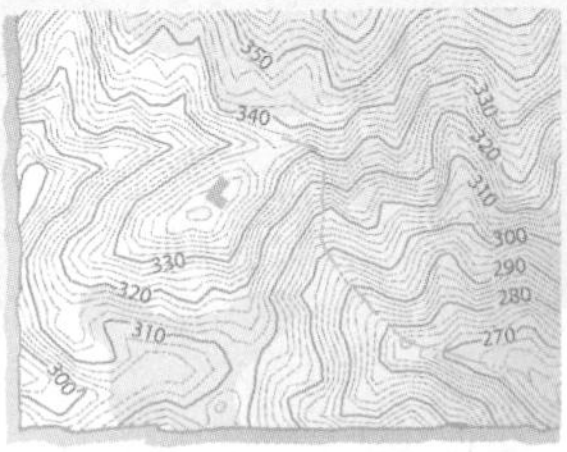

Chapter Reviews begin with a summary of the new mathematical ideas. Boldface terms can be found in the glossary at the back of the textbook. Exercises are similar to those that appear in the lessons and provide a cumulative assessment at the chapter level. **Assessing What You've Learned** helps students capture their learning experiences and identify gaps in understanding.

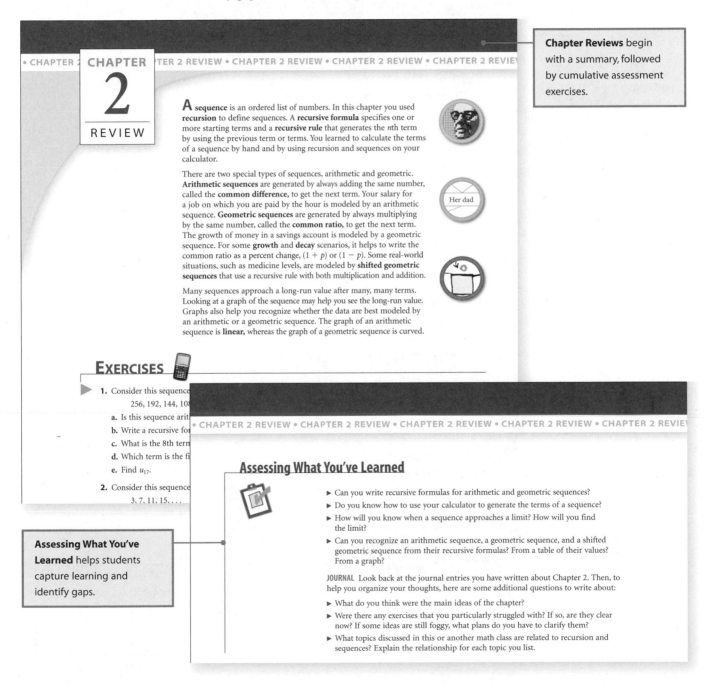

• CHAPTER 2 REVIEW • CHAPTER 2 REVIEW • CHAPTER 2 REVIEW • CHAPTER 2 REVIEW

CHAPTER 2 REVIEW

A **sequence** is an ordered list of numbers. In this chapter you used **recursion** to define sequences. A **recursive formula** specifies one or more starting terms and a **recursive rule** that generates the *n*th term by using the previous term or terms. You learned to calculate the terms of a sequence by hand and by using recursion and sequences on your calculator.

There are two special types of sequences, arithmetic and geometric. **Arithmetic sequences** are generated by always adding the same number, called the **common difference,** to get the next term. Your salary for a job on which you are paid by the hour is modeled by an arithmetic sequence. **Geometric sequences** are generated by always multiplying by the same number, called the **common ratio,** to get the next term. The growth of money in a savings account is modeled by a geometric sequence. For some **growth** and **decay** scenarios, it helps to write the common ratio as a percent change, $(1 + p)$ or $(1 - p)$. Some real-world situations, such as medicine levels, are modeled by **shifted geometric sequences** that use a recursive rule with both multiplication and addition.

Many sequences approach a long-run value after many, many terms. Looking at a graph of the sequence may help you see the long-run value. Graphs also help you recognize whether the data are best modeled by an arithmetic or a geometric sequence. The graph of an arithmetic sequence is **linear,** whereas the graph of a geometric sequence is curved.

Chapter Reviews begin with a summary, followed by cumulative assessment exercises.

EXERCISES

1. Consider this sequence
   256, 192, 144, 10
   a. Is this sequence arit
   b. Write a recursive fo
   c. What is the 8th ter
   d. Which term is the fi
   e. Find u_{17}.
2. Consider this sequence
   3, 7, 11, 15, . . .

• CHAPTER 2 REVIEW • CHAPTER 2 REVIEW • CHAPTER 2 REVIEW • CHAPTER 2 REVIEW • CHAPTER 2 REVIE

Assessing What You've Learned

- Can you write recursive formulas for arithmetic and geometric sequences?
- Do you know how to use your calculator to generate the terms of a sequence?
- How will you know when a sequence approaches a limit? How will you find the limit?
- Can you recognize an arithmetic sequence, a geometric sequence, and a shifted geometric sequence from their recursive formulas? From a table of their values? From a graph?

JOURNAL Look back at the journal entries you have written about Chapter 2. Then, to help you organize your thoughts, here are some additional questions to write about:

- What do you think were the main ideas of the chapter?
- Were there any exercises that you particularly struggled with? If so, are they clear now? If some ideas are still foggy, what plans do you have to clarify them?
- What topics discussed in this or another math class are related to recursion and sequences? Explain the relationship for each topic you list.

Assessing What You've Learned helps students capture learning and identify gaps.

Technology

Technology is an integral part of students' discoveries. The textbook was created with technology in mind, so the use of the tools is a seamless part of the pedagogy, not an afterthought. Throughout the book, solutions to examples are presented with **graphing calculator** screens, so students become accustomed to using the tool in their Investigations.

Students also use the graphing calculator to complete homework assignments. **Calculator Notes** provide step-by-step instructions for using the most popular graphing calculators. These instructions are referenced in the text and are included on the Student CD provided in the back of the text.

Graphing calculator screens are shown in the solutions.

Check this model by graphing the original data and the sequence on your calculator. The graph shows that this model fits only one data point—it does not decay fast enough. [▶ See **Calculator Note 2B** to learn about sequences on your calculator and **Calculator Note 2C** to learn about graphing sequences.◀]

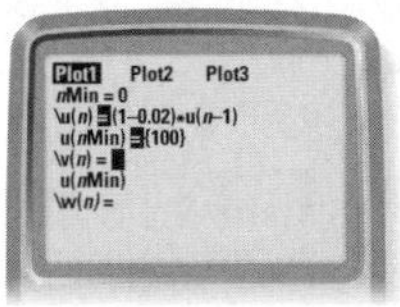

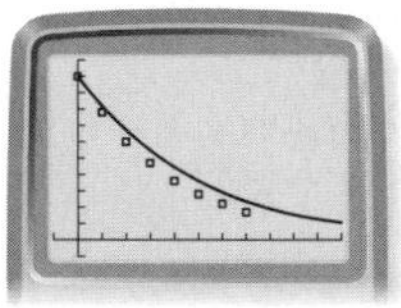

[−10, 110, 10, −10, 110, 10]

Calculator Notes, referenced in the text, are found on the student CD. These offer step-by-step graphing calculator instructions.

Note 1H• Random Numbers

There are several ways to generate a list of random numbers within an interval.

Random Integers

To find a random integer between 1 and 20, on the Home screen press MATH and arrow to PRB. Select 5:randInt(and enter 1,20), then press ENTER. If you want five random numbers, either press ENTER five times, or enter randInt(1,20,5) and press ENTER. If you ask for more numbers than show on one line of the screen, you can scroll to see the rest of the list. Or you can press 2nd [RCL] 2nd [ANS] ENTER to see the entire list on the screen.

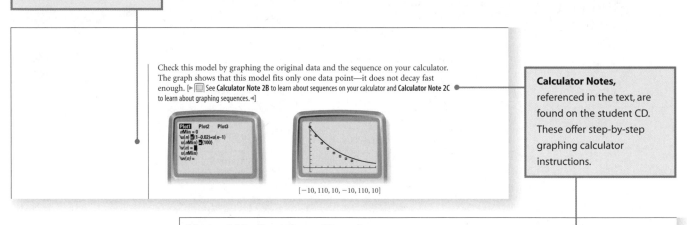

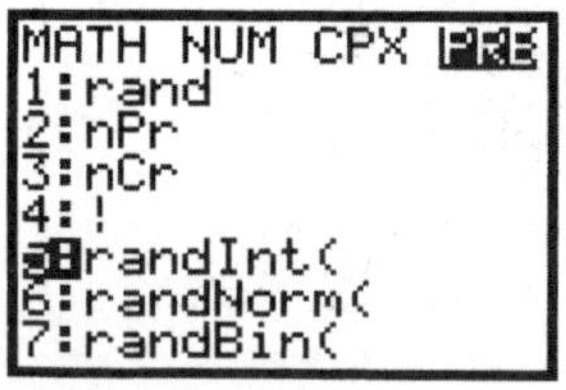

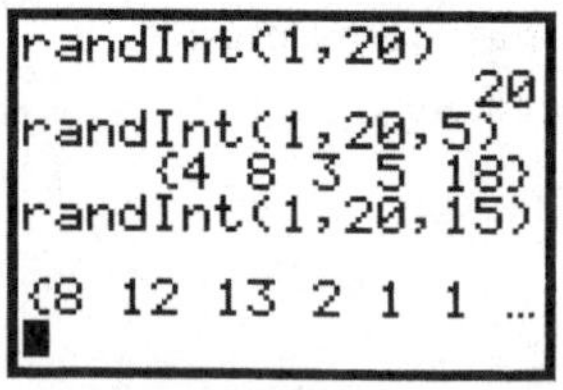

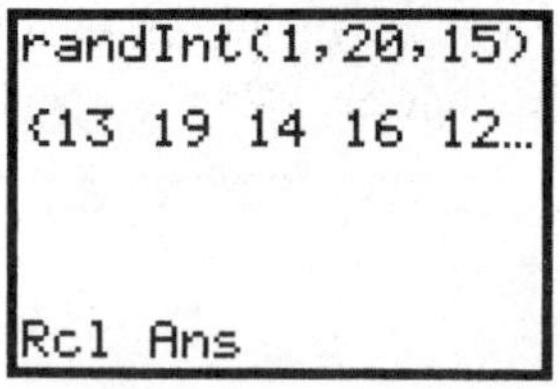

Recognizing that many students will use **Microsoft® Excel** in the workplace, optional spreadsheet Explorations have been incorporated in the text. Using Excel, students apply what they've learned in a technological environment to help them become comfortable in that environment. Students first work with an interactive spreadsheet, accessed at the Web-based **Resource Center** or on their CD, then build their own spreadsheets in a connected Activity.

Explorations with Microsoft® Excel allow students to explore concepts with dynamic templates, then build their own spreadsheets.

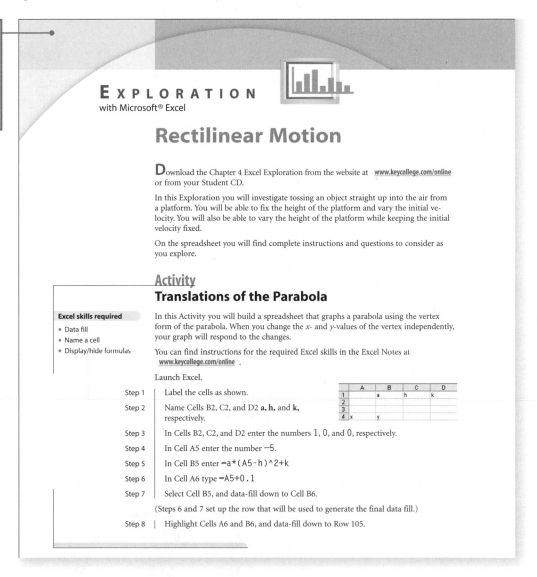

EXPLORATION
with Microsoft® Excel

Rectilinear Motion

Download the Chapter 4 Excel Exploration from the website at www.keycollege.com/online or from your Student CD.

In this Exploration you will investigate tossing an object straight up into the air from a platform. You will be able to fix the height of the platform and vary the initial velocity. You will also be able to vary the height of the platform while keeping the initial velocity fixed.

On the spreadsheet you will find complete instructions and questions to consider as you explore.

Activity
Translations of the Parabola

Excel skills required
- Data fill
- Name a cell
- Display/hide formulas

In this Activity you will build a spreadsheet that graphs a parabola using the vertex form of the parabola. When you change the *x*- and *y*-values of the vertex independently, your graph will respond to the changes.

You can find instructions for the required Excel skills in the Excel Notes at www.keycollege.com/online .

Launch Excel.

Step 1 Label the cells as shown.

	A	B	C	D
1		a	h	k
2				
3				
4	x	y		

Step 2 Name Cells B2, C2, and D2 **a, h,** and **k,** respectively.

Step 3 In Cells B2, C2, and D2 enter the numbers 1, 0, and 0, respectively.

Step 4 In Cell A5 enter the number −5.

Step 5 In Cell B5 enter `=a*(A5-h)^2+k`

Step 6 In Cell A6 type `=A5+0.1`

Step 7 Select Cell B5, and data-fill down to Cell B6.

(Steps 6 and 7 set up the row that will be used to generate the final data fill.)

Step 8 Highlight Cells A6 and B6, and data-fill down to Row 105.

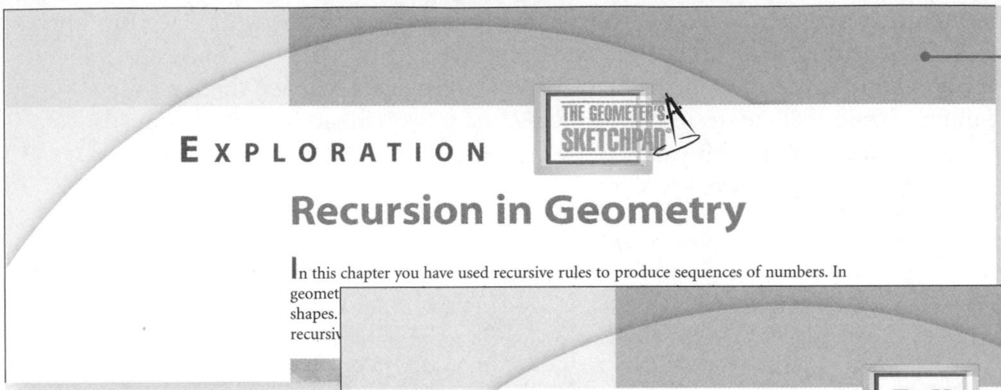

EXPLORATION

Recursion in Geometry

In this chapter you have used recursive rules to produce sequences of numbers. In
geomet
shapes.
recursi

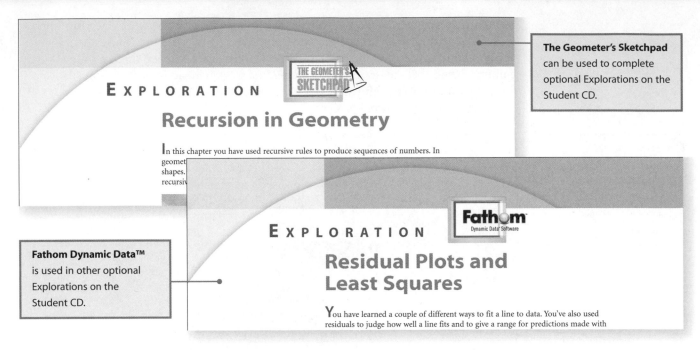

The Geometer's Sketchpad can be used to complete optional Explorations on the Student CD.

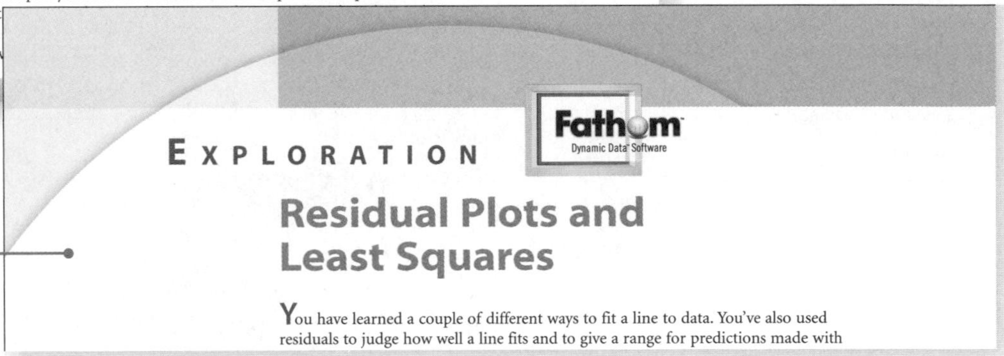

EXPLORATION

Residual Plots and Least Squares

You have learned a couple of different ways to fit a line to data. You've also used residuals to judge how well a line fits and to give a range for predictions made with

Fathom Dynamic Data™ is used in other optional Explorations on the Student CD.

Finally, if students have access to geometry or statistical software, they can investigate mathematical concepts in new ways. Created with the use of **The Geometer's Sketchpad®** and **Fathom Dynamic Data™** in mind, Explorations on the Student CD give students opportunities to interact with the concepts they're learning.

Summary

Through the investigations, exercises, and examples in *Investigating College Algebra with Technology,* students explore interesting problems and generalize concepts. And if a student forgets concepts, formulas, or procedures, they can always re-create them—because they developed them themselves the first time. Students will find that this discovery approach allows them to form a meaningful understanding of college algebra topics. In addition to the integrated graphing calculator exercises and optional Microsoft® Excel Explorations in the book, the Student CD has more Explorations using The Geometer's Sketchpad® and Fathom Dynamic Data™ software. Its many features make *Investigating College Algebra with Technology* an innovative and exciting book as well as a coherent and streamlined teaching and learning tool.

If you are a student, we hope that what you learn using *Investigating College Algebra with Technology* as your guide will serve you throughout your lifetime. If you are an instructor, we believe that this textbook will help create a classroom atmosphere where problem solving is integral to instruction. The Instructor Resources for this book—which include lesson-by-lesson pedagogical notes and suggested teaching strategies—will help you create the appropriate classroom structure to facilitate this approach. We look forward to hearing about your experiences.

To the Student

You're about to embark on an exciting mathematical journey. The goal of your trip is to develop the essential algebraic tools and mathematical power to participate fully as a productive citizen in a changing world. By using technology on this journey, you will naturally make the necessary connections for using algebra to analyze the data found in the world around you.

Success in algebra is a recognized gateway to many varied career opportunities. Furthermore, important decision-making situations will confront you in life, and your ability to integrate algebra and data analysis can help you make informed decisions. You will acquire skills that evolve and adapt to new situations. You'll need to interpret numerical information and use it as a basis for making decisions. And you'll need to find ways to solve problems that arise in real life, not just in textbooks.

With your instructor as an additional guide, you'll learn algebra by actively doing mathematics. You'll make sense of important algebraic concepts, learn essential algebraic skills, and discover how to use algebra. This requires a far bigger commitment than just studying worked-out examples or waiting for the instructor to show you.

Your personal involvement is critical to success as you work through **Investigations.** Keeping your measurements, data, and calculations accurate will make your task easier and the concepts clearer in the long run. Whether you work in groups or discuss your individual outcomes with the class, you can learn from your fellow students. Communicating with your classmates will strengthen your understanding of the mathematical concepts.

The approach of this textbook will lead you to explore ideas and ponder questions. Read the textbook carefully—with paper, pencil, and calculator close at hand—and take good notes. Concepts and problems you have encountered before can help you solve new problems. Work through the **Examples** and answer the questions that are asked along the way. At the end of each lesson, the first set of Exercises has you practice the basic skills you have just learned. Other exercises require a great deal of thought. Don't give up. Problem solving and mathematical thinking takes practice just as it takes practice to excel in playing a sport or a musical instrument.

Every chapter ends with **Assessing What You've Learned,** which suggests ways to reflect on what you have learned. Just as your instructor and this book are your guides, your notes can be a log of your travels through college algebra. You might also want to keep a journal of your personal impressions along the way. Record your successes and the challenges you need to overcome.

Your calculator and computer help you explore new ideas and answer questions. Throughout the text you will see references to **Calculator Notes** contained on the CD that accompanies your textbook. These step-by-step graphing calculator instructions in everything from basic skills to more advanced shortcuts will help you make effective use of the graphing calculator. Using your graphing calculator, and optionally Microsoft® Excel, you will be able to manipulate large amounts of data quickly so that you can see the overall picture. Calculator Programs and Interactive Spreadsheets with

Excel are available on your CD. Also included on your CD are optional exercises using geometry and statistical software such as The Geometer's Sketchpad® and Fathom Dynamic Data™.

You should expect challenges, hard work, and occasional frustration. Yet, as you gain algebra skills, you'll overcome obstacles and be rewarded with a deeper understanding of mathematics, an increased confidence in your own problem-solving abilities, and the opportunity to be creative. From time to time, look back and reflect on where you have been. We hope that your journey through *Investigating College Algebra with Technology* will be a meaningful and rewarding experience.

And now it is time to begin. You are about to discover some pretty fascinating things.

Student Support Materials

Packaged in the back of your text, the Student CD contains many resources that will help make your algebra studies effective and exciting. These include:

- **Prerequisite Review** of algebra fundamentals, with exercises and selected solutions
- **Calculator Notes** with detailed instructions for using various types of graphing calculators
- **Excel Notes** with instructions that will help you with the optional Microsoft® Excel Explorations
- **Explorations** using The Geometer's Sketchpad® and Fathom Dynamic Data™ software (optional)

The **Student Study and Solutions Companion** contains study hints and solutions to select homework exercises, along with self-quiz exercises and solutions.

The **Resource Center** at www.keycollege.com/InvestigatingCA has a wealth of innovative educational tools, including Dynamic Algebra Demonstrations created with JavaSketchpad™; prerequisite skill reviews referenced at key points in the text; and Connections that link your algebra study to the real world.

To the Instructor

The mathematics we teach in colleges and universities has been evolving over the last few decades. Our workplaces are changing, and technology is present everywhere. Technology is not just a collection of tools that make work easier—it fundamentally changes the work we do. That's why the algebra you find in this book won't look quite like the algebra you are accustomed to seeing in other textbooks. There are some new topics that are now possible to explore with technology and some standard topics that can be approached in new ways.

This is why the authors and the Key College Publishing editorial team created *Investigating College Algebra with Technology.* Through the Investigations and Explorations, students will encounter interesting problems and generalize concepts. As you progress through this book, you'll see that graphing calculators, and optimally Microsoft® Excel and other technologies, are used to explore patterns and to make, test, and generalize conjectures. This discovery approach allows students to form a conceptual understanding of College Algebra that will serve them well in their future studies and in life in general. *Investigating College Algebra with Technology* empowers students, giving them confidence that they can understand mathematics and solve new problems on their own.

The American Mathematical Association of Two-Year Colleges (AMATYC) sets standards in its *Crossroads for Mathematics* calling for meaningful and relevant mathematics, mathematics taught as a laboratory discipline, and the effective use of technology[1]. The Investigations in *Investigating College Algebra with Technology* enable students to approach mathematics as a laboratory science and actively discover mathematics. Technology is used effectively throughout the book, as students use graphing calculators and optionally Microsoft® Excel to discover mathematics through Investigations and Explorations. Optional activities on the Student CD use Fathom Statistical Software and the Geometer's Sketchpad. Several exercises in each chapter require the use of the graphing calculator or Excel.

The Mathematical Association of America's Committee on Undergraduate Programs in Mathematics recommends that, in addition to using technology, students experience varied applications of mathematics, and have opportunities to solve problems, reason critically, exhibit persistence, and test conjectures[2]. Each lesson of *Investigating College Algebra* includes a Reason and Apply section. These problems often show practical uses for the mathematics being learned. Many of the problems require students to think critically to choose the best approach to solving the problem and demonstrate persistence through trying multiple approaches. Some lessons include Mini-Investigations within the exercise sets where students can make and test conjectures.

This textbook supports a student-centered learning environment, allowing effective use of an instructor's preparatory time. If you normally write your own discovery activities, then we anticipate that having Investigations built into the textbook will allow you to spend more time focusing on your students' learning and less time on classroom

[1] *http://www.amatyc.org/Crossroads/CrsrdsXS.html*
[2] *http://www.maa.org/cupm/execsumm.pdf*

preparation. Typical student answers to the Investigations are included in the *Instructor Resources,* as are solutions to all the problems in the textbook, since many of these exercises take some time to complete.

The student support package is also designed to let you spend more classroom time helping students understand college algebra and less time on tasks such as demonstrating calculator usage and reviewing high school algebra.

If we have succeeded, this textbook not only will help your students learn College Algebra, it will help you give your students an appreciation that mathematics can explain the world around them and the confidence to explore their world using mathematics.

Instructor Support Materials

The **Instructor Resources,** available in print, on CD, and on the Web, provide lesson-by-lesson support for *Investigating College Algebra with Technology.* Included in this package are:

- detailed teaching strategies based on the authors' experience and extensive classroom testing
- complete solutions to exercises and investigations
- in-class demonstrations using The Geometer's Sketchpad® and Fathom Dynamic Data™ software

The **Test Generator,** developed especially for use with *Interactive College Algebra with Technology,* is extremely versatile. It features essay and fill-in-the-blank as well as multiple-choice and true–false questions; dynamic test questions in which variables may be changed to create unique assessments; the option to write new test questions; and complete, ready-to-print quizzes and tests for each textbook chapter.

Visit the ***Investigating College Algebra with Technology Resource Center*** at www.keycollege.com/InvestigatingCA .

Acknowledgments

A project of this magnitude involves a vast number of individuals to guide it to successful completion. The authors and publisher wish to thank the following reviewers for their invaluable feedback:

Linda Bolte, Eastern Washington University
Jodi Cotten, Westchester Community College, New York
Julie DePree University of New Mexico-Valencia
Mark Firmin, University of New Orleans
Mehri Hagar, Moorpark College, California
Christopher N. Hay-Jahans, University of Alaska Southeast
Judy B. Kidd, James Madison University, Virginia
Bonnie Oppenheimer, Mississippi University for Women
David Platt, Front Range Community College, Colorado
Pamela G. Powell, University of Texas at Austin
Jeff Reitz, Garrett College, Maryland
Nicole Sifford, Three Rivers Community College, Missouri
Pamela Smith, Fort Lewis College, Colorado
Peggy Tibbs, Weatherford College, Texas
Kevin Wheeler, Three Rivers Community College, Missouri
Martha J. Zimmerman, University of Louisville

And the following consultants and others from whose expertise we benefited:

Christian Aviles-Scott, Consultant in Educational Mathematics
Jim Bohan, Technology Consultant
Larry Copes, Consultant in Educational Mathematics
Jane Ries Cushman, Consultant in Educational Mathematics
George Rhys, College of the Canyons, California
Mariano Rodrigues, Rhode Island College
Charles Vonder Embse, Central Michigan University

The authors are also indebted to many people at Key College Publishing. We appreciate the vision and support of Richard Bonacci and Mike Simpson in bringing an investigative algebra textbook to the college market. Steve Rasmussen's commitment to quality innovative mathematics publishing made this project possible. The project took shape through the creative skills of the designer, Marilyn Perry, and our production editor, Laura Ryan. We were very fortunate to have a wonderful development editor, Erika Shaffer, who helped make this work a joy.

We would like to especially acknowledge the important work of Jerald Murdock and Ellen and Eric Kamischke in integrating data analysis and technology in the teaching of algebra, and making this approach accessible to students.

We also owe thanks to our colleagues and students at Aquinas College and Henry Ford Community College. They help keep us grounded in the realities of the college mathematics classroom. Finally, thanks to our families, who remind us there is life outside the mathematics classroom.

CHAPTER

1 Problem Solving

Dr. Wayne Daniel, a retired physicist and puzzle expert, designed this interlocking wooden puzzle. On the outside is an icosahedron, inside that a dodecahedron, inside that a cube, inside that a tetrahedron, and at the core, a tiny tetrahedron. Each form is itself a puzzle that must be assembled from interlocking pieces. When dreaming up a puzzle, says Dr. Daniel, "There are two problems to be solved: how to design it geometrically, then how to actually make it." It's an art that combines the hands and the mind.

A Puzzle Finally Makes the "Cosmic Figures" Fit, by Margaret Wertheim, *The New York Times,* May 10, 2005

OBJECTIVES

In this chapter you will

- solve problems both on your own and as a member of a group
- practice strategies for organizing information before you solve a problem
- use pictures, diagrams, and graphs as problem-solving tools
- learn a four-step process for solving problems with symbolic algebra
- create, interpret, and compare graphs of data sets
- calculate numerical measures that will help you understand and interpret data sets
- make conclusions about a data set and compare it to other data sets based on graphs and numerical values

LESSON

1.1 Pictures, Graphs, and Diagrams

As information becomes ever more quantitative and as society relies increasingly on computers and the data they produce, an innumerate citizen today is as vulnerable as the illiterate peasant of Gutenberg's time.

LYNN ARTHUR STEEN

MORE BACKGROUND

Before you begin this lesson, you should be familiar with these concepts:

- coordinate graphing
- slope
- operations with negative numbers

A short review of these concepts can be found at www.keycollege.com/online .

Many problems in this textbook ask you to look at situations in new and different ways. Often, the answer to a problem—or even the first steps toward solving the problem—will not be immediately obvious. You may even feel a little panicky if you read a problem and do not know immediately how to begin to solve it. This chapter offers some strategies to help you get started as you approach new problems.

Problem solving is a set of skills that develop with practice. We know that people think about problems and collect their thoughts in different ways. For some people, drawing a picture is a good way to begin to think about a new problem. Talking about the problem also helps many people to collect their thoughts. As you gain more experience and work collaboratively with others, you'll see many ways to approach problems, and you'll learn more ways to get started. Although some of the problems in this chapter are fictitious, they give you the chance to practice skills you will use throughout this book and in other fields of study.

This first lesson focuses on using a sketch, graph, or diagram to help solve a problem.

EXAMPLE A

The Candy Dish Problem

In a large family there is a candy dish filled with individually wrapped peppermint candies. Sometimes, during the night, family members sneak downstairs to grab handfuls of candy. One night, dinner was not very filling, so everyone in the family wanted a late-night snack.

Shortly after the family had gone upstairs to bed, the father went back downstairs and grabbed one-sixth of the pieces of candy in the dish. Then, the mother went down and took one-fifth of the candies remaining. A short while later, the oldest daughter came down and took one-fourth of the remaining candies. Then the oldest son came down and took one-third of the remaining pieces. Still later, the second son took half the remaining candy. Finally, the youngest daughter came downstairs and found only three pieces of candy in the dish.

How many pieces of candy were originally in the dish?

▶ *Solution*

How might you begin to solve this problem? First you need to understand it. A sketch or model is useful.

When you first read the problem, you may have had no idea how to begin to solve it. Your first impulse may have been to write an equation. However, the sketch may suggest other, simpler strategies.

Using pictures is one way to visualize a problem. Sometimes it helps to use actual objects and act out the problem. For instance, you could use pennies to represent the pieces of candy in Example A. When you act out a problem, record positions and quantities on paper as you solve the problem so you can recall your steps.

One way to solve this problem is to start with the fact that the youngest daughter finds three candies and move backward. If the youngest son took half the candy and left three pieces, then he must have found six candies. This in turn tells us that the oldest son must have found nine candies, since he took one-third and left six. Using similar reasoning, we can work our way up to find out how much candy the father found in the dish.

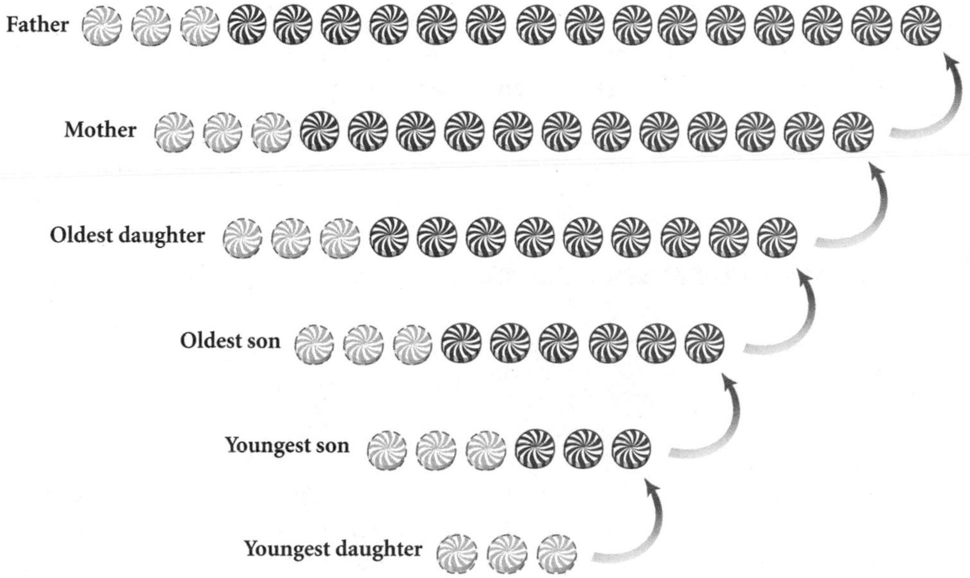

Problem solving often requires a group effort. Because different people have different approaches to solving problems, working in groups gives you the opportunity to hear and see different strategies. Sometimes group members can divide the work according to each person's strengths and expertise; at other times it's more productive for everyone to do the same task and then compare results. Each time you work in a group, decide how to assign tasks so that each person has a productive role. The next investigation will give you a chance to work in a group and practice some problem-solving strategies.

Paul Erdos

MathematicsConnection Paul Erdos (1913–1996), a prominent Hungarian mathematician, was famous as a solver of problems. He was also famous for collaborating with others in his problem solving. This collaboration resulted in a phenomenally successful career during which Erdos was credited with authorship of about 1500 academic papers. Mathematicians often identify themselves with *Erdos numbers,* which indicate an author's degree of separation from the work of Paul Erdos. For example, those who co-authored papers with Erdos have an Erdos number of one. Mathematicians who co-author a paper with someone who has an Erdos number of one achieve an Erdos number of two. If you collaborate with someone with an Erdos number of two, you will then have an Erdos number of three. A list of mathematicians with Erdos numbers of one or two is maintained by Professor Jerry Grossman of Oakland University in Rochester, Michigan. In all, Erdos co-authored papers with 502 people, more successful collaborations than any other mathematician has yet achieved. Nowadays, collaboration in mathematics is the norm. The old image of the mathematician toiling alone is no longer valid, as more than half of all published mathematical papers have two or more co-authors.

Investigation
Camel Crossing the Desert

A camel rests beside a pile of 3000 bananas at the edge of a 1000-mile-wide desert. He plans to travel across the desert, transporting as many bananas as possible to the other side. He can carry as many as 1000 bananas at any given time, but he must eat one banana every mile.

What is the maximum number of bananas the camel can transport across the desert? How does he do it? Work as a group, and prepare a written or visual solution that includes a description of your group's problem-solving strategies.

ScienceConnection In real life camels eat plant vegetation, not bananas, and can go only a few days without water. The camel's humps store fat, not water. Camels do not store any more water than other mammals do—the water is simply absorbed more slowly in their stomachs.

Pictures as problem-solving tools in mathematics are not limited to diagrams like those in Example A. Coordinate graphs are some of the most important problem-solving pictures in mathematics.

EXAMPLE B A line passes through the point (4, 7) and has slope $\frac{3}{5}$. Find another point on the same line.

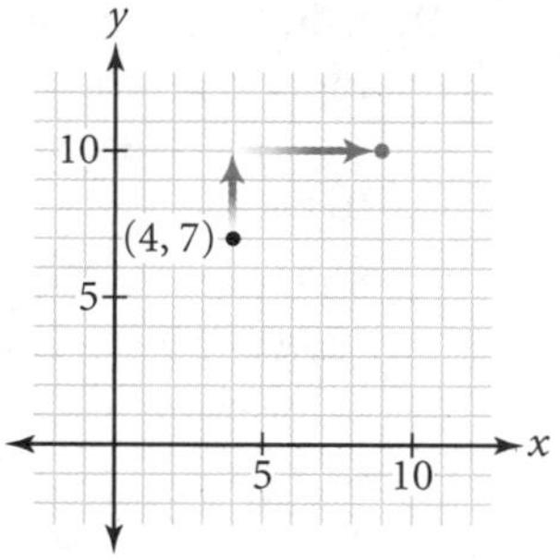

► Solution You could use a formula for slope and solve for an unknown point, but drawing a graph may be simpler.

Plot the point (4, 7). Recall that slope is $\frac{\text{change in } y}{\text{change in } x}$, and move from (4, 7) according to the slope, $\frac{3}{5}$. This means that as x increases by 5, y increases by 3. One possible point, (9, 10), is shown.

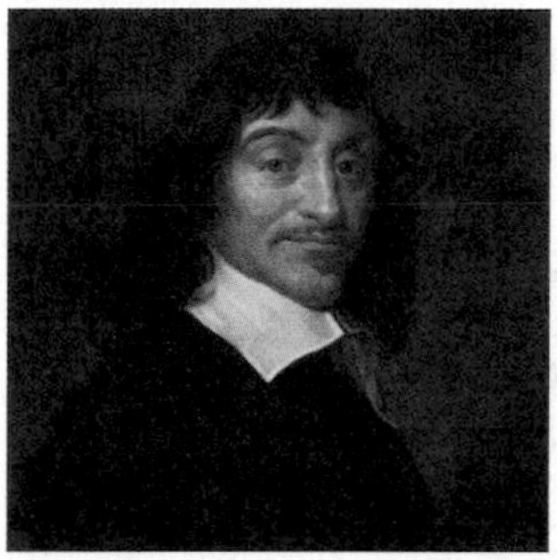
René Descarte

MathematicsConnection Coordinate graphs are also called *Cartesian graphs* after the French mathematician and philosopher René Descartes (1596–1650). Descartes was not the first to use coordinate graphs, but he was the first to publish his work using two-dimensional graphs with a horizontal axis, a vertical axis, and an origin. Descartes' goal was to apply algebra to geometry, which today is called *analytic geometry.* Analytic geometry in turn laid the foundations for much of modern mathematics, including calculus.

Although pictures and diagrams are the focus of this lesson, problem solving requires a variety of strategies. As you work on the exercises, don't limit yourself. You are always welcome to use any and all of the strategies that you know.

EXERCISES

► Practice Your Skills

1. Here is a problem similar to that in Example A. However, this problem has an additional character, the grandfather.

In a large family there is a candy dish filled with individually wrapped peppermint candies. Sometimes, during the night, family members sneak downstairs to grab handfuls of candy. One night, dinner was not very filling, so everyone in the family was quite hungry.

Before leaving the kitchen, the grandfather took one-seventh of the pieces of candy in the dish. Shortly after the family had gone upstairs to bed, the father went back downstairs and grabbed one-sixth of the pieces of candy in the dish. Then, the mother went down and took one-fifth of the candies remaining. A short while later, the oldest daughter came down and took one-fourth of the remaining candies. Then, the oldest son came down and took one-third of the remaining pieces. Still later, the second son took half the remaining candy. Finally, the youngest child came downstairs and found only _?_ pieces of candy in the dish. How many pieces of candy were originally in the dish?

a. Fill in the blank with a number that will allow this problem to have a whole-number solution, that is, one in which no pieces of candy must be broken into parts.

b. Give two more examples of numbers that could be the amount of candy the youngest daughter finds. How will changing the numbers at the end change the solution to the problem? Explain your answer.

c. What if the family also had a grandmother, and she went first, taking one-eighth of the candy? Would your number from 1a still work? Explain why or why not.

2. Symbolic representations can be very helpful in analyzing general situations or in answering more general questions, such as, "What is the relationship between the amount of candy at the end and the amount that the father takes?" Use the situation from Question 1 to create symbolic representations as specified here.

   a. Let the variable n represent the number of pieces of candy the youngest daughter finds. Write a symbolic expression using the variable n for the corresponding number of pieces of candy the father finds.

   b. Let the variable p represent the number of pieces of candy the grandfather finds. Give a symbolic expression using the variable p for the corresponding number of pieces of candy the youngest daughter finds.

   c. Sketch a graph to represent the relationship between p and n. Use the x-axis (horizontal axis) for the number of pieces the grandfather finds and the y-axis (vertical axis) for the number of pieces the youngest daughter finds.

   d. What is the slope of the graph?

   e. Explain why the value of the slope makes sense in this situation.

3. Find the slope of each line.

   a.

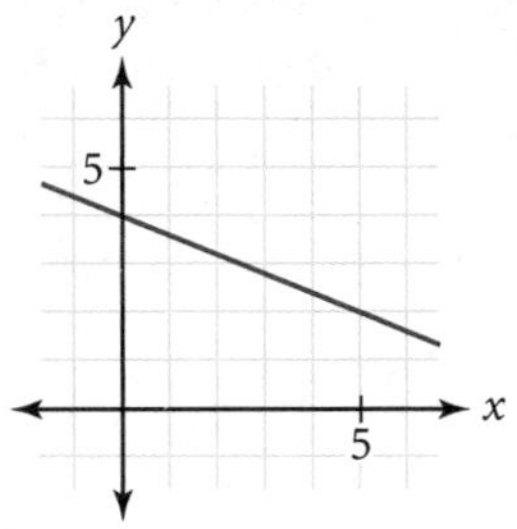

   b.

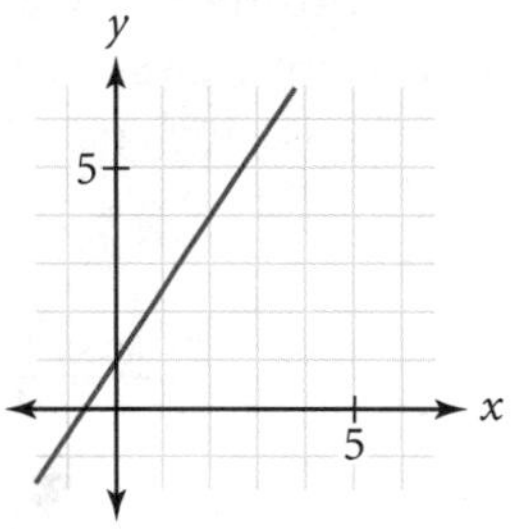

4. A line passes through the point (4, 7) and has slope $\frac{3}{5}$. Find two more points on the line other than (9, 10), which was found in Example B.

5. First, explain how this graph helps you solve these proportion problems, and then solve them.

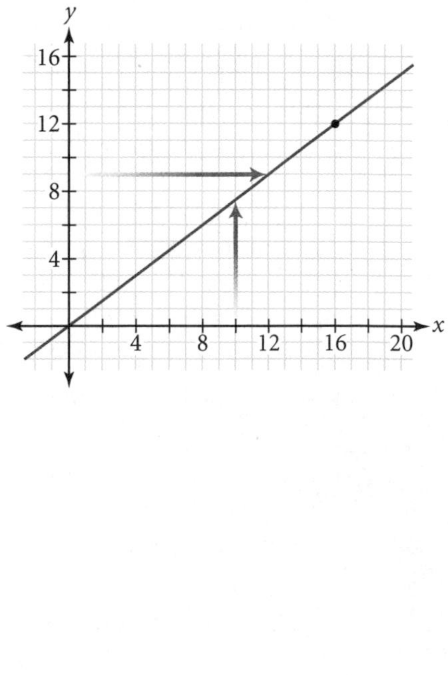

   a. $\frac{12}{16} = \frac{9}{a}$

   b. $\frac{12}{16} = \frac{b}{10}$

6. The following problem appears in *Problems for the Quickening of the Mind*, a collection of problems compiled by the mathematician Alcuin of York (735–804 C.E.). Describe a strategy for solving this problem, but do not actually solve the problem.

   A wolf, a goat, and a cabbage must be moved across a river in a boat that can hold only one besides the ferryman. The goat may not be left unattended with the cabbage. Similarly, the wolf may not be left alone with the goat. How can the ferryman carry them across so that the goat shall not eat the cabbage, nor the wolf the goat?

Reason and Apply

7. Find the slope of the line that passes through each pair of points.

a. (2, 5) and (7, 10) **b.** (3, −1) and (8, 7)

c. (−2, 3) and (2, −6) **d.** (3, 3) and (−5, −2)

8. For each scenario draw and label a diagram. Do not actually solve the problem.

a. A 25-foot ladder leans against the wall of a building. The ladder's foot is 10 feet from the wall. How high does the ladder reach?

b. A cylindrical tank with diameter 60 cm and length 150 cm rests on its side. The fluid in the tank leaks from a valve that is 20 cm off the ground. When no more fluid leaks out, what is the volume of the remaining fluid?

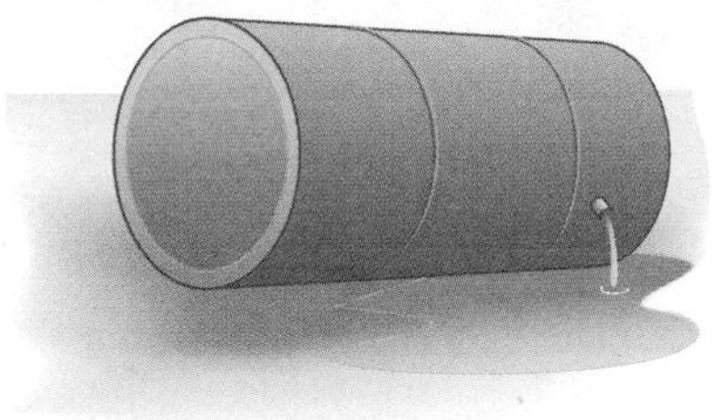

c. Five sales representatives each send letters to every other representative. How many letters do they send?

9. Use this graph to estimate these conversions of Fahrenheit temperatures to degrees Celsius and vice versa.

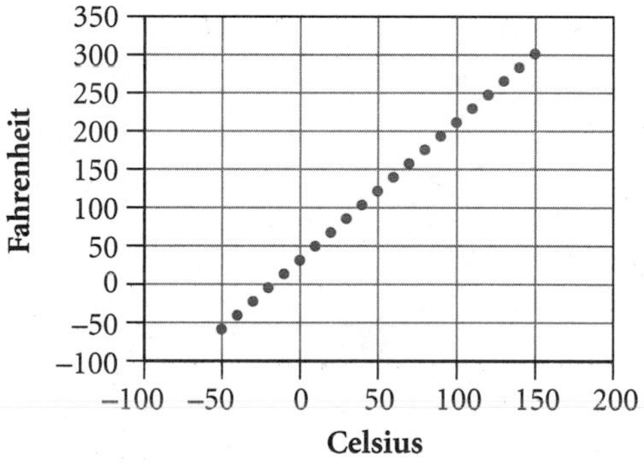

a. 0°C = _?_ F

b. 10°C = _?_ F

c. 212°F = _?_ C

d. −13°F = _?_ C

e. What is the slope of this line? Explain the real-world meaning of the slope.

10. You can use diagrams to represent algebraic expressions. Explain how this rectangle diagram demonstrates that $(x + 2)(x + 3)$ is equivalent to $x^2 + 2x + 3x + 6$.

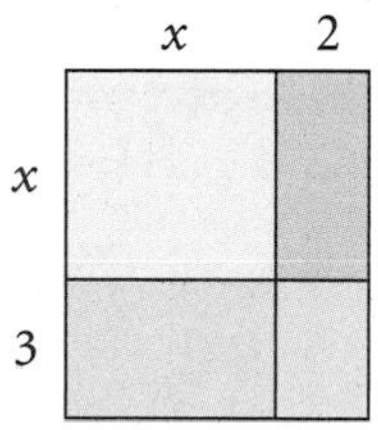

11. Draw a rectangle diagram to represent each product. Use the diagrams to expand each product.

a. $(x + 4)(x + 7)$ **b.** $(x + 5)^2$

c. $(x + 2)(y + 6)$ **d.** $(x + 3)(x - 1)$

12. Solve the problem in Exercise 6. Explain your solution.

PROJECT

UNIT CONVERSIONS

In Exercise 9 of Lesson 1.1, you used a graph to convert Fahrenheit temperatures to Celsius and vice versa. You can create similar unit-conversion graphs for other units of measurement.

Select a pair of units that measure the same characteristic, such as centimeters and inches for length, or fluid ounces and milliliters for volume. (Be sure to select units for which you can collect data.) Measure several items using both units, and make a graph. Find a line that fits your data, calculate the slope, and explain the real-world meaning of the slope.

Research the actual conversion factor for the two units you selected. You can do some of your research with the Internet links at www.keycollege.com/online . How accurate is your graphical model?

Your project should include:

- a table of the data you collected
- a graph of your data with a line of fit
- an explanation of the slope of your line
- a comparison of your model to the actual unit-conversion factor

LESSON

1.2

Symbolic Representation

A mathematician is a machine for turning coffee into theorems.

PAUL ERDOS

MORE BACKGROUND

Before you begin this lesson, you should be familiar with these concepts and processes:

- absolute value
- solving a system of three equations with three variables
- using the distributive property of multiplication over addition

A short review of these concepts can be found at www.keycollege.com/online .

The customer-service team had planned to double the number of calls they answered on the second day, but they exceeded that by three dozen. Seventy-five dozen customer-service calls in two days set a new record.

You can translate the preceding paragraph into an algebraic equation. Although you don't know how many calls were answered each day, an equation will help you figure it out.

For many years problems like this were solved without equations. Then, around the 17th century, the development of symbolic algebra made writing equations and finding solutions much simpler. Verbal statements could be translated into symbols by representing unknown quantities with letters, called variables, and converting the rest of the sentence into numbers and operations. (You will actually translate this telephone-call problem into algebraic notation when you do the exercises.)

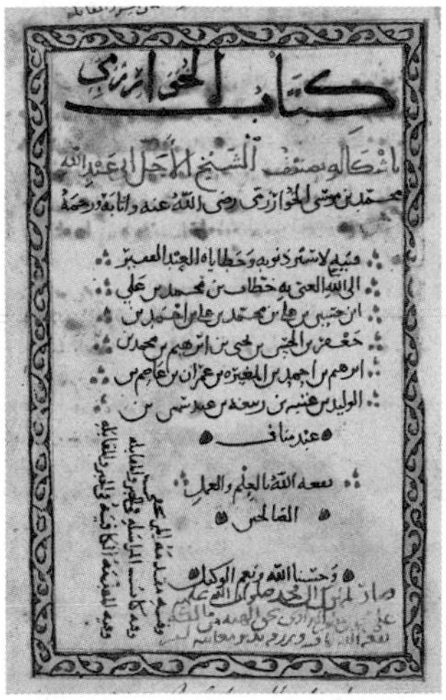

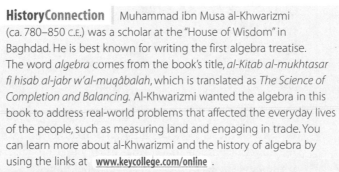

HistoryConnection Muhammad ibn Musa al-Khwarizmi (ca. 780–850 C.E.) was a scholar at the "House of Wisdom" in Baghdad. He is best known for writing the first algebra treatise. The word *algebra* comes from the book's title, *al-Kitab al-mukhtasar fi hisab al-jabr w'al-muqâbalah*, which is translated as *The Science of Completion and Balancing.* Al-Khwarizmi wanted the algebra in this book to address real-world problems that affected the everyday lives of the people, such as measuring land and engaging in trade. You can learn more about al-Khwarizmi and the history of algebra by using the links at www.keycollege.com/online .

This postage stamp depicts al-Khwarizmi and was issued in the former Soviet Union. At left is the title page of al-Khwarizmi's treatise on algebra.

Language is very complex and subtle, designed for general descriptions and qualitative communication. For quantitative communication, the ability to translate words into symbols can be a very helpful problem-solving skill. The symbols stand for numbers that vary or remain constant, that are given in the problem, or that are unknown. In applications the numbers quantify data such as time, weight, or position.

EXAMPLE A

Jason is using his computer to design small square stickers. He prints the stickers on uncut label paper. He is considering changing the dimensions of his stickers. If he halves the side lengths of the stickers, how will it affect the number of stickers that will fit on each page of label paper?

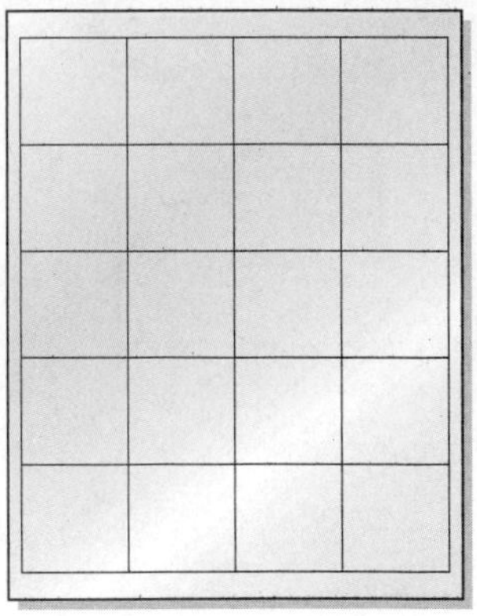

▶ Solution

First, list the unknown quantities, and assign a variable to each.

Let s represent the original side length of each sticker.

Because the side lengths of the stickers are being halved, let $\frac{s}{2}$ represent the side length of each new sticker.

The amount of space that a sticker takes up is the *area* of the sticker. Let A represent the original area taken up by one sticker.

Let B represent the area taken up by each new sticker.

Second, write equations for the problem.

To find how much space the old stickers took up, use the formula for the area of a square, with a side length of s:

$$A = s^2$$

The area of each new sticker, with side lengths of $\frac{s}{2}$, can be found using the same formula

$$B = \left(\frac{s}{2}\right)^2$$

Squaring the expression inside the parentheses shows that

$$B = \frac{s^2}{4}$$

Last, interpret your solution. Because $B = \frac{s^2}{4}$ and $A = s^2$, we know that B is one-fourth as large as A. If Jason halves the dimensions, then each sticker will take up one-fourth as much space. That means that he will be able to fit four times as many of the new stickers on each page of label paper.

EXAMPLE B

Evelyn needs four liters of 30% antiseptic solution. She has two other solutions on hand: three liters of 28% solution and two liters of 40% solution. Explain how she can mix the two solutions to get two liters of 30% solution.

▶ Solution

Although other approaches, such as trial-and-error, might be used to solve this problem, symbolic algebra provides a very efficient method. First, list the unknown quantities and assign a variable to each.

The unknown quantities are the amounts of each solution that will be mixed. Let x stand for the amount of 28% solution. For the amount of 40% solution, it would be possible either to use a new variable or to use what is known about the relationship between the amounts of solutions. Because we know that in the end there will be four liters of solution, it makes sense to use the expression $4 - x$ for the amount of 40% solution.

An antiseptic solution consists of the active ingredient—the concentrated antiseptic—and inert ingredients, such as water. The result of the mixing will be a four-liter solution that includes 30% concentrated antiseptic. That means that the amount of actual antiseptic in the final solution is 30% of four liters, or (.3)(4).

This information allows us to write an equation to describe the relationships.

$$.28x + .4(4 - x) = (.3)(4)$$

Next, solve for x and $4 - x$:

$$.28x + 1.6 - .4x = 1.2$$

$$x = \frac{10}{3} \text{ and } 4 - x = \frac{2}{3}$$

Now that we have solved for x and $4 - x$, how do we interpret this answer? Because x stands for the amount of 28% solution, we need $\frac{10}{3}$ of a liter of 28% solution. Unfortunately, in the original problem it was stated that Evelyn had only three liters of 28% solution, so there is no way she can combine these solutions to get the desired outcome.

Although symbolic algebra seems to have failed in Example B, it did help you recognize that no correct answer existed. If a problem has a mathematical answer, it usually can be obtained with symbolic algebra. Therefore, as you approach problems of a descriptive nature, it is often helpful to translate the problem into variables, expressions, and equations.

EXAMPLE C

Three friends went to the gym to work out. None of the friends would tell how much weight he could bench-press, but each hinted at his friends' bench-press weights. Chen said that Juan and Lou averaged 168 lb. Juan said that Chen bench-pressed 9 lbs more than Lou. Lou said that eight times Juan's amount equals seven times Chen's amount. Find how much each friend could bench-press.

► Solution

First, list the unknown quantities, and assign a variable to each.

Let C represent Chen's weight.
Let J represent Juan's weight.
Let L represent Lou's weight.

Second, write equations from the problem.

$\frac{J + L}{2} = 168$	Chen's statement translated into an algebraic equation. Call this Equation 1.
$C - L = 9$	Juan's statement as Equation 2.
$8J = 7C$	Lou's statement as Equation 3.

Third, solve the equations to find the values of the variables.

$J + L = 336$	Multiply both sides of Equation 1 by 2.
$C - L = 9$	Equation 2.
$J + C = 345$	Add the equations.
$7J + 7C = 2415$	Multiply both sides of the sum by 7.
$7J + 8J = 2415$	Equation 3 allows you to substitute $8J$ for $7C$.
$15J = 2415$	Add like terms.
$J = 161$	Divide both sides by 15.
$8(161) = 7C$	Substitute 161 for J in Equation 3.
$C = 184$	Solve for C.
$184 - L = 9$	Substitute 184 for C in Equation 2.
$L = 175$	Solve for L.

Last, interpret your solution. Chen bench-presses 184 lb, Juan bench-presses 161 lb, and Lou bench-presses 175 lb.

Notice that the previous examples used a four-step algebra solution process. The next investigation gives you a chance to try these four steps on your own or with a group.

Investigation
Problems, Problems, Problems

Select one or more of the problems listed here. First, use a sketch or other model to make sense of the problem. Then, use these four steps to find an algebraic solution.

Step 1 List the unknown quantities, and assign a variable to each.

Step 2 Write one or more equations that relate the unknown quantities to conditions of the problem.

Step 3 Solve the equations to find the value of each variable.

Step 4 Interpret your solution according to the context of the problem.

When you finish, write a paragraph answering these questions: Which of the four algebraic solution steps was hardest for you? Why?

Problem 1

When Adam and his sister Megan arrive at a basketball game, they see that there is 1 player for every 4 fans. Right behind them come 30 more hometown fans, and Megan notices that the ratio is now 2 hometown fans to each visiting fan. However, behind the extra hometown fans come 30 more visiting fans. There are now 4 visiting fans for every 3 hometown fans. What is the final ratio of players to fans?

Problem 2

Abdul and William decide to start a house-painting business. So far, they have done some preliminary work, going door to door and making phone calls. Abdul has spent

10 hours getting the business started, and William has spent 6 hours. Celia finds out what they are doing and wants to join their business venture. She is willing to contribute \$120 to repay Abdul and William for their time. To be fair, how much money should she give each of them?

Problem 3

A caterer claims that a birthday cake will serve either 20 children or 15 adults. At Tina's party there are 12 children and 7 adults. Is there enough cake?

Before you attempt to solve a problem, first create a sketch or other model of the situation. Then, if algebra seems to be an appropriate tool for solving the problem, use the four algebraic solution steps to help you organize information, find a solution, and interpret the final answer. As you do the exercises in this lesson, refer back to these four steps and practice using them.

EXERCISES

Practice Your Skills

1. Explain what you would do to change the first equation to the second.

a. $a + 12 = 47$
$a = 35$

b. $5b = 24$
$b = 4.8$

c. $-18 + c = 28$
$c = 46$

d. $\frac{d}{-3} = 4.5$
$d = -13.5$

2. Which equation would help you solve this problem?

Each member of the committee made 3 copies of the letter to the senator. Adding these to the 5 original letters, there is now a total of 32 letters. How many members does the committee have?

a. $5 + c = 32$ **b.** $3 + 5c = 32$ **c.** $5 + 3c = 32$

3. Solve each equation.

a. $5 + c = 32$ **b.** $3 + 5c = 32$ **c.** $5 + 3c = 32$

4. Which equation would help you solve this problem?

The customer-service team had planned to double the number of calls they answered on the second day, but they exceeded that by three dozen. Seventy-five dozen customer-service calls in two days set a new record.

a. $x + 3 = 75$ **b.** $x + 2x + 3 = 75$ **c.** $2x + 3 = 75$

5. Solve each equation.

a. $x + 3 = 75$ **b.** $x + 2x + 3 = 75$ **c.** $2x + 3 = 75$

Reason and Apply

6. Here are a problem and three related equations.

Anita buys 6 large beads and 20 small beads to make a necklace. Iwanda buys 4 large and 25 small beads for her necklace. Jill selects 8 large and 16 small beads to make an ankle bracelet. Without tax, Anita pays $2.70, and Iwanda pays $2.85. How much will Jill pay?

$6L + 20S = 270$ (Equation 1)

$4L + 25S = 285$ (Equation 2)

$8L + 16S = J$ (Equation 3)

a. What do the variables L, S, and J represent?

b. What are the units of L, S, and J?

c. What does Equation 1 represent?

7. Follow these steps to solve Exercise 6.

a. Multiply Equation 1 by negative two.

b. Multiply Equation 2 by three.

c. Add the resulting equations from 7a and b.

d. Solve the equation in 7c for S. Interpret the real-world meaning of this solution.

e. Use the value of S to find the value of L. Interpret the real-world meaning of the value of L.

f. Use the values of S and L to find the value of J. Interpret this solution.

8. The following problem appears in *Liber Abaci* (1202), or *Book of Calculations*, by the Italian mathematician Leonardo Fibonacci (1170–1250).

If *A* gets from *B* seven *denarii* [ancient Roman coins], then *A*'s sum is fivefold *B*'s. If *B* gets from *A* five *denarii*, then *B*'s sum is sevenfold *A*'s. How much has each?

$a + 7 = 5\,(b - 7)$ (Equation 1)

$b + 5 = 7\,(a - 5)$ (Equation 2)

a. What does a represent?

b. What does b represent?

c. Explain Equation 1 with words.

d. Explain Equation 2 with words.

9. Use Equations 1 and 2 from Exercise 8.

a. Explain how to get

$b + 5 = 7((5b - 42) - 5)$

b. Solve the equation in 9a for b.

c. Use your answer from 9b to find the value of a.

d. Use the context of Exercise 8 to interpret the values of a and b.

10. According to Mrs. Randolph's will, each of her great-grandchildren who lived in Georgia received \$700 more than each of her great-grandchildren who lived in Florida. In all, \$206,100 was divided among 36 great-grandchildren. The Georgian great-grandchildren decide that the will wasn't really fair, so they each contribute \$175 to be divided among the Floridian great-grandchildren. If all great-grandchildren now have equal shares, how many great-grandchildren live in Georgia?

a. List the unknown quantities, and assign a variable to each.

b. Write one or more equations that relate the variables to conditions of the problem.

c. Solve the equations to find the value of each variable.

d. Interpret the value of each variable according to the context of the problem.

Review

The following exercises test your understanding of prerequisite material for college algebra.

11. Rewrite each expression without parentheses.

a. $3(x + 7)$ **b.** $-2(6 - n)$ **c.** $x(4 - x)$

12. Rewrite each expression without parentheses.

a. $2x(x + 3)$ **b.** $-3y + 4(1 - y)$ **c.** $z(5 - z)$

13. Rewrite each expression without parentheses.

a. $3a(a + b)$ **b.** $-2(6 - n)$ **c.** $x - 4(9 - x)$

14. Substitute the given value of the variable(s) in each expression, and evaluate.

a. $47 + 3x$ when $x = 17$

b. $29 - 34n + 14m$ when $n = -1$ and $m = -24$

15. Substitute the given value of the variable(s) in each expression, and evaluate.

a. $5 + x^2$ when $x = -3$

b. $n - 5m$ when $n = 3$ and $m = -2$

16. Substitute the given value of the variable(s) in each expression and evaluate.

a. $-5x + 3y$ when $x = -2$ and $y = -4$

b. $(a + b)^2 - b^2$ when $a = 1$ and $b = 4$

LESSON

1.3 Organizing Information

MORE BACKGROUND

Before you begin this section, you should be familiar with these concepts:

- finding the volume of a rectangular solid (box)
- properties of exponents

A short review of these concepts can be found at www.keycollege.com/online.

Information is the oxygen of the modern age.

RONALD REAGAN

If one and a half chickens lay one and a half eggs in one and a half days, then how long does it take six monkeys to make nine omelets?

What sort of problem-solving strategy can you apply to a crazy problem like this? You could draw a picture or make a diagram. You could assign variables to all sorts of unknown quantities. But do you really have enough information to solve the problem? Sometimes the best strategy is to begin by organizing what you know and what you want to know. With the information organized, you may then find a way to get to the solution.

There are different methods of organizing information. One valuable technique uses the units in the problem to show how the information fits together.

EXAMPLE A

To qualify for the Labor Day 400 auto race, each driver must complete two laps of the track at an average speed of 100 mi/h. Due to some problems at the start, Naomi averages only 50 mi/h on her first lap. How fast must she go on the second lap to qualify for the race?

Sort the information into two categories: what you know and what you might need to know.

Assign variables to the quantities that you don't know.

Know	Need to Know
Speed of first lap: 50 mi/h	Speed of second lap (in mi/h): s
Average speed of both laps: 100 mi/h	Length of each lap (in mi): l
	Time for first lap (in h): t_1
	Time for second lap (in h): t_2

Next, look at the units to find connections between the pieces of information. Speed is measured in miles per hour and is therefore calculated by dividing distance by time, so you can write these equations:

$$50 \text{ mi/h} = \frac{l \text{ mi}}{t_1 \text{ h}}$$ The speed, distance, and time for the first lap

$$s \text{ mi/h} = \frac{l \text{ mi}}{t_2 \text{ h}}$$ The second lap

$$100 \text{ mi/h} = \frac{2l \text{ mi}}{(t_1 + t_2) \text{ h}}$$ The average speed needed for both laps to qualify

You can write the first and third equations as

$$t_1 \text{ h} = \frac{l \text{ mi}}{50 \text{ mi/h}} = \frac{l}{50} \text{h}$$

$$(t_1 + t_2) \text{ h} = \frac{2l \text{ mi}}{100 \text{ mi/h}} = \frac{l}{50} \text{h}$$

This means that the time for the first lap, t_1, and the time for both laps together, $(t_1 + t_2)$, are the same, and $t_2 = 0$. There is no time at all to complete the second lap.

It is not possible for Naomi to qualify for the race.

In Example A it was easy to find relationships between the known and unknown quantities because the units were miles, hours, and miles per hour. Sometimes you will need to use *common knowledge*—information that is not given in the problem—to make the intermediate connections between the units.

EXAMPLE B

How many seconds are in a calendar year?

First, identify what you know and what you want to know.

Know	Need to Know
1 year	Number of seconds

It may seem as though you don't have enough information, but consider these commonly known facts:

1 year = 365 days (non-leap year)
1 day = 24 hours
1 hour = 60 minutes
1 minute = 60 seconds

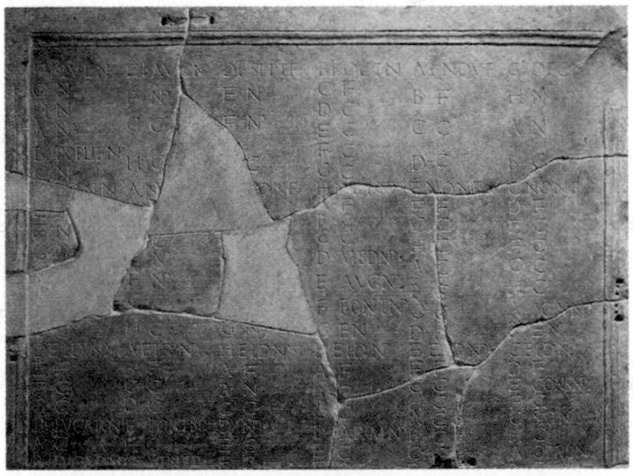

This fragment of an ancient Roman calendar shows months, days, and special events. You can learn how to read Roman calendars, or *fasti,* with an Internet link at www.keycollege.com/online .

You can write each equality as a fraction and multiply the chain of fractions such that the units are reduced to seconds.

$$1\,\cancel{\text{year}} \cdot \frac{365\,\cancel{\text{days}}}{1\,\cancel{\text{year}}} \cdot \frac{24\,\cancel{\text{hours}}}{1\,\cancel{\text{day}}} \cdot \frac{60\,\cancel{\text{minutes}}}{1\,\cancel{\text{hour}}} \cdot \frac{60\text{ seconds}}{1\,\cancel{\text{minute}}} = 31{,}536{,}000\text{ seconds}$$

There are 31,536,000 seconds in a non-leap-year calendar year.

Some problems overwhelm you with information. Identifying and categorizing what you know is always a good way to start organizing information.

EXAMPLE C

Lab assistant Jerry Thule has just finished cleaning a messy lab table and is putting the equipment back on the table when he reads a note telling him not to disturb the positions of three water samples. Not recalling the positions of the three samples, he finds these facts in the lab notes:

The water that is highest in sulfur was on one end.

The water that is highest in iron is in the Erlenmeyer flask.

The water taken from the spring was not next to the water in the bottle.

The water that is highest in calcium was to the left of the water taken from the lake.

The water in the Erlenmeyer flask, the water taken from the well, and the water that is highest in sulfur are three distinct samples.

The water in the round flask is not highest in calcium.

Determine which water sample goes where.

▶ Solution

Organize the facts into categories. (This is the first step in actually determining which sample goes where. You will finish the problem when you do the exercises.) Information is given about the types of containers, the sources of the water, the chemicals found in the samples, and the positions of the samples on the table. You can find three options for each category:

Containers: round flask (C1), Erlenmeyer flask (C2), bottle (C3)

Sources: spring (S1), lake (S2), well (S3)

Chemicals: sulfur (Ch1) , iron (Ch2), calcium (Ch3)

Positions: left (P1), center (P2), right (P3)

Now that the information is categorized, you can make an organized list of all possible situations. One quick way to make an organized list is with a tree diagram like the one on the next page:

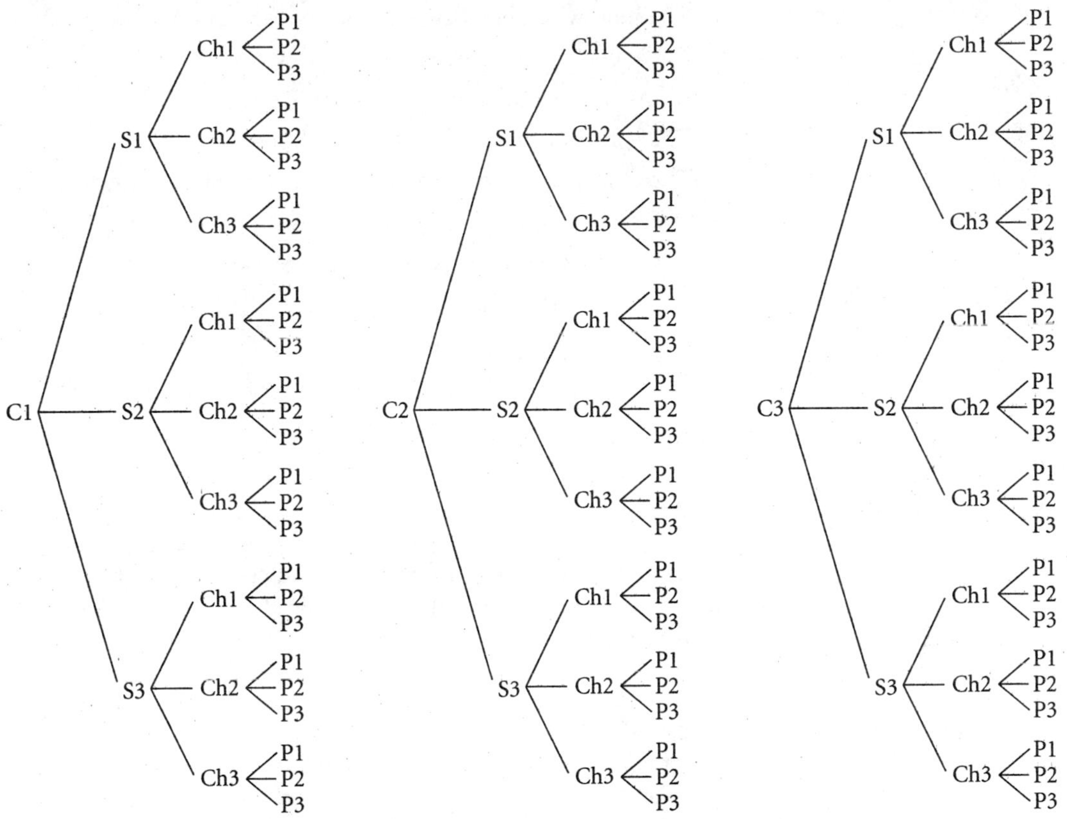

By looking at the tree, you can see that there are 81 possible combinations of container, source, chemical, and position. Although there are many possible combinations, it will be relatively easy to use the clues to eliminate those that will not work.

You will finish this problem in Exercise 6.

Use this investigation as an opportunity to practice categorizing and organizing information as you did in Example C.

Investigation
Who Owns the Zebra?

There are five houses on one side of Birch Street, each a different color. The homeowners each drive a different kind of car, and each has a different kind of pet. The owners all read different newspapers, and each grows a different plant in the garden.

- The family with the station wagon lives in the red house.
- The owner of the SUV has a dog.
- The family with the van reads the *Gazette.*
- The green house is immediately to the left of the white house.
- The *Chronicle* is delivered to the green house.

- The man who plants zucchini keeps birds.
- The owners of the yellow house grow corn.
- In the middle house they read the *Times.*
- The compact car parks at the first house.
- The family that grows eggplant lives in the house next to the house with cats.
- The people who live in the house next to the house where they have a horse, grow corn.
- The woman who grows beets receives the *Daily News.*
- The owner of the sports car grows okra.
- The family with the compact car lives next to the blue house.
- They read the *Bulletin* in the house next to the house where they grow eggplant.

Who owns the zebra?

Organizing the known information and clarifying what you need to find out is a very useful strategy. Whether it involves simply keeping track of units or sorting out masses of information, an organized plan is essential to finding a solution efficiently.

ScienceConnection Piecing together clues to understand a bigger picture was a major problem-solving strategy used in the Human Genome Project. In 1990 researchers from around the world began working cooperatively to map and sequence the human genome—the complete set of more than 3 billion human DNA base pairs. In April 2003, despite the overwhelming amount of information that had had to be organized, the International Human Genome Sequencing Consortium announced that the Human Genome Project was complete. By understanding the organization and function of human DNA, the researchers hope to improve human health and create guidelines for the ethical use of genetic knowledge.

EXERCISES

Practice Your Skills

1. Use units to help you find the missing information.

a. How many seconds would it take to travel 15 ft at a rate of 3.5 ft/s (feet per second)?

b. How many centimeters are in 25 ft? There are exactly 2.54 cm/in. (centimeters per inch) and 12 in./ft (inches per foot.)

c. How many miles could you drive on 15 gal of gasoline in a car that got 32 mi/gal?

d. How many hours would it take to drive 424 mi at an average speed of 65 mi/h?

e. After traveling 4.5 h at 65 mi/h, how far have you driven?

f. A car's gas tank holds 10.2 gal. The car has an estimated highway mileage of 37 mi/gal. How far can the car go on one tank of gas?

2. Emily and Alejandro are part of a math marathon team the members of which take turns solving math problems for four hours each day. On Monday Emily worked for three hours and Alejandro for one hour. On Tuesday Emily worked for two hours and Alejandro for two hours. On Monday they collectively solved 139 problems, and on Tuesday they solved 130 problems. Find the average problem-solving rates for Emily and for Alejandro.

a. Identify the unknown quantities and assign variables. What are the units of each variable?

b. What does the equation $3e + 1a = 139$ represent?

c. Write an equation for Tuesday.

d. Which of these ordered pairs (e, a) is a solution to the problem?

i. (37, 28) **ii.** (31, 38) **iii.** (28, 31) **iv.** (27, 23)

e. Interpret the solution from 2d according to the context of the problem.

3. To qualify for the Labor Day 400 auto race, each driver must complete two laps of the track at an average speed of 100 mi/hr. Benjamin averages only 75 mi/hr on his first lap. How fast must he go on the second lap to qualify for the race?

4. **APPLICATION** Alyse earns $15.40 per hour, and she earns time and a half for working after 8 P.M. Last week she worked 35 hours and earned $600.60. How many hours did she work after 8 P.M.?

5. The dimensions used to measure length, area, and volume are related by multiplication and division. Find the information for these rectangular boxes. Include the units in your solution.

a. A box has volume 486 in.^{3} and height 9 in. Find the area of the base.

b. A box has base area 3.60 m^2 and height 0.40 m. Find the volume.

c. A box has base area 2.40 ft^2 and volume 2.88 ft^3. Find the height.

d. A box has volume 12,960 cm^3, height 18 cm, and length 30 cm. Find the width.

6. Lab assistant Jerry Thule from Example C has just finished cleaning a messy lab table and is putting the equipment back on the table when he reads a note telling him not to disturb the positions of three water samples. Not knowing the correct positions of the three samples, he finds these facts in the lab notes:

- The water that is highest in sulfur was on one end.
- The water that is highest in iron is in the Erlenmeyer flask.
- The water taken from the spring was not next to the water in the bottle.
- The water that is highest in calcium was to the left of the water taken from the lake.
- The water in the Erlenmeyer flask, the water taken from the well, and the water that is highest in sulfur are three distinct samples.
- The water in the round flask is not highest in calcium.

Determine which water sample goes where. Identify each sample by its container, source, chemical, and position.

7. Al and Belinda are taking their three children, Cassie, David, and Emily, on a long car trip in their minivan. The van seats two in the front seats, two in the middle row of seats, and three in the back row of seats. They want to decide the best seating arrangement for the family.

- The only two licensed drivers are Al and Belinda. Therefore, one of them must sit in the driver's seat.
- Cassie sometimes gets queasy when she sits in the very back of the van. Therefore, Al and Belinda will not let her ride in a third-row seat.
- Because the front seats have air bags, only people over the age of 12 may ride in the front row. Cassie and David are both under 12.
- On the last car trip, David and Emily quarreled incessantly. Therefore, their parents will not let them sit next to each other this time.
- Emily wants to use the family's laptop computer to finish her homework. Because the battery does not work properly, she needs to sit near a power outlet. The van has two power outlets, one near the front and the other near the back row of seats.

Use this information to make an organized list of all possible seating arrangements. How many acceptable seating combinations does the family have?

8. **APPLICATION** Paul can paint the area of a 12-by-8-ft wall in 15 minutes. China can paint the same area in 20 minutes.

a. Fill in the blanks to complete the table.

Minutes	Area Paul Paints	Area China Paints
1		
2		
4		
8		

b. Extend the table in 8a to determine how long it would take the two of them together to paint the area of the wall.

c. In which equation does t represent how long it would take Paul and China to paint the area of the wall together? Explain your choice.

Equation i. $\frac{15t}{96} + \frac{20t}{96} = \frac{1}{96}$

Equation ii. $(96)15t + (96)20t = 96$

Equation iii. $\frac{96t}{15} + \frac{96t}{20} = 96$

d. Solve all three equations in 8c.

e. How does the first solution to the problem, from 8b, compare with the solution to the problem obtained by solving the appropriate equation?

9. Kiane has a photograph of her four cats sitting in a row. The cats are different ages, and each cat has its own favorite toy and favorite sleeping spot.

- Rocky and the 10-year-old cat would never sit next to each other for a photo.
- The cat that sleeps in the blue chair and the cat that plays with the rubber mouse are the two oldest cats.
- The cat that plays with the silk rose is the third cat in the photo.
- Sadie and the cat on Sadie's left in the photo never sleep on the furniture.

- The 8-year-old cat sleeps on the floor.
- The cat that sleeps on the sofa eats the same kind of food as the 13-year-old cat.
- Pascal likes to chase the 5-year-old cat.
- If you add the ages of the cat that sleeps in a box and the one that plays with a stuffed toy, you get the age of Winks.
- The cat that sleeps on the blue chair likes to hide the catnip ball that belongs to one of the cats sitting next to the ball in the photo.

Who plays with the catnip ball?

10. Joel is 16 years old. His cousin Rachel is 12.

a. What is the difference in their ages?

b. What is the ratio of Joel's age to Rachel's age?

c. In eight years, what will be the difference in their ages?

d. In eight years, what will be the ratio of Joel's age to Rachel's age?

11. Draw a rectangle diagram to represent each product. Use the diagrams to expand each product.

a. $(x + 1)(x + 5)$

b. $(x + 3)^2$

c. $(x + 3)(x - 3)$

IMPROVING YOUR REASONING SKILLS

Internet Access

A nationwide Internet service provider advertises "1000 hours free for 45 days" for new customers. How many hours per day do you need to be online to use 1000 hours in 45 days? Do you think it is ethical for the Internet service provider to make this offer?

LESSON

1.4

Measures of Central Tendency and Box Plots

Everybody gets so much information all day long that they lose their common sense.

GERTRUDE STEIN

In the computer age it sometimes seems as though we are bombarded with information. Newspapers, magazines, the Internet, the evening news, commercials, government bulletins, and sports publications are loaded with data and statistics. As an informed citizen, you need to be able to interpret this information to make intelligent decisions.

In the first part of this chapter, you worked on problem solving and then on analyzing information. We will now broaden this to include analyzing large amounts of information, generally called *data sets.* You will be asked to graph data sets in several different ways. You will also study some numerical measures that will help you better understand what a data set tells you. Each numerical measure can be called a **statistic.** A collection of measures, or the mathematical study of data collection and analysis, is called **statistics.**

HistoryConnection One of the earliest large-scale statistical surveys was that of the 14th-century Hawaiian king Umi. He is said to have collected all his people on a small plain, afterward called the Plain of Numbering, and to have asked each to deposit a stone in an area encircling the temple on that plain. The stones were placed in piles according to district, and the piles were located in the direction of the districts. The result showed the relative sizes of the districts' populations.

EXAMPLE A

The table below shows the round-trip commuting distance of 30 randomly chosen students at a regional state university. How far does a typical student at this university commute to class?

Student	Gender	Commuting distance (mi)	Student	Gender	Commuting distance (mi)	Student	Gender	Commuting distance (mi)
1	Female	0	11	Male	5	21	Female	10
2	Female	0	12	Female	6	22	Male	10
3	Male	0	13	Female	6	23	Female	12
4	Female	0	14	Male	6	24	Female	12
5	Male	1	15	Male	7	25	Male	14
6	Male	2	16	Male	7	26	Male	15
7	Male	2	17	Female	7	27	Male	19
8	Female	4	18	Male	8	28	Female	20
9	Male	4	19	Male	9	29	Male	30
10	Female	5	20	Male	10	30	Female	54

▶ Solution

There are three statistics you could use to describe a typical item from a list of numerical data: the mean, the median, or the mode.

The **mean** is 9.5 miles—the sum of data values, 285 miles, divided by the number of values, 30.

The **median** is 7 miles, the middle value when the data are arranged in order.

0, 0, 0, 0, 1, 2, 2, 4, 4, 5, 5, 6, 6, 6, 7, 7, 7, 8, 9, 10, 10, 10, 12, 12, 14, 15, 19, 20, 30, 54

$$\frac{7+7}{2} = 7$$

Because there are an even number of values, the median is the mean of the two middle values.

The **mode** is 0 miles, the distance that occurs most frequently.

[▶ See **Calculator Note 1E** to learn how to enter these data into your calculator. See **Calculator Note 1J** to calculate the mean and median.◀]

You can justify using any of these three statistics as a typical commuting distance. If you want to present a statistic that implies that students travel long distances, you might want to use the mean because it is higher than the median or mode, due to one very large data value called an *outlier.* When a data set includes one value that is far from the rest, the median is often more representative of the data.

The data set in Example A does not include every student at the university, so it may or may not tell much about all students' commuting distances. If this sample included only one class, then the time the class is taught might be important. A night class might include more part-time students, who would be less likely to live on campus. In general, an important first step in analyzing any data sample is to determine whether it represents the larger population.

If you assume that these data are from a random sample of *all* students, then you can draw some general conclusions about commuting distances to this university. The three values commonly used to describe a typical data value—the mean, the median, and the mode—are called **measures of central tendency.**

In statistics the mean is often referred to by the symbol $\bar{x}$ (pronounced "x bar"). Another symbol, Σ (capital *sigma*), is used to indicate the sum of the data values. For example, $\sum_{i=1}^{5} x_i$ means $x_1 + x_2 + x_3 + x_4 + x_5$, where x_1, x_2, x_3, x_4, and x_5 are the individual data values. So the mean of n data values is given by

$$\bar{x} = \frac{\sum_{i=1}^{n} x_i}{n}$$

The sum of all data values.

The number of data values.

Summarizing a data set with a single "typical" number or statistic is an incomplete picture. Sharing the entire data set is not usually informative either. A good description of the data set includes not only a measure of central tendency but the spread and shape of the data as well. Such a description is often made with a set of summary values or a graph.

The **box plot** (or **box-and-whisker plot**) is a visual tool for analyzing information about a data set. This is the box plot of all of the commuting data from Example A. [▶ See **Calculator Note 1K** to learn how to create a box plot on your calculator.◀]

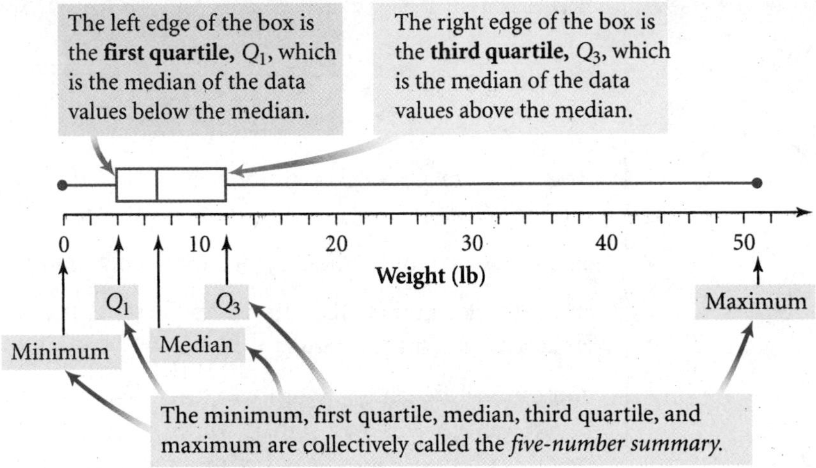
The left edge of the box is the first quartile, Q_1, which is the median of the data values below the median.
The right edge of the box is the third quartile, Q_3, which is the median of the data values above the median.
0
10
20
30
40
50
Weight (lb)
Q_1
Q_3
Maximum
Minimum
Median
The minimum, first quartile, median, third quartile, and maximum are collectively called the five-number summary.

The lines emanating from the "box" are called "whiskers." They identify the minimum and maximum values of the data. The difference between the maximum and minimum is the **range** of the data.

EXAMPLE B

Use the box plot above to analyze the commuting-distance data.

a. What percentage of the data values is represented by the lower whisker?

b. What are the values for the first quartile, the median, and the third quartile?

c. What is the five-number summary for this data set?

▶ Solution

Read the data values from the box plot at the five-number summary points, and use the definition of quartile.

a. One quarter, or 25%, of the data values are represented by the lower whisker. As a matter of fact, one quarter of the data values are represented by the upper whisker, one quarter are represented by the upper part of the box, and one quarter are represented by the lower part of the box.

b. There are 30 values, so after they have been arranged in order, the first quartile is the 8th value, or 4 miles; the median is the mean of the 15th and 16th values, or 7 miles; and the third quartile is the 23rd value, or 12 miles.

c. The five-number summary is 0, 4, 7, 12, 54.

Statisticians often talk about the *shape* of a data set. Shape describes how the data are distributed relative to the position of the measure of central tendency. A **symmetric** data set is balanced, or nearly so, at the center. Note that it does not have to be exactly equal on both sides to be called symmetric. **Skewed** data are spread more to one side of the center than to the other side. You will learn more about spread in the next lesson. For now, a box plot can be a good indicator of shape because the median and the range are clearly visible.

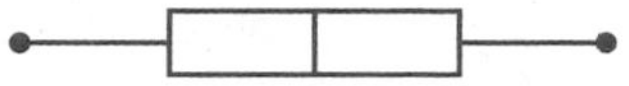

This box plot shows a symmetric data set.

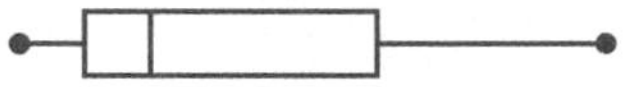

Skewed right implies that the data are spread more to the right of the center than to the left.

This data set is skewed left.

Investigation
Pulse Rates

You will need

- a watch or clock with a second hand

Pulse rate is often used as a measure of good physical conditioning. In this investigation you will practice making box plots, comparing box plots, and drawing some conclusions about pulse rates.

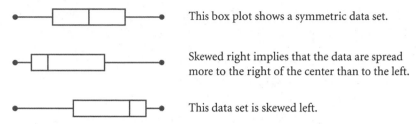

Step 1 Measure and record your resting pulse for 15 seconds. Multiply this value by 4 to get the number of beats per minute. Pool data from the entire class.

Step 2 Exercise for 2 minutes by doing jumping jacks or running in place. Afterward, measure and record your exercise pulse rate. Pool your data.

Step 3 Order each set of data. Calculate the five-number summaries for your class's resting pulse rates and your own exercise pulse rates.

Step 4 Prepare a box plot of the resting pulse rates and a box plot of the exercise pulse rates. Determine a range suitable for displaying both of these graphs on a single axis.

Step 5 Draw a conclusion about pulse rates by comparing these two graphs. Be sure to compare not only centers but also spreads and shapes.

How could a physician or a personal trainer use your results to determine whether a client is in good physical condition?

The box plot is a convenient way to compare two data sets. Not only can you readily compare the medians, but you can also see whether the two sets are distributed in the same manner.

EXERCISES

Practice Your Skills

1. Find the mean, median, and mode of each data set.

a. The time for pizza delivery (min): {28, 31, 26, 35, 26}

b. Yearly rainfall (cm): {11.5, 17.4, 20.3, 18.5, 17.4, 19.0}

c. The cost of a small box of popcorn at movie theaters ($): {2.75, 3.00, 2.50, 1.50, 1.50, 2.00, 2.50, 3.00}

d. The number of pets per household: {3, 2, 1, 0, 3, 4, 1}

2. A data set has a mean of 12 days; the median is 14 days; and there are three values in the data set.

a. What is the sum of all three values?

b. What is the one value you know?

c. Invent a data set that includes the statistics given.

d. Invent a different data set that also includes the statistics given.

e. Suppose the data may only include whole-number values. How many different data sets are possible that have the statistics given?

3. Approximate the values of the five-number summary for this box plot. Give the full name of each value.

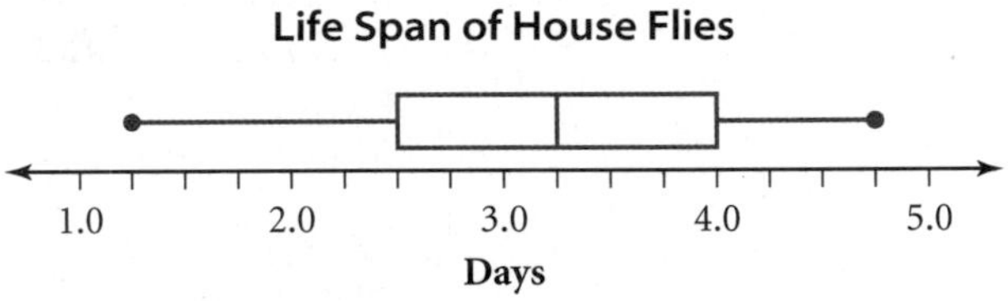

4. Match this data set to one of the four box plots.

Licks to the center of a lollipop:
{470, 510, 547, 558, 561, 574, 593}

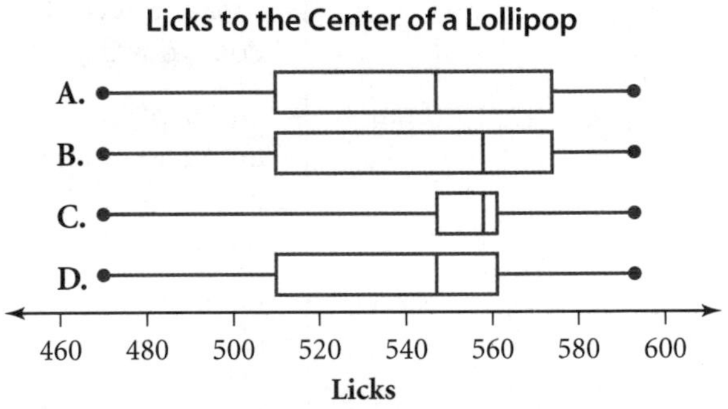

5. Match this box plot to one of the data sets for the number of minutes on the phone spent by 13 customer-service representatives in a given hour.

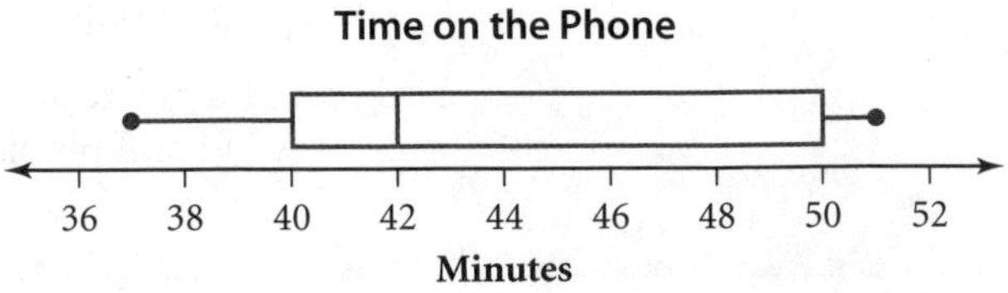

a. {36, 37, 38, 39, 40, 41, 42, 43, 44, 46, 48, 50, 52}

b. {37, 40, 42, 50, 51, 51, 51, 51, 51, 51, 51, 51, 51}

c. {36, 37, 40, 40, 40, 42, 42, 43, 44, 50, 50, 51, 52}

d. {37, 39, 40, 40, 40, 41, 42, 43, 44, 49, 51, 51, 51}

Reason and Apply

6. The lists below are the recorded scores on semester assignments for two students.

Connie: {82, 86, 82, 84, 85, 84, 85}

Oscar: {72, 94, 76, 96, 90, 76, 84}

Find the mean and median of each set of scores, and explain why they do not tell the whole story about the differences between Connie's and Oscar's scores.

7. These box plots represent Connie's and Oscar's scores from Exercise 6.

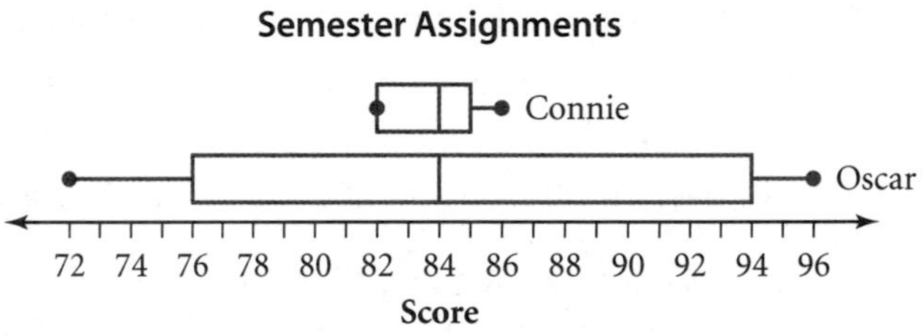

Write a paragraph describing the information pictured in the box plots. Use the box plots to help you draw some statistical conclusions. Include in your description answers to such questions as:

What does it mean that the second box plot is longer?

Where is the left whisker of the top box plot?

What does it mean when the median isn't in the middle of the box?

What does it mean when the left whisker is longer than the right whisker?

8. Homer Mueller has played in the minor leagues for 11 years. His home-run totals in order for those years are 56, 62, 49, 65, 58, 52, 68, 72, 25, 51, and 64.

a. Construct a box plot showing Homer's data.

b. Give the five-number summary.

c. Find $\bar{x}$.

d. How many home runs would Homer need to hit next season to achieve a 12-year mean of 60?

9. The *interquartile range (IQR)* is the difference between the first and third quartiles, or the length of the box in a box plot.

a. Look at the box plots in Exercise 7. What are the range and *IQR* for Connie? For Oscar?

b. Find the range and *IQR* for your box plot from Exercise 8.

10. Invent a data set with seven values and a mean of 12.

11. Invent a data set with seven values and

a. a mode of 70 and a median of 65

b. a mode of 60 and a median of 65

12. Invent a data set with seven values that creates this box plot.

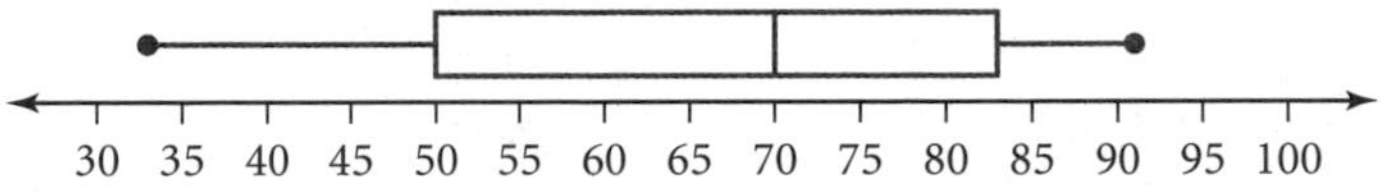

13. Refer to the commuting data listed at the beginning of this lesson. Separate the data by gender.

a. Compute the mean commuting distance for the males and the mean commuting distance for the females.

b. Calculate the median for each gender.

c. Compare the mean and median values for each gender. Which is the greater value, mean or median, in each set?

14. Use the commuting-distance data separated by gender from Exercise 13.

a. Create two box plots—one for each gender. Put both box plots on the same axis.

b. Write a brief statement based on the information in your graph, analyzing these two groups. In your statement, use the vocabulary introduced in this lesson.

c. Based on the box plots, explain why the mean or median may have been greater in 13c.

ScienceConnection One of the first people to study the density of nitrogen was the English scientist Lord Rayleigh (1842–1919, born John William Strutt). Working with fairly small samples, he noticed that the density of nitrogen produced by chemical compounds was different from the density of nitrogen produced by the atmosphere. On the supposition that the air-derived gas was heavier than the "chemical" nitrogen, he conjectured the existence of an unknown ingredient in the atmosphere. In 1894 Lord Rayleigh isolated the unknown ingredient, the colorless, tasteless, and odorless gas called argon. In 1904 Lord Rayleigh was awarded the Nobel Prize in Physics for his discovery. You can learn more about Rayleigh's work by using the links at www.keycollege.com/online.

Lord Rayleigh

15. **APPLICATION** Lord Rayleigh was one of the early pioneers in the study of the density of nitrogen. (Read the Science Connection.) The following are data that he collected. Lord Rayleigh's measurements first appeared in *Proceedings of the Royal Society* (London, 55, 1894, pp. 340–344). The units are the mass in grams of nitrogen filling a certain flask under a specified temperature and pressure.

a. Calculate the five-number summary for each set of data.

b. On the same axis construct a box plot for each set of data.

c. Describe any similarities and differences in the shapes of the box plots.

Do the box plots support Lord Rayleigh's conjecture?

Mass of Nitrogen Produced from Chemical Compounds (g)

2.30143	2.29890	2.29816
2.30182	2.29869	2.29940
2.29849	2.29889	2.30074
2.30054		

Mass of Nitrogen Produced from the Atmosphere (g)

2.31017	2.30986	2.31010
2.31001	2.31024	2.31010
2.31028	2.31163	2.30956

LESSON

1.5

Measures of Spread

Out on the edge you see all kinds of things you can't see from the center.

KURT VONNEGUT

If you ask several individuals to estimate the number of people in a crowd, their estimates will usually differ. The mean or median would measure the central tendency of the estimates, but neither of these statistics tell how widely peoples' estimates differ. Measuring variability, or **spread,** in numerical data allows a more complete description than merely stating a measure of central tendency. In this lesson you will investigate different ways to measure and describe variability.

On August 28, 1963, approximately 250,000 people gathered at the Lincoln Memorial in support of civil rights legislation. It was here that Martin Luther King, Jr., gave his "I have a dream" speech. The March on Washington was the largest recorded gathering of people anywhere to that date.

Investigation
A Good Design

You will need

- a rubber band
- a ruler
- a measuring tape
- paper
- books

In a well-designed experiment you should be able to follow a specific procedure and get very similar results every time you perform the experiment. In this investigation you will attempt to control the setup of an experiment in order to limit the variability of your results.

Experiment 1: Rubber-Band Launch

In this experiment you'll use a ruler to launch a rubber band. Select the height and angle of your launch and the length of your stretch, and determine any other factors that might affect your results. Launch the rubber band into an area clear of obstructions (especially other people). Record the horizontal distance of the flight. Repeat this procedure as precisely as you can with the same rubber band, the same launch setup, and the same stretch another seven or eight times.

Experiment 2: Rolling Ball

In this experiment you'll roll a ball of paper down a ramp and off the edge of your desk. Build your ramp from books, notebooks, or a pad of paper. Select the height and slope of your ramp and the distance from the edge of your desk, and determine any other factors that might affect your results. Make a ball by crumpling a piece of paper, and roll it down the ramp. Record the horizontal distance to the spot where the ball hits the floor. Repeat this procedure with the same ball, the same ramp setup, and the same release another seven or eight times.

Step 1 Using your data from Experiment 1 or 2, calculate the mean distance for your trials.

Step 2 On average, how much do your data values differ from the mean? How does the variability in your results relate to how controlled your setup was? Determine a way to calculate a *single* value that tells how consistently your group repeated the procedure. Write a formula to calculate your statistic.

Step 3 A value known as the *standard deviation* helps measure the spread of data away from the mean. Use your calculator to find this value for your data. [► **See Calculator Note 1J** to learn how to figure the standard deviation using your calculator. ◄]

Step 4 Repeat the experiment, and collect another set of data from seven or eight trials. For your new set of data, calculate the same statistic you used before and the standard deviation. How do the results of your experiments compare? Do you observe any pattern, over the two sets of data, between your statistic and the standard deviations that you found using your calculator?

To measure the spread of data, it is typical to start by finding the mean. Next, find the *deviation*, or directed distance, from each data value to the mean. When you compare the results from two groups in the last investigation, you may find that one group's mean is 200 cm and another group's mean is 300 cm. However, if their deviations are similar, then they performed the experiment equally well. The deviations let you compare the spread independent of the mean.

Consider two groups that each do the rubber-band launch seven times.

Group A Distance (cm): {182, 186, 182, 184, 185, 184, 185}
Group B Distance (cm): {152, 194, 166, 216, 200, 176, 184}

The mean for each group is 184. The individual deviations, $x_i - \bar{x}$, for each data value, x_i, are given in the table below.

Group A			Group B		
Data value	**Distance**	**Deviation**	**Data value**	**Distance**	**Deviation**
x_1	182	$182 - 184 = \mathbf{-2}$	x_1	152	$152 - 184 = \mathbf{-32}$
x_2	186	$186 - 184 = \mathbf{2}$	x_2	194	$194 - 184 = \mathbf{10}$
x_3	182	$182 - 184 = \mathbf{-2}$	x_3	166	$166 - 184 = \mathbf{-18}$
x_4	184	$184 - 184 = \mathbf{0}$	x_4	216	$216 - 184 = \mathbf{32}$
x_5	185	$185 - 184 = \mathbf{1}$	x_5	200	$200 - 184 = \mathbf{16}$
x_6	184	$184 - 184 = \mathbf{0}$	x_6	176	$176 - 184 = \mathbf{-8}$
x_7	185	$185 - 184 = \mathbf{1}$	x_7	184	$184 - 184 = \mathbf{0}$

These deviations show more variation in Group B's distances than in Group A's distances. That might imply that Group B's experiment was not designed well enough to produce consistent results.

Is there a way to combine the deviations into a single value that reflects the spread in a data set? Finding the average of all the deviations is a natural choice. What happens when you add all the deviations in the table above? If you think of the mean as a balance point in a data set, then it makes sense that the directed distances above and below the mean will balance out. The deviation sum for both Group A and Group B is zero.

Group A's deviation sum: $-2 + 2 - 2 + 0 + 1 + 0 + 1 = 0$
Group B's deviation sum: $-32 + 10 - 18 + 32 + 16 - 8 + 0 = 0$

For the sum of the deviations to be useful, you need to eliminate the effect of the different signs. This might be done by adding the absolute values of all the deviations or by squaring the deviations. Statisticians have found that the results are generally more consistent when the deviations are made into positive numbers by squaring them. The table below shows the square of each deviation.

	Group A			Group B		
	Distance	Deviation	$(\text{Deviation})^2$	Distance	Deviation	$(\text{Deviation})^2$
	182	−2	**4**	152	−32	**1024**
	186	2	**4**	194	10	**100**
	182	−2	**4**	166	−18	**324**
	184	0	**0**	216	32	**1024**
	185	1	**1**	200	16	**256**
	184	0	**0**	176	−8	**64**
	185	1	**1**	184	0	**0**
	$sum = \mathbf{14}$			$sum = \mathbf{2792}$		
Variance	$\frac{sum}{6} = \mathbf{2.333}$			$\frac{sum}{6} = \mathbf{465.3}$		
Standard deviation	$\sqrt{\frac{sum}{6}} \approx 1.5$			$\sqrt{\frac{sum}{6}} \approx 21.6$		

After the deviations are squared, the sum is no longer zero. We can compute the average of the squared deviations by dividing the sum of the squares by the number of deviations in the set. The sum of the squares of the deviations, divided by the number of values, is called the **variance** of the data. The square root of the variance is called the **standard deviation** of the data. The standard deviation provides one way to judge the "average difference" between data values and the mean. It is a measure of how the data are spread around the *mean*. This process is used to compute the standard deviation of a population. The most commonly used symbol for the standard deviation of a population is the lowercase Greek letter sigma, σ.

Did you notice that in the table above, the sum was divided by 6 instead of 7? The reason is that the sample was quite small. When calculating the standard deviation of a small sample, it is customary to divide by a number that is one less than the number of deviations. The letter s is used to designate the standard deviation of a small sample. The result of dividing the sum of the deviations by this slightly smaller number is that the computed value of the standard deviation will be slightly increased. This compensates for the fact that the sum of the deviations of a small sample is usually an underestimate of the true spread of a larger population's data. In this text we will use relatively small data sets. Therefore, we will use the computation for the sample standard deviation, s. The graphing calculator display below shows both forms of the standard deviation, s and σ, for the data in Group A.

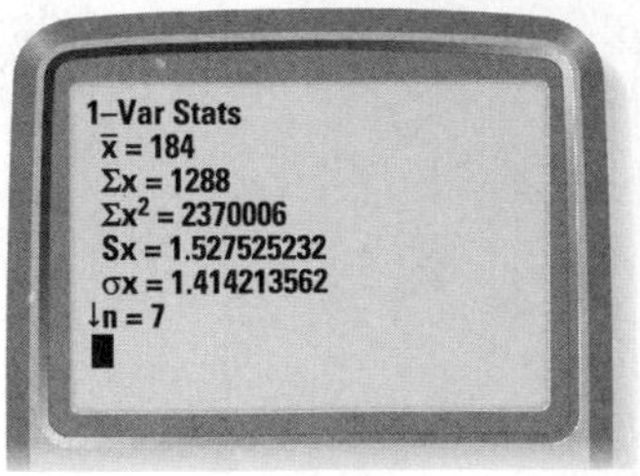

Standard Deviation

The standard deviation, s, is a measure of the spread of a sample data set.

$$s = \sqrt{\frac{\sum_{i=1}^{n}(x_i - \bar{x})^2}{n - 1}}$$

where x_i represents the individual data values, n is the number of values, and $\bar{x}$ is the mean.

The units of the standard deviation are the same as the units of the data. The larger standard deviation for Group B indicates that their distances generally lie much farther from the mean than do Group A's. A large value for the standard deviation tells you that the data values are not tightly clustered around the mean. As a general rule, a set with more data near the mean will have less spread and a smaller standard deviation.

Investigation
Standard Deviation Experiment

You will need

- a calculator

You can test the difference experimentally between the standard deviation of a population, σ, and the standard deviation of a small sample, s. On the next page is a set of 500 numbers.

The mean of this set of numbers is 486.726. The standard deviation of the population is approximately 286.7. This calculation was made by using the formula

$$\sigma = \sqrt{\frac{\sum_{i=1}^{n}(x_i - \bar{x})^2}{n}}$$

For this activity, each student or group of students will find the standard deviation of a small sample of the population consisting of eight numbers. Before you begin, discuss how each student or group will randomly choose eight numbers from the whole population. For this experiment, use the population standard deviation formula, σ above, to calculate the standard deviations.

a. After all groups have computed their σ-values, find the mean of the class standard deviation calculations. How did the mean of the standard deviations, σ, computed from the small samples compare to the standard deviation of the whole population of data?

b. The sample standard deviation formula, s, is used to correct the systematic bias we observe when taking the standard deviation using a small number of data points. How would dividing by a smaller number, $n - 1$, have changed the results of your standard deviation calculation?

800	544	120	663	228	902	75	725	318	121	886	207	462	518	550	128	600	727	446	729
623	89	849	340	675	11	52	992	529	630	632	736	896	986	218	419	686	942	324	28
64	211	152	744	959	614	166	523	87	345	604	881	731	8	9	433	556	706	634	662
583	385	458	542	700	788	781	56	987	894	482	378	827	640	904	828	943	103	94	347
127	990	52	473	164	656	799	592	681	755	301	916	229	787	27	59	275	249	330	206
473	299	583	782	174	451	222	90	516	952	107	185	466	250	455	914	136	594	977	62
397	373	76	163	903	151	69	314	256	510	890	911	534	178	973	122	738	446	136	454
755	443	933	283	118	456	795	71	377	546	187	501	274	526	298	31	73	52	683	667
834	417	50	666	572	259	168	831	510	143	687	268	425	736	405	163	499	662	858	262
135	875	241	875	834	473	72	160	679	445	615	243	908	461	399	456	990	246	386	189
413	713	84	744	747	899	171	876	330	800	81	766	601	22	409	530	442	265	500	238
357	713	986	15	569	341	614	194	576	840	30	496	931	520	309	121	238	852	396	594
14	558	614	740	950	853	30	660	290	100	602	778	144	647	500	522	793	654	700	372
209	262	307	568	45	512	845	629	350	170	813	452	641	326	502	118	31	38	500	556
340	558	877	795	296	926	324	881	555	394	423	311	130	745	332	929	336	433	369	803
910	296	751	940	137	793	136	990	364	207	348	319	352	866	550	485	402	893	912	375
66	865	486	414	737	942	520	578	985	88	960	369	49	381	957	508	145	891	319	716
12	899	170	81	87	637	973	291	125	602	222	350	815	950	194	287	433	818	527	874
490	686	718	677	878	789	57	487	787	32	662	557	736	253	420	438	37	486	94	705
354	246	836	236	129	501	208	192	761	92	184	385	348	50	296	769	500	463	313	992
88	900	434	903	784	658	249	867	326	952	115	392	481	474	392	263	881	615	225	549
650	383	389	861	190	461	802	780	223	387	106	713	780	71	836	973	901	379	950	182
583	428	770	288	796	508	265	116	434	640	187	474	901	378	56	584	461	200	502	7
846	348	565	576	730	615	286	633	30	48	63	842	655	798	447	570	350	952	813	801
50	566	994	427	767	904	719	183	241	982	857	826	224	210	881	436	422	157	23	384

When you make a box plot, you have a visual representation of how data are spread around the *median.* The range (the distance between the whisker endpoints) and the interquartile range (the length of the box) are measures of spread around the median. **Outliers** are data values that differ significantly from the majority of the data. The exact definition of an outlier may vary according to different textbooks. The **interquartile range (*IQR*)** is the difference between the first and third quartiles, or the length of the box in a box plot. Some statisticians identify outliers in a box plot as values that are more than $1.5 \cdot IQR$ from either end of the box.

EXAMPLE

Since 1975 Congress has mandated that automakers meet fuel economy standards. These Corporate Average Fuel Economy (CAFE) standards required the average mileage rate of all 2004-model passenger cars in each manufacturer's fleet to be

27.5 miles per gallon (mi/g). For all pickups, minivans, and sport utility vehicles, the average fleet mileage had to be 20.7 mi/g.

The legislation required average fuel economy standards, but individual vehicles were allowed mileage rates that deviated widely from the mean. This table gives the combined city/highway miles-per-gallon data for some vehicles sold in the United States in 2004.

Combined City/Highway Environmental Protection Agency Rated Miles per Gallon (2004)

Dodge Caravan Minivan (2WD)	22	GMC Safari (2WD)	18
Ford Ranger Pickup (2WD)	19	Honda Insight (hybrid)	63
Ford Focus (manual)	31	Honda Accord (manual)	29
Honda Civic (hybrid, automatic)	48	Chevrolet Impala	25
Toyota Prius (hybrid)	55	Mazda B2300 Pickup	24
Cadillac Escalade (AWD)	15	Kia Sedona Minivan	18
Chevrolet Astro Van	15	Ferrari 575 M Maranello	12
Dodge Ram Pickup (4WD)	13	Dodge Neon SRT–4/SX 2.0 (manual)	32
Volkswagen Jetta	41	BMW 545i	19
Toyota Echo (manual)	38	Mini Cooper (manual)	32

(U.S Environmental Protection Agency)

a. Calculate the mean and the standard deviation of this sample. What do the statistics tell you about the spread of fuel economy rates?

b. Make a box plot of the data. Identify any outliers using the $1.5 \cdot IQR$ definition.

▶ *Solution*

a. The mean miles-per-gallon rating for all vehicles listed is 28.45 mi/g. The standard deviation is approximately 14.33. So, the data are quite spread out.

b. The five-number summary is 12, 18, 24.5, 35, 63.

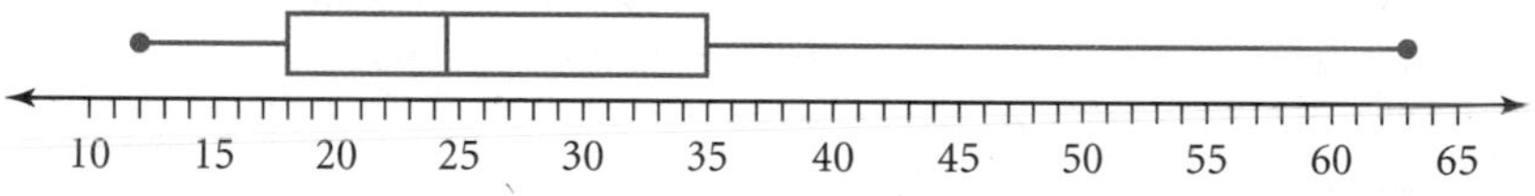

The interquartile range is 17. To be an outlier, a data value must be $1.5 \cdot 17$, or 25.5, away from an end of the box. That means it must be less than $(18.0 - 25.5)$, or -7.5 (impossible for mi/g!), or more than $(35 + 25.5)$, or 60.5. There is one mileage rate, 63, that meets this requirement, so it is an outlier. [▶ Revisit **Calculator Note 1K** to learn how to make a box plot that shows outliers on your calculator.◀]

As you work the exercises, you may notice that finding the measure of central tendency is usually an integral step in measuring the spread. To calculate the standard deviation, you first need to calculate the mean. The interquartile range relies on the first and third quartiles, which in turn rely on the median.

EXERCISES

Practice Your Skills

1. Given the data set {41, 55, 48, 44},

a. Find the mean.

b. Find the deviation from the mean for each value.

c. Find the sample standard deviation, *s*, of the data set.

2. The lengths in minutes of nine music CDs are 45, 63, 74, 69, 72, 53, 72, 73, and 50.

a. Find the mean.

b. Find the deviation from the mean for each value.

c. Find the standard deviation of the data set.

d. What are the units of the mean, the deviation, and the standard deviation?

3. In a controlled experiment 11 bean plants were grown from seeds. After two weeks, the heights in centimeters of the plants were 9, 10, 10, 13, 13, 14, 15, 16, 17, 19, and 21.

a. Find the five-number summary.

b. Find the range and *IQR*.

c. What are the units of the range and *IQR*?

4. To monitor weight, a cookie manufacturer samples jumbo chocolate chip cookies as they come off the production line. The weight in grams of 11 cookies were 22, 30, 27, 35, 32, 28, 18, 22, 25, 30, and 28.

a. Find the five-number summary.

b. Find the range and *IQR*.

c. What are the units of the range and *IQR*?

5. Invent a data set with seven data values such that the mean and the median are both 84, the range is 23, and the interquartile range is 12.

Reason and Apply

6. **APPLICATION** The mean diameter of a Purdy Goode Compact Disc is 12.0 cm with a standard deviation of 0.012 cm. No CDs can be shipped that differ from the mean by more than one standard deviation. How would the company's quality-control engineer use those statistics?

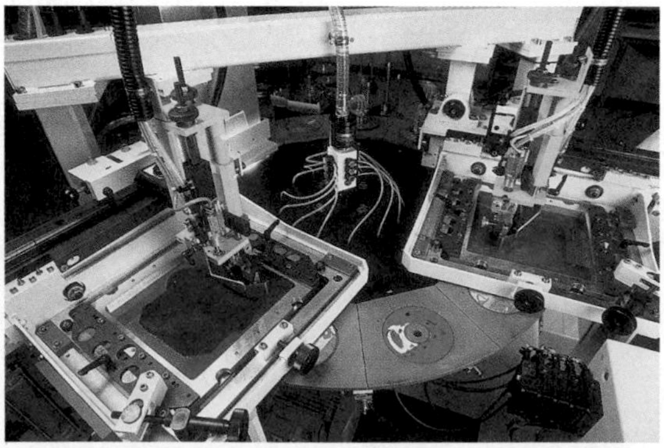

This automated machine paints labels on compact discs.

7. Some statisticians define outliers as data values that are more than $2s$ from the mean, where s is the sample standard deviation. Use this method to identify any outliers in the fuel economy rates from the table on page 37. How do the outliers found by standard deviation compare to the outliers found by interquartile range?

8. Find the standard deviation and *IQR* of the combined commuting-distance data (including both males and females) from Example A in Lesson 1.4. Which of these two values is larger? Will this value always be larger? Explain your reasoning, and find or create another data set that supports your answer.

9. Two data sets have the same range and *IQR*, but the first is symmetric, and the second is skewed left.

a. Sketch two box plots that satisfy the conditions for the two sets.

b. Would you guess that the standard deviation of the skewed data set is less than, more than, or the same as the first? Explain your reasoning.

c. Invent data sets of seven values that satisfy the conditions.

d. Find the standard deviations of the two sets. Do the standard deviations support your answer to 9b?

10. Students collected eight length measurements during a mathematics lab. The mean measurement was 46.3 cm, and the deviations of *seven* individual measurements were 0.8 cm, −0.4 cm, 1.6 cm, 1.1 cm, −1.2 cm, −0.3 cm, −1.0 cm, and −0.6 cm.

a. What were the original eight measurements collected?

b. Find the standard deviation of the original measurements.

c. Which measurements are more than one standard deviation above or below the mean?

11. APPLICATION The students in four classes recorded their resting pulse rates. The class means and standard deviations are given in the table.

Meeting time	Mean	Standard deviation
8 A.M.	79.4	3.2
11 A.M.	74.6	5.6
4 P.M.	78.2	4.1
7 P.M.	80.2	7.6

a. In which course section were students' pulse rates most alike? How can you tell?

b. Can you tell which class is attended by students with the fastest pulses? Why or why not?

12. The temperatures during one year for two cities at 12:00 noon on the 15th day of each month are given here in degrees Fahrenheit.

Temperatures on the 15th of Each Month (°F)

Month	Juneau, Alaska	New York City, New York	Month	Juneau, Alaska	New York City, New York
January	24	31	July	56	76
February	28	33	August	55	75
March	33	41	September	49	68
April	40	51	October	42	57
May	47	60	November	32	47
June	53	69	December	27	37

a. Find the mean temperature and standard deviation for each city.

b. Draw a box plot for each city. Find each median and interquartile range.

c. Which city has more consistent temperatures? Justify your conclusion.

d. Is a better measure of the spread given by the interquartile range or by the standard deviation? Justify your conclusion.

13. APPLICATION The musical society sells bagels every weekday morning to raise funds for its activities. The numbers of bagels sold each day in a five-week period are shown here.

65	76	100	67	44
147	82	94	92	79
158	77	62	85	71
69	88	80	63	75
62	68	71	73	74

a. Find the median and the interquartile range for this data set.

b. Find the mean and the standard deviation.

c. Draw a box plot for this data set. Use the $1.5 \cdot IQR$ rule to name any numbers that are outliers.

d. Remove the outliers from the data set, and draw another box plot.

e. After removing the outliers, recalculate the median and the interquartile range, and the mean and the standard deviation.

f. Which is more affected by outliers, the mean or the median? The standard deviation or the interquartile range? Explain your answers.

14. Refer to the data in Exercise 13.

a. Suppose each bagel sold yields a net profit of 28¢. Draw a box plot for the profit on bagel sales generated each day. Find the median and interquartile range, and the mean and standard deviation. Compare your income statistics with the original statistics describing the numbers of bagels sold. How are the two sets of statistics and the graphs related?

b. Use your findings from 14a to predict the net profit statistics if the net profit per package were 35¢.

c. Suppose the treasurer's audit finds that 20 fewer bagels were sold each day than originally reported. Find the median and interquartile range, and the mean and standard deviation. Describe a process you could use to find the corrected results for all the information requested in Exercise 13.

d. Use your findings from 14c to predict the statistical results if instead 10 fewer bagels were sold per day than originally reported.

15. APPLICATION Matt Decovsky wants to buy a 160W CD player for his car at an online auction site. Before bidding, he decides to do some research on the selling price of recently sold CD players. His search comes up with these 20 prices.

$74.00	$102.50	$64.57	$74.00
$82.87	$73.01	$77.00	$71.00
$71.01	$112.50	$86.00	$102.50
$76.00	$56.00	$135.50	$66.00
$71.00	$76.00	$51.00	$88.00

a. Find the mean, median, and mode.

b. Draw a box plot of the data. Describe its shape.

c. Find the IQR, and determine whether there are any outliers.

d. While looking over the items that were sold, Matt realizes that the outlier stereos contain features that don't interest him. If he removes any outliers, which measure will be least affected, the mean or the median? Explain.

e. Sketch a new box plot using the 18 data points. How does this help you support your answer in 15d?

Review

16. (Lesson 1.4) Create a data set that has 12 points, a median of 40, a mode of 5, and a mean of 36.

17. (Lesson 1.4) The data sets below give the weights in pounds of the offensive and defensive teams of the 2002 Super Bowl Champion New England Patriots. *Source:* www.nfl.com

Offensive players' weights (lb): {190, 305, 310, 320, 315, 322, 255, 190, 220, 245, 230}

Defensive players' weights (lb): {280, 305, 280, 270, 250, 253, 245, 199, 196, 207, 218}

a. Find the mean and median weights of each team.

b. Prepare a box plot of each data set. Use the box plots to make general observations about the differences between the two teams.

EXPLORATION
with Microsoft® Excel

Percentiles

Download the Chapter 1 Excel Exploration from the website at www.keycollege.com/online or from your Student CD.

This interactive spreadsheet contains a table of grouped data. Beneath the data table is a cell in which to enter the desired percentile. When you enter different values of the desired percentile, the spreadsheet will point you to the data range in which that percentile falls. It will also show you the calculation of the percentile score.

On the spreadsheet you will find complete instructions and questions to consider as you explore.

Activity
Tennis Elimination

Excel skills required

- Name a cell
- Adjust column width
- Enter a logical formula
- Data fill

1,025 tennis players show up for a tournament. They will play elimination rounds. If a player loses, he or she is eliminated. Ties will be broken with "sudden death" so that all matches will produce a winner. If there is an odd number of players in a given round, one player will be chosen at random to draw a bye (that is, to sit out the round and advance automatically to the next round). How many matches must be scheduled in order for the champion to emerge?

You can find instructions for the required Excel skills in the Excel Notes at www.keycollege.com/online or on your Student CD.

Launch Excel.

Step 1 Start by entering labels for the things you will need.

a. Into Cell A1 type #PLAYERS, and then press ENTER. (Always press ENTER after you type text or data into a cell.)

b. Similarly, enter the other labels as shown here.

	A	B	C	D
1	# Players			# Matches
2	Remaining	Eliminations	Byes	

Step 2 Click in Cell E1, and type =SUM(B3:B50). This cell will show the total number of matches in all rounds.

Step 3 Click in Cell B1, then give the cell a name. Do this by clicking in the upper left corner, where at first you see the cell reference "B1." In that space type PLAYERCT. This is a more powerful way to work in Excel—by identifying with a name the numbers you want to explore. In this example, in any other place where you want the number in B1 to appear, you can reference PLAYERCT, and the number in that cell will be used. An item of this type—one that can change but is fixed for the problem—is called a *parameter.*

At this stage you have

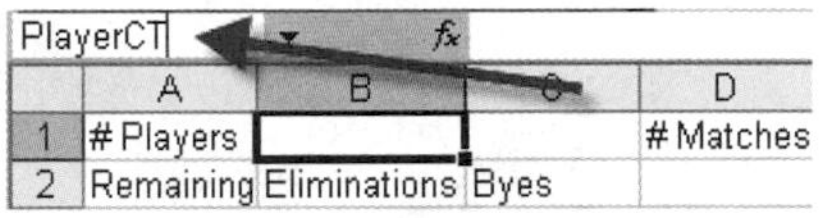

Step 4 Click in Cell B1, and type `1025`. (In Excel, pressing the = before a number is optional, but it is required when you enter a formula or calculation of any kind.)

Step 5 Click in Cell A3, and type `=PLAYERCT`.

The number 1025 will automatically appear in Cell A3. If you change Cell B1 to any other number, A3 will change with it. If you decide to experiment with this, change it back to 1025 when you are done.

Step 6 Click in Cell B3. Type `=INT(A3/2)`. INT is a built-in Excel function that rounds down to the nearest integer less than or equal to the number. For example, INT(23.75) = 23. Explain why the use of this function here gives you the number of matches that must be played in that round. The number 512 will appear if 1025 is the original number.

Step 7 If Cell A3 is odd, then one person will draw a bye. So, in Cell C3 you will enter a logical formula based on another useful Excel function, ISODD. Click in Cell C3, and type `=IF(ISODD(A3),1,0)`. If the number in Cell A3 is even, there will be a 0 in Cell C3. With 1025 as a starting number, the number that shows up in C3 will be 1.

Step 8 You need to calculate the number of people remaining to play and enter that answer in Cell A4. Rather than doing the calculations yourself, you can use Excel to calculate it. Click in Cell A4, and type `=B3+C3`.

Explain why this answers the question, "What is the number of people in the next round?"

Step 9 The "rules" in Cells B3 and C3 can be copied to other cells, thanks to Excel's *relative copying*. In Steps 9 and 10, you will use the power of relative copying to fill a number of cells. This process is called *data fill.* Highlight Cell B3. Click and drag the mouse so that both Cells B3 and C3 are highlighted. The screen will look like this:

	A	B	C
1	# Players	1025	
2	Remaining	Eliminations	Byes
3	1025	512	1

Step 10 Notice the small black square in the lower corner of the highlighted area.

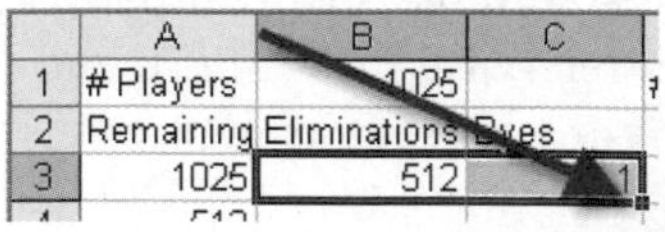

Click on that dot, drag downward to cover Cells B4 and C4, and then let go.

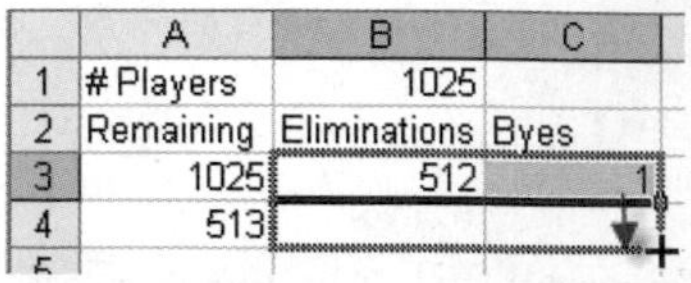

You have just completed a *data fill.* When you click out of Cells B4 and C4, you will get the numbers for the next round.

At this point you can answer the question "How many matches are there in the first two rounds?"

Now that the groundwork is done, you can data-fill the entire row to project all the rounds. Highlight Cells A4 to C4, then perform a data fill in as many rows as you like. If you fill so that there are 10 to 15 rows, you can answer quite a few questions.

Questions

1. The original question was, "How many matches must be scheduled in order for the champion to emerge?" What is the answer when you start with 1025 players?
2. What is the answer with 249 players?
3. What is the answer with 5,125 players?
4. Is there a pattern?
5. Suppose only one player shows up. Do you need the spreadsheet to answer the question?
6. Suppose there are only two players? Three players? Is there a pattern? Is this the same pattern as previously discussed? Can you explain why this pattern should give you the answer to this question?

CHAPTER 1 REVIEW

This chapter introduced various strategies for analyzing mathematical situations. Problem-solving strategies can help you get started even when you have no idea where to begin. There is no single way to solve a problem. Different people prefer different problem-solving strategies, yet not all strategies can be applied to all problems.

Here are some of the strategies available to you. Organize the information that is given, or that you figure out in the course of your work, into a list or table. Draw a picture, graph, or diagram, and label it to illustrate information you are given and what you are trying to find. Special types of diagrams, such as tree diagrams and coordinate graphs, can help you organize information. Make a physical representation of the problem. Look for patterns in numbers or units of measure. Be sure that all your measures use the same system of units and that you compare quantities with the same unit. Eliminate some possibilities. If you know what the answer cannot be, you are partway there. Solve sub-problems that present themselves as part of the problem context, or solve a simpler problem by substituting easier numbers or by looking at a special case. Don't forget to use algebra! Assign variables to unknown quantities, and write expressions for related quantities. Translate verbal statements into equations, and solve the equations. Work backward from the solution to the problem. Use guess-and-check, adjusting each successive guess by the result of your previous guess. Finally but most importantly, do not panic—just start working. Even if your first ideas do not result in a solution, you will still probably learn more about the problem as you try different strategies.

Sets of data, such as temperatures, the sugar content of breakfast cereals, and the distance you commute daily, can be analyzed and pictured using tools discussed in this chapter. The **mean, median,** and **mode** are **measures of central tendency**. They tell you about typical values for the data set. But a measure of central tendency alone does not tell the whole story. You also need to look at the **spread** of the data values. The **variance** and **standard deviation** help you determine spread about the mean, and the **interquartile range** (*IQR*) helps you determine spread about the median. These measures of spread are also frequently used to identify **outliers** in the data set.

One way to display a data set visually is with a **box plot.** A box plot shows the median of the data set, the **range** of the entire set, and the interquartile range between the **first quartile** and the **third quartile.** A box plot does not show individual data values but may help you see whether the data set is **symmetric** or **skewed.**

By using a combination of **statistics** and graphs, you can better understand the meaning and implications of a data set. Careful analysis of data sets helps you solve problems that form general conclusions about the past and predict the future.

EXERCISES

1. This graph shows the relationship between distance driven and gasoline consumed for two cars going 60 miles per hour (mi/hr).

a. How far can Car A drive on 7 gallons of gasoline?

b. How much gasoline is needed for Car B to drive 342 miles?

c. Which car can drive farther on 8.5 gallons of gas? How much farther?

d. What is the slope of each line? Explain the real-world meaning of each slope.

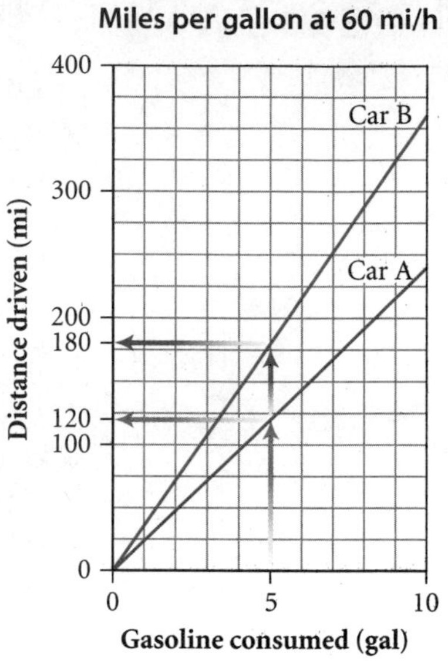

2. You are given a 3-liter bucket, a 5-liter bucket, and an unlimited supply of water. Describe or illustrate a procedure that will result in exactly 4 liters of water in the 5-liter bucket.

3. Draw a rectangle diagram to represent each product. Use the diagrams to expand each product.

a. $(x + 3)(x + 4)$ **b.** $(2x)(x + 3)$

c. $(x + 6)(x + 2)$ **d.** $(x - 4)(2x - 1)$

4. Solve each equation, and check your answers by substituting them for the unknowns in the original equation.

a. $3(x + 5) + 2 = 26$

b. $3.75 - 1.5(y + 4.5) = .75$

5. In 5a and 5b, translate each verbal statement into a symbolic expression. Combine the expressions to solve 5c.

a. Six more than twice a number

b. Five times three less than a number

c. Six more than twice a number is five times three less than the number. Find the number.

6. **APPLICATION** Keisha is moving to a new apartment 12 miles from her old one. She rents a small truck from You-Do-It Truck Rental for $19.95 per day plus $0.35 per mile. Keisha hopes she can complete the move in five trips all on the same day. She estimates that she will drive the truck another 10 miles for pickup and return.

a. Write an expression that represents the cost of a one-day rental with any number of miles.

b. How much will Keisha pay if she does complete the move in five trips?

c. How much will Keisha save for each trip she does *not* take?

7. Amy belongs to the Kent Valley Puzzle Club. At the first meeting she introduces herself by saying that it is her birthday and that three times her age three years ago is twice as much as her age six years from now. How old is Amy?

8. Every night Scott empties his pocket change into a bowl on his dresser. By Saturday he has 47 coins totaling \$5.02. He notices that the number of pennies is the same as the number of quarters and that the sum of the number of pennies and quarters is one more than the sum of the number of nickels and dimes. How many of each coin does Scott have?

9. Toby has only a balance scale, a single 40-gram mass, and a stack of white blocks and red blocks. (Assume that all white blocks have the same mass and all red blocks have the same mass.) Toby discovers that four white blocks and one red block balance two white blocks, two red blocks, and the 40-gram mass. He also finds that five white blocks and two red blocks balance one white block and five red blocks.

a. List the unknown quantities, and assign a variable to each.

b. Translate Toby's discoveries into equations.

c. Solve the equations to find a value for each variable.

d. Interpret the solution according to the context of the problem.

10. The height of a golf ball in flight is given by $h = -16t^2 + 48t$, where h represents the height in feet above the ground, and t represents the time in seconds since the ball was hit.

a. Fill in the table, showing the height of the ball at various times.

b. Explain the real-world meaning of each of the values in the table.

Time (t)	Height (h)
0 seconds	
1 second	
2 seconds	
3 seconds	

11. (Prerequisite skill: review of exponents) Rewrite each expression using the properties of exponents such that the variable appears only once.

a. $(4x^{-2})(x)$ **b.** $\dfrac{4x^2}{8x^3}$ **c.** $(x^3)^5$

12. (Prerequisite skill: review of exponents) Consider the equation $y = 2^x$. Fill in the missing values for x and y in the table.

x	y
−2	
0	
3	
	32

13. Use units to help you find the missing information.

a. How many ounces are in 5 gallons? (There are 8 ounces per cup, 4 cups per quart, and 4 quarts per gallon.)

b. How many meters are in 1 mile? (There are 2.54 centimeters per inch, 12 inches per foot, 5280 feet per mile, and 100 centimeters per meter.)

14. Bethany Rogers works for a temporary-staffing agency. On Wednesday she is sent to fill in for an absent sales department assistant. She finds out that all three sales representatives have important meetings with clients, but she uncovers only these clues:

Mr. Birmingham is a sales representative, although he is not meeting with Mr. Green.

Miss Hunt is the client who will be meeting in the lunchroom.

Mr. Green is the client with a 9 A.M. appointment.

Mr. Mendoza is the sales representative who is meeting in the conference room.

Ms. Phoung is a client, but she will not be in the 3 P.M. meeting.

Mrs. Plum is a sales representative but not the one meeting at 12:00 noon.

The client with the 9 A.M. appointment is not meeting in the convention hall.

Help Bethany figure out which sales representative is meeting with which client, where, and when.

15. Penny calculates that the deviations from the mean for a data set of eight values are 0, −40, −78, −71, 33, 36, 42, and 91.

a. How do you know that at least one of the deviations is incorrect?

b. If it turns out that 33 is the only incorrect deviation, what should it be?

c. Use the corrected deviations to find the actual data values, the standard deviation, the median, and the interquartile range if the mean is

i. 747 **ii.** 850

16. The Los Angeles Lakers won the 2002 NBA Championship. This table gives the total points scored by each player during that season:

Points Scored by Los Angeles Lakers Players (2001–2002 Season)

Player	Total points scored	Player	Total points scored
Kobe Bryant	2019	Samaki Walker	460
Shaquille O'Neal	1822	Stanislav Medvedenko	331
Derek Fisher	786	Mitch Richmond	260
Rick Fox	645	Brian Shaw	169
Devean George	581	Mark Madsen	167
Robert Horry	550	Jelani McCoy	26
Lindsey Hunter	473	Mike Penberthy	5

(*www.nba.com*)

a. Find the mean, median, and mode of this data set.

b. Find the five-number summary.

c. Draw a box plot of the data. Describe the shape of the data set.

d. Calculate the interquartile range.

e. Identify any outliers. Use the 1.5 · *IQR* definition for outliers.

Kobe Bryant and Shaquille O'Neal

17. Below is a list of Academy Award–winners in the Best Actress and Best Actor categories and each person's age at the time he or she received the award.

Academy Award Winners 1970–2003

Year	Best actress in a leading role, age	Best actor in a leading role, age
1970	Glenda Jackson, 34	George C. Scott, 43
1971	Jane Fonda, 34	Gene Hackman, 42
1972	Liza Minnelli, 26	Marlon Brando, 48
1973	Glenda Jackson, 37	Jack Lemmon, 48
1974	Ellen Burstyn, 42	Art Carney, 56
1975	Louise Fletcher, 41	Jack Nicholson, 38
1976	Faye Dunaway, 35	Peter Finch, 60
1977	Diane Keaton, 31	Richard Dreyfuss, 30
1978	Jane Fonda, 41	Jon Voight, 40
1979	Sally Field, 33	Dustin Hoffman, 42
1980	Sissy Spacek, 30	Robert De Niro, 37
1981	Katharine Hepburn, 74	Henry Fonda, 76
1982	Meryl Streep, 33	Ben Kingsley, 39
1983	Shirley MacLaine, 49	Robert Duvall, 52
1984	Sally Field, 38	F. Murray Abraham, 45
1985	Geraldine Page, 61	William Hurt, 35
1986	Marlee Matlin, 21	Paul Newman, 61
1987	Cher, 41	Michael Douglas, 43
1988	Jodie Foster, 26	Dustin Hoffman, 51
1989	Jessica Tandy, 81	Daniel Day-Lewis, 32
1990	Kathy Bates, 42	Jeremy Irons, 42
1991	Jodie Foster, 29	Anthony Hopkins, 54
1992	Emma Thompson, 33	Al Pacino, 52
1993	Holly Hunter, 35	Tom Hanks, 37
1994	Jessica Lange, 45	Tom Hanks, 38
1995	Susan Sarandon, 49	Nicolas Cage, 31
1996	Frances McDormand, 39	Geoffrey Rush, 45
1997	Helen Hunt, 34	Jack Nicholson, 60
1998	Gwyneth Paltrow, 26	Roberto Benigni, 46
1999	Hillary Swank, 26	Kevin Spacey, 40
2000	Julia Roberts, 33	Russell Crowe, 36
2001	Halle Berry, 35	Denzel Washington, 47
2002	Nicole Kidman, 35	Adrien Brody, 29
2003	Charlize Theron, 28	Sean Penn, 43

(*www.imdb.com*)

a. Find the mean age and the median age of the Best Actress winners.

b. Find the mean age and the median age of the Best Actor winners.

c. On the same axis draw two box plots, one for the age of Best Actress winners and the other for the age of Best Actor winners.

d. Use your box plot to predict which data set has the greater standard deviation. Explain your reasoning. Then calculate the standard deviations to check your prediction.

At the 74th annual Academy Awards, Halle Berry became the first African American to win Best Actress. Denzel Washington became the second African American to win Best Actor.

18. APPLICATION Following the 1998 Academy Awards ceremony, Best Actress nominee Fernanda Montenegro (age 70) said that Gwyneth Paltrow (age 26) had won for *Shakespeare In Love* because she was younger. This comment caused Pace University undergraduate Michael Gilberg and Professor Terence Hines to test in their statistics class the proposition that younger women and older men are more likely to receive Academy Awards than are older women and younger men. Their study was published in the March 2000 issue of the *Journal of Psychological Reports.*

Assume that you are working with Gilberg and Hines. Use your statistics and graphs from Exercise 17 to confirm or refute the theory that younger women and older men are more likely to win these awards. Explain your conclusions.

19. Which box plot has the greater standard deviation? Explain your reasoning.

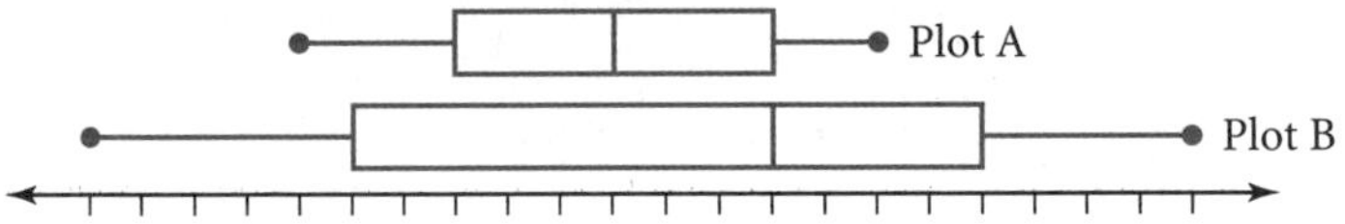

20. Consider these box plots. Group A conducted the rubber-band launch experiment 30 times, and Group B conducted the experiment 25 times.

a. How many data values are represented in each whisker of each box plot?

b. Which data set has the greater standard deviation? Explain how you know.

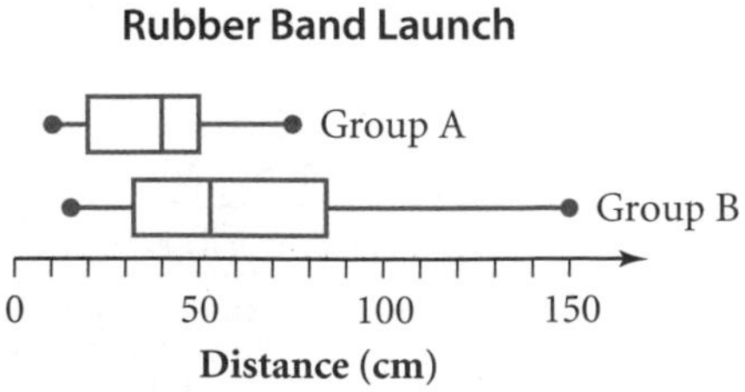

21. Invent two data sets, each with seven values, such that Set A has the greater standard deviation, and Set B has the greater interquartile range.

22. The table shown here contains the recorded extreme temperatures in degrees Fahrenheit for each of the seven continents.

a. Find the mean and standard deviation of the high temperatures.

b. Find the mean and standard deviation of the low temperatures.

Highest and Lowest Recorded Temperatures

Continent	High (°F)	Low (°F)
Africa	136	−11
Antarctica	59	−129
Asia	129	−90
Australia	128	−9
Europe	122	−67
North America	134	−87
South America	120	−27

(*www.infoplease.com/ipa/A0001375.html, www.infoplease.com/ipa/A0001377.html*)

23. Which temperatures, if any, are outliers in each set of data? Use the $2 \cdot IQR$ definition of outliers.

24. For a 20-day period, Ignacio kept a log of the amounts of time he spent doing homework and working.

Day	1	2	3	4	5	6	7	8	9	10
Homework (min)	20	50	200	55	275	230	230	115	285	140
Work (min)	390	150	75	360	125	150	450	200	175	280

Day	11	12	13	14	15	16	17	18	19	20
Homework (min)	325	290	260	190	190	195	225	135	205	220
Work (min)	60	25	475	135	190	250	50	210	300	170

a. Draw two box plots, one showing the amount of time spent doing homework and one showing the amount of time working. Which distribution has the greater spread?

b. Calculate the median and interquartile range, and the mean and standard deviation, for both homework and work. Which measure of spread best represents the data?

Assessing What You've Learned

- When presented with an unfamiliar problem, where will you start? Do you have a repertoire of strategies to help you get started? For example, some students always begin by drawing a picture. Just the act of putting a pencil to paper often helps begin the process of problem solving.
- What do you do when your first problem-solving attempt does not work? Do you have a repertoire of problem-solving strategies, such as working backward, creating a simpler problem, or using symbolic representation?
- Are you comfortable with entering large sets of data on your calculator and using the calculator to find statistical measures such as mean, median, mode, quartiles, and standard deviation?

- Do you know how to use the statistical measures listed above to analyze sets of data?
- Can you create and interpret box plots both by hand and with a graphing calculator?

Now that you have finished the first chapter, it's time to take stock of what you've learned. The ideas in this chapter are among the most important in the course and will be used throughout this book. Below are some general strategies for success.

JOURNAL Keeping a journal helps you to learn as you reflect on the course work. Informal notes about the mathematics you are learning and about areas of confusion or frustration help you decide when to seek assistance from your professor and what questions to ask. Keeping a journal is a good way to collect these informal notes, and if you write in it regularly, you'll track the progress of your understanding throughout the course. Write in such a way that you wouldn't mind having your professor read directly from the journal. Many successful mathematics students have found that small composition books with graph-paper pages are convenient for recording ideas and sketching graphs, tables, and diagrams. If you write in your journal each day after class, it will become a useful tool to use in reviewing for exams.

To help you organize your thoughts, here are some additional questions you might start writing about:

- What do you think were the main ideas of the chapter?
- Were there any exercises that you particularly struggled with? If so, are they clearer now? If some ideas are still foggy, what plans do you have to clarify them?
- How has your notion of algebra changed since you finished your last algebra course? Do you have particular expectations about what you will learn in this course? If so, what are they?
- As you worked on problem solving, what did you learn about your own strengths and weaknesses as a problem-solver? Do you consider yourself well organized? Do you have a systematic approach? What goals do you have for your own problem-solving skills?

FORM A STUDY GROUP Many people learn mathematics more effectively when they talk about the problems. A study group allows you to consolidate your own understanding as you discuss problems with others. You may also benefit from the regular discipline of getting together with your group at scheduled times to do homework or review for tests.

CHAPTER

2 Patterns and Recursion

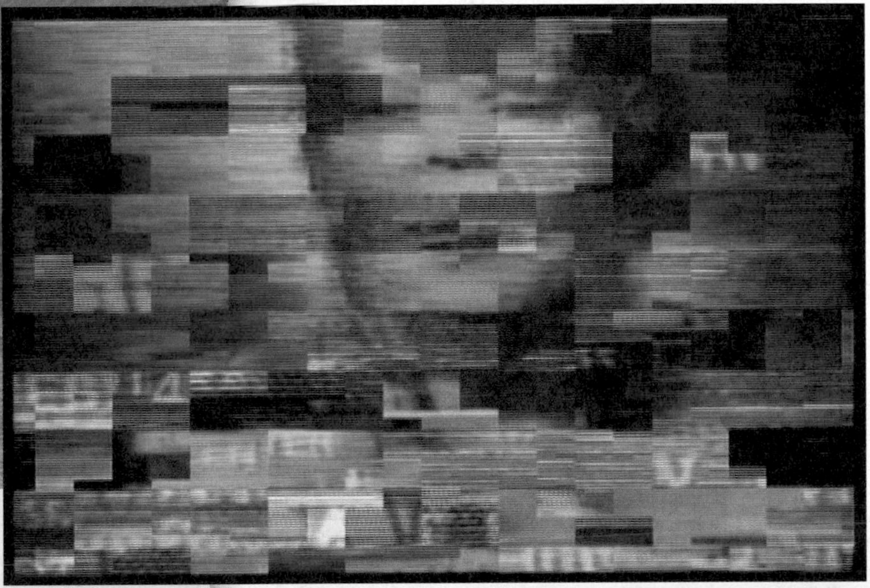

To create the video piece *Residual Light,* experimental video artist Anthony Discenza (American, b 1967) recorded 3 hours of commercial television by filming the TV screen while continuously channel surfing. This 3-hour sample was then compressed in stages by recording and re-recording the material on analog and digital tape while controlling the speed of the playback. Through this recursive process, the original 3 hours was gradually reduced to just 3 minutes. This 3-minute sequence was then slowed back down, resulting in a 20-minute loop.

OBJECTIVES

In this chapter you will

- recognize and visualize mathematical patterns called *sequences*
- write recursive definitions for sequences
- display sequences with graphs
- investigate what happens to sequences in the long run

LESSON 2.1

Recursively Defined Sequences

For every pattern that appears, a mathematician feels he ought to know why it appears.

W. W. SAWYER

Look around! You are surrounded by patterns and influenced by how you perceive them. You have learned to recognize visual patterns in floor tiles, window panes, tree leaves, and flower petals. In every discipline, people discover, observe, recreate, explain, generalize, and use patterns. Artists and architects use patterns that are attractive or practical. Scientists and manufacturing engineers follow patterns and predictable processes that ensure quality, accuracy, and uniformity. Mathematicians frequently encounter patterns in numbers and shapes and use them to connect familiar mathematical models to the real world or to expand their understanding of existing mathematical models.

You can discover and explain many mathematical patterns by thinking about recursion. **Recursion** is a process in which each step of a pattern is dependent on the step or steps that preceded it. It is often easy to define a pattern recursively, and a recursive definition reveals a lot about the properties of the pattern.

Scientists use patterns and repetition to conduct experiments, gather data, and analyze results.

The arches in the Santa Maria Novella cathedral in Florence, Italy, show an artistic use of repeated patterns.

EXAMPLE A

A square table seats 4 people. Two square tables pushed together seat 6 people. Three tables pushed together seat 8 people. How many people can sit at 10 tables arranged in a straight line? How many tables arranged in a straight line are needed to seat 32 people? Write a recursive definition to find the number of people who can sit at any linear arrangement of square tables.

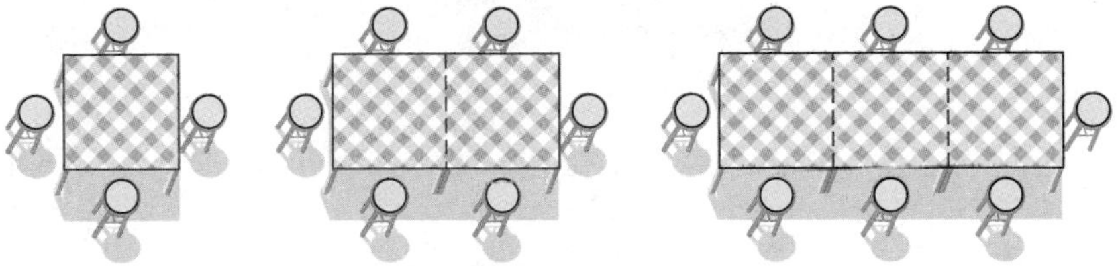

► *Solution*

If you sketch the arrangements of 4 tables and 5 tables, you'll notice that when you add another table, you seat 2 more people than in the previous arrangement. If you then put this information into tablular form, a clear pattern is revealed. You can continue this pattern to find that 10 tables seat 22 people.

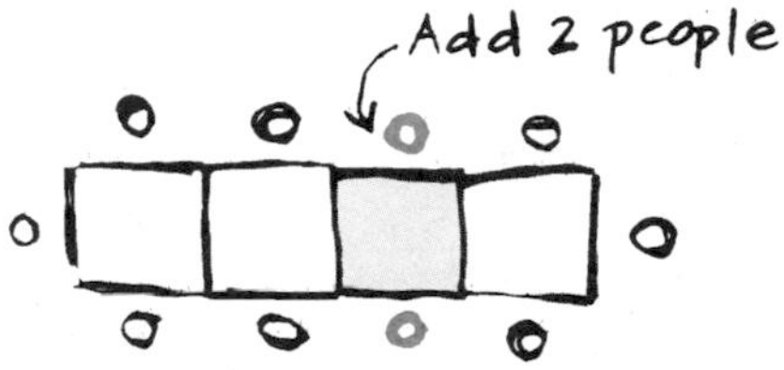

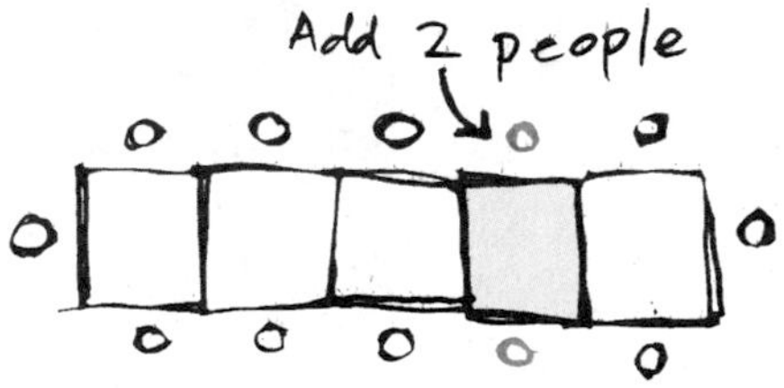

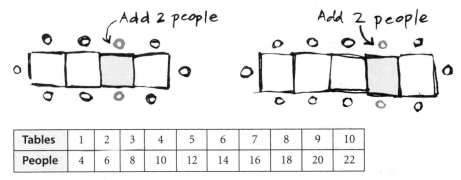

Tables	1	2	3	4	5	6	7	8	9	10
People	4	6	8	10	12	14	16	18	20	22

This graph shows the same information by plotting the points (1, 4), (2, 6), (3, 8), and so on. The graph too reveals a clear pattern. You can extend the graph to find that 15 tables are needed for 32 people.

You can use recursion—repeatedly adding 2 to the previous number of people—to find more numbers in the pattern or more points on the graph. You can write a recursive definition that gives the starting value and the mathematical operations required to find each subsequent value. For this example you could write

number of people at 1 table = 4

number of people at n tables = number of people at $(n - 1)$ tables + 2

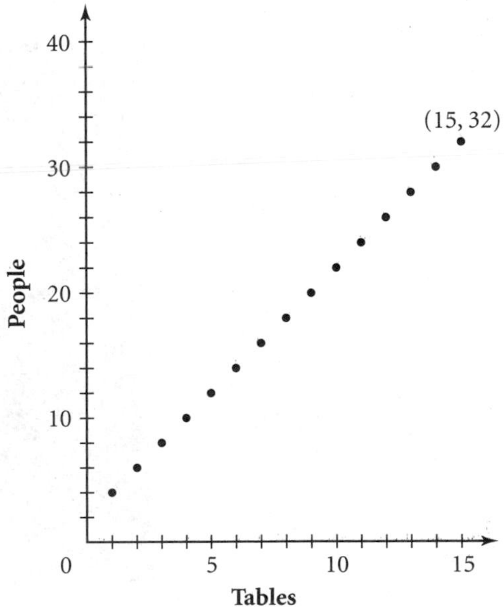

This definition summarizes how to use recursion to find the number of people who can sit at any linear arrangement of tables. For example, to find the number of people who can sit at 10 tables, take the number of people at 9 tables and add 2, or $20 + 2 = 22$.

A **sequence** is an ordered list of numbers. The table and graph in Example A represent the sequence

4, 6, 8, 10, 12, . . .

Each number in the sequence is called a *term.* The first term, u_1 (pronounced "u sub one"), is 4. The second term, u_2, is 6, and so on.

The *n*th term, u_n, is called the *general term* of the sequence. The general term is used to write the recursive definition algebraically as a **recursive formula.** The formula must specify one (or more) starting terms and a **recursive rule** that defines the *n*th term in relation to the previous term (or terms). The sequence 4, 6, 8, 10, 12, . . . is generated with this recursive formula:

$u_1 = 4$

$u_n = u_{n-1} + 2$ where $n \geq 2$

Because the starting value is $u_1 = 4$, the recursive rule $u_n = u_{n-1} + 2$ is first used to find u_2. This is clarified by saying that n must be greater than or equal to 2 to use the recursive rule.

This means that *the first term is 4,* and *each subsequent term is equal to the previous term plus 2.* Notice that each term, u_n, is defined in relation to the previous term, u_{n-1}. xample, the 10th term relies on the 9th term, or $u_{10} = u_9 + 2$.

EXAMPLE B

A concert hall has 59 seats in Row 1, 63 seats in Row 2, 67 seats in Row 3, and so on. The concert hall has 35 rows of seats. Write a recursive formula to find the number of seats in each row. How many seats are in Row 4? Which row has 95 seats?

An opera house in Sumter, South Carolina.

▶ Solution

First, it helps to organize the information as a table.

Row	1	2	3	4	...
Seats	59	63	67		...

Every recursive formula requires a starting term. Here, the starting term is 59, the number of seats in Row 1. That is, $u_1 = 59$.

This sequence also appears to contain a common difference between successive terms: 63 is 4 more than 59, and 67 is 4 more than 63. Use this information to write the recursive rule for the nth term, $u_n = u_{n-1} + 4$.

Therefore, this recursive formula generates the sequence representing the number of seats in each row:

$$u_1 = 59$$
$$u_n = u_{n-1} + 4 \text{ where } n \geq 2$$

You can use this recursive formula to calculate the number of seats in each row.
[▶ **See Calculator Note 2A** to learn how to perform recursion on your calculator. ◀]

u_1	59	The starting term is 59.
u_2	63	Substitute 59 for u_1. Add 4.
u_3	67	Substitute 63 for u_2. Add 4.
u_4	71	Continue, using recursion.

Because $u_4 = 71$, there are 71 seats in Row 4. If you continue the recursion process, you will find that $u_{10} = 95$, or that Row 10 has 95 seats.

In Examples A and B, the terms of the sequence are related in that each term is equal to the previous term plus a constant. This type of sequence is called an **arithmetic sequence.**

Arithmetic Sequence

An **arithmetic sequence** is a sequence in which each term is equal to the previous term plus a constant. This constant is called the **common difference.** If d is the common difference, the recursive rule for the sequence has the form

$$u_n = u_{n-1} + d$$

The key to identifying an arithmetic sequence is to recognize the common difference. If you are given a few terms and need to write a recursive formula, first try subtracting consecutive terms. If $u_n - u_{n-1}$ is constant for each pair of terms, then you know that your recursive rule must be a rule for an arithmetic sequence.

Investigation
Monitoring Inventory

CareerConnection In 2004, economists who held bachelor's degrees and went to work for the U.S. government could earn starting salaries of as much as $60,000 per year. Those with specialized knowledge, such as familiarity with Chinese culture, could earn considerably more. *Source: usajobs.opm.gov* Many colleges require substantial mathematics, including statistics and calculus for economics degrees.

Heater King, Inc., has purchased the parts to make 2000 water heaters. Each day, the workers assemble 50 water heaters from the available parts. The company has agreed to supply MegaDepot with 40 water heaters per day and Smalle Shoppe with 10 water heaters per day. MegaDepot currently has 470 water heaters in stock, and Smalle Shoppe has none. The management team at Heater King, Inc., needs a way to monitor inventory and demand.

Step 1 As a group, model what happens to the number of water heaters not made, the inventory at MegaDepot, and the inventory at Smalle Shoppe. Keep track of your daily results in a table like this one. [▶ See **Calculator Note 2A** for different ways to perform recursion on your calculator.◀]

Day	Water heaters not made	MegaDepot	Smalle Shoppe
1	2000	470	0

Step 2 Use your table from Step 1 to answer these questions:

a. In how many days will MegaDepot have a number of water heaters greater than or equal to the number of water heaters left to be made?

b. The assembly-line machinery malfunctions on the first day MegaDepot has a number of water heaters greater than twice the number of water heaters not made. How many water heaters does Smalle Shoppe have on the day assembly stops?

Step 3 Write a short summary of how your group modeled inventory and found the answers to the questions in Step 2. Compare your group's methods to the methods of other groups.

The sequences in Example A, Example B, and the Investigation are arithmetic sequences. Example C introduces a different kind of sequence that is still defined recursively.

EXAMPLE C

MathematicsConnection The Sierpiński triangle is named after the Polish mathematician Waclaw Sierpiński (1882–1969). He was most interested in number theory, set theory, and topology, three branches of mathematics that study the relations and properties of sets of numbers or points. Sierpiński was highly involved in the development of mathematics in Poland between World War I and World War II. He published 724 papers and 50 books in his lifetime. He introduced his famous triangle pattern in a 1915 paper, in which he described its properties.

This stamp, part of Poland's 1982 "Mathematicians" series, portrays Waclaw Sierpiński

The geometric pattern below is created recursively. If you continue the pattern endlessly, you create a *fractal* called a Sierpiński triangle. In a fractal the structure of any part is similar to the structure of the entire pattern. How many red triangles are there at Stage 20? How are the triangles at Stage 20 similar to the triangles at earlier stages?

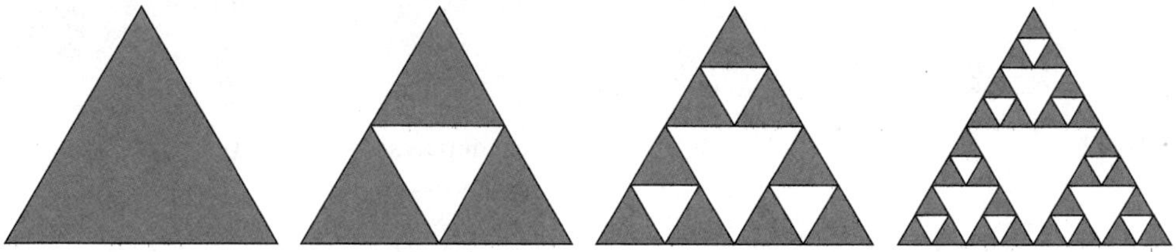

Stage 0 Stage 1 Stage 2 Stage 3

▸ Solution

Count the number of red triangles at each stage, and write a sequence.

1, 3, 9, 27, . . .

The starting term, 1, represents the number of triangles at Stage 0. In this case, $u_1 = 1$.

Starting with the secondary term, each term of the sequence multiplies the previous term by 3: so 3 is 3 times 1, 9 is 3 times 3, and 27 is 3 times 9. Use this information to write the recursive rule and complete your recursive formula.

$u_1 = 1$

$u_n = 3 \cdot u_{n-1}$ where $n \geq 2$

Using the recursive rule 20 times, you will find that $u_{21} = 3486784401$. There are more than 3 billion triangles at Stage 20!

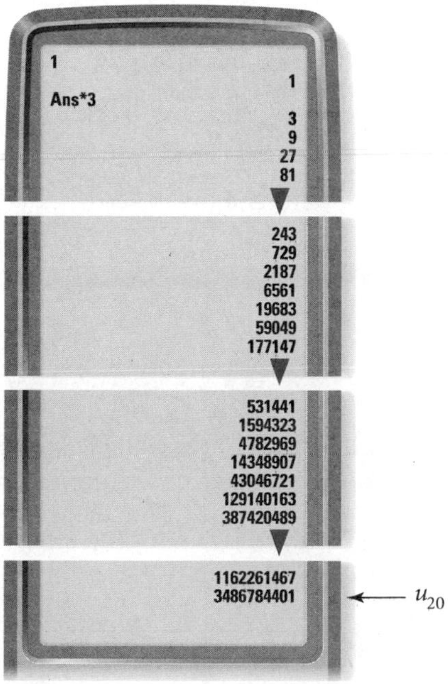

In Example C, consecutive terms of the sequence are related by multiplication. The common multiple relating the terms is called the *common ratio.* This type of sequence is called a *geometric sequence.*

Guitar feedback is a real-world example of recursion. When the amplifier is turned up high enough, the sound is picked up by the guitar and amplified again and again, creating a feedback loop. Jimi Hendrix (1942–1970), a pioneer in the use of feedback and distortion in rock music, is one of the legendary guitar players of the 1960s.

Geometric Sequence

A **geometric sequence** is a sequence in which each term is equal to the previous term multiplied by a constant, the **common ratio.** If r is the common ratio, the recursive rule for the sequence has the form

$$u_n = r \cdot u_{n-1}$$

A geometric sequence is identified by dividing consecutive terms. If $\frac{u_n}{u_{n-1}}$ is constant for each pair of terms, then you know you must write a recursive rule for a geometric sequence.

Arithmetic and geometric sequences are the most basic sequences because the recursive rule for each uses only one operation, addition in the case of arithmetic sequences and multiplication in the case of geometric sequences. Recognizing these basic operations will help you easily identify sequences and write recursive formulas.

EXERCISES

Practice Your Skills

1. Write the first 4 terms of each sequence.

a. $u_1 = 20$
$u_n = u_{n-1} + 6$ where $n \geq 2$

b. $u_1 = 47$
$u_n = u_{n-1} - 3$ where $n \geq 2$

c. $u_0 = 32$
$u_n = 1.5 \cdot u_{n-1}$ where $n \geq 1$

d. $u_1 = -18$
$u_n = u_{n-1} + 4.3$ where $n \geq 2$

2. Identify each sequence in Exercise 1 as arithmetic or geometric. State the common difference or the common ratio for each.

3. Write a recursive formula, and use it to find the missing table values.

n	1	2	3	4	5	. . .	?
u_n	40	36.55	33.1	29.65	?	. . .	12.4

4. Write a recursive formula to generate an arithmetic sequence with a first term 6 and a common difference 3.2. Find the 10th term.

5. Write a recursive formula to generate each sequence. Then find the indicated term.

a. 2, 6, 10, 14, . . . Find the 15th term.

b. 10, 5, 0, −5, . . . Find the 12th term.

c. 0.4, 0.04, 0.004, 0.0004, . . . Find the 10th term.

d. −2, −8, −14, −20, −26, . . . Find the 30th term.

e. 1.56, 4.85, 8.14, 11.43, . . . Find the 14th term.

f. −6.24, −4.03, −1.82, 0.39, . . . Find the 20th term.

HistoryConnection Rózsa Péter (1905–1977), a Hungarian mathematician, was a leading contributor to the theory of recursion and the first person to propose the study of recursion in its own right. Her interest in communicating mathematics to nonmathematicians led her to write *Playing with Infinity: Mathematical Explorations and Excursions.* Another of her books, *Recursive Functions in Computer Theory,* describes the connections between recursion and computer languages. In an interview she described recursion this way:

> The Latin technical term "recursion" refers to a certain kind of stepping backwards in the sequence of natural numbers, which necessarily ends after a finite number of steps. With the use of such recursions the values of even the most complicated functions used in number theory can be calculated in a finite number of steps.

Rózsa Péter

6. Write a recursive formula for the sequence graphed at right. Find the 46th term.

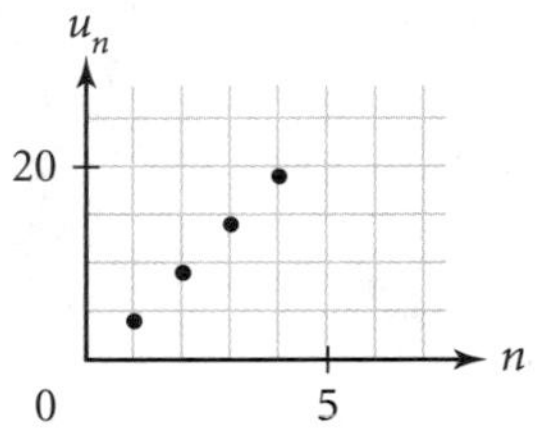

Reason and Apply

7. Write a recursive formula that you can use to find the number of segments, u_n, for Figure n of this arithmetic pattern. Use your formula to complete the table.

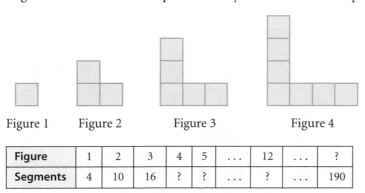

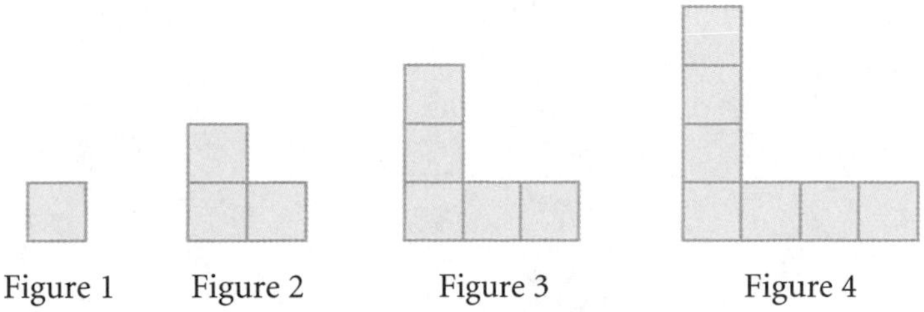

Figure	1	2	3	4	5	. . .	12	. . .	?
Segments	4	10	16	?	?	. . .	?	. . .	190

8. A 50-gallon (gal) bathtub contains 20 gal of water and is filling at a rate of 2.4 gal/min. You check the tub every minute on the minute.

a. Suppose that the drain is closed. When will you discover that the water is flowing over the top?

b. Now, suppose that the bathtub contains 20 gal of water and is filling at a rate of 2.4 gal/min, but the drain is open, and water drains at a rate of 3.1 gal/min. When will you discover that the tub is empty?

c. Write a recursive formula that you can use to find the water level at any minute due to both the 2.4-gal/min rate of filling and the 3.1 gal/min rate of draining.

9. A car leaves town heading west at 57 km/h.

a. How far will the car travel in 7 h?

b. A second car leaves town 2 h after the first car, but the second car is traveling at 72 km/h. To the nearest hour, when will the second car pass the first?

10. **APPLICATION** Inspector 47 at the Zap Battery plant keeps a record of the number of defective AA batteries she finds each day on an assembly line. Although the number of faulty batteries do not make an exact sequence, she estimates an arithmetic sequence.

a. Write a recursive formula for an arithmetic sequence that estimates the number of defective batteries. Explain your reasoning.

b. Predict the number of batteries that will be found defective on the 10th day if nothing is done to improve the process.

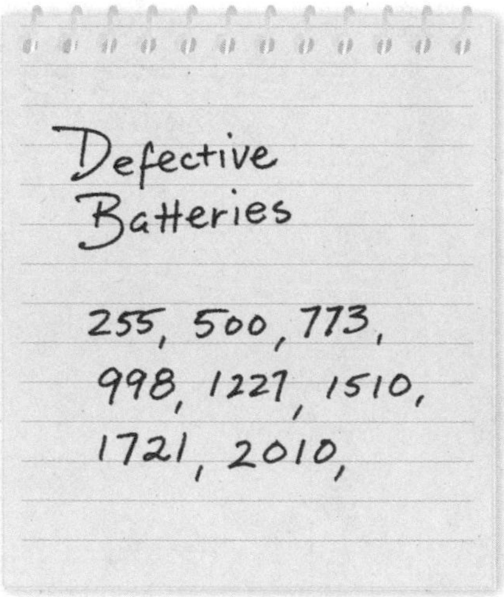

11. During the week of February 14, the owner of Nickel's Appliances stocks hundreds of red, heart-shaped vacuum cleaners. The next week he still has hundreds of red, heart-shaped vacuum cleaners. He tells his manager, "Discount the price 25% each week until they are gone."

a. On February 14 the vacuum cleaners are priced at \$80. What is the price of a vacuum cleaner during the 2nd week?

b. What is the price during the 4th week?

c. When will the vacuum cleaners sell for less than \$10?

12. Taoufik picks up his homework paper from the puddle into which it has fallen. Sadly he reads the first problem and finds that the arithmetic sequence is a blur except for two terms.

a. What is the common difference? How did you find it?

b. What are the missing terms?

c. What is the answer to Taoufik's homework problem?

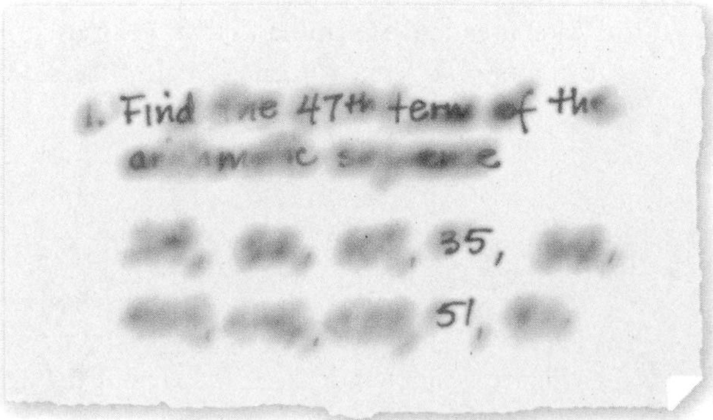

Review

13. (Lessons 1.4 and 1.5) Dr. Tesla determines the means of his student's scores to calculate their grades. Students who average above 90% earn a grade of A. Those who earn more than 80% but less than 90% earn a grade of B. The table below gives the scores for two students, Maria and Syed, on the six tests given this semester.

	Test 1	Test 2	Test 3	Test 4	Test 5	Test 6
Maria	92	94	54	100	90	90
Syed	85	88	86	88	89	84

a. What grade did each student earn?

b. Do you think this system is fair to both students? Using mathematical terms you learned in Chapter 1, explain why or why not.

c. How could Maria make an argument using standard deviation and outliers (and one really bad week) to make the case that she deserves a higher grade?

LESSON

2.2 Modeling Growth and Decay

MORE BACKGROUND

Before you begin this lesson, you should be familiar with this concept:

- Using the distributive property to factor expressions

A short review of this concept can be found at www.keycollege.com/online .

Each sequence you generated in Lesson 2.1 was either an arithmetic sequence with a recursive rule in the form $u_n = u_{n-1} + d$ or a geometric sequence with a recursive rule in the form $u_n = r \cdot u_{n-1}$. Arithmetic sequences have a common difference between terms, whereas geometric sequences have a common ratio. You will have to consider each sequence carefully to decide whether it is arithmetic or geometric.

So far you have used u_1 as the first term of each sequence. In some situations (like the one in the next Investigation), it is more meaningful to treat the first term as the zero term, or u_0. The subscript 0 is often used in situations involving time. The zero term represents the starting value before any change occurs. In many problems you can decide whether it would be better to begin at u_0 or at u_1.

EXAMPLE A

An automobile depreciates, or loses value, as it gets older. Different automobiles depreciate at different rates. Suppose that a particular automobile loses $\frac{1}{5}$ of its value each year. Write a recursive formula to find the value of this car when it is 6 years old if it cost $23,999 when it was new.

ConsumerConnection The *Kelley Blue Book,* first compiled in 1926 by Les Kelley, annually publishes the standard values of every vehicle on the market. Many people who want to know the value of an automobile will ask what its "Blue Book" value is. The *Kelley Blue Book* calculates the value of a car by accounting for its make, model, year, mileage, location, and condition.

▶ Solution

Each year, the car will be worth $\frac{4}{5}$ of its previous year's value, so the recursive sequence is geometric. It is convenient to start with $u_0 = 23999$ to represent the value of the car when it was new so that u_1 will represent the value after one year and so on. The recursive rule that generates the sequence of annual values is

$u_0 = 23999$	Starting value.
$u_n = 0.8 \cdot u_{n-1}$ where $n \geq 1$	$\frac{4}{5}$ is 0.8.

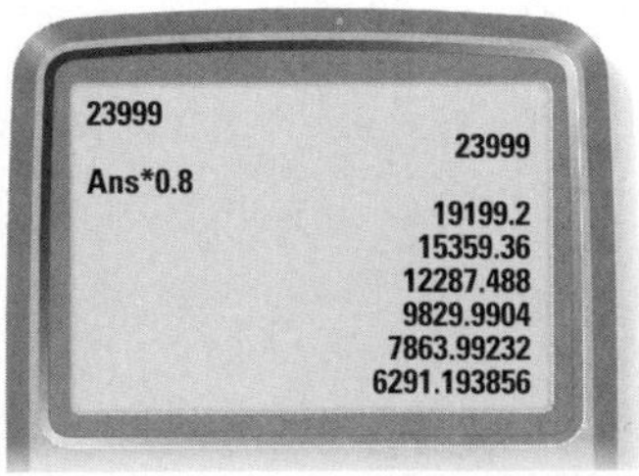

Use this rule to find the 6th term.

After 6 years the car is worth $6,291.19.

Investigation
Looking for the Rebound

You will need

- a ball
- a motion sensor

When you drop a ball, the rebound height becomes smaller after each bounce. In this investigation you will write a recursive formula for the height of a real ball as it bounces.

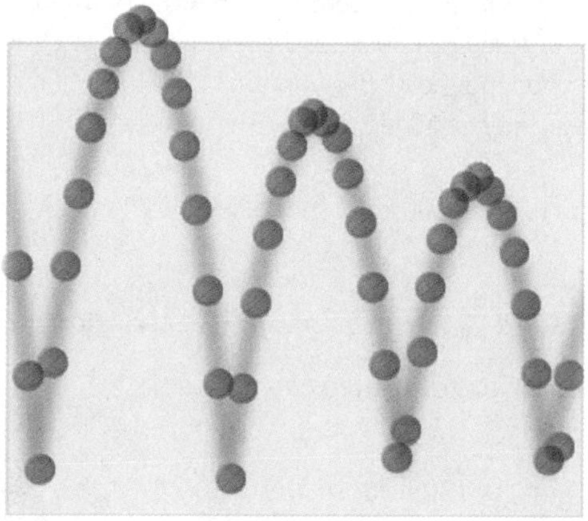

Procedure Note

1. Hold the motion sensor above the ball.
2. Press the trigger, then release the ball.
3. If the ball drifts, try to follow it and maintain the same height with the motion sensor.
4. If you do not capture at least six good consecutive bounces, repeat the procedure.

Step 1 Set up your calculator and motion sensor to collect bouncing ball data. [▸ See **Calculator Note 2D** for a calculator program to help you gather data for this investigation.◂]

Step 2 The data transferred to your calculator is in the form (x, y), where x is the time since you pressed the trigger and y is the height of the ball. Trace the data graphed by your calculator to find the initial height and the rebound height after each bounce. Record your data in a table.

Step 3 Graph your data table on paper and on your calculator. [▸ See **Calculator Note 2C.**◂]

Step 4 Compute the rebound ratios for consecutive bounces.

$$\text{rebound ratio} = \frac{\textit{rebound height}}{\textit{previous rebound height}}$$

Step 5 Decide on a single value that best represents the rebound ratio for your ball. Use this ratio to write a recursive formula that models your sequence of *rebound height* data, and use it to generate the first six terms.

Step 6 Compare your experimental data to the terms generated by your recursive formula. How close are they? Describe some of the factors that might affect this experiment. For example, how might the formula change if you used a different kind of ball?

You may find it easier to think of the common ratio as the whole (1) plus or minus a percent change. In place of r, you can write $(1 + p)$ or $(1 - p)$. The car example involved a 20% $\left(\frac{1}{5}\right)$ loss, so the common ratio could be written as $(1 - 0.20)$. Your bouncing ball may have had a ratio of 0.75, which you can write as $(1 - 0.25)$, or a 25% loss per bounce. These are examples of **decay,** or geometric sequences that decrease. The next example is one of **growth,** or a geometric sequence that increases.

EXAMPLE B Gloria deposits \$2000 into a bank account that pays 7% annual interest, compounded annually. This means the bank pays her 7% of her account balance as interest at the end of each year, and she leaves the original amount and the interest in the account. Find the number of years it will take for the original deposit to double in value.

> **Economics Connection** Interest is a charge that you pay for borrowing money or that the bank pays you to keep the money in your bank account. Simple interest is a percentage paid on the total amount (*principal*) over a period of time. If you leave the interest in the account, in the next time period you will receive interest on both the principal and the interest that were in your account. This is called *compounded interest* because you are receiving interest on the interest. Most banks compound the interest monthly or daily, resulting in higher annual percentage yields (APY), or total interest added to your account each year.

▶ Solution The balance starts at \$2000 and increases by 7% each year.

$u_0 = 2000$

$u_n = (1 + 0.07)u_{n-1}$ where $n \geq 1$ The recursive rule that represents 7% growth.

Use your calculator to compute year-end balances recursively.

The 11th term, u_{11}, is 4209.70, so the investment balance will more than double in 11 years.

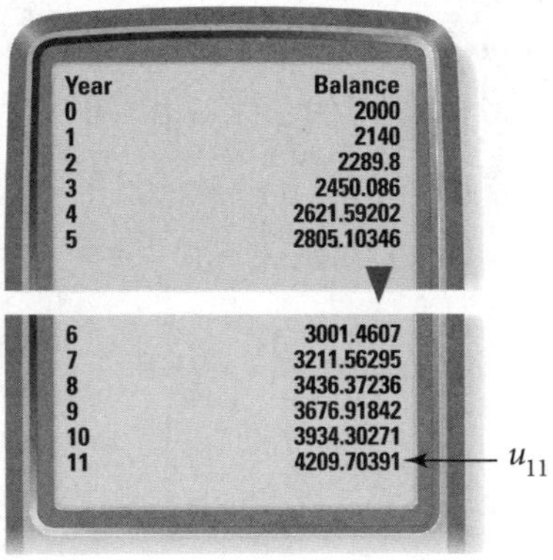

Year Balance
0 2000
1 2140
2 2289.8
3 2450.086
4 2621.59202
5 2805.10346
6 3001.4607
7 3211.56295
8 3436.37236
9 3676.91842
10 3934.30271
11 4209.70391
u_{11}

Spreadsheets are useful tools for calculating recursive sequences.

Compounding interest has many applications in everyday life. The interest on both savings accounts and loans is almost always compounded, often leading to surprising results. For example, a graph of the account balance in the previous example looks like this:

Leaving just \$2000 in the bank at a good interest rate for 11 years can double your money. In another 6 years, the money will triple. Some banks will compound the interest monthly. You can write the common ratio as $\left(1 + \frac{0.07}{12}\right)$ to represent $\frac{1}{12}$ of the interest, compounding monthly. How would you change this rule to show that the interest is compounded 52 times a year?

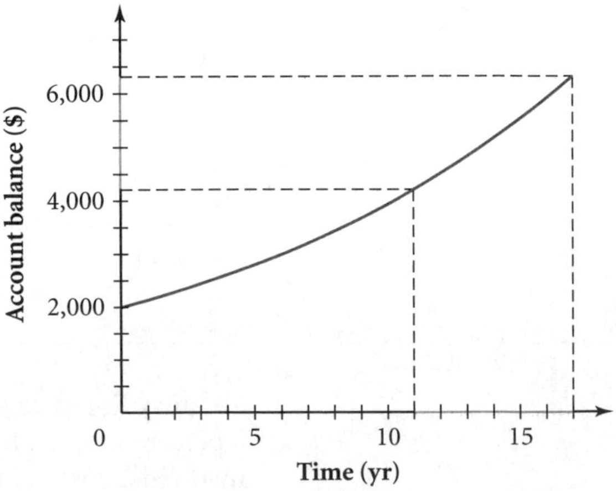

EXERCISES

Practice Your Skills

1. Find the common ratio for each sequence.

a. 100, 150, 225, 337.5, 506.25 . . .

b. 73.4375, 29.375, 11.75, 4.7, 1.88 . . .

c. 80.00, 82.40, 84.87, 87.42, 90.04 . . .

d. 208.00, 191.36, 176.05, 161.97 . . .

2. Identify each sequence in Exercise 1 as growth or decay. Give the percent change of each.

3. Write a recursive formula for each sequence in Exercise 1, and find the 10th term. Use u_1 for the first term given.

4. Match each recursive rule to a graph. Explain your reasoning.

a. $u_1 = 10$

$u_n = (1 - 0.25) \cdot u_{n-1}$ where $n \geq 2$

b. $u_1 = 10$

$u_n = (1 + 0.25) \cdot u_{n-1}$ where $n \geq 2$

c. $u_1 = 10$

$u_n = 1 \cdot u_{n-1}$ where $n \geq 2$

i.

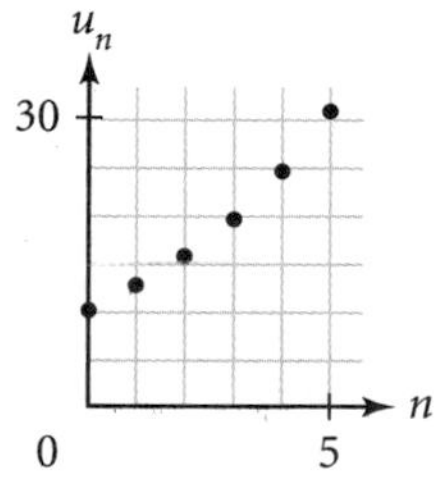

ii.

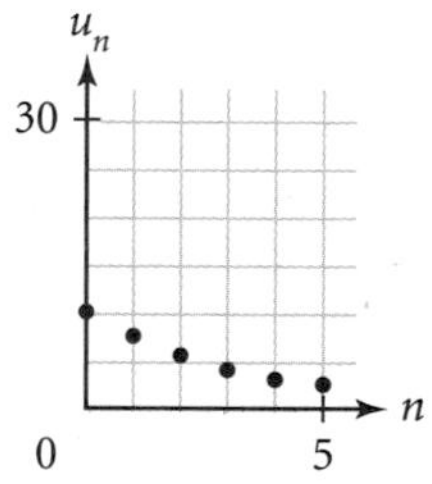

iii.

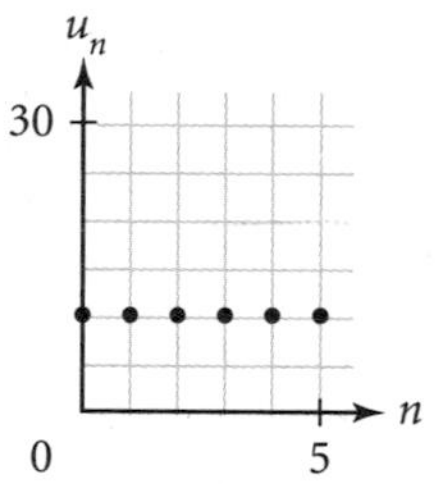

5. Factor these expressions so that the variable appears only once.

a. $u_{n-1} + 0.07u_{n-1}$

b. $A - 0.18A$

c. $x + 0.08125x$

d. $2u_{n-1} - 0.85u_{n-1}$

Reason and Apply

6. Suppose the recursive formulae $u_0 = 100$ and $u_n = (1 - 0.20)u_{n-1}$, where $n \geq 1$, models a bouncing ball. Give real-world meanings for the numbers 100 and 0.20.

7. Suppose the initial height from which a rubber ball drops is 100 in. The rebound heights to the nearest inch are 80, 64, 51, 41, ...

a. What is the rebound ratio for this ball?

b. What is the height of the 10th rebound?

c. After how many bounces will the rebound height be less than 1 in.? Less than 0.1 in.?

8. Suppose the recursive formula $u_{2003} = 250000$ and $u_n = (1 - 0.025)u_{n-1}$, where $n \geq 2004$, describes an investment made in the year 2003. Give real-world meanings for the numbers 250,000 and 0.025.

9. **APPLICATION** A small company with 12 employees is growing at a rate of 20% a year. If its business keeps growing at the same rate, the company will need to hire more employees to keep up with the growth.

a. How many people should the company plan to hire in each of the next 5 years?

b. How many employees will the company have in 5 years?

EconomicsConnection In labor negotiations both the union and the management use recursion and spreadsheets to determine wage growth. The union is interested in seeing that wages keep pace with inflation, the management in seeing that wages do not outpace growth in the company's income.

10. **APPLICATION** The table below shows investment balances over time.

Elapsed time n (yr)	0	1	2	3	. . .
Balance u_n (\$)	2000	2170	2354.45	2554.58	. . .

a. Write a recursive formula that generates the balances in the table.

b. What is the annual interest rate (APY)?

c. In how many years will the original deposit triple in value?

11. Suppose Jill's biological family tree looks like the diagram below. You can model recursively the number of people in each generation.

a. Make a table showing the number of Jill's ancestors in each of the past five generations. Use u_0 to represent Jill's generation.

b. Describe how to find the number of ancestors within a generation if the number of ancestors in the preceding generation is known. Write a recursive formula.

c. Find the number of the term of this sequence that is closest to 1 billion. What is the real-world meaning of this answer?

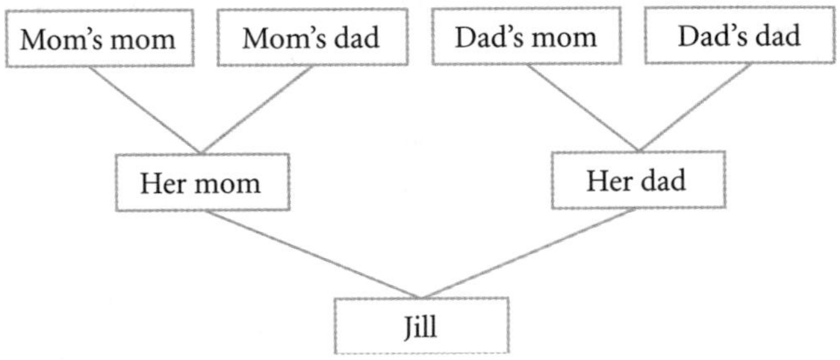

d. If a new generation is born every 25 years, approximately when did Jill have 1 billion living ancestors?

e. Your answer to 11c assumes there are no duplicates, that is, no common ancestors on Jill's mother's or her father's sides of the family. Look up Earth's population for the year you found in 11d. You will find helpful links at www.keycollege.com/online . Write a few sentences describing any problems you have with the assumption that Jill's parents had no common ancestors.

12. **APPLICATION** Carbon dating is used to find the ages of remains of once-living things. Carbon-14 is found naturally in all living things, and it decays slowly after their deaths. About 88.55% of it remains after any 1000-year period.

Let 100%, or 1, be the beginning amount of carbon-14. At what point will less than 5% remain? Write the recursive formula you used.

13. **APPLICATION** Suppose that $500 is deposited into an account that earns 6.5% annual interest and that no more deposits and no withdrawals are made.

a. If the interest is compounded monthly, what is the monthly rate?

b. What is the balance after 1 month?

c. What is the balance after 1 year?

d. What is the balance after 29 months?

14. **APPLICATION** Between 1970 and 2000, the population of Grand Traverse County, Michigan, grew from 39,175 to 77,654.

a. Find the percent increase over the 30-year period.

b. What do you think the *annual* growth rate was during this period?

c. Check your answer to 14b by using a recursive formula. Do you get 77,654 people after 30 years? Explain why your recursive formula may not work.

d. Use guess-and-check to find a growth rate, to the nearest 0.1% (or 0.001), that comes closest to producing the 30-year growth in Grand Traverse County.

e. Use your answer to 14d to estimate the population in 1985. How does this compare with the average of the populations of 1970 and 2000? Why is this so?

15. Taufiq looked at the second problem in his wet notes that had fallen into a puddle.

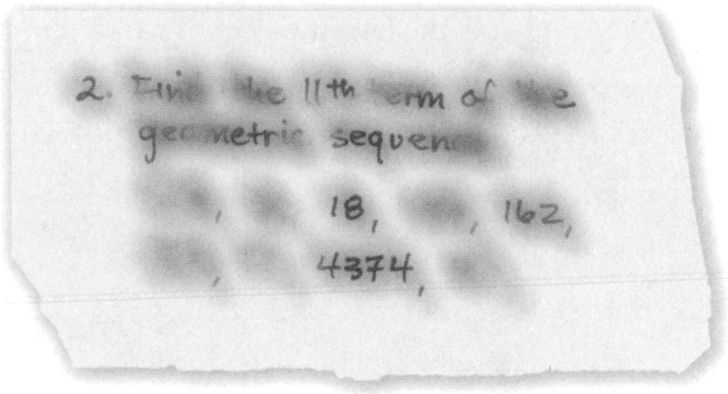

a. What is the common ratio? How did you find it?

b. What are the missing terms?

c. Solve the problem.

16. **APPLICATION** Suppose the signal strength in a fiber-optic cable diminishes by 15% every 10 miles.

a. What percent of the original signal is left after a stretch of 10 miles?

b. Create a table of signal strength in 10-mile intervals, and make a graph of the sequence.

c. If the phone company plans to boost the signal just before it falls to 1%, how far apart should the company place its booster stations?

TechnologyConnection Fiber-optic technology uses light pulses to transmit information from one transmitter to another over fiber lines made of silica (glass). Fiber-optic strands are used in telephone wires, cable television lines, power systems, and other communications. These strands operate on the principle of total internal reflection, which means that the light pulses cannot escape from the glass tube and instead bounce information from transmitter to transmitter.

Review

17. (Lesson 2.1) Consider the sequence 180, 173, 166, 159, ...

a. Write a recursive formula. Use $u_1 = 180$.

b. What is u_{10}?

c. What is the first term with a negative value?

18. (Lesson 2.1) Write a recursive formula for each sequence. Then find the indicated term.

a. 3, 9, 15, 21, . . .	Find the 35th term.
b. 6, 2, −2, −6, . . .	Find the 11th term.
c. 2, 1, 0.5, 0.25, . . .	Find the 20th term.
d. 100, 107, 114.49, 122.5043	Find the 6th term.

19. (Lesson 2.1) In the first lesson of the book, the candy dish problem is introduced. Suppose a dish is full of candy at the beginning of an evening. During the night, Dad sneaks downstairs and takes 25% of the candy. Later, Mom sneaks downstairs and eats 25% of the candy. Next, the brother eats 25% of the candy. Finally, the sister comes downstairs and eats the rest of the candy.

a. Let u_0 represent the amount of candy at the start of the evening. Write a recursive formula to describe the amount of candy left in the dish after each family member has taken 25%.

b. Supposing that no one ate only part of a piece of candy, find a possible value for u_3. What does u_3 represent in terms of the problem?

c. What is the fewest pieces of candy the sister could have eaten?

d. What is the fewest pieces of candy the sister could have eaten if instead of eating 25% of the candy each of the other family members ate 20%?

LESSON

2.3

A First Look at Limits

MORE BACKGROUND

Before you begin this lesson, you should be familiar with this concept:

- the Pythagorean Theorem

A short review of this concept can be found at: www.keycollege.com/online .

In three words I can sum up everything I know about life: It goes on.

ROBERT FROST

Increasing arithmetic and geometric sequences, such as the number of new triangles at each stage in a Sierpiński triangle or the periodic balance of money earning interest in the bank, have terms that get larger and larger. But can a tree continue to grow taller year after year? Can people continue to build taller and taller buildings, run faster and faster, and jump higher and higher, or are there limits?

Decreasing sequences may also have limits. For example, the temperature of a cup of hot cocoa as it cools, taken at one-minute intervals, produces a sequence that approaches the temperature of the room. In the long run, the hot cocoa will be at room temperature.

In the next Investigation you will explore what happens to a sequence in the long run.

How large can a tree grow? It depends partly on environmental factors, such as disease and climate. Like humans, trees have mechanisms that slow their growth as they age. (Unlike humans, however, a tree may not reach maturity until 100 years after it starts growing.) The tallest tree alive is a redwood in Montgomery Woods State Reserve, California, that is over 367 feet—five stories taller than the Statue of Liberty.

The women's world record for the fastest time in the 100-m dash has decreased by about 3 s in 66 years. Marie Mejzlíková of Czechoslovakia set the record at 13.6 s in 1922, and Florence Griffith-Joyner of the United States set it at 10.49 s in 1988. In the 1998 article entitled "How Good Can We Get?" Jonas Mureika predicts that the ultimate performance by a woman in the 100-m dash will be 10.15 s.

Investigation
Doses of Medicine

You will need

- a bowl
- a supply of water
- a supply of tinted liquid
- measuring cups graduated in milliliters (mL)
- a sink or waste bucket

Our kidneys continuously filter our blood, removing impurities. Doctors take this into account when prescribing the dosage and frequency of medicine.

In this investigation you will simulate what happens in the body when a patient takes medicine. To represent the blood in a patient's body, use a bowl containing a total of 1 liter (L) of liquid. Start with 16 milliliters (mL) of tinted liquid to represent a dose of medicine in the blood, and use clear water for the rest.

Science Connection To find out how well a person's kidneys function, physicians perform a creatinine clearance test. Creatine is a substance that the body makes as it converts food to energy through the process of metabolism. Creatine is further broken down into creatinine, which the kidneys release and expel through the bladder. Doctors use this formula to estimate a person's ideal creatinine clearance:

$$\frac{(\textit{body weight}) \cdot (140 - \textit{age}) \cdot (0.85 \textit{ if female})}{72 \cdot (\textit{stable creatinine})}$$

Step 1 Suppose a patient's kidneys filter out 25% of this medicine each day. To simulate this, remove $\frac{1}{4}$, or 250 mL, of the mixture from the bowl, and replace it with 250 mL of clear water to represent filtered blood. Make a table like the one below, and record the amount of medication in the blood over several days. Repeat the blood/medicine simulation for each day.

Day	Amount of medicine (mL)
0	16
1	
2	
3	

Step 2 Write a recursive formula that generates the sequence in your table.

Step 3 How many days will pass before there is less than 1 mL of medicine in the blood?

Step 4 Is the medicine ever completely removed from the blood? Why or why not?

Step 5 Sketch a graph, and describe what happens in the long run.

Often a single dose of medicine is not enough. Doctors prescribe regular doses to produce and maintain a sufficiently high level of medicine in the body. You will now modify your simulation to see what happens when a patient takes medicine daily over a period of time.

Step 6 Start over with 1 L of liquid. Again, all of the liquid is clear water, representing the blood, except for 16 mL of tinted liquid that represents the initial dose of medicine. Each day 250 mL of liquid are removed and replaced with 234 mL of clear water and 16 mL of tinted liquid to represent a new dose of medicine. Complete another table like the one in Step 1, recording the amount of medicine in the blood over several days.

Step 7 Write a recursive formula that generates this sequence.

Step 8 Do the contents of the bowl ever turn into pure medicine? Why or why not?

Step 9 Sketch a graph, and explain what happens to the level of medicine in the blood after many days.

The elimination of medicine from the human body is a real-world example of a dynamic, or changing, system. A contaminated lake and its cleanup processes are another. Dynamic systems often reach a point of stability in the long run. The quantity associated with that stability, such as the number of milliliters of medicine, is called a *limit.* In mathematics, we say that the sequence of numbers associated with the system approaches that limit. The ability to predict limits is very important for analyzing these situations. The long-run value helps you estimate limits.

Each of the sequences in the investigation approached a different long-run value. The first sequence approached zero. The second sequence was *shifted,* and it approached a nonzero value. A **shifted geometric sequence** includes an added term in the recursive rule. Let's look at another example of a shifted geometric sequence.

EnvironmentalConnection On April 25, 1986, a nuclear reactor in Chernobyl exploded, spreading radioactive elements over a large portion of Eastern Europe. Some areas near Chernobyl were contaminated at 40 times the level considered safe by the Soviet government. The decay of radioactive elements can be modeled by a geometric series with a limit value of 0. Lesson 5.1 includes an Investigation modeling radioactive decay.

EXAMPLE

Antonio and Deanna are working at the community swimming pool for the summer. They need to provide a "shock" treatment of 450 g of dry chlorine to prevent the growth of algae in the pool. After the initial treatment, they add 45 g of chlorine each day. Each day, the sun burns off 15% of the chlorine. Find the amount of chlorine in the pool after 1 day, 2 days, and 3 days. Create a graph that shows the chlorine level after several days and in the long run.

▶ *Solution*

The starting value is given as 450. This amount decays by 15% a day, but 45 g are also added each day. The amount remaining after each day is generated by the rule $u_n = (1 - 0.15)u_{n-1} + 45$, or $u_n = 0.85u_{n-1} + 45$. Use this rule to find the chlorine level in the long run.

$u_0 = 450$ The initial shock treatment.

$u_1 = 0.85(450) + 45 = 427.5$ The amount after 1 day.

$u_2 = 0.85(427.5) + 45 \approx 408.4$ The amount after 2 days.

$u_3 = 0.85(408.4) + 45 \approx 392.1$ The amount after 3 days.

To find the long-run value representing the amount of chlorine, you can continue to evaluate terms until the value stops changing, or see where the graph levels off. From the graph, the long-run value appears to be 300 g of chlorine.

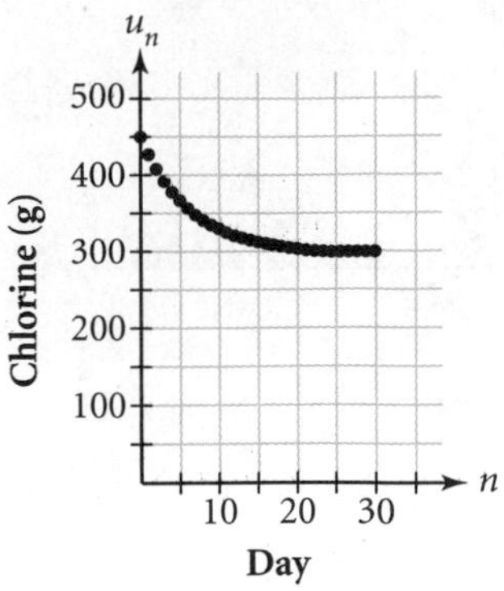

You can also use algebra to find the value of the terms as they level off. If you assume that terms stop changing, then you can set the value of the next term equal to the value of the previous term and solve the equation.

$u_n = 0.85u_{n-1} + 45$	Recursive rule.
$c = 0.85c + 45$	Assign the same variable to u_n and u_{n-1}.
$0.15c = 45$	Subtract 0.85c from both sides.
$c = 300$	Divide both sides by 0.15.

The amount of chlorine will level off at 300 g, which agrees with the long-run value estimated from the graph.

The study of limits is an important part of calculus, the mathematics of change. Understanding limits mathematically will help you work with other real-world applications in biology, chemistry, physics, and social science.

EXERCISES

Practice Your Skills

1. Find the value of u_1, u_2, and u_3. Identify the type of sequence (arithmetic, geometric, or shifted geometric) and state whether it is increasing or decreasing.

a. $u_0 = 16$
$u_n = (1 - 0.05)u_{n-1} + 16$ where $n \geq 1$

b. $u_0 = 800$
$u_n = (1 - 0.05)u_{n-1} + 16$ where $n \geq 1$

c. $u_0 = 50$
$u_n = (1 - 0.10)u_{n-1}$ where $n \geq 1$

d. $u_0 = 40$
$u_n = (1 - 0.50)u_{n-1} + 20$ where $n \geq 1$

2. Solve each equation.

a. $a = 210 + 0.75a$

b. $b = 0.75b + 300$

c. $c = 210 + c$

d. $d = 0.75d$

3. Find the long-run value for each sequence in Exercise 1.

4. Write a recursive formula for each sequence.

a. 200.00, 216.00, 233.28, 251.94 . . .

b. 0, 10, 15, 17.5, 18.75

Reason and Apply

5. The Osbornes have a small swimming pool and are giving it a chlorine treatment. The recursive formula below gives the daily amount of chlorine in the pool.

$u_0 = 300$

$u_n = (1 - 0.15)u_{n-1} + 30$ where $n \geq 1$

a. Explain the real-world meanings of the values 300, 0.15, and 30 in this formula.

b. Describe what happens to the chlorine level over the long run.

6. APPLICATION On October 1, 2005, Sal invests $24,000 in a bank account that earns 3.4% annually, compounded monthly. A month later, he withdraws $100 and continues to withdraw $100 on the first day of every month thereafter.

a. Write a recursive formula for this problem.

b. List the first five terms of this sequence of balances, starting with the initial investment.

c. What is the meaning of the value of u_4?

d. What is the balance on October 2, 2006? On October 2, 2008?

7. APPLICATION Consider the bank account in Exercise 6.

a. What happens to the balance if the same interest and withdrawal patterns continue for a long time? Does the balance ever level off?

b. What monthly withdrawal amount would maintain a constant balance of $24,000 in the long run?

8. APPLICATION The Forever Green Nursery owns 7000 white pine trees. The nursery plans to sell 12% of its trees and plant 600 new ones each year.

a. Find the number of pine trees owned by the nursery after 10 years.

b. Find the number of pine trees owned by the nursery after many years, and explain what is occurring.

c. What equation can you solve to find the number of trees in the long run?

d. Try different starting totals in place of the 7000 trees. Describe any changes to the long-run value.

e. In the 5th year, a disease destroys many of the nursery's trees. How does the solution change?

9. APPLICATION Jack takes a capsule containing 20 milligrams (mg) of a prescribed allergy medicine early in the morning. By the same time a day later, 25% of the medicine has been eliminated from his body. Jack doesn't take any more medicine, and his body continues to eliminate 25% of the remaining medicine each day. Write a recursive formula for the daily amount of this medicine in Jack's body. How long will it be before there is less than 1 mg of the medicine remaining in his body?

10. Consider the last part of the Investigation "Doses of Medicine." If you double the amount of medicine taken each time from 16 mL to 32 mL but continue to filter only 250 mL of liquid, will the limit of the concentration be doubled? Write a convincing argument for your position.

11. **APPLICATION** An anti-asthmatic drug has a half-life of about 9 hours. This means that 9 hours is the length of time it takes for the amount of drug present in a person's blood to decrease to half that amount.

a. What does this mean for the amount of this drug in a person's bloodstream if the initial concentration is 16 mg/L?

b. Create a graph of (*elapsed time, drug concentration*) using 9-hour increments on the time axis.

c. What dosage of this drug should a person take every 9 hours to maintain a balance of 16 mg/L?

12. Suppose square *ABCD* with side length 8 in. is cut from paper. Another square, *EFGH,* is placed with its corners at the midpoints of *ABCD,* as shown. A third square is placed with its corners at midpoints of *EFGH,* and so on.

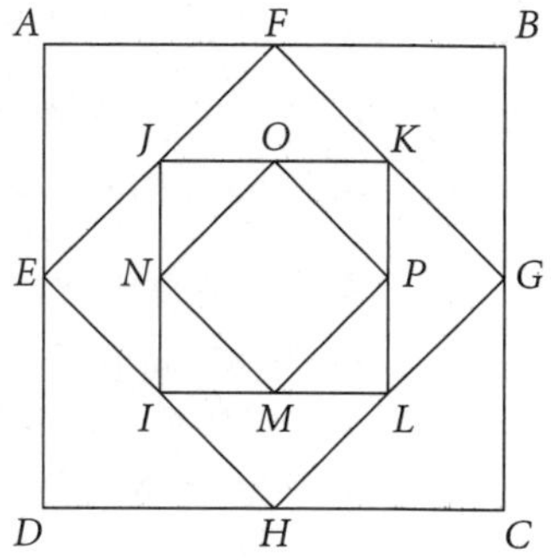

a. What is the perimeter of the 9th square?

b. What is the area of the 9th square?

c. What happens to the ratio of perimeter to area as the squares get smaller?

Review

13. (Lesson 2.1) Assume that two terms of a sequence are $u_3 = 16$ and $u_4 = 128$.

a. Find u_2 and u_5 if the sequence is arithmetic.

b. Find u_2 and u_5 if the sequence is geometric.

14. (Lesson 2.2) Suppose that a park biologist estimates the moose population in a national park over a 4-year period of mild winters. She makes this table.

Year	Estimated number of moose
1998	760
1999	835
2000	920
2001	1010

a. Write a recursive formula that approximately models the growth in the moose population over this 4-year period.

b. The winter of 2002 was particularly severe, and the park biologist has predicted a decline of 10% to 15% in the moose herd. What is the range of moose population she predicts for 2002?

15. (Lesson 2.2) If a rubber ball rebounds back to 97% of its height with each bounce, how many times will it bounce before it rebounds to half its original height?

16. (Lesson 1.4) Livonia Gymnastics Academy's Level 5 team earned the following scores in floor exercise at the 2004 Splitz Splash meet:

8.35	8.15	8.35	7.9	8.2	8.55	8.65	8.875	9.3	8.6
8.9	8.25	8.775	8.95	8.85	8.85	9	8.7	8.7	

Draw a box plot for the data, and give the data's five-number summary.

LESSON

2.4

Graphing Sequences

Mathematics is the lantern by which what before was dimly visible now looms up in firm, bold outlines.

IRVING FISHER

By looking for numerical patterns, you can write a recursive formula that generates a sequence of numbers quickly and efficiently. You can also use graphs to help you identify patterns in a sequence. The general shape of the graph of a sequence's terms indicates the type of sequence needed to generate the terms.

This graph is a visual representation of the first five terms of the arithmetic sequence generated by the recursive formula $u_1 = 1$ and $u_n = u_{n-1} + 2$ where $n \geq 2$. This graph in particular appears to be **linear,** that is, the points appear to lie on a line. The common difference, $d = 2$, makes each new point rise 2 units above the previous point.

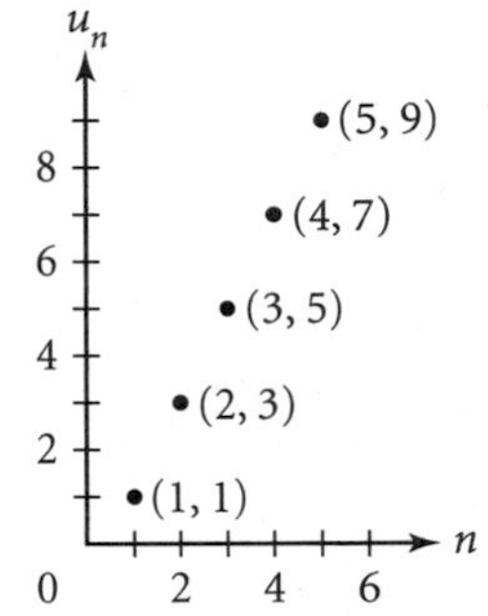

Graphs of sequences are examples of *discrete graphs,* or graphs made of isolated points. It is incorrect to connect those isolated points with a continuous line or curve, because the term number, n, must be a whole number.

This graph might make you think that all arithmetic sequences have linear graphs. The investigation will help you test that idea as well as make other generalizations.

Investigation

Match Them Up

Below are 18 different representations of sequences. Match each table with a recursive formula and a graph that represent the same sequence. As you do the matching, think about similarities and differences between the sequences and how those similarities and differences affect the tables, formulas, and graphs.

1.

n	u_n
0	8
1	4
3	1
6	0.125
9	0.015625

2.

n	u_n
0	0.5
1	1
2	2
3	4
4	8

3.

n	u_n
0	−2
1	1
2	2.5
4	3.625
5	3.8125

4.

n	u_n
0	−2
2	2
5	8
7	12

5.

n	u_n
0	8
1	6
3	2
5	−2
7	−6

6.

n	u_n
0	−4
1	−4
2	−4
4	−4
8	−4

a. $u_0 = 8$

$u_n = u_{n-1} - 2$ where $n \geq 1$

b. $u_0 = 8$

$u_n = 0.5u_{n-1}$ where $n \geq 1$

c. $u_0 = 0.5$

$u_n = 2u_{n-1}$ where $n \geq 1$

d. $u_0 = -8$

$u_n = u_{n-1} + 2$ where $n \geq 1$

e. $u_0 = -4$

$u_n = u_{n-1}$ where $n \geq 1$

f. $u_0 = -2$

$u_n = 0.5u_{n-1} + 2$ where $n \geq 1$

i.

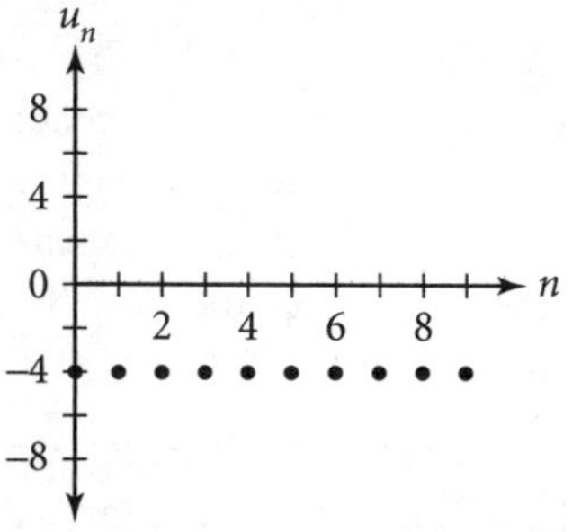

ii.

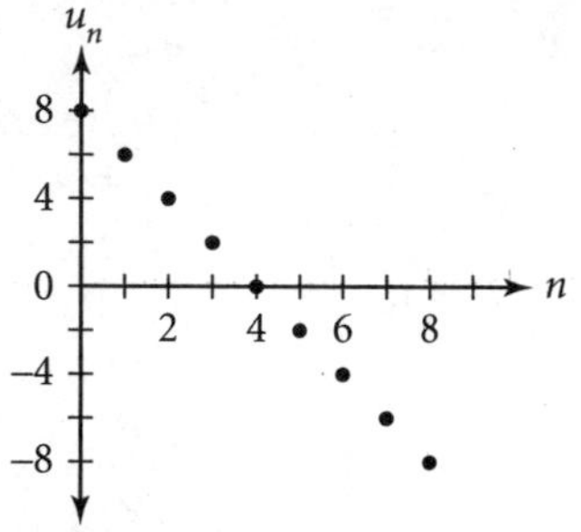

iii.

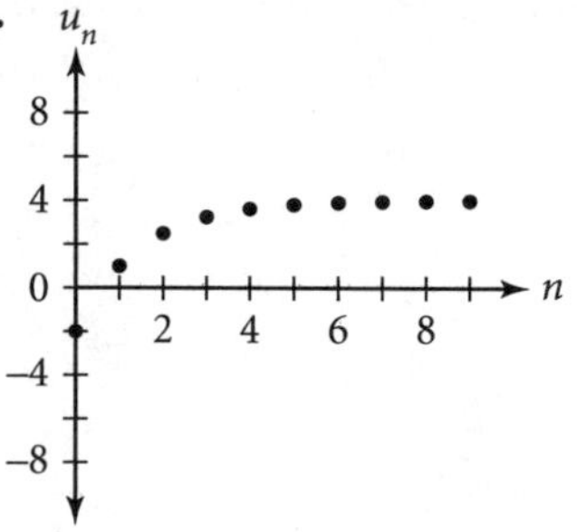

iv.

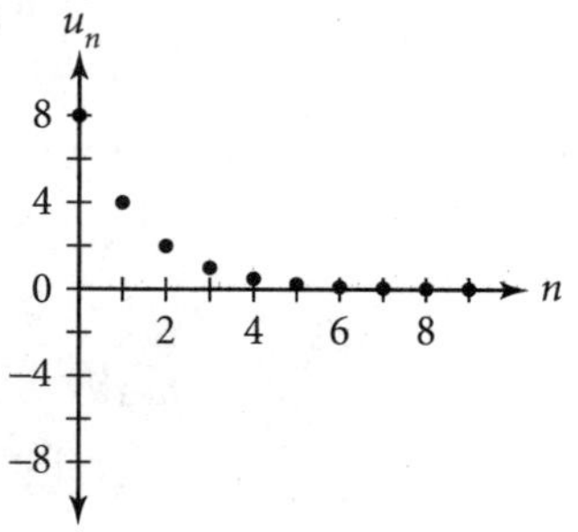

v.

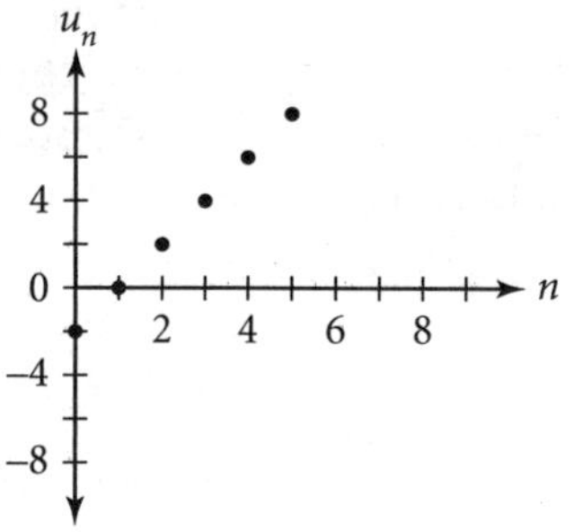

vi.

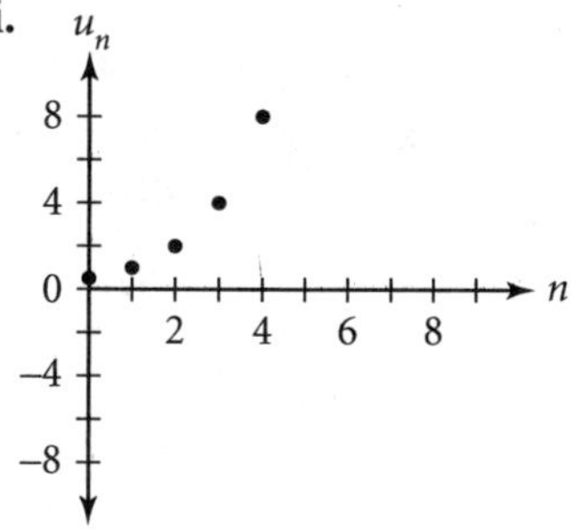

Write a paragraph that summarizes the relationships between different types of sequences, different types of recursive formulas, and different types of graphs. What generalizations can you make?

Meteorology (weather forecasting) is one career that relies on mathematical modeling. Meteorologists use computers and sophisticated models to monitor changes in the atmosphere. Trends in the data can help predict the trajectory and severity of impending storms, such as hurricanes.

The general shape of the graph of a sequence allows you to recognize whether the sequence is arithmetic or geometric. Even if the graph represents data that are not generated by a sequence, you may be able to find a sequence that is a *model,* or a close fit, for the data. The more details you can identify from the graph, the better you will be at fitting a model.

EXAMPLE

In deep water, divers find it darker the deeper they go. The data here give the percent of surface light intensity that remains at depth n ft in a particular body of water.

Depth (ft)	0	10	20	30	40	50	60	70
Light (%)	100	78	60	47	36	28	22	17

Find a sequence model that approximately fits the data.

▶ Solution

A graph of the data shows a decreasing, curved pattern. It is not linear, so an arithmetic sequence is not a good model. A geometric sequence with a long-run value of 0 will be a better choice.

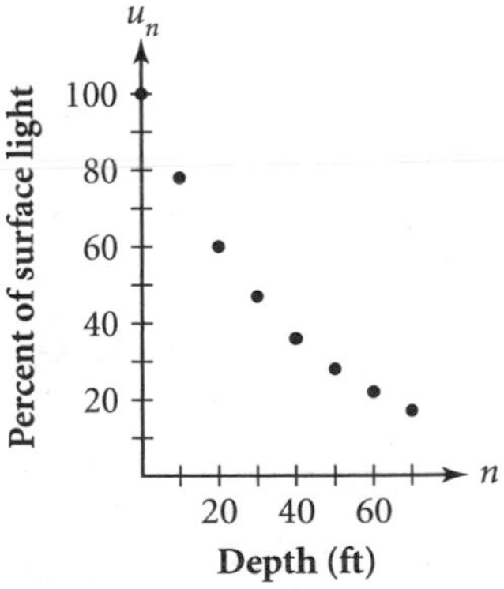

The starting value at depth 0 ft is 100% light intensity, so use $u_0 = 100$. The recursive rule should have the form $u_n = (1 - p)u_{n-1}$, but the data are not given for every foot, so you cannot immediately find a common ratio. The ratios between the given values are all approximately 0.77, or $(1 - 0.23)$. Because the light intensity decreases at a rate of 0.23 every 10 feet, it must decrease at a smaller rate every foot. A starting guess of 0.02 gives the model

$$u_0 = 100$$
$$u_n = (1 - 0.02)u_{n-1} \text{ where } n \geq 1$$

Check this model by graphing the original data and the sequence on your calculator. The graph shows that this model fits only one data point—it does not decay fast enough. [► See **Calculator Note 2B** to learn about sequences on your calculator and **Calculator Note 2C** to learn about graphing sequences.◄]

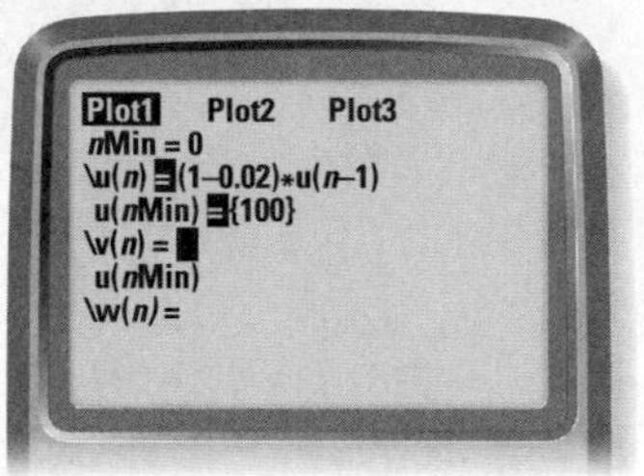

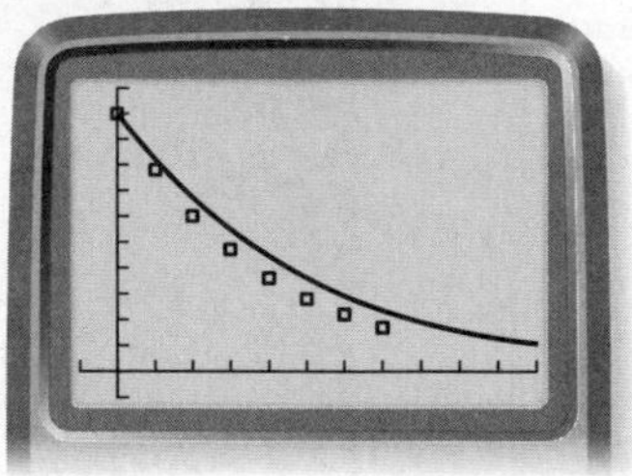

[−10, 110, 10, −10, 110, 10]

Experiment by increasing the rate of decay. With some trial and error you can find a model that better fits the data.

$u_0 = 100$

$u_n = (1 - 0.025)u_{n-1}$ where $n \geq 1$

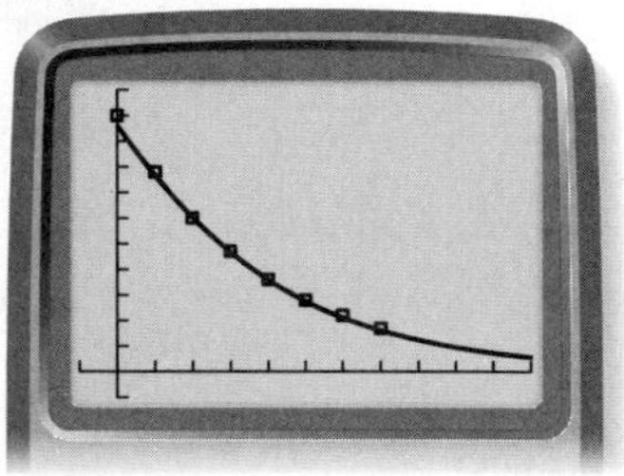

[−10, 110, 10, −10, 110, 10]

Once you have a good sequence model, you can use your calculator to find specific terms or to make a table of terms. For example, the value of u_{43} means that at depth 43 ft, approximately 34% of surface light intensity remains.

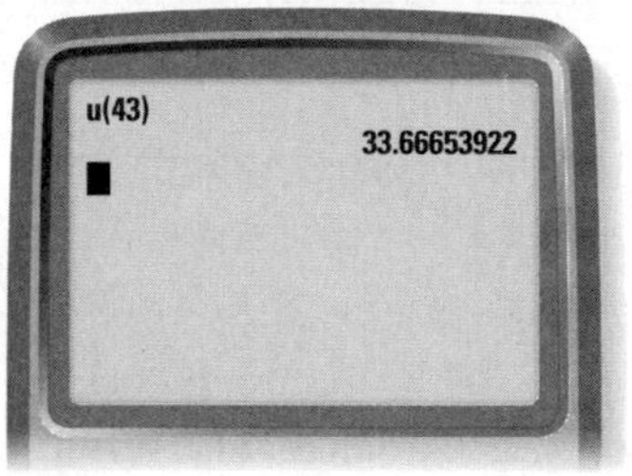

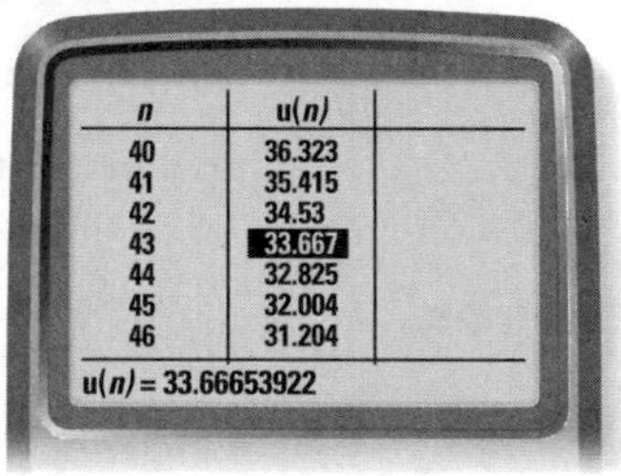

As you see in the calculator screens in the previous example, some calculators use $u(0)$, $u(1)$, $u(2)$, . . . , $u(n-1)$, and $u(n)$ instead of the subscripted notation u_0, u_1, u_2, . . . , u_v, and u_n. Be aware that $u(5)$ means u_5, not u multiplied by 5. You may also see other variables, such as a_n or v_n, used for recursive formulas in other textbooks. It is important to be able to make sense of these equivalent mathematical notations and to be flexible when reading other people's work.

Alertness also pays off when you are working with graphs. Graphs help you understand, explain, and visualize the mathematics of situations. When you make a graph or look at a graph, try to find connections between the graph and the mathematics used to create it. Consider the variables and units used on each axis and the smallest and largest values of those variables. Sometimes this will be clear and obvious, but sometimes you will need to look at the graph in a new way to see the connections.

EXERCISES

Practice Your Skills

1. Suppose you are going to graph the specified terms of these four sequences. For each sequence, what minimum and maximum values of n and u_n would you use on the axes to get a good graph?

a.

n	0	1	2	3	4	5	6	7	8	9
u_n	2.5	4	5.5	7	8.5	10	11.5	13	14.5	16

b. The first 20 terms of the sequence generated by

$u_0 = 400$

$u_n = (1 - .18)u_{n-1}$ where $n \geq 1$

c. The first 30 terms of the sequence generated by

$u_0 = 25$

$u_n = u_{n-1} - 7$ where $n \geq 1$

d. The first 70 terms of the sequence generated by

$u_0 = 15$

$u_n = (1 + .08)u_{n-1}$ where $n \geq 1$

2. Identify each graph as a representation of an arithmetic sequence, a geometric sequence, or a shifted geometric sequence. Use an informed guess to write a recursive formula for each.

a.

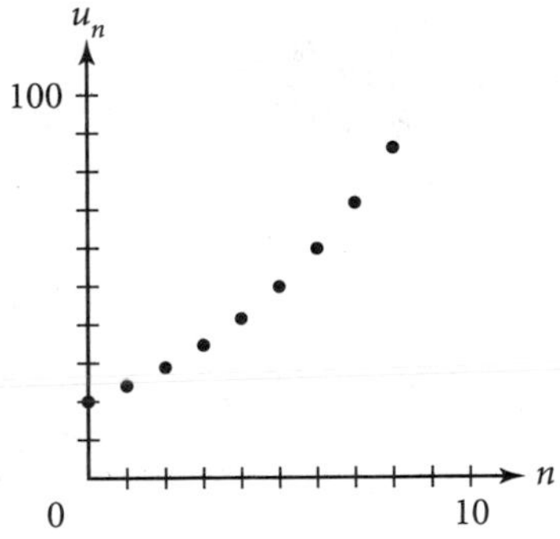

b.

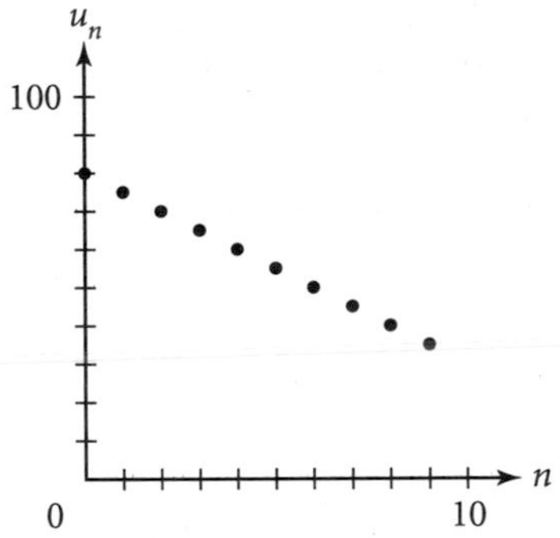

c.

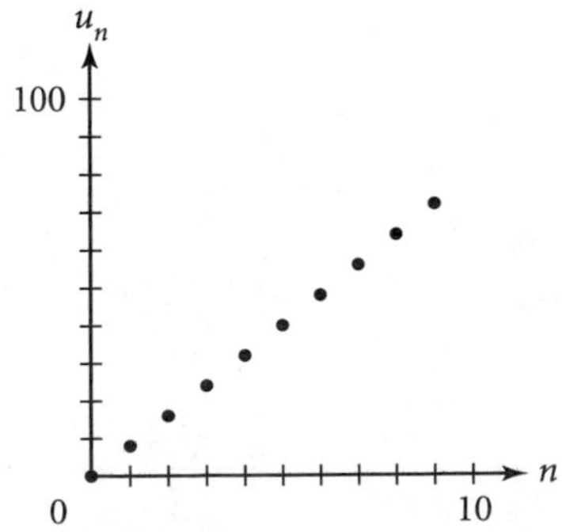

d.

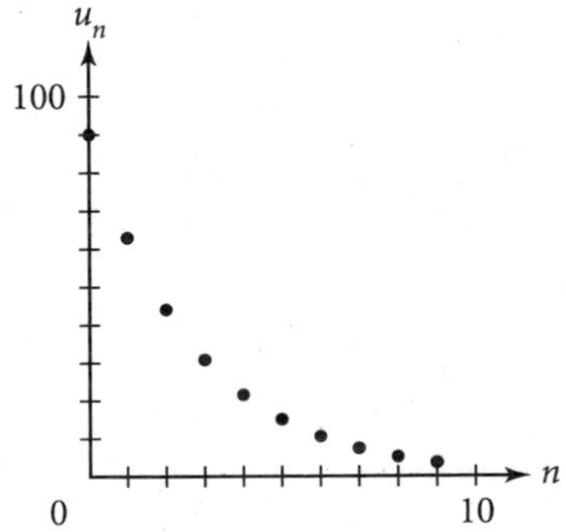

3. Imagine the graphs of the sequences generated by these recursive formulas. Describe each graph using exactly three of these terms: arithmetic, decreasing, geometric, increasing, linear, nonlinear, shifted geometric.

a. $u_0 = 450$

$u_n = a \cdot u_{n-1}$ where $n \geq 1$ and $0 < a < 1$

b. $u_0 = 450$

$u_n = b + u_{n-1}$ where $n \geq 1$ and $b < 0$

c. $u_0 = 450$

$u_n = u_{n-1} \cdot c$ where $n \geq 1$ and $c > 1$

d. $u_0 = 450$

$u_n = u_{n-1} + d$ where $n \geq 1$ and $d > 0$

Reason and Apply

4. Consider the recursive rule $u_n = 0.75u_{n-1} + 210$.

a. What is the long-run value of any shifted geometric sequence that is generated by this recursive rule?

b. Sketch the graph of a sequence that is generated by this recursive rule and has a starting value

i. below the long-run value

ii. above the long-run value

iii. at the long-run value

c. Write a short paragraph describing how the long-run value and the starting value of each shifted geometric sequence in 4b influence the general look of the graph.

5. Match each recursive formula with the graph of the same sequence. Give your reason for each choice.

a. $u_0 = 20$

$u_n = u_{n-1} + d$ where $n \geq 1$

b. $u_0 = 20$

$u_n = r \cdot u_{n-1}$ where $n \geq 1$

c. $u_0 = 20$

$u_n = r \cdot u_{n-1} + d$ where $n \geq 1$

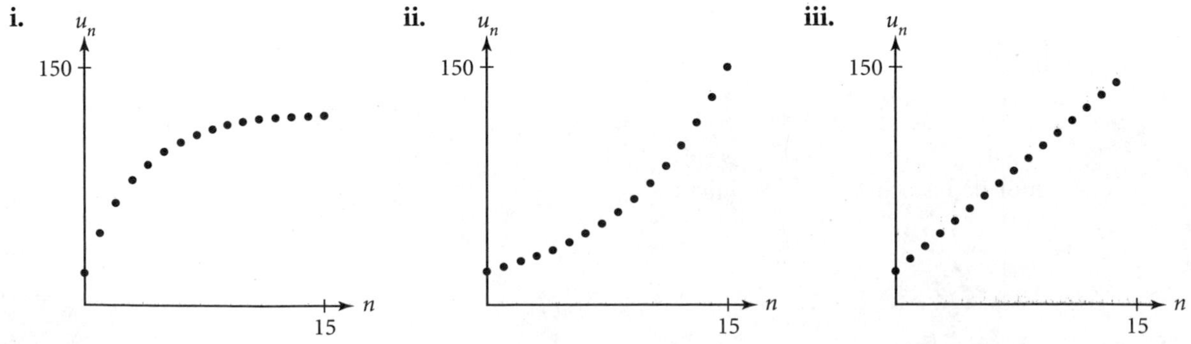

6. Consider the geometric sequence 18, −13.5, 10.125, −7.59375, ...

a. Write a recursive formula that generates this sequence.

b. Sketch a graph of the sequence. Describe how the graph is similar to other graphs that you have seen and also how it is unique.

c. What is the long-run value?

7. Your friend calls on the phone, and the conversation goes like this:

Friend: What does the graph of an arithmetic sequence look like?

You: Be more specific.

Friend: Why? Don't they all look the same?

You: Yes and no.

Explain to your friend how the graphs of arithmetic sequences are similar and how they are different.

8. Your friend calls back and asks about geometric sequences. Explain how the graphs of geometric sequences are similar and how they are different.

9. APPLICATION The Forever Green Nursery has 7000 white pine trees. The nursery plans to sell 12% of its trees and plant 600 new ones each year.

a. Make a graph that shows the number of trees at the nursery over the next 20 years.

b. Use the graph to estimate the number of trees in the long run. How does your estimate compare to the long-run value you found in Exercise 8, Lesson 2.3?

10. APPLICATION The Bayside Community Water District has decided to add fluoride to the drinking water. Board member Evelyn King does research and finds that the ideal concentration of fluoride in drinking water is between 1.00 mg/L and 2.00 mg/L. If the concentration gets as high as 4.00 mg/L, people may suffer health problems, such as bone disease or damage to developing teeth. If the concentration is less than 1.00 mg/L, it is too low to promote dental health. Ms. King supposes that 15% of the fluoride present in the water supply is consumed during a period of one day. Create a graph to help her analyze each of these scenarios.

a. If the fluoride content begins at 3.00 mg/L, and no additional fluoride is added, how long will it be before the concentration is too low to promote dental health?

b. If the fluoride content begins at 3.00 mg/L, and 0.50 mg/L is added daily, will the concentration increase or decrease? What is the long-run value? Explain your reasoning.

c. Suppose the fluoride content begins at 3.00 mg/L, and 0.10 mg/L is added daily. Describe what happens.

d. The Water District board members vote that there should be an initial treatment of 3.00 mg/L but that the long-run fluoride content should be 1.50 mg/L. How much fluoride needs to be added daily for the fluoride content to stabilize at 1.50 mg/L?

11. APPLICATION As the air temperature gets warmer, snowy tree crickets chirp faster. You can actually use a snowy tree cricket's rate of chirping per minute as a close approximation of the temperature in degrees Fahrenheit. Use a graph to find a sequence model that approximately fits this data.

Snowy Tree Crickets' Rate of Chirping

Temperature (°F)	50	55	60	65	70	75	80
Rate (chirps/min.)	40	60	80	100	120	140	160

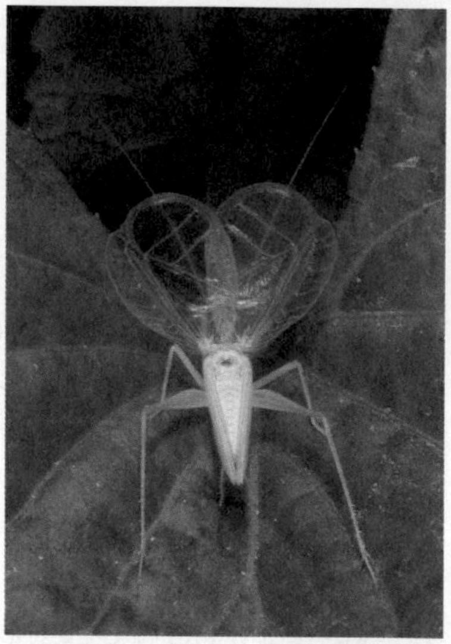

Snowy tree crickets are about 0.7 in. long and pale green; they live in shrubs and bushes. Only male crickets chirp, and they have different chirps for different activities, such as mating and fighting. All species of crickets chirp by rubbing their wings together.

12. APPLICATION This table gives the estimated population of Peru from 1950 to 2000. Use a graph to find a sequence model that approximately fits these data.

Population of Peru

Year	Population (millions)
1950	7.6
1960	9.9
1970	13.2
1980	17.3
1990	21.8
1996	24.5
2000	26.2

(U.S. Bureau of the Census, International Data Base)

Review

13. (Lesson 2.1) Consider the recursive formula

$u_0 = 450$

$u_n = 0.75u_{n-1} + 210$ where $n \geq 1$

a. Find u_1, u_2, u_3, u_4, and u_5.

b. How can you calculate backward from the value of u_1 to u_0? In general, what operations can you perform on any term to find the value of the previous term?

c. Write a recursive formula that generates the values of u_5 to u_0 backward.

14. (Lesson 2.3) For 14a–c, find the long-run value of the sequence generated by the recursive formula.

a. $u_0 = 50$

$u_n = (1 - 0.30)u_{n-1} + 10$ where $n \geq 1$

b. $u_0 = 50$

$u_n = (1 - 0.30)u_{n-1} + 20$ where $n \geq 1$

c. $u_0 = 50$

$u_n = (1 - 0.30)u_{n-1} + 30$ where $n \geq 1$

d. Use any patterns you notice in your answers to 14a–c to find the long-run value of the sequence generated by

$$u_0 = 50$$
$$u_n = (1 - 0.30)u_{n-1} + 70 \text{ where } n \geq 1$$

15. **APPLICATION** (Lesson 1.4) On warranty cards, manufacturers often ask about a consumer's income level. A manufacturer of DVD recorders finds that the lowest income reported by a consumer is \$20,000, the first quartile is \$40,000, the median income is \$70,000, the third quartile is \$125,000, and the highest income is \$825,000.

a. Do you think the mean or median is higher? Explain.

b. The median income earned by U.S. households in 2002 was between \$40,000 and \$45,000. How does this compare with the incomes of purchasers of the DVD recorder? Can you explain this?

LESSON

2.5

Loans and Investments

In life you will face many financial situations, which may include car loans, checking accounts, credit cards, long-term investments, life insurance, retirement accounts, and home mortgages. You will need to make intelligent choices about your money and about whom to trust. Fortunately, much of the mathematics is no more complicated than the recursive rule $u_n = r \cdot u_{n-1} + d$.

Investigation
Life's Big Expenditures

In this Investigation you will use recursion to explore loan balances and payment options. Your calculator is a helpful tool for trying different sequence models.

Part 1

You plan to borrow $22,000 over 5 yr (60 mo) to purchase a new car. The bank charges interest at an annual rate of 7.9%, compounded monthly. Part of each monthly payment is applied to the interest, and the remainder reduces the starting balance, or principal.

Step 1 What is the *monthly* interest rate? What is the first month's interest on the $22,000? If you make a payment of $300 at the end of the first month, then what is the remaining balance?

Step 2 Record the balances for the first 6 months with monthly payments of $300. How many months would it take to pay off the loan?

Step 3 Experiment with other values for the monthly payment. What monthly payment allows you to pay off the loan in exactly 60 months without underpaying?

Step 4 How much do you actually pay for the car using the monthly payment you found in Step 3? (*Hint:* The last payment should be a little less than the other 59 payments.)

Part 2

Use the techniques you discovered in Part 1 to find the monthly payment for a 30-year home mortgage of $146,000 with an annual interest rate of 7.25%, compounded monthly. How much do you actually pay for the house?

ConsumerConnection In the last quarter of the year 2000, the National Association of Home Builders reported that the median price of a home in the United States was $151,000. Mortgage lenders usually require monthly payments to be no more than 29% of a family's monthly income, which means that fewer than 60% of families could afford to buy a home in 2000. The most affordable place to buy a house was Des Moines, Iowa, where people with median incomes could afford 88.9% of the homes sold (median home price of $107,000). In contrast, the least affordable houses were in San Francisco, California, where only 6.1% of people could afford the median home price of $530,000.

Investments are mathematically similar to loans. With an investment, deposits are added on a regular basis so that your balance increases.

EXAMPLE

Gwen's employer offers an investment plan that invests a portion of each paycheck before taxes are deducted. Gwen gets paid every week. The plan has a fixed annual interest rate of 4.75%, compounded weekly, and she decides to contribute $10 each week. What will Gwen's balance be in 5 years?

▶ Solution

Gwen's starting balance is $10. Each week the previous balance is multiplied by $\left(1 + \frac{0.0475}{52}\right)$, and Gwen adds another $10. A recursive formula that generates the balance is

$$u_0 = 10$$

$$u_n = \left(1 + \frac{0.0475}{52}\right)u_{n-1} + 10 \text{ where } n \geq 1$$

There are 52 weeks in a year and 260 weeks in 5 years. The value of u_{260} shows that the balance after 5 years is $2945.89.

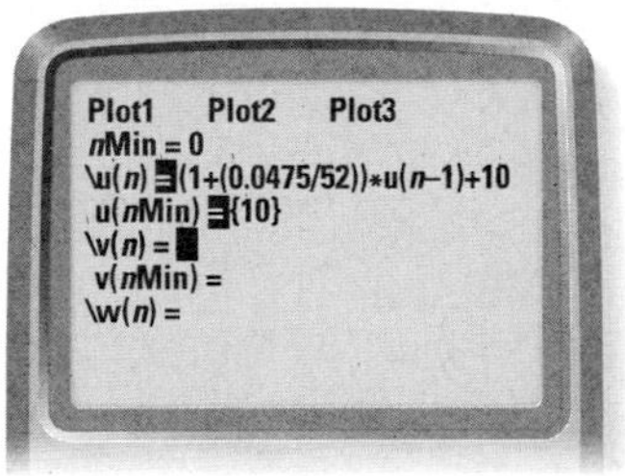

As you work on each exercise, look for these important pieces of information: the principal, the deposit or payment amount, the annual interest rate, and the frequency with which interest is compounded. Using recursion, you will be able to solve many financial problems with these values.

EXERCISES

▶ Practice Your Skills

1. Assume that the sequence generated by $u_0 = 450$ and $u_n = (1 + 0.039)\,u_{n-1} + 50$ where $n \geq 1$ represents a financial situation, and n is measured in years.

a. Is this a loan or an investment? Explain your reasoning.

b. What is the principal?

c. What is the deposit or payment amount?

d. What is the annual interest rate?

e. What is the frequency with which interest is compounded?

2. Answer the questions in Exercise 1a–e for the sequence generated by $u_0 = 500$ and $u_n = \left(1 + \frac{0.04}{4}\right)u_{n-1} - 25$ where $n \geq 1$.

3. Find the first month's interest on a $32,000 loan at an annual interest rate of

a. 4.9%

b. 5.9%

c. 6.9%

d. 7.9%

4. Write a recursive formula for each financial situation.

a. You borrow $10,000 at an annual interest rate of 10%, compounded monthly, and each payment is $300.

b. You buy $7000 worth of furniture on a credit card with an annual interest rate of 18.75%, compounded monthly. You plan to pay $250 each month.

c. You invest $8000 at 6%, compounded quarterly, and you deposit $500 every three months. (Quarterly means four times per year.)

d. You enroll in an investment plan that deducts $100 from your monthly paycheck and deposits it into an account with an annual interest rate of 7%, compounded monthly.

Reason and Apply

5. **APPLICATION** Find the balance after 5 years if $500 is deposited into an account with an annual interest rate of 3.25%, compounded monthly.

6. **APPLICATION** Consider a $1000 investment at an annual interest rate of 6.5%, compounded quarterly. Find the balance after

a. 10 years

b. 20 years

c. 30 years

7. **APPLICATION** Find the balance of a $1000 investment after 10 years at an annual interest rate of 6.5% when compounded

a. annually

b. monthly

c. daily (compound 360 times each year)

d. After the same length of time, how will the balances compare of investments compounded annually, monthly, and daily?

8. **APPLICATION** Beau and Shaleah each get a $1000 bonus at work and decide to invest it. Beau puts his money into an account that earns an annual interest rate of 6.5%, compounded yearly. He also decides to deposit $1200 each year. Shaleah finds an account that earns 6.5%, compounded monthly, and decides to deposit $100 each month.

a. Compare the amounts of money that Beau and Shaleah deposit each year. Describe any differences or similarities.

b. Compare the balances of Beau's and Shaleah's accounts over several years. Describe any differences or similarities.

9. **APPLICATION** Regis deposits \$5000 into an account for his 10-year-old child. The account has an annual interest rate of 8.5%, compounded monthly.

a. What regular monthly deposit amount is needed to make the account worth \$1 million by the time the child is 55 years old?

b. Sketch a graph of the increasing balances.

10. **APPLICATION** Cici purchased \$2000 worth of merchandise with her credit card this past month. Then she was unexpectedly laid off. She decides to make no more purchases with the card and to make only the minimum payment of \$40 each month. Her annual interest rate is 18%, compounded monthly.

a. Find the balance on the credit card over the next 6 months.

b. When will Cici pay off the total balance on her credit card?

c. What is the total amount paid for the \$2000?

11. **APPLICATION** Megan Flanigan is a loan officer with L. B. Mortgage Company. She is offering a loan of \$60,000 to a borrower at 7.6% annual interest, compounded monthly.

a. What should she tell the borrower the monthly payment will be if the loan must be paid off in 25 years?

b. Sketch a graph that shows the unpaid balance over time.

Review

12. (Lesson 2.1) Is this sequence arithmetic, geometric, shifted geometric, or something else? Explain why.

$\frac{1}{1}, \frac{1}{2}, \frac{1}{3}, \frac{1}{4}, \ldots$

13. (Lesson 2.3) Consider the geometric sequence generated by

$$u_0 = 4$$
$$u_n = 0.7u_{n-1} \text{ where } n \geq 1$$

a. What is the long-run value?

b. What is the long-run value if the common ratio is changed to 1.3?

c. What is the long-run value if the common ratio is changed to 1?

14. (Lesson 2.1) Write a recursive formula for each sequence.

a. 1, 2, 4, 8, . . .

b. 1, 5, 9, 13, . . .

c. 1, −3, −7, −11, . . .

d. 1, 0.5, 0.25, 0.125, . . .

E X P L O R A T I O N

Refining the Growth Model

Until now, you have assumed that the rate of growth or decay remains constant over time. The terms generated by the recursive formulas you have used to model arithmetic or geometric growth increase infinitely in the long run.

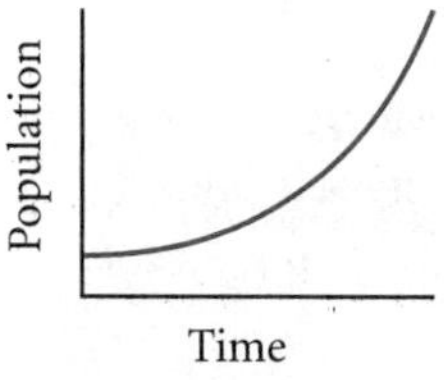

Geometric growth

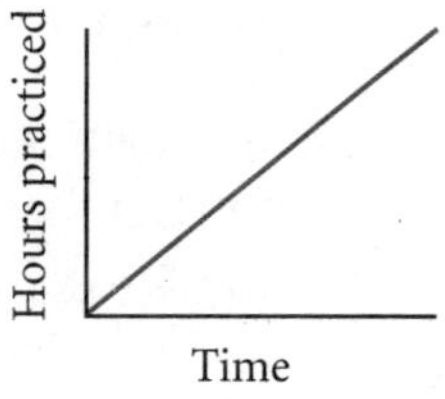

Arithmetic growth

However, environmental factors rarely support unlimited growth. Because of space and resource limitations, or competition among individuals or species, an environment usually supports a population only up to a limiting value.

EnvironmentalConnection Isle Royale in Lake Superior had no moose or wolves until the early 1900s, when a few moose swam to the island. For decades the moose had no predators, and their population grew logistically, limited only by food supply. (A few harsh winters also had an effect). In 1949 a rare ice bridge formed to Ontario, and a wolf pack crossed over to the island, changing the situation, as now the moose had a predator.

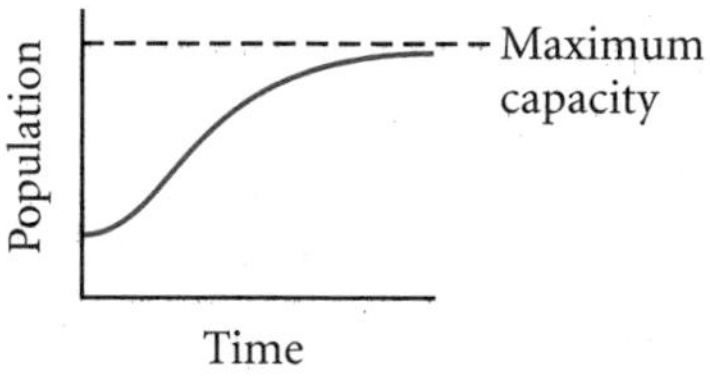

The graph above shows a *logistic function* with the population leveling off at the maximum capacity. If a population's growth is modeled by a logistic function, the growth rate is not constant but rather changes as the population changes.

Logistic functions are used to model more than just population. Researchers in many fields apply logistic functions to the study of such things as the spread of disease or consumer buying patterns. In this Investigation you will see an example of logistic growth.

Investigation
Cornering the Market

A company has invented a new gadget, and everyone who hears about it wants one! The company hasn't started advertising the new gadget yet, but the news is spreading fast. Simulate this situation with your class to see what happens when a popular new product enters the market.

Each person in your class will be assigned a unique ID number between 1 and the number of students in your class. At time zero only one person has bought the gadget, and at the end of every time period each person who has one tells another person about it, and that person goes out and buys one (unless he or she already has one).

Step 1 For about ten time periods each person who knows about the gadget generates a random ID number and tells that person, who immediately goes out and buys one. [▶ See **Calculator Note 1H** to learn how to generate random numbers.◀] Create a table like the one here and record the total number of people who have the gadget.

Time period	Number of people who own the gadget
0	1
1	
2	
3	

Step 2 Enter your data into lists L1 and L2. Make a scatter plot of your data, and sketch it on your paper. Describe your scatter plot. Explain why the number of people who own the gadget doesn't double each time period.

Step 3 Calculate the ratios of consecutive terms in list L2, and enter these in list L3. These ratios show you the rate at which the number of people who own the gadget grows each time period.

Step 4 In this activity the growth rate depends on the number of people who own the gadget. Shorten L1 and L2 by deleting the last value in each list so that all three lists are the same length. Make a scatter plot of (L2, L3), and sketch this graph on your paper. What happens to the growth rate as the number of people who own the gadget increases?

The net growth at each step depends on the previous population size u_{n-1}. So the net growth is a function of the population. This changes the simple growth model of $u_n = u_{n-1}(1 + p)$, in which the rate, p, is a constant, to a growth model with a variable rate

$$u_n = u_{n-1}\left(1 + p \cdot \left(1 - \frac{u_{n-1}}{L}\right)\right)$$

where p is the unrestricted growth rate, L is the limiting capacity or maximum population, and $p \cdot \left(1 - \frac{u_{n-1}}{L}\right)$ is the net growth rate.

HistoryConnection At the end of the 18th century, data on population growth were not available, and social scientists generally agreed with the theory of English economist Thomas Malthus (1766–1834), that a population always increases exponentially. In the 19th century, Belgian mathematician Pierre François Verhulst (1804–1849) and Belgian social statistician Adolphe Quételet (1796–1874) formulated the net growth rate expression $p \cdot \left(1 - \frac{u_{n-1}}{L}\right)$ for a population model. Quételet believed that checks on population growth needed to be accounted for in a more systematic manner than Malthus had described. Verhulst was able to incorporate the changes in growth rate into a mathematical model.

Step 5 What is the maximum population, L, for the gadget-buying scenario? What is u_0? What is the unrestricted growth rate, p? What does $\frac{u_{n-1}}{L} u_{n-1}$ represent? Write the recursive formula for this logistic function.

Step 6 Create list L4, defined as $1 - \frac{L2}{L}$, and plot (L2, L4) with (L2, L3). Add these new points in a new color to your sketch from Step 4.

Step 7 Turn off these scatter plots, and check how well your recursive formula from Step 5 models your original data (L1, L2).

A "closed" environment creates clear limitations on space and resources and calls for a logistic function model.

EXAMPLE

Suppose the unrestricted growth rate of a deer population on a small island is 12% annually, but the island's maximum capacity is 2000 deer. The current deer population is 300.

a. What net growth rate can you expect for next year?

b. What will the deer population be in 1 year?

c. Make a graph of the population over the next 50 years.

▶ Solution

a. There is a maximum capacity, so the population can be modeled with a logistic function. The net growth rate is $p\left(1 - \frac{u_{n-1}}{L}\right)$, where p is the unrestricted growth rate and L is the limiting value, in this case the island's maximum capacity. Because $p = 12\%$ and $L = 2000$ deer, the net growth rate will be

$$0.12\left(1 - \frac{u_{n-1}}{2000}\right)$$

Using $u_0 = 300$, this gives a growth rate of $0.12\left(1 - \frac{300}{2000}\right) = 0.102$, or 10.2%, for the first year.

b. The recursive rule for this logistic function is

$$u_n = u_{n-1}\left(1 + 0.12\left(1 - \frac{u_{n-1}}{2000}\right)\right)$$

Using $u_0 = 300$,

$$u_n = 300\left(1 + 0.12\left(1 - \frac{300}{2000}\right)\right)$$
$$= 300(1 + 0.102)$$
$$= 330.6$$

So the deer population after 1 year will be approximately 331.

c. The graph shows the population as it grows toward the maximum capacity of 2000 deer.

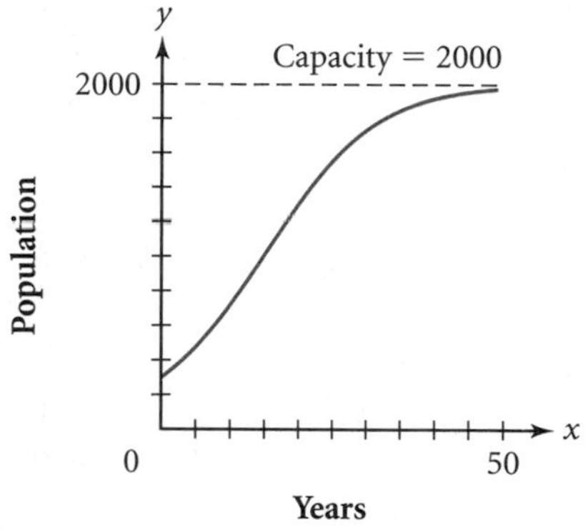

Now, try your hand at applying logistic function models with these questions.

Questions

1. Bacteria grown in a culture dish are provided with plenty of food but a limited amount of growing space. Eventually the population will become overcrowded, even though there is plenty of food. Initially, the bacteria grow at an unrestricted rate of 125%. The starting population is 50, and the capacity of the dish is 5000. Find the net growth rate and the population at the end of each week for 7 weeks.

2. A large field provides enough food to feed 500 healthy rabbits. When food and space are unlimited, the population growth rate of the rabbits is 20%, or 0.20. The population that can be supported is 500 rabbits. Complete each statement using the choices less than 0, greater than 0, 0, or close to 0.20.

   a. When the population is fewer than 500, the net growth rate will be __?__.

   b. When the population is more than 500, the net growth rate will be __?__.

   c. When the population is very small, the net rate will have a value near __?__.

   d. When the population is 500, the net rate will have a value of __?__.

3. Suppose the recursive rule $d_n = d_{n-1}\left(1 + 0.35\left(1 - \frac{d_{n-1}}{750}\right)\right)$ will give the number of daisies growing in the median strip of a highway each year. Presently there are about 100 daisies. Write a paragraph or two explaining what will happen. Explain and support your reasoning.

EXPLORATION
with Microsoft® Excel

Arithmetic and Geometric Growth

Download the Chapter 2 Excel Exploration from the website at www.keycollege.com/online or from your Student CD.

This Exploration will allow you to study two types of sequences defined recursively. Arithmetic sequences grow (or shrink) by adding a fixed number to the previous term, whereas geometric sequences grow (or shrink) based on multiplying the previous term by a fixed number.

On the spreadsheet you will find complete instructions and questions to consider as you explore.

Activity
Savings

Excel skills required

- Name a cell
- Data fill
- View formulas

Zoeigh was lucky. Her grandparents started a savings account for her when she was born. They started her off with $5000. Her grandparents placed the money in an account in which it would be held until her 21st birthday. Each month, interest will be calculated on the amount present, and added to the account. As you've learned, this is called compound interest. Banks report interest on an annual basis. If the interest is compounded monthly, then the monthly interest rate is $\frac{1}{12}$ of the APR. You will set up a flexible spreadsheet so that Zoeigh's parents can explore possibilities for this account.

You can find instructions for the required Excel skills in the Excel Notes at www.keycollege.com/online or on your Student CD.

Launch Excel.

Step 1 Label the cells as shown.

	A	B	C	D
1				Beginning Amount
2				Annual Percentage Rate
3				
4	Month	Balance	Interest	

Step 2 Name Cell C1 AMOUNT. If you completed the Excel Activity in Chapter 1, you've already named a cell, but we'll go through the steps again here.

a. Click inside Cell C1. Note that the cell address appears near the upper left corner.

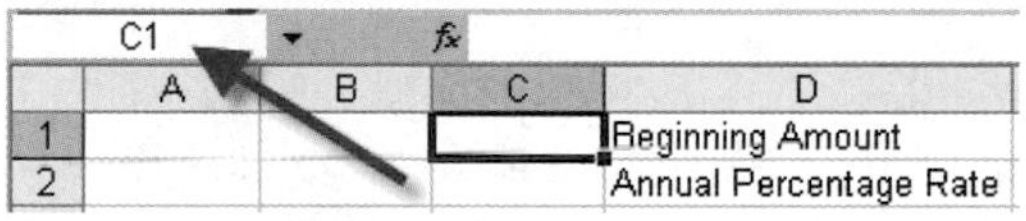

b. Click where you see the cell address, and change that by typing AMOUNT. (Remember that you must press ENTER to name the cell.)

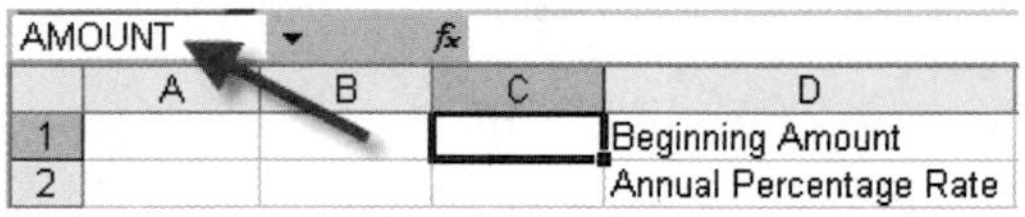

Step 3 In Cell C1 enter the number by typing 5000, and click another cell.

Step 4 Name Cell C2 APR.

Step 5 In Cell C2 enter the annual percentage rate (APR) by typing 6%.

Step 6 In Cell A5, type 1.

Step 7 In Cell B5 type =AMOUNT.

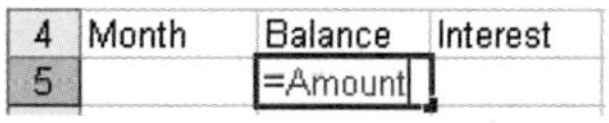

Step 8 In Cell C5 type =B5*(APR/12).

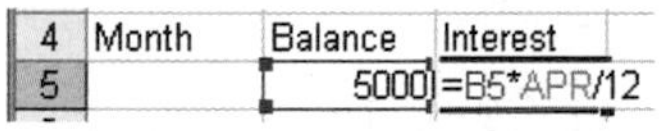

When you press ENTER, the number 25 will appear. This represents $25.00. This is the first month's interest if the annual rate is 6%. With the annual rate at 6%, the monthly, or periodic, rate is $\frac{1}{12} \cdot 6\% = 0.5\%$, and $5000 \cdot 0.5\% = 25$.

Step 9 Now we'll format cells B5 and C5 as currency.

a. Highlight Cells B5 and C5 together

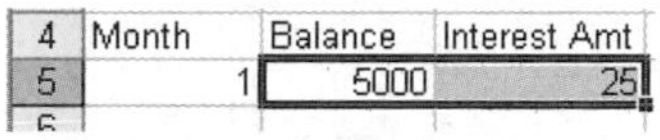

b. Use Format/Cells/Number/Currency to display these cells as currency.

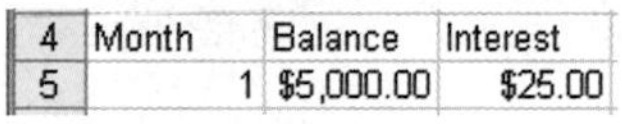

Step 10 Now you are ready to create the row that sets up the iterative relations.

a. In Cell A6 enter =A5+1.

b. In Cell B6 enter =B5+C5.

c. In Cell C6 enter =B6*(APR/12).

Thus far, you have

4	Month	Balance	Interest
5	1	=AMOUNT	=B5*APR/12
6	=A5+1	=B5+C5	=B6*APR/12

Step 11 To view this in your own spreadsheet, press CTRL+` (the accent can be found to the left of the 1 key). In this mode you see the formulas and not the values. This is a toggle, so when you press it again, you will get the numbers back:

4	Month	Balance	Interest
5	1	$5,000.00	$25.00
6	2	$5,025.00	$25.13

Step 12 You are now ready to data-fill.

a. Highlight Cells A6 to C6.

4	Month	Balance	Interest
5	1	$5,000.00	$25.00
6	2	$5,025.00	$25.13
7			

b. Notice the small black square in the lower right corner. You can click and drag the square down to fill the cells beneath as far as you want to go. For now, fill down to Row 16.

4	Month	Balance	Interest
5	1	$5,000.00	$25.00
6	2	$5,025.00	$25.13
7			
8			
9			
10			
11			

When you let go, the cells will fill with the numbers you need.

Step 13 Use CTRL+` to view the formulas.

4	Month	Balance	Interest
5	1	=AMOUNT	=B5*APR/12
6	=A5+1	=B5+C5	=B6*APR/12
7	=A6+1	=B6+C6	=B7*APR/12
8	=A7+1	=B7+C7	=B8*APR/12
9	=A8+1	=B8+C8	=B9*APR/12
10	=A9+1	=B9+C9	=B10*APR/12
11	=A10+1	=B10+C10	=B11*APR/12
12	=A11+1	=B11+C11	=B12*APR/12
13	=A12+1	=B12+C12	=B13*APR/12
14	=A13+1	=B13+C13	=B14*APR/12
15	=A14+1	=B14+C14	=B15*APR/12
16	=A15+1	=B15+C15	=B16*APR/12

Notice that APR is common to all cells in the Interest column. Remember that when you name a cell and refer to it this way, Excel treats that as an "absolute reference." In general math terms, this is called a *parameter*—it is arbitrary but fixed for the duration of the exercise. The reference to Column B, of course, varies in each formula in the Interest column; the "B" reference is a variable.

This is an iterative process in which the current step depends on two things in the previous step.

See if you can figure out how much money would have to be put into the account to yield $6000 at the end of a year. This may take quite a few tries. If you work in pairs, discuss this with your teammate and determine a strategy for finding an answer.

Step 14 Use CTRL+` to toggle back to normal view.

Questions

1. If the $5000 is in an account with an annual rate of 6%, how much will be in the account after 1 year?
2. If the $5000 is in an account with an annual rate of 4%, how much will be in the account after 1 year?
3. Highlight Row 16, and fill the data downward so you can see the balances develop for 20 years. How much money will the account contain in 20 years if you start with $5000 at an annual rate of 6%?
4. How much money will the account contain in 20 years if on Zoeigh's 10th birthday Grandma gives another $5000?
5. Suppose you start with $5000, and each year you add $500. How would you adjust the spreadsheet to represent this?
6. Which earns more money, adding $5000 every 10 years or adding $400 every year, assuming the same interest rate?

Activity
Loans

This Activity is very similar to the previous one, except that instead of adding on each line, you will subtract.

You will create a spreadsheet that will amortize a loan for a given monthly payment and annual interest rate.

Step 1 Inputs should include the amount of the loan (name that cell `LOANAMOUNT`), the APR (name that cell `APR`), and the planned payment (name that cell `PAYMENT`). Use cells in Rows 1–3 for these.

Step 2 Label some columns using Row 5: "Balance" (Column A), "Interest" (Column B), "Principal" (Column C).

Step 3 Enter =LOANAMOUNT in A6.

Step 4 In B6 enter a formula to calculate the interest earned in one month on the balance in A6.

Step 5 In C6 enter a formula to calculate the portion of the payment that is not applied to interest.

Step 6 In A7 enter a formula to calculate the new balance after the interest and payment are applied.

Step 7 Data-fill the contents of B6 and C6 to Row 7. Then data-fill the contents of Cells A7 to C7 as far as you need to for a zero balance.

Questions

1. How long will it take to pay off an $8000 credit card balance at 18% if you pay $160 per month?

2. Paying more will reduce the term of a loan. How long will it take to pay off an $8000 credit card balance at 18% if you pay $320 per month? (Actually calculate your answer—don't assume it is half the answer to Question 1).

3. Negotiating a lower rate will reduce the term of a loan. How long will it take to pay off an $8000 credit card balance at 9% if you pay $160 per month?

4. Some companies allow you to reduce payments for a time. How long will it take to pay off an $8000 credit card balance at 18% if you pay $100 per month?

CHAPTER 2 REVIEW

A **sequence** is an ordered list of numbers. In this chapter you used **recursion** to define sequences. A **recursive formula** specifies one or more starting terms and a **recursive rule** that generates the nth term by using the previous term or terms. You learned to calculate the terms of a sequence by hand and by using recursion and sequences on your calculator.

There are two special types of sequences, arithmetic and geometric. **Arithmetic sequences** are generated by always adding the same number, called the **common difference,** to get the next term. Your salary for a job on which you are paid by the hour is modeled by an arithmetic sequence. **Geometric sequences** are generated by always multiplying by the same number, called the **common ratio,** to get the next term. The growth of money in a savings account is modeled by a geometric sequence. For some **growth** and **decay** scenarios, it helps to write the common ratio as a percent change, $(1 + p)$ or $(1 - p)$. Some real-world situations, such as medicine levels, are modeled by **shifted geometric sequences** that use a recursive rule with both multiplication and addition.

Many sequences approach a long-run value after many, many terms. Looking at a graph of the sequence may help you see the long-run value. Graphs also help you recognize whether the data are best modeled by an arithmetic or a geometric sequence. The graph of an arithmetic sequence is **linear,** whereas the graph of a geometric sequence is curved.

EXERCISES

1. Consider this sequence:

256, 192, 144, 108, . . .

a. Is this sequence arithmetic or geometric?

b. Write a recursive formula that generates the sequence. Use u_1 for the starting term.

c. What is the 8th term?

d. Which term is the first to have a value less than 20?

e. Find u_{17}.

2. Consider this sequence:

3, 7, 11, 15, . . .

a. Is this sequence arithmetic or geometric?

b. Write a recursive formula that generates the sequence. Use u_1 for the starting term.

c. What is the 128th term?

d. Which term has the value 159?

e. Find u_{20}.

3. List the first five terms of each sequence. For each set of terms, what minimum and maximum values of n and u_n would you use on the axes to make a good graph?

a. $u_1 = -3$

$u_n = u_{n-1} + 1.5$ where $n \geq 2$

b. $u_2 = 3$

$u_n = 3u_{n-1} - 2$ where $n \geq 2$

4. **APPLICATION** A large barrel contains 12.4 gal of oil 18 minutes after its drain is opened. How many gallons of oil were in the barrel to start, if it drains at a rate of 4.2 gal/min?

5. **APPLICATION** The atmospheric pressure is 14.7 pounds per square inch (psi) at sea level. An increase in altitude of 1 mi produces a 20% decrease in the atmospheric pressure. Mountain climbers use this relationship to determine whether or not they can safely climb a mountain and to periodically calculate their altitude after they begin climbing.

a. Write a recursive formula that generates a sequence that represents the atmospheric pressure at different altitudes.

b. Sketch a graph that shows the relationship between altitude and atmospheric pressure.

c. What is the atmospheric pressure when the altitude is 7 mi?

d. At what altitude does the atmospheric pressure drop below 1.5 psi?

EnvironmentalConnection Humans of any age or physical condition can become ill when they experience extreme changes in atmospheric pressure in a short span of time. Atmospheric pressure changes the amount of oxygen a person is able to inhale, which in turn causes the buildup of fluid in the lungs or brain. Someone who travels from a low-altitude city like Akron, Ohio, to a high-altitude city like Aspen, Colorado, and immediately climbs the summit of a mountain to ski could get a headache, become nauseated, or even fall seriously ill.

6. **APPLICATION** The enrollment at a college is currently 5678. The board of administrators estimates that each year from now on, the school will graduate 24% of its students and admit 1250 new students. What will the enrollment be during the 6th year? What will the enrollment be in the long run? Sketch a graph of the enrollment over 15 years.

7. **APPLICATION** You deposit $500 into a bank account that has an annual interest rate of 5.5%, compounded quarterly.

a. How much money will you have after 5 years if you never deposit more money?

b. How much money will you have after 5 years if you deposit an additional $150 every 3 months after the initial $500?

8. **APPLICATION** Oliver wants to buy a small vacation cabin and needs to borrow $80,000. What monthly payment is necessary to pay off the mortgage in 30 years if the annual interest rate is 8.9%, compounded monthly?

9. Match each recursive formula with the graph of the same sequence. Give your reason for each choice.

a. $u_0 = 5$

$u_n = u_{n-1} + 1$ where $n \geq 1$

b. $u_0 = 1$

$u_n = (1 + 0.5)u_{n-1}$ where $n \geq 1$

c. $u_0 = 5$

$u_n = (1 - 0.5)u_{n-1}$ where $n \geq 1$

d. $u_0 = 5$

$u_n = u_{n-1} - 1$ where $n \geq 1$

i.

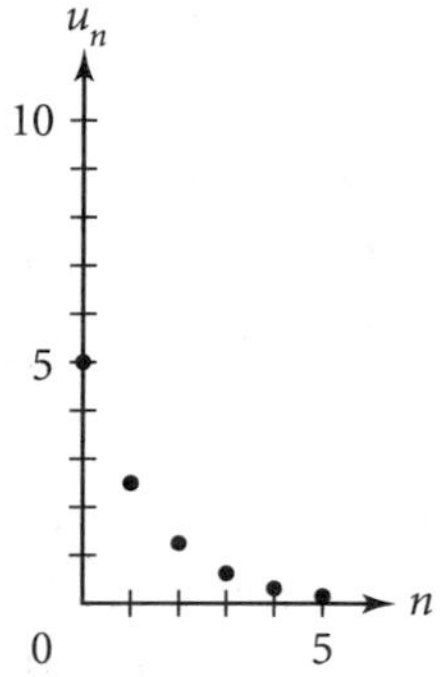

ii.

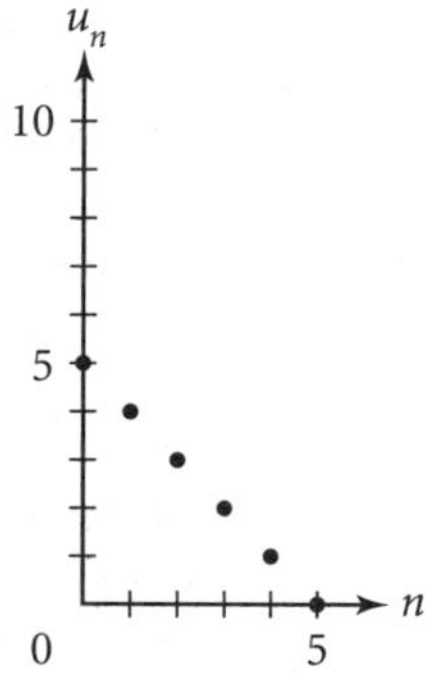

iii.

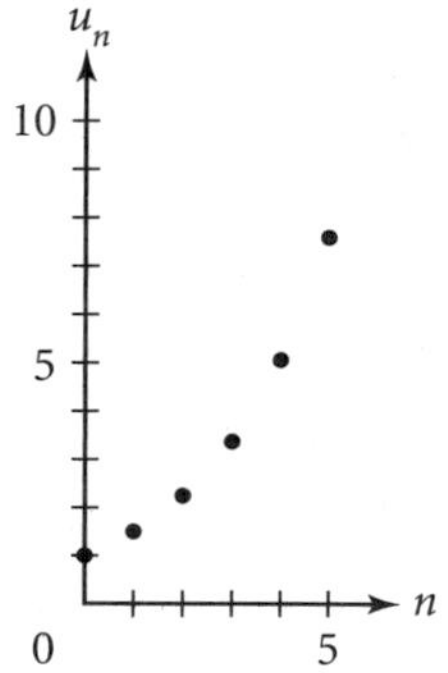

iv.

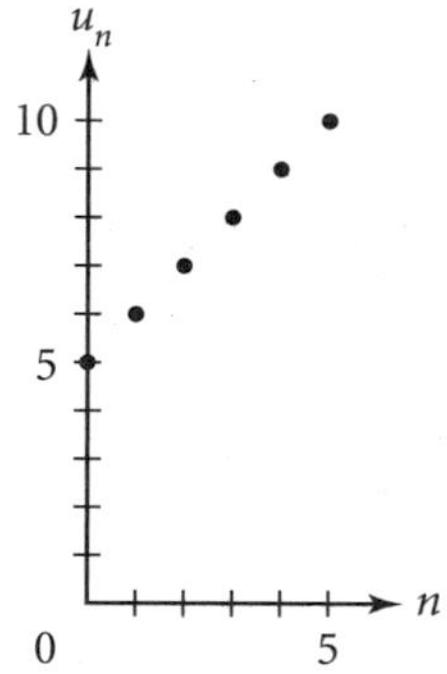

10. This table gives the consumer price index for medical care from 1970 to 2000. Use a graph to find a sequence model that approximately fits this data.

U.S. Consumer Price Index for Medical Care

Year	Consumer price index
1970	34.0
1975	47.5
1980	74.9
1985	113.5
1990	162.8
1995	220.5
2000	260.8

(U.S. Department of Labor, Bureau of Labor Statistics)

EconomicsConnection The consumer price index (CPI) is a measure of the average change over time in the prices paid by urban consumers for goods and services, such as food, clothing, shelter, fuel, and health care. Each month, the Bureau of Labor Statistics obtains price information for 80,000 items in order to adjust the index. For each category of goods and services, the price of specific items in the period 1982–1984 is assigned an index of 100, and the index for all subsequent years is given in relation to this reference period. For example, an index of 130 means the price of an item increased 30% over its 1982–1984 price. The CPI is used as an economic indicator and as a means of adjusting dollar values. You can learn more about the CPI by using the links at **www.keycollege.com/online** .

Assessing What You've Learned

- Can you write recursive formulas for arithmetic and geometric sequences?
- Do you know how to use your calculator to generate the terms of a sequence?
- How will you know when a sequence approaches a limit? How will you find the limit?
- Can you recognize an arithmetic sequence, a geometric sequence, and a shifted geometric sequence from their recursive formulas? From a table of their values? From a graph?

JOURNAL Look back at the journal entries you have written about Chapter 2. Then, to help you organize your thoughts, here are some additional questions to write about:

- What do you think were the main ideas of the chapter?
- Were there any exercises that you particularly struggled with? If so, are they clear now? If some ideas are still foggy, what plans do you have to clarify them?
- What topics discussed in this or another math class are related to recursion and sequences? Explain the relationship for each topic you list.

CHAPTER

3 Linear Models and Systems

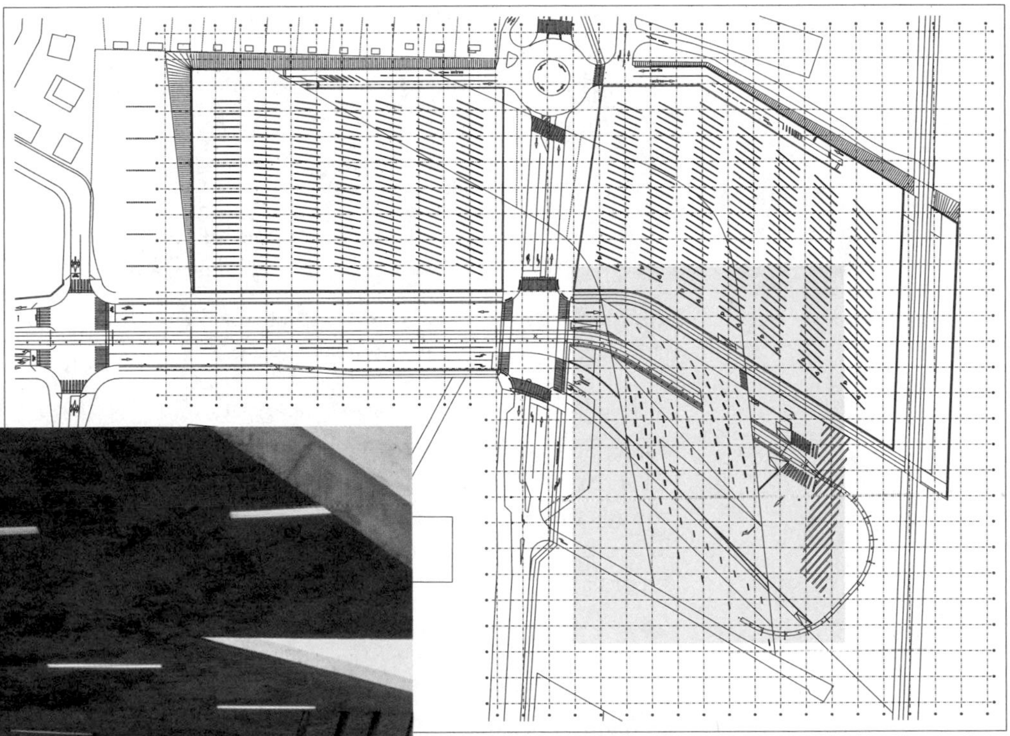

Iraqi-British architect Zaha Hadid (b 1950) designed this commuter train station on the outskirts of Strasbourg, France. She worked with the concept of overlapping fields and lines to represent the interacting patterns of moving cars, trains, bicycles, and pedestrians. The roof of the waiting area in the station is punctured by angled lines that let sunlight shine on the floor, shifting throughout the day. You can see these lines in the photograph, which shows the region of the architectural plan highlighted in blue.

OBJECTIVES

In this chapter you will

- review linear equations in intercept form
- explore connections between arithmetic sequences and linear equations
- find lines of fit for data sets that are approximately linear
- solve systems of linear equations

LESSON

3.1

Linear Equations

You can solve many rate problems by using recursion.

Matias wants to call his aunt in Chile on her birthday. He learned that placing the call costs \$0.23 and that each minute he talks costs \$0.11. How much would it cost to talk for 30 minutes?

Last year I went fishing with Salvador Dali. He was using a dotted line. He caught every other fish.

STEVEN WRIGHT (b. 1955)

You can calculate the cost of Matias's phone call with the recursive formula

$$u_0 = 0.23$$
$$u_n = u_{n-1} + 0.11$$

To find the cost of a 30-minute phone call, calculate the first 30 terms as shown on the calculator screen.

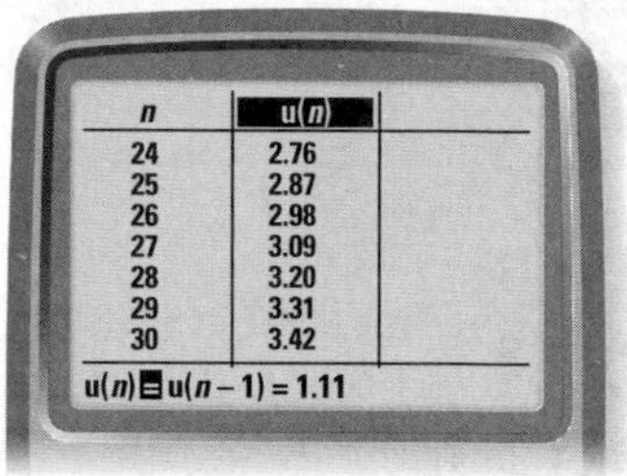

n
u(n)
24 2.76
25 2.87
26 2.98
27 3.09
28 3.20
29 3.31
30 3.42
u(n) u(n – 1) = 1.11

You or Matias can also find the cost of a 30-minute call by using the linear equation

$$y = 0.23 + 0.11x$$

where x is the length of the phone call in minutes and y is the cost in dollars. If the phone company always rounds the length of the call upward to the nearest whole minute, then the costs become a sequence of discrete points, and you can write the relationship as an explicit formula,

$$u_n = 0.23 + 0.11n$$

where n is the length of the phone call in minutes and u_n is the cost in dollars.

An **explicit formula** gives a direct relationship between two discrete quantities. How does the explicit formula differ from the recursive formula? How would you use each one to calculate the cost of a 15-minute call or an n-minute call?

In this lesson you will write and use explicit formulas for arithmetic sequences. You will also write linear equations for lines through the discrete points of arithmetic sequences.

EXAMPLE A

Find an explicit formula for the recursively defined arithmetic sequence

$$u_0 = 2$$
$$u_n = u_{n-1} + 6 \text{ where } n \geq 1$$

a. Use the explicit formula to find u_{22}.

b. Find the value of n so that $u_n = 86$.

▶ Solution

Look for a pattern in the sequence.

u_0 2

u_1 $8 = 2 + 6 = 2 + 6 \cdot 1$

u_2 $14 = 2 + 6 + 6 = 2 + 6 \cdot 2$

u_3 $20 = 2 + 6 + 6 + 6 = 2 + 6 \cdot 3$

Notice that the common difference (or rate of change) between the terms is 6. You start with 2 and just keep adding another 6. That means each term is equivalent to 2 plus 6 times the term number. In general, when you write the formula for a sequence, you use n to represent the number of the term and u_n to represent the term itself.

Term value = Initial value + Rate · Term number

$$u_n = 2 + 6 \cdot n$$

a. You can use the explicit formula to find u_{22} without calculating all of the previous terms. By substituting 22 for n, you get $u_{22} = 2 + 6 \cdot 22$. So, u_{22} equals 134. [▶ **See Calculator Note 2A.**◀]

b. To find n so that $u_n = 86$, substitute 86 for u_n in the formula $u_n = 2 + 6n$.

$86 = 2 + 6n$ Substitute 86 for u_n

$14 = n$ Solve for n

So, 86 is the 14th term.

You graphed sequences of points (n, u_n) in Chapter 1. The variable n stands for a term number, so it is a whole number: 0, 1, 2, 3, ... So, using different values for n will produce a set of discrete points. This graph shows the arithmetic sequence from Example A.

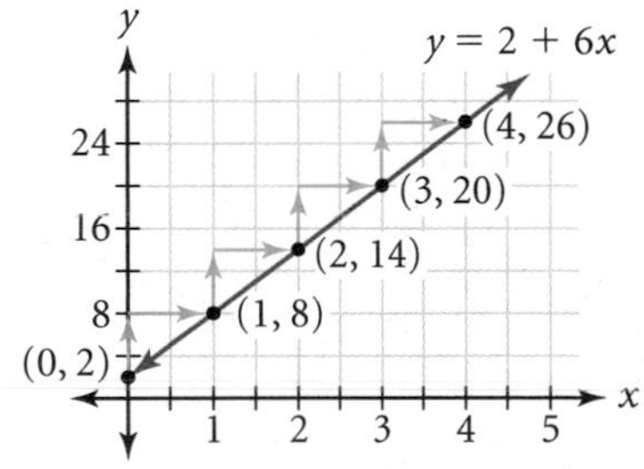

When n increases by 1, u_n increases by 6, the common difference. In terms of x and y, the number 6 is the change in the y-value, or function value, that corresponds to a unit change (a change of 1) in the x-value. So, the points representing the sequence lie on a line with a slope of 6. In general, the common difference, or rate of change, between consecutive terms of an arithmetic sequence is the **slope** of the line through those points.

The pair (0, 2) names the starting value 2, which is the y-intercept. Using the intercept form of a linear equation, you can now write the equation of the line through the points of the sequence as $y = 2 + 6x$, or $y = 6x + 2$.

In this course you will use x and y to write linear equations. You will use n and u_n to write both recursive formulas and explicit formulas for sequences of discrete points.

In the Investigation you will focus on this relationship between the formula for an arithmetic sequence and the equation of the line through the points representing the sequence.

Investigation
Match Point

Below are three recursive formulas, three graphs, and three linear equations.

1. $u_0 = 4$
$u_n = u_{n-1} - 1$ where $n \geq 1$

2. $u_0 = 2$
$u_n = u_{n-1} + 5$ where $n \geq 1$

3. $u_0 = -4$
$u_n = u_{n-1} + 3$ where $n \geq 1$

a.

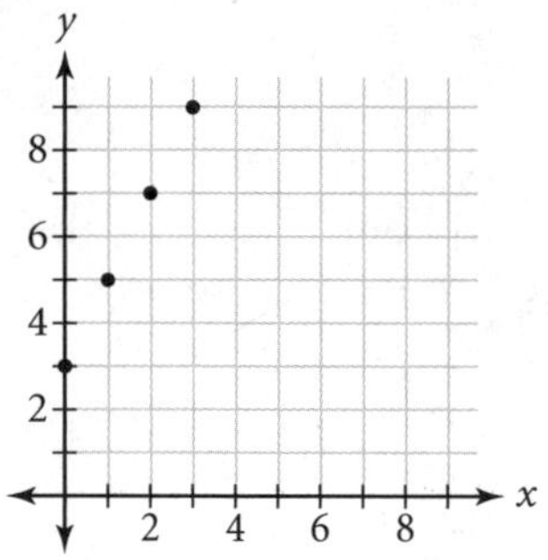

b.

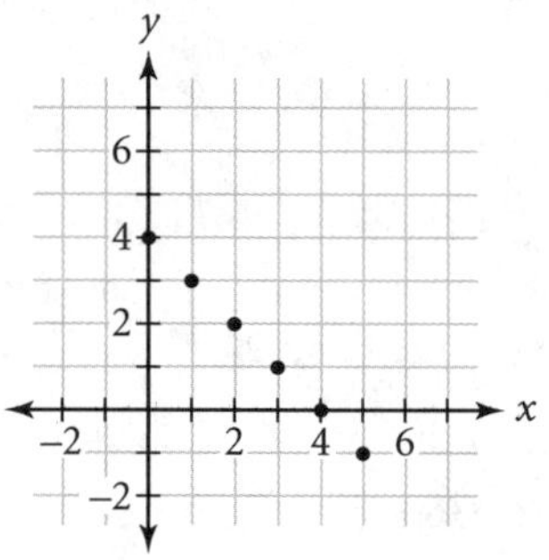

c.

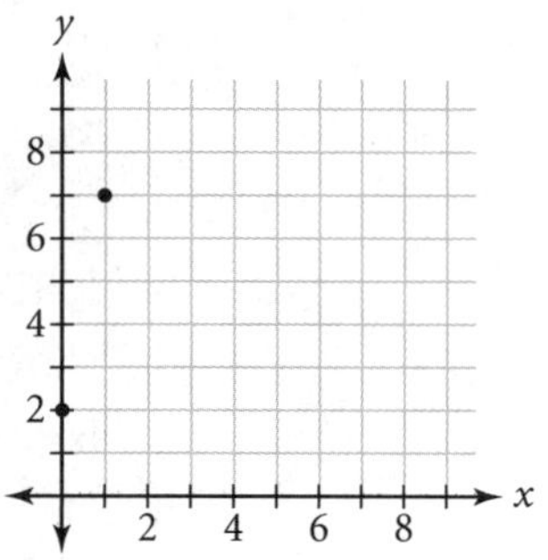

i. $y = -4 + 3x$ **ii.** $y = 4 + x$ **iii.** $y = 2 + 5x$

Step 1 Your task is to match the recursive formulas, graphs, and linear equations that go together. (Not all of the appropriate matches are listed. If the recursive rule, graph, or equation is missing, you will need to create it.)

Step 2 Write a brief statement relating the starting value and common difference of an arithmetic sequence to the corresponding equation $y = a + bx$.

Step 3 Are points (n, u_n) of an arithmetic sequence always collinear (on the same line)? Write a brief statement supporting your answer.

EXAMPLE B Retta typically spends \$2 a day on lunch. After today's lunch she notices that she has \$17 left. She thinks of this sequence to model her daily cash balance:

$u_1 = 17$
$u_n = u_{n-1} - 2 \quad$ where $n > 1$

a. Find the explicit formula that represents her daily cash balance and the equation of the line through the points of this sequence.

b. How useful is this formula for predicting the amount of money Retta will have each day?

▶ Solution

Use the common difference and starting term to write the explicit formula.

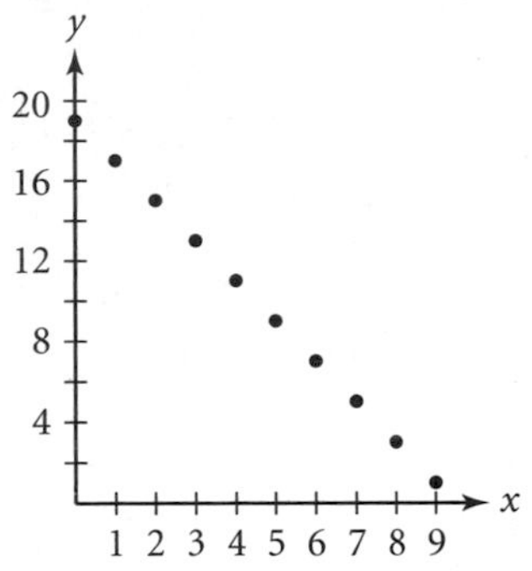

a. Each term is 2 less than the previous term, so the common difference of the arithmetic sequence and the slope of the line are both -2.

The term u_1 is 17, so the previous term, u_0, or the y-intercept, is 19.

The explicit formula for the arithmetic sequence is $u_n = 19 - 2 \cdot n$, and the equation of the line containing these points is $y = 19 - 2x$.

b. We don't know whether Retta has any other expenses, when she receives her paycheck, or whether she buys lunch on the weekend. The formula could be valid for eight more days (u_9) until she has \$1 left, as long as she gets no more money and spends only \$2 per day.

For both sequences and equations, it is important to consider the conditions for which the relationship is valid. For example, a phone company usually rounds the length of a phone call upward to determine the charges, so the relationship between the length of the call and the cost of the call is valid only for a call length that is a positive whole number. Also, the portion of the line to the left of the y-axis, where x is negative, is part of the mathematical model but has no relevance for the phone-call scenario.

In the Investigation you found the connection between the recursive formula for a sequence and the intercept form of the equation of a line.

EXERCISES

Practice Your Skills

1. Consider the sequence

$u_0 = 18$

$u_n = u_{n-1} - 3$ where $n \geq 1$

a. Graph the sequence.

b. What is the slope of the line that contains the points? How is that related to the common difference of the sequence?

c. What is the y-intercept of the line that contains these points? How is it related to the sequence?

d. Write the equation of the line that contains these points.

2. Refer to the graph of the sequence.

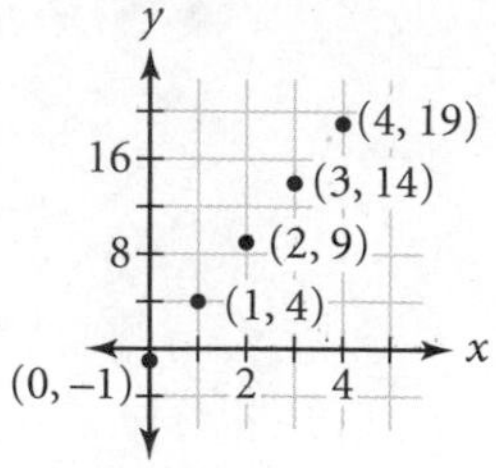

a. Write a recursive formula for the sequence. What is the common difference? What is the value of u_0?

b. What is the slope of the line through the points? What is the y-intercept?

c. Write the equation of the line that contains these points.

3. Write the equation of the line that passes through the points of an arithmetic sequence in which u_0 is equal to 7 and 3 is added to get each new term.

4. Write a recursive formula for a sequence whose points lie on the line $y = 6 - 0.5x$.

5. Find the slope of each line.

a. $y = 2 + 1.7x$ **b.** $y = x + 5$ **c.** $y = 12 - 4.5x$ **d.** $y = 12$

Reason and Apply

6. An arithmetic sequence has a starting term, u_0, of 6.3 and a common difference of 2.5.

a. Write an explicit formula for the sequence.

b. Use the formula to figure out which term is 78.8.

7. Suppose you drive through Macon, Georgia (which is 82 mi from Atlanta), on your way to Savannah, Georgia, at a steady 54 mi/h.

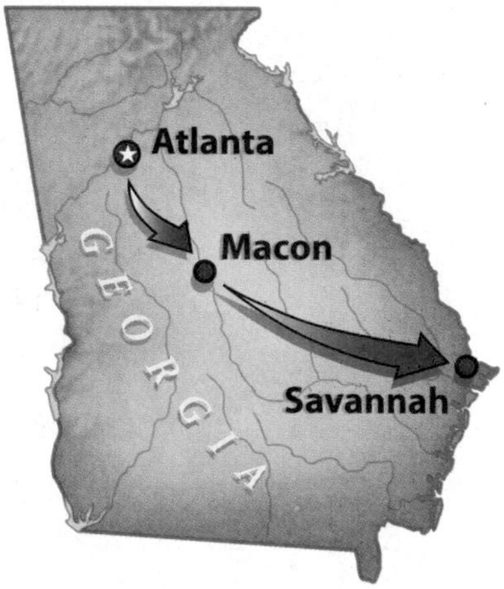

a. What is your distance from Atlanta two hours after you leave Macon?

b. Write an equation that represents your distance y from Atlanta x hours after leaving Macon.

c. Graph the equation.

d. Does this equation model an arithmetic sequence? Why or why not?

8. **APPLICATION** Melissa and Roy both sell cars at the same dealership and have to meet the same weekly profit goals. Last week, Roy sold only three cars, and he was below his goal by \$2050. Melissa sold seven cars, and she beat her goal by \$1550. Assume that the profit is related to the number of cars sold and that the profit is approximately the same for each car they sell.

a. Use a graph to find a few terms of this sequence of profits.

b. What is the real-world meaning of the common difference?

c. What is the explicit formula relating profit to the number of cars sold? What is the linear equation of the model?

d. What is the real-world meaning of the horizontal and vertical intercepts?

e. How many more cars must Roy sell to be within \$500 of his goal?

9. The points on this graph represent the first five terms of an arithmetic sequence. The height of each point is its distance from the x-axis, or the value of the 2nd coordinate of the point.

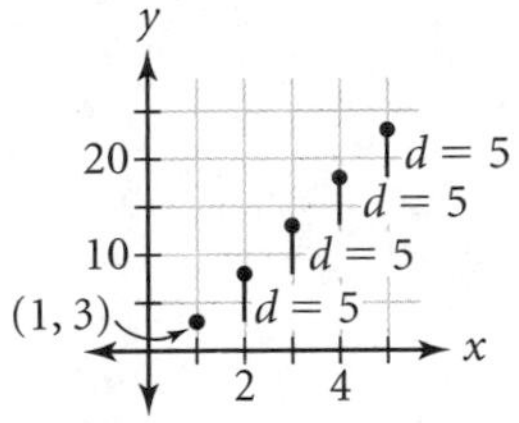

a. Find u_0, the 2nd coordinate of the point preceding those given.

b. How many common differences (d's) do you need to get from the height of $(0, u_0)$ to the height of $(5, u_5)$?

c. How many d's do you need to get from the height of $(0, u_0)$ to the height of $(50, u_{50})$?

d. Explain why the height from the x-axis to $(50, u_{50})$ can be found using the equation $u_{50} = u_0 + 50d$.

e. In general, for an arithmetic sequence the explicit formula is $u_n =$ ___?___.

10. A gardener planted a new variety of ornamental grass and kept a record of its height for the first two weeks of growth.

Time (days)	0	3	7	10	14
Height (cm)	4.2	6.3	9.1	11.2	14

a. How much does the grass grow each day?

b. Write an explicit formula that gives the height of the grass after n days.

c. How long will it take for the grass to be 28 cm tall?

11. Follow parts 11a–c to find four missing terms in an arithmetic sequence from 7 to 27.

a. Name two points on the graph of this sequence: (___?___, 7) and (___?___, 27).

b. Plot the two points you named in 11a, and find the slope of the line connecting the points.

c. Use the slope to find the missing terms.

d. Plot all the points, and write the equation of the line that contains them.

12. **APPLICATION** If an object is dropped, then it will fall about 16 ft during the first second. In each second that follows, the object falls 32 ft farther than in the previous second.

a. Write a recursive formula to describe free fall.

b. Find an explicit formula to describe free fall.

c. How far will the object fall during the 10th second?

d. How fast will the object be moving in feet per second (ft/s)?

e. During which second will the object fall 400 feet?

HistoryConnection Leonardo da Vinci was able to discover the formula for the velocity (directional speed) of a freely falling object by looking at a sequence. He let drops of water fall at equally spaced time intervals between two boards covered with blotting paper. When a spring mechanism was disengaged, the boards clapped together. By measuring the distances between successive blots and noting that these distances increased arithmetically, da Vinci discovered the formula $v = gt$, where v is the velocity of the object, t is the time since it was released, and g is a constant that represents any object's downward acceleration due to the force of gravity.

Review

13. APPLICATION (Lesson 2.1) Suppose a company offers a new employee a starting salary of \$33,400 with annual raises of \$1500, or a starting salary of \$32,900 with raises of \$2000 each year. At what point does one choice earn more than the other for the employee? Explain.

14. (Lesson 2.3) Suppose that you add 300 mL of water to an evaporating dish at the start of each day, and each day 40% of the water in the dish evaporates. Write and solve an equation that computes the long-run water level in the dish. Will a 1-L dish do the job? Explain why or why not.

15. APPLICATION (Lesson 1.4) Five stores in Tulsa, Oklahoma, sell the same model of graphing calculator for \$89.95, \$93.49, \$109.39, \$93.49, and \$97.69. What are the median price, the mean price, and the standard deviation? If these stores are representative of all stores in the Tulsa area, of what importance is it to a consumer to know the median, mean, and standard deviation? Which is probably more helpful, the median or the mean?

IMPROVING YOUR REASONING SKILLS

Sequential Slopes

Here's a sequence that generates coordinate points. What is the slope between any two points of this sequence?

$(x_0, y_0) = (0, 0)$

$(u_n, y_n) = (x_{n-1} + 2, y_{n-1} + 3)$ where $n \geq 1$

Now, match each of these recursive rules to the slope between points.

a. $(x_n, y_n) = (x_{n-1} + 2, y_{n-1} + 2)$

b. $(x_n, y_n) = (x_{n-1} + 1, y_{n-1} + 3)$

c. $(x_n, y_n) = (x_{n-1} + 3, y_{n-1} - 4)$

d. $(x_n, y_n) = (x_{n-1} - 2, y_{n-1} + 10)$

e. $(x_n, y_n) = (x_{n-1} + 1, y_{n-1})$

f. $(x_n, y_n) = (x_{n-1} + 9, y_{n-1} + 3)$

i. 0 **ii.** 1 **iii.** 3 **iv.** -5 **v.** $\frac{1}{3}$ **vi.** $\frac{-4}{3}$

In general, how do these recursive rules determine the slope between points?

LESSON

3.2

Revisiting Slope

There is more to life than increasing its speed.

MOHANDAS GANDHI

Suppose you are taking a long trip in your car. At 5 P.M. you notice that the odometer reads 45,623 miles. At 9 P.M. you notice that it reads 45,831. You find your average speed during that time period by dividing the difference in distance traveled by the difference in time elapsed.

$$\text{Average speed} = \frac{45831 \text{ mi} - 45623 \text{ mi}}{9 \text{ h} - 5 \text{ h}} = \frac{208 \text{ mi}}{4 \text{ h}} = 52 \text{ mi/h}$$

You can also write the rate 52 mi/h as the ratio $\frac{52 \text{ mi}}{1 \text{ h}}$. If you graph the information as points of the form (*time, distance*), the slope of the line connecting the two points is $\frac{52}{1}$, which also tells you the average speed.

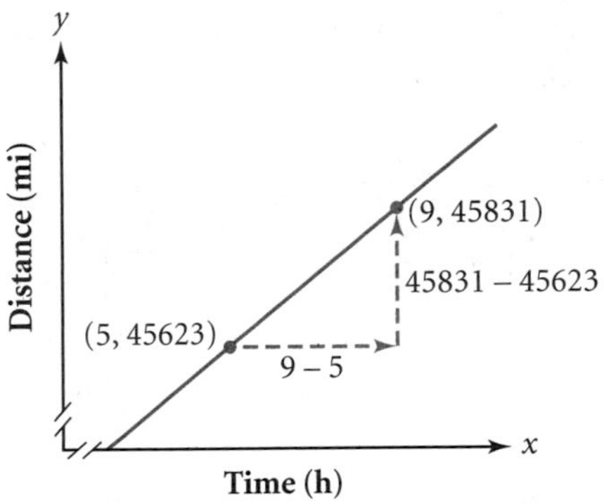

Slope

The formula for the slope of the line between two points, (x_1, y_1) and (x_2, y_2), is

$$\text{slope} = \frac{y_2 - y_1}{x_2 - x_1}$$

where $x_2 \neq x_1$.

Table 1

x	y
3	5
4	8
5	11
6	14
7	17

Table 2

x	y
−2	100
0	90
2	80
4	70
6	60

Table 3

x	y
2	7
3	7
4	7
5	7
6	7

Table 4

x	y
2	5
2	6
2	7
2	8
2	9

The slope will be the same for any two points selected on the line. In other words, a line has only one slope. In Table 1, for instance, the difference between the y-coordinates is three times the difference between the corresponding x-coordinates. In Table 2, no matter which two points you choose, you can calculate that the slope is -5.

Two points on a line can have the same y-value; in that case, the slope of the line is 0. Table 3 lists some points that are on a line with points sharing the same y-value. Using the points (2, 7) and (3, 7) from the table, we can confirm that the slope is 0.

$$\frac{7-7}{3-2}=0$$

If the points had the same x-value, the denominator would be 0, and the slope would be undefined. So, the definition of slope specifies that the points cannot have the same x-value. Table 4 gives some points from a line that have identical x-values. What would the graph of this line look like?

Recall that *slope* is another word for the steepness or rate of change of a line. If a linear equation is in intercept form, then the slope of the line is the coefficient of x.

Intercept Form of the Equation of a Line

You can write the equation of a line as

$$y = a + bx$$

where a is the y-intercept and b is the slope of the line.

Slope is often represented by the letter m. However, we will use the letter b in linear equations, as in the intercept form $y = a + bx$.

When you use real-world data, choosing different pairs of points will result in choosing lines with slightly different slopes. However, if the data are nearly linear, these slopes should not differ greatly.

Investigation
Balloon Blastoff

You will need

- paper
- tape
- a balloon
- straw
- string
- a motion sensor

In this Investigation you will launch a rocket and use your sensor data to calculate the rocket's speed. Then you will write an equation for the rocket's distance as a function of time. Choose one person to be the monitor and one person to be the launch controller.

Procedure Note

1. Make a rocket of paper and tape. Design your rocket so that it can hold an inflated balloon and be taped to a drinking straw threaded on a string.
2. Tape your rocket to the straw on the string.
3. Inflate a balloon, but do not tie off the end. The launch controller should insert it into your rocket and hold it closed.

Step 1 The monitor holds the sensor about 3 meters farther down the string and counts down to blastoff. When the monitor presses the trigger on the sensor and says, "Blast off," the launch controller releases the balloon. [▶ See **Calculator Note 3B.**◀]

Step 2 Upload the data from the sensor to each calculator in the group.

Step 3 Graph the data with time as the independent variable. What are the domain and range of your data? Explain.

Step 4 Select four points that you think will give the most accurate slope for the data. Indicate the points you selected on a graph of these data, and explain why you chose them. Use these points in pairs to calculate slopes. This should give six values for the slope.

Step 5 Are all six slope values that you calculated in Step 4 the same? Why or why not? Find the mean, median, and mode of your slopes. With your group decide what value best represents the slope of your data. Explain why you chose this value.

Step 6 What is the real-world meaning of the slope, and how is this related to the speed of your rocket?

Step 7 Write an equation for the rocket's distance x seconds after it is released, and explain what each part of the equation means.

Step 8 Graph your line, and label it "Our rocket." Imagine each of these scenarios. On the same graph, sketch lines to represent each of these rockets.

a. A rocket that was released at the same time as yours but traveled at 75% the speed of your rocket. Label this line "Slower rocket."

b. A rocket that was released 2 seconds before yours and traveled at the same speed as yours. Label this line "Same speed."

c. A rocket that was released 2 seconds after yours but got caught on the string and did not go anywhere. Label this line "No movement."

Write the equation for each of the imaginary rockets in Step 8. Label each of the lines on your graph.

In many cases when you try to model the correct steepness and trend of the points, you may have some difficulty deciding which points to use. In general, select two points that are far apart to minimize the error. Disregard data points that you think might represent measurement errors.

When you begin to analyze a relationship between two variables, you must decide which variable you will express in terms of the other. When one variable depends on the other variable, it is called the **dependent variable.** The other variable is called the **independent variable.** Time is usually considered an independent variable.

EXAMPLE

Daron's car uses 0.05 gal of gas for every mile it is driven. He starts out with a full tank, 16.4 gal.

a. Identify the independent and dependent variables.

b. Write a linear equation in intercept form to model this situation.

c. How much gas will be left in Daron's tank after he has driven 175 mi?

d. How far can he travel before he has less than 2 gal remaining?

▶ ***Solution***

The two variables are the distance Daron has driven and the amount of gas in his tank.

a. The amount of gasoline remaining in Daron's tank depends on the number of miles he has driven. This means the amount of gasoline is the dependent variable, and distance is the independent variable. So, you will use x for the distance (in miles) and y for the amount of gasoline (in gallons).

b. Daron starts out with $y = 16.4$ gal, and this amount decreases by 0.05 gal/mi. The equation is $y = 16.4 - 0.05x$.

c. You know the x-value is 175 mi. You can substitute 175 for x and solve for y.

$$\begin{aligned} y &= 16.4 - 0.05 \cdot 175 \\ &= 7.65 \end{aligned}$$

He will have 7.65 gal remaining.

d. You know the y-value is 2.0 gal. You can substitute 2.0 for y and solve for x.

$$\begin{aligned} 2.0 &= 16.4 - 0.05x \\ -14.4 &= -0.05x \\ 288 &= x \end{aligned}$$

Once he has traveled more than 288 mi, he will have less than 2 gal in his tank.

EXERCISES

Practice Your Skills

1. Find the slope of the line containing each pair of points.

a. $(3, -4)$ and $(7, 2)$

b. $(5, 3)$ and $(2, 5)$

c. $(-0.02, 3.2)$ and $(0.08, -2.3)$

2. Find the slope of each line.

a. $y = 3x - 2$

b. $y = 4.2 - 2.8x$

c. $y = 5(3x - 3) + 2$

d. $y - 2.4x = 5$

e. $4.7x + 3.2y = 12.9$

f. $\frac{2}{3}y = \frac{2}{3}x + \frac{1}{2}$

3. Solve each equation.

a. Solve $y = 4.7 + 3.2x$ for y if $x = 3$.

b. Solve $y = -2.5 + 1.6x$ for x if $y = 8$.

c. Solve $y = a - 0.2x$ for a if $x = 1000$ and $y = -224$.

d. Solve $y = 250 + bx$ for b if $x = 960$ and $y = 10$.

4. Find the equations of both lines in each graph.

a.

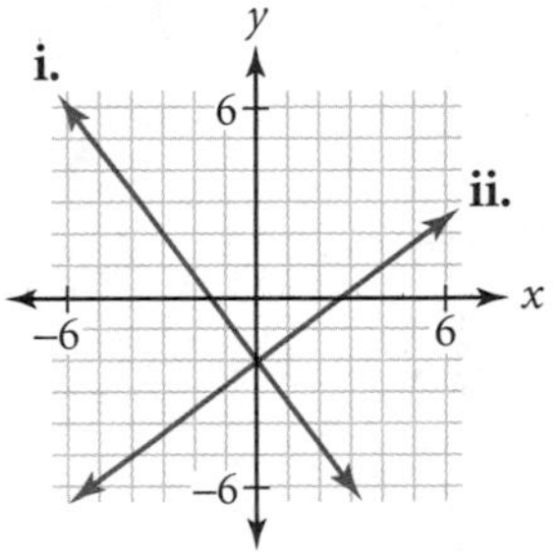

b.

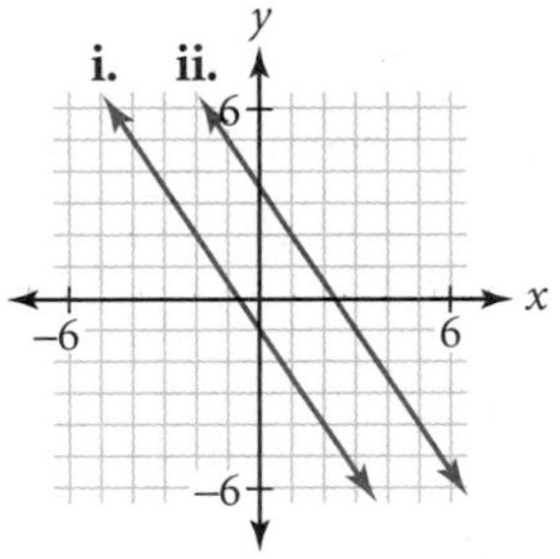

5. Consider the equations and graphs of Exercise 4.

a. What do the equations in 4a have in common? What do you notice about their graphs?

b. What do the equations in 4b have in common? What do you notice about their graphs?

Reason and Apply

6. **APPLICATION** Layton measures the voltage across different numbers of batteries placed end to end. He records his data in a table.

Number of batteries	1	2	3	4	5	6	7	8
Voltage (volts)	1.43	2.94	4.32	5.88	7.39	8.82	10.27	11.70

a. Find the slope of the line through these data. Be sure to include units with your answer.

b. Which points did you use and why? What is the real-world meaning of this slope?

c. Does it make sense that the y-intercept of the line is 0? Explain why or why not.

7. Use this graph to determine the speed of a balloon rocket. The independent variable is time in seconds (s), and the dependent variable is distance from the sensor in meters (m).

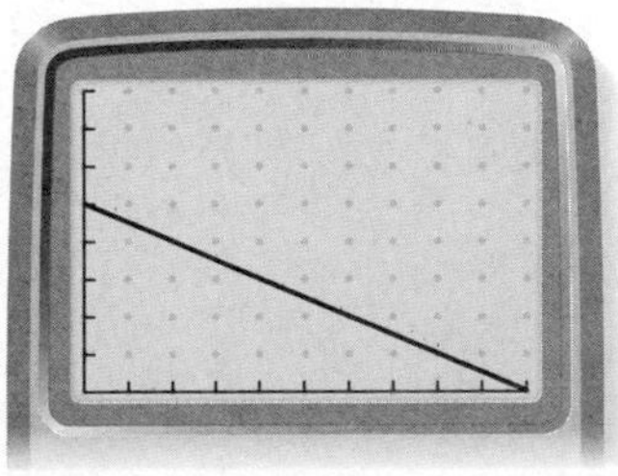

[0, 10, 1, 0, 8, 1]

8. This graph shows the relationship between the height of some high-rise buildings and the number of stories in them. The line drawn to fit the data has the equation $y = 13.02x + 20.5$.

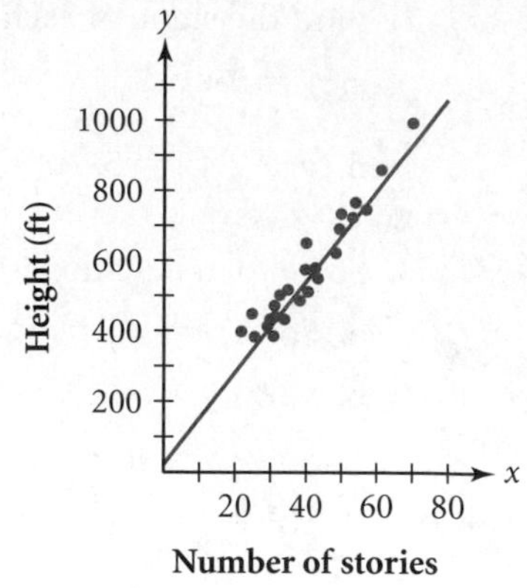

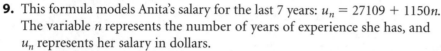

a. Estimate the slope. What is the meaning of the slope?

b. Estimate the y-intercept. What is the meaning of the y-intercept?

c. Explain why some of the points lie above the line and some lie below.

d. According to the graph, what are the domain and range of this relationship?

9. This formula models Anita's salary for the last 7 years: $u_n = 27109 + 1150n$. The variable n represents the number of years of experience she has, and u_n represents her salary in dollars.

a. What did she earn in the 5th year? What did she earn in her 1st year? (Think carefully about what n represents.)

b. What is the rate of change of her salary?

c. What is the 1st year in which Anita's salary is more than $45,000? Set up an equation and solve it.

d. How would Anita's salary be different in the year you found in 9c if she earned a 3% raise each year instead of a $1150 raise?

10. APPLICATION This table shows how long it took to lay tile for hallways of different lengths.

Length of hallway (ft)	3.5	9.5	17.5	4.0	12.0	8.0
Time (min)	85	175	295	92	212	153

a. What is the independent variable? Why? Graph the data.

b. Find the slope of the line through these data. What is the real-world meaning of this slope?

c. Which points did you use and why?

d. Find the y-intercept of the line. What is the real-world meaning of this value?

11. APPLICATION The manager of a concert hall records the total number of tickets sold and total sales income, or revenue, for each event. Two different ticket prices are offered.

Total tickets	448	601	297	533	523	493	320
Total revenue ($)	3357.00	4495.50	2011.50	3784.50	3334.50	3604.50	2353.50

a. Find the slope of the line through these data. What is the real-world meaning of this slope?

b. Which points did you use and why?

12. APPLICATION How much does air weigh? The following table gives the weight of a cubic foot of dry air at the same pressure at various temperatures in degrees Fahrenheit.

Temp. (°F)	0	12	32	52	82	112	152	192	212
Weight (lb)	0.0864	0.0842	0.0807	0.0776	0.0733	0.0694	0.0646	0.0609	0.0591

a. Make a scatter plot of the data. [► See **Calculator Note 1F.**◄]

b. What is the approximate slope of the line through the data? Include units in your answer.

c. Describe the real-world meaning of the slope.

RecreationConnection Hot air rises, cool air sinks. As air is heated, it becomes less dense and lighter than the cooler air surrounding it. This simple law of nature is the principle behind hot-air ballooning. By heating the air in the balloon envelope, maintaining its temperature, or letting it cool, the balloon's pilot is able to climb higher, fly levelly, or descend. Joseph and Jacques Montgolfier developed the first hot-air balloon in 1783, inspired by the rising of a shirt that was drying above a fire.

Review

13. (Lessons 1.4 and 1.5) Charlotte and Emily measured the pulse rates of everyone in their nursing class in beats per minute (beats/min) and collected this set of data.

{62, 68, 68, 70, 74, 66, 82, 74, 76, 72, 70, 68, 80, 60, 84, 72, 66, 78, 70, 68, 66, 82, 76, 66, 66, 80}

a. What is the mean pulse rate for the class?

b. What is the standard deviation? What does this tell you?

14. (Lesson 2.4) Each of these two graphs was generated by a recursive rule in the form $u_0 = a$ and $u_n = (1 + r)u_{n-1} + p$ where $n \geq 1$. Describe the parameters a, r, and p that produce each graph. (There are two answers to 14b.)

a.

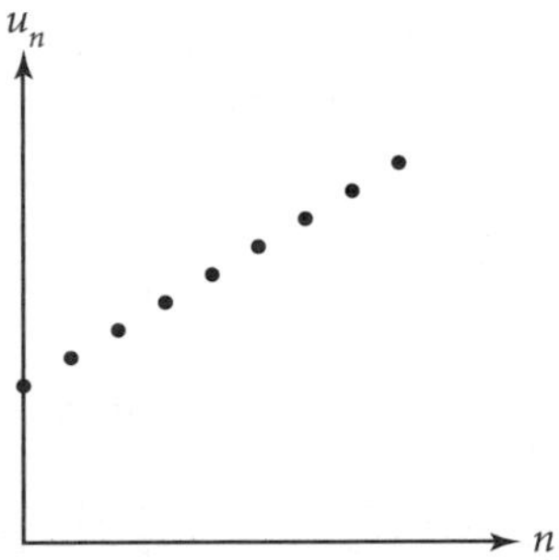

b.

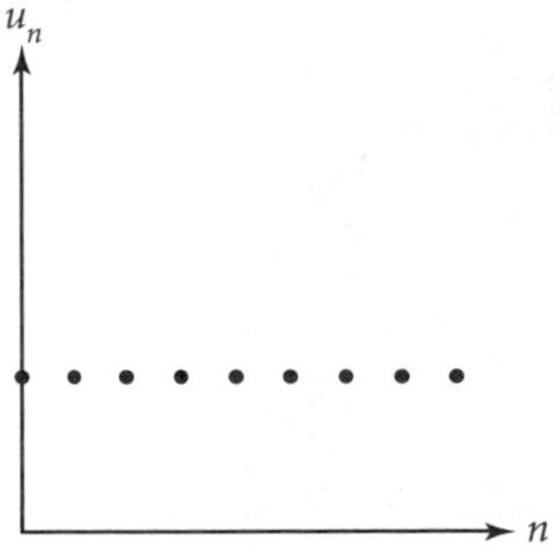

15. (Lesson 3.1) Study each graph, and complete the table by filling in > 0, < 0, or $= 0$ for a and b in the equation $y = a + bx$.

i.

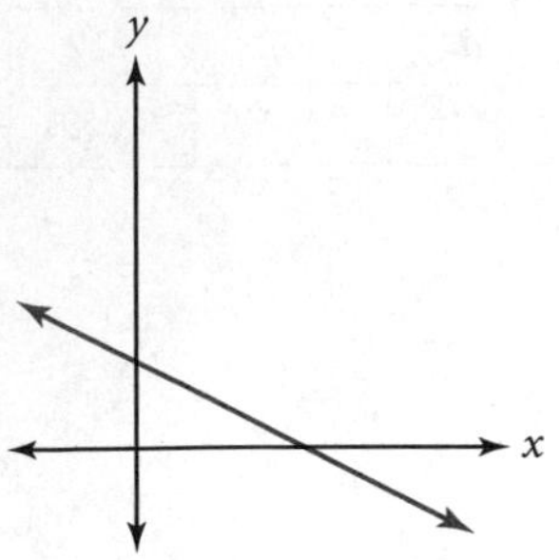

ii.

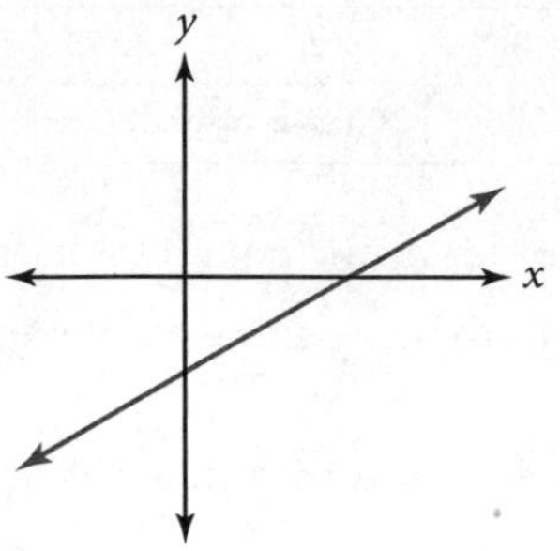

iii.

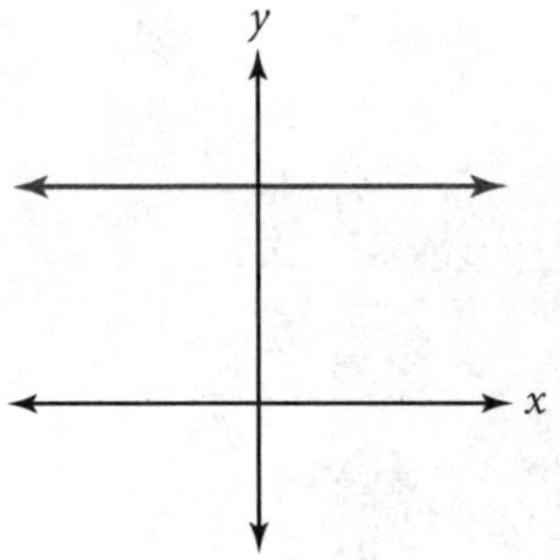

iv.

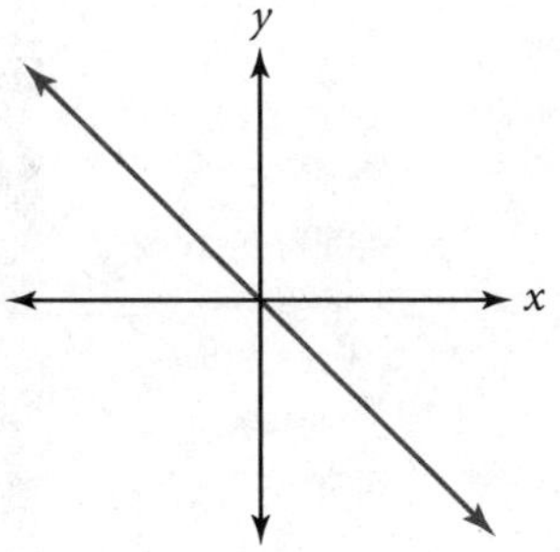

	i	ii	iii	iv
a				
b				

LESSON

3.3

Fitting a Line to Data

MORE BACKGROUND

Before beginning this lesson, you should be familiar with this concept:

- using function notation

A short review of this concept can be found at: www.keycollege.com/online .

Out of intense complexities, intense simplicities emerge.

WINSTON CHURCHILL

Investigation

Height and Arm Span

You will need

- a tape measure
- a graphing calculator

Sometimes you can find a nearly linear relationship in data.

Height	Arm span

Step 1 For each student in your class, measure the student's height and the student's arm span in inches. To measure arm span, have the student stand with arms held straight out at shoulder height. Measure from fingertip to fingertip. Record the data in a table like the one below.

Step 2 Create a scatter plot of your data, using height as the independent variable.

Step 3 Now try to enter equations in your calculator with graphs that best represent the data in the scatter plot. [► See **Calculator Note 3A** for instructions on how to graph equations on a calculator. ◄] Decide which of your graphs best represents the data.

Step 4 Compare your equation and graph with those of several classmates. Which do you think best represents the points on the scatter plot? What criteria did you use to decide which was best?

All the points of an arithmetic sequence lie on a line. When you collect data and make a graph, sometimes the data will appear to have a linear relationship. However, the points will rarely lie on a single line. Usually they will be scattered, and it is up to you to determine a reasonable location for the line that summarizes or gives the trend of the data set. A line that fits the data reasonably well is called a **line of fit.**

There is no single list of rules that will give the best line of fit in every instance, but you can use these guidelines to obtain a reasonably good fit.

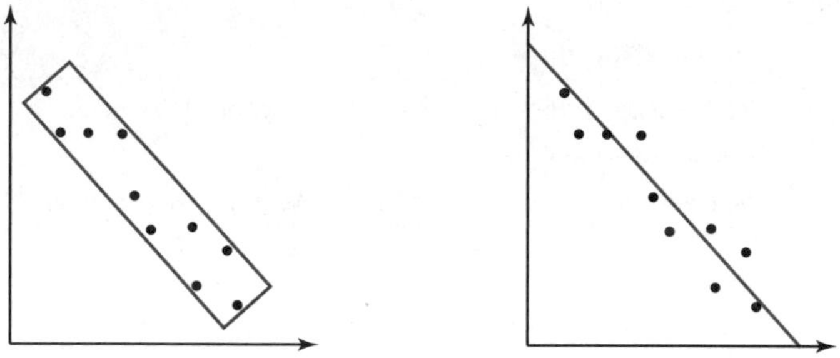

Finding a Line of Fit

1. Determine the direction of the points. The longer side of the smallest rectangle that contains most of the points shows the general direction of the line.
2. The line should divide the points equally. Draw the line so that there are about as many points above the line as below the line. The line has nearly the same slope as the longer sides of the rectangle.
3. The points above the line should not be concentrated at one end, nor should the points below the line.
4. The line should have a y-intercept that makes sense. Consider whether the y-intercept should be the origin.

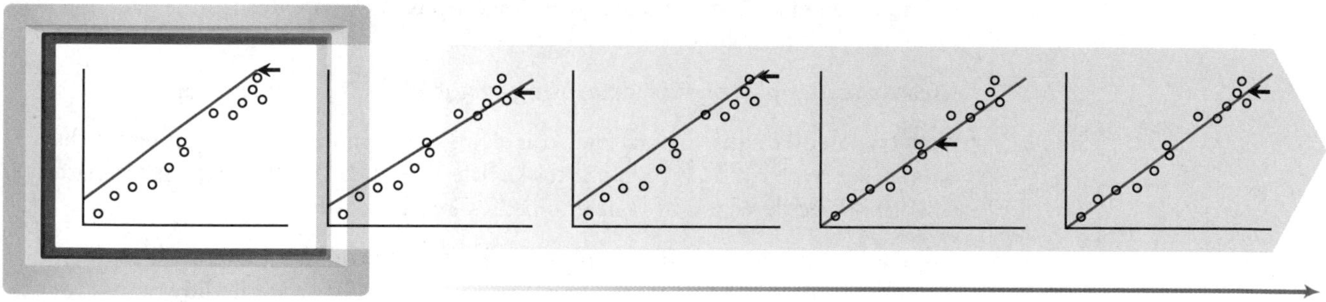

Once you have drawn a line of fit for your data, you can write an equation that expresses the relationship.

There are several ways to write the equation of a line. One way is to find the slope and the y-intercept and write the intercept form of the equation. Once you find an equation, you can then use this equation to predict points for which data are not available.

EXAMPLE

On a barren lava field on top of the Mauna Loa volcano in Hawaii, scientists have been monitoring the concentration of CO_2 (carbon dioxide) in the atmosphere continuously since 1959. This site is favorable because it is relatively isolated from vegetation and human activities that produce CO_2. The concentrations for 17 different years, measured in parts per million (ppm), are shown in this table.

Year	CO_2 (ppm)	Year	CO_2 (ppm)
1980	337.84	1990	353.50
1981	339.06	1993	356.63
1982	340.57	1994	358.34
1983	341.20	1995	359.98
1984	343.52	1996	362.09
1986	346.11	1997	363.23
1987	347.84	1998	365.38
1988	350.25	1999	368.24
1989	352.60		

(*Carbon Dioxide Information and Analysis Center*)

a. Find a line of fit to summarize the data.

b. Predict the concentration of CO_2 in the atmosphere in the year 2050.

ScienceConnection Mauna Loa, which means "long mountain," has an area of 2035 mi^2, covers half of the island of Hawaii, and is Earth's largest active volcano. This active volcano's last eruption was in 1984. Volcanologists routinely monitor Mauna Loa for signs of eruption and specify hazardous areas of the mountain.

▶ *Solution*

A graph of the data shows a linear pattern.

a. You can draw a line that seems to fit the trend of the data.

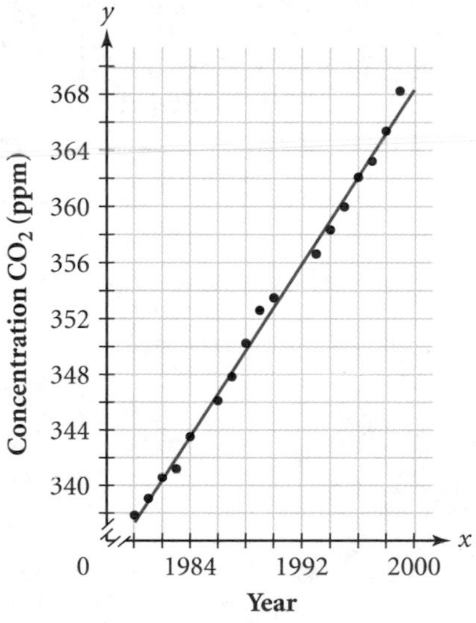

The line shown on the graph is a good fit because many of the points lie on or near the line, the data points above and below the line are roughly equal in number, and they are evenly distributed on both sides along the line. We do not know what the y-intercept should be, because we do not know the concentration in the year 0; we have no reason to believe that the line passes through (0, 0).

Next, you need to find an equation for this line of fit. The y-intercept is not easily available, but you can choose two points to determine the slope. Note that the points do not need to be data points. You might choose the points (1984, 343.5) and (1994, 359.0). They are far enough apart, and their coordinates are easy to find on the graph.

You can use these points to find the slope.

$$\text{slope} = \frac{359.0 - 343.5}{1994 - 1984} = \frac{15.5}{10} = 1.55$$

This means the concentration of CO_2 in the atmosphere has been increasing at a rate of about 1.55 ppm each year.

You can then use trial and error to find a good candidate for the y-intercept. Pick a y-intercept, graph the line, and then adjust your prediction for the y-intercept until you get a satisfactory line.

$y = -2732 + 1.55x$ is one possibility for the line of best fit.

b. You can substitute 2050 for x and solve for y to predict the CO_2 concentration in the year 2050.

$$y = -2732 + 1.55\,(2050)$$
$$= 445.6$$

The CO_2 concentration is about 446 ppm.

Investigation
The Wave

You will need

- a stop watch or a watch with a second hand

Sometimes at sporting events people in the audience stand up quickly in succession with their arms upraised and then sit down again. The continuous rolling motion that this creates through the crowd is called "the wave." You and your class will investigate how long it takes groups of different sizes to do the wave.

Step 1 Using groups of different sizes, determine the time each group takes to complete the wave. Collect at least nine pieces of data of the form (*number of people, time*), and record them in a table.

Step 2 Plot the points, and find the equation of a reasonable line of fit. Write a paragraph about your results. Be sure to include this information:

- ▶ What is the slope of your line, and what is its real-world meaning?
- ▶ What are the x- and y-intercepts of your line, and what are their real-world meanings?
- ▶ What is the x-intercept of your equation, and what is its real-world meaning?
- ▶ What is a reasonable domain for this equation? Why?

Step 3 Can you use your line of fit to predict how long it would take to complete the wave if everyone in your college participated? Everyone in a large stadium? Explain why or why not.

Keep these data and the equation. You will use them in a later lesson.

Finding a value between other values given in a data set is called **interpolation.** Using a model to extend beyond the first or last data points is called **extrapolation.** Referring to the Mauna Loa data in the example, how would you use interpolation to estimate the CO_2 levels in 1960? In 1991?

EXERCISES

Practice Your Skills

1. Write an equation of each line shown.

a.

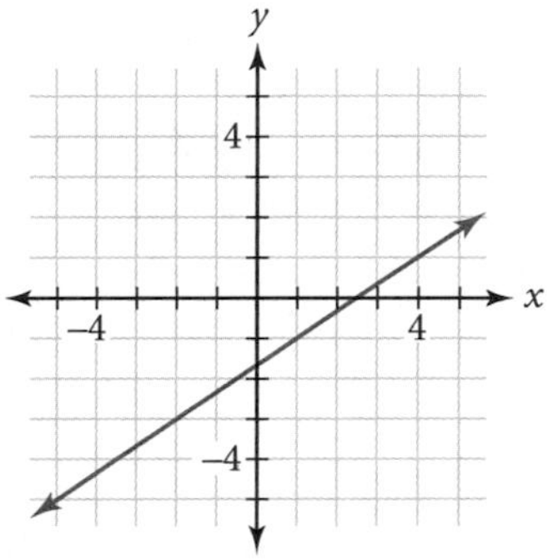

b.

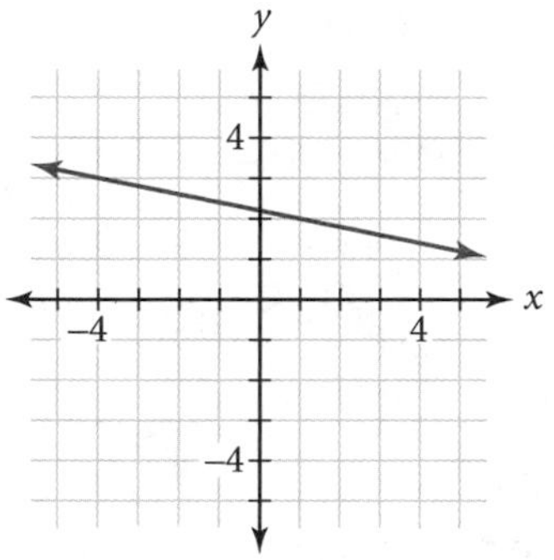

2. Write an equation of each line described.

a. Slope $\frac{2}{3}$ passing through $(5, -7)$

b. Slope -4 passing through $(1, 6)$

c. Parallel to $y = -2 + 3x$ passing through $(-2, 8)$

d. Parallel to $y = -4 - \frac{3}{5}(x + 1)$ passing through $(-4, 11)$

3. Solve each equation. Use function notation in your answer.

a. Solve $f(n) = 23 + 2(n - 7)$ for $f(n)$ if $n = 11$.

b. Solve $f(t) = -47 - 4(t + 6)$ for t if $f(t) = 95$.

c. Solve $f(x) = 56 - 6(x - 10)$ for x if $f(x) = 107$.

4. Consider the line $y = 5$.

a. Graph this line, and identify two points on it.

b. What is the slope of this line?

c. Write an equation of the line that contains the points $(3, -4)$ and $(-2, -4)$.

d. Write three statements about horizontal lines and their equations.

5. Consider the line $x = -3$.

a. Graph it, and identify two points on it.

b. What is the slope of this line?

c. Write an equation of the line that contains the points $(3, 5)$ and $(3, 1)$.

d. Write three statements about vertical lines and their equations.

Reason and Apply

6. Look at each graph below, and choose the *one* with the line that best satisfies the guidelines on page 120. For each of the other graphs, explain which guidelines the line violates.

a.

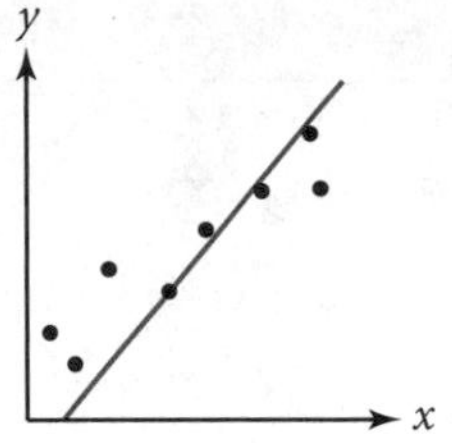

b.

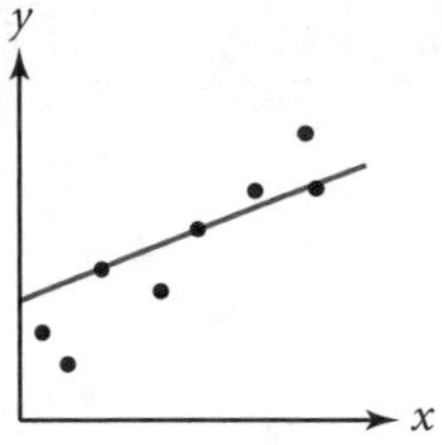

c.

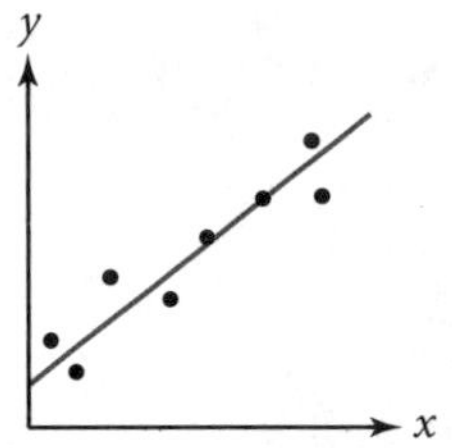

d.

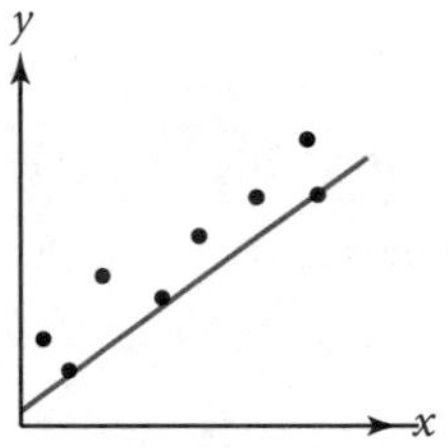

7. For each graph below, lay your ruler along your best estimate of the line of fit. Estimate the y-intercept and the coordinates of one other point on the line. Write an equation in intercept form for the line of fit.

a.

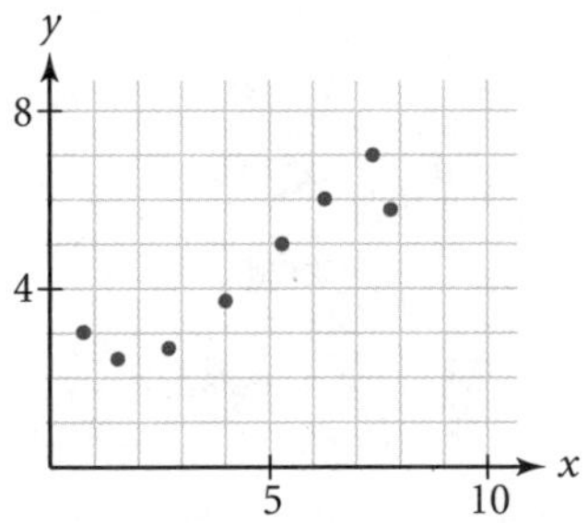

b.

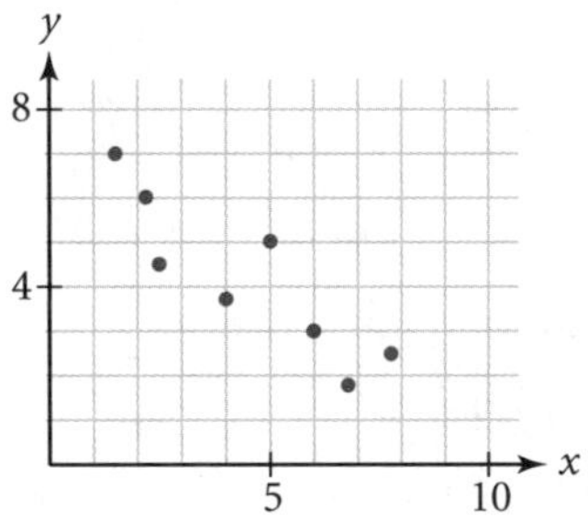

c.

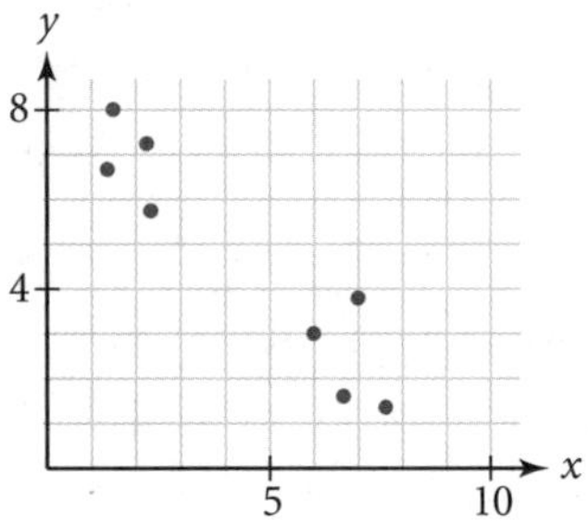

8. **APPLICATION** A photography studio offers several packages to families posing for portraits.

Number of pictures	44	31	24	15
Total cost ($)	189.00	159.00	129.00	109.00

a. Plot the data, and find an equation of a line of fit. Explain the real-world meaning of the slope of this line.

b. Find the y-intercept of your line of fit. Explain the real-world meaning of the y-intercept.

c. If the studio offers a 75-print package, what do you think it should charge?

d. How many prints do you think the studio should include in the package for a special priced at $89.99?

9. Use height as the independent variable and length of forearm as the dependent variable for the data collected from nine students.

Height	Forearm
185.9	48.5
172	44.5
155.0	41.0
191.5	50.5
162.0	43.0
164.3	42.5
177.5	47.0
180.0	48.0
179.5	47.5

a. Name a good graphing window for your scatter plot.

b. Write a linear equation that models the data.

c. Write a sentence describing the real-world meaning of the slope of your line.

d. Write a sentence describing the real-world meaning of the y-intercept. Explain why this doesn't make sense and how you might correct it.

e. Use your equation to estimate the height of a student with a 50-cm forearm and to estimate the length of a forearm of a student 158 cm tall.

10. APPLICATION This data set was collected by a college psychology class to determine the effects of sleep deprivation on students' ability to solve problems. Ten participants went 8, 12, 16, 20, or 24 hours without sleep and then completed a set of simple addition problems. The number of addition errors was recorded.

Hours without sleep	8	8	12	12	16	16	20	20	24	24
Number of errors	8	6	6	10	8	14	14	12	16	12

a. Create a scatter plot of the data.

b. Write an equation of a line that fits the data, and sketch it on your graph.

c. Based on your model, how many errors will a person make if she or he hasn't slept in 22 hours?

d. In 10c did you use interpolation or extrapolation? Explain.

Review

11. (Algebra) Rewrite each expression without parentheses, combining exponents whenever possible.

a. $3(x^2y)^3$ **b.** $(a^3b^4)(a^{-2}b^3)$ **c.** $\dfrac{12x^2y^8}{(2x^3y)^3}$

12. (Lesson 2.3) Write the first four terms of this sequence, and describe its long-run behavior.

$u_1 = 56$

$u_n = \dfrac{u_{n-1}}{2} + 4$ where $n \geq 2$

13. (Lesson 1.4) Given the data set {20, 12, 15, 17, 21, 15, 30, 16, 14},

a. Find the median.

b. Add as few elements as possible to the set in order to make 19.5 the median.

14. (Lesson 3.1) You start 8 meters from a marker and walk toward it at the rate of 0.5 m/s.

a. Write a recursive rule that gives your distance from the marker after each second.

b. Write an explicit function that allows you to find your distance from the marker at any time.

c. Interpret the real-world meaning of a negative value for the function in 14a or 14b.

PROJECT

LIFE INSURANCE

Term insurance policies that provide level premiums for 10 or 20 years are popular because of their low cost relative to other insurance products. In this project you will gather quotes for different dollar amounts of life insurance.

1. Go to an insurance company's website. You will need to check life insurance costs. If you wish to protect your privacy, make up data about your age and gender. Be sure to be consistent for each estimate.
2. Find out the cost of $100,000 of life insurance. Select a term policy that provides a level premium for 10 years.
3. Find out the cost of $200,000 of life insurance. Select a term policy that provides a level premium for 10 years.
4. Find the equation of a line that passes through those two data points. Let x be the amount of insurance you want, and let y be the annual cost.
5. Use your equation to predict the cost of $500,000 in insurance.
6. Check the insurer's website again. How much do they actually charge for a $500,000 policy?
7. Is the price of insurance linear (in terms of how much you want to buy)?
8. If it was linear in 7, what is the y-intercept in dollars? If the y-intercept is not $0, why not? What does the y-intercept represent?

LESSON

3.4

Linear Systems

The number of tickets sold for a fund-raiser such as a spaghetti dinner helps determine the financial success of the event. Income from ticket sales can be less than, equal to, or greater than expenses. The break-even value is the intersection of the expense function and the line $y = x$. This equation models all the situations where the expenses, y, are equal to income, x. In this lesson you will focus on mathematical situations involving two or more equations or conditions that must be satisfied at the same time. A set of two or more equations that share the same variables and are solved or studied simultaneously is called a **system of equations.**

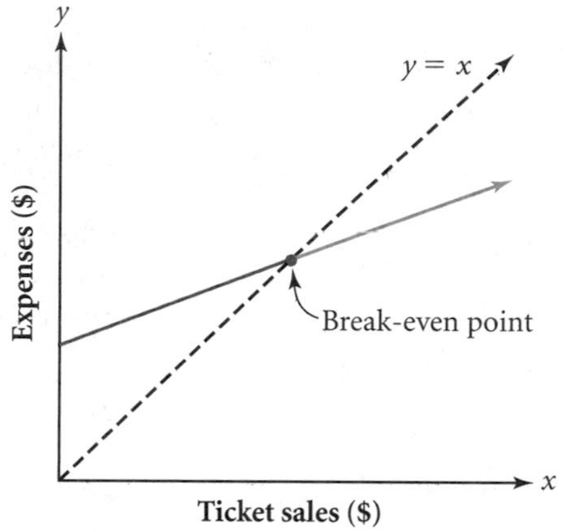

EXAMPLE A

Minh and Daniel are starting a business together, and they need to decide between long-distance phone carriers. One company offers the Phrequent Phoner Plan, which costs \$20 for the first 200 minutes and \$17 for each 200 minutes after that. A competing company offers the Small Business Plan, which costs \$50 for the first 200 minutes and \$11 for each additional 200-minute block. Which plan should Minh and Daniel choose?

▶ Solution

The plan they choose depends on the nature of their business and how many minutes of calling time they need each month. Because the Phrequent Phoner Plan costs less for the first 200 minutes, it is obviously the better plan if they do not make many calls. However, if they make a lot of calls, the Small Business Plan probably will be cheaper because the cost is less for additional minutes. There is a usage for which both plans cost the same. To find it, let x represent the number of 200-minute blocks purchased, and let y represent the cost in dollars.

A cost equation that models the Phrequent Phoner Plan is

$$y = 20 + 17(x - 1)$$

A cost equation for the Small Business Plan is

$$y = 50 + 11(x - 1)$$

A graph of these equations shows that the Phrequent Phoner Plan cost is less than the Small Business Plan cost until the lines intersect. You may be able to estimate the coordinates of this point from the graph.

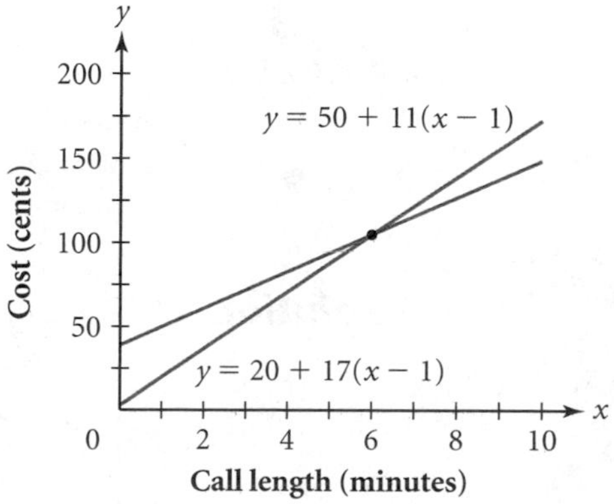

Or you can look in a table to find an answer. The point of intersection at (6, 105) tells you that six 200-minute blocks will cost $105 with either plan.

If Minh and Daniel think that their average monthly usage will be less than 1200 minutes, they should choose the Phrequent Phoner Plan. Otherwise, the Small Business Plan is the better option.

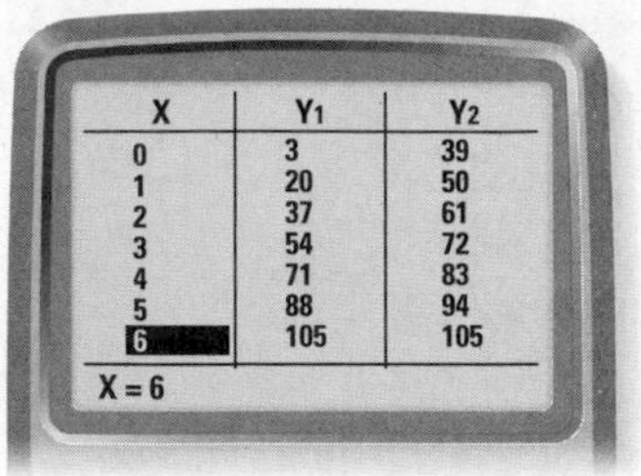

In Example B we will see that when you have the equations of both lines in a linear system, you can use those to solve the system. Example A and the following Investigation show that you can also use a table or graph to see a trend in the data estimate solutions.

Investigation
Population Trends

The table below gives the populations of San Jose, California, and Detroit, Michigan. Study the data with your group, then follow the steps to make some predictions.

Populations

Year	1950	1960	1970	1980	1990	2000
San Jose	92,280	204,196	459,913	629,442	782,248	894,943
Detroit	1,849,568	1,670,144	1,514,063	1,203,339	1,027,974	951,270

(*World Almanac and Book of Facts 2002*)

Step 1 If the current trends continue, when will San Jose have a larger population than Detroit? What will the two populations be at that time?

Step 2 Show the method you used to make this prediction. Choose a different method to check your answer.

There are different methods for finding the exact coordinates of an intersection point by solving systems of equations. One method is illustrated in the next example.

EXAMPLE B Justine and her little brother Evan are running a race. Because Evan is younger, Justine gives him a 50-foot head start. Evan runs at 12.5 ft/s, and Justine runs at 14.3 ft/s. How far will they run before Justine passes Evan? What distance should Justine mark for a close race?

▶ ***Solution*** You can compare the time and distance that each person runs. Because distance, d, depends on time, t, you can write these equations:

$d = rt$ Distance equals the rate, or speed, multiplied by time.

$d = 14.3t$ Justine's distance equation.

$d = 50 + 12.5t$ Evan's distance equation.

Graphing these two equations shows that Justine will eventually catch up to Evan and pass him if the race is long enough. At that moment, they will be the same distance from the start at the same time. You can estimate this point from the graph or scroll down until you find the answer in the table. You can also solve the system of equations.

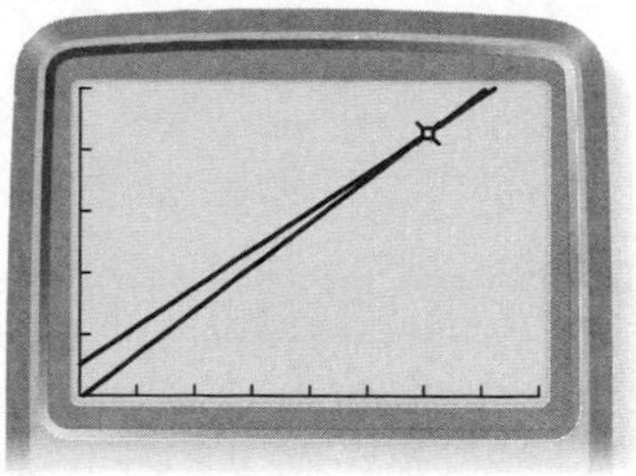

[0, 40, 5, 0, 500, 100]

Because both equations represent distances, and you want to know when those distances are equal, you can set the equations equal to each other and solve for the time, t, when the distances are equal.

$14.3t = 50 + 12.5t$	The right side of Justine's equation equals the right side of Evan's equation because they are equal to the same distance d.
$1.8t = 50$	Subtract $12.5t$ from both sides.
$t = \frac{50}{1.8} \approx 27.8$	Divide both sides by 1.8.

So, Justine will pass Evan after 27.8 seconds. Now you can substitute this value in either equation to find their distances from the starting line when Justine passes Evan.

$$d = 14.3t = 14.3 \cdot \frac{50}{1.8} \approx 397.2$$

If Justine marks a 400-ft distance for the race, she will win, but it will be a close race.

The method of solving a system demonstrated in Example B uses one form of substitution. In this case you substituted one expression for distance in place of the distance, d, in the other equation, resulting in an equation with only one variable, t. When the two equations are written in intercept form, substitution is a straightforward method for finding an exact solution. The solution to a system of equations with two variables is a pair of values that satisfies both equations.

It is possible for a system of equations with two variables to have no solution.

Consider the system

$$y = 3x - 1$$
$$y = 6x - 3(x + 2)$$

Substituting for y in the second equation,

$$3x - 1 = 6x - 3(x + 2)$$

Distributing:

$$3x - 1 = 6x - 3x - 6$$
$$3x - 1 = 3x - 6$$

Subtracting $3x$ from both sides:

$-1 = -6$ FALSE!!!

The original system has no solution. Graphing both lines in the original problem, you can see that they are parallel and do not intersect.

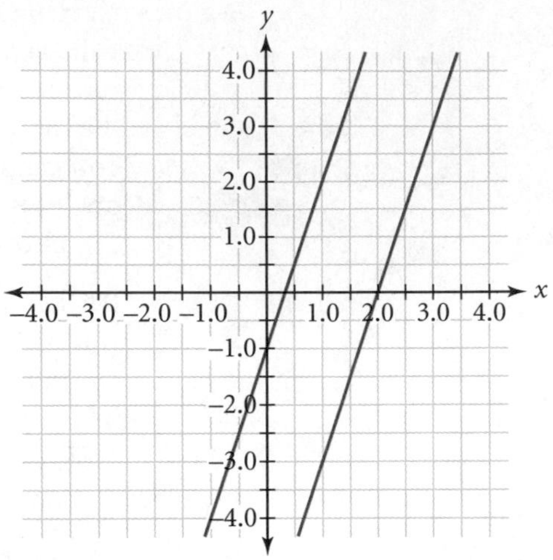

EXERCISES

Practice Your Skills

1. Use a table to find the point of intersection for each pair of linear equations.

a. $\begin{cases} y = 3x - 17 \\ y = -2x - 8 \end{cases}$ **b.** $\begin{cases} y = 28 - 3(x - 5) \\ y = 6 + 7x \end{cases}$ **c.** $\begin{cases} y = 7 - 3x \\ y = 3x - 6(x + 2) \end{cases}$

2. Write a system of equations that has (2, 7.5) as its solution.

3. Write the equation of the line perpendicular to $y = 4 - 2.5x$ and passing through the point (1, 5).

4. Solve each equation.

a. $4 - 2.5(x - 6) = 3 + 7x$ **b.** $11.5 + 4.1t = 6 + 3.2(t - 4)$

5. Use substitution to find the point (x, y) where each pair of lines intersect. Use a graph or table to verify your answer.

a. $\begin{cases} y = -2 + 3(x - 7) \\ y = 10 - 5x \end{cases}$ **b.** $\begin{cases} y = 0.23x + 9 \\ y = 4 - 1.35x \end{cases}$

c. $\begin{cases} y = -1.5x + 7 \\ y = -3x + 14 \end{cases}$ **d.** $\begin{cases} y = 4x + 9 \\ y = 2x + 2(x + 9) \end{cases}$

Reason and Apply

6. The equations $s_1 = 18 + 0.4m$ and $s_2 = 11.2 + 0.54m$ give the lengths of two different springs in centimeters, s_1 and s_2, as mass amounts in grams, m, are separately added to each.

a. When are the springs the same length?

b. When is one spring at least 10 cm longer than the other?

c. Write a statement comparing the two springs.

7. **APPLICATION** A group of retirees forms a small business, the Kangaroo Company, to market dolls at craft shows. This graph shows the Kangaroo Company's production costs and revenue for the dolls it sells at craft shows. Use the graph to estimate the answers to the questions.

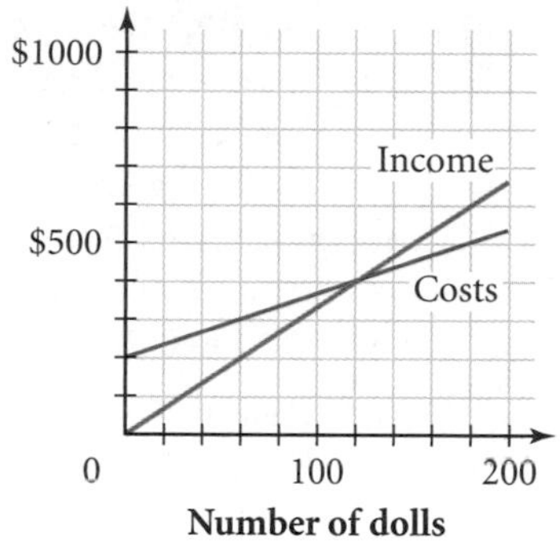

a. If 25 dolls are sold, will the company earn a profit? Describe how you can use the graph to answer this question.

b. If the company sells 200 dolls, will it earn a profit? If so, approximately how much?

c. How many dolls must the company sell to break even? How do you know?

8. **APPLICATION** Winning times for men and women in the 1500-m Olympic speed skating event are given below.

1500-m Olympic Speed Skating

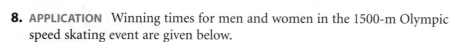

Year	1964	1968	1972	1976	1980	1984	1988	1992	1994	1998
Men	2:10.3	2:03.4	2:02.96	1:59.38	1:55.44	1:58.36	1:52.06	1:54.81	1:51.29	1:47.87
Women	2:22.6	2:22.4	2:20.85	2:16.58	2:10.95	2:03.42	2:00.68	2:05.87	2:02.19	1:57.58

a. Analyze the data and predict when the winning times for men and women will be the same if the current trends continue.

b. How reasonable do you think your prediction is? Explain your reasoning.

c. Predict the winning times for the 2002 Winter Olympics. How close are your predictions to the actual results?

d. Is it appropriate to use a linear model for these data? Why?

9. **APPLICATION** The 4 Line Company charges $0.44 to connect, plus 4 cents for every minute including the first minute. The 6 Saver Company charges 6 cents per minute, with no connection fee. Both companies calculate their charges so that a call of exactly 3 min will cost the same as a call of 3.25 min or 3.9 min, and there is no increase in cost until you have been connected for 4 min. Increases are calculated after each additional minute. A function that models this situation is the *greatest integer function*, $[x]$. It removes the fractional part of the number and gives only the integer part. [► See **Calculator Note 3D** to learn how to use the greatest integer function on your calculator. ◄]

a. Use the greatest integer function to write cost equations for the two companies.

b. Graph the two equations.

c. Now determine when each plan is the most desirable. Explain your reasoning.

10. Write a system of equations to model each situation, and solve for the appropriate values of the variables.

a. The perimeter of a rectangle is 44 cm. Its length is 2 cm more than twice its width.

b. The perimeter of an isosceles triangle is 40 cm. The base length is 2 cm less than the length of a leg of the triangle.

c. The Fahrenheit reading on a dual thermometer is 0.4 degree less than three times the Celsius reading. (Hint: $F = 1.8C + 32$ is a formula that converts Celsius temperatures to Fahrenheit temperatures.)

11. **APPLICATION** At an excavation site, anthropologists use various clues to discover the identity of skeletal remains. For instance, when a skeleton is found, an anthropologist uses the lengths of certain bones to calculate the height of the living person. The humerus is the single large bone that extends from the elbow to the shoulder socket. The following formulas, attributed to the work of Mildred Trotter and G. C. Gleser, can be used to predict a male's height, m, and a female's height, f, knowing the length, h, of the humerus: $m = 3.08h + 70.45$ and $f = 3.36h + 57.97$. All measurements are in centimeters.

a. Sketch the two lines on the same graph in an appropriately labeled window.

b. If a humerus were found and it measured 42 cm, how tall would the person be if the bone was determined to have come from a male? From a female?

c. For what length humerus would the male and female heights be the same?

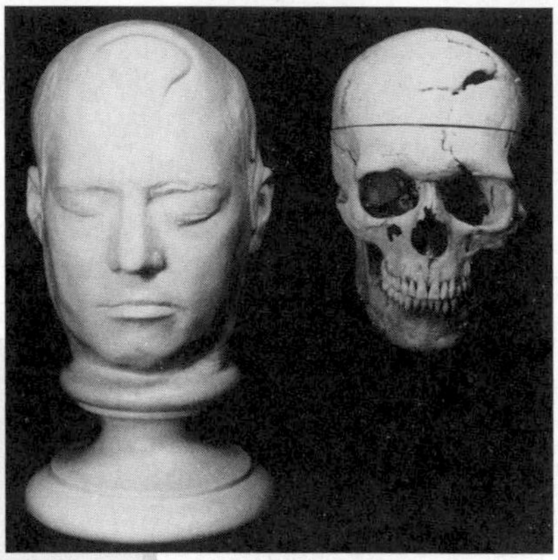

Science Connection Forensic scientists can use these equations and other information to help identify human remains. By studying bones and bone fragments, forensic scientists can gather a wealth of information on age, sex, ancestry, height, and diet. A forensic scientist needs a bachelor's degree that includes many science and mathematics courses.

Review

12. (Algebra) Use a graph or table to find the point (x, y) at which each pair of lines intersects.

a. $y = 44.61 + 3.2x$

$y = 5.6 - 5.1x$

b. $y = \frac{2}{3}x - 3$

$y = -\frac{5}{6}x + 7$

13. (Algebra) Use the model $y = -2732 + 1.55x$ from the example in Lesson 3.3.

a. Predict the concentration of carbon dioxide (in ppm) in the year 1984.

b. Use the model to predict the concentration of carbon dioxide (in ppm) in the year 2010.

c. According to this model, when will the carbon dioxide level be double the preindustrial level of 280 ppm?

14. (Lesson 1.4) The following are six quotes to insure a 2003 Ford Taurus for one year.

$1140 $985 $1450 $945 $1080 $1295

a. What is the mean quote?

b. What is the standard deviation of the quotes?

c. Do any quotes seem to be outliers? Explain.

LESSON

3.5

Substitution and Elimination

A solution to a system of equations in two variables is a pair of values that satisfies both equations and represents the intersection of their graphs. In Lesson 3.4, you reviewed solving a system of equations using substitution when both equations are in intercept form. Suppose you want to solve a system, and one or both of the equations are not in intercept form. You can rearrange them in intercept form, or you can use a different method.

If one equation is in intercept form, you can still use substitution.

EXAMPLE A

Solve this system for x and y.

$$\begin{cases} y = 15 + 8x \\ -10x - 5y = -30 \end{cases}$$

► Solution

You can rearrange the second equation into intercept form and substitute the right side of one equation for y in the other equation. Or you can simply substitute the right side of the first equation for y in the second equation.

$-10x - 5y = -30$	Original form of the second equation.
$-10x - 5(15 + 8x) = -30$	Substitute the right side of the first equation for y.
$-10x - 75 - 40x = -30$	Distribute -5.
$-50x = 45$	Add 75 to both sides, and combine like terms.
$x = -0.9$	Divide both sides by -50.

Now that you know the value of x, you can substitute it in either equation to find the value of y.

$y = 15 + 8(-0.9)$	Substitute -0.9 for x in the first equation.
$y = 7.8$	Multiply, and combine like terms.

Write your solution as an ordered pair. The solution to this system is $(-0.9, 7.8)$.

The substitution method relies on the **substitution property,** according to which if $a = b$, then a may be replaced by b in an algebraic expression. Substitution is a powerful mathematical tool that allows you to rewrite expressions and equations in forms that are easier to use and solve. Notice that substituting an expression for y, as you did in Example A, eliminates y from the equation, allowing you to solve a single equation for a single variable, x.

A third method for solving a system of equations is the **elimination method.** The elimination method uses the addition property of equality: if $a = b$, then $a + c = b + c$. In other words, if you add equal quantities to both sides of an equation, the equation is still true. If necessary, you can also use the multiplication property of equality, according to which if $a = b$, then $ac = bc$, or, if you multiply both sides of an equation by equal quantities, then the equation is still true.

EXAMPLE B

Solve these systems for x and y.

a. $\begin{cases} 4x + 3y = 14 \\ 3x - 3y = 13 \end{cases}$ **b.** $\begin{cases} -3x + 5y = 6 \\ 2x + y = 6 \end{cases}$

▶ Solution

Because none of these equations is in intercept form, it is probably easier to solve the systems using the elimination method.

a. You can solve the system without changing either equation to intercept form by adding the two equations.

$4x + 3y = 14$	Original equations.
$3x - 3y = 13$	Add equal sides to equal sides.
$7x = 27$	The variable y is eliminated.
$x = \frac{27}{7}$	Solve for x.
$4\left(\frac{27}{7}\right) + 3y = 14$	Substitute $\frac{27}{7}$ for x in either equation.
$y = -\frac{10}{21}$	Rearrange, and calculate y.

The solution to this system is $\left(\frac{27}{7}, -\frac{10}{21}\right)$. You can substitute the coordinates in both equations to check that the point is a solution for both.

$$4\left(\frac{27}{7}\right) + 3\left(-\frac{10}{21}\right) = \frac{108}{7} - \frac{30}{21} = \frac{294}{21} = 14$$

$$3\left(\frac{27}{7}\right) - 3\left(-\frac{10}{21}\right) = \frac{81}{7} + \frac{30}{21} = \frac{273}{21} = 13$$

b. Adding the equations as they are written will not eliminate either of the variables. You need to multiply one or both equations by some value so that if you add the equations together, one of the variables will be eliminated. The easiest choice is to multiply the second equation by -5 and then add it to the first equation.

$-3x + 5y = 6$	$\rightarrow$	$-3x + 5y = 6$	Original form of the first equation.
$-5(2x + y) = -5(6)$	$\rightarrow$	$-10x - 5y = -30$	Multiply both sides of the second equation by -5.
		$-13x = -24$	Add the equations.

This eliminates the y-variable and gives $x = \frac{24}{13}$. Substituting this x-value in either of the original equations gives the y-value. Or you can use the same process to eliminate the x-variable.

$$\begin{aligned} -3x + 5y = 6 \quad &\rightarrow \quad -6x + 10y = 12 \quad \text{Multiply both sides by 2.} \\ 2x + y = 6 \quad &\rightarrow \quad \underline{6x + 3y = 18} \quad \text{Multiply both sides by 3.} \\ &\qquad\qquad\quad 13y = 30 \\ &\qquad\qquad\qquad y = \frac{30}{13} \end{aligned}$$

The solution to this system is $\left(\frac{24}{13}, \frac{30}{13}\right)$. You can use your calculator to verify the solution.

It would take a lot of effort to solve this last system using a table. If you had used one of the substitution methods to solve the systems in Example B, you would have had to work with fractions or carry a lot of decimal values for accuracy. To solve these systems, the easiest method to use is the elimination method.

Investigation
It All Adds Up

In this Investigation you may discover that some interesting things happen when you multiply both sides of an equation by the same number or add equations. Work with a partner, and follow the steps below.

Step 1 Graph each of these equations on the same coordinate axes. Where do the lines appear to intersect?

$$\begin{cases} 7x + 2y = -3 & \text{(Equation 1)} \\ 3x + 4y = 5 & \text{(Equation 2)} \end{cases}$$

Step 2 One partner needs to select any number, M, and the other partner a different number, N. The first partner multiplies Equation 1 by the number M he or she chose, while the other partner multiplies Equation 2 by the number N he or she chose.

Step 3 Add the two new equations to form Equation 3. Graph Equation 3 on the same coordinate axis as the two original lines, and describe the location of this new line in relation to the original lines.

Step 4 Next, one partner multiplies the original Equation 1 by $M = -3$, while the other partner multiplies the original Equation 2 by $N = 7$.

Step 5 Add the two new equations from Step 4 to form Equation 4. Graph Equation 4, and describe the location of the line representing Equation 4 in relation to the two original lines.

Step 6 How does Equation 4 differ from the other equations?

Step 7 Repeat this process with different systems of equations, and make a conjecture based on your observations. Summarize your findings.

The elimination method explored in the investigation uses this combination technique to eliminate one of the variables. Solving for the variable that remains gives you the x- or y-coordinate of the point of intersection.

EXERCISES

Practice Your Skills

1. Solve each equation for the specified variable.

a. $w - r = 11$, for w

b. $2p + 3h = 18$, for h

c. $w - r = 11$, for r

d. $2p + 3h = 18$, for p

2. Multiply both sides of each equation by the given value. What is the relationship between the graphs of the new equation and the original equation?

a. $j + 5k = 8$, by -3

b. $2p + 3h = 18$, by 5

c. $6f - 4g = 22$, by 0.5

d. $\frac{5}{6}a + \frac{3}{4}b = \frac{7}{2}$, by 12

3. Add each pair of equations. What is the relationship between the graphs of the new equation and the original pair?

a. $\begin{cases} 3x - 4y = 7 \\ 2x + 2y = 5 \end{cases}$

b. $\begin{cases} 5x - 7y = 3 \\ -5x + 3y = 5 \end{cases}$

4. Graph each system, and find an approximate solution. Then choose a method, and find the exact solution. List each solution as an ordered pair.

a. $\begin{cases} y = 3.2x + 44.61 \\ y = -5.1x + 5.60 \end{cases}$

b. $\begin{cases} y = \frac{2}{3}x - 3 \\ y = -\frac{5}{6}x + 7 \end{cases}$

c. $\begin{cases} y = 4.7x + 25.1 \\ 3.1x + 2y = 8.2 \end{cases}$

d. $\begin{cases} -6x - 7y = 20 \\ -5x + 4y = -5 \end{cases}$

e. $\begin{cases} 2.1x + 3.6y = 7 \\ -6.3x + y = 8.2 \end{cases}$

5. Solve each system of equations.

a. $\begin{cases} 5.2x + 3.6y = 7 \\ -5.2x + 2y = 8.2 \end{cases}$

b. $\begin{cases} \frac{1}{4}x - \frac{2}{5}y = 3 \\ \frac{3}{8}x + \frac{2}{5}y = 2 \end{cases}$

c. $\begin{cases} 4x + 9y = 12 \\ 3x - 8y = 10 \end{cases}$

d. $\begin{cases} s = 7 - 3n \\ 7n + 2s = 40 \end{cases}$

e. $\begin{cases} f = 3d + 5 \\ 10d - 4f = 6 \end{cases}$

f. $\begin{cases} \frac{1}{4}x - \frac{4}{5}y = 7 \\ \frac{3}{4}x + \frac{2}{5}y = 2 \end{cases}$

g. $\begin{cases} 3x - 6y = 8 \\ y = \frac{1}{2}x + 4 \end{cases}$

h. $\begin{cases} 2x + 3y = 4 \\ 1.2x + 1.8y = 2.6 \end{cases}$

Reason and Apply

6. Solve each problem.

a. If $4x + y = 6$, then what is $(4x + y - 3)^2$?

b. If $4x + 3y = 14$ and $3x - 3y = 13$, what is $7x$?

7. Solve each system of equations.

a. $\begin{cases} 3x + 2y = 7 \\ -5x + 4y = 6 \end{cases}$ b. $\begin{cases} y = x^2 - 4 \\ y = -2x^2 + 2 \end{cases}$ c. $\begin{cases} y = x^2 + 3 \\ y = -x^2 + 1 \end{cases}$

8. The formula for converting Fahrenheit to Celsius is $C = \frac{5}{9}(F - 32)$. What temperature on the Fahrenheit scale is three times the equivalent temperature on the Celsius scale?

9. **APPLICATION** Ellen must decide between two cameras. The first camera costs \$47.00 and uses two alkaline AA batteries. The second camera costs \$59.00 and uses one \$4.95 lithium battery. She plans to use the camera often enough to replace the AA batteries probably six times a year at a cost of \$11.50 a year. The lithium battery, however, will last an entire year.

a. Write an equation to represent the overall expense of each camera.

b. Which camera is less expensive in the short term? In the long term? When will the overall cost of the less expensive camera equal the overall cost of the other camera?

c. Carefully describe three different ways to verify your solution.

10. Write a system of two equations that has a solution of $(-1.4, 3.6)$.

11. The two sequences below have one term that is the same. Determine which term this is, and find its value.

$u_1 = 12$
$u_n = u_{n-1} + 0.3$ where $n \geq 2$

$v_1 = 15$
$v_n = v_{n-1} + 0.2$ where $n \geq 2$

12. Formulas play an important part in many fields of mathematics and science. You can create a new formula using substitution to combine formulas.

a. Using the formulas $A = s^2$ and $d = s\sqrt{2}$, write a formula for A in terms of d.

b. Using the formulas $P = IE$ and $E = IR$, write a formula for P in terms of I and R.

c. Using the formulas $A = \pi r^2$ and $C = 2\pi r$, write a formula for A in terms of C.

13. **APPLICATION** A support bar will be in equilibrium (balanced) at the fulcrum, O, if $m_1x + m_2y = m_3z$, where m_1, m_2, and m_3 represent masses and x, y, and z represent the distance of the masses to the fulcrum. Draw a diagram for each question, and calculate the answer.

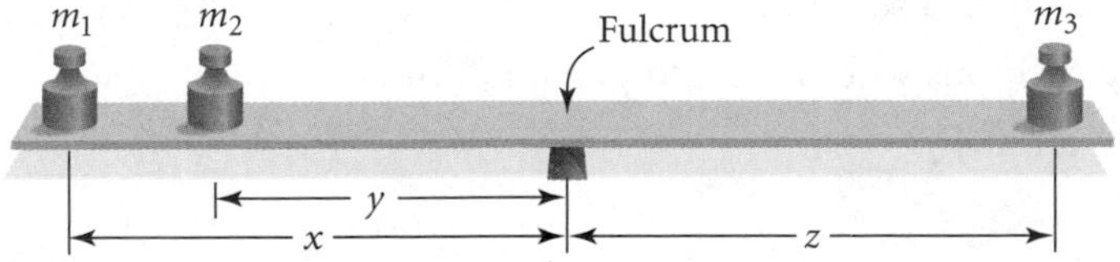

a. A 40-in. bar is in equilibrium when weights of 6 lb and 9 lb are hung from the ends. Find the position of the fulcrum.

b. While in the park, Michael and his two sons, Justin and Alden, play on a 16-ft seesaw. Michael, who weighs 150 lb, sits at the edge of one end while Justin and Alden move to the other side and try to balance. The seesaw balances with Justin at the edge of the other end and Alden 3 ft from him. After some additional experimentation the seesaw balances once again with Alden at the edge and Justin 5.6 ft from the fulcrum. How much does each boy weigh?

Review

14. (Lesson 3.2) Consider the equation $3x + 2y - 7 = 0$.

a. Solve the equation for y.

b. Graph this equation.

c. What is the slope?

d. What is the y-intercept?

e. Write an equation for a line that is perpendicular to this one and has the same y-intercept. Graph this equation.

15. APPLICATION (Lesson 3.3) The table gives the mean price of a gallon of gasoline in the United States from 1950 through 2000.

Year	1950	1960	1970	1980	1990	2000
Price ($)	0.27	0.31	0.36	1.19	1.15	1.56

(*2002 New York Times Almanac*, p. 365)

a. Make a scatter plot of the data.

b. Find a line that models the data.

c. Assuming that the same trend continues, predict the price of gas in 2010.

16. APPLICATION (Lesson 1.4) The table shows the normal monthly precipitation in inches for Pittsburgh, Pennsylvania, and Portland, Oregon.

Month	J	F	M	A	M	J	J	A	S	O	N	D
Pittsburgh	2.5	2.4	3.4	3.1	3.6	3.7	3.8	3.2	3.0	2.4	2.9	2.9
Portland	5.4	3.9	3.6	2.4	2.1	1.5	0.7	1.1	1.8	2.7	5.3	6.1

(*2002 New York Times Almanac*, p. 469)

a. Display the data in two box plots on the same screen.

b. Give the five-number summary of each data set.

c. Describe the differences in living conditions with respect to precipitation.

d. Which city generally has more rain annually?

17. (Lessons 2.1 and 3.1) Consider these three sequences.

i. 243, −324, 432, −576, . . .

ii. 22, 26, 31, 37, 44, . . .

iii. 24, 25.75, 27.5, 29.25, 31, . . .

a. Find the next two terms in each sequence.

b. Identify each sequence as arithmetic, geometric, or other.

c. If a sequence is arithmetic or geometric, write a recursive routine to generate the sequence.

d. If a sequence is arithmetic, give an explicit formula that generates the sequence.

EXPLORATION
with Microsoft® Excel

Finding Trends

Download the Chapter 3 Excel Exploration from the website at www.keycollege.com/online or from your Student CD.

In this exploration, you will analyze ordered pairs of data and try to find trends.

On the spreadsheet you will find complete instructions and questions to consider as you explore.

Activity
Cramer's Rule

Cramer's Rule for 2-by-2 systems makes it very easy to solve and think about systems of equations in two variables.

Simply put, if you have a system of equations

$$ax + by = e$$
$$cx + dy = f$$

then you can solve this system directly using the formulas

$$x = \frac{de - bf}{ad - bc} \quad \text{and} \quad y = \frac{af - ce}{ad - bc}$$

It is easy to program a spreadsheet to do this.

You will create a spreadsheet that will solve systems of equations.

Step 1 To represent the system of equations $\begin{cases} 2x + 5y = 9 \\ 3x + 2y = 8 \end{cases}$, enter cell contents as follows:

	A	B	C	D	E	F
1	2	5	9		x	y
2	3	2	8			

Step 2 Enter the formulas as below in Cells E2 and F2.

E	F
x	y
=(B2*C1-B1*C2)/(A1*B2-B1*A2)	=(A1*C2-A2*C1)/(A1*B2-B1*A2)

This should result in the values 2 and 1, respectively.

Verify that (2, 1) is a solution of the original system.

Questions

1. Change the cells as follows:

	A	B	C
1	3	-7	1
2	2	4	18

What system of equations does this represent? What does your spreadsheet report as the solution? Verify that the solution is correct.

2. Change the cells as follows:

	A	B	C
1	5	1	1
2	3	7	18

What values would you place in C1 and C2 so that the solution will turn out to be $(1, -1)$? Verify that by entering those values into C1 and C2. What system of equations does this represent?

3. Change the cells as follows:

	A	B	C
1	2	5	1
2	6	15	18

What system of equations does this represent? What error is reported? Explain why the spreadsheet gives an error message. What is the solution to the system?

4. Change the cells as follows:

	A	B	C
1	-3	2	5
2	6	-4	-10

What system of equations does this represent? What error is reported? Explain why the spreadsheet gives an error message. What is the solution to the system?

5. How would you represent the system of equations $\begin{cases} 5x + 2y = 8 \\ 3x - 4y = 6 \end{cases}$ in the spreadsheet? What does the spreadsheet report as the solution?

6. How would you represent the system of equations $\begin{cases} 7x - 3y = 12 \\ 3x + 8y = 10 \end{cases}$ in the spreadsheet? What does the spreadsheet report as the solution?

CHAPTER 3 REVIEW

In this chapter you analyzed sets of two-variable data. A plot of two-variable data may be linear, or it may appear nearly linear over a short domain. If it is linear, you can find a **line of fit** to model the data.

One variable, the **dependent variable,** will depend on the other, the **independent variable.** You can write an **explicit formula** for this line and use it to **interpolate** or **extrapolate** points for which data are not available.

You can also solve a **system of equations** to find the intersection of two lines by graphing the line using the **substitution property** or by using the **elimination method.**

To write the equation of a line, you can use its **slope** and y-intercept to write the equation in intercept form.

You learned two methods for finding a line of fit: estimating a line of fit by looking at the trend of the data and fitting a line based on certain criteria, and using the calculator regression feature.

EXERCISES

1. Find the slope of the line containing the points (16, 1300) and (−22, 3250).

2. Consider the line $y = -5.02 + 23.45x$.
   a. What is the slope of this line?
   b. Write the equation for a line that is parallel to this line.
   c. Write the equation for a line that is perpendicular to this line.

3. Find the point on each line where y is equal to 740.0.
   a. $y = 16.8x + 405$
   b. $y = -7.4 + 4.3(x - 3.2)$

4. Consider the system of equations

$$\begin{cases} y = 3.2x - 4 & \text{(Equation 1)} \\ y = 3.1x - 3 & \text{(Equation 2)} \end{cases}$$

   a. Substitute the y-value from Equation 1 in Equation 2 to obtain a new equation. Solve the new equation for x.
   b. Subtract the expression on the right side of Equation 2 from the expression on the right side of Equation 1. Set the difference equal to y to get another equation. Graph this equation and find its x-intercept.
   c. Explain why the solutions to 4a and 4b are the same.

5. The graphs below show four different lines of fit for the same set of data. For each graph, decide whether or not the line is a good line of fit, and explain your decision.

a.

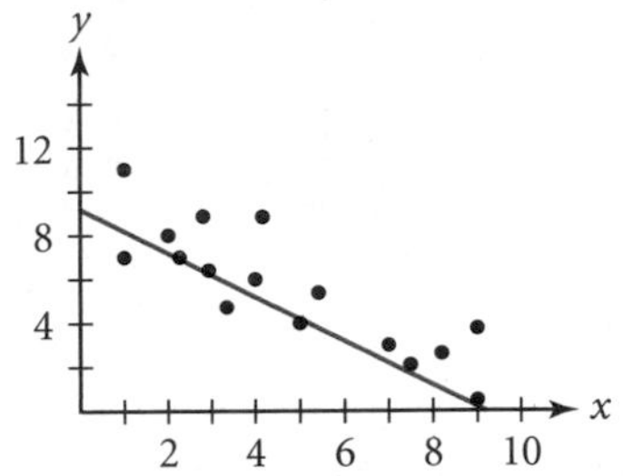

b.

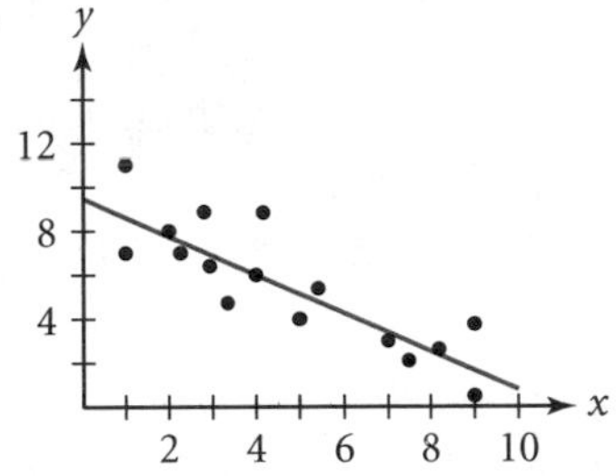

c.

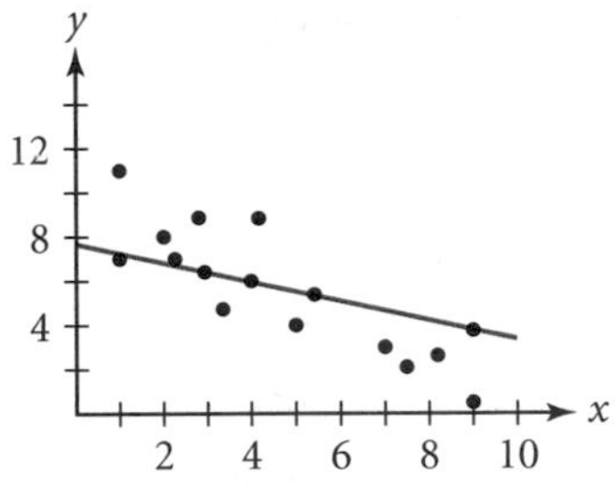

6. Use substitution to solve each system.

a. $\begin{cases} y = 6.2x + 18.4 \\ y = -2.1x + 7.40 \end{cases}$

b. $\begin{cases} y = \frac{3}{4}x - 1 \\ y = \frac{7}{10}x + \frac{2}{5}y = 8 \end{cases}$

c. $\begin{cases} 2x - 4y - 9 \\ -x + 2y = 3 \end{cases}$

d. $\begin{cases} 3x + 2y = 4 \\ -3x + 5y = 3 \end{cases}$

7. Find the point (or points) where each pair of lines intersects.

a. $\begin{cases} 5x - 4y = 5 \\ 2x + 10y = 2 \end{cases}$

b. $\begin{cases} y = \frac{1}{4}(x - 8) + 5 \\ y = 0.25x + 3 \end{cases}$

c. $\begin{cases} \frac{3}{5}x - \frac{2}{5}y = 3 \\ 0.6x - 0.4y = -3 \end{cases}$

ArchitectureConnection In 1173 C.E. the Tower of Pisa was built on soft ground. Ever since, it has been leaning to one side as it sinks into the soil. This eight-story, 117-foot tower was built with a foundation only seven feet deep. The tower was completed in the mid-1300s, even though it started to lean after the first three stories were completed. The tower's structure consists of a cylindrical body, arches, and columns. Rhombuses and rectangles decorate the surface.

The Tower of Pisa in Pisa, Italy

8. Read the Architecture Connection. The table lists the amount of the tower's lean, measured in millimeters, for 13 different years.

a. Make a scatter plot of the data. Sketch this graph on your paper.

b. Find a line modeling the data.

Tower of Pisa

Year	Lean
1975	2964.2
1976	2964.4
1977	2965.6
1978	2966.7
1979	2967.3
1980	2968.8
1981	2969.6
1982	2969.8
1983	2971.3
1984	2971.7
1985	2972.5
1986	2974.2
1987	2975.7

c. What is the slope of the line? Interpret the slope in the context of the problem.

d. Find the amount of lean predicted by your equation for 1992 (the year work was started to secure the foundation).

9. The ratio of the weight of an object on Mercury to its weight on Earth is 0.378.

a. Explain why you can use the equation $m = 0.378e$ to model the weight of an object on Mercury.

b. How much would a 160-lb person weigh on Mercury?

c. The ratios for the Moon and Jupiter are 0.166 and 2.364 respectively. The equations $Y1 = 0.378x$, $Y2 = 1x$, $Y3 = 0.166x$, and $Y4 = 2.364x$ are graphed here. Match each planet with its graph and equation.

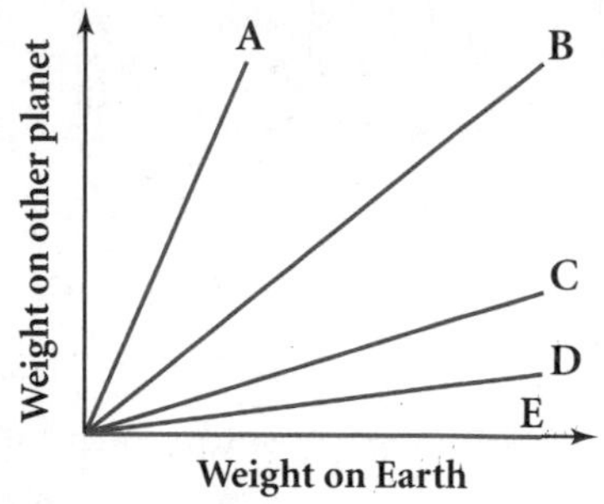

ScienceConnection Planning a trip to outer space? The following table gives the ratio of an object's weight on each of the planets and the Moon to its weight on Earth.

Mercury	Venus	Earth	Moon	Mars	Jupiter	Saturn	Uranus	Neptune	Pluto
0.378	0.907	1	0.166	0.377	2.364	0.916	0.889	1.125	0.670

You can calculate your weight on each of these celestial bodies. Some of these calculations will make you seem like a lightweight, and others will make you seem heavy! But, of course, your body will still be the same. Your *mass* won't change. Your *weight* on other planets depends on your mass, *m*, the planet's mass, *M*, and your distance, *r*, from the center of the planet.

10. The 4th term of an arithmetic sequence is 64. The 54th term is −61. Find the 23rd term.

MIXED REVIEW

11. (Lesson 2.4) State whether each recursive formula defines a sequence that is arithmetic, geometric, shifted geometric, or none of these. State whether a graph of the sequence would be linear or curved. Then list the first five terms of the sequence.

a. $u_1 = 4$ and $u_n = 3u_{n-1}$ where $n \geq 2$

b. $u_0 = 20$ and $u_n = 2u_{n-1} + 7$ where $n \geq 1$

12. **APPLICATION** (Lesson 2.2) You receive a bonus of \$500 and deposit it into a savings account on June 15. The account has an annual interest rate of 5.9%, compounded annually.

a. Write a recursive formula for this problem.

b. List the first three terms of the sequence.

c. What is the meaning of the value of u_3?

d. How much money will you have in your account when you retire 35 years later?

e. If you deposit an additional \$100 in your account each year on June 15, how much will you have in savings after 35 years?

13. APPLICATION (Lesson 2.2) An Internet website gives the current world population and projects the population for future years. Its projections for the number of people on Earth on January 1 in the years 2005 through 2009 are given in the table.

a. Find a recursive formula to model the population growth. What kind of sequence is this?

b. Predict the population on January 1, 2010.

c. In what year will the world's population exceed 10 billion?

d. Is this a realistic model that could predict world population in the next millennium? Explain.

Population Projections

Year	World population
2005	6,486,915,022
2006	6,581,251,218
2007	6,676,959,301
2008	6,774,059,223
2009	6,872,843,077

(*www.ibiblio.org*)

14. APPLICATION (Lesson 2.2) The table gives the population of Wayne County, Michigan, from 1960 to 2000.

a. Find a recursive model for the population of Wayne County, Michigan.

b. What sort of sequence did you use?

c. How does this compare to the model of the world's population in Exercise 13?

Year	Population
1960	2,666,297
1970	2,666,751
1980	2,337,891
1990	2,111,687
2000	2,061,162

(*U.S. Census Bureau*)

15. APPLICATION (Lesson 2.3) Jonah must take an antibiotic every 12 hours. Each pill is 25 milligrams (mg), and after every 12 hours 50% of the drug in his body remains. What is the amount of antibiotic in his body over the first 2 days? What amount will there be in his body in the long run?

16. (Lesson 1.4) Create a box plot that has this five-number summary: 5, 7, 12, 13, 1.

a. Are the data skewed left, skewed right, or symmetric?

b. What is the median of the data?

c. What is the *IQR*?

d. What percentage of data values are above 12? Above 13? Below 5?

17. (Lessons 2.1, 3.1) Consider the arithmetic sequence 6, 13, 20, 27, 34 . . .

a. Write a recursive formula that describes this sequence.

b. Write an explicit formula for this sequence.

c. What is the slope of your equation in 19b? What relationship does this have to the arithmetic sequence?

d. Determine the value of the 32nd term. Is it easier to use your formula from 19a or 17b for this?

18. (Lessons 2.1, 3.1) For an arithmetic sequence $u_1 = 12$ and $u_{10} = 52.5$,

a. What is the common difference of the sequence?

b. Find the equation of the line through the points (1, 12) and (10, 52.5).

c. What is the relationship between 18a and 18b?

19. (Lesson 1.5) The table shows high school dropout rates reported by states in the period 1998–1999. Note that data are unavailable for some states.

a. What are the mean, median, mode, and standard deviation of the data?

b. Do any states lie more than 2 standard deviations above or below the mean?

High School Dropout Rates, 1998–1999

State	Rate (%)	State	Rate (%)
Alabama	4.4	Montana	4.5
Alaska	5.3	Nebraska	4.2
Arizona	8.4	Nevada	7.9
Arkansas	6.0	New Jersey	3.1
Connecticut	3.3	New Mexico	7.0
Delaware	4.1	North Dakota	2.4
District of Columbia	8.2	Ohio	3.9
Georgia	7.4	Oklahoma	5.2
Idaho	6.9	Oregon	6.5
Illinois	6.5	Pennsylvania	3.8
Iowa	2.5	Rhode Island	4.5
Kentucky	4.9	South Dakota	4.5
Louisiana	10.0	Tennessee	4.6
Maine	3.3	Utah	4.7
Maryland	4.4	Vermont	4.6
Massachusetts	3.6	Virginia	4.5
Minnesota	4.5	West Virginia	4.9
Mississippi	5.2	Wisconsin	2.6
Missouri	4.8	Wyoming	5.7

(*http://necs.ed.gov/pubs2002/2002114.pdf*)

20. (Lessons 3.4 and 3.5) Use an appropriate method to solve each system of equations.

a. $\begin{cases} 2.1x - 3y = 4 \\ 5x + 3y = 7 \end{cases}$

b. $\begin{cases} y = \frac{1}{3}x + 5 \\ y = -3x + \frac{1}{2} \end{cases}$

c. $\begin{cases} 3x + 4y = 12 \\ 2x - 6y = 5 \end{cases}$

Assessing What You've Learned

- Can you identify the slope and y-intercept of a linear equation? How are slopes and common differences related?
- Can you tell what the slope represents in an equation modeling a real-world situation?
- How do you use your calculator to find a line of best fit?
- What techniques do you know for solving systems of linear equations? What is the key step in each technique?
- How do you know when a linear system has no solution?

JOURNAL Look back at the journal entries you have written for Chapter 3. To help you organize your thoughts, here are some additional questions to write about.

- What do you think were the main ideas of the chapter?
- Were there any exercises that you particularly struggled with? If so, are they clear now? If some ideas are still foggy, what plans do you have to clarify them?
- From the chapter, select an exercise that you could not fully solve. Write out the problem and as much of the solution as possible. Then clearly explain what is keeping you from solving the problem. Be as specific as you can.
- How are arithmetic sequences and linear equations similar? Give an example that would be better modeled with an arithmetic sequence. Give an example that would be better modeled with a linear equation.

CHAPTER

4 Functions, Relations, and Transformations

Starbucks, Seattle: *Compression,* 1998 by Benjamin Edwards. No matter what you think of corporate franchises, they are a ubiquitous marker of the modern cityscape: modular spaces for any corner, any mid-block storefront; logos transformed infinitely, blown up on windows and banners, shrunken onto a million paper cups; a single vision in innumerable and essentially unchanging iterations.

OBJECTIVES

In this chapter you will

- interpret graphs of functions and relations
- review function notation
- learn about the linear, quadratic, square root, absolute value, and semicircle families of functions
- apply transformations—translations, reflections, stretches, and shrinks—to the graphs of functions and relations
- transform functions to model real-world data

LESSON

4.1

Interpreting Graphs

Fashion is something that goes in one year and out the other.

DENISE KLAHN

In this chapter you will learn more about functions. Functions make algebra useful in the real world. Once you become familiar with a variety of function families, you can examine real data and decide whether they are modeled by a particular type of function. Finding the function family that models a given situation then enables you to use algebra to make predictions and solve real-life problems. In addition to explaining how to identify function families, this chapter will equip you to use transformations to model more effectively a variety of situations.

A picture can be worth a thousand words—if you can interpret the picture. In this lesson you will investigate the relationship between real-world situations and graphs that represent them.

What is the real-world meaning of this graph, which shows the relationship between the number of customers getting haircuts each week and the price charged for each haircut?

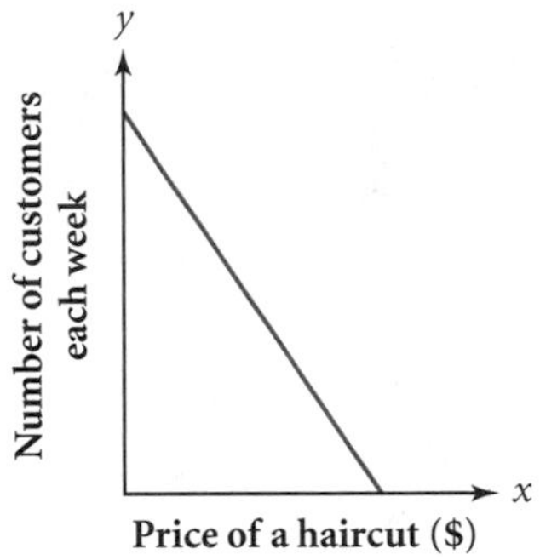

The number of customers depends on the price of the haircut. Therefore, we say that the number of customers is the *dependent variable.* The price in dollars is the *independent variable.* As the price increases, the number of haircuts decreases linearly. As you would expect, fewer people are willing to pay a high price; a lower price attracts more customers. The slope indicates the number of haircuts lost for each dollar increase. The x-intercept represents the haircut price that is too high for anyone. The y-intercept indicates the number of haircuts when they are free.

EXAMPLE

As a fringe benefit, Washtenaw Computer Services provides employees with free soft drinks from a vending machine. However, the employees are complaining that the soft drink machine is frequently empty. A committee is appointed to study this problem. The committee records the number of cans in the soda machine at various times during a typical weekday and makes a graph.

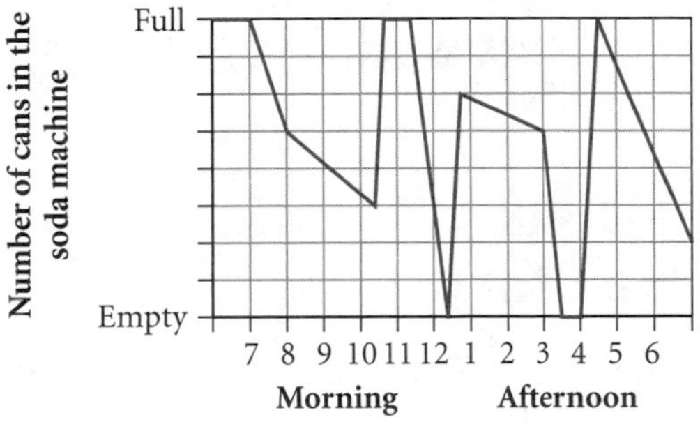

a. Based on the graph, at what times is soda consumed most rapidly?

b. When is the machine refilled? How can you tell?

c. When is the machine empty? How can you tell?

d. What do you think the committee will recommend to solve the problem?

▶ Solution

Each horizontal segment indicates a time interval during which soda does not sell. Negative slopes represent consumption of soda, and positive slopes show when the soda machine is refilled.

a. The most rapid consumption is pictured by the steep, negative slopes from 11:30 A.M. to 12:30 P.M., and from 3:00 to 3:30 P.M.

b. The machine is completely refilled during the night, again at 10:30 A.M., and again at 4:30 P.M. The machine is also refilled at 12:30 P.M. but only to 75% capacity.

c. The machine is empty from 3:30 to 4:00 P.M. and briefly at about 12:30 P.M.

d. The committee might recommend refilling the machine once more around 2:00 or 3:00 P.M. to solve the problem of frequent emptiness. Refilling the machine completely at 12:30 P.M. may also help solve the problem.

Although the committee members in the example are interested in solving a problem related to soda consumption, they could also use the graph to answer other questions about the company: When do employees begin to arrive in the morning? What time are morning staff meetings? When is lunch? When do most employees leave for the day?

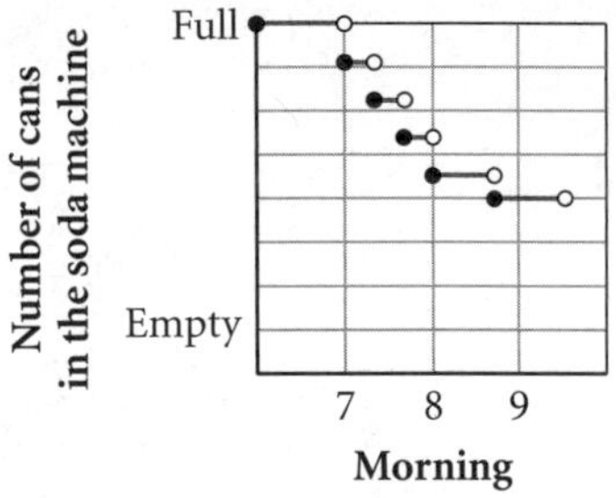

Both the graph of haircut customers and the graph in the example are shown as continuous graphs. In other words, there are x and y values for every possible number in the section of the graph that is shown. This is in contrast to discrete graphs, in which only certain values are shown. In reality, the quantity of soda in the machine can take on only discrete values, because the number of cans must be a whole number. The graph might be drawn more accurately with a series of short horizontal segments, as shown here. The price of a haircut and the number of haircuts can also take on only discrete values because prices are usually given as an exact number of dollars and cents. This graph might be more accurately drawn with separate points. However, in both cases a continuous "graph sketch" makes it easier to see the trends and patterns.

Investigation
Graph a Story

Every graph tells a story. Make a graph to go with the story in Part 1. Then invent your own story to go with the graph in Part 2.

Part 1

Sketch a graph that reflects all the information given in this story.

"It was a dark and stormy night. Before the torrents of rain came, the bucket was empty. The rain subsided at daybreak. The bucket remained untouched through the morning until Old Dog Trey arrived as thirsty as a dog. The sun shone brightly through the afternoon. Then Billy, the kid next door, arrived. He noticed two plugs in the side of the bucket. One of them was about a quarter of the way up, and the second one was near the bottom. As fast as you could blink an eye, he pulled out the plugs and ran away."

Part 2

This graph tells a story. It could be a story about a lake, a bathtub, or whatever you imagine. Spend some time with your group discussing the information contained in the graph. Write a story that conveys all of this information, including when and how the rates of change increase or decrease.

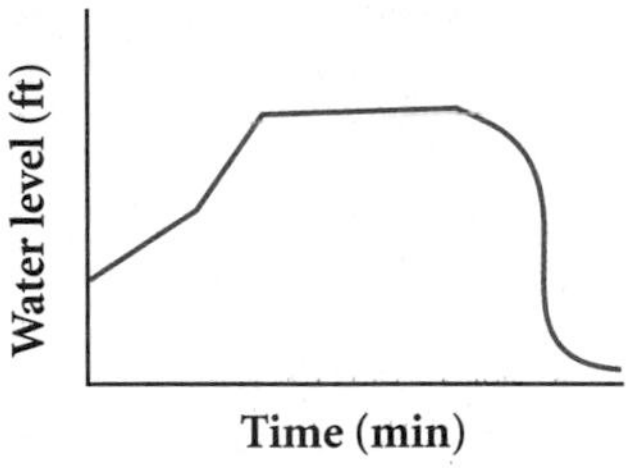

As you interpret data and graphs that show a relationship between two variables, you must always decide which is the independent variable and which is the dependent variable. You should also consider whether the variables are discrete or continuous. That is, are only certain values for x and y relevant to the problem, or does it make sense for all inputs and outputs to be defined?

Science Connection Contour maps are a way to represent altitude graphically. Each line marks all of the points that are the same height in feet above sea level. Using the distance between two contour lines, you can calculate the rate of change in altitude. These maps are used by hikers, forest firefighters, and scientists.

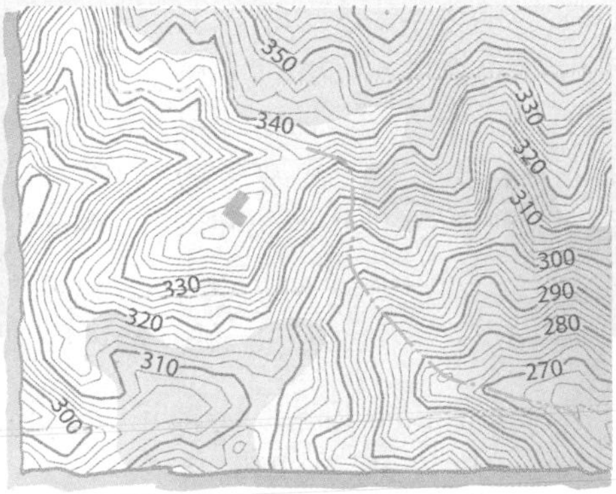

EXERCISES

Practice Your Skills

1. Sketch a graph to match each description.

a. Increasing throughout, first slowly and then at a faster rate

b. Decreasing slowly, then more and more rapidly, then suddenly becoming constant

c. Alternately increasing and decreasing without any sudden changes in rate

2. For each graph write a description like those in Exercise 1.

a. 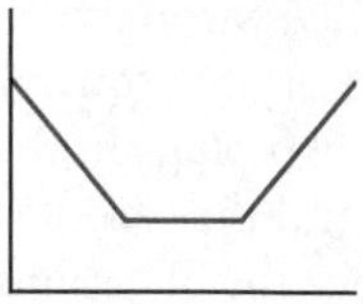

b. 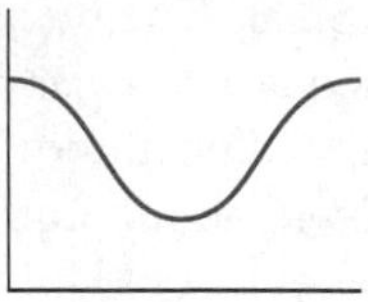

c. 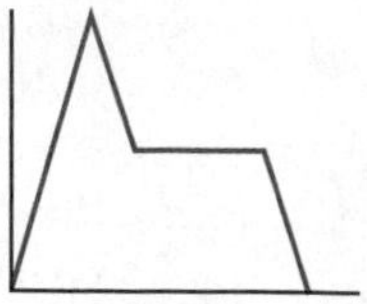

3. Match a description to each graph.

a. 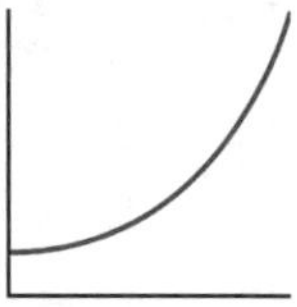

b. 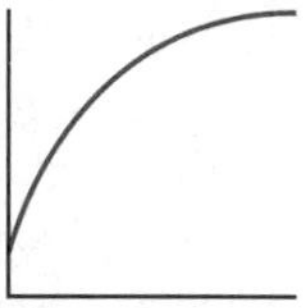

c. 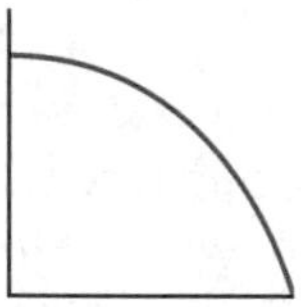

d. 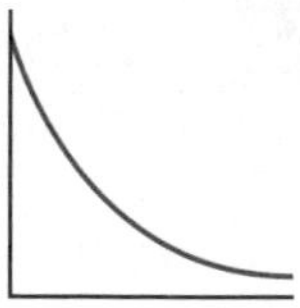

A. Increasing at an increasing rate

B. Decreasing at a decreasing rate

C. Increasing at a decreasing rate

D. Decreasing at an increasing rate

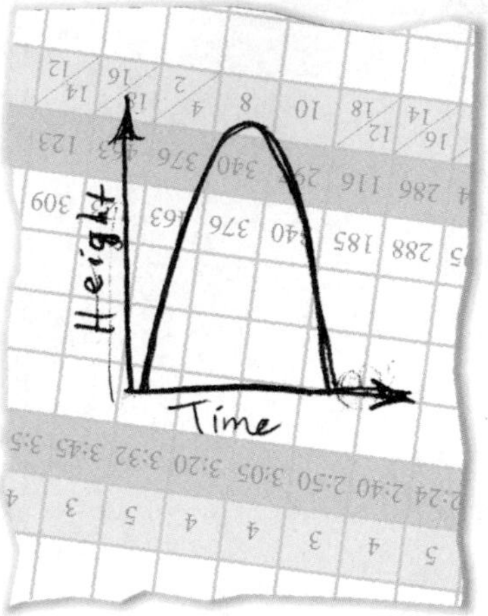

Reason and Apply

4. Harold's concentration often wanders from his golf game to the mathematics involved in it. His scorecard frequently contains mathematical doodles and graphs.

a. What is a real-world meaning of this graph, which was found on one of his recent scorecards?

b. What units might he be using?

c. Describe a realistic domain and range for this graph.

d. Does this graph show how far the ball traveled? Explain.

5. Make up a story to go with the graph at right. Be sure to interpret the x- and y-intercepts.

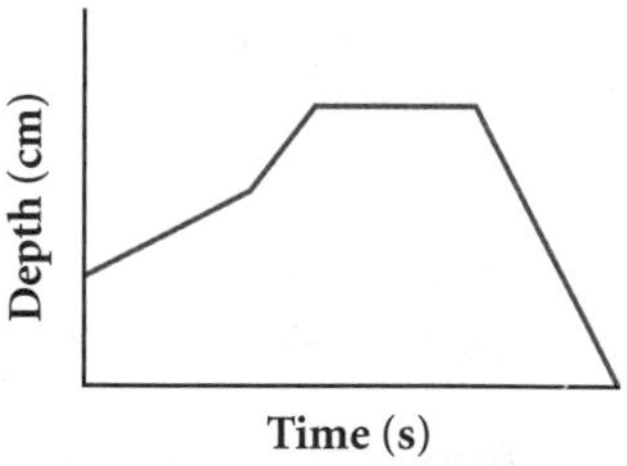

6. Sketch what you think is a reasonable graph for each relationship described. In each situation, identify the variables, and label your axes accordingly.

a. The height of a basketball during the first ten seconds of a game

b. The distance it takes to brake a car to a full stop, compared to the car's speed when the brakes are first applied

c. The temperature of an iced drink as it sits on a table for a long period of time

d. The speed of a falling acorn after a squirrel drops it from the top of an oak tree

e. Your height above the ground as you ride a Ferris wheel

7. Sketch what you think is a reasonable graph for each relationship described. In each situation, identify the variables, and label your axes accordingly. In each situation, will the graph be continuous, or will it be a collection of discrete points or pieces? Explain your answer.

a. The amount of money you have in a savings account that is compounded annually over a period of several years. Assume no additional deposits are made.

b. The same amount of money that you started with in 7a, hidden under your mattress over the same period of several years

c. An adult's shoe size compared to the adult's foot length

d. Your distance from Detroit during a flight from Detroit to Newark if your plane is forced to circle the airport in a holding pattern when you approach Newark

e. The daily maximum temperature of a town over a one-month period

8. Charlie and Emily decided to hike to a picnic spot along a trail in a large state park. They began walking rather slowly. After the first half mile, they realized that they had forgotten the dessert, so they quickly jogged back to the car. They then walked at an even, medium speed for the next two miles. They stopped for an hour to eat, then turned around and walked slowly home.

Sketch two graphs to go with this story.

a. Use time as the independent variable and speed as the dependent variable.

b. Use time as the independent variable and distance from the car as the dependent variable.

9. For each of the following, describe a relationship of your own, and draw a graph to go with it.

a. As the x-value increases, the y-value decreases. The y-value decreases quickly at first, then less quickly as the x-value gets larger.

b. As the x-value increases, the y-value increases at a steady rate, then suddenly stops increasing.

c. As the x-value increases, the y-value increases at first, then decreases rapidly.

10. Car A and Car B are at the starting line of a race. At the green light they both accelerate to 60 mi/h in 1 min. The graph here represents their velocities in relation to time.

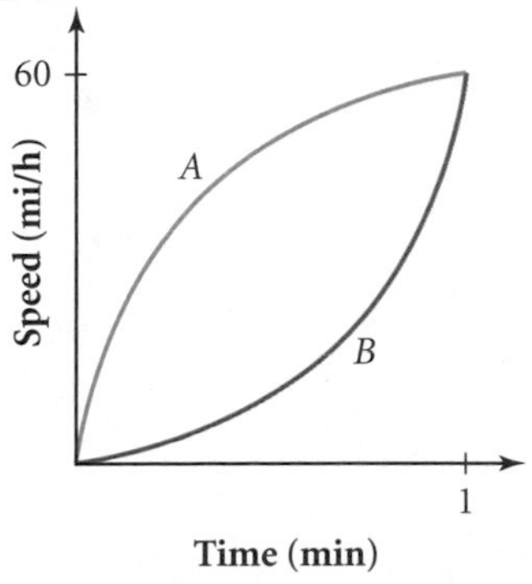

a. Describe the rate of change for each car.

b. After 1 min, which car is in the lead? Explain your reasoning.

Review

11. (Lesson 3.2) Write an equation for the function that fits each situation. Remember to use x to represent the independent variable and y to represent the dependent variable.

a. The length of a rope is 1.70 m, and it decreases by 0.12 m for every knot that is tied in it.

b. When you join a CD club, you get the first eight CDs for \$7.00. After that, your bill increases by \$9.50 for each additional CD you purchase.

12. **APPLICATION** (Lesson 3.4) Albert starts a business reproducing high-quality copies of pictures. It costs \$155 to prepare the picture and then \$15 to make each print. Albert plans to sell each print for \$27.

a. Write a cost equation and graph it.

b. Write an income equation, and graph it on the same set of axes.

c. How many prints does Albert need to sell before he makes a profit?

13. **APPLICATION** (Lesson 2.3) Suppose you have a \$200,000 home loan with an annual interest rate of 6.5%, compounded monthly.

a. If you pay \$1200 per month, what balance remains after 20 years?

b. If you pay \$1400 per month, what balance remains after 20 years?

c. If you pay \$1500 per month, what balance remains after 20 years?

d. Make an observation about the answers to 12a–c.

LESSON

4.2

Function Notation

She had not understood mathematics until he had explained to her that it was the symbolic language of relationships. "And relationships," he had told her, "contained the essential meaning of life."

PEARL S. BUCK
THE GODDESS ABIDES, 1972

Rachel's parents keep track of her height as she gets older. They plot these values on a graph and connect the points with a smooth curve. For every age you choose on the x-axis, there is only one height that pairs with it on the y-axis. That is, Rachel is only one height at any time during her life.

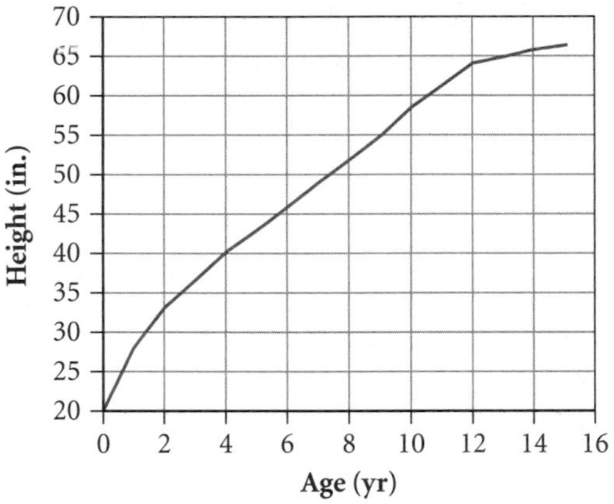

A **relation** is any relationship between two variables. A **function** is a special type of relation. In order for a relation to be a function, for every input (independent variable), there is exactly one output (dependent variable). If x is the independent variable, a function pairs at most one y per x. At any given point in time (the input), Rachel is one particular height (the output). Therefore, you can say that Rachel's height is a function of her age.

Graphs of functions can be continuous, such as the graph of Rachel's height. Or they can be made up of discrete points, such as a graph of the maximum temperatures for each day of a month. Although real-world data often have an identifiable pattern, a function does not necessarily need to have a rule that connects the two variables. The definition of a function requires only that there be a definite y-value for each x-value. The tables below show other relations that are functions:

x	y
3	5
4	5
5	31
6	8
7	19
8	5

x	y
−1	5
0	5
1	5
2	5
3	5
4	5

The reason we can say that the relations in the two tables are both functions is that for every value of the independent variable, there is a definite value for the dependent variable. For each of the functions, if you were given a value of x, you then also would be able to find the value of y. In other words, for each input there is exactly one corresponding output.

If there were more than one y-value for a corresponding x-value, then you could not use the value of x to predict the corresponding y-value. You may remember the **vertical line test** from previous mathematics classes. It helps you determine whether or not a graph represents a function. If a vertical line crosses the graph more than once, then an x-value has two or more corresponding y-values, and the relation is not a function. If no vertical line crosses the graph more than once, then the relation is a function. Take a minute to think about how you could apply this technique to the graph of Rachel's height and the graph in the next example.

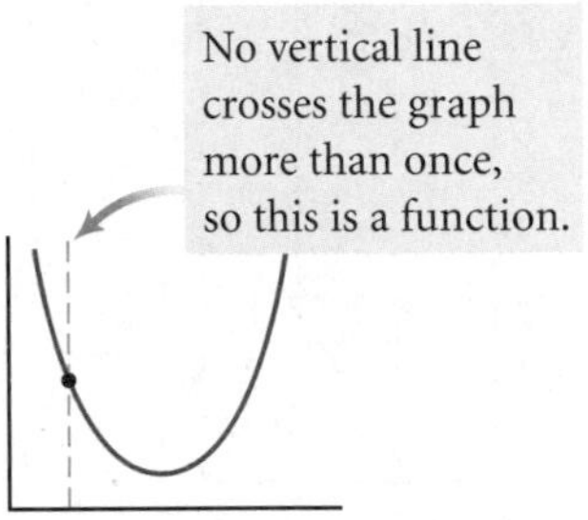
No vertical line crosses the graph more than once, so this is a function.
Function

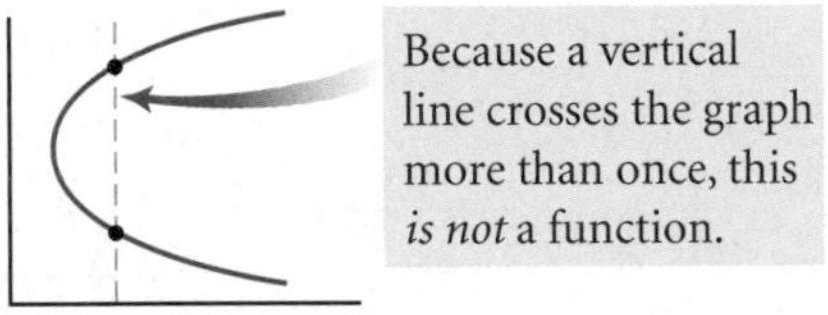
Because a vertical line crosses the graph more than once, this *is not* a function.
Not a function

Function notation emphasizes the dependent relationship between the variables that are used in a function. The notation $y = f(x)$ indicates that values of the dependent variable, y, are explicitly defined in terms of the independent variable, x, by the function f. The expression $y = f(x)$ is read, "y equals f of x."

EXAMPLE

Function f is defined by the equation $f(x) = \frac{2x+5}{x-3}$. Function g is defined by the graph shown.

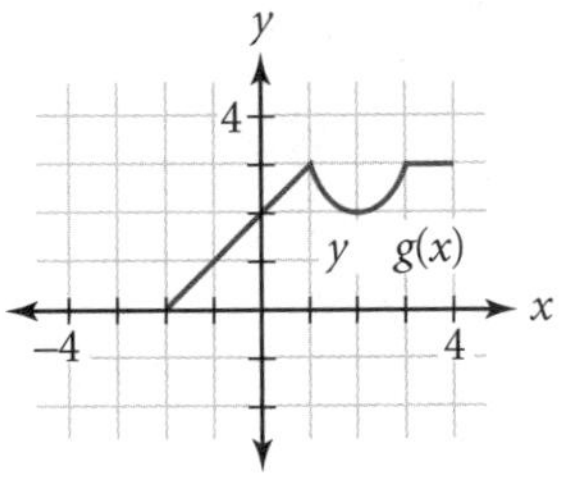
y
4
y g(x)
x
−4
4

Find these values.

a. $f(8)$

b. $f(-7)$

c. $g(1)$

d. $g(-2)$

► Solution

When a function is defined by an equation, simply replace each x with the x-value and evaluate.

a. $f(x) = \dfrac{2x+5}{x-3}$

$$f(8) = \frac{2 \cdot 8 + 5}{8 - 3} = \frac{21}{5} = 4.2$$

b. $f(-7) = \dfrac{2 \cdot -7 + 5}{-7 - 3} = \dfrac{-9}{-10} = 0.9$

You can check your work with your calculator. [► See **Calculator Note 4A** to learn about evaluating functions. ◄]

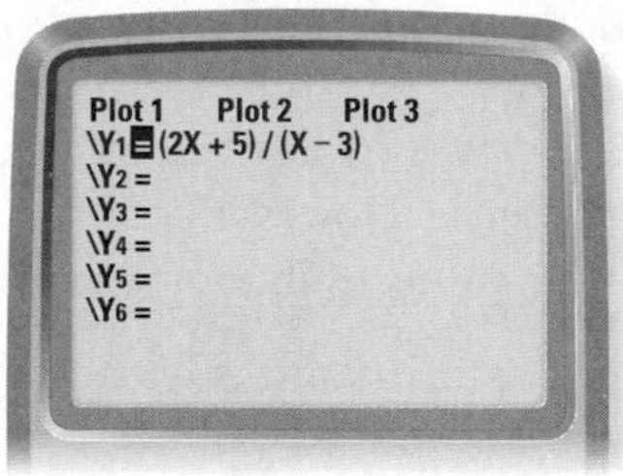

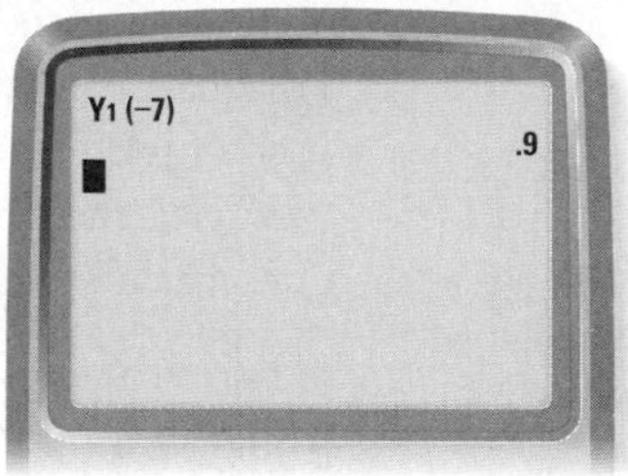

c. The notation $y = g(x)$ tells you that the values of y are explicitly defined in terms of x by the graph of the function g. To find $g(1)$, locate the value of y when x is 1. The point $(1, 3)$ on the graph means that $g(1) = 3$.

d. The point $(-2, 0)$ on the graph means that $g(-2) = 0$.

In the Investigation you will practice identifying functions and using function notation. As you do so, keep notes about how you can identify functions in different forms.

Investigation
To Be or Not to Be

Below are nine representations of relations.

a.

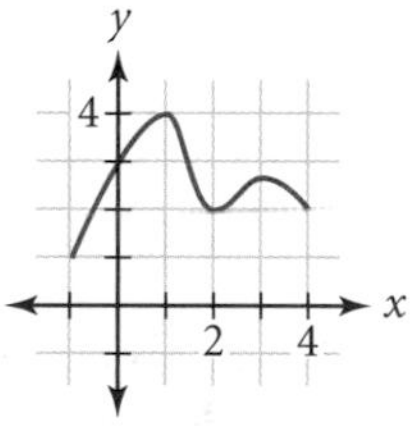

b.

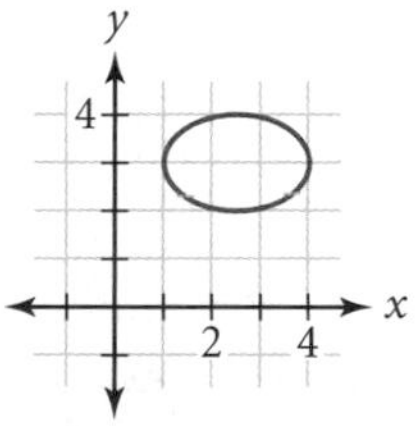

c.

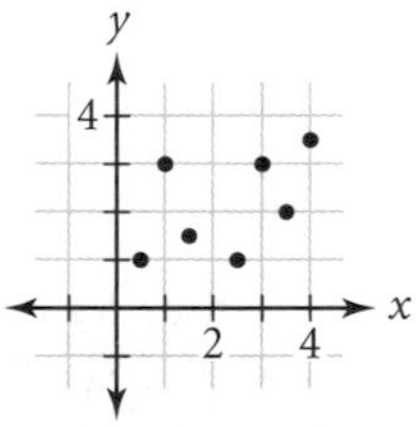

d.

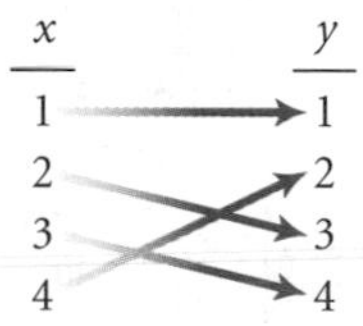

e.

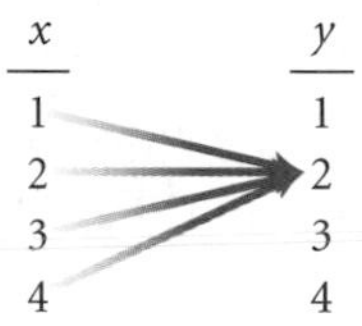

f.

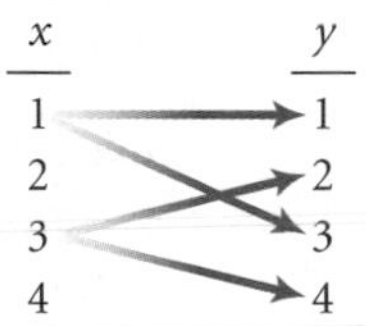

g. Independent variable: the age of each student in your class
Dependent variable: the height of each student

h. Independent variable: an automobile in the state of Kentucky
Dependent variable: that automobile's license plate number

i. Independent variable: the day of the year
Dependent variable: the time of sunset

Step 1 Identify each relation that is also a function. For each relation that is not a function, explain why it is not.

Step 2 For each function in parts a–f, find the y-value when $x = 2$, and find the x-value(s) when $y = 3$. Write each answer in function notation using the letter of the subpart as the function name, for example, $y = d(x)$ for part d.

When you use function notation to refer to a function, you can use any letters you like. For example, you might use $y = h(x)$ if the function represented height, or $y = p(x)$ if the function represented population. Often, in describing real-world situations you would use letters that make sense. However, to avoid confusion, avoid using the dependent variable as the function name, as in $y = y(x)$. Choose freely, but choose wisely.

When looking at real-world data, often it is hard to decide whether or not there is a functional relationship. For example, if you measured the height of every student in your class and the weight of his or her backpack, you might collect a data set in which each student height is paired with only one backpack weight. But does that mean no two students of the same height could have backpacks of different weights? Does it mean you shouldn't try to model the situation with a function?

PROJECT

STEP FUNCTIONS

The graph at right is an example of a *step function.* The open circles indicate points that are not included in the graph. For example, the value of $f(3)$ is 5, not 2. The places where the graph "jumps" are called *discontinuities.*

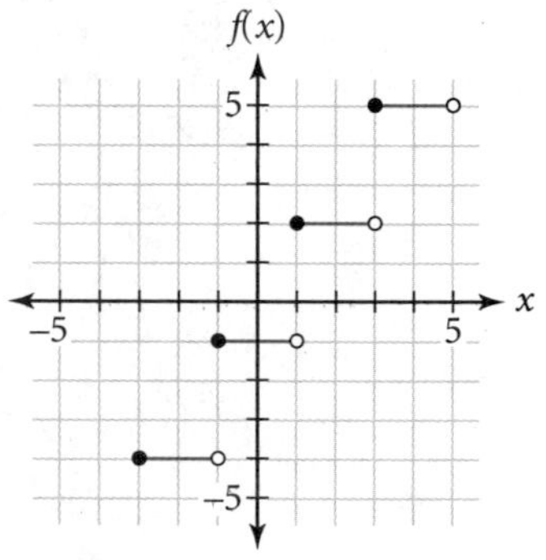

In previous mathematics courses you may have seen one of the best-known step functions—the *greatest integer function.* The output value, $f(x)$, is created by rounding the input value, x, downward to a whole number. For example, if $x = 1.6$, then $f(x) = 1$.

Another well-known example of a step function is an Internal Revenue Service income tax table, such as the one shown here. You may also find step functions as you explore phone service plans or other consumer charges.

If line 42 (taxable income) is—		And you are—			
At least	But less than	Single	Married filing jointly *	Married filing sepa-rately	Head of a house-hold
		Your tax is—			
35,000					
35,000	**35,050**	5,494	4,539	5,494	4,744
35,050	**35,100**	5,506	4,546	5,506	4,751
35,100	**35,150**	5,519	4,554	5,519	4,759
35,150	**35,200**	5,531	4,561	5,531	4,766
35,200	**35,250**	5,544	4,569	5,544	4,774
35,250	**35,300**	5,556	4,576	5,556	4,781
35,300	**35,350**	5,569	4,584	5,569	4,789
35,350	**35,400**	5,581	4,591	5,581	4,796

Do further research on step functions. Prepare a report or class presentation describing a real-world example. Describe the step function in three ways:

- in words
- as a graph
- as a table

EXERCISES

Practice Your Skills

1. For each graph below tell whether or not it represents a function, and state why or why not.

a.

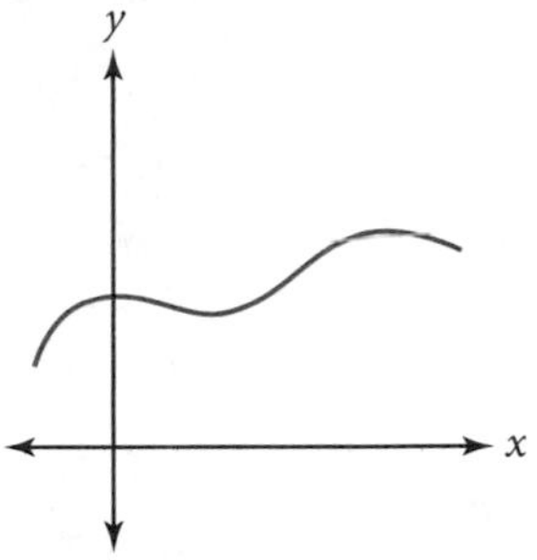

b.

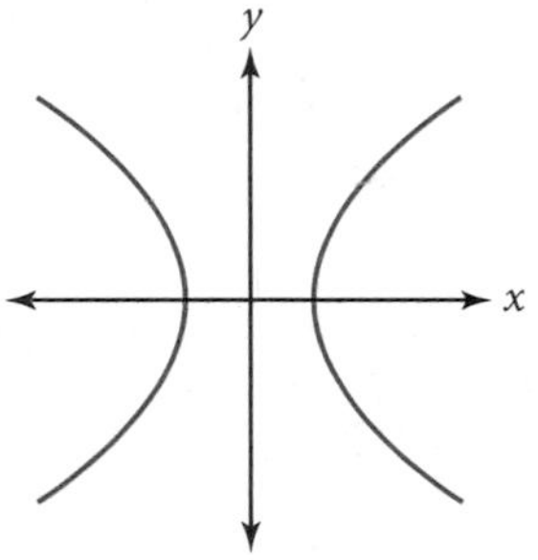

c.

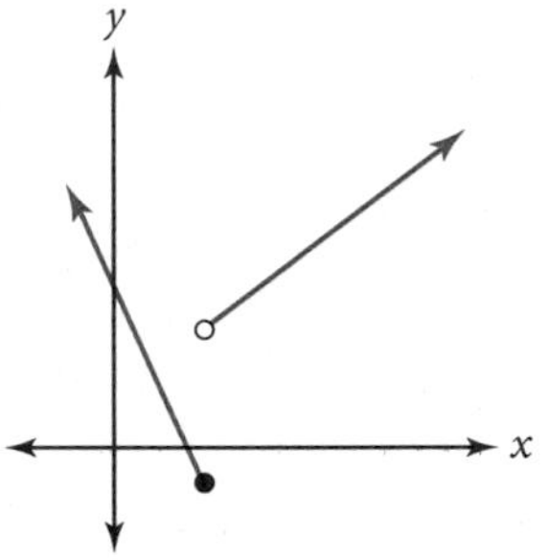

2. Use the functions $f(x) = 3x - 4$ and $g(x) = x^2 + 2$ to find these values.

a. $f(7)$ **b.** $g(5)$ **c.** $f(25)$ **d.** $g(-3)$ **e.** x when $f(x) = 7$

3. Miguel works at an appliance store. He is paid \$8.25 an hour and works eight hours a day. In addition, he earns a 3% commission on all items he sells. Let x represent the total dollar value of the appliances that Miguel sells, and let the function m represent Miguel's daily earnings as a function of x. Which function describes how much Miguel earns in a day?

a. $m(x) = 8.25 + 0.03x$ **b.** $m(x) = 66 + 0.03x$

c. $m(x) = 8.25 + 3x$ **d.** $m(x) = 66 + 3x$

4. Use the graph shown here to find each value. Each answer will be an integer from 1 to 26. Relate the answer to the position of the letters of the alphabet (1 = A, 2 = B, etc.), and fill in the name of a famous mathematician.

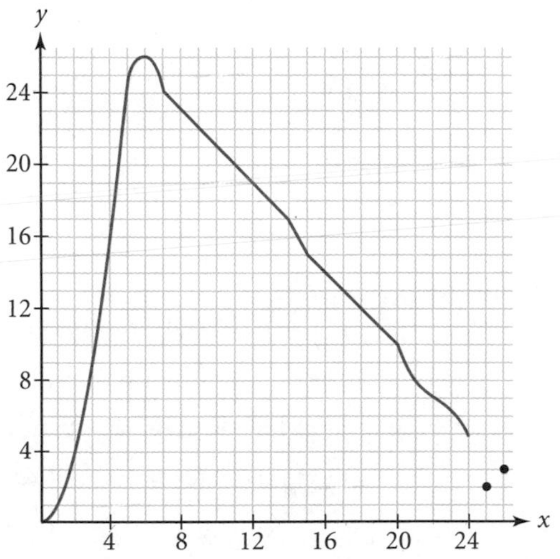

a. $f(13)$ **b.** $f(25) + f(26)$

c. $2f(22)$ **d.** $\dfrac{f(3) + 11}{\sqrt{f(3 + 1)}}$

e. $\dfrac{f(1 + 4)}{f(1) + 4} - \dfrac{1}{4}\left(\dfrac{4}{f(1)}\right)$ **f.** x when $f(x + 1) = 26$

g. $\sqrt[3]{f(21) + f(14)}$ **h.** x when $2f(x + 3) = 52$

i. x when $f(2x) = 4$ **j.** $f(f(2) + f(3))$

k. $f(9) - f(25)$ **l.** $f(f(5) - f(1))$

m. $f(4 \cdot 6) + f(4 \cdot 4)$

___	___	___	___		___	___	___	___	___	___	___	___	___
a	b	c	d		e	f	g	h	i	j	k	l	m

5. Identify the independent variable in each relation. Is the relation a function?

a. The price of a graphing calculator and the sales tax you pay

b. The amount of money in your savings account and the time it has been in the account

c. The amount your hair has grown since the time of your last haircut

d. The amount of gasoline in your car's fuel tank and how far you have driven since your last fill-up

Reason and Apply

6. Sketch a reasonable graph for each relation described in Exercise 5. In each situation, identify the variables, and label your axes accordingly.

7. Suppose $f(x) = 25 - 0.6x$.

a. Draw a graph of this function.

b. What is $f(7)$?

c. Identify the point $(7, f(7))$ by marking it on your graph.

d. Find the value of x when $f(x) = 27.4$. Mark this point on your graph.

e Give an example of a real-world situation that might be modeled by this equation.

8. Identify the domain and range of the function of f in the graph shown here.

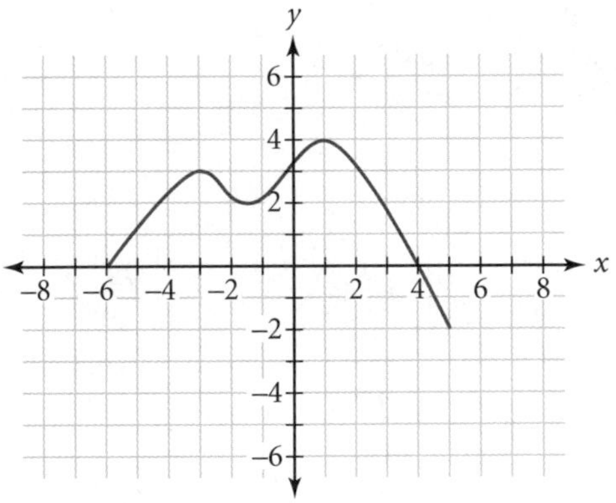

9. Sketch a graph of each function.

a. $y = f(x)$ has a domain of all real numbers and range $f(x) \leq 0$.

b. $y = g(x)$ has domain $x > 0$ and a range of all real numbers.

c. $y = h(x)$ has a domain of all real numbers and range $h(x) = 3$.

10. Consider the function $f(x) = 3(x + 1)^2 - 4$.

a. Find $f(5)$. **b.** Find $f(n)$. **c.** Find $f(x + 2)$.

d. Use your calculator to graph $y = f(x)$ and $y = f(x + 2)$ on the same axes. How do the graphs compare?

11. Ken walks toward and then away from a wall. Is the graph of his distance from the wall a function of elapsed time? Why or why not?

12. **APPLICATION** The length of a pendulum in inches, L, is a function of its period, or the length of time it takes to swing back and forth, in seconds, t. The function is defined by the formula $L = 9.78t^2$.

a. Find the length of a pendulum if its period is four seconds.

b. In the lobby of the General Assembly Building of the United Nations in New York City is a large pendulum that has a 200-lb bob on a 75-ft wire. What is its period?

The pendulum at the United Nations building was a gift from Queen Juliana of the Netherlands in 1955.

13. **APPLICATION** The number of diagonals of a polygon, d, is a function of the number of sides of the polygon, n, and is given by the formula $d = \frac{n(n-3)}{2}$.

a. Find the number of diagonals in a dodecagon (a 12-sided polygon).

b. How many sides would a polygon have if it contained 170 diagonals?

14. The sum of the degree measures of the interior angles of a polygon is a function of the number of sides of the polygon, n. This table shows the degree sums of several polygons.

Number of sides	Sum of interior angle measures
3	180°
4	360°
5	540°
6	720°

a. Find the sum of the degree measures in a dodecagon (a 12-sided polygon).

b. How many sides does a polygon have if its interior angle sum is 18,000°?

LanguageConnection You probably have noticed that some words, like *biannual, triplex,* and *quadrant,* have prefixes that indicate a number. Knowing the meaning of a prefix can help you determine the meaning of a word. The word *polygon* comes from the Greek *poly* (many) and *gonia* (angle). The Greek word for *angle* is also related to the Greek word for "knee," as angles can be thought of as the "knees" of polygons. Many mathematical words use the following Greek prefixes.

1 mono	4 tetra	7 hepta	10 deca
2 di	5 penta	8 octa	20 icosa
3 tri	6 hexa	9 ennea	100 hecto or hecato

A polyhedron is a three-dimensional shape with many faces. What is the name of this shape?

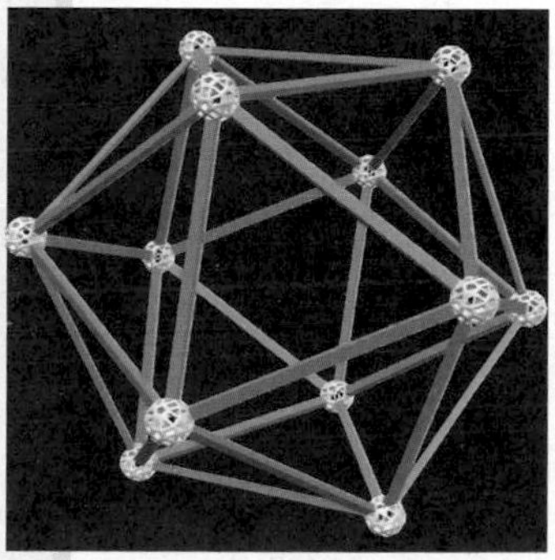

15. Create graphs depicting the water heights as each bottle is filled with water at a constant rate.

a.

b.

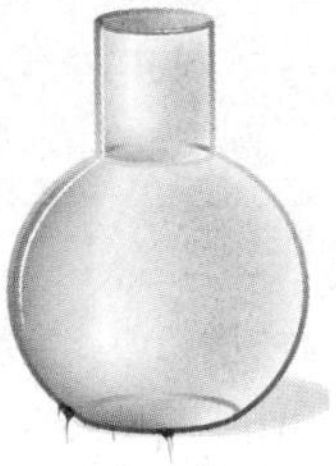

c.

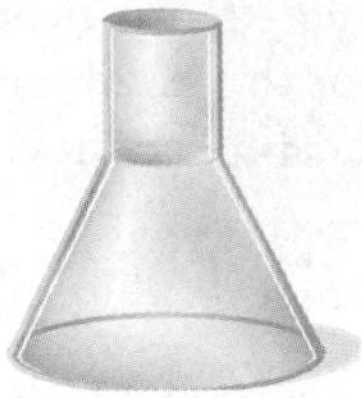

Review

16. (Lesson 1.4). The box plot represents the amount that 32 children collected for the United Nations Children's Fund (UNICEF). Estimate the total amount of money raised. Explain your reasoning.

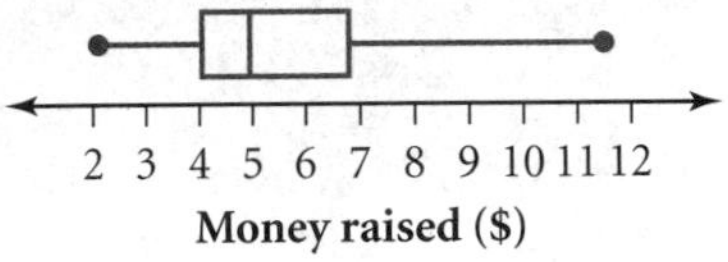

17. (Lesson 3.4) Given the graph at right, find the intersection of lines ℓ_1 and ℓ_2.

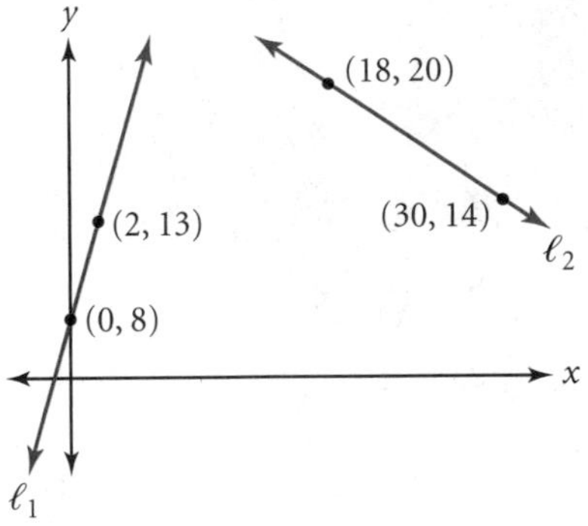

LESSON

4.3

Lines in Motion

This Investigation will help you see how moving the graph of a line affects its equation. Moving a graph horizontally or vertically is called a **translation.** The discoveries you make about translations of lines will also apply to the graphs of other functions.

Investigation
Movin' Around

You will need

- two motion sensors

In this Investigation you will explore what happens to the equation of a linear function when you translate the graph of the line. You'll then use your discoveries to interpret data.

Part 1

Graph the lines in each step, and look for patterns.

Step 1 On graph paper, graph the line $y = 2x$, and then draw a line parallel to it but 3 units higher. What is the equation of this new line?

Step 2 On the same set of axes, draw a line parallel to the line $y = 2x$ but shifted down 4 units. What is the equation of this line?

Step 3 On a new set of axes, graph the line $y = \frac{1}{2}x$. Mark the point where the line passes through the origin. Plot another point right 3 units and up 4 units from the origin, and draw a line through this point parallel to the original line. Write at least two equations of the new line.

Step 4 What happens if you move every point on $y = \frac{1}{2}x$ to a new point up 1 unit and right 2 units? Complete the table, then write an equation for this new line. What do you notice?

f*(*x*) = *x

Original function		**Translated function (Up 1 and right 2)**	
x	$f(x)$	x	y
0	0	2	1
1		3	
2	1	4	
3		5	
4	2	6	

Now try moving every point up 2 units and right 2 units. What do you notice? Explain why the results of the two translations are different.

Step 5 In general, what is the effect of translating a line on its equation?

Part 2

Your class will now use motion sensors to create a function and a translated copy of that function. [▶ See **Calculator Note 4B** for instructions on collecting and retrieving data from two motion sensors.◀] This activity will require three volunteers. In the steps of the activity, the volunteers are called A, B, and C.

Step 1 At 0 seconds: C begins to walk slowly toward the motion sensors, and A begins to collect data.

About 2 seconds:	B begins to collect data.
About 5 seconds:	C begins to walk backward.
About 10 seconds:	A's sensor stops.
About 12 seconds:	B's sensor stops, and C stops walking.

Step 2 After collecting the data, connect a calculator to each of the two sensors, and retrieve the data. Be sure to keep track of which sensor was in front and which was behind.

Step 3 Graph both sets of data on the same screen. Record a sketch of what you see, and answer these questions:

a. How are the two graphs related?

b. If A's graph is $y = f(x)$, what equation describes B's graph? Explain how you determined this equation.

c. In general, if the graph of $y = f(x)$ is translated horizontally h units and vertically k units, what is the equation of this translated function?

If you know the effects of translations, you can write an equation that translates any function on a graph. No matter the shape of a function $y = f(x)$, the graph of $y = f(x - 3) + 2$ will look just the same as $y = f(x)$, but it will be translated up 2 units and right 3 units. Understanding this relationship will enable you to graph functions and write equations for graphs more easily.

Translation of a Function

A *translation* moves a graph horizontally or vertically or both.

Given the graph of $y = f(x)$, the graph of

$$y = f(x - h) + k$$

is a translation horizontally h units and vertically k units.

LanguageConnection The word *translation* can refer to the act of converting one language to another. As in mathematics, *translation* of foreign languages is an attempt to keep meanings parallel. Direct substitution of words often loses the nuances and subtleties of meaning of the original text. The subtleties involved in the art and craft of translation have inspired the emergence of translation studies programs in universities throughout the world.

In a translation every point (x_1, y_1) is mapped to a new point, $(x_1 + h, y_1 + k)$. This new point is called the *image* of the original point. In $y = y_1 + b(x - x_1)$, the point at (0, 0) is translated to the new point at (x_1, y_1). In fact, every point is translated horizontally x_1 units and vertically y_1 units.

EXAMPLE

Describe how the graph of $f(x) = 4 + 2(x - 3)$ is a translation of the graph of $f(x) = 2x$.

► Solution

The graph of $f(x) = 4 + 2(x - 3)$ passes through the point (3, 4). Consider this point to be the translated image of (0, 0) on $f(x) = 2x$. It is translated right 3 units and up 4 units from its original location, so the graph of $f(x) = 4 + 2(x - 3)$ is simply the graph of $f(x) = 2x$ translated right 3 units and up 4 units.

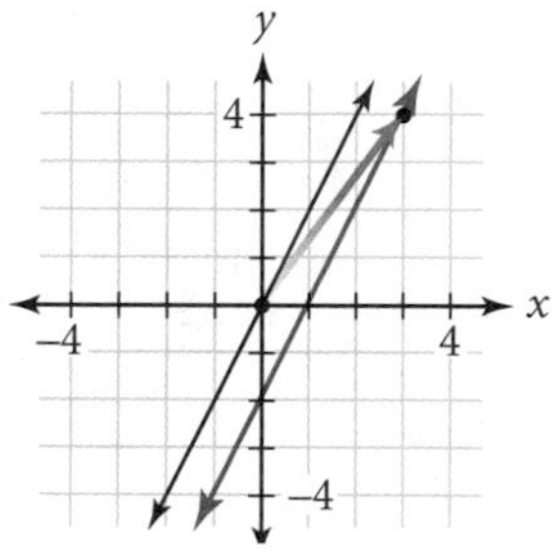

Note that you can distribute and combine like terms in $f(x) = 4 + 2(x - 3)$ to get $f(x) = -2 + 2x$. The fact that these two equations are equivalent means that translating the graph of $f(x) = 2x$ right 3 units and up 4 units is equivalent to translating the line down 2 units. In the graph in the example, this appears to be true.

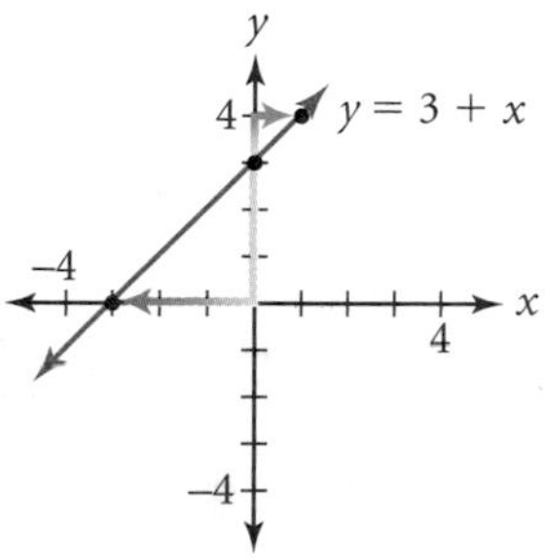

If you imagine translating a line in a plane, some translations will give you the same line you started with. For example, if you start with the line $y = 3 + x$ and translate every point up 1 unit and right 1 unit, you will map the line onto itself. If you translate every point on the line down 3 units and left 3 units, you also map the line onto itself.

There are infinitely many translations that will map a line onto itself. Similarly, there are infinitely many translations that map a line onto another line, m_1. To map the line $y = 3 + x$ onto the line m_1 shown below it, you could translate every point down 2 units and right 1 unit, or you could translate every point down 3 units.

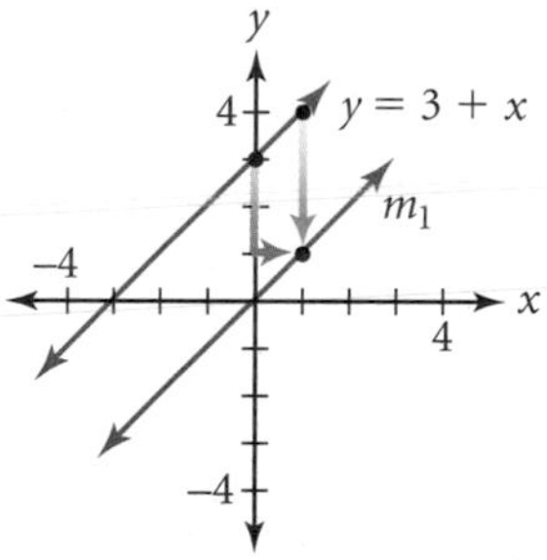

In the next few lessons you will see how to translate and otherwise transform other functions.

EXERCISES

Practice Your Skills

1. The graph of the line $y = \left(\frac{2}{3}\right)x$ is translated right 5 units and down 3 units. What is the equation of the new line?

2. How does the graph of $y = f(x - 3)$ compare with the graph of $y = f(x)$?

3. If $f(x) = -2x$, find

a. $f(x + 3)$ **b.** $-3 + f(x - 2)$ **c.** $5 + f(x + 1)$

4. Consider the line that passes through the points $(-5.2, 3.18)$ and $(1.4, -4.4)$, as shown.

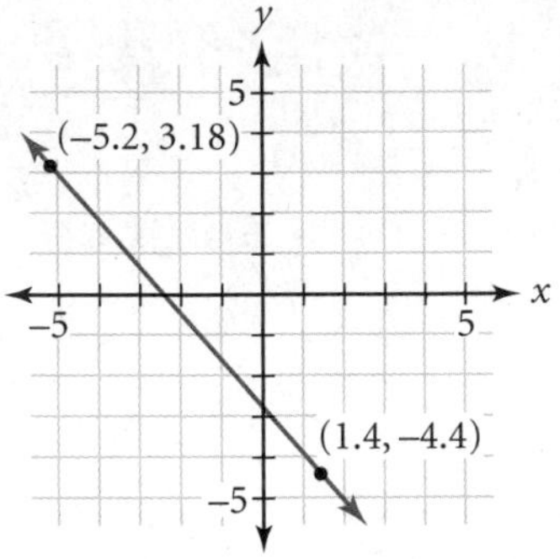

a. Find an equation of the line.

b. Write an equation of the parallel line that is 2 units above this line.

5. Write an equation of each line.

a. The line $y = 4.7x$ translated down 3 units

b. The line $y = 2.8x$ translated right 2 units

c. The line $y = x$ translated up 4 units and left 1.5 units

Reason and Apply

6. The graph of $y = f(x)$ is shown at right. Write an equation for each of the graphs in 6a–d.

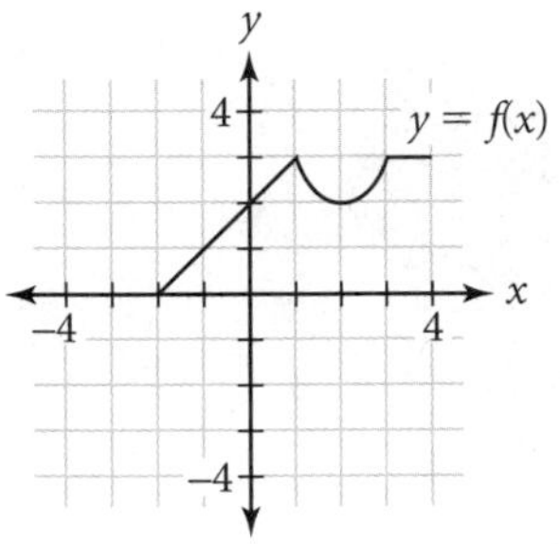

a.

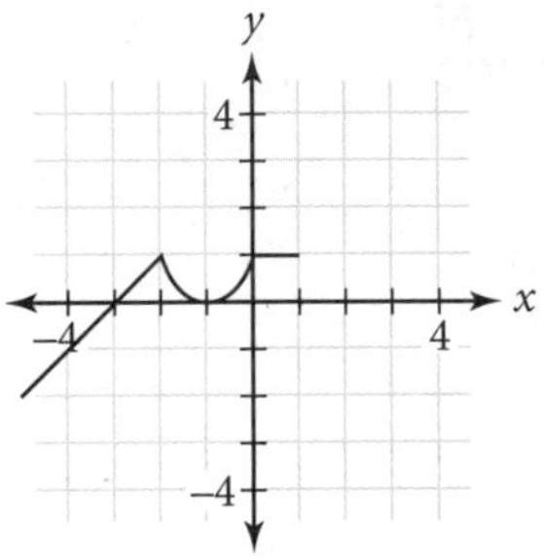

b.

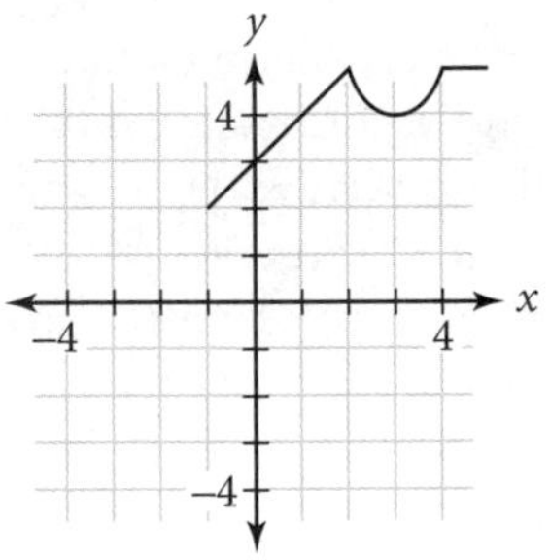

c.

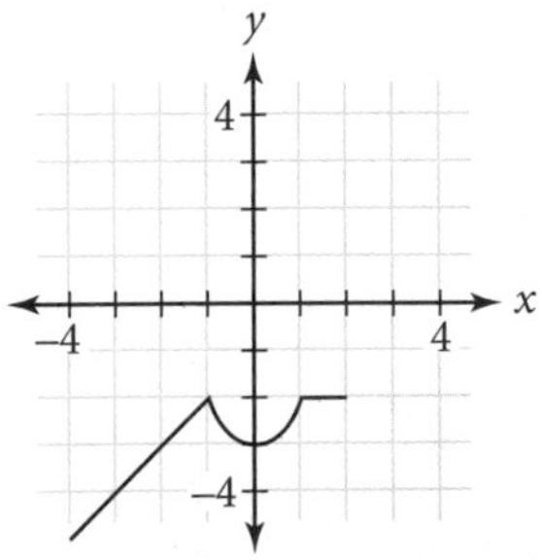

d.

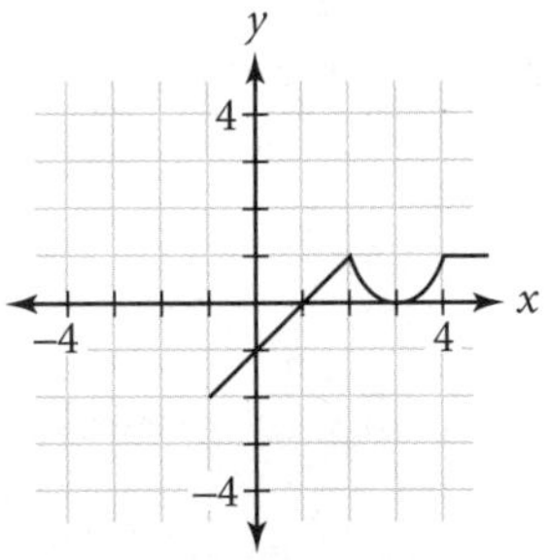

7. As knots are tied in a length of rope, the rope will become effectively shorter because each knot takes up rope length. Jeannette and Keegan collect data about the length of a piece of rope as more and more knots are tied in it. The equation that fits their data is $y = 102 - 6.3x$, where x represents the number of knots, and y represents the length of the rope in centimeters. Mitch has a piece of rope cut from the same source but of a different starting length. He ties his knots in exactly the same way as Jeannette and

Keegan. Unfortunately, he loses his data and can remember only that his rope was 47 cm long after he had tied 3 knots. Which of the following equations describes Mitch's rope?

a. $y = 47 - 6.3\,(x - 3)$ **b.** $y = 102 - 47x$ **c.** $y = 102 - 6.3\,(x - 47)$

d. $y = 47 - 3\,(6.3)$ **e.** $y = 47 + (x - 3)$

8. Rachel, Pete, and Brian perform Part 2 of the Investigation in this lesson. Rachel walks while Pete and Brian hold the motion sensors. They create this unusual graph:

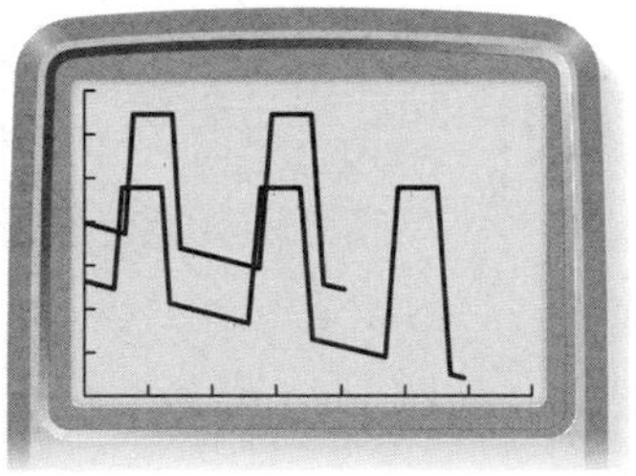

a. The lower curve is made from the data collected by Pete's motion sensor. Where was Brian standing, and when did he start his motion sensor to create the upper curve?

b. If Pete's curve is the graph of $y = f(x)$, what equation represents Brian's curve?

9. **APPLICATION** Kari's assignment in her computer programming course is to simulate the motion of an airplane by repeatedly translating it across the screen. The coordinate system in the software program is shown below with the origin, (0, 0), in the upper left corner. In this program, coordinates to the right and *down* are positive. The starting position of the airplane is (1000, 500), and Kari would like the airplane to end at (9000, 4000). She thinks that moving the airplane in 15 equal steps will model the motion well.

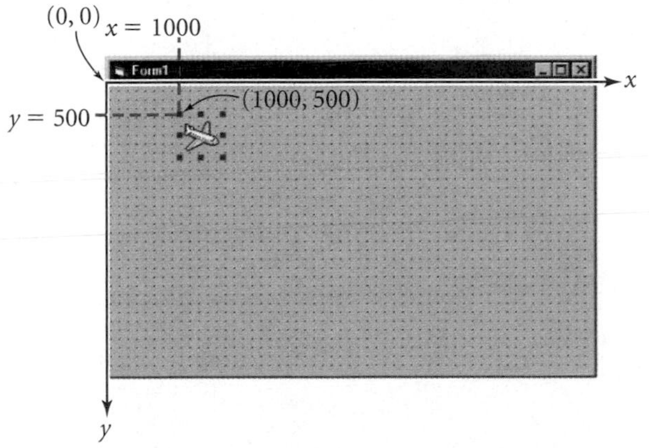

a. What should be the airplane's first position after (1000, 500)?

b. If the airplane's position at any time is given by (x, y), what is the next position in terms of x and y?

c. If the plane moves down 175 units and right 400 units in each step, how many steps will it take to reach the final position of (9000, 4000)?

ArtConnection Animation is the simulation of movement produced by rapidly displaying a sequence of related images. Flip books are an old-fashioned method of animation. Altering the picture slightly from page to page and flipping rapidly through the book create the special effect of movement. The same concept is used to portray motion in cartoons and other animated films. An animated feature film may consist of more than 65,000 pictures. Computer-generated animation has largely replaced hand-drawn animation, drastically reducing the amount of labor needed to produce an animated motion picture.

Since the mid-1990s Macromedia Flash animation has given websites striking visual effects.

© 2002 Eun-Ha Paek. Stills from "L'Feaux Episode 7" on www.MilkyElephant.com

10. ***Mini-Investigation*** Linear equations can also be written in standard form:

$$ax + by + c = 0$$

a. Identify the values of a, b, and c for each of these equations in standard form:

i. $4x + 3y + 12 = 0$

ii. $-x + y + 5 = 0$

iii. $7x - y + 1 = 0$

iv. $-2x + 4y - 2 = 0$

v. $2y + 10 = 0$

vi. $3x - 6 = 0$

b. Solve the standard form, $ax + by + c = 0$, for y. The result should be an equivalent equation in intercept form. What is the y-intercept? What is the slope?

c. Use what you learned from 10b to find the y-intercept and slope of each of the equations in 10a.

d. The graph of $4x + 3y + 12 = 0$ is translated as described below. Write an equation in standard form for each of the translated graphs.

i. A translation right 2 units

ii. A translation left 5 units

iii. A translation up 4 units

iv. A translation down 1 unit

v. A translation right 1 unit and down 3 units

vi. A translation up 2 units and left 2 units

e. In general, if the graph of $ax + by + c = 0$ is translated horizontally h units and vertically k units, what is the equation of the translated line?

Review

11. (Lesson 1.4) Suppose that your basketball team's scores in the first four games of the season were 86 points, 73 points, 76 points, and 90 points.

a. Write a function that gives the mean score in terms of the fifth-game score.

b. What will be your team's mean score if the fifth-game score is 79 points?

c. What score will give a five-game average of 84 points?

12. **APPLICATION** (Lesson 3.2) The Internal Revenue Service has approved ten-year linear depreciation as one method of determining the value of business property. This means that the value declines to zero over a ten-year period. Suppose a piece of business equipment costs \$12,500 and is depreciated over a ten-year period. Here is a sketch of the linear function that represents this depreciation:

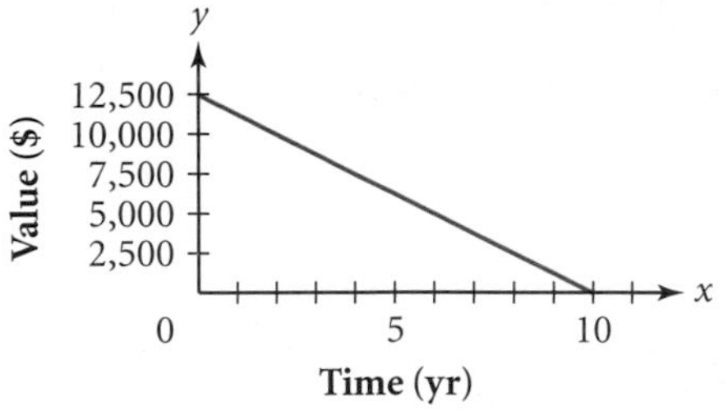

a. What is the y-intercept? Give the real-world meaning of this value.

b. What is the x-intercept? Give the real-world meaning of this value.

c. What is the slope? Give the real-world meaning of the slope.

d. Write an equation that describes the value of the equipment during the ten-year period.

e. When is the equipment worth \$6500?

LESSON

4.4 Translations and the Quadratic Family

Call it a clan, call it a network, call it a tribe, call it a family. Whatever you call it, whoever you are, you need one.

JANE HOWARD

In the previous lesson you looked at translations of the graphs of linear functions. Translations can occur in other settings as well. For instance, what would the graph look like if the professor decided to add five points to each of the scores?

What translation will map the black triangle on the left onto its red image on the right?

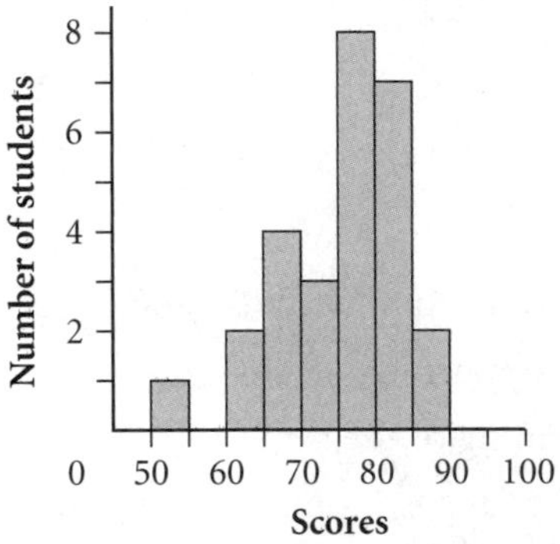

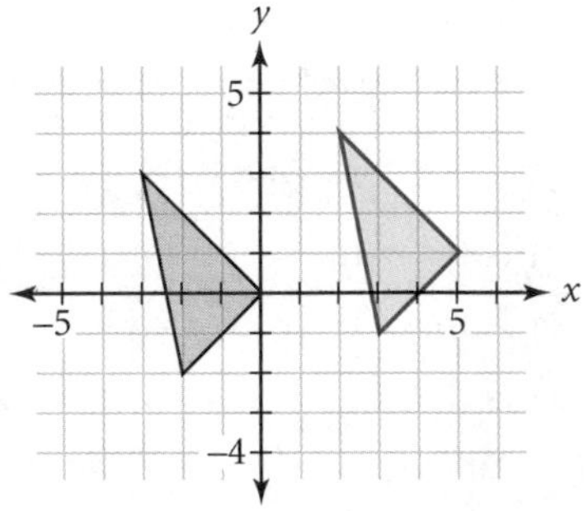

Translations are also a natural feature of the real world, including the world of art. Music can be transposed from one key to another. Melodies are often translated by a certain interval within a composition.

MusicConnection When a song is in a key that is difficult to sing or play, it can be translated, or transposed, into an easier key. To transpose music means to change the pitch of each note without changing the relationships between the notes. Musicians have several techniques for transposing music, and because these techniques are mathematically based, computer programs have been written that can do it as well.

In mathematics a change in the size or position of a figure or graph is called a **transformation.** Translations are one type of transformation. You may recall other types of transformations, such as reflections, dilations, stretches, shrinks, and rotations, from other mathematics classes.

In this lesson you will experiment with translations of the graph of the function $y = x^2$. The special shape of this graph is called a *parabola.* Parabolas always have a *line of symmetry* that passes through the vertex.

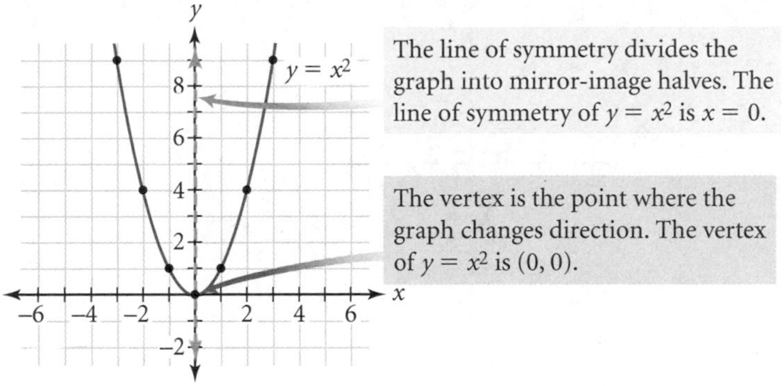
y
y = x²
8
6
4
2
−2
x
−6
−4
−2
2
4
6
The line of symmetry divides the graph into mirror-image halves. The line of symmetry of y = x² is x = 0.
The vertex is the point where the graph changes direction. The vertex of y = x² is (0, 0).

EngineeringConnection Several types of bridge designs involve the use of the parabolic arch. The cable of a suspension bridge the total weight of which is uniformly distributed over the length of the span forms a parabola.

The function $y = x^2$ is a building-block function, or **parent function.** By transforming the graph of a parent function, you can create infinitely many new functions, or a **family of functions.** The function $y = x^2$ and all functions created from transformations of its graph are called *quadratic functions.* Although Greek numeric prefixes were listed in Lesson 4.2, not all mathematical words have only Greek roots. The term *quadratic* uses the Latin prefix *quad,* which means "four." This is because the highest power of x is x squared, and squares have four sides.

Quadratic functions are very useful, as you will discover throughout this book. You can use functions in the quadratic family to model the height of a projectile from the ground as a function of time, or the area of a square as a function of the length of its side.

This lesson focuses on writing the quadratic equation of a parabola after a translation and on graphing a parabola given its equation. You will see that locating the vertex is fundamental to your success in understanding and using parabolas to solve problems.

Investigation
Make My Graph

> *Procedure Note*
>
> For this investigation use a "friendly" calculator window with a factor of 2. [▶ See **Calculator Note 4C** to learn about friendly windows.◀] Enter the parent function $y = x^2$ into Y1. Enter the equation for the transformation in Y2, and graph both Y1 and Y2 to check your work.

Step 1 Each graph below shows in black the graph of the parent function $y = x^2$. Find a quadratic equation that produces the congruent red parabola. (*Congruent* figures have the same shape and size.) Recall what you learned about translations of the graphs of linear equations in Lesson 4.3, and try to apply it here.

a.

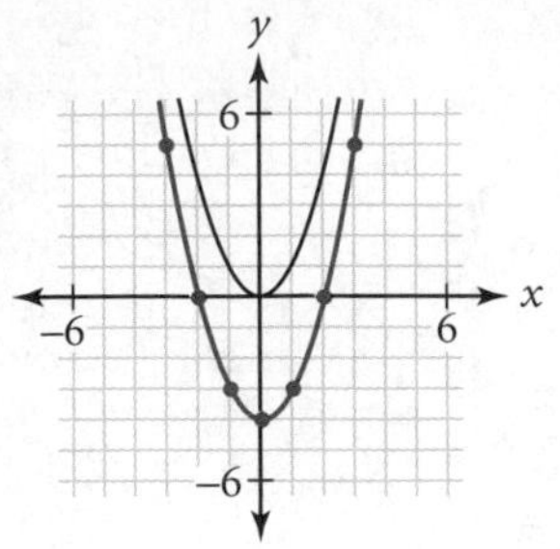

b.

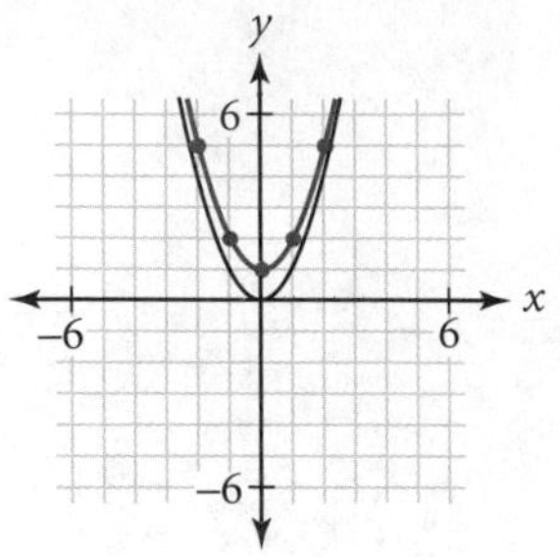

c.

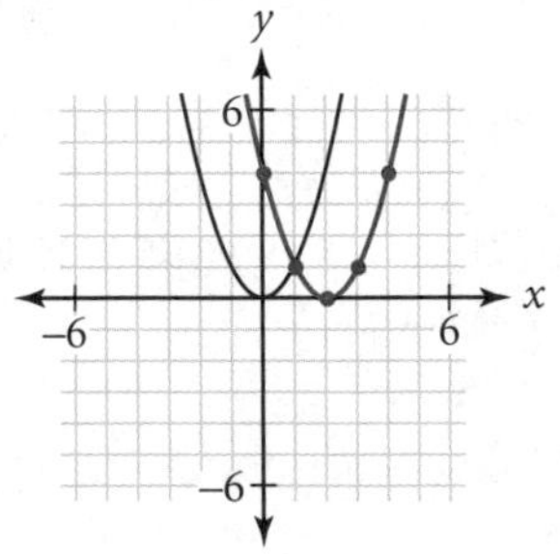

d.

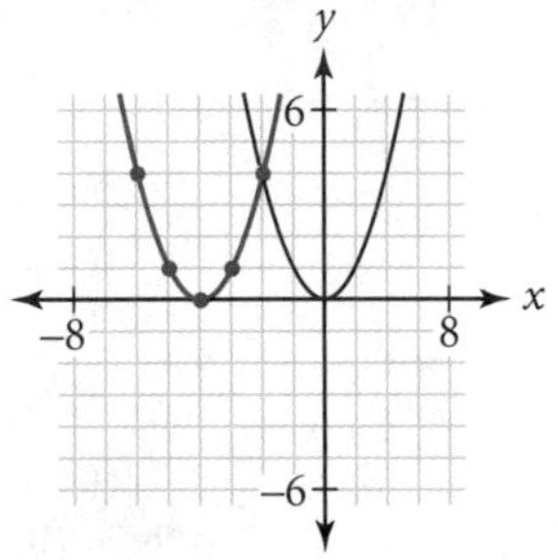

e.

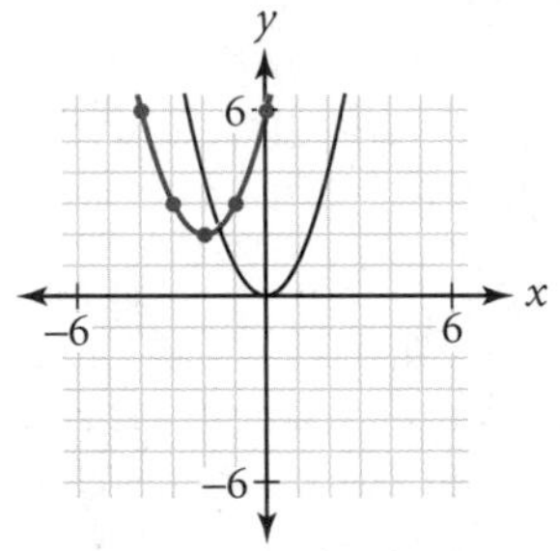

f.

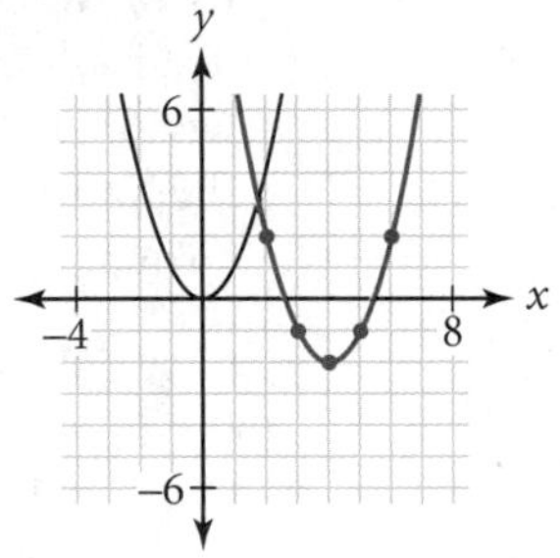

Step 2 Write a few sentences describing any connections you discovered between the graphs of the translated parabolas, the equation for the translated parabola, and the equation of the parent function.

Step 3 In general, what will be the equation of a parabola if the graph of $y = x^2$ is translated horizontally h units and vertically k units?

The next example shows one simple application involving parabolas and translations of parabolas. In later chapters you will discover many applications of this important mathematical curve.

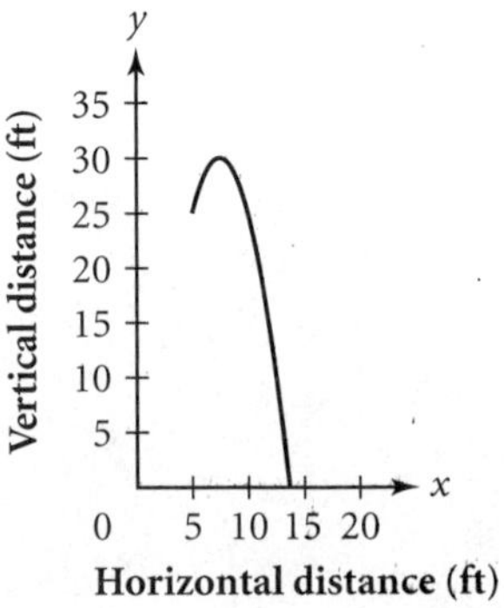

EXAMPLE

This graph shows a portion of a parabola. It represents a diver's position (horizontal and vertical distance) from the edge of a pool as he dives from a 5-ft-long board 25 ft above the water.

a. Sketch a graph of the diver's position if he dives from a 10-ft-long board 10 ft above the water. (Assume that he leaves the board at the same angle and with the same force.)

b. In the scenario described in part a, what will be the diver's position when he reaches his maximum height?

▶ *Solution*

First, make sure that you can interpret the graph. The point (5, 25) represents the moment when the diver leaves the board, which is 5 ft long and 25 ft high. The vertex, (7.5, 30), represents the position at which the diver's height is at a maximum, or 30 ft; it is also the point at which the diver's motion changes from upward to downward. The x-intercept, approximately (13.6, 0), indicates that the diver hits the water approximately 13.6 ft from the edge of the pool.

a. If the length of the board increases from 5 ft to 10 ft, then the parabola translates right 5 units. If the height of the board decreases from 25 ft to 10 ft, then the parabola translates down 15 units. If you define the original parabola as the graph of $y = f(x)$, then the function of the new graph is $y = f(x - 5) - 15$.

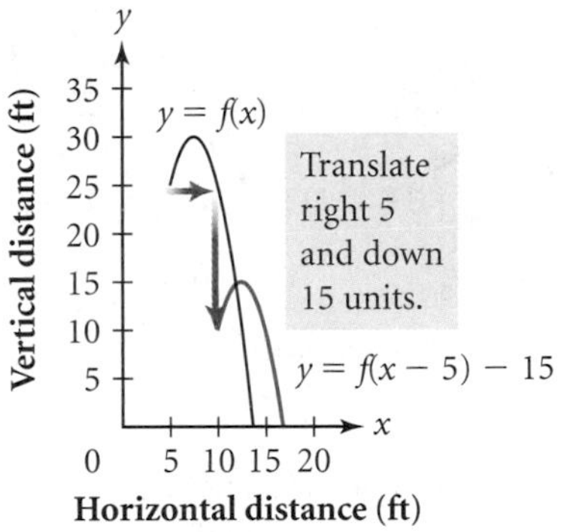

b. As with every point on the graph, the vertex translates right 5 units and down 15 units. The new vertex is $(7.5 + 5, 30 - 15)$, or $(12.5, 15)$. This means that when the diver's horizontal distance from the edge of the pool is 12.5 ft, he reaches his maximum height of 15 ft.

The translations you investigated with linear functions and functions in general work the same way with quadratic functions. If you translate the graph of $y = x^2$ horizontally h units and vertically k units, then the equation of the translated parabola is $y = (x - h)^2 + k$. You may also see this equation written as $y = k + (x - h)^2$ or $y - k = (x - h)^2$. When you translate horizontally, you can think of it as replacing x in the equation with $(x - h)$. Likewise, a vertical translation replaces y with $(y - k)$.

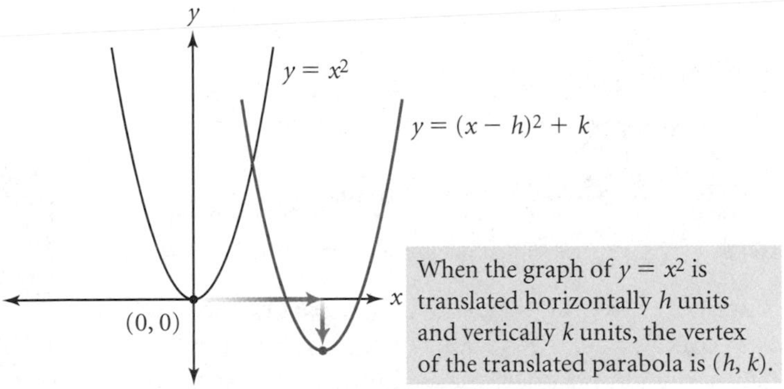

It is important to notice that the vertex of the translated parabola is (h, k). That's why finding the vertex is fundamental to determining translations of parabolas. In every function you learn, there will be key points to locate. Finding the relationships between these points and the corresponding points in the parent function will enable you to write equations more easily.

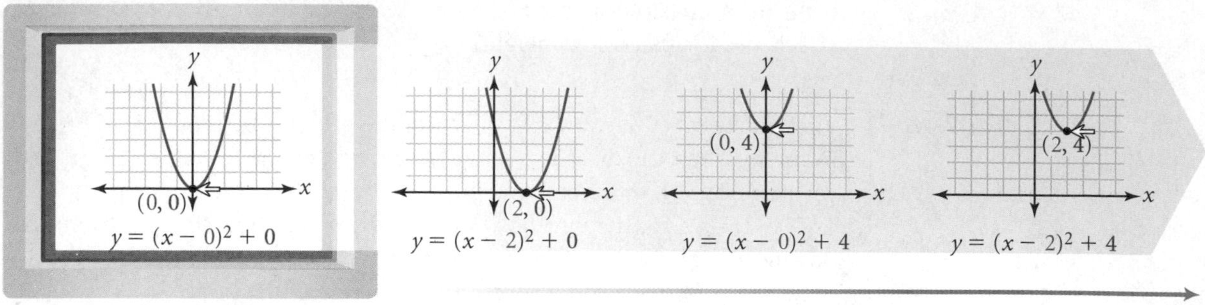

EXERCISES

Practice Your Skills

1. Write an equation of each parabola at right. Each parabola is a translation of the graph of the parent function $y = x^2$.

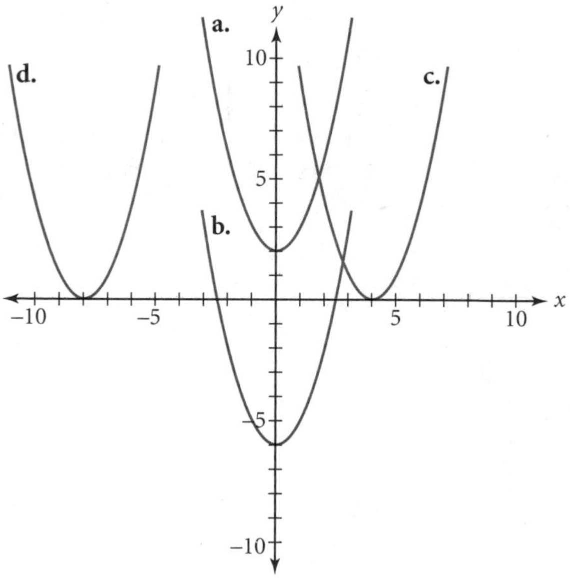

2. Each parabola described is congruent to the graph of $y = x^2$. Write an equation of each parabola and sketch its graph.

a. The parabola is translated down 5 units.

b. The parabola is translated up 3 units.

c. The parabola is translated right 3 units.

d. The parabola is translated left 4 units.

IMPROVING YOUR REASONING SKILLS

Translating Stars

The group of stars known as the Big Dipper, which is part of the constellation Ursa Major, contains stars at various distances from the Earth. Imagine translating the Big Dipper that you see in the night sky to a new position. Would all of the stars need to be moved the same distance? Why or why not?

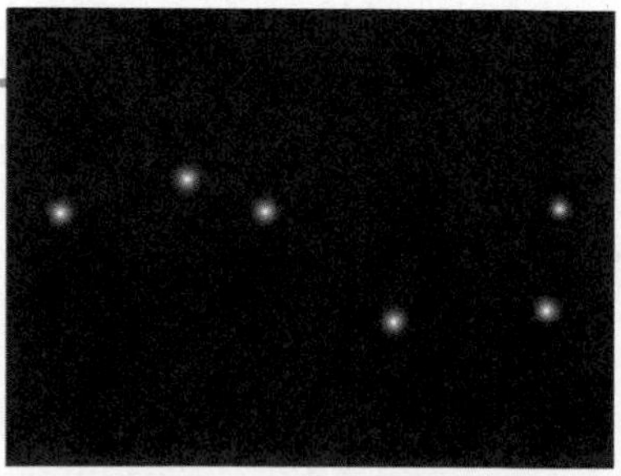

3. If $f(x) = x^2$, then the graph of each equation below is a parabola. Describe the location of the parabola relative to the graph of $f(x) = x^2$.

a. $y = f(x) - 3$ b. $y = f(x) + 41$ c. $y = f(x - 2)$ d. $y = f(x + 4)$

4. Describe what happens to the graph of $y = x^2$ in these situations.

a. x is replaced with $(x - 3)$.
b. x is replaced with $(x + 3)$.
c. y is replaced with $(y - 2)$.
d. y is replaced with $(y + 2)$.

5. Solve.

a. $x^2 = 4$
b. $x^2 + 3 = 19$

Reason and Apply

6. Write an equation of each parabola.

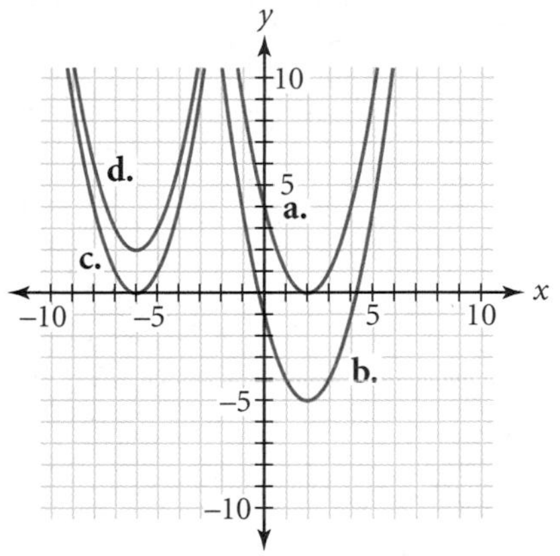

7. The red parabola is the image of the graph of $y = x^2$ after a translation right 5 units and down 3 units.

a. Write an equation of the red parabola.
b. Where is the vertex of the red parabola?
c. What are the coordinates of the other four points if they are 1 or 2 horizontal units from the vertex? How are the coordinates of each point on the black parabola related to the coordinates of the corresponding point on the red parabola?
d. What is the length of segment b? Of segment c?

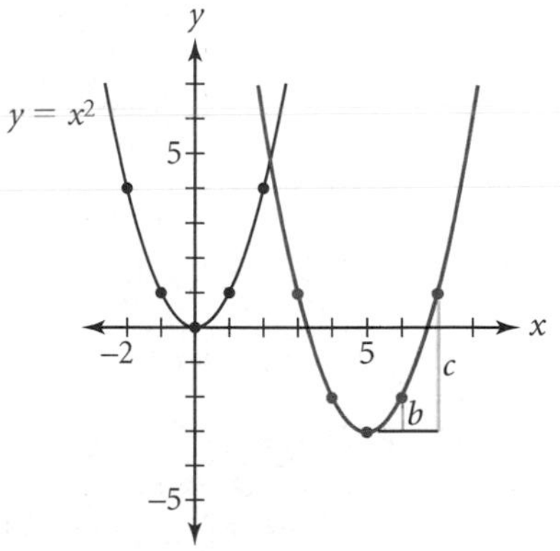

8. **APPLICATION** This table of values compares the number of teams in a Scrabble™ league and the number of games required for each team to play every other team twice (once at home and once away from home).

Number of teams (x)	0	1	2	3	...
Number of games (y)	0	0	2	6	...

a. Continue the table to include 10 teams.

b. Plot each point, and describe the graph produced.

c. Write an explicit function of this graph.

d. Use your function to find how many games are required if there are 30 teams.

9. Given the graph of $y = f(x)$, draw a graph of each of these related functions.

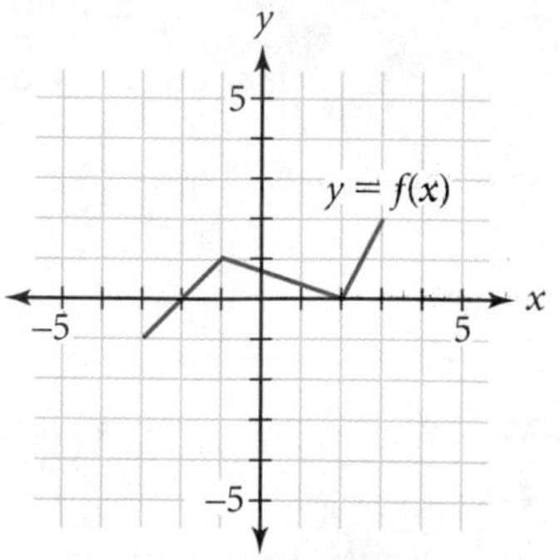

a. $y = f(x + 2)$

b. $y = f(x - 1) - 3$

10. Solve.

a. $3 + (x - 5)^2 = 19$

b. $(x + 3)^2 = 49$

c. $5 - (x - 1) = -22$

d. $-15 + (x + 6)^2 = -7$

11. This histogram shows the students' scores on a recent quiz in Professor Noah's class. Sketch what the graph will look like if Professor Noah

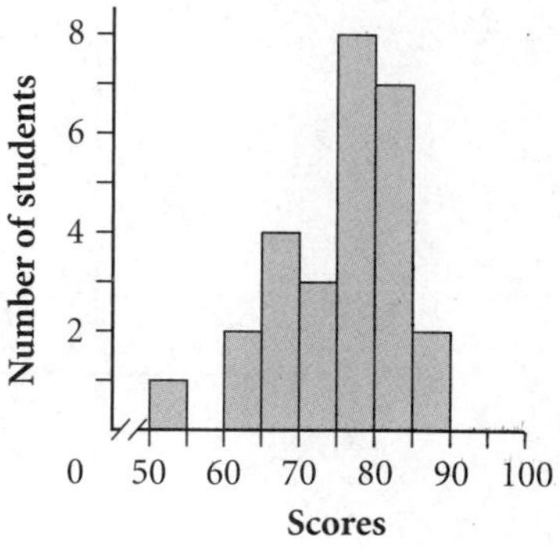

a. Adds five points to everyone's score

b. Subtracts ten points from everyone's score

Review

12. (Lesson 2.1) Match each recursive formula with the equation of the line that contains the sequence of points, n, u_n, generated by the formula.

a. $u_0 = -8 \quad u_n = u_{(n-1)} + 3$ where $n \geq 1$

b. $u_1 = 3 \quad u_0 = u_{(n-1)} - 8$ where $n \geq 2$

A. $y = 3x - 11$

B. $y = 3x - 8$

C. $y = 11 - 8x$

D. $y = -8x + 3$

13. (Lesson 4.1) A car drives at a constant speed along the road pictured at right from point A to point X. Sketch a graph showing the straight-line distance between the car and point X. Mark points A, B, C, D, E, and X on your graph.

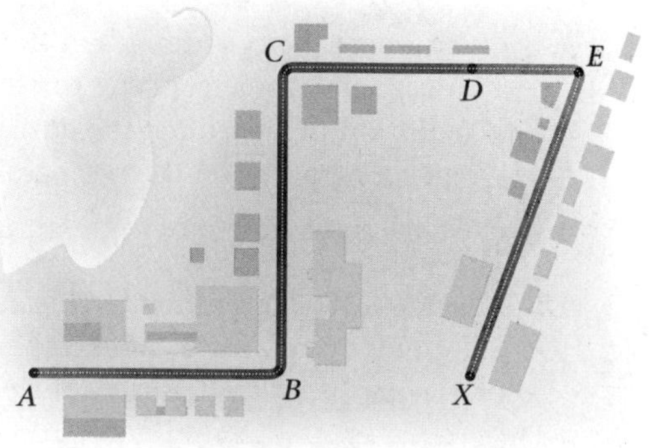

14. **APPLICATION** (Lesson 3.4) You need to rent a car for one day. Mertz Rental charges \$32 per day plus \$0.10 per mile. Saver Rental charges \$24 per day plus \$0.18 per mile. Luxury Rental charges \$51 per day with unlimited mileage.

a. Write a cost equation of each rental agency's rates.

b. Graph the three equations on the same pair of axes.

c. Explain which rental agency offers the cheapest alternative under various circumstances.

LESSON

4.5

Reflections and the Square Root Family

The *square root function,* $y = \sqrt{x}$, is another parent function you can use to illustrate transformations. Recall that the domain of a function is the set of all possible values for x (the independent variable). The range is the set of all possible values for y (the dependent variable). From the graph shown here, what are the domain and range of $f(x) = \sqrt{x}$? If you can graph $y = \sqrt{x}$ on your calculator, you can trace to show that $\sqrt{3}$ is approximately 1.732. What is the approximate value of $\sqrt{8}$? How would you use the graph to find $\sqrt{31}$? What happens when you try to trace for values of $x < 0$?

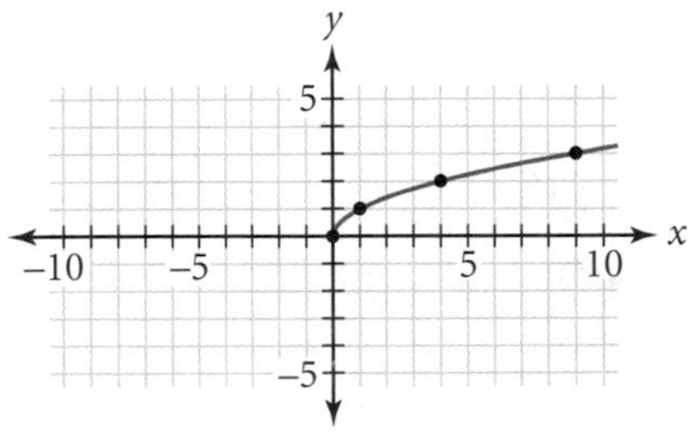

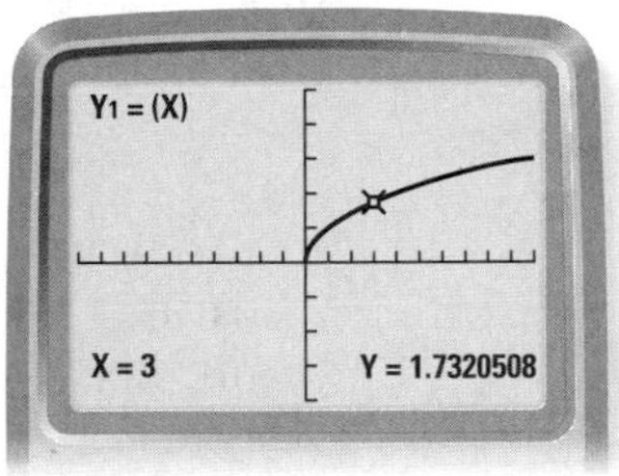

[−9.4, 9.4, 1, −6.2, 6.2, 1]

Investigation
Take a Moment to Reflect

Procedure Note

For this Investigation use a friendly window with a factor of 2.

In this Investigation you first will work with linear functions to discover how to create a new transformation—a **reflection.** Then you will apply reflections to quadratic functions and square root functions.

Step 1 Enter $y = 0.5x + 2$ into Y1, and graph it on your calculator. Then enter the equation Y2 $= -$Y1(x) and graph it. [▶ See **Calculator Note 4D** to learn how to use Y1 in the equation of Y2.◀]

a. Write the equation for Y2 in terms of x. How does the graph of Y2 compare with the graph of Y1?

b. Change Y1 to $y = -2x - 4$, and repeat the instructions in Step 1a.

c. Change Y1 to $y = x^2 + 1$ and repeat.

d. In general, how are the graphs of $y = f(x)$ and $y = -f(x)$ related?

Step 2 Enter $y = 0.5x + 2$ into Y1. Enter the equation Y2 = Y1$(-x)$, and graph both Y1 and Y2.

a. Write the equation for Y2 in terms of x. How does the graph of Y2 compare with the graph of Y1?

b. Change Y1 to $y = -2x - 4$, and repeat the instructions in Step 2a.

c. Change Y1 to $y = x^2 + 1$ and repeat. Explain what happened.

d. Change Y1 to $y = (x - 3)^2 + 2$ and repeat.

e. In general, how are the graphs of $y = f(x)$ and $y = f(-x)$ related?

Step 3 Enter $y = \sqrt{x}$ into Y1, and graph it on your calculator.

a. Predict what the graphs of $Y2 = -Y1(x)$ and $Y2 = Y1(-x)$ will look like. Use your calculator to verify your predictions. Write equations of both of these functions in terms of x.

b. Predict what the graph of $Y2 = -Y1(-x)$ will look like. Use your calculator to verify your prediction.

c. Do you notice that the graph of the square root function looks like half of a parabola oriented horizontally? Why isn't it an entire parabola? What function would you graph to complete the bottom half of the parabola?

In algebra, reflections may be across the x- or y-axis or across some other defined line, such as $y = x$, or $x = 3$. Rules for reflecting over the x- or y-axis are summarized in the box.

Reflection of a Function

A *reflection* is a transformation that flips a graph across a line, creating a mirror image.

Given the graph of $y = f(x)$, the graph of

$$y = f(-x)$$

is a reflection across the y-axis, and the graph of

$$y = -f(x)$$

is a reflection across the x-axis.

Because the graph of the square root function looks like half a parabola, it's easy to see the effects of reflections. The square root family also has many real-world applications, such as finding the time it takes a falling object to reach the ground. This example shows you how to apply a square root function.

ScienceConnection Obsidian, a natural volcanic glass, was a popular material for tools and weapons in prehistoric times because it makes a very sharp edge. In 1960 scientists Irving Friedman and Robert L. Smith discovered that obsidian absorbs moisture at a slow, predictable rate and that measuring the thickness of the layer of moisture with a high-power microscope helps determine its age. The age of prehistoric artifacts is predicted by a square root function similar to $d = \sqrt{5t}$, where t is time in thousands of years and d is the thickness of the layer of moisture in microns (millionths of a meter).

EXAMPLE

Objects fall to the ground because of gravity. When an object is dropped from an initial height of d meters (m), the height, h, after t seconds is given by the quadratic function $h = -4.9t^2 + d$. If an object is dropped from a height of 1000 m, how long does it take for the object to reach a height of 750 m? 500 m? How long will it take the object to hit the ground?

ScienceConnection Isaac Newton formulated the theory of gravitation in the 1680s, building on the work of Galileo and Kepler. Gravity is the force of attraction that exists between all objects. In general, larger objects pull smaller objects toward them. The force of gravity keeps objects on the surface of a planet, and it keeps objects in orbit around a planet or the Sun. You may have seen pictures of astronauts "floating" in space or on the surface of the Moon. This is because gravity in space and on the Moon is much weaker than near Earth.

▶ Solution

The height of an object dropped from a height of 1000 m is given by the function $h = -4.9t^2 + 1000$. You want to know t for various values of h, so first solve this equation for t.

$$h = -4.9t^2 + 1000$$ Original equation.

$$h - 1000 = -4.9t^2$$ Subtract 1000 from both sides.

$$\frac{h - 1000}{-4.9} = t^2$$ Divide both sides by -4.9.

$$\pm\sqrt{\frac{h - 1000}{-4.9}} = t$$ Take the square root of both sides.

$$t = \sqrt{\frac{h - 1000}{-4.9}}$$ Because it doesn't make sense to have a negative value for time, you use only the positive root.

To find when the height is 750 meters, substitute 750 for h.

$$t = \sqrt{\frac{750 - 1000}{-4.9}}$$

$$t \approx 7.143$$

The height of the object is 750 m after approximately 7 s.

A similar substitution shows that the height of the object is 500 m after approximately 10 s.

$$t \approx 10.102$$

The object hits the ground when its height is 0 m. That occurs after approximately 14 s.

$$t \approx 14.286$$

From the example you may notice that square root functions play an important part in solving quadratic functions. However, unlike the example you cannot always eliminate the negative root. You'll have to let the context of a problem dictate when to use the positive root, the negative root, or both.

EXERCISES

Practice Your Skills

1. Each graph is a transformation of the graph of the parent function $y = \sqrt{x}$. Write an equation of each graph.

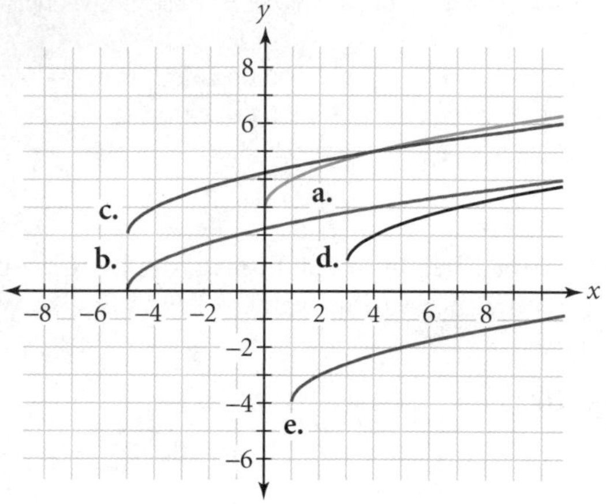

2. Describe what happens to the graph of $y = \sqrt{x}$ in the following situations.

a. x is replaced with $(x - 3)$.

b. x is replaced with $(x + 3)$.

c. y is replaced with $(y - 2)$.

d. y is replaced with $(y + 2)$.

IMPROVING YOUR GEOMETRY SKILLS

Lines in Motion Revisited

Imagine that the graph of any line $y = a + bx$ is translated 2 units in a direction perpendicular to its slope. What horizontal and vertical translation would be equivalent to this translation? What are the values of h and k? What is the linear equation of the image? You may want to use your calculator or geometry software to experiment with some specific linear equations before you try to generalize for h and k.

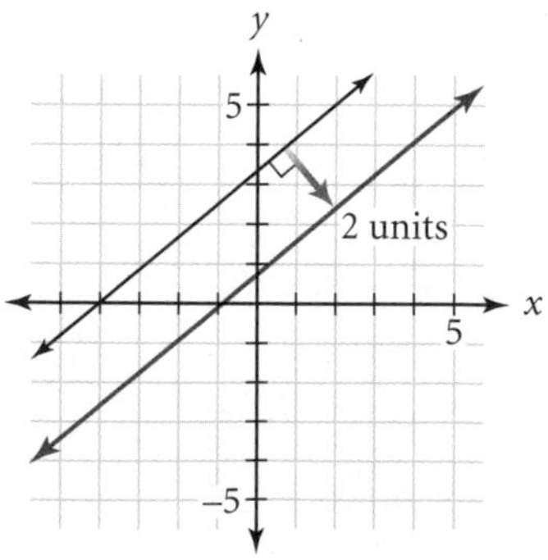

3. Each curve (right) is a transformation of the graph of the parent function $y = \sqrt{x}$. Write an equation of each curve.

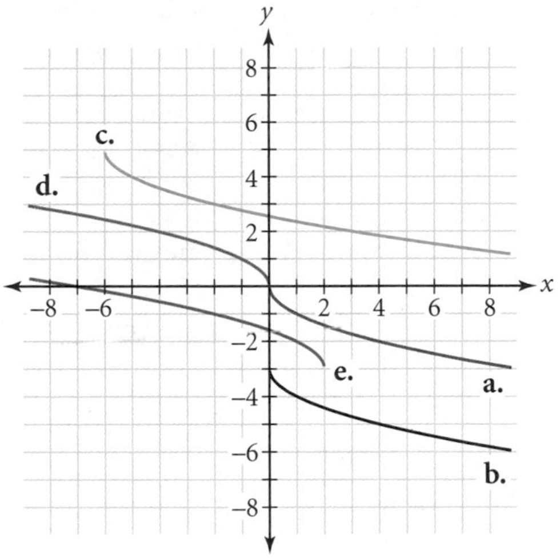

4. Given the graph below of $y = f(x)$, draw a graph of each of these related functions.

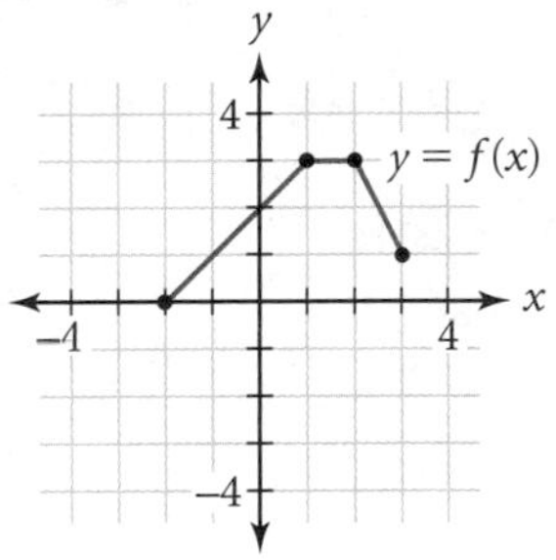

a. $y = f(-x)$ **b.** $y = -f(x)$ **c.** $y = -f(-x)$

Reason and Apply

5. Consider the parent function $f(x) = \sqrt{x}$.

a. Name three pairs of integer coordinates that are on the graph of $y = f(x + 4) - 2$.

b. Write $y = f(x + 4) - 2$ using a *radical,* or square root symbol, and graph it.

c. Write $y = -f(x - 2) + 3$ using a radical, and graph it.

6. Consider this parabola:

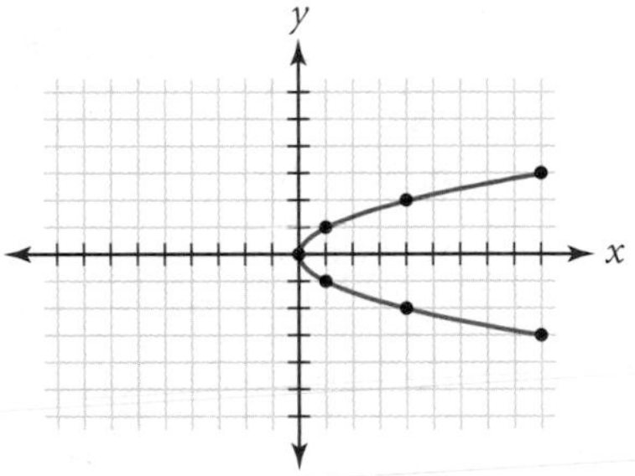

a. Graph the parabola on your calculator. What two functions did you use?

b. Combine both functions from 6a using ± notation to create a single equation. Square both sides of the equation. What is the result?

7. Refer to the two parabolas shown.

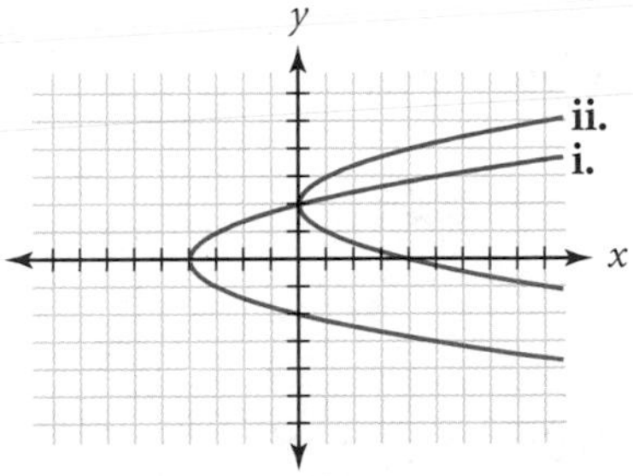

a. Explain why neither graph represents a function.

b. Write a single equation of each parabola using ± notation.

c. Square both sides of each equation in 7b. What is the resulting equation of each parabola?

8. Write the equation of each parabola. Each parabola is a transformation of the graph of the parent function $y = x^2$.

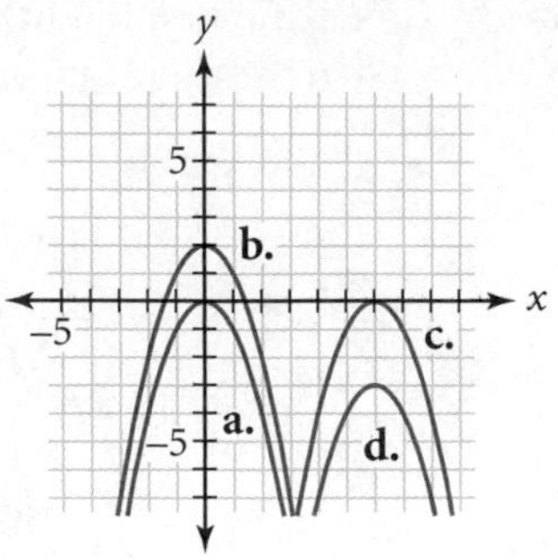

9. Write the equation of a parabola that is congruent to the graph of $y = -(x + 3)^2 + 4$ but translated right 5 units and down 2 units.

10. As Jake and Arthur travel together from Detroit to Chicago, each makes a graph relating time and distance. Jake, who lives in Detroit and keeps his watch on Detroit time, graphs his distance from Detroit. Arthur, who lives in Chicago and keeps his watch on Chicago time (an hour earlier than Detroit's), graphs his distance from Chicago. They both use the time of day for their x-axes. The distance between Detroit and Chicago is 250 miles.

a. Sketch what you think each graph might look like.

b. If Jake's graph is described by the function $y = f(x)$, what function describes Arthur's graph?

c. If Arthur's graph is described by the function $y = g(x)$, what function describes Jake's graph?

11. APPLICATION Police measure the lengths of skid marks to determine the initial speed of a vehicle before the brakes were applied. Many variables, such as the type of road surface, whether or not the car has an antilock braking system (ABS), and weather conditions, play an important role in determining the speed. For cars without ABS, the formula used to determine the initial speed is $S = 5.5\sqrt{D \cdot f}$, where S is the speed in miles per hour, D is the average length of the skid marks in feet, and f is a constant called the "drag factor." At a particular accident scene assume it is known that the road surface has a drag factor of 0.7.

a. Write an equation that will determine the initial speed on this road as a function of the length of skid marks.

b. Sketch a graph of this function.

c. If the average length of the skid marks is 60 ft, estimate the initial speed of the car when the brakes were applied.

d. Solve your equation from 11a for D. What can you determine using this equation?

e. Graph your equation from 11a. What shape is it?

f. If you traveled on this road at a speed of 65 mi/h and suddenly slammed on your brakes, how long would your skid marks be?

Review

12. (Lesson 4.2) Identify each relation that is also a function. For each relation that is not a function, explain why it is not.

a. Independent variable: city
Dependent variable: area code

b. Independent variable: any pair of whole numbers
Dependent variable: their greatest common factor

c. Independent variable: any pair of fractions
Dependent variable: their least common denominator

d. Independent variable: the day of the year
Dependent variable: the time of sunrise

13. Solve for x. Solving square root equations often results in *extraneous solutions,* or answers that don't check in the original equation, so be sure to check your work.

a. $3 + \sqrt{x - 4} = 20$

b. $\sqrt{x + 7} = -3$

c. $4 - (x - 2)^2 = -21$

d. $5 - \sqrt{-(x + 4)} = 2$

14. (Lesson 4.4) Find the equation of the parabola with vertex $(-6, 4)$ and containing the point $(-5, 5)$.

15. (Lesson 3.1) The graph of the line ℓ_1 is shown here.

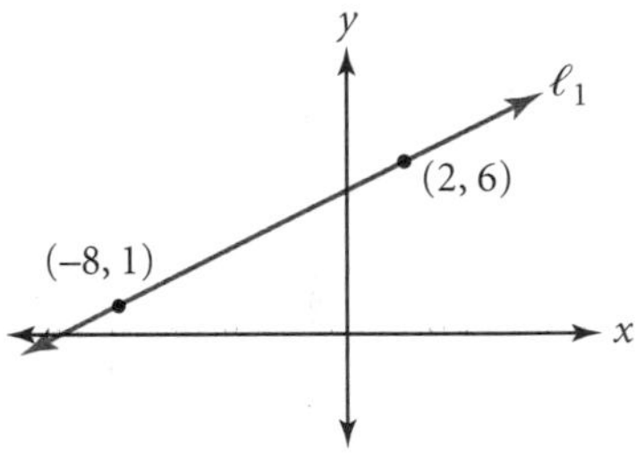

a. Write the equation of the line ℓ_1.

b. The line ℓ_2 is the image of the line ℓ_1 translated right 8 units. Sketch the line ℓ_2, and write its equation in a way that emphasizes the horizontal translation.

c. The line ℓ_2 also can be thought of as the image of the line ℓ_1 after a vertical translation. Write the equation of the line ℓ_2 in a way that emphasizes the vertical translation.

d. Show that the equations in 15b and 15c are equivalent.

16. (Lesson 1.4) Consider this data set:

{37, 40, 36, 37, 37, 49, 39, 47, 40, 38, 35, 46, 43, 40, 47, 49, 70, 65, 50, 73}

a. Give the five-number summary.

b. Display the data in a box plot.

c. Find the interquartile range.

d. Identify any outliers, based on the interquartile range.

LESSON

4.6 Stretches and Shrinks and the Absolute Value Family

A mind that is stretched by a new experience can never go back to its old dimensions.

OLIVER WENDELL HOLMES

Hao and Dayita ride the subway each day. They live on the same east-west subway route. Hao lives 7.4 miles west of the college, and Dayita lives 5.2 miles east of the college. This information is shown on the number line below.

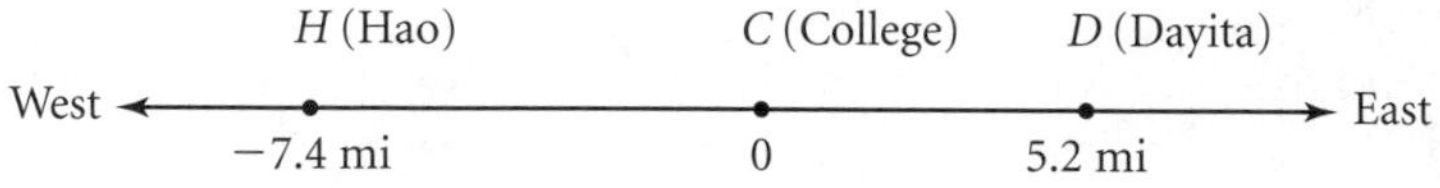

The distance between two points is always positive. However, if you calculate Hao's distance from the college, or *HC*, by subtracting his starting position from his ending position, you get a negative value:

$$-7.4 - 0 = -7.4$$

In order to make the distance positive, you use the absolute value function, which makes any input positive or zero. For example, the absolute value of -3 is 3, or = 3. For Hao's distance from the college, use the absolute value function to calculate

$$HC = |-7.4 - 0| = |-7.4| = 7.4$$

What is the distance from *D* to *H*? What is the distance from *H* to *D*?

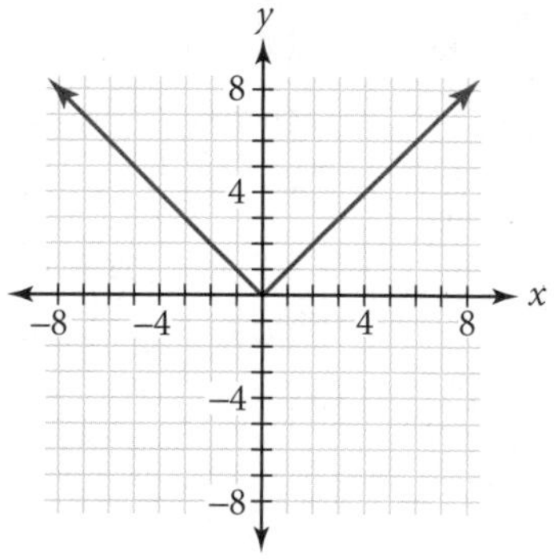

In this lesson you will explore transformations of the graph of the parent function $y = |x|$. [► See **Calculator Note 4E** to learn how to use and graph the absolute value function. ◄] In this lesson you will write and use equations of the form $y = a\left|\frac{x-h}{b}\right| + k$. What you have learned about translating and reflecting other graphs will apply to this function as well. You will also learn about transformations that stretch and shrink a graph.

EXAMPLE A

Graph the function with each of these functions. How does the graph of each function change the original graph?

a. $y = 2|x|$

b. $y = \left|\frac{x}{3}\right|$

c. $y = 2\left|\frac{x}{3}\right|$

► Solution

In the graph of each function, the vertex remains at the origin. Notice, however, how the points (1, 1) and (−2, 2) on the parent function are mapped to a new location.

a. Every point on the graph of $y = 2|x|$ has a y-coordinate that is 2 times the y-coordinate of the corresponding point on the parent function. You say the graph of $y = 2|x|$ is "a vertical stretch by a factor of 2."

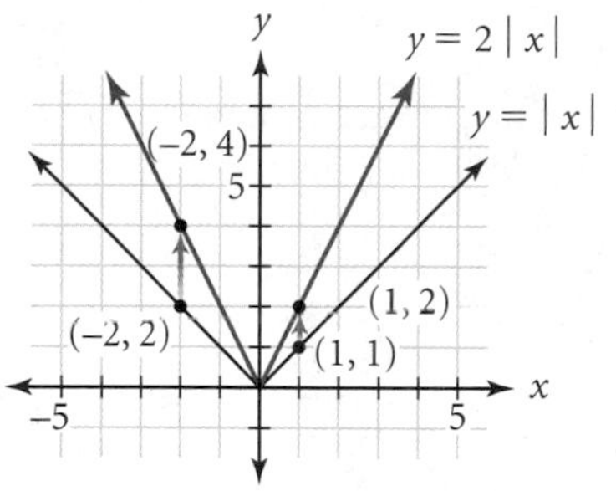

b. Replacing x with $\frac{x}{3}$ multiplies the x-coordinates by a factor of 3. The graph of $y = \frac{x}{3}$ is a horizontal stretch by a factor of 3.

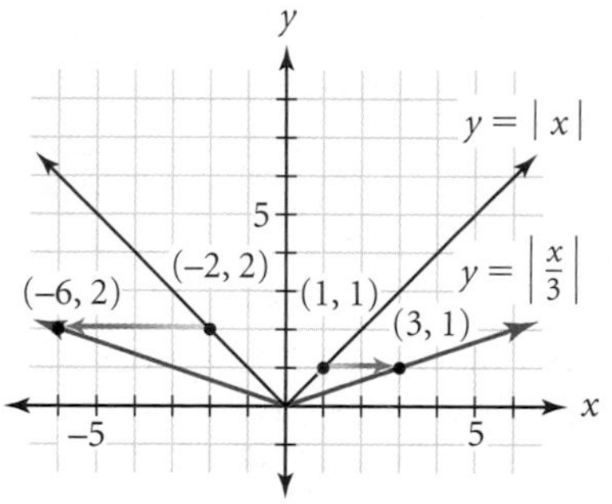

c. The combination of multiplying the parent function by 2 and dividing x by 3 results in a vertical stretch by a factor of 2 *and* a horizontal stretch by a factor of 3.

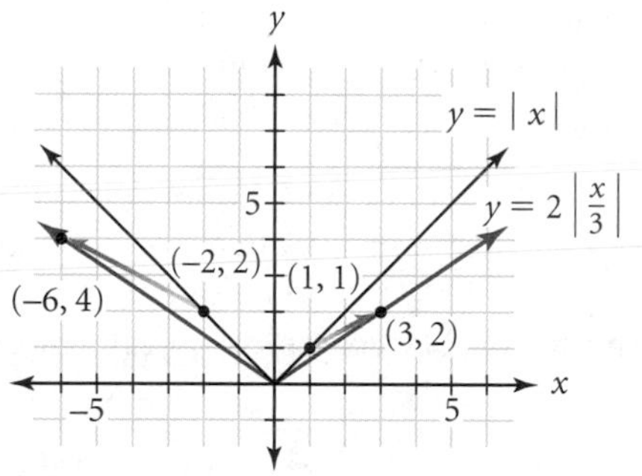

Translations and reflections are *rigid transformations*—they produce an image that is congruent to the original figure. Stretches and shrinks are *nonrigid transformations*—the image is not congruent to the original figure (unless you use a factor of 1 or -1). If you stretch or shrink a figure by the same factor both vertically and horizontally, then the image and the original figure will be similar. If you stretch or shrink by different vertical and horizontal scale factors, then the image and the original figure will not be similar.

Using what you know about translations, reflections, and stretches, you can fit functions to data by locating only a few key points. For quadratic, square root, and absolute value functions, first locate the vertex of the graph. Then use any other point to find the factors by which to stretch or shrink the image.

EXAMPLE B

These data are from one bounce of a ball. Find an equation that fits the data over this domain.

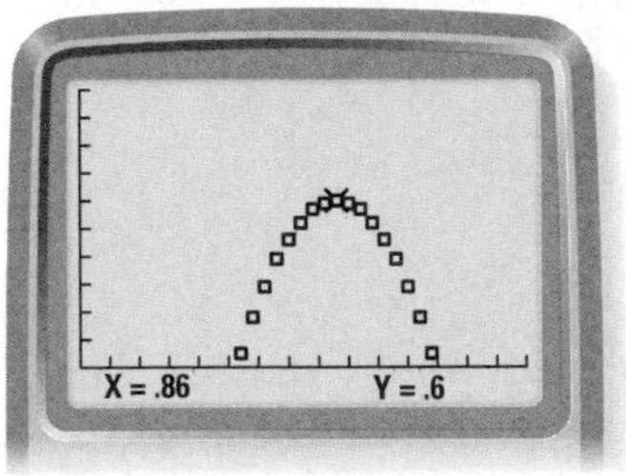

Time (s) x	Height (m) y
0.54	0.05
0.58	0.18
0.62	0.29
0.66	0.39
0.70	0.46
0.74	0.52
0.78	0.57
0.82	0.59
0.86	0.60
0.90	0.59
0.94	0.57
0.98	0.52
1.02	0.46
1.06	0.39
1.10	0.29
1.14	0.18
1.18	0.05

► ***Solution***

Graph the data on your calculator. The graph appears to be a parabola. However, the parent function $y = x^2$ has been reflected, translated, and possibly stretched or shrunk. Start by determining the translation. The vertex has been translated from (0, 0) to (0.86, 0.60). This is enough information for you to write the equation in the form $y = (x - h)^2$, or $y = (x - 0.86)^2 + 0.60$. If you think of replacing x with $(x - 0.86)$ and y with $(y - 0.60)$, you could also write the equivalent equation $y - 0.6 = (x - 0.86)^2$.

The graph of $y = (x - 0.86)^2 + 0.60$ does not fit the data. The function still needs to be reflected and, as you can see from the graph, shrunk. Both of these transformations can be accomplished together.

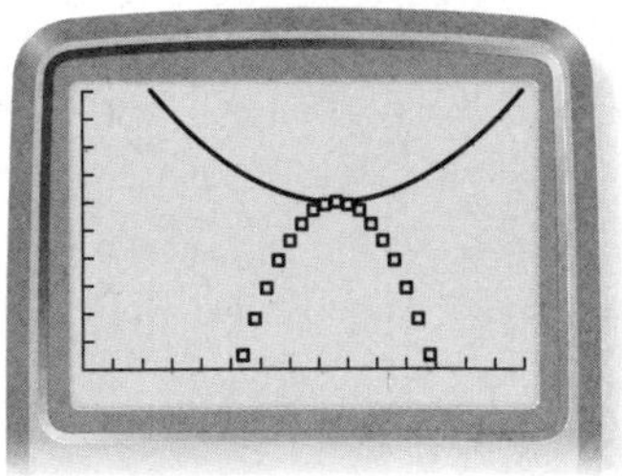

Select one other data point to determine the scale factors, a and b. You can use any other point, but you will get a better fit if you choose one that is not too close to the vertex. For example, you can choose the data point (1.14, 0.18).

Assume this data point is the image of the point (1, 1) in the parent parabola $y = x^2$. In the graph of $y = x^2$, (1, 1) is 1 unit away from the vertex (0, 0) both horizontally and vertically. The data point is $1.14 - 0.86$, or 0.28, unit away from the x-coordinate of the vertex, and $0.18 - 0.60$, or -0.42, unit away from the y coordinate of the vertex. So, the horizontal scale factor is 0.28, and the vertical scale factor is -0.42. The negative vertical scale factor also produces a reflection across the x-axis.

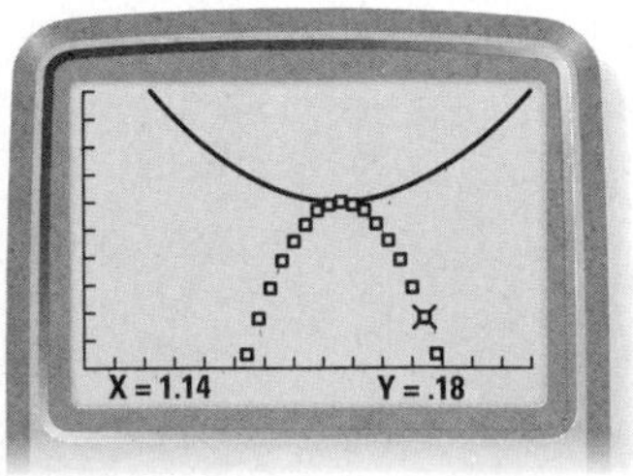

Combine these scale factors with the translations to get the final equation

$$\frac{y-0.6}{-0.42}=\left(\frac{x-0.86}{0.28}\right)^2 \text{ or } y=-0.42\left(\frac{x-0.86}{0.28}\right)^2+0.6$$

This model fits the data nicely.

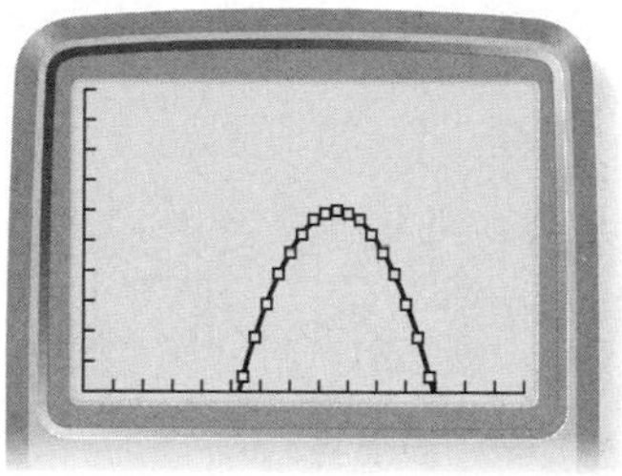

The same procedure works with the other functions you have studied so far. As you continue to add new functions to your mathematical repertoire, you will find that what you have learned about function transformations continues to apply.

Investigation
The Pendulum

You will need

- string
- a small weight
- a stopwatch or a watch with a second hand

Procedure Note

1. Tie the weight at one end of a length of string to make a pendulum. Firmly hold the other end of the string, or tie it to something, so that the weight hangs freely.
2. Measure the length of the pendulum from the center of the weight to the point where the string is held.
3. Pull the weight to one side, and release it so that it swings back and forth in a short arc, about 10° to 20°. Time ten complete swings (forward and back is one swing).
4. The *period* of your pendulum is the time of one complete swing (forward and back). Find the period by dividing the elapsed time by 10.

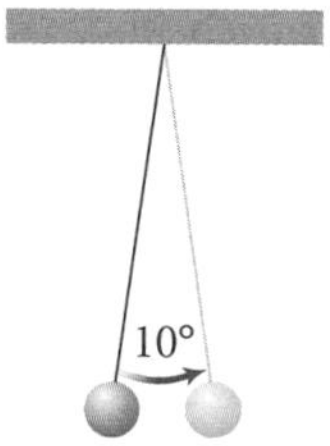

Galileo Galilei (1564–1642) made many contributions to our understanding of gravity, the physics of falling objects, and the orbits of the planets. One of his famous experiments involved the periodic motion of a pendulum. In this Investigation you will carry out the same experiment and find a function to model the data.

Step 1 Follow the Procedure Note to find the period of your pendulum. Repeat the experiment for several different string lengths, and complete a table of values. Use a variety of short, medium, and long string lengths.

Step 2 Graph the data, using *length* as the independent variable. The vertex is at the origin, (0, 0). Why do you suppose it is there?

Step 3 What is the shape of the graph? What do you suppose is the parent function?

Step 4 Divide up the points, and have each person in your group find the horizontal or vertical stretch or shrink from the parent function. Apply these transformations to find an equation to fit the data.

Step 5 Compare the collection of equations from your group. Which points are the best to use to fit the curve? Why do these points work better than others?

In the exercises you will use techniques you discovered in this lesson. Remember that replacing y with $\frac{y}{a}$ stretches a graph by a factor of a vertically. Replacing x with $\frac{x}{b}$ stretches a graph by a factor of b horizontally. When graphing a function, you should do stretches and shrinks before translations to avoid the vertex moving around.

Stretch or Shrink of a Function

A **stretch** and a **shrink** are transformations that expand or compress a graph either horizontally or vertically.

Given the graph of $y = f(x)$, the graph of

$$\frac{y}{a} = f(x) \text{ or } y = af(x)$$

is a vertical stretch or shrink by a factor of a. When $a > 1$, it is a stretch; when $0 < a < 1$, it is a shrink. When $a < 0$, a reflection across the x-axis is also involved.

Given the graph of $y = f(x)$, the graph of

$$y = f\left(\frac{x}{b}\right) \text{ or } y = f\left(\frac{1}{b} \cdot x\right)$$

is a horizontal stretch or shrink by a factor of b. When $b > 1$, it is a stretch; when $0 < b < 1$, it is a shrink. When $b < 0$, a reflection across the y-axis is also involved.

EXERCISES

Practice Your Skills

1. Each graph is a transformation of the graph of one of the parent functions you've studied. Write an equation for each graph.

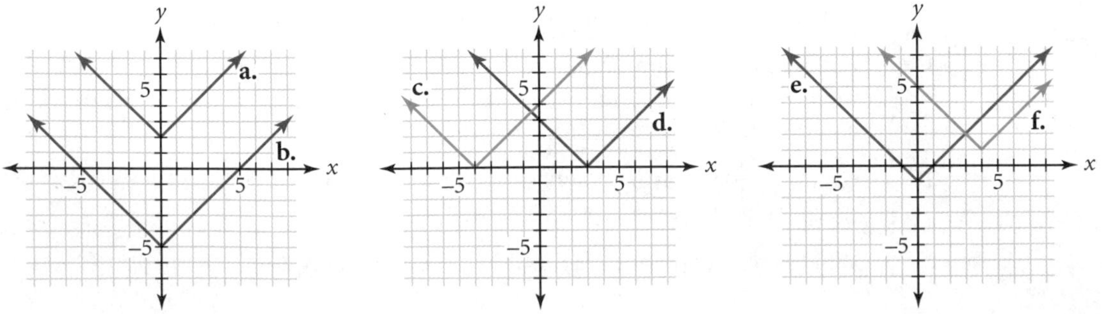

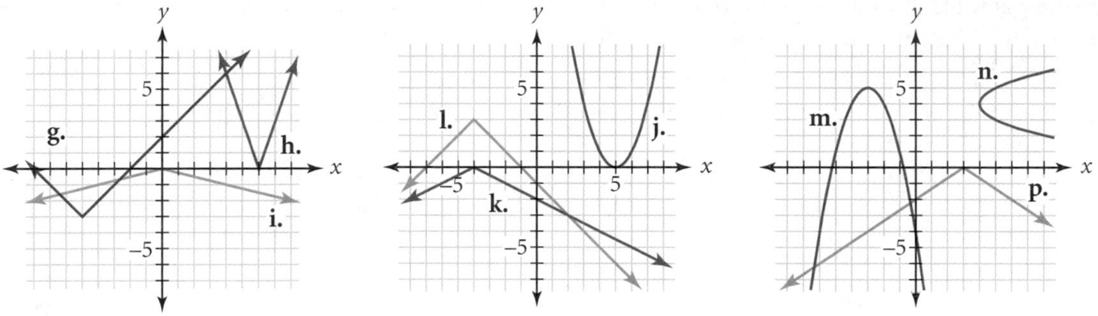

2. Describe what happens to the graph of $y = f(x)$ in these situations.

a. x is replaced with $\frac{x}{3}$.

b. x is replaced with $-x$.

c. y is replaced with $\frac{y}{2}$.

d. y is replaced with $-y$.

3. Solve each equation for y.

a. $y + 3 = 2(x - 5)^2$

b. $\frac{y + 5}{2} = \left|\frac{x + 1}{3}\right|$

c. $\frac{y + 7}{-2} = \sqrt{\frac{x - 6}{-3}}$

Reason and Apply

4. Choose a few different values for a. What can you conclude about $y = a|x|$ and $y = |ax|$? Are they the same function?

5. The graph at right shows how to solve the equation $|x - 4| = 3$ graphically. The equations $y = |x - 4|$ and $y = 3$ are graphed on the same coordinate axes.

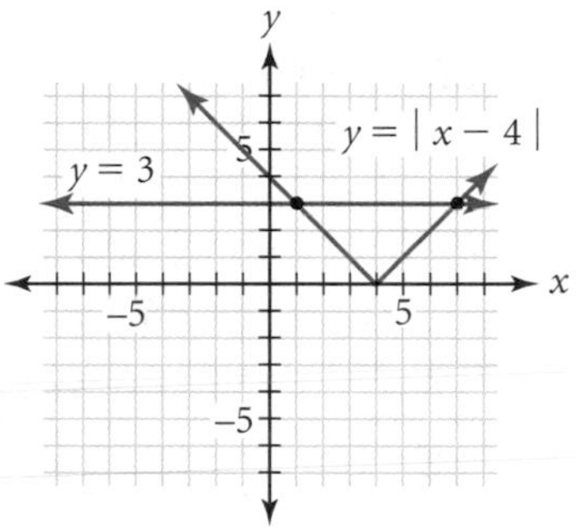

a. What is the x-coordinate of each point of intersection? What x-values are solutions of the equation $|x - 4| = 3$?

b. Solve the equation $|x + 3| = 5$ graphically.

6. **APPLICATION** You can use a single radio receiver to find the distance to a transmitter by measuring the strength of the signal, but you cannot determine the direction from which the signal is being broadcast. These distances are measured with a receiver as you drive east along a straight road. Find a model that fits the data. Where do you think the transmitter might be located?

Miles traveled	0	4	8	12	16	20	24	28	32	36
Distance to object	18.4	14.4	10.5	6.6	3.0	2.6	6.0	9.9	13.8	17.8

7. Assume you know that the vertex of a parabola is (5, 4).

a. If the parabola is stretched vertically by a factor of 2 in relation to the graph of $y = x^2$, what are the coordinates of the point 1 unit to the right of the vertex?

b. If the parabola is stretched horizontally by a factor of 3 in relation to the graph of $y = x^2$, what are the coordinates of the points 1 unit above the vertex?

c. If the parabola is stretched vertically by a factor of 2 and horizontally by a factor of 3, name two points that are symmetric with respect to the vertex.

8. Given the parent function $y = x^2$, describe the transformations represented by the function $\frac{y-2}{3} = \left(\frac{x+7}{4}\right)^2$. Sketch a graph of the transformed parabola.

9. A parabola has vertex (4.7, 5) and passes through the point (2.8, 9).

a. What is an equation of the axis of symmetry for this parabola?

b. What is an equation of this parabola?

c. Is this the only parabola passing through this vertex and point? Explain. Sketch a graph to support your answer.

10. Sketch a graph of each of these equations.

a. $\frac{y-2}{3} = (x-1)^2$ **b.** $\left(\frac{y+1}{2}\right)^2 = \frac{x-2}{3}$ **c.** $\frac{y-2}{2} = \left|\frac{x+1}{3}\right|$

11. Given the graph of $y = f(x)$, draw graphs of these related functions.

a. $\frac{y}{-2} = f(x)$

b. $y = f\left(\frac{x-3}{2}\right)$

c. $\frac{y+1}{1/2} = f(x+1)$

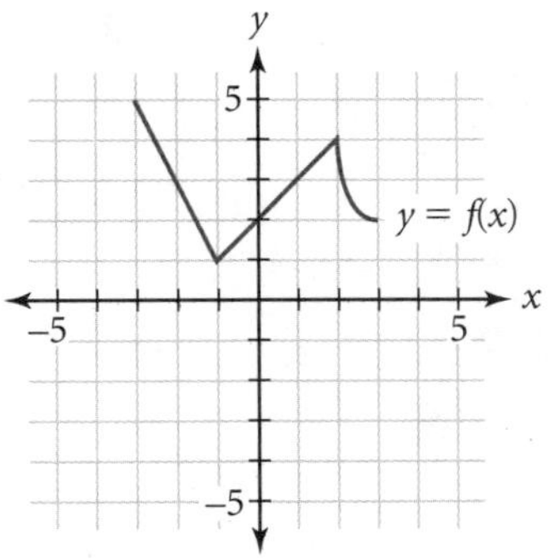

12. **APPLICATION** A chemistry class gathers these data on the conductivity of a base solution as acid is added to it.

Graph the data, and use transformations to find a model to fit the data.

Acid volume (mL) x	Conductivity (μS/cm) y
0	4152.95
1	3140.97
2	2100.34
3	1126.55
4	162.299
5	1212.47
6	2358.11
7	3417.83
8	4429.81

Review

13. (Lesson 1.5) A panel of art fair judges rates 20 exhibits as follows:

Exhibit number	1	2	3	4	5	6	7	8	9	10	11	12	13	14	15	16	17	18	19	20
Rating	79	81	94	92	68	79	71	83	89	92	85	88	86	83	89	90	92	77	84	73

The judges decide that the top rating should be 100, so they add 6 points to each rating.

a. What are the mean and the standard deviation of the ratings before the 6 points are added?

b. What are the mean and the standard deviation of the ratings after 6 points are added?

c. What do you notice about the change in the mean? In the standard deviation?

14. **APPLICATION** (Lesson 3.1) This table shows the percentage of U.S. households with computers in various years.

Year	1995	1996	1997	1998	1999	2000
Households (%)	31.7	35.5	39.2	42.6	48.2	53.0

(*2002 New York Times Almanac*, p. 798)

a. Plot these data on an x-y axis.

b. Is a linear model a good model for this situation? Explain your reasoning.

15. (Lesson 3.2) Sketch a graph of a function that has the following characteristics.

a. domain: $x \geq 0$
range: $f(x) \geq 0$
linear and increasing

b. domain: $-10 \leq x \leq 10$
range: $-3 < f(x) \leq 3$
nonlinear and increasing

c. domain: $x \geq 0$
range: $-2 < f(x) \leq 10$
increasing, then decreasing, then increasing, and then decreasing

LESSON

4.7 Transformations and the Circle Family

Many times the best way, in fact the only way, to learn is through mistakes. A fear of making mistakes can bring individuals to a standstill, to a dead center.

GEORGE BROWN

You have explored several functions and relations and transformed them in a plane. You know that a horizontal translation occurs when x is replaced with $(x - h)$ and that a vertical translation occurs when y is replaced with $(y - k)$. You have reflected graphs across the y-axis by replacing x with $(-x)$ and across the x-axis by replacing y with $(-y)$. You have also stretched and shrunk a function vertically by replacing y with $\frac{y}{a}$ and horizontally by replacing x with $\frac{x}{b}$.

In this lesson you will stretch and shrink a graph of a relation that is not a function and discover how to create the equation of a new shape.

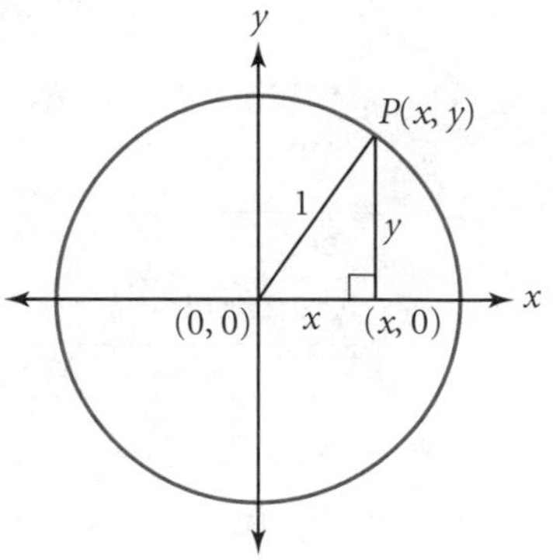

You will start by investigating the circle. A *unit circle* has a radius of 1 unit. Suppose P is any point on a unit circle with center at the origin. Draw the slope triangle between the origin and point P.

You can derive the equation of a circle from this diagram by using the Pythagorean Theorem. The legs of the right triangle have lengths x and y, and the length of the hypotenuse is 1 unit, so its equation is $x^2 + y^2 = 1$.

The Equation of a Unit Circle

$x^2 + y^2 = 1$ is the equation of a *unit circle* with center (0, 0).

What are the domain and the range of this circle? If a value, such as 0.5, is substituted for x, what are the output values of y? Is this the graph of a function? Why or why not?

In order to draw the graph of a circle on your calculator, you need to solve the equation $x^2 + y^2 = 1$ for y. When you do this, you get two equations, $y = +\sqrt{1 - x^2}$ and $y = -\sqrt{1 - x^2}$. Each of these is a function. You will have to graph both of them to get the complete circle.

You can transform a circle to get an ellipse. An *ellipse* is a stretched or shrunken circle.

EXAMPLE A

What is the equation of this ellipse?

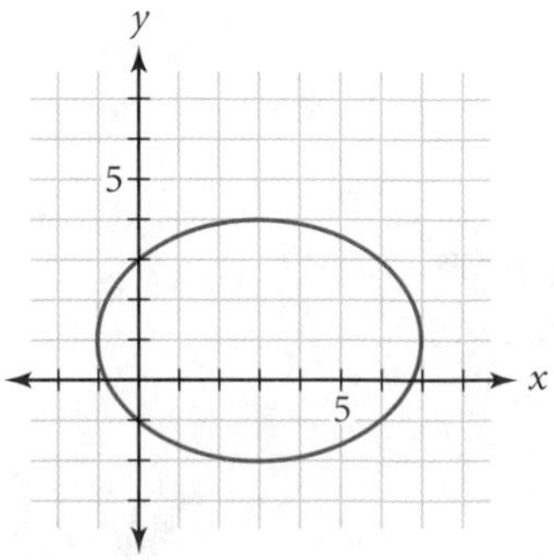

▶ Solution

The original unit circle has been translated and stretched both vertically and horizontally. It has been stretched horizontally by a factor of 4 and vertically by a factor of 3, and the new center is at (3, 1), so the equation changes like this:

Original unit circle.	$x^2 + y^2 = 1$
Stretch horizontally by a factor of 4 (replace x with $\frac{x}{4}$).	$\left(\frac{x}{4}\right)^2 + y^2 = 1$
Stretch vertically by a factor of 3 (replace y with $\frac{y}{3}$).	$\left(\frac{x}{4}\right)^2 + \left(\frac{y}{3}\right)^2 = 1$
Translate to new center at (3, 1).	$\left(\frac{x-3}{4}\right)^2 + \left(\frac{y-1}{3}\right)^2 = 1$

To enter this equation into your calculator to check your answer, you need to solve for y.

$$\left(\frac{y-1}{3}\right)^2 = 1 - \left(\frac{x-3}{4}\right)^2 \quad \text{Subtract } \left(\tfrac{x-3}{4}\right)^2 \text{ from both sides.}$$

$$\frac{y-1}{3} = \pm\sqrt{1 - \left(\frac{x-3}{4}\right)} \quad \text{Take the square root of both sides.}$$

$$y = 1 \pm 3\sqrt{1 - \left(\frac{x-3}{4}\right)^2} \quad \text{Multiply both sides by 3, then add 1.}$$

It will take two equations to graph this on your calculator. By graphing both of these equations, you can draw the complete ellipse and verify your answer.

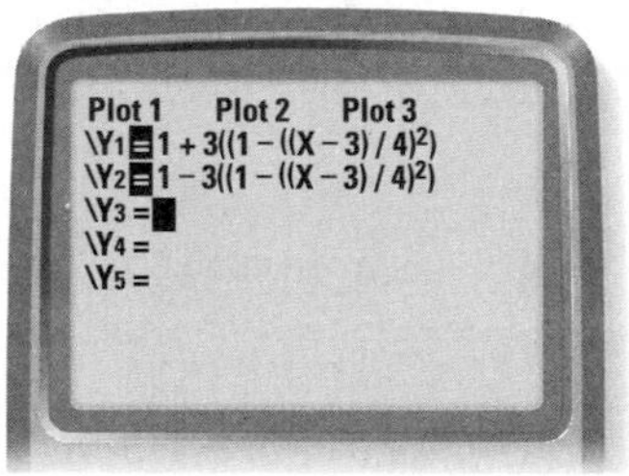

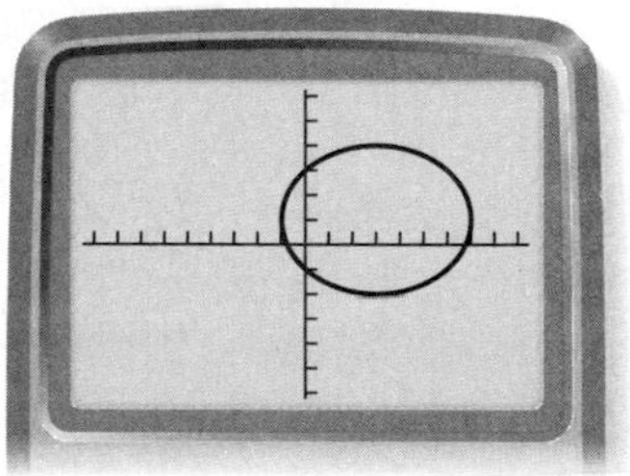

[−9.4, 9.4, 1, −6.2, 6.2, 1]

Investigation
Modeling an Ellipse

You will need

- the worksheet "Modeling an Ellipse," available at *www.keycollege.com/online.*

Choose one of the ellipses from the worksheet. Use your ruler carefully to place axes on the ellipse, and scale your axes in centimeters. Find the equation to model your ellipse. Graph your equation on your calculator, and verify that it creates an ellipse with the same dimensions as those on the worksheet.

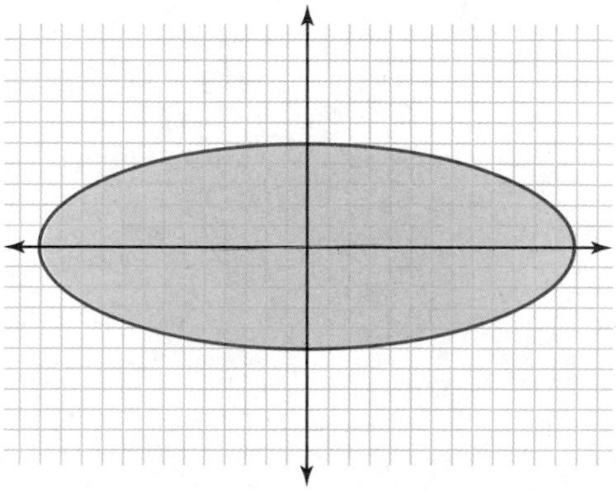

Equations for transformations of relations such as circles and ellipses are sometimes easier to work with in the general form before you solve them for y. You need to solve for y to enter the equations into your calculator. If you start with a function such as the top half of the unit circle, $f(x) = \sqrt{1 - x^2}$, you can transform it in the same way you transformed any other function, but it may be a little messier to deal with.

EXAMPLE B If $f(x) = \sqrt{1 - x^2}$, write $g(x) = 2f(3(x - 2)) + 1$ as a simple function in terms of x, and sketch a graph of this new function.

▶ Solution In $g(x) = 2f(3(x - 2)) + 1$, note that $f(x)$ is the parent function, x has been replaced with $3(x - 2)$, and $f(3(x - 2))$ is then multiplied by 2 and 1 is added. You can rewrite the function $g(x)$ as

$$g(x) = 2\sqrt{(1 - (3(x - 2))^2} + 1 \text{ or } g(x) = 2\sqrt{1 - \left(\frac{x - 2}{1/3}\right)^2} + 1.$$

This indicates that $f(x)$, a semicircle, has been horizontally shrunk by a factor of $\frac{1}{3}$, vertically stretched by a factor of 2, then shifted right 2 units and up 1 unit. The resulting transformed semicircle is graphed here. What are the coordinates of the right endpoint of the graph? Describe how the original semicircle's right endpoint of (1, 0) was mapped to this new location.

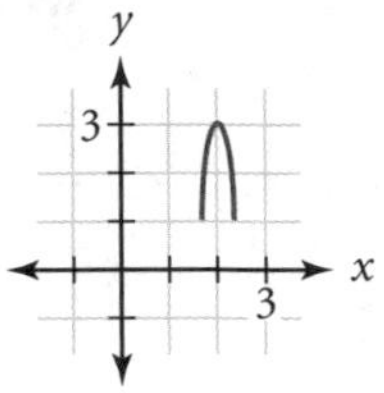

You have now learned to translate, reflect, stretch, and shrink functions and relations. These transformations are the same for all equations.

Transformations of Functions and Relations

Translations

The graph of $y = k + f(x - h)$ translates the graph of $y = f(x)$ h units horizontally and k units vertically.

or

Replacing x with $(x - h)$ translates the graph h units horizontally. Replacing y with $(y - k)$ translates the graph k units vertically.

Reflections

The graph of $y = f(-x)$ is a reflection of the graph of $y = f(x)$ across the y-axis.
The graph of $y = -f(x)$ is a reflection of the graph of $y = f(x)$ across the x-axis.

or

Replacing x with $-x$ reflects the graph across the y-axis. Replacing y with $-y$ reflects the graph across the x-axis.

Stretches and Shrinks

The graph of $y = af\left(\frac{x}{b}\right)$ is a stretch or shrink of the graph of $y = f(x)$ by a vertical scale factor of a and by a horizontal scale of b.

or

Replacing x with $\frac{x}{b}$ stretches or shrinks the graph by a horizontal factor of b.
Replacing y with $\frac{y}{a}$ stretches or shrinks the graph by a vertical factor of a.

EXERCISES

Practice Your Skills

Verify your work by graphing each equation on your calculator. Be sure to use a friendly graphing window.

1. Each equation represents a single transformation. Copy and complete this chart.

Equation	Transformation (translation, reflection, stretch, or shrink)	Orientation (horizontal, vertical)	Direction (up, down, left, right, across *x*-axis, across *y*-axis)	Amount
$y + 3 = x^2$	Translation	Vertical	Down	3
$-y = \|x\|$				
$y = \sqrt{\frac{x}{4}}$				
$\frac{y}{0.4} = x^2$				
$y = \|x - 2\|$				
$y = \sqrt{-x}$				

2. The equation $y = \sqrt{(1 - x^2)}$ is the equation of the top half of the unit circle with center at (0, 0) shown on the left. What is the equation of the top half of an ellipse shown on the right?

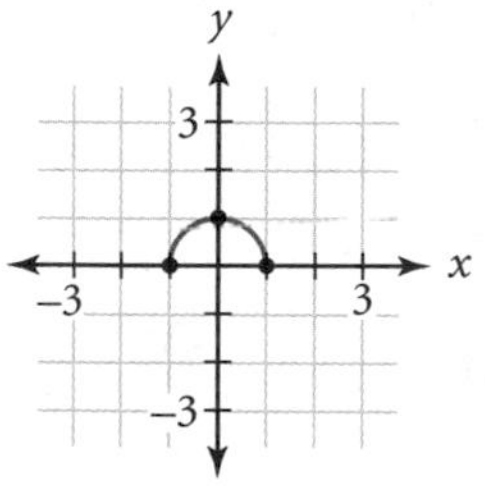

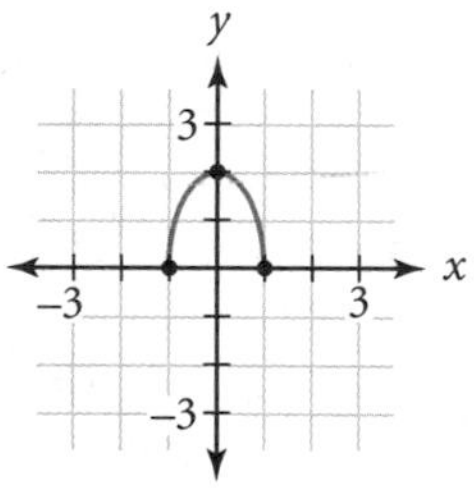

3. Use $f(x) = \sqrt{(1 - x^2)}$ to graph each of the transformations below.

a. $g(x) = -f(x)$

b. $h(x) = -2f(x)$

c. $j(x) = -3 + 2f(x)$

4. Write an equation in $y=$ form of each graph.

a.

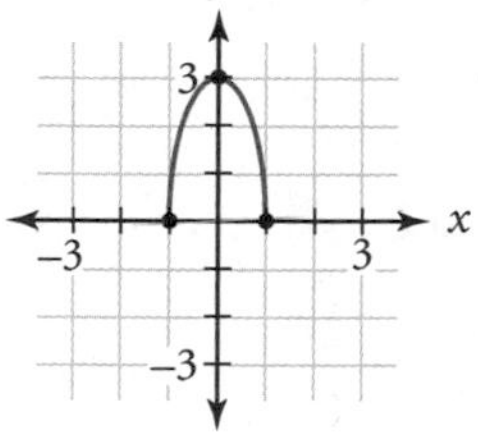

b.

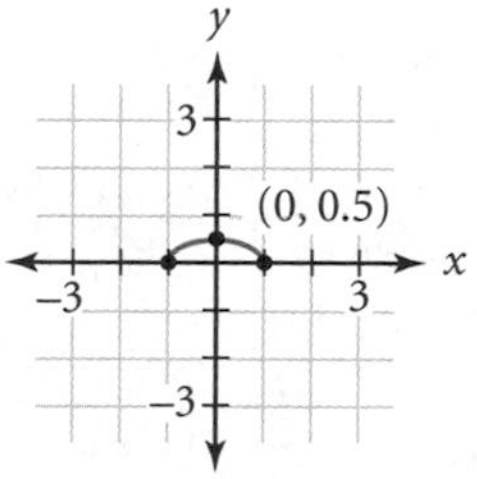

c.

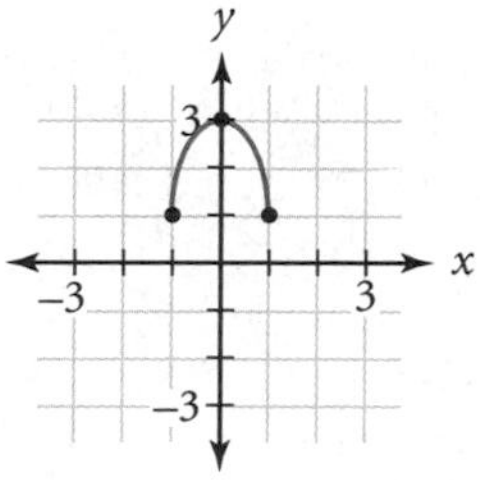

d. 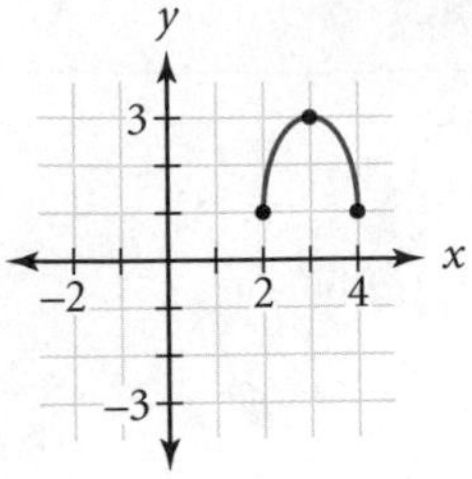

e. 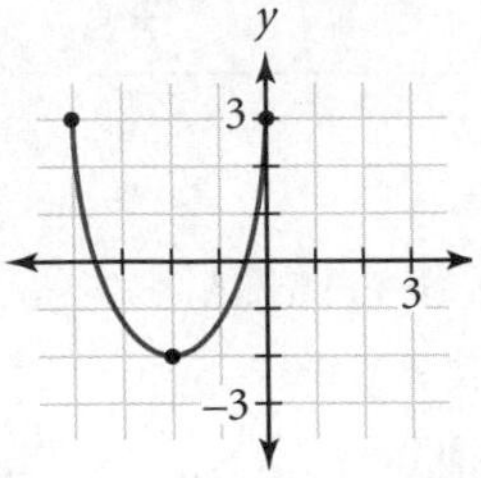

f. 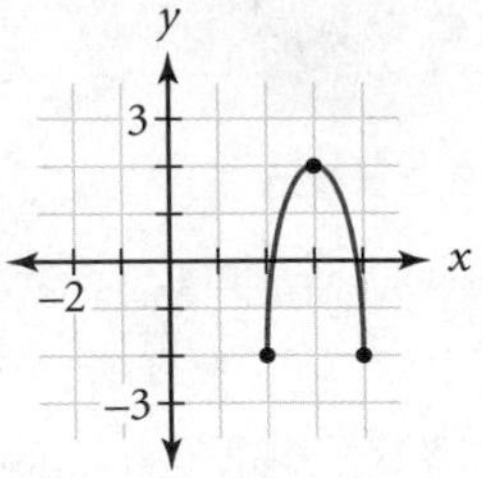

5. Write an equation and draw a graph of each transformation of the unit circle.

a. Replace y with $(y - 2)$.

b. Replace x with $(x + 3)$.

c. Replace y with $\frac{y}{2}$.

d. Replace x with $\frac{x}{2}$.

Reason and Apply

6. In the ellipse shown, the x-coordinate of each point on a unit circle has been multiplied by a factor of 3.

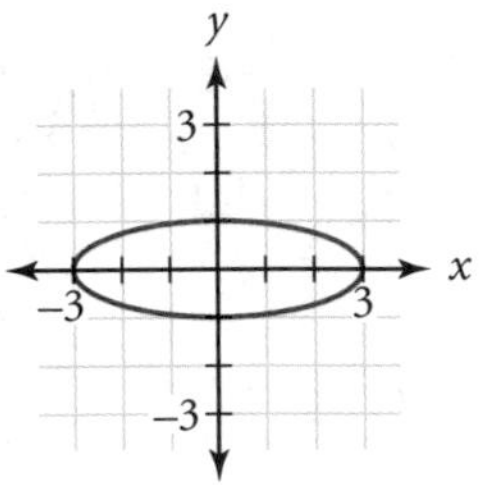

a. Write the equation of this ellipse.

b. What term did you substitute for x in the parent equation?

c. If $f(x)$ is the top half of a unit circle, then what is the equation of the top half of the ellipse, $g(x)$, in terms of $f(x)$?

ScienceConnection The first satellite, *Sputnik I*, was launched by Russia in 1957. Its name means "fellow traveler." Satellites have been used to aid in navigation, communication, research, and military reconnaissance, and today they are an indispensable part of our daily lives. They help us communicate by telephone and television, gather weather data, and provide global positioning for our GPS devices. The job the satellite is intended to do will determine the type of orbit into which it is placed.

A satellite in geosynchronous orbit (from *geo*, meaning "Earth" and *synchronous*, meaning "occurring at the same time") is located approximately 22,300 miles from Earth, above the equator. At that distance it takes the satellite a full 24 hours to circle the planet, and because it takes Earth 24 hours to make one rotation on its axis, the satellite and Earth move together. A satellite in geosynchronous orbit always stays directly over the same spot on Earth; its orbital path is circular. Because we always know where such satellites are, satellite dish antennae can be aimed in the right direction.

A satellite may also be placed in an elliptical orbit. A satellite in this orbit takes about 12 hours to circle the planet. Elliptical orbits move in a north-south (or *polar*) direction, as opposed to geosynchronous orbits, which move from east to west. Satellites in elliptical orbits cover areas of the Earth that are not covered by geosynchronous satellites.

Polar orbits are used for viewing the Earth's surface. As a satellite orbits in a north-south direction, Earth spins beneath it in an east-west direction. As a result, a satellite in polar orbit eventually can scan the entire surface of the Earth. For this reason, satellites that monitor the global environment are almost always in polar orbit.

7. For the circle shown at right, write an equation that generates each transformation.

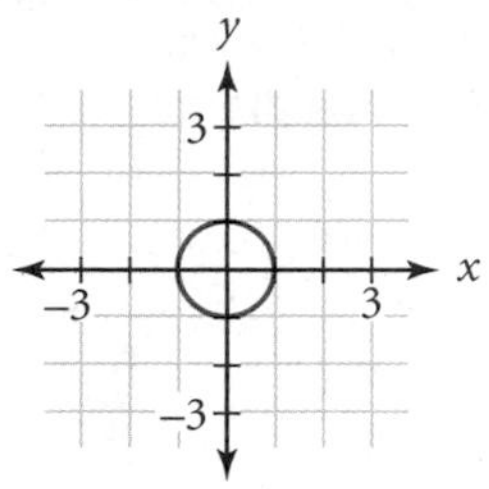

a. Each y-value is half the original y-value.

b. Each x-value is half the original x-value.

c. Each y-value is half the original y-value, and each x-value is twice the original x-value.

8. Consider the ellipse shown below.

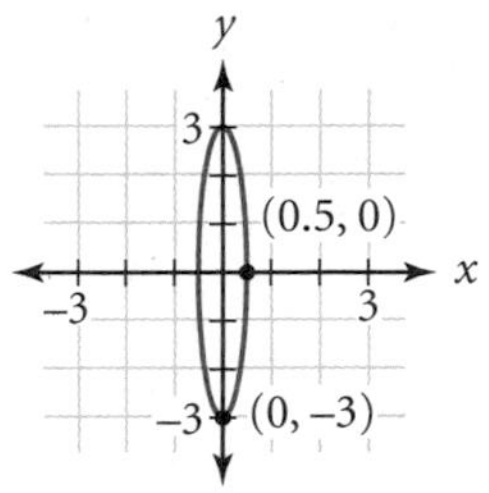

a. Write two functions that you could use to graph this ellipse.

b. Use the $\pm$ symbol to write one equation that could replace the two equations in 8a.

c. Write another equation for the ellipse by squaring both sides of the equation in 8b.

9. In the window [−0.5, 1.5, 0.5, −0.5, 1.5, 0.5], graph $y_1 = x$, $y_2 = x^2$, $y_3 = \sqrt{x}$, $y_4 = |x|$, and $y_5 = \sqrt{1 - x^2}$.

a. Four of the five functions intersect in the same two points. What are the coordinates of these points?

b. Use the intersection points to draw a box that just encloses the right half of the 5th function, a semicircle. How do the points you found in 9a relate to the dimensions of the box?

c. Use a small friendly window. Solve for y, and graph:

$$\frac{y}{2} = \frac{x}{4} \qquad \frac{y}{2} = \left(\frac{x}{4}\right)^2 \qquad \frac{y}{2} = \sqrt{\frac{x}{4}} \qquad \frac{y}{2} = \left|\frac{x}{4}\right| \qquad \frac{y}{2} = \sqrt{1 - \left(\frac{x}{4}\right)^2}$$

Where do the first four functions intersect?

d. Use the two points from 9c to draw a box that encloses just the right half of the transformed semicircle. How do the points relate to the dimensions of the box?

e. Use a larger friendly window. Solve for y, and graph

$$\frac{y-3}{2} = \frac{x-1}{4} \qquad \frac{y-3}{2} = \left(\frac{x-1}{4}\right)^2 \qquad \frac{y-3}{2} = \sqrt{\frac{x-1}{4}} \qquad \frac{y-3}{2} = \left|\frac{x-1}{4}\right|$$

$$\frac{y-3}{2} = \sqrt{1 - \left(\frac{x-1}{4}\right)^2}$$

Where do the first four functions intersect?

f. What are the dimensions of a box that encloses the right half of the semicircle? How do these dimensions relate to the two points in 9e?

g. In each set of functions, one of the points of intersection located the center of the transformed semicircle. How did the other point relate to the shape of the modified semicircle?

Review

10. (Lesson 4.6) Refer to Exercise 13 in Lesson 4.6. After the judges raised the rating scores by adding the same number to each score, one of the judges suggested that perhaps they should instead *multiply* the original scores by a factor that would make the highest score equal 100. They decide to try this method.

Exhibit number	1	2	3	4	5	6	7	8	9	10
Rating	68	71	73	77	79	79	81	83	83	84

Exhibit number	11	12	13	14	15	16	17	18	19	20
Rating	85	86	88	89	89	90	92	92	92	94

a. By what scale factor should they multiply the highest score, 94, to get 100?

b. Use this same multiplier to alter all of the scores, and record the altered scores in a table.

c. What are the mean and the standard deviation of the original scores? Of the altered scores?

d. Plot the original and altered scores on the same graph. Describe what happened to the scores visually. How does this explain what happened to the mean and the standard deviation?

e. Which method do you think the judges should use? Explain your reasoning.

11. (Lesson 2.2) Find the next three terms in this sequence: 16, 40, 100, 250, . . .

12. APPLICATION (Lesson 2.3) Throughout her career, Professor Blake has been making contributions to her individual retirement account. At the time of her retirement, she is 65 years old and has \$250,000 in her account. Suppose the money in her account is earning 4.5% interest, compounded monthly. She plans to withdraw \$2,000 per month from the account to support herself during retirement. She will make the first withdrawal exactly one month after she retires.

a. Write a recursive formula for this problem, showing how her account balance changes each month.

b. List the first five terms of this sequence of balances, starting with her retirement.

c. What is the meaning of the value of u_4?

d. What is the balance one year (12 months) after she retires? What is the balance ten years after she retires?

13. APPLICATION (Lesson 2.3) Consider the retirement account in Exercise 12.

a. What happens to the balance if the same interest and withdrawal patterns continue for a long time? Does the balance ever level off?

b. What monthly withdrawal amount would maintain a constant balance of \$250,000 in the long run?

14. (Lesson 4.5) Consider the line $f(x) = 3x + 1$.

a. Write the equation, in $y=$ form, of the image of $y = 3x + 1$ after a reflection across the x-axis. Graph both lines on the same axes.

b. Write the equation, in $y=$ form, of the image of $y = 3x + 1$ after a reflection across the y-axis. Graph both lines on the same axes.

c. Write the equation, in $y=$ form, of the image of $y = 3x + 1$ after a reflection across the x-axis and then across the y-axis. Graph both lines on the same axes.

d. How does the image line in 14c relate to the original line?

15. (Lesson 1.4) This table shows passenger activity in the world's 30 busiest airports in 2000.

a. Estimate the total number of passengers who used the 30 airports. Explain any assumptions you make.

b. Estimate the mean usage among the 30 airports in 2000.

c. Estimate the five-number summary values. Explain any assumptions you make.

d. Display the data in a box plot. [► See **Calculator Note 1K** ◄]

Number of passengers (in millions)	Number of airports
$25 \le p < 30$	5
$30 \le p < 35$	8
$35 \le p < 40$	8
$40 \le p < 45$	1
$45 \le p < 50$	2
$50 \le p < 55$	1
$55 \le p < 60$	2
$65 \le p < 70$	1
$70 \le p < 75$	1
$80 \le p < 85$	1

(*2002 New York Times Almanac*, p. 413)

Compositions of Functions

Sometimes you'll need two or more functions to answer a question or analyze a problem. Suppose an offshore oil well is leaking. Graph A shows the radius r of the spreading oil slick growing as a function of time t, so $r = f(t)$. Graph B shows the area a of the circular oil slick as a function of its radius r, so $a = g(r)$. Time is measured in hours, the radius is measured in kilometers, and the area is measured in square kilometers.

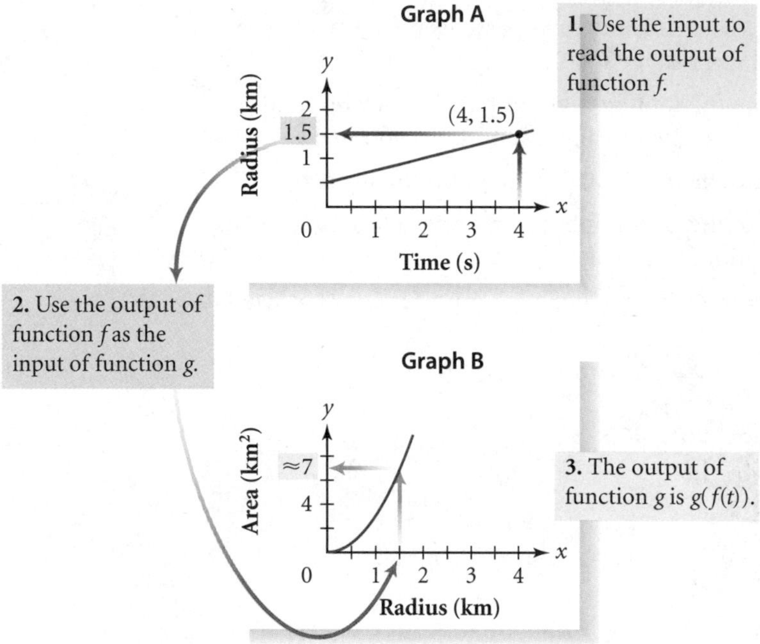

If you want to find the area of the oil slick after 4 hours, use function f on Graph A to find that when t equals 4 h, r equals 1.5 km. Next, using function g on Graph B, you find that when r equals 1.5 km, a is approximately 7 km^2.

Input t (hours)	Output $f(t)$ (km)
0	0.50
1	0.75
2	1.00
3	1.25
4	1.50

Input $f(t)$ (km)	Output $g(f(t))$ (km^2)
0.50	1
0.75	2
1.00	3
1.25	5
1.50	7

You used the graphs of two different functions, f and g, to find the solution that after 4 h, the oil slick has area 7 km^2. You actually used the output from one function, f, as the input in the other function, g. This is an example of the **composition** of two functions to form a new functional relationship between area and time, that is, $a = g(f(t))$. The symbol $g(f(t))$, read "g of f of t," is a composition of the two functions f and g.

The composition $g(f(t))$ gives the final outcome when an x-value is substituted in the "inner" function, f, and its output value, $f(t)$, is then substituted as input into the "outer" function, g.

EXAMPLE A

Consider these functions:

$$f(x) = \frac{3x}{4} - 3 \quad \text{and} \quad g(x) = |x|$$

What will $g(f(x))$ look like?

▶ Solution

Function f is the inner function, and function g is the outer function. Use equations and tables to identify the output of f, and use it as the input of g.

Find several $f(x)$ output values.

x	f(x)
−2	−4.5
0	−3
2	−1.5
4	0
6	1.5
8	3

Use the $f(x)$ output values as the input of $g(x)$.

f(x)	g(f(x))
−4.5	4.5
−3	3
−1.5	1.5
0	0
1.5	1.5
3	3

Match the input of the inner function, f, with the output of the outer function, g, and plot the graph.

x	g(f(x))
−2	4.5
0	3
2	1.5
4	0
6	1.5
8	3

The solution is the composition graph at far right. All the function values of f, whether positive or negative, give positive output values under the rule of g, the absolute value function. So, the part of the graph of function f showing negative output values is reflected across the x-axis in this composition.

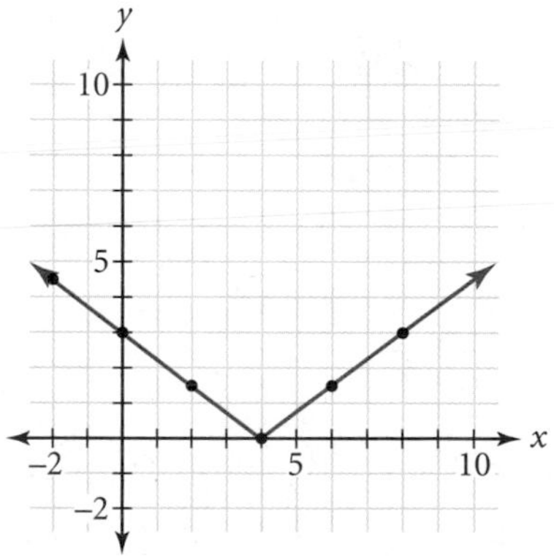

You can use what you know about transformations to get the specific equation for $y = g(f(x))$. Use the parent function $y = |x|$, apply a horizontal stretch of 4/3, and translate the vertex right 3 units. You get the graph of $y = \left|\frac{3x}{4} - 3\right|$. You have actually been composing functions as you transformed graphs using two or more steps.

Investigation
Looking Up

You will need

- a small mirror
- one or more tape measures or metersticks

Part 1

Procedure Note

1. Place the mirror flat on the floor about 0.5 m (20 in.) from a wall.
2. Use tape to attach tape measures or metersticks up the wall to a height of 1.5 to 2 m.

First, you'll establish a relationship between your distance from the mirror and what you can see in it.

Step 1 Stand a short distance from the mirror, and look down into it. Move slightly left or right until you can see the tape measure on the wall reflected in the mirror.

Step 2 Have a group member slide his or her finger up the wall to help locate a height mark, h, that is reflected in the mirror. Record h as elsewhere and the distance, d, from your toe to the center of the mirror.

Step 3 Change your distance from the mirror and repeat Step 2. Collect several pairs of data in the form (d, h). Make sure that some distances from the mirror are short and some are long.

Step 4 Use transformations to fit the function $h(d) = \frac{1}{d}$ to your data so that it gives the approximate height, h for any distance, d.

Part 2

Now you'll combine your work from Part 1 with the scenario of a timed walk toward and away from the mirror.

Step 5 Suppose this table gives you your position at 1-s intervals:

Time (s)	0	1	2	3	4	5	6	7
Distance to mirror (cm)	163	112	74	47	33	31	40	62

Use one of the families of functions from this chapter to fit these data. Call this function $d(t)$. It should give the distance from the mirror for seconds 0 to 7.

Step 6 Use your two functions to answer these questions.

a. How high up the wall can you see when you are 47 cm from the mirror?

b. Where are you at 1.3 s?

c. How high up the wall can you see at 3.4 s?

Step 7 Change each expression into words relating to the context of this Investigation, and find an answer. Show the steps you needed to evaluate each expression.

a. $h(60)$ **b.** $d(5.1)$ **c.** $h(d(2.8))$

Step 8 Find a single function, $H(t)$, that does the work of $h(d(t))$. Show that $H(2.8)$ gives the same answer as Step 7c above.

Don't confuse a composition of functions with multiplying two functions. Composing functions requires you to replace the independent variable in one function with the output value of the other function. Therefore, $f(g(x)) \neq g(f(x))$, except for certain functions. Except in special cases, function composition is generally not commutative. In other words, the order in which you do the compositions makes a difference. You can compose a function with itself.

EXAMPLE B

Suppose $A(x) = \left(1 + \frac{0.07}{12}\right)x - 250$ gives the balance of a loan at 7% in the month after a \$250 payment. In the equation, x is the current balance, and $A(x)$ is the next balance. Translate these expressions into words, and find their values.

a. $A(15000)$

b. $A(A(20000))$

c. $A(A(A(18000)))$

d. $A(A(x))$

▸ Solution

Each expression builds from the inside out.

a. $A(15000)$ asks, "What is the loan balance after one monthly payment if the starting balance is \$15,000?" Substituting 15000 for x in the given equation, you get $A(15000) = \left(1 + \frac{0.07}{12}\right)15000 - 250 = 15087.50 - 250 = 14837.50$, or \$14,837.50.

b. $A(A(20000))$ asks, "What is the loan balance after two monthly payments if the starting balance is \$20,000?" Substitute 20000 for x in the given equation. You get 19866.67 and use it as input in the given equation. That is, $A(A(20000)) = A(19866.67) = 19732.56$, or \$19,732.56.

c. $A(A(A(18000)))$ asks, "What is the loan balance after three monthly payments if the starting balance is \$18,000?" Working from the inner expression outward, you get $A(A(A(18000)) = A(A(17855)) = A(17709.15) = 17562.46$, or \$17,562.46.

d. $A(A(x))$ asks, "What is the loan balance after two monthly payments if the starting balance is x?"

$A(A(x)) = A\left(A\left(1 + \frac{0.07}{12}\right)x - 250\right)$	Use the given equation to substitute $\left(1 + \frac{0.07}{12}\right)x - 250$ for $A(x)$.
$= \left(1 + \frac{0.07}{12}\right)\left[\left(1 + \frac{0.07}{12}\right)x - 250\right] - 250$	Use the output, $\left[\left(1 + \frac{0.07}{12}\right)x - 250\right]$ in place of the input, x.
$= 1.005833(1.005833x - 250) - 250$	Simplify the fractions.
$= 1.0117x - 251.458 - 250$	Apply the distributive law.
$= 1.0117x - 501.458$	Simplify.

PROJECT

BOOLEAN GRAPHS

In his book *An Investigation into the Laws of Thought* (1854), the English mathematician George Boole (1815–1864) approached logic in a system that reduced it to simple algebra. In his system, later called *Boolean algebra* or *symbolic algebra,* expressions are combined using *and* (multiplication), *or* (addition), and *not* (negative) and then interpreted as true (1) or false (0). Today, Boolean algebra plays a fundamental role in the design, construction, and programming of computers. You can learn more about Boolean algebra with the Internet links at www.keycollege.com/online.

An example of a Boolean expression is $x \leq 5$. In this case, if x is 10, the expression is false and is assigned a value of 0. If x is 3, then the expression is true, and it is assigned a value of 1. You can use Boolean expressions to limit the domain of a function when graphing on your calculator. For example, the graph of $Y1 = (x + 4)/(x \leq 5)$ does not exist for values of x greater than 5, because your calculator would be dividing by 0. [► See **Calculator Note 4F** to learn more about graphing functions with Boolean expressions. ◄]

You can use your calculator to draw this car by entering the following short program.

```
PROGRAM:CAR
ClrDraw
DrawF 1/(X≥1)(X≤9)
DrawF (1.2√(X-1)+1)/(X≤3.5)
DrawF (1.2√(-(X-9))+1)/(X≥6.5)
DrawF (-0.5(X-5)-+4)/(X≥3.5)(X≤6.5)
DrawF -√(1-(X-2.5)⁻)+1
DrawF -√(1-(X-7.5)⁻)+1
DrawF (abs(X-5.5)+2)/(X≥5.2)(X≤5.8)
```

Write your own program that uses functions, transformations, and Boolean expressions to draw your initial. Your project should include

- a screen capture or sketch of your drawing
- the functions you used to create your drawing

EXERCISES

Practice Your Skills

1. Graph A shows a swimmer's speed as a function of time. Graph B shows the swimmer's oxygen consumption as a function of her speed. Time is measured in seconds, speed in meters per second (m/s) and oxygen consumption in liters per minute (L/min). Use the graphs to estimate the values.

Graph A

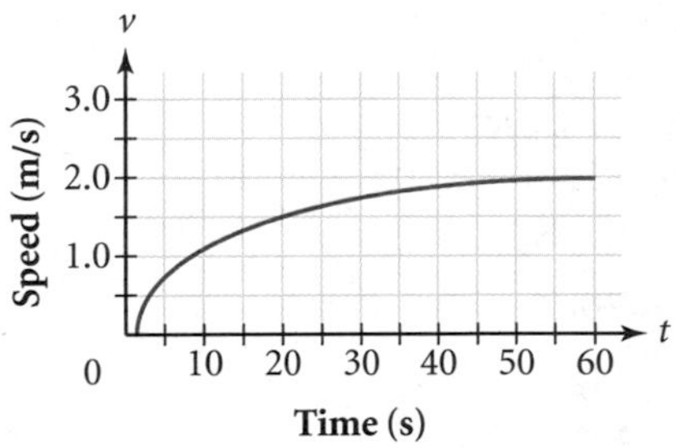

Graph B

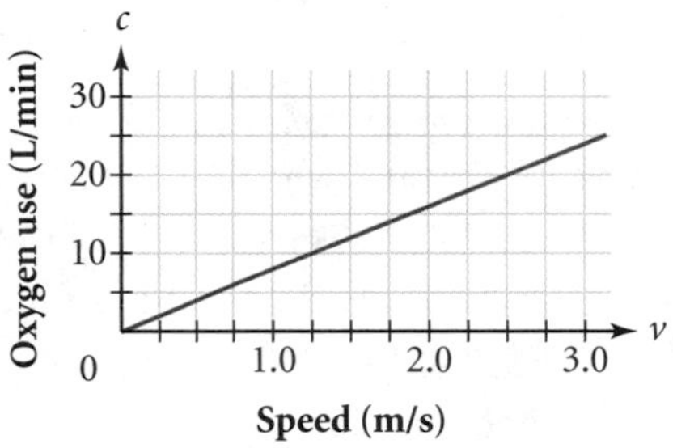

a. The swimmer's speed after 20 s of swimming

b. The swimmer's oxygen consumption at a swimming speed of 1.5 m/s

c. The swimmer's oxygen consumption after 40 s of swimming

2. Given that $f(x) = 3 + \sqrt{x + 5}$ and $g(x) = 2 + (x - 1)^2$, evaluate these expressions:

a. $f(4)$

b. $f(g(4))$

c. $g(-1)$

d. $g(f(-1))$

3. Suppose $g = \{(1, 2), (-2, 4), (5, 5), (6, -2)\}$ and $f = \{(0, -2), (4, 1), (3, 5), (5, 0)\}$.

a. Find $g(f(4))$.

b. Find $f(g(-2))$.

c. Find $f(g(f(3)))$.

Reason and Apply

4. Consider the graph at right.

a. Write an equation of this graph in $y=$ form.

b. Write two functions, f and g, such that the figure is the graph of $f(g(x))$.

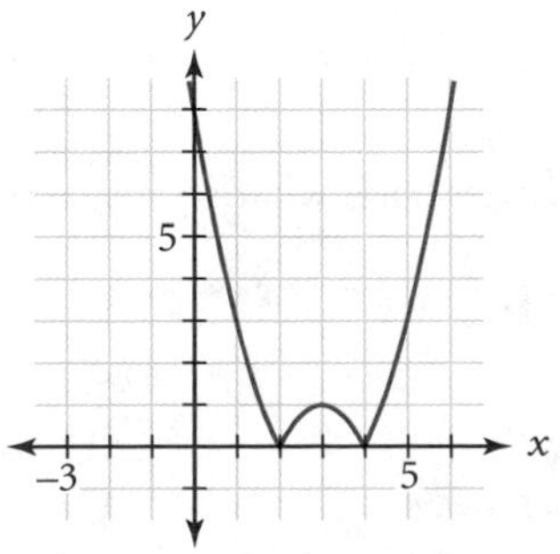

5. Suppose $g = \{(1, 2), (-2, 4), (5, 5), (6, -2)\}$ and $f = \{(2, 1), (4, -2), (5, 5), (-2, 6)\}$.

a. Find $g(f(2))$.

b. Find $f(g(6))$.

c. Use the numbers in the domains of g and f to complete the table below. Describe what is happening.

x	$g(x)$	$f(x)$	$f(g(x))$	$g(f(x))$
−2				
1				
2				
4				
5				
6				

6. **APPLICATION** A, B, and C are gauges with different linear measurement scales. When A measures 12, B measures 13, and when A measures 36, B measures 29. When B measures 20, C measures 57, and when B measures 32, C measures 84.

a. Sketch separate graphs for readings of B as a function of A and for readings of C as a function of B. Label the axes.

b. If A reads 12, what does C read?

c. Write a function with B as the dependent variable and A as the independent variable.

d. Write a function with C as the dependent variable and B as the independent variable.

e. Write a function with C readings depending on A readings.

7. A graph of the function $g(x)$ is shown at right. Draw a graph of each related function, $h(x)$, given below.

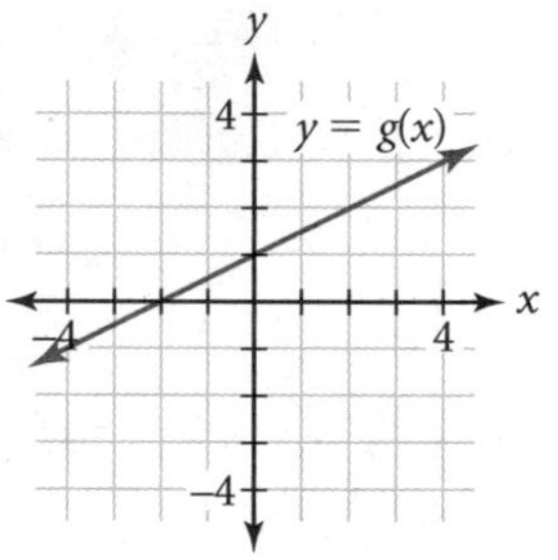

a. $h(x) = \sqrt{g(x)}$

b. $h(x) = |g(x)|$

c. $(g(x))^2$

8. The two lines pictured here are $f(x) = 2x - 1$ and $g(x) = \frac{1}{2}x + \frac{1}{2}$. Begin by making an accurate copy of the graph. Solve each problem both graphically and numerically.

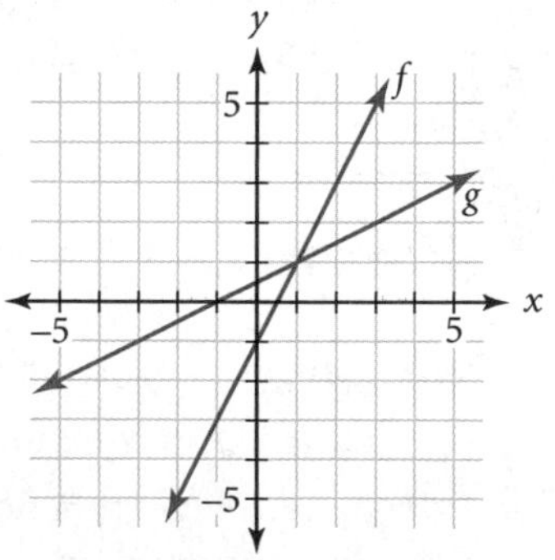

a. Find $g(f(2))$.

b. Find $f(g(-1))$.

c. Create a table showing five x-values in the domain of f, and find $g(f(x))$.

d. Create a table showing five x-values in the domain of g, and find $f(g(x))$.

e. Carefully describe what seems to be happening in these compositions.

9. Suppose $f(x) = -x^2 + 2x + 3$ and $g(x) = (x - 2)^2$. Find each value below.

a. $f(g(3))$ **b.** $f(g(2))$ **c.** $g(f(0.5))$ **d.** $g(f(1))$

e. $f(g(x))$. Simplify to remove all parentheses. [► See **Calculator Note 4A** to use your calculator to check the answer. ◄]

f. $g(f(x))$. Simplify to remove all parentheses.

10. Aaron and Davis need to write the equation that will produce this graph:

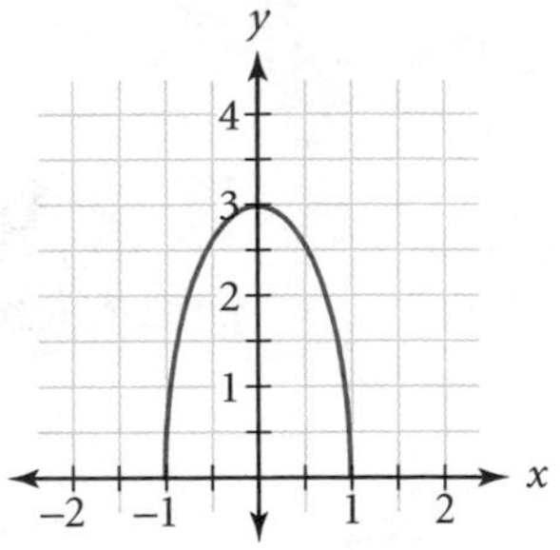

Aaron: "This is impossible! How are we supposed to know if the parent function is a parabola or a semicircle? If we don't know the parent function, there is no way to write the equation."

Davis: "Don't panic yet. I am sure we can determine its parent function if we study the graph carefully."

Do you agree with Davis? Explain completely and, if possible, write the equation of the graph.

11. **APPLICATION** Jen and Priya are mall employees. They decide to go out to the Hamburger Shack for lunch. They each have a 50-cent coupon from the Sunday newspaper for the Super-Duper-Deluxe \$5.49 Value Meal. In addition, if they show their employee ID cards, they'll get a 10% discount. Jen's server rang up the order as Value Meal, coupon, and then ID discount. Priya's server rang it up as Value Meal, ID discount, and then coupon.

a. How much did each woman pay?

b. Write a function, $C(x)$, that will deduct 50 cents from a price, x.

c. Write a function, $D(x)$, that will take 10% off a price, x.

d. Find $C(D(x))$.

e. Which server used $C(D(x))$ to calculate the price of the meal?

f. Is there a price for the Value Meal that would result in both women paying the same price? If so, what is it?

Review

12. (Lesson 3.1) Solve these equations for x.

a. $\sqrt{x-4} = 3$

b. $(3 - \sqrt{x+2})^2 = 4$

c. $3 - \sqrt{x} = 5$

d. $3 + 5\sqrt{1 + 2x^2} = 13$

13. (Lesson 4.6) $x^2 + y^2 = 1$ is the equation of a unit circle.

a. Apply a horizontal stretch by a factor of 3 and a vertical stretch by a factor of 3, and write the equation that results.

b. Sketch the graph. Label the intercepts.

14. (Lesson 3.5) If $4x - 2y = 24$ and $2x + 3y = 12$, what is $6x$?

15. (Lesson 4.4) Suppose $f(x) = x^2$ is the parent equation of a parabola. Translate f 3 units to the right and 5 units up, and call the image f'.

a. Give the equation for $f'(x)$.

b. What is the vertex of the graph of $f'(x)$?

c. Give the coordinates of the image point with an x-coordinate that is 2 units to the right of the vertex.

E X P L O R A T I O N
with Microsoft® Excel

Rectilinear Motion

Download the Chapter 4 Excel Exploration from the website at www.keycollege.com/online or from your Student CD.

In this Exploration you will investigate tossing an object straight up into the air from a platform. You will be able to fix the height of the platform and vary the initial velocity. You will also be able to vary the height of the platform while keeping the initial velocity fixed.

On the spreadsheet you will find complete instructions and questions to consider as you explore.

Activity
Translations of the Parabola

Excel skills required

- Data fill
- Name a cell
- Display/hide formulas

In this Activity you will build a spreadsheet that graphs a parabola using the vertex form of the parabola. When you change the *x*- and *y*-values of the vertex independently, your graph will respond to the changes.

You can find instructions for the required Excel skills in the Excel Notes at www.keycollege.com/online.

Launch Excel.

	A	B	C	D
1		a	h	k
2				
3				
4	x	y		

Step 1 Label the cells as shown.

Step 2 Name Cells B2, C2, and D2 **a, h,** and **k,** respectively.

Step 3 In Cells B2, C2, and D2 enter the numbers `1`, `0`, and `0`, respectively.

Step 4 In Cell A5 enter the number `−5`.

Step 5 In Cell B5 enter `=a*(A5-h)^2+k`

Step 6 In Cell A6 type `=A5+0.1`

Step 7 Select Cell B5, and data-fill down to Cell B6.

(Steps 6 and 7 set up the row that will be used to generate the final data fill.)

Step 8 Highlight Cells A6 and B6, and data-fill down to Row 105.

Questions

1. Inspect the table. Using CTRL+` (the accent can be found to the left of the 1 key), display formulas and study Column B. Note that this represents $y = a(x - h)^2 + k$. Using CTRL+`, toggle back to normal view. Viewing the column, you will see that when Excel performed the data fill, the parameters a, h, and k did not change, but the cell reference to the left did change. The parameters a, h, and k remain fixed to allow you to explore the impact of any one of them.

   a. With $a = 1$, $h = 0$, and $k = 0$, which equation is represented?

   b. With $a = 1$, $h = 1$, and $k = 0$, which equation is represented?

   c. With $a = 1$, $h = 0$, and $k = 1$, which equation is represented?

   d. With $a = 1$, $h = 1$, and $k = 1$, which equation is represented?

2. Highlight the block of cells from A5 to B105. Using Insert/Chart, you will now insert a chart. Using the Standard Types tab, select XY (Scatter). Under Chart sub-type, select the scatter plot showing data points connected by smoothed lines.

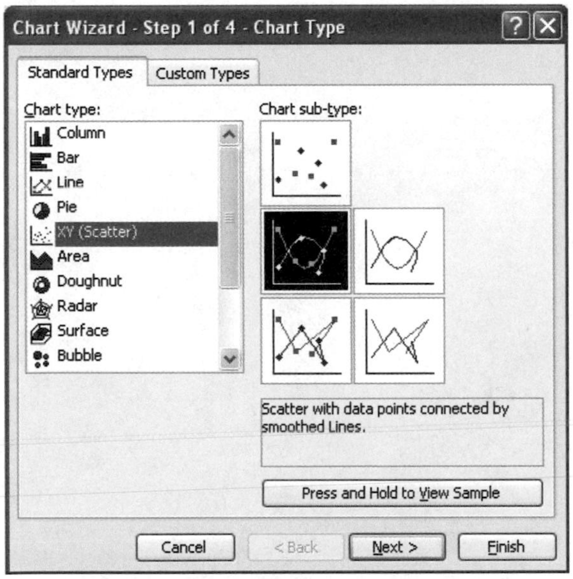

Click the **Finish** button.

Click on the series box 

, and press the Delete button.

Double-click any point on the graph. Select the Patterns tab. On the right of the resulting box, set the marker at None. What are the coordinates of the lowest point on the graph?

3. Now, make the following changes in your table, and answer the associated question.

a. Change the values in Cells C2 and D2 to 1 and 0, respectively. What are the coordinates of the lowest point on the resulting graph?

b. Change the values in Cells C2 and D2 to 0 and 1, respectively. What are the coordinates of the lowest point on the resulting graph?

c. Change the values in Cells C2 and D2 to 1 and 1, respectively. What are the coordinates of the lowest point on the resulting graph?

d. Change the values in Cells C2 and D2 to 2 and 2, respectively. What are the coordinates of the lowest point on the resulting graph?

e. Change the value in Cell B2 to -1. What happens? Compare this to your original. What does it have in common with the previous graph?

CHAPTER 4 REVIEW

This chapter introduced the concept of a **function** and reviewed **function notation.** You saw real-world situations represented by rules, sets, functions, graphs, and, most important, equations.

You learned to distinguish between functions and other **relations** using either the definition of a function—at most one y-value per x-value—or the **vertical line test.** You also classified functions according to their **function families.**

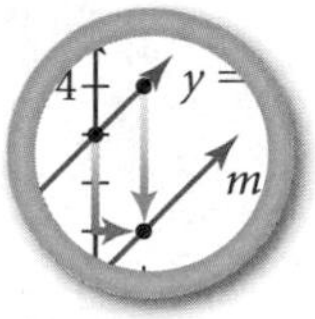

This chapter also introduced several **transformations,** including **translations, reflections,** and vertical and horizontal **stretches** and **shrinks.** You learned how to transform the graphs of **parent functions** to create several families of functions—linear, quadratic, square root, absolute value, and semicircle. For example, if you stretch the graph of the parent function $y = x^2$ by a factor of 3 vertically and by a factor of 2 horizontally, and translate it right 1 unit and up 4 units, then you get the graph of the function $y = 3\left(\frac{x-1}{3}\right)^2 + 4$.

Finally, you looked at the **composition** of functions. Many times, solving a problem involves two or more related functions. You can find the value of a composition of functions by using algebraic or numeric methods or by graphing.

EXERCISES

1. Sketch a graph that shows the relationship between the time in seconds after you start microwaving a bag of popcorn and the number of pops per second. Describe in words what your graph shows.

2. Use these three functions to find each value:

   $f(x) = -2x + 7$

   $g(x) = x^2 - 2$

   $h(x) = (x + 1)^2$

   a. $f(4)$
   b. $g(-3)$
   c. $h(x + 2) - 3$
   d. $f(g(3))$
   e. $g(h(-2))$
   f. $h(f(-1))$
   g. $f(g(a))$
   h. $g(f(a))$
   i. $h(f(a))$

3. The graph of $y = f(x)$ is shown here. Sketch the graph of each of these functions:

a. $y = f(x) - 3$ **b.** $y = f(x - 3)$

c. $y = 3f(x)$ **d.** $y = f(-x)$

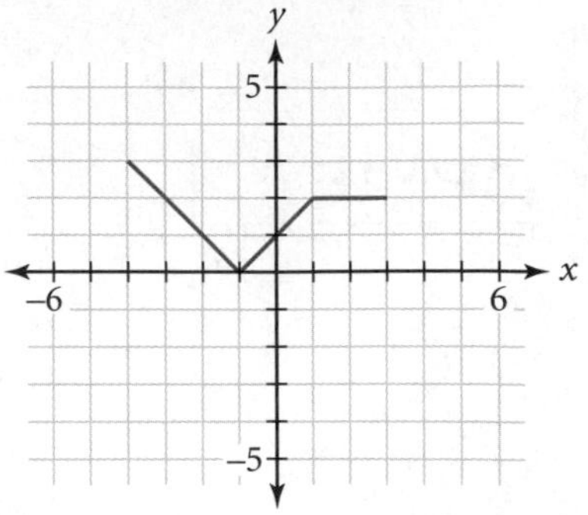

4. Assume you know the graph of $y = f(x)$. Describe the transformations, in order, that would give you the graphs of these functions:

a. $y = f(x + 2) - 3$

b. $\dfrac{y - 1}{-1} = f\left(\dfrac{x}{2}\right) + 1$

c. $y = 2f\left(\dfrac{x - 1}{0.5}\right) + 3$

5. The graph of $y = f(x)$ is shown here. Use what you know about transformations to sketch these related functions:

a. $y - 1 = f(x - 2)$ **b.** $\dfrac{y + 3}{2} = f(x + 1)$

c. $y = f(-x) + 1$ **d.** $y + 2 = f\left(\dfrac{x}{2}\right)$

e. $y = -f(x - 3) + 1$ **f.** $\dfrac{y - 2}{-2} = f\left(\dfrac{x - 1}{1.5}\right)$

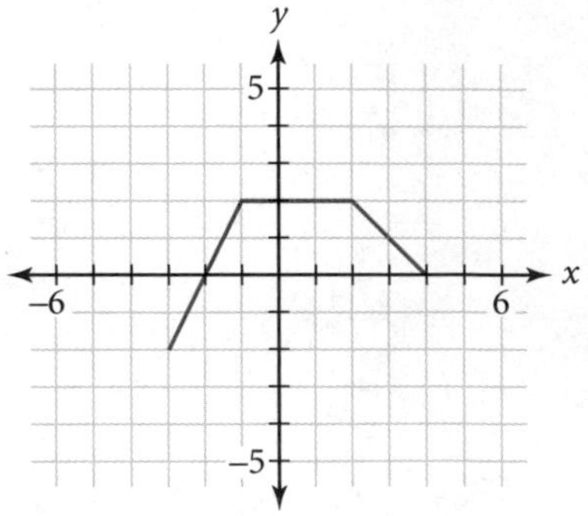

6. For each graph, name the parent function, and write an equation of the graph.

a.

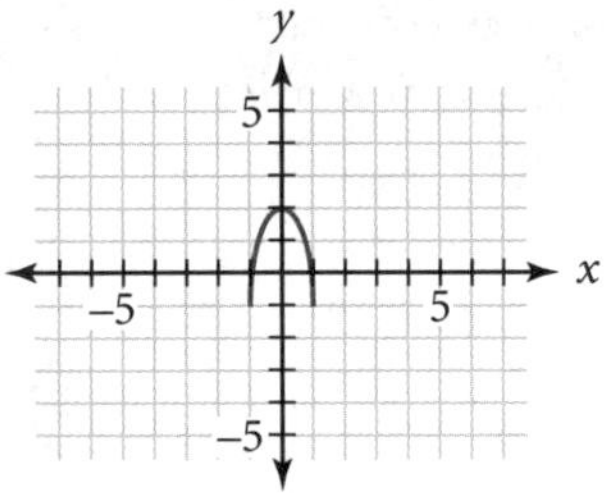

b.

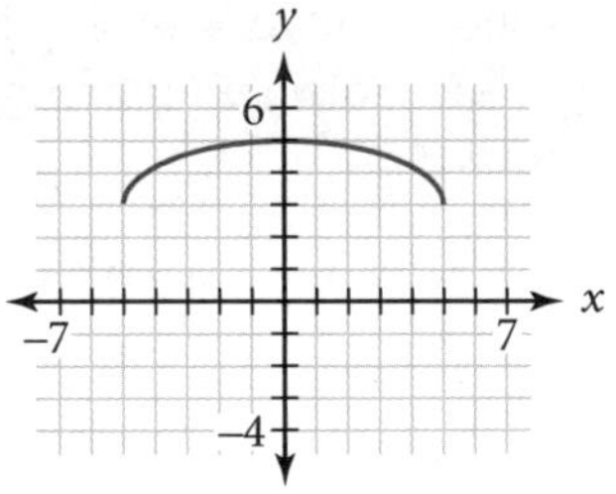

c.

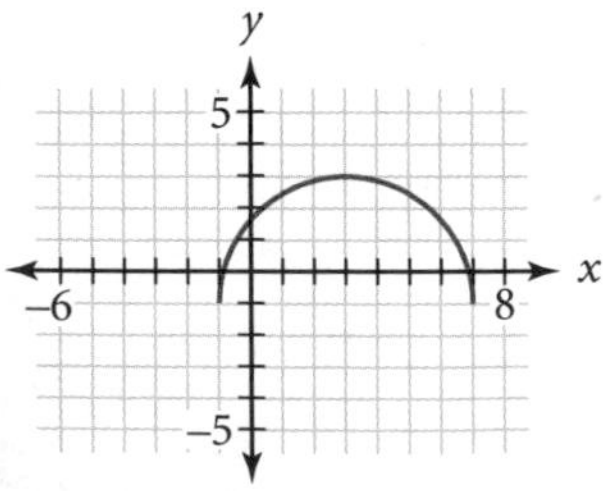

d.

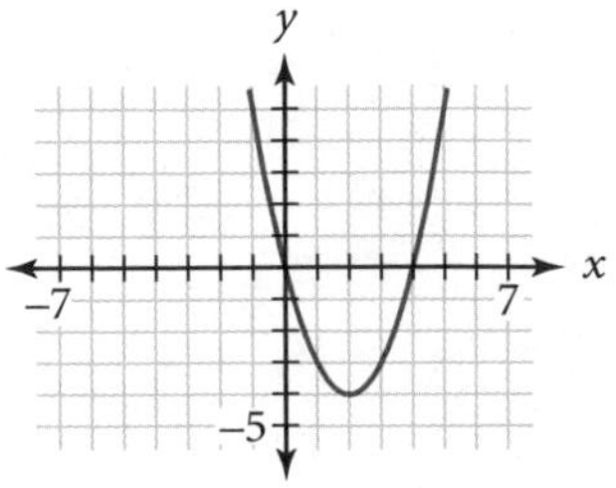

e.

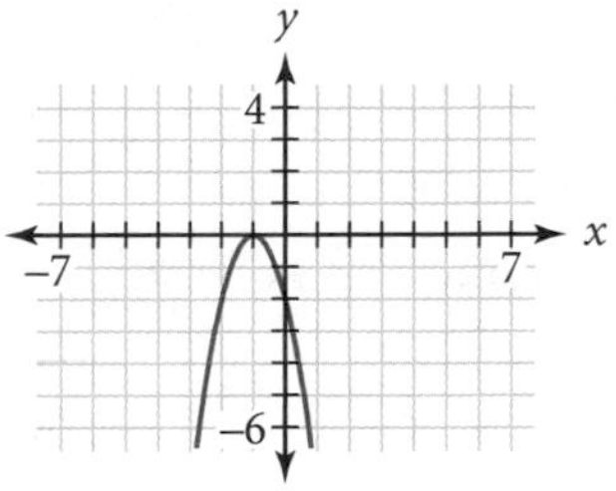

f.

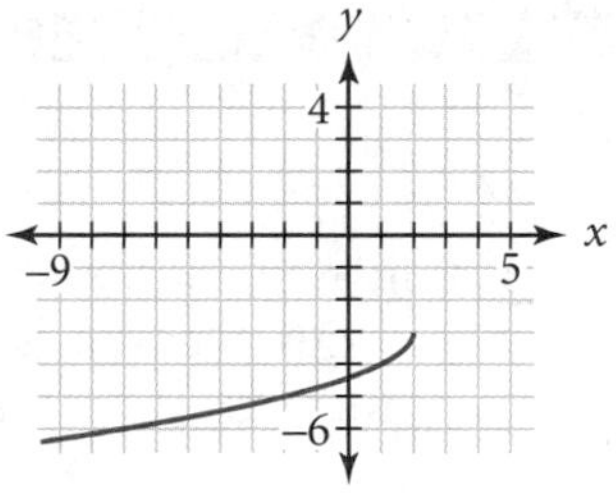

g.

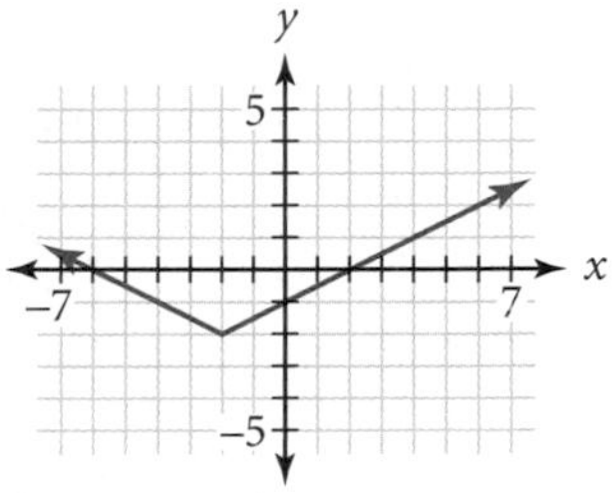

h.

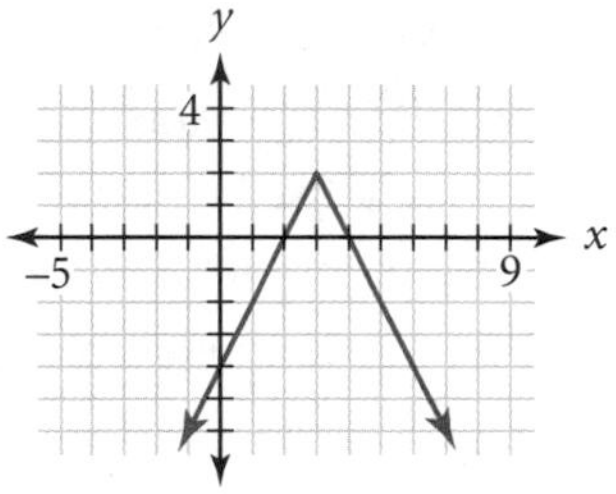

7. Solve for y.

a. $2x - 3y = 6$ **b.** $(y + 1)^2 - 3 = x$ **c.** $\sqrt{1 - y^2} + 2 = x$

8. Solve for x.

a. $4\sqrt{x - 2} = 10$

b. $\left(\frac{x}{-3}\right)^2 = 5$

c. $\frac{x - 3}{2} = 4$

d. $3\sqrt{1 + \left(\frac{x}{5}\right)^2} = 2$

9. APPLICATION The Acme Bus Company has a daily ridership of 18,000 passengers and charges \$1.00 per ride. The company wants to raise the fare yet keep its revenue as large as possible. (The revenue is found by multiplying the number of passengers by the fare charged.) From previous fare increases, the company estimates that for each increase of \$0.10, it will lose 1000 riders.

a. Complete this table.

Fare (\$) x	1.00	1.10	1.20	1.30	1.40	1.50	1.60	1.70	1.80
Number of passengers	18000								
Revenue (\$) y	18000								

b. Make a graph of the the revenue (y) versus fare charged (x). You should recognize the graph as a parabola.

c. What are the coordinates of the vertex of the parabola? Explain the meaning of each coordinate of the vertex.

d. Find a quadratic function that models these data. Use your model to find

i. The revenue if the fare is \$2.00.

ii. The fare(s) that make no revenue (\$0).

Assessing What You've Learned

- Can you sketch a graph to match a description of a situation?
- Can you identify whether a graph, a table of data values, or a description of a situation represents a function?
- Do you know the difference between an independent variable and a dependent variable?
- What are step functions?
- What are translations, reflections, stretches, and shrinks? How does applying one of these transformations change the graph of a function?
- Can you use transformations to solve problems that involves circles or ellipses?
- Do you know how to interpret and work with function compositions?

JOURNAL Look back at the journal entries you have written about Chapter 4. To help you organize your thoughts, here are some additional questions to write about.

- What do you think were the main ideas of the chapter?
- Were there any exercises that you particularly struggled with? If so, are they clear now? If some ideas are still foggy, what plans do you have to clarify them?

Organize your notes on each type of parent function and each type of transformation discussed in this chapter. Review how each transformation affects the graph of a function or relation and how the equation of the function or relation changes. You might want to create a large chart with rows for each type of transformation and columns for each type of parent function; don't forget to include a column for the general function, $y = f(x)$.

CHAPTER

5 Exponential, Power, and Logarithmic Functions

Art can take on living forms. To create *Tree Mountain,* conceptualized by artist Agnes Denes, 11,000 people planted 11,000 trees on a human-made mountain in a former gravel pit in Finland. The trees were planted in a mathematical pattern similar to a golden spiral, imitating the arrangement of seeds in a sunflower.

All living things grow and eventually decay. Both growth and decay can be modeled using exponential functions.

Tree Mountain—A Living Time Capsule—11,000 People, 11,000 Trees, 400 Years 1992–1996, Ylöjärvi, Finland (420 × 270 × 28 meters) © Anges Denes.

OBJECTIVES

In this chapter you will

- write explicit equations for geometric sequences
- use exponential functions to model real-world growth and decay scenarios
- review the properties of exponents and the meaning of fractional exponents
- find the inverse of a function
- apply logarithms, the inverses of exponential functions

LESSON

5.1

The Exponential Function

"Obvious" is the most dangerous word in mathematics.

ERIC TEMPLE BELL

In Chapter 2 you used sequences and recursive rules to model the geometric growth and decay of money, populations, and other phenomena. Recursive formulas generate only values, such as the amount of money after one year or two years, or the population in a certain year. But usually growth and decay happen continuously. In this lesson you will focus on finding explicit formulas for these patterns, which will allow you to model situations involving continuous growth and decay or to find discrete points without using recursion.

Science Connection Atoms that have more neutrons than protons in their nuclei are unstable. Their nuclei can split apart, emitting radiation and resulting in a more stable atom. This process is called *radioactive decay.* The time it takes for half the atoms in a radioactive sample to decay is called the *half-life,* and the half-life is specific to the element. For instance, the half-life of carbon-14 is 5,700 years, whereas the half-life of uranium-238 is 4.5 billion years.

Submerged for two years in a storage tank in La Hague, France, this radioactive waste glows blue. The blue light is known as the "Cherenkov glow."

Investigation
Radioactive Decay

You will need

- one die per person or a graphing calculator with a random number generator

Procedure Note

1. All members of the class should stand up except the recorder. The recorder counts and records the number standing at each stage.
2. Each person rolls a die, and those with ones on their cubes sit down.
3. Wait for the recorder to count and record the number of people standing.
4. Repeat Steps 2 and 3 until fewer than 3 students are standing.

This Investigation is a simulation of radioactive decay. Each person will need a standard six-sided die. [▶ See **Calculator Note 5A** to simulate this with your calculator instead. ◀] Each standing person represents a radioactive atom in a sample. The people who sit down at each stage represent the atoms that have undergone radioactive decay.

Step 1 Follow the Procedure Note to obtain data in the form (stage, number standing).

Step 2 Write a recursive formula that fits your data.

Step 3 Write an expression to calculate the 8th term, using only u_0 and the ratio.

Step 4 | Write an expression that calculates the nth term, using only u_0 and the ratio.

Step 5 | What was the half-life of this sample?

Step 6 | Write a paragraph explaining how this activity simulates the life of a radioactive sample.

In Chapter 3 you learned how to find the equation of the line that passes through the points of an arithmetic sequence. In this lesson you will find the equation of a curve that passes through the points of a geometric sequence.

You probably recognized the geometric decay model in the Investigation. As you recall from Chapter 2, geometric decay is nonlinear. At each step the previous term is multiplied by a common ratio. So, the nth term has been multiplied by the common ratio n times. Exponents are used to represent a number that appears as a factor n times, so exponential functions are used to model geometric growth. An **exponential function** is a continuous function with a variable as the exponent and is used to model growth or decay.

EXAMPLE A

Most automobiles depreciate as they get older. Suppose that an automobile that originally cost \$14,000 depreciates by one-fifth of its value every year.

a. What is the value of this automobile after two and a half years?

b. When is this automobile worth half its initial value?

▶ Solution

a. The recursive formula gives automobile values only after one year, two years, three years, and so on.

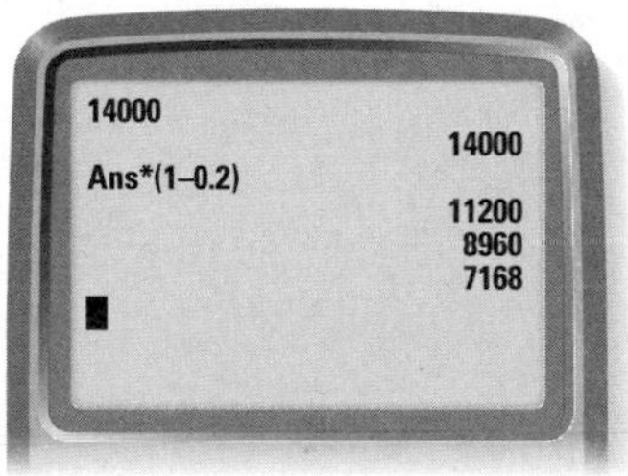

The value decreases by $\frac{1}{5}$, or 0.2, each year. We can build a table.

Year	Value
0	\$14000
1	\$14000 − .2(\$14000) = \$11200
2	\$11200 − .2(\$11200) = \$8960
3	\$8960 − 0.2(\$8960) = \$7168

The distributive property tells us that $x - 0.2x = 0.8x$ for any number x. Also, notice the pattern in the table:

Year	Value
0	\$14000
1	\$11200
2	\$8960
3	\$7168

× 0.8
× 0.8
× 0.8

So, to find the value after a given number of years, n, multiply by 0.8 n times. The explicit formula is $u_n = 14000(0.8)^n$.

Year	Value
0	$\$14000(0.8)^0 = \$14000$
1	$\$14000(0.8)^1 = \$11200$
2	$\$14000(0.8)^2 = \$8960$
3	$\$14000(0.8)^3 = \$7168$

The equation of the continuous function through the points of this sequence is

$$y = 14000(0.8)^x$$

You can use the continuous function to find the value of the car at any point. To find the value after $2\frac{1}{2}$ years, substitute $2\frac{1}{2}$ for x.

$$y = 14000(0.8)^{2.5} = \$8,014.07$$

It makes sense that the automobile's value after 2.5 years should be between the values for u_2 and u_3, \$8,960 and \$7,168. Because this is not a linear function, finding the value halfway between these two values does not give an accurate value for the car after $2\frac{1}{2}$ years.

b. To find when the automobile is worth half its initial value, substitute 7000 for y and find x.

$y = 14000(0.8)^x$	Original equation.
$7000 = 14000(0.8)^x$	Substitute 7000 for y.
$0.5 = (0.8)^x$	Divide both sides by 14000.

You can experiment with different exponents to find one that produces a value very close to 0.5. The value of $(0.8)^{3.10628372}$ is very close to 0.5. This means that the value of the car is \$7,000, or half of its original value after 3.10628372 years (about 3 years and 39 days). This is the "half-life" of the value of the automobile, or the amount of time needed for the value to decrease to half its original amount.

An equation of the form $y = ab^x$ is an exponential function, where the coefficient a is the y-intercept and the base b is the growth rate. Exponential growth and decay are both modeled by the same parent function. Growth is modeled by a base that is greater than one, and decay is modeled by a base that is between zero and one. In general, the larger the base, the faster the growth; the closer the base to zero, the faster the decay.

Decay can be expressed as a percent decrease. For example, the car in Example A also could be said to decrease in value by 20% per year. To reflect that idea, the base 0.80 also can be expressed as $1 - 0.20$. Growth can be expressed as a percent increase as well. If a classic car were to rise in value by 5% per year, it would have a base of $1 + 0.05$, or 1.05.

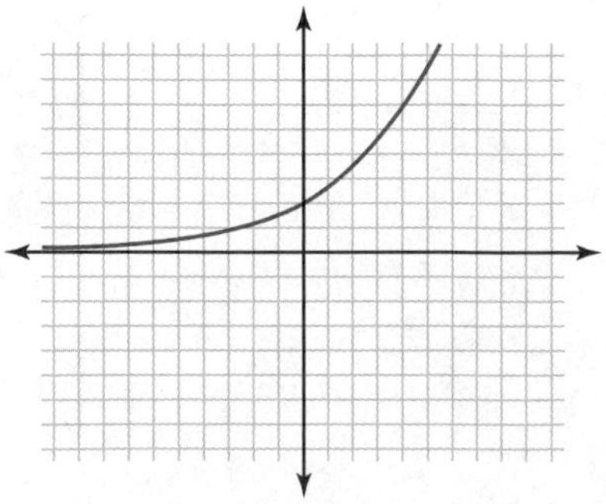
An exponential graph with $a > 0$ and $b > 1$.

Because the value of an exponential function with $a > 0$ never reaches zero, it has range $y > 0$.

All exponential growth curves have what is called a **doubling time,** just as decay has a half-life. This time is independent of the value of the function. In other words, if the ratio is constant, it takes just as long to double \$1,000 to \$2,000 as it takes to double \$5,000 to \$10,000.

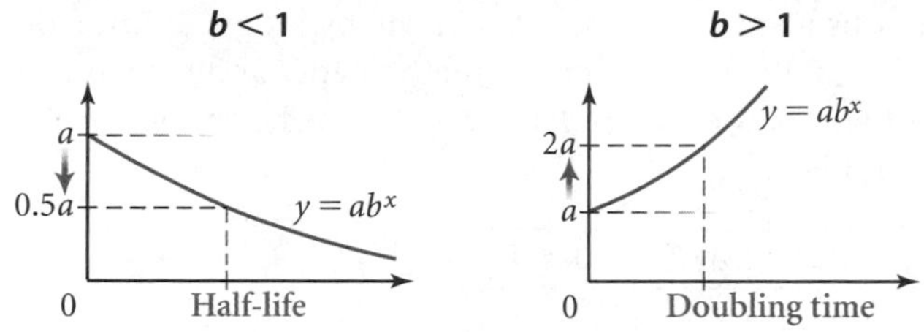

EXERCISES

Practice Your Skills

1. Evaluate the functions at the given value.

a. $f(x) = 4.753(0.9421)^x,\ x = 5$

b. $g(h) = 238(1.37)^h,\ h = 14$

c. $h(t) = 47.3(0.835)^t + 22.3,\ t = 24$

d. $j(x) = 225(1.0825)^{x-3},\ x = 37$

2. Record three terms of the sequence, and then write an explicit function for the sequence.

$u_0 = 16$
$u_n = 0.75u_{n-1}$

$u_0 = 24$
$u_n = 1.5u_{n-1}$

3. Evaluate the function for $x = 0$, 1, and 2, and then write a recursive formula for the pattern.

a. $f(x) = 125(0.6)^x$

b. $f(x) = 3(2)^x$

4. Find the growth rate of each exponential function. Then state the percent increase or decrease.

x	y
3	48
4	36

x	y
0	54
1	72

x	y
41	50
42	47

x	y
41	47
42	50

Reason and Apply

5. In 1991 the population of the People's Republic of China was 1.151 billion with a percent increase of 1.5% annually.

a. Write a recursive formula that models this growth.

b. Create a table.

c. Write an explicit equation that models this growth. Choose two data points, and show that your equation works.

The actual population of China in the year 2001 was 1,273,111,290. How does this compare with the value predicted by your equation? What does this tell us?

Year	Population (in billions)
1991	
1992	
1993	
1994	
1995	
1996	
1997	
1998	
1999	
2000	

6. Jack planted a mysterious bean just outside his kitchen window. It immediately sprouted 2.56 cm above the ground. Being a student of mathematics and the sciences, Jack kept a careful log of the growth of the sprout. He measured the height of the plant each day at 8:00 A.M. and recorded this data.

Day 0	Day 1	Day 2	Day 3	Day 4
2.56 cm	6.4 cm	16 cm	40 cm	100 cm

a. Write an explicit formula for this pattern. If the pattern were to continue, what would be the heights on the 5th and 6th days?

b. Jack's younger brothers measured the plant at 8:00 P.M. on the evening of the 3rd day and found it to be about 63.25 cm tall. Show how this value can be found mathematically. (You may need to experiment with your calculator.)

c. Find the height of the sprout at noon on the 6th day.

d. Find the doubling time for this plant.

e. Experiment with the equation to find the day and time (to the nearest hour) when the stalk reached its final height of 1 km.

7. ***Mini-Investigation*** Graph the equations of 7a–d simultaneously on your calculator.

a. $y_1 = 1.5^x$

b. $y_2 = 2^x$

c. $y_3 = 3^x$

d. $y_4 = 4^x$

e. How do the graphs compare? What points (if any) do they have in common?

f. Predict what the graph of $y = 6^x$ will look like. Use your calculator to verify your prediction.

8. Each of the red curves is a transformation of the graph of $y = 2^x$, shown in black. Write an equation of each red curve.

a.

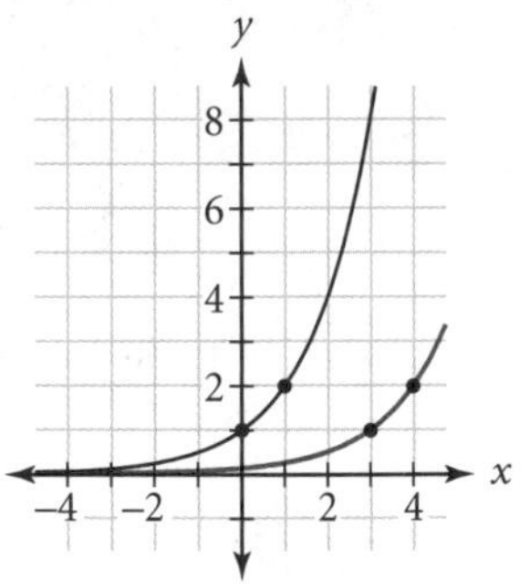

b.

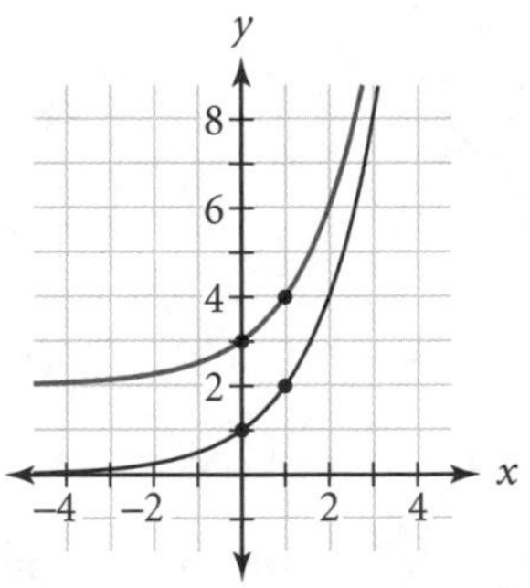

c.

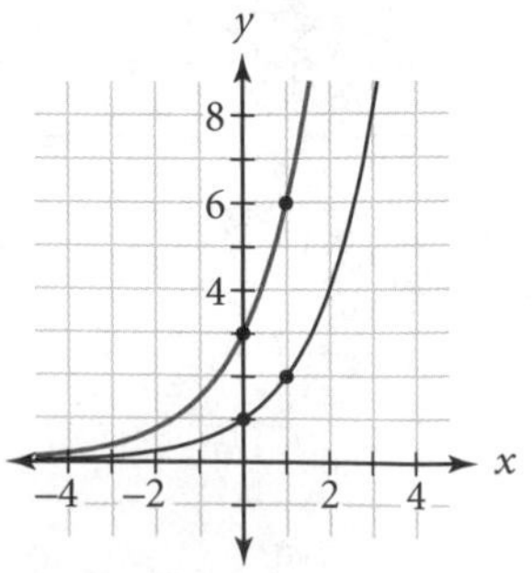

d.

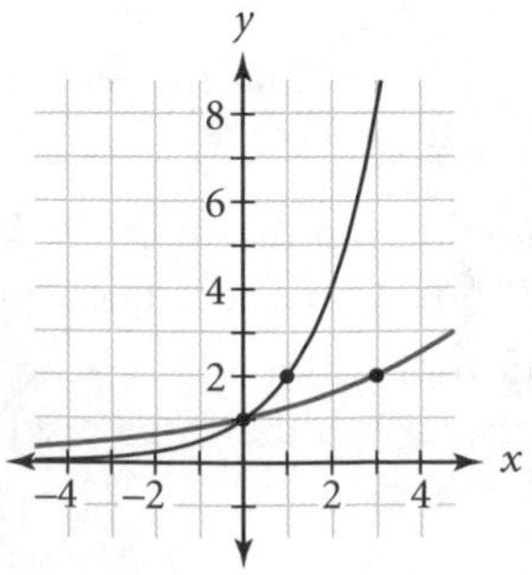

9. ***Mini-Investigation*** Graph the equations of 9a–d simultaneously on your calculator.

a. $y_1 = 0.2^x$

b. $y_2 = 0.3^x$

c. $y_3 = 0.5^x$

d. $y_4 = 0.8^x$

e. How do the graphs compare? What points (if any) do they have in common?

f. Predict what the graph of $y = 0.1^x$ will look like. Use your calculator to verify your prediction.

10. Each of the red curves is a transformation of the graph $y = 0.5^x$, shown in black. Write an equation of each red curve.

a.

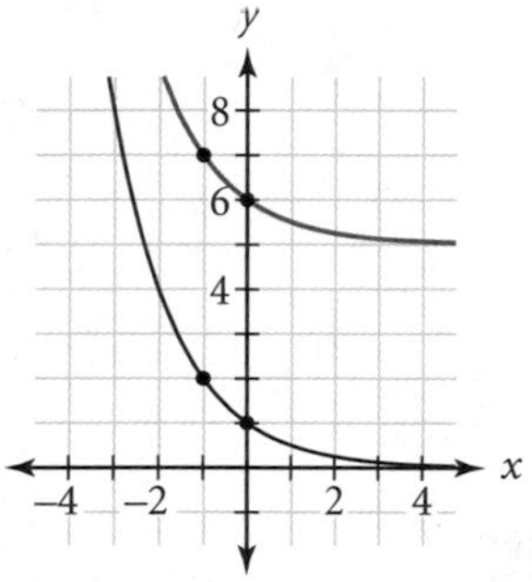

b.

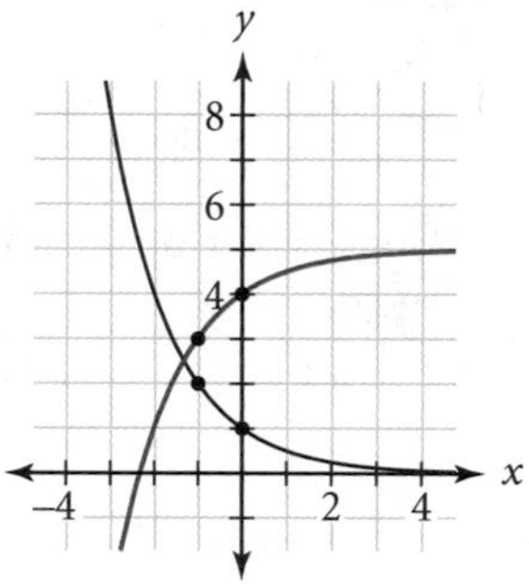

c.

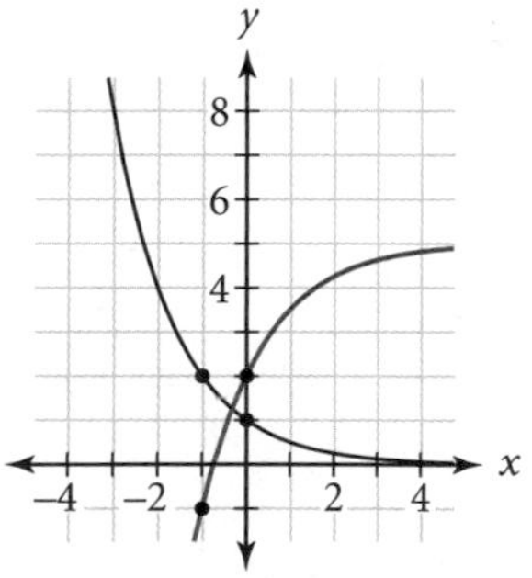

d.

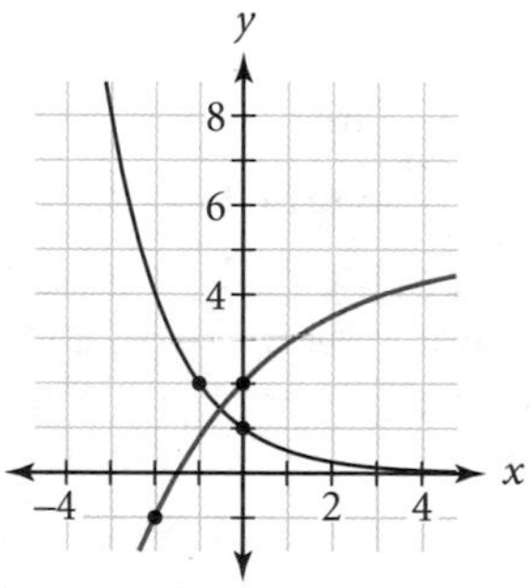

11. The exponential form, $y = ab^x$, is convenient when you know the y-intercept. Start with $f(0) = 30$ and $f(1) = 27$.

a. Find the common ratio.

b. Write the function $f(x)$.

c. Sketch a graph of $f(x)$ and $g(x) = f(x - 4)$ on the same axis.

d. What is the value of $g(4)$?

e. Write an equation of $g(x)$ that does not use the intercept.

Review

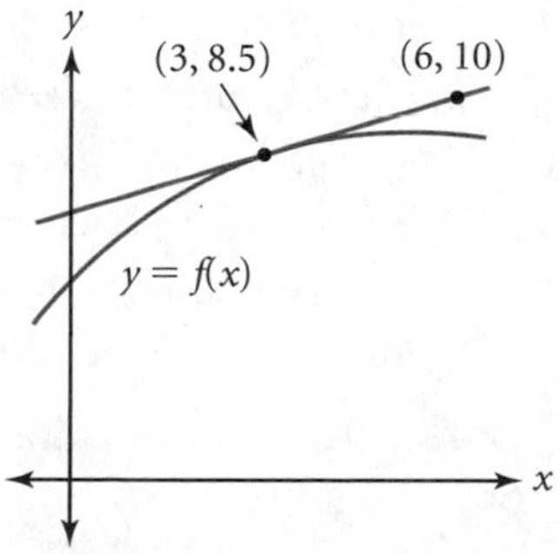

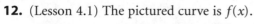

12. (Lesson 4.1) The pictured curve is $f(x)$.

a. Complete the missing values to make a true statement. $f(\underline{\ ?\ }) = \underline{\ ?\ }$.

b. Find the equation of the pictured line.

13. (Lesson 4.1) Janell starts 10 m from a motion sensor and walks at 2 m/s toward the sensor. When she is 3 m from the sensor, she instantly turns around and walks at the same speed back to the starting point.

a. Sketch a graph of the function that models Janell's walk.

b. Give the domain and range of the function.

c. Write an equation of the function.

14. ***Mini-Investigation*** (Lesson 4.8) Make up two linear functions, $f(x)$ and $g(x)$. Enter $f(x)$ in Y1 and $g(x)$ in Y2. Enter $f(g(x))$ in Y3 as Y1(Y2) and $g(f(x))$ in Y4 as Y2(Y1). Display the graphs of Y3 and Y4. Change $f(x)$ and/or $g(x)$ and again graph Y3 and Y4.

a. Describe the relationship between $f(g(x))$ and $g(f(x))$.

b. Explain the relationship algebraically .

15. ***Spreadsheet*** (Lesson 2.2) Use a spreadsheet to construct a table with 100 values modeled by the equation $y = 5(1.5)^x$.

a. Place the x values 0 to 49 in Column A, starting in Cell A1.

b. Enter 5 in cell B1.

c. Complete this formula and enter it in Cell B2: = B1 * ____. Use the data fill feature of your spreadsheet to extend the formula to Cell B49.

0	5
1	7.5
2	11.25
3	16.875

d. In column C, show y-values for an exponential function with an initial value of 1000 and a growth rate of 1.3.

e. For which x-value is Column C larger? Why?

f. Extension: According to popular legend Dutch traders purchased the island of Manhattan in 1626 for goods worth $24. Suppose someone invested $24 at 6% interest, compounded annually, in 1626. What would that be worth in 2006?

PROJECT

THE RULE OF 72

The rule of 72 says that the time it takes for an investment to double at a given annual percent increase (interest rate) can be found by dividing 72 by the interest rate. For example, the rule of 72 predicts that at a 6% interest rate an investment should double in $\frac{72}{6} = 12$ years.

a. Create a table with three columns, headed Interest Rate, Rule of 72 Doubling Time, and Actual Doubling Time. Include 2% and 24% as entries under Interest Rate. Choose at least five interest rates between 2% and 24%. Calculate the values of the other two columns.

b. How accurate is the rule of 72?

c. Create another table, this time using the rule of 70. (Divide 70 instead of 72 by the interest rates you used before).

d. Notice that the rule of 70 is more accurate for many values than the rule of 72. Why do you think the rule of 72 is used?

LESSON

5.2

Properties of Exponents

Frequently, you will need to rewrite a mathematical expression in a different form to make the expression easier to understand or an equation easier to solve. Some properties of exponents you learned about in previous math courses can help you rewrite expressions. Recall that in an exponential expression, such as 4^3, the numeral 4 is called the *base,* and the numeral 3 is called the *exponent.* You can say that 4 is raised to the *power* of 3. If the exponent is a positive integer, you can write the expression in expanded form, for example $4^3 = 4 \cdot 4 \cdot 4$. Because 4^3 equals 64, you say that 64 is a *power of* 4.

Investigation
Exponent Rules!

In this Investigation you'll use expanded form to review and generalize the properties of exponents.

Step 1 Write each product in expanded form, and then rewrite it in exponential form.

a. $2^3 \cdot 2^4$ **b.** $x^5 \cdot x^{12}$ **c.** $10^2 \cdot 10^5$

Step 2 Generalize your results from Step 1.

$a^m \cdot a^n = \underline{\ ?\ }$

Step 3 Write the numerator and denominator of each quotient in expanded form. Reduce by eliminating common factors, and then rewrite in exponential form the factors that remain.

a. $\dfrac{4^5}{4^2}$ **b.** $\dfrac{x^8}{x^6}$ **c.** $\dfrac{(0.94)^{15}}{(0.94)^5}$

Step 4 Generalize your results from Step 3.

$\dfrac{a^m}{a^n} = \underline{\ ?\ }$

Step 5 Write each quotient in expanded form, reduce, and then rewrite in exponential form.

a. $\dfrac{2^3}{2^4}$ **b.** $\dfrac{4^5}{4^7}$ **c.** $\dfrac{x^3}{x^8}$

Step 6 Rewrite each quotient in Step 5 using the property you discovered in Step 4.

Step 7 Generalize your results from Steps 5 and 6.

$\dfrac{1}{a^n} = \underline{\ ?\ }$

Step 8 Write several expressions in the form $(a^n)^m$. Use what you've learned to expand each, and then rewrite the expression in exponential form. Generalize your results.

Step 9 Write several expressions in the form $(a \cdot b)^n$. Use what you've learned to expand each, and then rewrite the expression in exponential form. Generalize your results.

Step 10 Explain a way to show that $a^0 = 1$. Use all of the properties you have discovered. Try writing at least two exponential expressions that can be expanded or rewritten to support your explanation.

Here's a summary of the properties of exponents. You discovered many of these in the Investigation; a few you did not. Can you write an example of each property?

Properties of Exponents

For $a > 0$, $b \neq 0$, and all values of m and n, these properties are true:

Multiplication Property of Exponents

$$a^m \cdot a^n = a^{m+n}$$

Division Property of Exponents

$$\frac{a^m}{a^n} = a^{m-n}$$

Reciprocal Property of Exponents

$$a^{-n} = \frac{1}{a^n} \text{ and } a^n = \frac{1}{a^{-n}}$$

Exponents of Zero

$$a^0 = 1$$

Power of a Power Property

$$(a^n)^m = a^{mn}$$

Distributive Property of Exponentiation over Multiplication

$$(ab)^n = a^n b^n$$

Distributive Property of Exponentiation over Division

$$\left(\frac{a}{b}\right)^n = \frac{a^n}{b^n}$$

Power Property of Equality

If $a = b$, then $a^n = b^n$.

Common Base Property of Equality

If $a^n = a^m$, then $n = m$.

In Lesson 5.1 you learned to solve equations with a variable in the exponent by using a calculator to try various values of x. The properties of exponents allow you to solve these types of equations algebraically. In one special case you can rewrite both sides of the equation with a common base. This strategy is fundamental to solving the more general equations you'll see later in this chapter.

EXAMPLE

Solve for x.

a. $8^x = 4$

b. $27^x = \dfrac{1}{81}$

c. $\left(\dfrac{49}{9}\right)^2 = \left(\dfrac{3}{7}\right)^{\frac{3}{2}}$

▶ Solution

If you use the power of a power property to convert each side of the equation to a common base, then you can solve without a calculator.

a.

$8^x = 4$	
$(2^3)^x = 2^2$	$8 = 2^3$ and $4 = 2^2$
$2^{3x} = 2^2$	Use the power of a power property to rewrite $(2^3)x$ as 2^{3x}.
$3^x = 2$	Use the common base property of equality.
$x = \dfrac{2}{3}$	Divide.

b.

$27^x = \dfrac{1}{81}$	
$(3^3)^x = \dfrac{1}{3^4}$	$27 = 3^3$ and $81 = 3^4$
$3^{3x} = 3^{-4}$	Use the power of a power property and the reciprocal property.
$3x = -4$	Use the common base property of equality.
$x = \dfrac{-4}{3}$	Divide.

c.

$\left(\dfrac{49}{9}\right)^x = \left(\dfrac{3}{7}\right)^{\frac{3}{2}}$	
$\left(\dfrac{7^2}{3^2}\right)^{\frac{3}{2}} = \left(\dfrac{3}{7}\right)^{\frac{3}{2}}$	$49 = 7^2$ and $9 = 3^2$
$\left(\left(\dfrac{7}{3}\right)^2\right)^x = \left(\left(\dfrac{7}{3}\right)^{-1}\right)^{\frac{3}{2}}$	Use the distributive property of exponentiation over division and the reciprocal property.
$\left(\dfrac{7}{3}\right)^{2x} = \left(\dfrac{7}{3}\right)^{-\frac{3}{2}}$	Use the power of a power property.
$2x = \dfrac{-3}{2}$	Use the common base property of equality.
$x = \dfrac{-3}{4}$	Divide.

EXERCISES

Practice Your Skills

1. Rewrite each expression as a fraction without exponents. Verify that your answer is equivalent to the original expression by evaluating each on your calculator.

a. 5^{-3} **b.** -6^2 **c.** -3^{-4}

d. $(-12)^{-2}$ **e.** $\left(\frac{3}{4}\right)^{-2}$ **f.** $\left(\frac{2}{7}\right)^{-1}$

2. Rewrite each expression in the form a^n.

a. $a^8 \cdot a^{-3}$ **b.** $\frac{b^6}{b^2}$

c. $(c^4)^5$ **d.** $\frac{d^0}{e^{-3}}$

3. State whether each equation is true or false. If it is false, explain why.

a. $3^5 \cdot 4^2 = 12^7$ **b.** $100(1.06)^x = (106)^x$

c. $\frac{4^x}{4} = 1^x$ **d.** $\frac{6.6 \cdot 10^{12}}{8.8 \cdot 10^{-4}} = 7.5 \cdot 10^{15}$

4. Solve for x.

a. $3^x = \frac{1}{9}$

b. $\left(\frac{5}{3}\right)^x = \left(\frac{27}{125}\right)$

c. $\left(\frac{1}{3}\right)^x = 243$

d. $5 \cdot 3^x = 5$

Reason and Apply

5. Rewrite each expression in the form ax^n.

a. $x^6 \cdot x^6$ **b.** $4x^6 \cdot 2x^6$

c. $(-5x^3) \cdot (-2x^4)$ **d.** $\frac{72x^7}{6x^2}$

e. $\left(\frac{6x^5}{3x}\right)^3$ **f.** $\left(\frac{20x^7}{4x}\right)^{-2}$

6. You've seen that the power property of exponentiation over multiplication allows you to rewrite $(a \cdot b)^n$ as $a^n \cdot b^n$. Is there a power property of exponentiation over addition that allows you to rewrite $(a + b)^n$ as $a^n + b^n$? Write some numeric expressions in the form $(a + b)^n$ and evaluate them. Are your answers equivalent to $a^n + b^n$? Write a short paragraph that summarizes your findings.

7. Consider this sequence:

$7^2, 7^{2.25}, 7^{2.5}, 7^{2.75}, 7^3$

a. Use your calculator to evaluate each term in the sequence. Give your answers approximate to four decimal places.

b. Find the differences between the consecutive terms of the sequence. What do these differences tell you?

c. Find the ratios of the terms in 7a. What do these values tell you?

d. What observation can you make about decimal powers?

8. For a–d, graph the equations on your calculator.

a. $y = x^2$ **b.** $y = x^3$ **c.** $y = x^4$ **d.** $y = x^5$

e. How are the graphs similar? How are they different? What points (if any) do they have in common?

f. Predict what the graph of $y = x^6$ will look like. Use your calculator to verify your prediction.

g. Predict what the graph of $y = x^7$ will look like. Use your calculator to verify your prediction.

9. Each red curve is a transformation of the graph of $y = x^3$, shown in black. Write an equation of each red curve.

a.

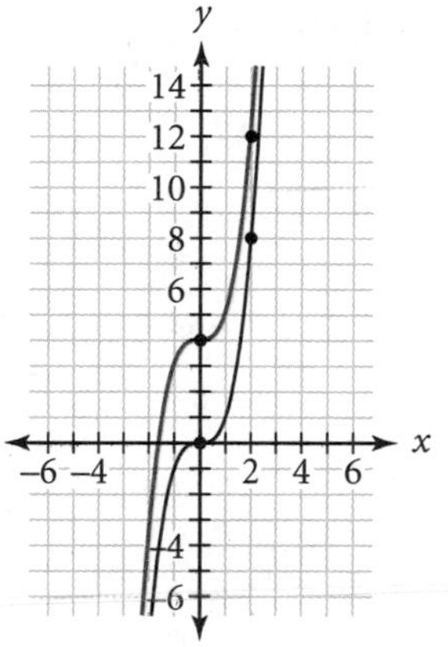

b.

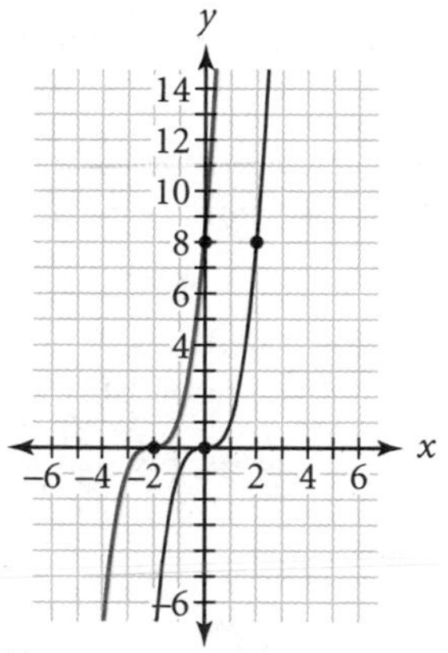

c.

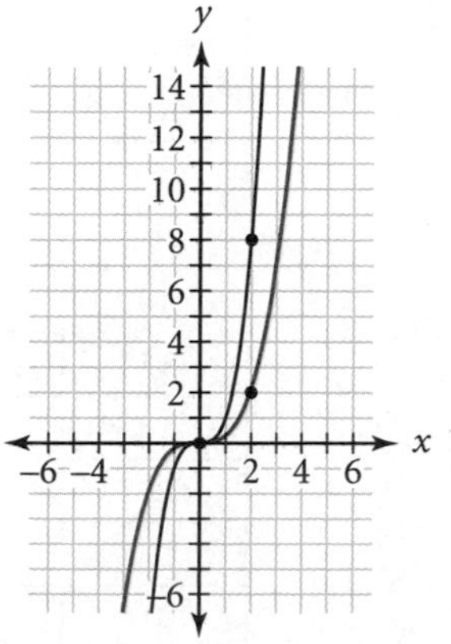

d.

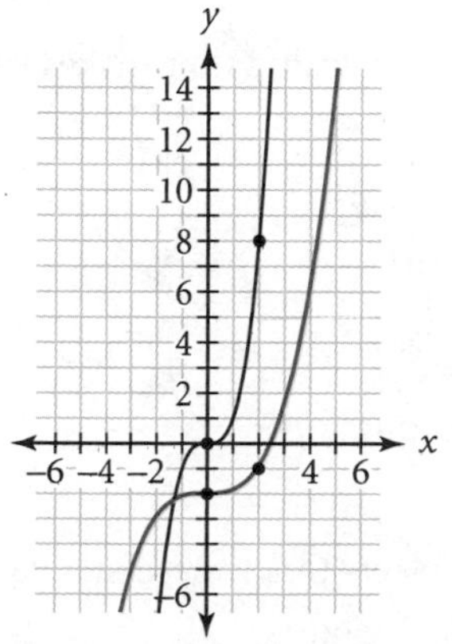

10. Consider the exponential equation $y = 47\,(0.9)^x$. Several points satisfying the equation are shown in this calculator table. Notice that when $x = 0$, $y = 47$.

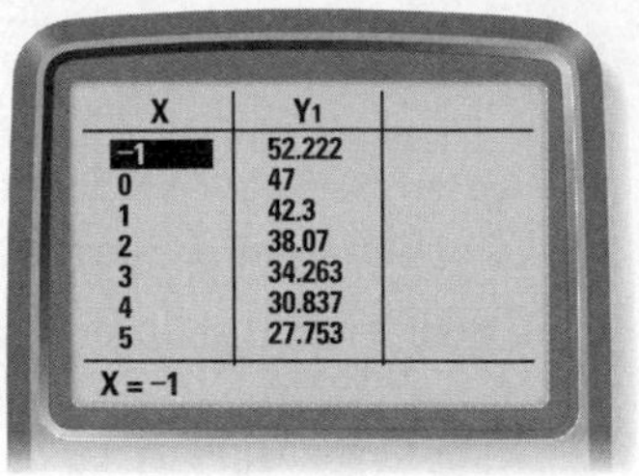

a. The expression $47(0.9)^x$ could be rewritten as $47(0.9)(0.9)^{x-1}$. Explain why this is true. Rewrite $47(0.9)(0.9)^{x-1}$ in the form $a \cdot b^{x-1}$.

b. The expression $47(0.9)^x$ could also be rewritten as $47(0.9)(0.9)(0.9)^{x-2}$. Rewrite $47(0.9)(0.9)(0.9)^{x-2}$ in the form $a \cdot b^{x-2}$.

c. Look for a connection between your answers to 10a and 10b and the values in the table. State a conjecture or general equation that generalizes your findings.

11. **APPLICATION** A radioactive sample was created in 1980. In 2002 a technician measured the radioactivity at 42.0 rads. One year later the radioactivity was 39.8 rads.

a. Find the ratio of 2002 radioactivity to 2003 radioactivity. Give an answer approximate to 4 decimal places.

b. Let 1980 correspond to $x = 0$. Let y be the sample's radioactivity in rads. Use the decay rate you calculated in 11a and the point (22, 42.0) to find the constant a in $y = ab^x$.

c. What was the radioactivity in 1980?

d. What equation models the sample's radioactivity?

e. Calculate the radioactivity in 2010 using the equation.

12. A ball bounced to a height of 30.0 cm on the 3rd bounce and to a height of 5.2 cm on the 6th bounce.

a. If you are using the equation $y = ab^x$ to model the ball's height y after x bounces, verify that you get the two equations $30.0 = ab^3$ and $5.2 = ab^6$ after substituting the points (3, 30.0) and (6, 5.2).

b. Divide both sides of one equation by the sides of the other equation to cancel a. This is similar to subtracting the equations when solving linear systems. Solve for b.

c. How high was the ball when it was dropped?

Review

13. (Lesson 5.1) Name the x-value that makes each equation true.

a. $37{,}000{,}000 = 3.7 \cdot 10^x$

b. $0.000801 = 8.01 \cdot 10^x$

c. $47{,}500 = 4.75 \cdot 10^x$

d. $0.0461 = x \cdot 10^{-2}$

14. (Algebra) Solve the equation $\dfrac{y+3}{2} = (x+4)^2$ for y. Then carefully graph it on your paper.

15. (Lesson 3.3) Paul collects this time-distance data for a remote-controlled car.

Time (s)	Distance (m)
5	0.8
8	1.7
8	1.6
10	1.9
15	3.3
18	3.4
22	4.1
24	4.6
31	6.4
32	6.2

a. Plot the points on a graph.

b. Find a line to fit the data.

LESSON

5.3

Fractional Exponents and Roots

In this lesson, you will investigate properties of **fractional,** or **rational, exponents.** You will see how they can be useful in solving exponential and power functions and in finding an exponential curve to model data.

The volume and surface area of a cube, such as this fountain in Osaka, Japan, are related by fractional exponents.

Investigation
Getting to the Root

In this Investigation you'll explore the relationship between x and $x^{1/2}$ and learn how to find the values of some fractional powers.

Step 1 Use your calculator to create a table for $y = x^{1/2}$ at integer values of x. When is $x^{1/2}$ a positive integer? Describe the relationship between x and $x^{1/2}$.

Step 2 Graph $Y1 = x^{1/2}$. This graph should look familiar. Make a conjecture about what other function is equivalent to $y = x^{1/2}$, enter your guess in Y2, and verify that the equations give the same y-value at each x-value.

Step 3 State what you have discovered about raising a number to the one-half power. Include an example with your statement.

Step 4 Clear the previous functions, and make a table for $f(x) = 25^x$ with $x = \frac{1}{2}, \frac{2}{2}, \frac{3}{2}$, and so on.

Step 5 Study your table, and explain any relationships you see. How would you find the value of $49^{3/2}$ without your calculator? Use the calculator to check your answer.

Step 6 How could you find the value of $27^{2/3}$ without a calculator? Verify your response, and then test your strategy on $8^{5/3}$. Check your answer.

Step 7 As a group, describe what it means to raise a number to a rational exponent, and generalize a procedure for simplifying $a^{m/n}$.

Fractional powers with a numerator of 1 indicate roots. For example, $x^{1/5} = \sqrt[5]{x}$, or the "5th root of x," and $x^{1/n} = \sqrt[n]{x}$, or the "nth root of x." Recall that the 5th root of x is the number that, multiplied together five times, gives you x. For fractional exponents with numerators other than one, such as $9^{3/2}$, the numerator is interpreted as the power to which to raise the root. That is, $9^{3/2} = (9^{1/2})^3$, or $(\sqrt{9})^3$.

Definition of Fractional Exponents

The power property of exponents states that $a^{m/n} = (a^{1/n})^m$ and $a^{m/n} = (a^m)^{1/n}$; therefore

$$a^{m/n} = (\sqrt[n]{a})^m \text{ or } \sqrt[n]{a^m} \text{ for } a \geq 0$$

EXAMPLE A

Rewrite with exponents, and solve for the variable.

a. $\sqrt[4]{a} = 14$

b. $\sqrt[9]{b^5} = 26$

c. $(\sqrt[3]{c})^8 = 47$

► Solution

Rewrite each expression with a rational exponent, then use properties of exponents to solve.

a.
$$\sqrt[4]{a} = 14$$
$$a^{1/4} = 14 \quad \text{Rewrite } \sqrt[4]{a} \text{ as } a^{1/4}.$$
$$(a^{1/4})^4 = 14^4 \quad \text{Raise both sides to a power of 4.}$$
$$a = 38412 \quad \text{Evaluate } 14^4.$$

b.
$$\sqrt[9]{b^5} = 26$$
$$b^{5/9} = 26 \quad \text{Rewrite } \sqrt[9]{b^5} \text{ as } b^{5/9}.$$
$$(b^{5/9})^{9/5} = 26^{9/5} \quad \text{Raise both sides to a power of } \tfrac{9}{5}.$$
$$b \approx 352.33 \quad \text{Approximate } 26^{9/5}.$$

c.
$$(\sqrt[3]{c})^8 = 47$$
$$c^{8/3} = 47 \quad \text{Rewrite } (\sqrt[3]{c})^8 \text{ as } c^{8/3}.$$

$$(c^{8/3})^{3/8} = 47^{3/8} \qquad \text{Raise both sides to a power of } \tfrac{3}{8}.$$

$$c \approx 4.237 \qquad \text{Approximate } 47^{3/8}.$$

There are actually two solutions to this equation because a negative number raised to the 8th power gives the same result as a positive number raised to the 8th power. The other solution is $x - 4.237$.

An exponential function in the form $y = ab^x$ has a variable as the exponent. A **power function,** in contrast, has a variable as the base.

Power Function

The equation

$$y = ax^n$$

where a and n are constants, is the parent of the family of **power functions.**

You will use different procedures to solve power equations. You must learn to recognize the difference between power equations and exponential equations.

EXAMPLE B

Solve for x.

a. $x^4 = 3000$

b. $6x^{2.5} = 90$

▶ *Solution*

To solve a power equation, use the power property of equality, and choose an exponent that will undo the exponent on x.

a.

$$x^4 = 3000$$

$$(x^4)^{1/4} = 3000^{1/4}$$ Use the power property of equality. Raising both sides to the power of $\frac{1}{4}$ "undoes" the power of 4 on x. (There is a second solution; see below part b).

$$x \approx 7.40$$ Use your calculator to approximate $3000^{1/4}$.

b.

$$6x^{2.5} = 90$$

$$x^{2.5} = 15$$ Divide both sides by 6.

$$(x^{2.5})^{1/2.5} = 15^{1/2.5}$$ Use the power property of equality, and choose the exponent $\frac{1}{2.5}$.

$$x \approx 2.95$$ Approximate $15^{1/2.5}$.

Solving equations symbolically is often no more than "undoing" the order of operations. To evaluate the expression $6x^{2.5}$ for a given a value of x, raise it to the power of 2.5, and then multiply the result by 6. Solving by "undoing" reverses this order. So, to solve $6x^{2.5} = 90$, divide by 6 and then raise the result to the power of $\frac{1}{2.5}$.

Generally, the properties of exponents apply only to exponents with positive bases. So, in Example B part a we found only one solution. Note that $x \approx -7.40$ is a second solution.

You learned in the last example that functions of the form $y = a \cdot x^n$ are power functions. A function like $y = \sqrt[9]{b^5}$ is included in the family of power functions because it can be rewritten with an exponent of $\frac{5}{9}$. All the transformations you discovered for parabolas and square root curves also apply to functions that can be written in the form $y = a \cdot x^n$.

You have seen that when $x = 0$, then $y = a$ in the general exponential equation $y = a \cdot b^x$. This means that a is the initial value of the function at time 0 (the y-intercept), and b is the growth or decay rate.

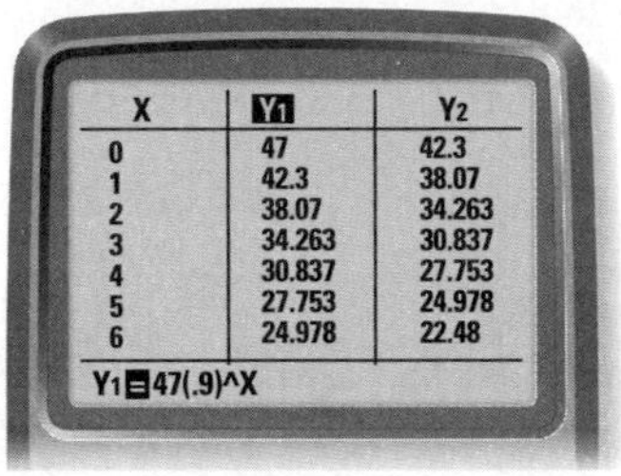

X	Y1	Y2
0	47	42.3
1	42.3	38.07
2	38.07	34.263
3	34.263	30.837
4	30.837	27.753
5	27.753	24.978
6	24.978	22.48

Y1=47(.9)^X

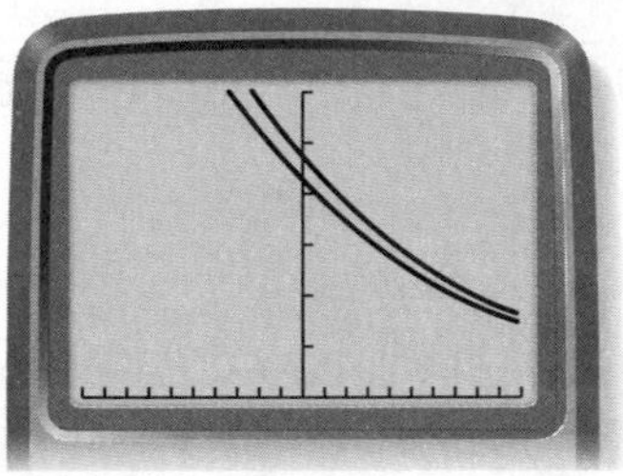

The table and graph above show the exponential functions $Y1 = f(x) = 47(0.9)^x$ and $Y2 = g(x) = 42.3(0.9)^x$. Both the table and graph indicate that if the $g(x)$ function is translated right 1 unit, it becomes the same as $f(x)$. So $f(x) = g(x - 1)$, or $f(x) = 42.3\ (0.9)^{(x-1)} = 47(0.9)^x$.

EXAMPLE C

Casey struck the bell in the university clock tower. Her sound probe, held nearby, measured the sound at 40 lb/in.^{2} when 4 seconds had elapsed and 4.7 lb/in.^{2} after 7 seconds had elapsed. Sound decays exponentially.

a. Name two points through which the exponential curve must pass.

b. Find an exponential equation that models these data.

c. How loud was the bell when it was struck (at 0 seconds)?

▶ ***Solution***

a. Time is the independent variable, and loudness is the dependent variable, so the two points are (4, 40) and (7, 4.7).

b. Start by putting each of the two points into an exponential equation, $y = y_1 \cdot b^x$.

$$40 = y_1 \cdot b^4 \text{ and } 4.7 = y_1 \cdot b^7$$

Substitute 4 and 7 for 7 for x.
Substitute 40 and 4.7 for y.

Note that you don't yet know what b is. Divide the two equations to find b.

$$\frac{4.7}{40} = \frac{y_1 b^7}{y_1 b^4}$$

Use division to combine the two equations into one.

$$0.1175 = b^3$$

Use properties of exponents to rewrite the right-hand side.

$$(0.1175)^{1/3} = (b^3)^{1/3}$$

Undo the cubing with a cube root.

$$b \approx 0.4898$$

Use the division property of exponents to simplify the right side.

Now substitute to find y_1.

$40 = y_1 \cdot b^4$

$40 = y_1 \cdot (0.4898)^4$

$40 = y_1(0.057551)$

$y_1 = 695.$

The equation is $y = 695(.4898)^x$.

c. To find the loudness at 0 seconds, use $y_1 = 695$ lb/in.^{2}.

Note that the base of 0.4898 is an approximation. You could use the exact value of $\left(\frac{4.7}{40}\right)^{1/3}$ if you needed more precision.

EXERCISES

Practice Your Skills

1. Match any expressions that are equivalent.

a. $\sqrt[5]{x^2}$ b. $x^{2.5}$

c. $\sqrt[3]{x}$ d. $x^{5/2}$

e. $x^{0.4}$ f. $\left(\frac{1}{x}\right)^{-3}$

g. $(\sqrt{x})^5$ h. x^3

i. $x^{1/3}$ j. $x^{2/5}$

2. Identify each function as a power function, an exponential function, or neither. (It may be translated, stretched, or reflected). Give a brief reason for your choice.

a. $f(x) = 17x^5$ b. $f(t) = t^3 + 5$

c. $g(v) = 200(1.03)^x$ d. $h(x) = 2x - 7$

e. $g(y) = 3\sqrt{y - 2}$ f. $f(t) = t^2 + 4t + 3$

g. $h(t) = \dfrac{12}{3^x}$ h. $g(w) = \dfrac{28}{x - 5}$

i. $f(y) = \dfrac{8}{y^4} + 1$ j. $g(x) = \dfrac{x^3 + 2}{1 - x}$

k. $h(w) = \sqrt[5]{4w^3}$

3. Rewrite each expression in the form b^x in which x is a fraction or a decimal.

a. $\sqrt[6]{a}$ b. $\sqrt[10]{b^8}$ c. $\dfrac{1}{\sqrt{c}}$ d. $(\sqrt[5]{d})^7$

4. Solve each equation for the variable, and show or explain the step(s).

a. $\sqrt[6]{a} = 4.2$ b. $\sqrt[10]{b^8} = 14.3$

c. $\dfrac{1}{\sqrt{c}} = 0.55$ d. $(\sqrt[5]{d})^7 = 23$

5. Solve for x. If answers are not exact, approximate to two decimal places.

a. $x^7 = 4000$ **b.** $x^{0.5} = 28$ **c.** $x^{-3} = 247$

d. $5x^{1/4} + 6 = 10.2$ **e.** $3x^{-2} = 2x^4$ **f.** $-2x^{1/2} + (9x)^{1/2} = -1$

Reason and Apply

6. Graph $y = x^{1/2}$, $y = x^{1/3}$, $y = x^{1/4}$, and $y = x^{1/5}$ in a standard window.

a. Describe any patterns you see.

b. What is the domain of each curve? Can you explain why?

c. Predict what the graph of $y = x^{1/7}$ will look like. Verify your conjecture.

7. Graph $y = x^{1/4}$, $y = x^{2/4}$, $y = x^{3/4}$, and $y = x^{4/4}$ in a standard window.

a. What patterns do you notice?

b. What do you predict the graph of $y = x^{5/4}$ will look like? Verify your prediction.

8. Compare your observations of the power functions in Exercises 6 and 7 to your previous work with exponential functions. How do the shapes of the curves differ?

9. Identify the following graphs as exponential functions, power functions, or neither.

a.

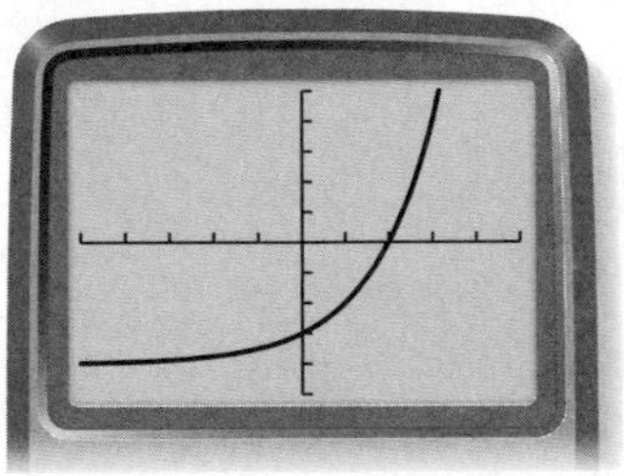

b.

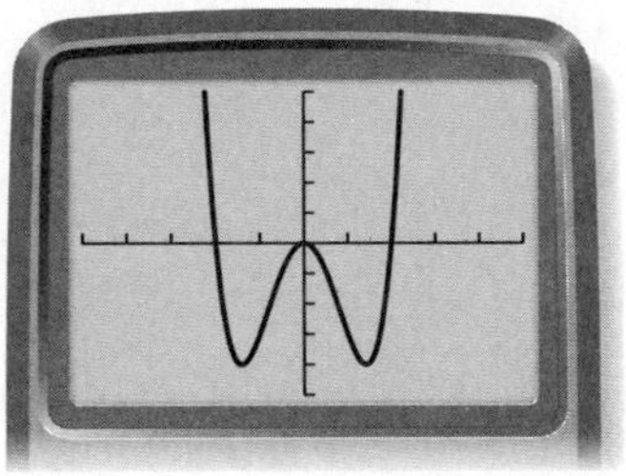

c.

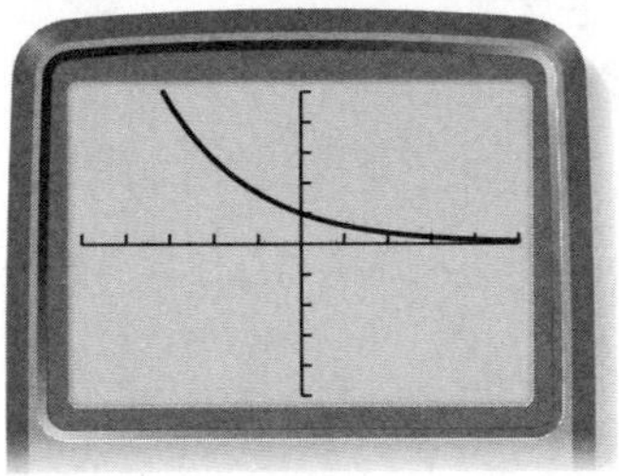

d.

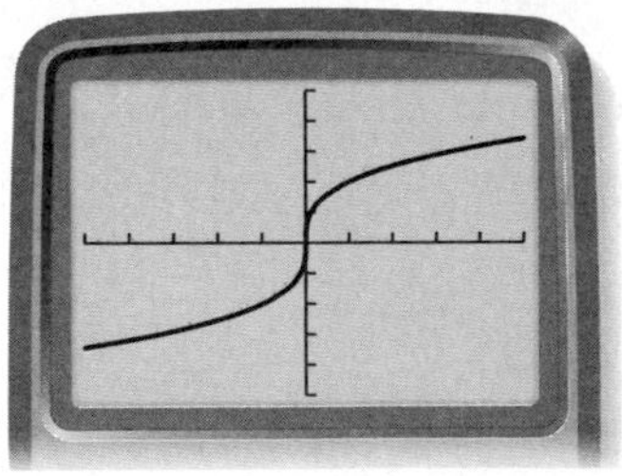

10. Each of these graphs is a transformation of the power function $y = x^{3/4}$. Write the equation for each curve.

a.

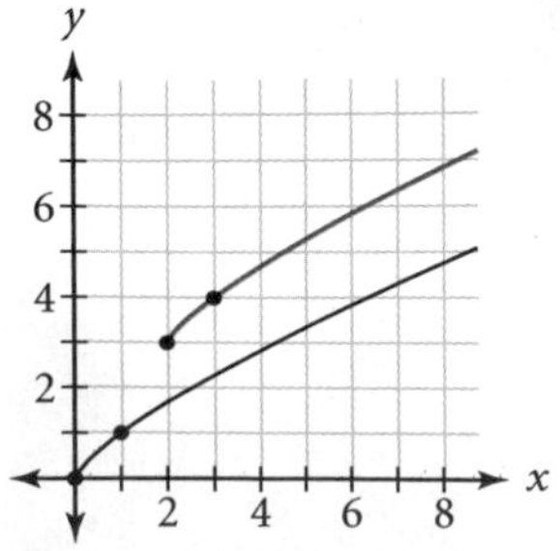

b.

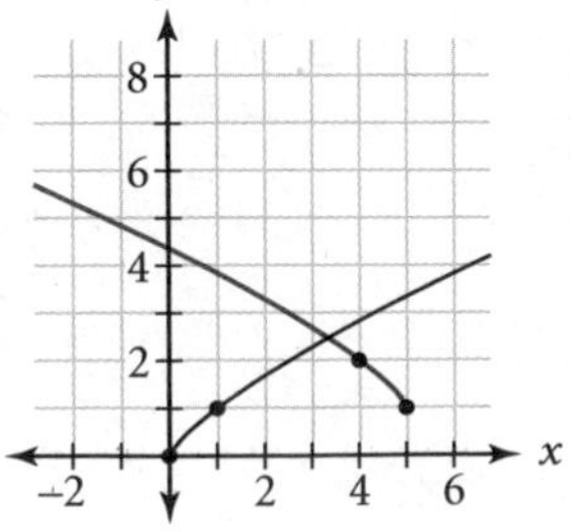

c.

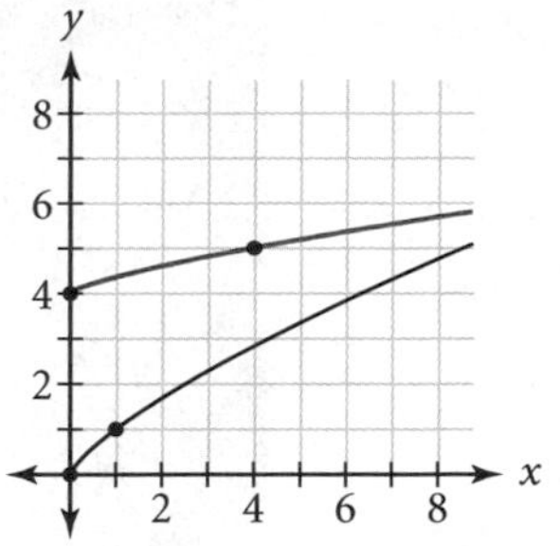

d.

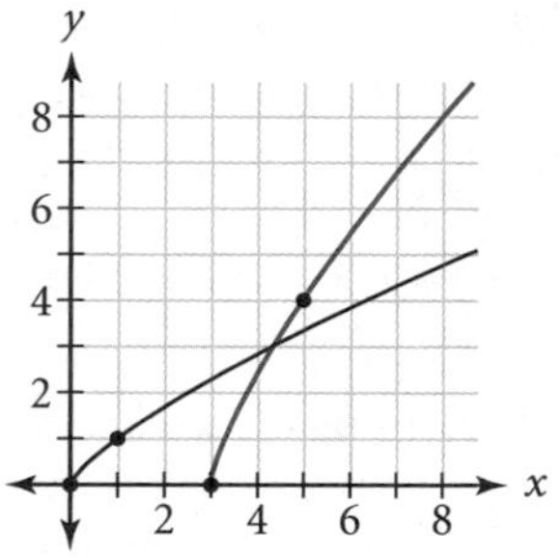

11. APPLICATION Dan placed three colored gels over the main spotlight in the theater so that the intensity of the light on stage was 900 W/cm^2. After he added two more gels, making a total of five over the light, the intensity on stage dropped to 600 W/cm^2. What will be the intensity of the light on stage with six gels over the spotlight, assuming that the intensity of light decays exponentially with the thickness of material covering it?

12. Solve each equation for x.

a. $9\sqrt[5]{x} + 4 = 17$ **b.** $\sqrt{5x^4} = 30$ **c.** $4\sqrt[3]{x^2} = \sqrt{35}$

13. APPLICATION Johannes Kepler discovered in 1609 that the mean orbital radius of a planet, measured in astronomical units (AU), is equal to the time of one revolution around the Sun (measured in years) raised to a power of $\frac{2}{3}$.

a. Venus has a "year" of 0.615 year. What is its radius?

b. Saturn has a radius of 9.541 AU. How long is its "year"?

c. Complete the table below:

Mercury	Venus	Earth	Mars	Jupiter	Saturn	Uranus	Neptune
0.387 AU			1.523 AU		9.542 AU		30.086AU
	0.615 yr	1.00 yr		11.861 yr		84.008 yr	

14. Invent a situation that can be modeled by the equation $y = 400(1 - 0.15)^x$. Explain the meaning of every number in this equation.

15. Write an equation of the form $y = ab^x$ that matches the data in this table. Use your equation to find the missing values.

x	y
0	
2	206.045
3	298.76525
5	
	1915.0058

16. APPLICATION Discovered by Robert Boyle in 1662, Boyle's law gives the relationship between pressure and volume of gas if temperature and amount are held constant. If the volume, V, of a container is increased, the pressure, P, decreases. If the volume of a container is decreased, the pressure increases. One way this rule can be written mathematically is $P = kV^{-1}$, where k is a constant.

a. Show that this formula is equivalent to $PV = k$.

b. If a gas occupies 12.3 L at a pressure of 40.0 mm Hg (millimeters of mercury), find the constant, k.

c. What is the volume when the pressure is increased to 60.0 mm Hg?

d. If the volume were 15 L, what would the pressure be?

Science Connection Scuba divers are trained in the effects of Boyle's law. As divers ascend, water pressure decreases, so the air in the lungs expands. It is relatively safe to make an emergency ascent from a depth of 60 feet, but you *must* exhale as you do so. If a diver were to hold his or her breath while ascending, the expanding oxygen in the lungs would cause rupturing air sacs, and bleeding in the lungs. Death would be likely.

A scuba diver swims beneath a coral reef in the Red Sea.

17. If an equation is linear, input values 1 unit apart have a common difference for their output values. Subtracting consecutive entries in a table always gives the same answer. If an equation is exponential, input values 1 unit apart have a common ratio for their output values. Dividing consecutive entries in a table always gives the same answer. Certain power functions can also be identified in a table.

a. Complete the table for $y = 3x^2$.

x	y	Difference between consecutive y-values	Difference of the differences
0	0		
1	3	3	
2	12	9	6
3	27	15	6
4	48		
5	75		

What do you notice?

b. Create a similar table for $y = 2x^3$. Do you see a pattern? If you add a column and take differences again, do you see a pattern?

c. Write a rule to determine whether a table of values comes from an equation of the form $y = ax^n$ where n is a whole number.

18. Determine whether the value in each table was generated by a power function, an exponential function, or neither.

a.

x	y
1	0.5
2	2
3	4.5
4	8
5	12.5
6	18

b.

x	y
−1	−3
0	−5
1	−3
2	−1
3	1
4	3

c.

x	y
3	29.7
4	70.4
5	137.5
6	237.6
7	377.3
8	563.2

d.

x	y
2	243
3	81
4	27
5	9
6	3
7	1

Review

19. (Lesson 5.1) Use properties of exponents to find an equivalent expression (if possible) in the form $a \cdot x^n$.

a. $(3x^3)^3$

b. $(2x^3)(2x^2)^3$

c. $\dfrac{6x^4}{30x^5}$

d. $(4x^2)(3x^2)^3$

e. $\dfrac{-72x^5y^5}{-4x^3y}$

f. $\dfrac{6x^3 - 3x^2}{3x}$

20. (Chapter 4) Write equations of these graphs.

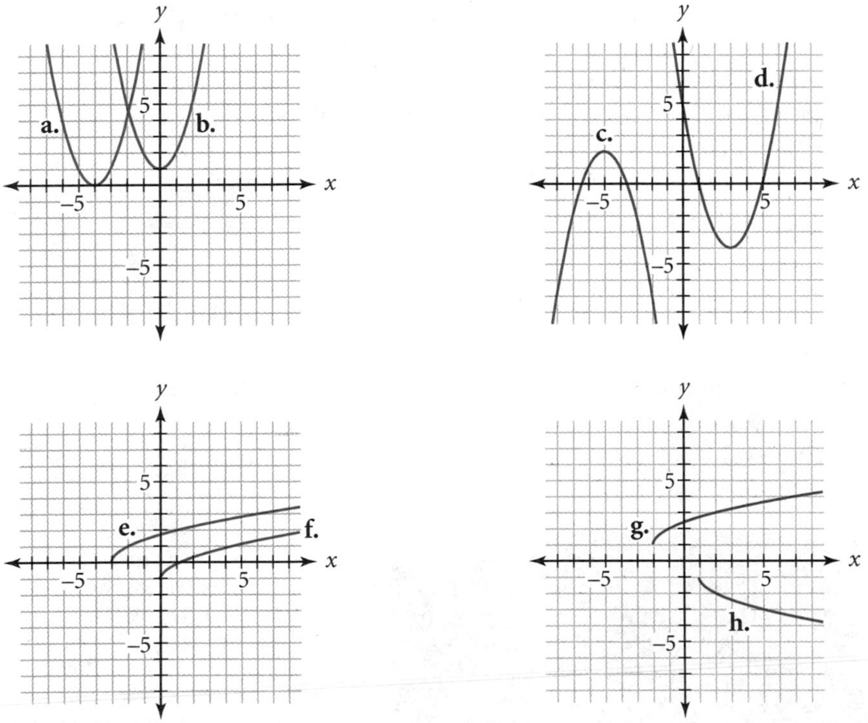

21. (Lesson 2.2) The town of Hamlin has a growing rat population. Eight summers ago there were 20 rat sightings, and the number has been increasing by about 20% each year.

a. Give a recursive rule that models the increasing rat population. Use the number of rats in year 1 as u_1.

b. About how many rat sightings do you predict for this year?

c. Define variables and write an equation that models the continuous growth of the rat population.

PROJECT

POWERS OF 10

How much longer does it take 1 billion seconds to pass than 1 million? How much taller are you than an ant? Are there more grains of sand on a beach than stars in the sky? You've studied scientific notation and exponents and how investments and other quantities change exponentially. But just how different are 10^9 and 10^{10}?

In this Project you'll identify as many objects as you can whose size or number represents a power of 10. Your objects can be related in some way, but they don't have to be. For instance, what is the area of your kitchen? Your house? The state you live in? The land on Earth?

First decide what you're going to measure: length, area, volume, speed, quantity, or any other unit of measure. Then find at least one object with a measurement on the order of each power of 10. Usually, when scientists describe something as "on the order of," they just look at the power of 10 without considering whether the number may be closer to the higher power. For example, 9.2×10^3 is on the order of 10^3. Decide for yourself how you'll deal with these numbers (does 80 count as 10^1 or 10^2?), but be consistent. You'll probably find some powers of 10 more easily than others. Are there any powers of 10 for which you can't find an object? Is there a largest power of 10 that you can find? A smallest power of 10?

Your project should include

- A list of the object or objects you found for each power of 10 and a source or calculation for each measurement. Try to include at least 15 powers of 10.
- An explanation of any powers of 10 you couldn't find and the largest and smallest values you found, if there are any. Don't forget negative powers.
- A visual aid or written explanation showing the different scales of your objects. If your objects are related, include an explanation of how they're related.

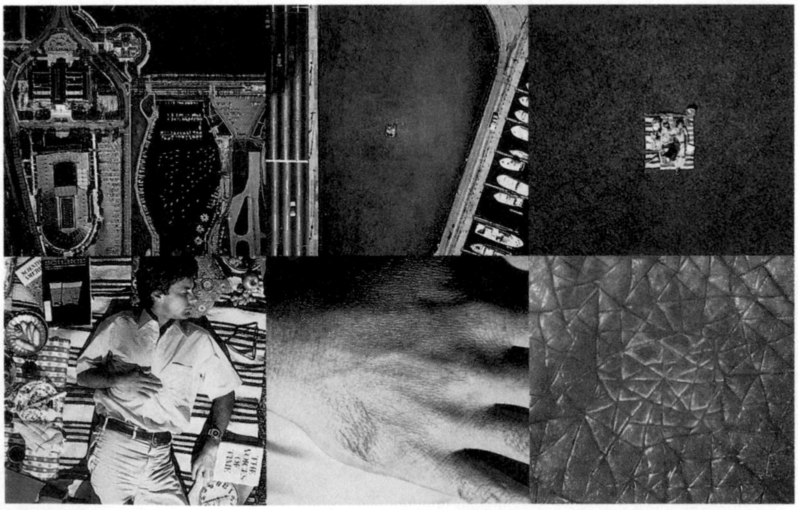

The film *Powers of Ten* (1977), by American designers Charles Eames (1907–1978) and Ray Eames (1912–1988), explores the vastness of the universe using the powers of ten. The film begins with a 1-meter-square image of a man in a Chicago park, which represents 10^0. Then the camera moves 10 times farther away each ten seconds until it reaches the end of the universe, representing 10^{25}. Then the camera zooms in so that the view is ultimately an atom inside the man, representing 10^{-18}. These stills from the film show images representing 10^3 to 10^{-2}.

LESSON

5.4

Applications of Power Equations

You have seen that many equations can be solved by undoing the order of operations. In Lesson 5.3 you applied this strategy to some simple power equations. The strategy also applies for more complex power equations that arise in application problems.

EXAMPLE A

Rita wants to invest \$500 in a savings account so that its doubling time will be eight years. What annual percentage rate is necessary for this to happen? (Assume the account is compounded annually.)

► ***Solution***

If the doubling time is eight years, the initial deposit of \$500 will double to \$1000. The interest rate, r, is unknown. Write an equation and solve for r.

$1000 = 500(1 + r)^8$	Original equation.
$2 = (1 + r)^8$	Undo the multiplication by 500 by dividing both sides by 500.
$2^{1/8} = ((1 + r)^8)^{1/8}$	Undo the power of 8 by raising both sides to the power of 1/8.
$2^{1/8} = 1 + r$	Use the properties of exponents.
$2^{1/8} - 1 = r$	Undo the addition of 1 by subtracting 1 from both sides.
$0.0905 \approx r$	Use a calculator to evaluate $2^{1/8} - 1$.

Rita will need an account with an annual percentage rate of approximately 9.05%.

You have also seen how the general form of an exponential equation can be used in real-world applications. The next example shows a more complex exponential application.

EXAMPLE B

A motion sensor is used to measure the distance between it and a pendulum. The motion sensor's "trigger" is used to collect the closest distance for every 10th swing. At rest, the pendulum hangs 1.25 m from the motion sensor. Find an equation that models these data.

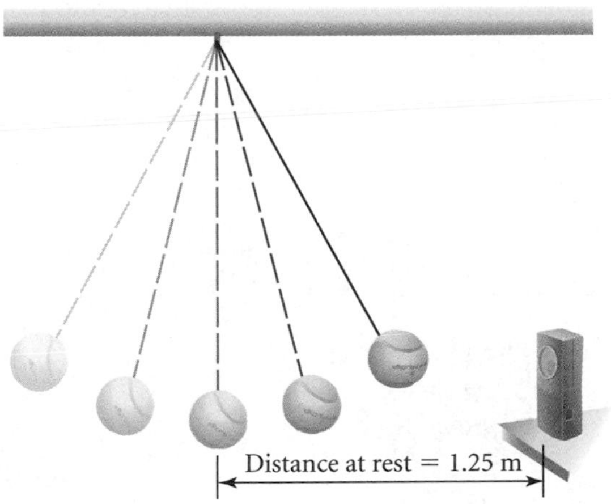

Swing number x	0	10	20	30	40	50	60
Closest distance (m) y	2.35	1.97	1.70	1.53	1.42	1.36	1.32

▶ Solution

Plot these data. The points form a curve, so the data are not linear. As the pendulum slowly stops swinging, these data approach a long-run value of 1.25 m, so the pattern is one of geometric decay. An exponential equation will best model these data.

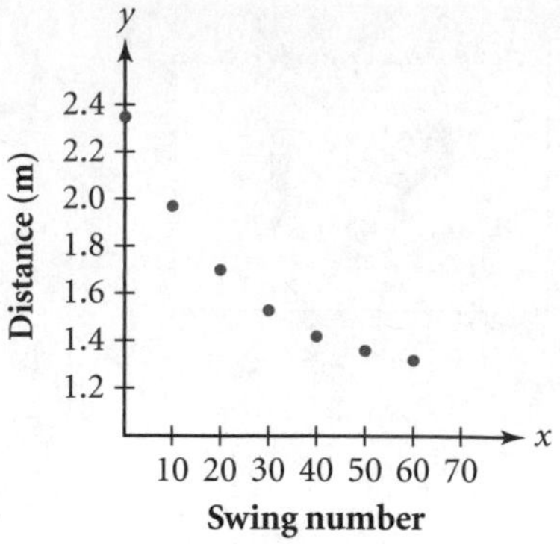

An exponential decay function $y = ab^x$ will approach a long-run value of 0. Since these data approach the long-run value of 1.25, the exponential function must be translated up 1.25 units. Therefore, the equation that models the pendulum will be $y = ab^x + 1.25$, or $y - 1.25 = ab^x$.

The table can be rewritten to show values of $y - 1.25$:

Swing number x	0	10	20	30	40	50	60
Closest distance (m) $y - 1.25$	1.10	0.92	0.45	0.28	0.17	0.11	0.07

This shows that $a = 1.10$.

Using the first and last pair of values for x and $y - 1.25$,

$$ab^{60} = 0.07$$

$$b^{60} \approx 0.063636 \qquad \text{Substitute for } a \text{ and divide.}$$

$$(b^{60})^{\frac{1}{60}} \approx (0.063636)^{\frac{1}{60}} \qquad \text{Undo raising to the 60th power by taking the 60th root.}$$

$$b \approx 0.9551$$

Therefore, the decay rate is approximately 0.9551, and the equation that approximately models the pendulum is $y = (1.10)(0.9551)^x$.

In the last example the exponential equation did not exactly match the pendulum movement. But when you work with real measurements, the best you will get are close values for the constant and a small variation without a pattern.

EXERCISES

▶ Practice Your Skills

1. Solve for x.

a. $x^5 = 50$ **b.** $\sqrt[3]{x} = 3.1$ **c.** $x^2 = -121$

2. Solve for x.

a. $x^{1/4} - 2 = 3$ **b.** $4x^7 - 6 = -2$ **c.** $33(x^{2/3} + 5) = 207$

d. $1450 = 800\left(1 + \frac{x}{12}\right)^{7.8}$ **e.** $14.2 = 222.1 \cdot x^{3.5}$

3. Rewrite each expression in the form ax^n.

a. $(27x^6)^{2/3}$ b. $(16x^8)^{3/4}$ c. $(36x^{-12})^{3/2}$

Reason and Apply

4. **APPLICATION** A sheet of translucent glass 1 mm thick is designed to reduce the intensity of light. If six sheets are placed together, then the outgoing light intensity is 50% of the incoming light intensity. What is the reduction rate of one sheet in this exponential relation?

5. Natalie performs a decay simulation using small colored candies with a letter printed on one side. She starts with 200 candies and pours them onto a plate. She removes all the candies on which the letter faces up, counts the remaining candies, and then repeats the experiment using the remaining candies. Below are her data for each stage.

Stage Number x	0	1	2	3	4	5	6
Candies remaining y	200	105	57	31	18	14	12

After Stage 6, she checks the remaining candies and finds that seven did not have a letter on either side.

a. Natalie uses an exponential equation, $y = ab^x$, to model her data. What must she do to the equation to account for the seven unmarked candies? Write the equation.

b. Find values of a and b.

6. According to the Consumer Price Index in July 2002 the average cost of a gallon of whole milk was $2.74. At the July 2002 rate of inflation for all items, a gallon will cost $3.41 in the year 2024. What was the rate of inflation for all items in July 2002?

7. **APPLICATION** There is a power relationship between the radius of an orbit and the time of one orbit for the moons of Saturn. (Eleven of Saturn's 22 moons are listed).

a. Make a scatter plot of these data.

b. Experiment with different values of a in the power equation $y = ax^2$ to find a good fit for the data.

c. Experiment with different values of a in the linear equation $y = ax$ to find a good fit for the data.

d. Experiment with different values of a in the power equation $y = ax^{1.5}$ to find a good fit for the data.

e. Which model, of those in 7b, 7c, or 7d, seems to best fit the data? Can you find an even better model by adjusting both a and b in the equation $y = ab^x$?

f. Use your model to find the orbit radius for Titan, which has an orbit time of 15.945 days.

g. Using your model, find the orbit time for Phoebe, which has an orbit radius of 12,952,000 km.

Moons of Saturn

Moon	Radius (100,000 km)	Orbit time (days)
Atlas	1.3767	0.602
Prometheus	1.3935	0.613
Pandora	1.4170	0.629
Epimetheus	1.5142	0.694
Janus	1.5147	0.695
Mimas	1.8554	0.942
Enceladus	2.3804	1.370
Tethys	2.9467	1.888
Dione	3.7742	2.737
Helene	3.7806	2.739
Rhea	5.2710	4.518

(*www.solarsystem.nasa.gov*)

8. **APPLICATION** relationship between the weight in tons, W, and the length in feet, L, of a sperm whale is given by the formula $W = 0.000137L^{3.18}$.

a. An average sperm whale is 62 ft long. What is its weight?

b. How long would a sperm whale be if it weighed 75 tons?

9. **APPLICATION** In order to estimate the height of an *Ailanthus ailtissima* tree, botanists have developed the formula $h = \frac{5}{3}d^{.8}$, where h is the height measured in meters and d is the diameter in centimeters.

a. If the height of an *Ailanthus ailtissima* tree is 18 m, find the diameter.

b. If the circumference of an *Ailanthus ailtissima* tree is 87 cm, estimate its height.

10. **APPLICATION** Fat reserves in birds are related to body mass by the formula $F = 0.033 \cdot M^{1.5}$, where F represents the mass in grams of the fat reserves and M represents the total body mass in grams.

a. How many grams of fat reserves would you expect in a 15-g warbler?

b. What percent of this warbler's body mass is fat?

ScienceConnection Allometry is the study of size relationships between different features of an organism as a consequence of growth. Many characteristics may vary greatly among different organisms, but within an organism there may exist a fairly constant relationship or very similar growth patterns during a particular stage of development. The study of these relationships produces mathematical models that can be used outside the laboratory.

Review

11. **APPLICATION** (Lesson 5.1) A sample of a particular radioactive material has been decaying for 5 years. After 2 years there were 6.0 g of radioactive material left. After 5 years there are 5.2 g left.

a. What is the rate of decay?

b. How much radioactive material was initially in the sample?

c. Find an equation to model the decay.

d. How much radioactive material will be left after 50 years (45 years from now)?

e. What is the half-life of this radioactive material?

12. **APPLICATION** (Lesson 1.4) In his geography class, Juan conjectures that more people live in cities that are warm (above 50°F) in the winter than in cities that are cold (below 32°F). To examine his conjecture, he collects the mean temperatures for January of 25 of the largest U.S. cities. These cities contain about 33.5 million people, or about 12% of the population in 2000.

31.8°, 56.0°, 21.4°, 51.4°, 31.2°, 52.3°, 56.8°, 44.0°, 50.4°, 23.4°, 26.0°, 48.5°, 53.2°, 27.1°, 49.1°, 32.7°, 39.6°, 18.7°, 29.6°, 35.2°, 37.1°, 44.2°, 39.1°, 29.5°, 40.5°

(*Time Almanac 2002*)

a. Construct a box plot of these data.

b. List the five-number summary.

c. What are the range and the interquartile range for these data?

d. Do the data support Juan's conjecture? Explain your reasoning.

LESSON 5.5

Building Inverses of Functions

Success is more a function of consistent common sense than it is of genius.

AN WANG

Alex and Sasha are sharing their graphs of the same set of data.

"I know my graph is right!" exclaims Alex. "I've checked and rechecked it. Yours must be wrong, Sasha."

Sasha disagrees. "I've entered these data in my calculator, and I made sure I entered the correct numbers."

The graphs are pictured below. Can you explain what is happening?

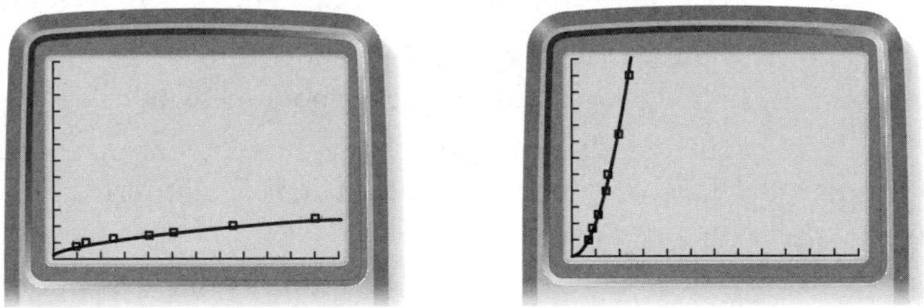

This lesson is about the **inverse** of a function—the exchange of the independent variable is with the dependent variable. Look again at Alex's and Sasha's graphs. If they labeled the axes, they might see that the only difference is their choice of independent variables. In some real-world situations, it makes sense for either of two related variables to be used as the independent variable. In the Investigation you will find some inverses and then discover how they relate to the original function.

Investigation
The Inverse

Consider these functions.

i. $f(x) = 6 + 3x$

ii. $f(x) = \sqrt{x + 4} - 3$

iii. $f(x) = (x - 2)^2 - 5$

iv. $f(x) = 2 + \sqrt{x + 5}$

v. $f(x) = \frac{1}{3}(x - 6)$

vi. $f(x) = \sqrt[3]{5x}$

Part 1

Find out how your calculator draws the inverse of a function.

a. Using your calculator, graph each function, and then draw its inverse by hand. (Switch the x- and y-coordinates of the function.) Sketch both the function and its inverse on your paper.

b. Give the coordinates of at least three points on the inverse.

c. Using what you learned in Chapter 4, find a function (or functions) to fit the inverse. Check your response by adding this graph to your calculator to see if it matches the inverse.

d. Record the equations of your function and its inverse in a table on your paper, and then repeat Part 1 with another function.

Part 2

a. Study the sketches you made of functions and their inverses. What observations can you make about the graphs of a function and its inverse?

b. Look at the graphs and equations of the pair i and v and of the pair iii and iv. What observations can you make about these pairs?

c. After studying the equations you wrote for the functions and their inverses, explain how you could find the equation of an inverse of a function without looking at its graph.

EXAMPLE A

A 589-mi flight from Washington, D.C., to Chicago took 118 min. Another flight, one of 1452 mi from Washington, D.C., to Denver, took 222 min. Model this relationship both as (time, distance) and as (distance, time). If a flight from Washington, D.C., to Seattle takes 323 min, what is the distance traveled? If the distance between Washington, D.C., and Miami is 910 mi, how long will a flight between these two cities take?

▶ Solution

If you know the time traveled and want to find the distance, then time is the independent variable, and the points are (118, 589) and (222, 1452). The slope is $\frac{1452 - 589}{222 - 118} = \frac{863 \text{ mi}}{104 \text{ min}} \approx 8.3$ mi/min. Using the first point to find the vertical intercept, you get $589 = b + \frac{863}{104}(118)$. So b, the vertical intercept, is $-\frac{40578}{104}$, and the equation is $d = -\frac{40578}{104} + \frac{863}{104}t$.

To find the distance between Washington, D.C., and Seattle, substitute 323 for *time.*

$$d = -\frac{40578}{104} + \frac{863}{104}(323)$$

The distance is 2290 mi.

If you know distance and want to find time, then distance is the independent variable. The two points then are (589, 118) and (1452, 222). This makes the slope $\frac{222 - 118}{1452 - 589} = \frac{104 \text{ min}}{863 \text{ mi}} \approx 0.12$ min/mi. Using the first point again, the equation is $t = \frac{40578}{863} + \frac{104}{863}d$.

To find the time of a flight from Washington, D.C., to Miami, substitute 910 for *distance.*

$$t = \frac{40578}{863} + \frac{104}{863}(910)$$

The flight will take 157 min.

You can also use the first equation for *distance*, and solve for t to get the second equation, for *time.*

$$d = -\frac{40578}{104} + \frac{863}{104}t \qquad \text{First equation.}$$

$$\frac{40578}{104} + d = \frac{863}{104}t \qquad \text{Add } \tfrac{40{,}570}{104} \text{ to both sides.}$$

$$\frac{40578}{863} + \frac{104}{863}d = t \qquad \text{Multiply both sides by } \tfrac{104}{863}.$$

Each of these equations is the inverse of the other. The choice of independent variable has been switched. Graph the two equations on your calculator. What do you notice?

In the Investigation you may have noticed that the inverse of a function is not necessarily a function. Recall from Chapter 4 that any set of points is called a **relation.** A relation may or may not be a function. When both an equation and its inverse are functions, the function is called a *one-to-one function.* When a function is one-to-one, any horizontal line touches its graph in at most one point.

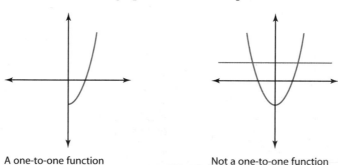

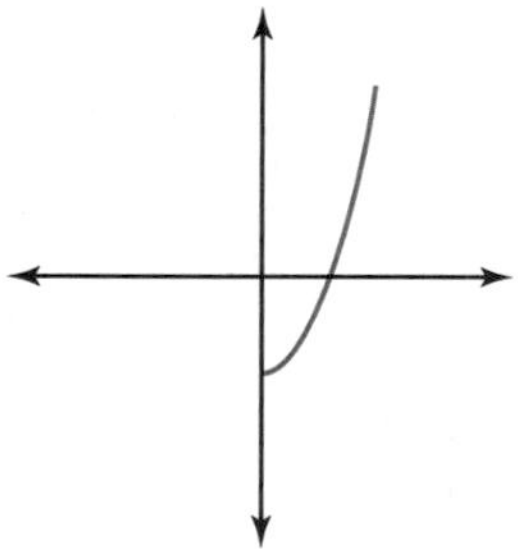

A one-to-one function

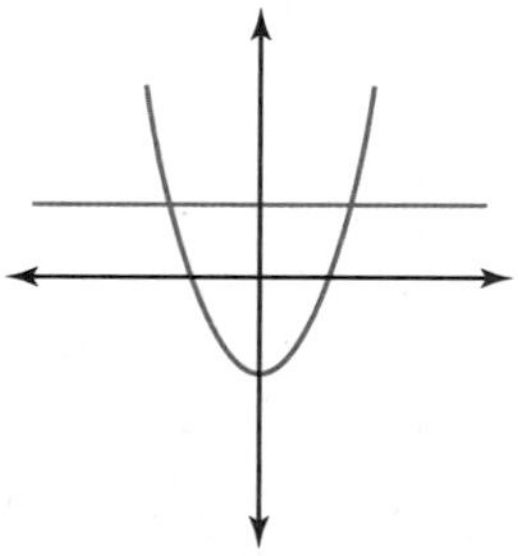

Not a one-to-one function

The Inverse of a Relation

The **inverse** of a relation is obtained by exchanging the x- and y-coordinates of all points or by exchanging the x and y variables in an equation.

The inverse of a one-to-one function $f(x)$ is written as $f^{-1}(x)$. Note that this notation is similar to the notation for an exponent of -1, but $f^{-1}(x)$ refers to the inverse function, not an exponent.

EXAMPLE B

a. Find the inverse function of $f(x) = 4 - 3x$. Then find $f(f^{-1}(x))$.

b. Find the inverse of $g(t) = \frac{5}{t}$. Then find $g(g^{-1}(t))$.

► Solution

a. The first step is to find the inverse. Exchange the independent and dependent variables. Then, solve for the new dependent variable.

$x = 4 - 3y$	Exchange x and y.
$x - 4 = -3y$	Subtract 4 from both sides.
$\dfrac{x-4}{-3} = y$	Divide by -3.
$f^{-1}(x) = \dfrac{x-4}{-3}$	Write in function notation.

The next step is to form the composite function.

$f(f^{-1}(x)) = 4 - 3\left(\dfrac{x-4}{-3}\right)$	Substitute $f^{-1}(x)$ for x in $f(x)$.

Let's see what happens when we distribute and remove some parentheses.

$$f(f^{-1}(x)) = 4 + (x - 4)$$

$$f(f^{-1}(x)) = x$$

What if you had found $f^{-1}(f(x))$ instead of $f(f^{-1}(x))$?

$f^{-1}(f(x)) = \dfrac{(4-3x)-4}{-3}$	Substitute $f(x)$ for x in $f^{-1}(x)$.
$f^{-1}(f(x)) = \dfrac{-3x}{-3}$	Combine like terms in the numerator.
$f^{-1}(f(x)) = x$	Divide.

b.

$t = \dfrac{5}{y}$	Exchange y and t.
$ty = 5$	Multiply both sides by y.
$y = \dfrac{5}{t}$	Divide both sides by t.
$g^{-1}(t) = \dfrac{5}{t}$	Write in function notation.
$g(g^{-1}(t)) = \dfrac{5}{5/t}$	Substitute $g^{-1}(t)$ for t.
$g(g^{-1}(t)) = 5\left(\dfrac{t}{5}\right) = t$	

Algebraically, for one-to-one function $f(f^{-1}(x)) = x$ and $f^{-1}(f(x)) = x$, that is, when you take the composite of a function and its inverse, you get the original value back. How does the graph of $y = x$ relate to the graphs of a function and its inverse? Look carefully at the graphs below to see the relationship of a function and its inverse.

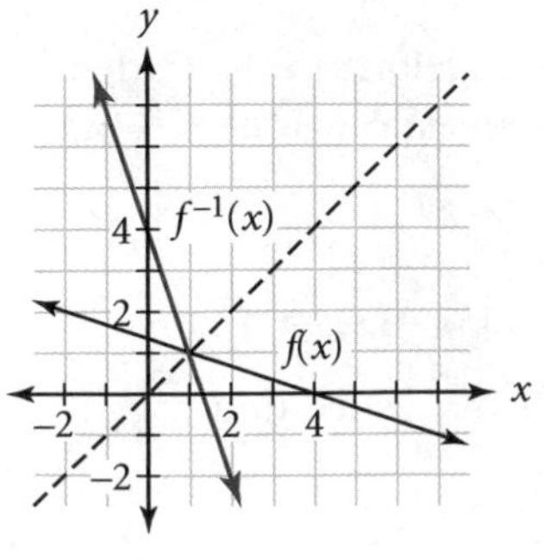

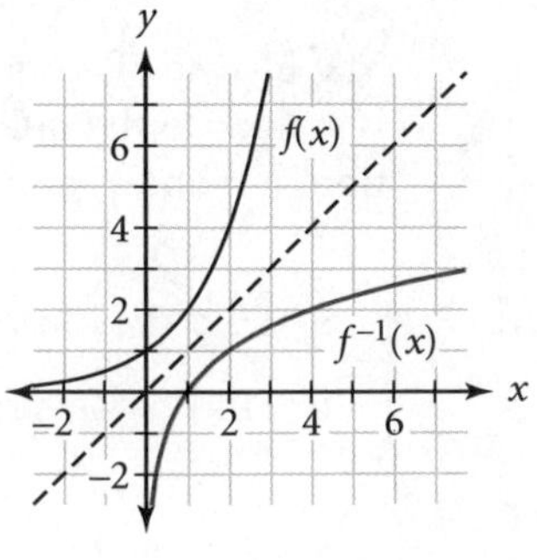

A turntable deejay, like Spooky shown here, applies special effects and mixing techniques to alter an original source of music. If you consider the original record to be one function and the effects or a second record to be another function, the music that the deejay creates is a composition of functions.

EXERCISES

Practice Your Skills

1. A function $f(x)$ contains the points $(-2, -3)$, $(0, -1)$, $(2, 2)$, and $(4, 6)$. Give the points known to be in the inverse of $f(x)$.

2. Given $g(t) = 5 + 2t$, find each value:

a. $g(2)$ **b.** $g^{-1}(9)$ **c.** $g^{-1}(20)$

3. Which one of graphs a, b, or c is the inverse of the graph at far right? Explain how you know.

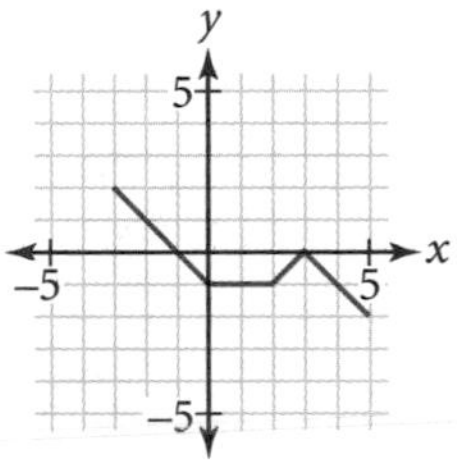

a.

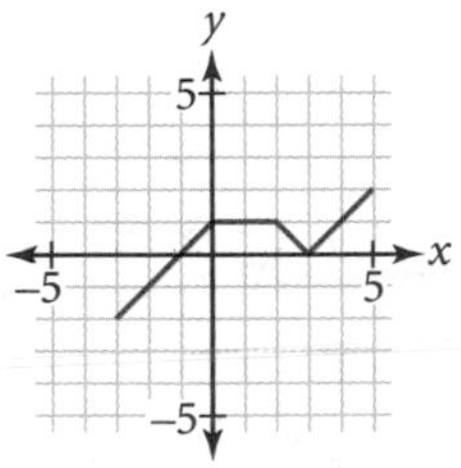

b.

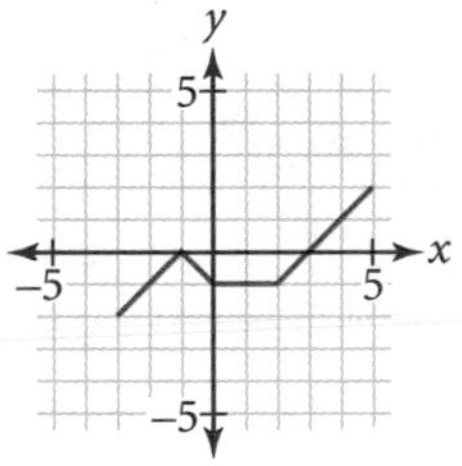

c.

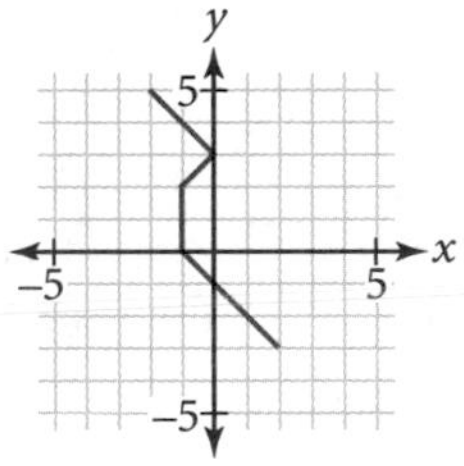

4. Match each function with its inverse.

a. $y = 6 - 2x$

b. $y = 2 - \frac{6}{x}$

c. $y = -6(x - 2)$

d. $y = \frac{-6}{x - 2}$

e. $y = \frac{-1}{2}(x - 6)$

f. $y = \frac{2}{x - 6}$

g. $y = 2 - \frac{1}{6}x$

h. $y = 6 + \frac{2}{x}$

Reason and Apply

5. Given the functions $f(x) = -4 + 0.5(x - 3)^2$ and $g(x) = 3 + \sqrt{2(x + 4)}$:

a. Find $f(7)$ and $g(4)$.

b. What does this imply?

c. Find $f(1)$ and $g(-2)$.

d. What does this imply?

e. Over what part of the domain of f are f and g inverse functions? (Suggestion: Graph both equations.)

6. Given $f(x) = 4 + (x - 2)^{\frac{3}{5}}$:

a. Solve for x when $f(x) = 12$.

b. Find $f^{-1}(x)$ symbolically.

c. How are solving for x and finding an inverse alike? How are they different?

7. Consider the equation $f(x) = 4 + (x - 2)^{\frac{3}{5}}$, given in Exercise 6.

a. Graph $\text{Y1} = 4 + (x - 2)^{\frac{3}{5}}$, and use your calculator to draw the inverse of Y1.

b. Graph the inverse function you found in Exercise 6b. How does it compare to the inverse drawn by your calculator?

c. How can you determine whether your answer to 6b is correct?

8. Write each function using $f(x)$ notation, then find its inverse. If the inverse is a function, write it using $f^{-1}(x)$ notation.

a. $y = 2x - 3$

b. $3x + 2y = 4$

c. $x^2 + 2y = 3$

9. For the functions given in a and b, find:

a. $f(x) = 6.34x - 140$

b. $f(c) = 1.8c + 32$

i. $f^{-1}(x)$

ii. $f(f^{-1}(15.75))$

iii. $f^{-1}(f(15.75))$

iv. $f(f^{-1}(x))$ and $f^{-1}(f(x))$

Note that the equation in 9b will convert Celsius temperatures to Fahrenheit temperatures. You will use either this function or its inverse in Exercise 10.

10. **APPLICATION** The data in the table describe the relationship between altitude and air temperature.

Feet	Meters	°F	°C
1,000	300	56	13
5,000	1,500	41	5
10,000	3,000	23	−5
15,000	4,500	5	−15
20,000	6,000	−15	−26
30,000	9,000	−47	−44
36,087	10,826	−69	−56

A caribou stands in front of Mt. McKinley in Denali National Park, Alaska. Mt. McKinley reaches an altitude of 6194 m above sea level.

a. Write a best-fit equation for $f(x)$ that describes the relationship (*altitude in meters, temperature in* °C). Use at least three decimal places in your answer.

b. Use your results from 10a to write the equation of $f^{-1}(x)$.

c. Write a best-fit equation of $g(x)$, describing (*altitude in feet, temperature in* °F).

d. Use the results of 10c to write the equation of $g^{-1}(x)$.

e. What would be the temperature in °F at the summit of Mt. McKinley, which is 6194 m high?

f. Write a composition of functions that would also provide the answer for 10e. You will use $f(c)$ or its inverse $c(f)$ from 9b.

HistoryConnection Anders Celsius (1701–1744) was a Swedish astronomer. His thermometric scale used the freezing and boiling temperatures of water as reference points; freezing corresponded to 100° and boiling to 0°. His colleagues at the Uppsala Observatory reversed his scale five years later, giving us the current version, on which freezing corresponds to 0° and boiling to 100°. This thermometer was known as the "Swedish thermometer" until the 1800s, when people started referring to it as the "Celsius thermometer." This temperature scale is also sometimes called centigrade.

11. On Celsius's original scale, freezing corresponded to 100° and boiling corresponded to 0°.

a. Write a formula that converts a temperature given by today's Celsius scale to the scale that Celsius invented. (See the History Connection).

b. Explain how you would convert a temperature given in degrees Fahrenheit to a temperature on the original scale that Celsius invented.

12. Here is the paper Alum turned in for a recent quiz in her mathematics class. If it is a four-point quiz, what is Alum's score? For each problem that Alum did not answer correctly, provide the correct answer, and explain it so that next time she will get it right!

QUIZ

1. Rewrite x^{-1}.

Answer: $\frac{1}{x}$

2. What does $f^{-1}(x)$ mean?

Answer: $\frac{1}{f(x)}$

3. Rewrite $9^{-1/5}$.

Answer: $\frac{1}{9^5}$

4. What number is 0^0 equal to?

Answer: 0

13. In looking over his water bills for the past year, Mr. Aviles saw that he was charged a basic monthly fee of \$7.18 and \$3.98 per thousand gallons (gal) used.

a. Write the monthly cost function in terms of the number of thousands of gallons used.

b. What is his monthly bill if he uses 8000 gal of water?

c. Write a function for the number of thousands of gallons used in terms of the cost.

d. If his monthly bill was \$54.94, how many gallons of water were used?

e. Show that the functions from 13a and 13c are inverses.

f. Mr. Aviles decides to fix his leaky faucets. He calculates that he is wasting 50 gal/d. About how much money will he save on his monthly bill?

g. A gallon is 231 cubic inches (in.^{3}). Find the dimensions of a rectangular container that will hold the contents of the water saved by Mr. Aviles in a month.

Environmental Connection Although about two-thirds of the world is covered with water, only 1% of Earth's water is available for drinking. Many of these freshwater sources are becoming increasingly polluted or affected by changing weather patterns. Increasing population and consumption and the building of dams and reservoirs are damaging ecosystems around the world. Conservation of water is critical.

Leaks account for nearly 12% of the average household's annual water consumption. About one in five toilets leaks at any given time, and that can waste more than 50 gal of water per day. Dripping sinks add up fast, too. A faucet that leaks one drop per second can lose as much as 30 gal per day.

Review

14. (Lesson 5.3) Rewrite the expression $125^{2/3}$ in as many different ways as you can.

15. (Lesson 5.1) Find an exponential function that contains the points (2, 12.6) and (5, 42.525).

16. (Lesson 5.1) Solve by rewriting with the same base.

a. $4^x = 8^3$

b. $3^{4x+1} = 9^x$

c. $2^{x-3} = \left(\frac{1}{4}\right)^x$

17. (Lesson 4.4) Give equations of two parabolas with vertex (3, 2) passing through the point (4, 5).

18. (Lesson 3.4) A rectangle has a perimeter of 155 in. Its length is 7 more than twice its width.

a. Write a system of equations to model the rectangle.

b. Solve the system to find the rectangle's dimensions.

LESSON

5.6

The Logarithmic Function

If all art aspires to the condition of music, all the sciences aspire to the condition of mathematics.

GEORGE SANTAYANA

You can model many phenomena, both natural and human made, with exponential functions. You have already used several methods to solve for x when it is contained in an exponent. In special, rare instances it is possible to solve by finding a common base. For example, finding the value of x that makes each of these equations true is straightforward because of your experience with exponents.

$$10^x = 1000 \qquad 3^x = 81 \qquad 4^x = \frac{1}{16}$$

Solving the equation $10^x = 47$ isn't as straightforward, because you probably don't know how to write 47 as a power of 10. You can, however, solve this equation by graphing $y = 10^x$ and $y = 47$ and finding the intersection—the solution to the system and the solution to $10^x = 47$. Take a minute to verify that $10^{1.6720979} \approx 47$.

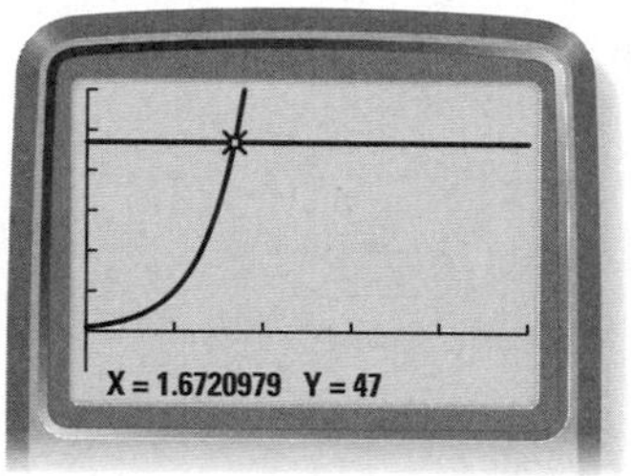

In the next Investigation you will discover an algebraic strategy to solve for x in an exponential equation. You'll use a new function called a *logarithm*, abbreviated *log*. Locate the LOG key on your calculator before going on.

Investigation
Exponents and Logarithms

In this Investigation you'll explore the connection between exponents on the base 10 and logarithms. This will help you understand what a logarithm is and how it works.

Step 1 Enter the equation $Y_1 = 10^x$ into your calculator. Make a table of values for Y1.

Step 2 Enter the equation Y2 $= \log(10^x)$ and compare the table values for Y1 and Y2. What observations can you make? Try starting your table at different values (including negative values), and use different decimal increment values.

Step 3 Based on your observations in Step 2, what are the values of the following expressions? Use the table to verify your answers.

a. $\log(10^{2.5})$

b. $\log(10^{-3.2})$

c. $\log(10^0)$

d. $\log(10^x)$

Step 4 Complete the following statements.

a. If $100 = 10^2$, then $\log 100 = \underline{\ ?\ }$.

b. If $400 \approx 10^{2.6021}$, then $\log \underline{\ ?\ } \approx \underline{\ ?\ }$.

c. If $\underline{\ ?\ } \approx 10^?$, then $\log 500 \approx \underline{\ ?\ }$.

d. If $y = 10^x$, then $\log \underline{\ ?\ } = \underline{\ ?\ }$

Step 5 Use logarithms to solve each equation for x. Check your answers.

a. $300 = 10^x$

b. $47 = 10^x$

c. $0.01 = 10^x$

d. $y = 10^x$

Step 6 Use a window similar to $[-4.7, 4.7, 1, -3.1, 3.1, 1]$ to investigate the graph of $y = \log x$. Is $\log x$ a function? What are the domain and range of $\log x$?

Step 7 Graph $y = 10^x$, and draw its inverse on the same set of axes. Now graph $y = \log x$. What observations can you make?

Step 8 If $f(x) = 10^x$, then what is $f^{-1}(x)$? What is $f(f^{-1}(x))$?

The expression $\log x$ is another way of expressing x as an exponent on the base 10. Ten is the *common base* for logarithms. So, $\log x$ is called a *common logarithm* and is shorthand for writing $\log_{10} x$. You read this as "the logarithm base 10 of x". $\log x$ is the exponent you place on 10 to get x.

EXAMPLE A

Solve $4 \cdot 10^x = 4650$.

▶ Solution

$4 \cdot 10^x = 4650$	Original equation.
$10^x = 1162.5$	Divide both sides by 4.
$x = \log_{10} 1162.5$	The logarithm base 10 of 1162.5 is the exponent you place on 10 to get 1162.5.
$x \approx 3.065392962$	Use the log key on your calculator to evaluate.

The general **logarithm function** is an exponent-producing function. A logarithm base b is the exponent you place on b to get x.

Definition of Logarithm

For $a > 0$ and $b > 0$, $\log_b a = x$ is the same as $a = b^x$.

The general logarithm function is dependent on the base of the exponential expression. The next example demonstrates how to use logarithms to solve exponential equations when the base is not 10.

EXAMPLE B

Solve $8^x = 256$.

▶ Solution

You know that $8^2 = 64$ and $8^3 = 512$, so x must be between 2 and 3 if $8^x = 256$. You can rewrite the equation as $x = \log_8 256$ by the definition of a logarithm. But the calculator doesn't have a built-in logarithm base 8 function. Have we hit a dead end?

One way to solve this equation is to rewrite each side of the equation $8^x = 256$ as a power with a base of 10.

$8^x = 256$	Original equation.
$(10^{0.9031})^x \approx 256$	Log $8 \approx 0.9031$, so $8 \approx 10^{0.9031}$.
$(10^{0.9031})^x \approx 10^{2.4082}$	Log $256 \approx 2.4082$, so $256 \approx 10^{2.4082}$.
$0.9031x \approx 2.4082$	Use the power of a power property of exponents and the common base property of exponents.
$x \approx \frac{2.4082}{.9031} \approx 2.6666$	Divide both sides by 0.9031.

Recall that $x = \log_8 256$, that 2.4082 was an approximation for log 256, and that 0.9031 was an approximation by log 8. The numerator and denominator of the last step above suggest a more direct way to solve $x = \log_8 256$.

$$x = \log_8 256 = \frac{\log 256}{\log 8} = \frac{8}{3} \approx 2.6667$$

The relationship at the end of Example B is called the *logarithm change-of-base property.* It enables you to solve problems involving logarithms with bases other than 10.

Logarithm Change-of-Base Property

$$\log_b a = \frac{\log a}{\log b}$$

where $a > 0$ and $b > 0$

This relationship holds because you can write any number using the inverse functions of logarithms and exponents. Composing functions that are inverses of each other produces an output value that is the same as the input. By definition the equation $10^x = 4$ is equivalent to $x = \log 4$. Substitution from the second equation in the first equation gives you $10^{\log 4} = 4$. More generally, $10^{\log x} = x$, which means that $y = 10^x$ and $y = \log x$ are inverses. This relationship allowed you to change from writing the solution in Example B from base 8 logarithms to the more convenient base 10.

EXAMPLE C

An initial deposit of \$500 is invested at 8.5%, compounded annually. How long will it take the balance to grow to \$800?

▶ *Solution*

Let x represent the number of years the investment is held. Use the general formula for exponential growth, $y = a(1 + r)^x$.

$500(1 + 0.085)^x = 800$	Growth formula for compounding interest.
$(1.085)^x = 1.6$	Divide both sides by 500.
$x = \log_{1.085} 1.6$	Use the definition of a logarithm.
$x = \frac{\log 1.6}{\log 1.085}$	Use the logarithm change-of-base property.
$x \approx 5.7613$	Evaluate.

It will take six years for the balance to grow to \$800.

EXERCISES

Practice Your Skills

1. Rewrite each logarithmic equation in exponential form using the definition of logarithms.

a. $\log 1000 = x$ **b.** $\log_5 625 = x$ **c.** $\log_7 \sqrt{7} = x$

d. $\log_8 2 = x$ **e.** $\log_5 \frac{1}{25} = x$ **f.** $\log_6 1 = x$

2. Solve each equation in Exercise 1 for x.

3. Rewrite each exponential equation in logarithmic form using the definition of logarithms. Then solve for x. (Give your answers rounded to four decimal places.)

a. $10^x = 0.001$ **b.** $5^x = 100$ **c.** $35^x = 8$

d. $0.4^x = 5$ **e.** $0.8^x = 0.03$ **f.** $17^x = 0.5$

4. Graph each equation. Write a sentence explaining how the graph compares to the graph of either $y = 10^x$ or $y = \log x$.

a. $y = \log (x + 2)$ **b.** $y = 3 \log x$ **c.** $y = -\log x - 2$

d. $y = 10^{x+2}$ **e.** $y = 3(10^x)$ **f.** $y = -(10^x) - 2$

Reason and Apply

5. Classify each statement as true or false. If false, change the second part to make the statement true.

a. If $6^x = 12$, then $x = \log_{12} 6$

b. If $\log_2 5 = x$, then $5^x = 2$

c. If $2 \cdot 3^x = 11$, then $x = \dfrac{\log 11}{2 \log 3}$

d. If $x = \dfrac{\log 7}{\log 3}$, then $x = \log_7 3$

6. The function $g(x) = 23(0.94)^x$ gives the Celsius temperature of a bowl of water x minutes after a large quantity of ice is added. After how many minutes will the water reach 5°C?

7. Assume the United States public debt can be estimated with the model $y = 0.051517(1.1306727)^x$, where x represents the number of years since 1900 and y represents the debt in billions of dollars.

a. According to the model, when did the debt pass $1 trillion?

b. According to the model, what is the annual growth rate of the public debt?

c. What is the doubling time for this growth model?

8. **APPLICATION** Carbon-14 is an isotope of carbon that is formed when radiation from the Sun strikes ordinary carbon dioxide in the atmosphere. Trees, which get their carbon dioxide from the air, contain small amounts of carbon-14. Once a tree is cut down, no more carbon-14 is formed, and the amount that is present begins to slowly decay. The half-life of the carbon-14 isotope is 5750 yr.

   a. Find an equation that models the percentage of carbon-14 in a sample of wood. (Assume that at time zero there is 100% and at time 5750 yr there is 50%.)

   b. A piece of wood is found to contain 48.37% of its original carbon-14. According to this information, approximately how long ago did the tree that it came from die? What assumptions are you making, and why is this answer approximate?

Fossilized wood can be found in Petrified Forest National Park. Some of the fossils are more than 200 million years old.

9. **APPLICATION** Crystal looks at an old radio dial and notices that the numbers are not evenly spaced. She hypothesizes that there is an exponential relationship involved. She tunes the radio to 88.7 FM. After six clicks of the tuning knob, she is listening to 92.9 FM.

   a. Write an exponential model. Let x represent the number of clicks, and let y represent the station number.

   b. Use the equation you have found to determine how many clicks Crystal should turn the dial to get from 88.7 FM to 106.3 FM.

Review

10. **APPLICATION** (Lesson 5.1) The C notes on a piano (C_1–C_8) are one octave apart. Their relative frequencies double from one C note to the next.

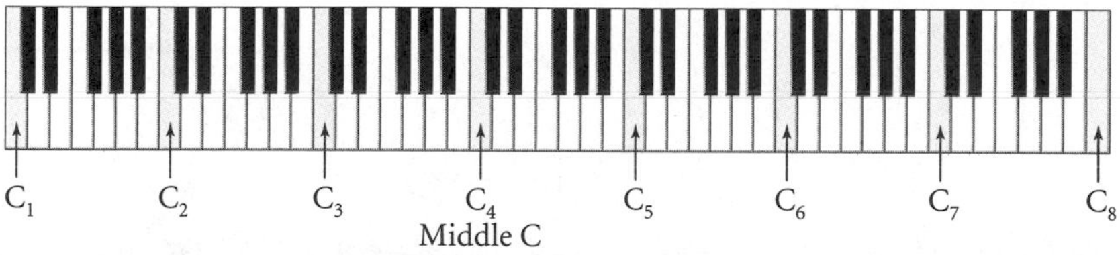

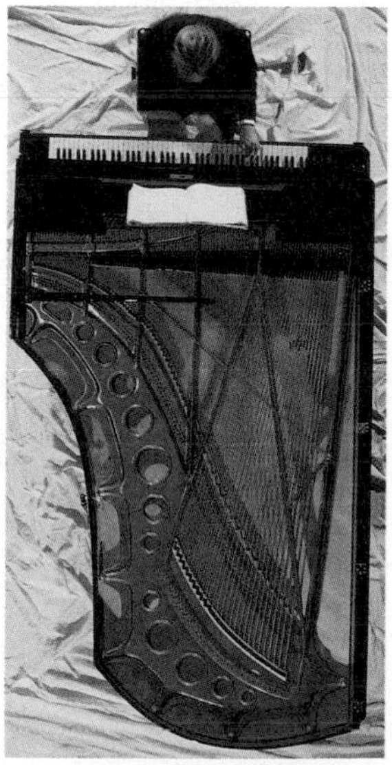

   a. If the frequency of middle C (C4) is 261.6 cycles per second, and the frequency of C5 is 523.2 cycles per second, find the frequencies of the other C notes.

   b. Even though the frequencies of the C notes form a discrete function, you can model it using a continuous explicit function. Write a function model of these notes.

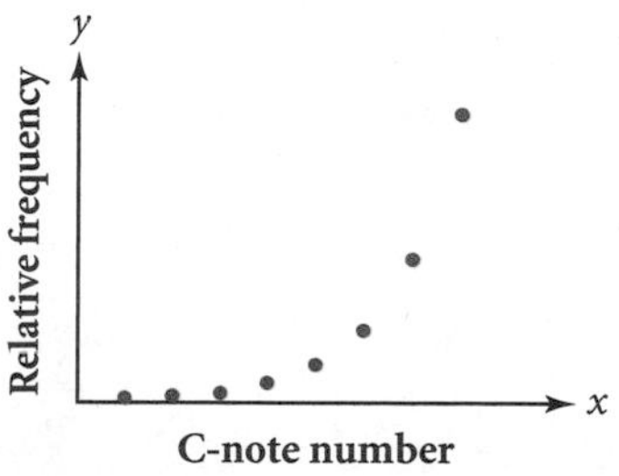

11. (Lesson 5.2) Solve each equation.

a. $(x - 2)^{2/3} = 49$

b. $3x^{2.4} - 5 = 16$

12. APPLICATION (Lesson 3.3) The number of railroad passengers has been increasing in the United States. The table below shows railroad ridership from 1988 to 2000.

Year x	1988	1989	1990	1991	1992	1993	1994	1995	1996	1997	1998	1999	2000
Passengers (millions) y	36.9	38.8	40.2	40.1	41.6	55.0	60.7	62.9	65.6	68.7	75.1	79.8	84.1

(*Time Almanac 2002*)

a. Plot the data, and find a regression line.

b. If the trend continues, what is a good estimate of the ridership in 2010?

13. (Chapter 4) In each case below use the graph and equation of the parent function to write an equation of the transformed image.

a.

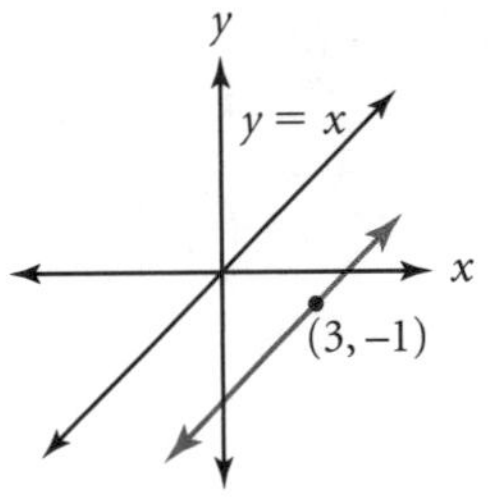

b.

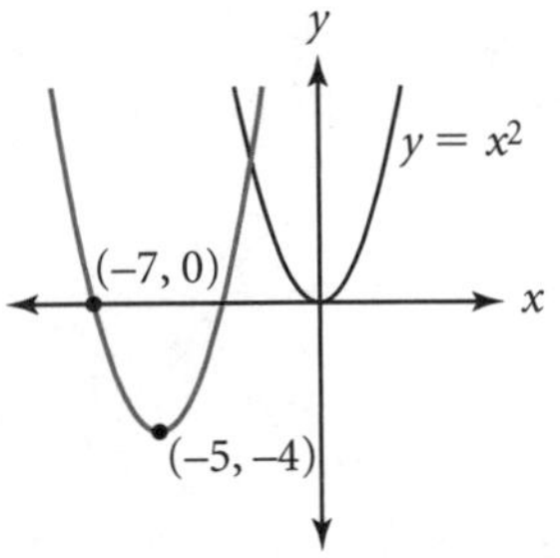

c.

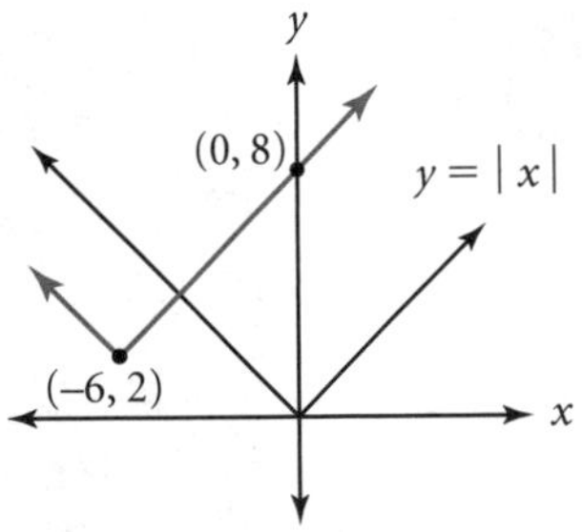

d.

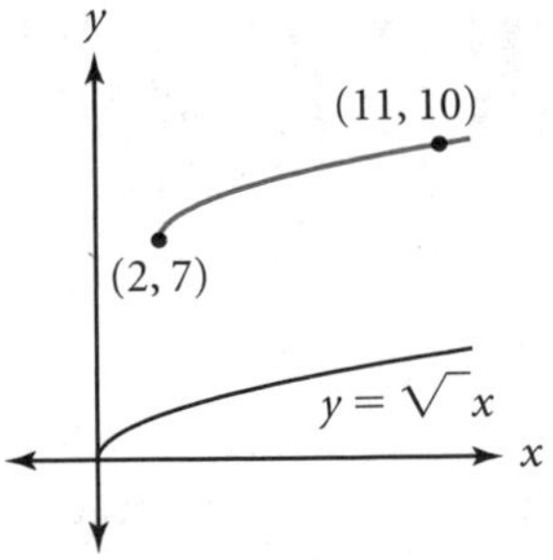

LESSON

5.7

Properties of Logarithms

Each problem that I solved became a rule which served afterwards to solve other problems.

RENÉ DESCARTES

Before machines and electronics were adapted for multiplication, division, and exponentiation, scientists spent long hours doing computations by hand. Early in the 17th century, Scottish mathematician John Napier (1550–1617) discovered a method that greatly reduced the time and difficulty of these calculations, using a table of numbers that he named *logarithms*. As you learned in Lesson 5.6, a logarithm is an exponent, $n = 10^{\log n}$, and you already know how to use the multiplication, division, and power properties of exponents. In the next example you will discover some shortcuts and simplifications.

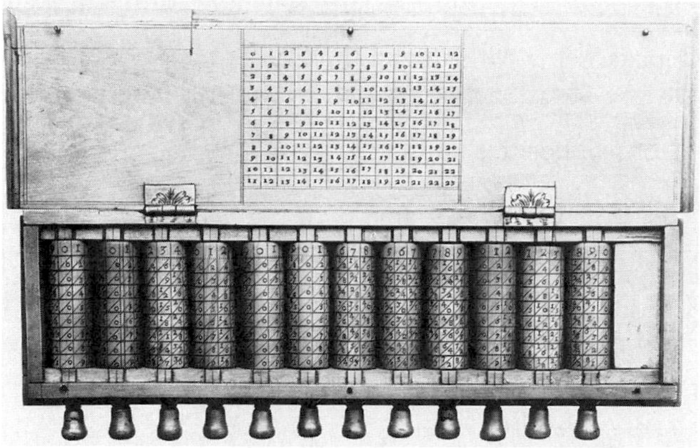

After inventing logarithms, John Napier designed a device for calculating with logarithms in 1617. Later called "Napier's bones," the device used multiplication tables carved on strips of wood or bone. The calculator shown here has an entire set of Napier's bones carved on each spindle. You can learn more about Napier's bones and early calculating devices at www.keycollege.com/online .

EXAMPLE A

a. Multiply 183.47 by 19.628 without using the multiplication key on your calculator.

b. Divide 183.47 by 19.628 without using the division key on your calculator.

c. Raise $4.7^{2.8}$ without using the exponentiation key on your calculator.

▶ Solution

a. Log 183.47 = 2.263565, and therefore $10^{2.263565} = 183.47$.

Log 19.628 = 1.292876, and therefore $10^{1.292876} = 19.628$.

$183.47 \cdot 19.628 = 10^{2.263565} \cdot 10^{1.292976} = 10^{2.263565 + 1.292976} = 10^{3.556441} \approx 3601.148$

b. $\dfrac{183.47}{19.628} = \dfrac{10^{2.263565}}{10^{1.292876}} = 10^{2.263565 - 1.292876} = 10^{0.970689} = 9.34736$

c. Log 4.7 ≈ 0.6721, and therefore $10^{0.6721} \approx 4.7$.

$4.7^{2.8} \approx (10^{0.6721})^{2.8} = 10^{0.6721 \cdot 2.8} = 10^{1.88188} \approx 18.34$

TechnologyConnection A few years after Napier's discovery, English mathematician William Oughtred (1574–1660) realized that sliding two logarithmic scales next to each other made calculations easier, and he invented the slide rule. Over the next three centuries, many people improved the slide rule, making it an indispensable tool for engineers and scientists, until computers and electronic calculators were invented and became available for use. For more on the history of computational machines, see the links at www.keycollege.com/online .

Investigation
Logarithm Rules!

Logarithms have properties related to properties of exponents. You will discover some of them in this Investigation.

Step 1 Choose two positive numbers, a and b. Calculate $\log a$, $\log b$, and $\log(ab)$. Try to see how $\log a$, $\log b$, and $\log(ab)$ are related to each other. Record your results.

Step 2 Repeat Step 1 with at least three other pairs of numbers, and record your results. Do you always notice the same relationship between $\log a$, $\log b$, and $\log(ab)$?

Step 3 Choose two positive numbers, a and b. Calculate $\log a$, $\log b$, and $\log\frac{a}{b}$. Try to see how $\log a$, $\log b$, and $\log\frac{a}{b}$ are related. Record your results.

Step 4 Repeat Step 3 with at least three other pairs of numbers, and record your results. Do you always notice the same relationship between $\log a$, $\log b$ and $\log\frac{a}{b}$?

Step 5 Choose a positive number, a.

Complete this table:

n	$\log a^n$	$n \cdot \log a$
1		
2		
3		
4		
5		

During the Investigation you probably discovered the following three properties of logarithms:

Properties of Logarithms

Product Property

$$\log_a xy = \log_a x + \log_a y$$

Quotient Property

$$\log_a \frac{x}{y} = \log_a x - \log_a y$$

Power Property

$$\log_a x^n = n \log_a x$$

You can also verify these properties using the rules of exponents. For instance, $\log xy = \log x + \log y$ because $10^{\log xy} = xy$, and $10^{\log x + \log y} = 10^{\log x}10^{\log y} = xy$ too.

EXAMPLE A

Use the properties of exponents to determine whether each statement is true or false.

a. $\log 2 + \log 9 = \log 18$

b. $\log 7 + \log 3 = \log 10$

c. $\log 81 = 4 \log 3$

d. $\log 12 - \log 3 = \log 4$

▶ ***Solution***

The statement in part a is true.

$\log 2 + \log 9$

$= \log 2 \cdot 9$ Apply the product property.

$= \log 18$

The statement in part b is false.

$\log 3 + \log 7 = \log 21$ because of the product property.

The statement in part c is true.

$\log 81$

$= \log 3^4$

$= 4 \log 3$ after applying the power property.

The statement in part d is true.

$\log 12 - \log 3$

$= \log \frac{12}{3}$ Apply the quotient property.

$= \log 4.$

You can also use your calculator to confirm that a, c, and d are true.

The power property can also be used as a second method to solve exponential equations.

EXAMPLE B

Solve each equation:

a. $3^x = 23$

$\log 3^x = \log 23$ Take the logarithm of both sides of the equation.

$x \log 3 = \log 23$ Apply the power property.

$x = \dfrac{\log 23}{\log 3}$ Solve for x by dividing by x's coefficient.

$x \approx 2.85$ Estimate using a calculator.

$3^{2.85} \approx 22.89.$

If we had made a more accurate estimate in the last step, we would see an output even closer to 23 when we checked the solution.

b. $7 + 3 \cdot 4^x = 25$

$3 \cdot 4^x = 18$ — Begin by isolating the base and exponent.

$4^x = 6$

$\log 4^x = \log 6$ — Take the logarithm of both sides of the equation.

$x \log 4 = \log 6$ — Apply the power property.

$x = \dfrac{\log 6}{\log 4}$ — Solve for x.

$x \approx 1.29$

In this chapter you have learned the properties of exponents and logarithms summarized below. You can use these properties to solve equations involving exponents. Remember to look carefully at the order of operations and then work step by step to undo each operation.

Properties of Exponents and Logarithms

Product Property

$(a^m)(a^n) = a^{(m+n)}$ or $\log_a xy = \log_a x + \log_a y$

Quotient Property

$\dfrac{a^m}{a^n} = a^{m-n}$ or $\log_a \dfrac{x}{y} = \log_a x - \log_a y$

Power Property

$(a^m)^n = a^{m-n}$ or $\log_a x^n = n \log_a x$

Definition of Logarithm

If $x = a^m$, then $\log_a x = m$.

Change-of-Base Property

$\log_a x = \dfrac{\log_b x}{\log_b a}$

Power of a Product Property

$(ab)^m = a^m b^m$

Definition of Fractional Exponents

$a^{\frac{m}{n}} = \sqrt[n]{a^m}$

Definition of Negative Exponents

$a^{-n} = \dfrac{1}{a^n}$

$\left(\dfrac{a}{b}\right)^{-n} = \left(\dfrac{b}{a}\right)^n$

EXERCISES

Practice Your Skills

1. Without looking back in the book, change the form of each expression below, using properties of logarithms or exponents. Name each property you used, either with the name used in the book or the name you gave the property.

a. $g^{(h+k)} =$ b. $\log s + \log t =$ c. $\left(\frac{f^w}{f^v}\right) =$ d. $\log \frac{h}{k} =$ e. $(j^s)^t =$

f. $\log b^g =$ g. $\sqrt[h]{k^m} =$ h. $\frac{\log_s t}{\log_s u} =$ i. $w^t w^s =$ j. $p^{-h} =$

2. Determine whether each equation is true or false. If false, rewrite one side of the equation to make it true. Check your answer on your calculator.

a. $\log 3 + \log 7 = \log 21$

b. $\log 5 + \log 3 = \log 8$

c. $\log 16 = 4 \log 2$

d. $\log 5 - \log 2 = \log 2.5$

e. $\log 9 - \log 3 = \log 6$

f. $\log \sqrt{7} = \log \frac{7}{2}$

g. $\log 35 = 5 \log 7$

h. $\log \frac{1}{4} = -\log 4$

i. $\frac{\log 3}{\log 4} = \log \frac{3}{4}$

j. $\log 64 = 1.5 \log 16$

3. Use the power property of logarithms to solve these equations. Show your steps.

a. $5.1^x = 247$

b. $17 + 1.25^x = 30$

c. $27(0.93^x) = 12$

d. $23 + 45(1.024^x) = 147$

Reason and Apply

4. This table lists the consecutive notes of an octave from one C note to the next C note. This scale is called a *chromatic scale,* and it increases in 12 steps, called *half-tones.* The frequencies associated with the consecutive notes form a geometric sequence, in which the frequency of the next C note of the octave is double the frequency of the preceding C note.

a. Find a function that will generate the frequencies.

b. Fill in the missing table values.

	Note	Frequency (Hz)
Do	C	261.6
	C#	
Re	D	
	D#	
Mi	E	
Fa	F	
	F#	
Sol	G	
	G#	
La	A	
	A#	
Ti	B	
Do	C	523.2

MusicConnection It is interesting to note that if an instrument is tuned to mathematically exact intervals, it will sound out of tune in a different key. With some adjustments, it will be a well-tempered scale, or a scale that is in tune for any key. However, not all music is based on an octave scale. Indian musical compositions are based on the *raga,* a structure of 5 to 7 notes. There are 72 *melas,* or parent scales, on which all ragas are based.

Anoushka Shankar plays the sitar in the tradition of classical Indian music.

5. The altitude of an airplane is calculated by measuring atmospheric pressure on the surface of the airplane. This pressure is exponentially related to the plane's height above the Earth's surface. At ground level the pressure is 14.7 pounds per square inch (abbreviated lb/in.^{2}, or psi). At an altitude of 2 mi, the pressure is reduced to 9.46 psi.

a. Use this information to write the exponential equation for altitude in miles in terms of air pressure.

b. What is the pressure at an altitude of 12,000 ft? (1 mi. = 5280 ft)

c. What is the altitude of an airplane if the atmospheric pressure is 3.65 psi?

ScienceConnection Air pressure is the weight of the atmosphere pushing down on objects in the atmosphere, including the Earth itself. Air pressure decreases with altitude because there is less air above you as you ascend. A barometer is an instrument that measures air pressure, usually in millibars or inches of mercury, both of which can be converted to lb/in^2, which is the weight of air pressing down on each square inch of surface. Although air pressure changes with weather conditions, it also drops about 1 in. of mercury for each 1000-ft altitude gain.

6. The half-life of carbon-14, which is used in dating archaeological finds, is 5,750 years.

a. Assume that 100% of the carbon-14 is present at time 0 years, or $x = 0$. Write the equation that expresses the percentage of carbon-14 remaining as a function of time.

b. In some bone fragments 25% of the carbon-14 remains. What is the approximate age of the bones?

c. In the movie *Raiders of the Lost Ark* (1981), a piece of the Ark of the Covenant found by Indiana Jones contained 62.45% of its carbon-14. When does this indicate the ark was constructed?

d. Coal is formed from trees that lived about 100 million years ago. Could carbon-14 dating be used to determine the age of a lump of coal? Explain your answer.

7. **APPLICATION** Carbon-11 decays at a rate of 3.5% per minute. Assume that 100% is present at time 0 min.

a. What percent remains after 1 min?

b. Write the equation that expresses the percent of carbon-11 remaining as a function of time.

c. What is the half-life of carbon-11?

d. Explain why carbon-11 is not used for dating archaeological finds.

Review

8. (Lesson 5.5) Draw the graph of a function whose inverse is not a function. Carefully describe what must be true about the graph of a function if its inverse is not a function.

9. (Lesson 5.1) Find an equation to fit each set of data.

a.

x	y
1	8
4	17
6	21
7	26

b.

x	y
0	2
3	54
4	162
6	1458

10. (Chapter 4) Describe how each function has been transformed from the parent function $y = 2^x$ or $y = \log x$. Then graph the function.

a. $y = -4 + 3(2^{x-1})$

b. $y = 2 - \log\left(\frac{x}{3}\right)$

11. (Lessons 1.4 and 1.5) Answer true or false. If the statement is false, explain why, or give a counterexample.

a. A grade of 86% is always better than being in the 86th percentile.

b. A mean is always greater than a median.

c. If the range of a set of data is 28, the difference between the maximum and the mean is 14.

d. If the standard deviation is 2.1 and the mean is 19.7, then the median must be between 17.6 and 21.8.

e. The mean of a box plot that is skewed left is to the left of the median.

12. (Lessons 3.1 and 3.2) A driver charges $14 per hour plus $20 for chauffeuring if one books directly with her. If she is booked through an agency, the agency charges 115% of what the driver charges plus $25.

a. Write a function to model the cost of hiring a driver directly. Identify the domain and range.

b. Write a function to model the agency's charge. Identify the domain and range.

c. Give a single function that a person can use to calculate the cost of using an agency to hire a driver for h hours.

LESSON

5.8 Applications of Logarithms

Drowning problems in an ocean of information is not the same as solving them.

RAY E. BROWN

In this lesson you will explore applications of the techniques and properties you discovered in the last lesson. You can use logarithms to rewrite and solve problems involving exponential and power functions that relate to the natural world as well as to life decisions. You will be better able to interpret information about investing and borrowing money, disposing of nuclear and toxic waste, interpreting chemical reaction rates, and managing natural resources if you have a good understanding of these functions and problem-solving techniques.

EXAMPLE A

Recall the pendulum example from Lesson 5.4. The equation $y - 1.25 = (1.10)(0.9557)^x$ gave the greatest distance from a motion sensor for each swing of the pendulum, based on the number of the swing. Use this equation to find the swing number when the distance was closest to 1.47 m. Explain each step.

▶ Solution

$y - 1.25 = (1.10)(0.9557)^x$	Original equation.
$0.22 = 1.47 - 1.25 = (1.10)(0.9557)^x$	Substitute 1.47 for y.
$0.2 = (0.9557)^x$	Divide both sides by 1.10.
$\log(0.2) = \log(0.9557^x)$	Take the logarithm of both sides.
$\log(0.2) = x\log(0.9557)$	Use the power property of logarithms.
$-0.6990 = -0.0197x$	Evaluate the logarithms.
$x = \dfrac{-0.6990}{-0.0197} = 35.48$	Divide both sides by -0.0197.

On the 35th swing the pendulum will be closest to 1.47 m from the motion sensor.

Like other operations on equations, you can take the logarithm of both sides, provided the value of the expressions is known to be positive. Recall that the domain of $y = \log x$ is $x > 0$, so you cannot find the logarithm of a negative number or 0. In the last problem we knew that both sides were equal to $+0.2$ before we took the logarithm of both sides.

In Chapter 3 we used linear regression to find models for linear data. Similarly, when data appear to grow or decay exponentially, we can use exponential regression to find a model.

EXAMPLE B

After Eva convinced the mill to treat its wastewater before returning it to the lake, she began to sample the lake water for toxin levels once every five weeks. Here are the data she collected.

Week #	0	5	10	15	20	25	30
Toxin (ppm)	349.0	130.2	75.4	58.1	54.2	52.7	52.1

Eva hoped that the level would be much closer to zero after this much time. Does she have evidence that the toxin is still getting into the lake? Find an equation that models this data that she can present to the mill to justify her conclusion.

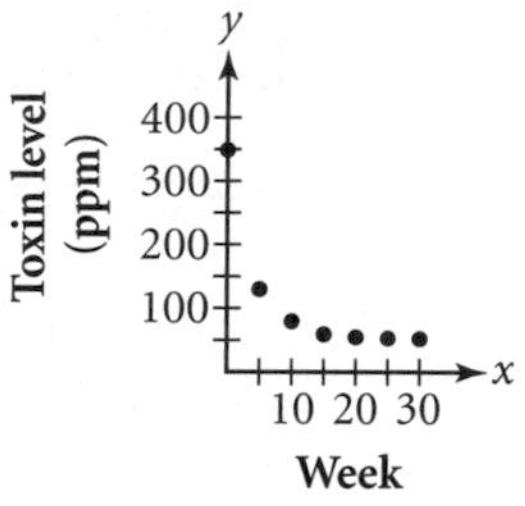

▶ *Solution*

The scatter plot of the data indicates exponential decay, so the model to which she must fit these data is $y = k + ab^x$, where k is the toxin level toward which the lake is dropping. If k is 0, then eventually the lake will be clean. If not, some toxins continue to be released. The table indicates that the toxin levels may be leveling off at 52 ppm. If k were 0, then the general equation would become $y = ab^x$. Subtract 52 from each toxin. Then take an exponential regression for the remaining data. [▶ Follow the procedure in **Calculator Note 5B.** ◀]

$Y = 302.4(0.774)^x$ is an equation modeling these data.

The equation that models the amount of toxin, T, in the lake after w weeks is $T = 52 + 302.4(0.774)^w$. If you graph this equation with the original data, you see that it fits quite well.

Investigation
Cooling

You will need

- a cup of hot water (optional)
- a temperature probe

In this Investigation you will find the relationship between the temperature of a cooling object and time.

Step 1 Connect a temperature probe to a data collector, and set it up to collect 60 data points over 10 min, or one data point every 10 sec. Heat the end of the probe by placing it in hot water or by holding it tightly in the palm of your hand. When it is hot, set the probe on a table so that the tip is not touching anything, and begin data collection. [▶ See **Calculator Note 5C.** ◀]

Step 2 While the data are being collected, draw a sketch of what you expect the graph of (*time, temperature*) data to look like as the temperature probe cools. Label the axes, and mark the scale on your graph. Did everyone in your group draw the same graph? Discuss any differences of opinion.

Step 3 Plot data of the form (*time, temperature*) on an appropriately scaled graph. Your graph should appear to be an exponential function. Study the graph and the data, and guess the temperature limit.

Step 4 Subtract this limit from your temperatures, and use the exponential regression feature of your calculator to find an equation modeling this modified data.

Step 5 Find the equation that models this data. Give real-world meaning to the values in the equation.

EXERCISES

Practice Your Skills

1. Prove that these statements are true by taking the logarithm of both sides, then using the properties of logarithms to re-express each side until you have two identical expressions.

a. $10^{n+p} = (10^n)(10^p)$

b. $\frac{10^d}{10^e} = 10^{d-e}$

2. Solve each equation by taking the logarithm of each side. Check your answers by substituting your answer for x.

a. $800 = 10^x$

b. $2048 = 2^x$

c. $16 = 0.5^x$

d. $478 = 18.5(10^x)$

e. $155 = 24.0(1.89^x)$

f. $0.0047 = 19.1(0.21x)$

3. Suppose you invest \$3,000 at 6.75% annual interest, compounded monthly. How long will it take to triple your money?

Reason and Apply

4. **APPLICATION** The length of time that milk (and many other perishable substances) will stay fresh varies with the storage temperature. Suppose that milk will keep for 146 hr in a refrigerator at 4°C. Milk that is left out in the kitchen at 22°C will keep for only 42 hr. Because bacteria grow exponentially, you can assume that freshness decays exponentially.

a. Write an equation that expresses the number of hours, h, that milk will keep in terms of the temperature, T.

b. Use your equation to predict how long milk will keep at 30°C and at 16°C.

c. If a container of milk soured after 147 hr, at what temperature was it kept?

d. Graph the relationship between hours and temperature using your equation from 4a and the five data points you have found.

e. What is a realistic domain for this relationship? Why?

Science Connection In 1860 Louis Pasteur developed a method of killing bacteria in fluids called *pasteurization*. It was first developed for use with wines but was quickly adapted for raw milk. Milk is heated not quite to its boiling point, which would affect its taste and nutritional value, but to 63°C (145°F) for 30 min or 72°C (161°F) for 15 s. This kills most, but not all, harmful bacteria. Refrigerating milk slows the growth of the remaining bacteria, but eventually the milk will spoil when there are too many bacteria for it to be healthy. The bacteria in milk change the lactose to lactic acid, which smells and tastes bad to humans.

Louis Pasteur

5. The equation $f(x) = \frac{12000}{1 + 499(1.09^{-x})}$ gives the total sales each day since the release of a new video game. Find each value, and give a real-world meaning.

a. $f(20)$

b. $f(80)$

c. x when $f(x) = 6000$

d. Show the steps to solve 5c symbolically.

e. Graph the equation in a window large enough for you to see the overall behavior of the curve. Use your graph to describe how the number of games sold each day changes. Does this model seem reasonable?

6. The intensity of sound, D, measured in decibels (dB) is given by the formula

$$D = 10 \log\left(\frac{I}{10^{-16}}\right)$$

where I is the power of the sound in watts per square centimeter (W/cm^2) and 10^{-16} W/cm^2 is the power of sound just below the threshold of hearing.

a. Find the number of decibels of a 10^{-13} W/cm^2 whisper.

b. Find the number of decibels in a normal conversation of $3.16 \cdot 10^{-10}$ W/cm^2.

c. Find the power of the sound (W/cm^2) experienced by the orchestra members seated in front of the brass section, measured at 107 dB.

d. How many times more powerful is a sound of 47 dB than a sound of 42 dB?

7. Find an equation relating the loudness of spoken words measured at the source and the maximum distance at which another person can recognize the speech.

a. Plot the data on your calculator, and make a rough sketch on your paper.

b. Notice that the graph of the data looks like the graph of a logarithmic function. Find an equation of the form $y = a + b\log x$ to model this data by experimenting with values of a and b.

c. Graph this new equation with the data. Does it seem to be a good model?

Loudness (dB)	Distance (m)
0.5	0.1
3.2	16.0
5.3	20.4
16.8	30.5
35.8	37.0
84.2	44.5
120.0	47.6
170.0	50.6

8. A glass of ice-cold water is left in a room at a temperature of 74°F, and the temperature of the water is read at the time intervals listed.

Time (min)	0	2	4	6	8	10	12	14	16	18	20	22	24	26	28	30
Temp. (°F)	5	14	22	29	35	40	45	49	52	55	57	60	61	63	64	66

a. Plot the data using an appropriate window. Make a rough sketch of this graph.

b. Find an exponential model given temperature as a function of time.

9. When Quinn starts treating her pool for the season, she begins with a shock treatment of 4 gal of chlorine. Every 24 hr, 15% of the chlorine evaporates. The next morning, she adds 1 quart ($\frac{1}{4}$ gal) of chlorine to the pool, and she continues to do so each morning.

a. How much chlorine is there in the pool after one day (after she adds the first daily quart of chlorine)? After two days? After three days?

b. Use recursive notation to write a formula for this pattern.

c. Use the formula from 9b to make a table of values, and sketch a graph of 20 terms.

d. Find an explicit model that fits the data.

Review

10. (Lesson 5.7) The function $y = 20\log\left(\frac{x}{0.00002}\right)$ measures the intensity of sound measured in decibels (dB), where x is the pressure created by the sound on the eardrum measured in Pascals (Pa).

a. How many decibels is the sound of the hum of a refrigerator that causes 0.00356 Pa of pressure on the eardrum?

b. A noise causing 20 Pa of pressure on the eardrum brings severe pain to most people. How many decibels is this?

c. Write the inverse function.

d. What is the pressure on the eardrum caused by a 90-dB sound?

11. (Lesson 5.1)

a. Find an exponential function that passes through the points (4, 18) and (10, 144).

b. Find a logarithmic function that passes through the points (18, 4) and (144, 10).

12. (Lesson 3.2) The Highland Fish Company is starting a new line of frozen fish sticks. It will cost $19,000 to set up the production line, and it will cost $1.75 per pound to buy and process the fish. Highland Fish will sell the final product at a wholesale cost of $1.92 per pound.

a. Write a cost function and an income function for Highland Fish Company's new venture.

b. Graph both functions on the same axes over a domain $0 \le x \le 1{,}000{,}000$.

c. How many pounds of fish sticks will HFC have to produce before they start making a profit on the new venture?

d. What profit do they stand to make on the first 500,000 pounds of fish?

13. (Lesson 4.7) Sketch the graph of $(4(x + 5))^2 + \left(\frac{y - 8}{2}\right)^2 = 1$. Give coordinates of appropriate defining points.

14. (Lesson 5.3) Solve each equation for x.

a. $x^5 = 3418$ **b.** $(x - 5.1)^4 = 256$ **c.** $7.3x^6 + 14.4 = 69.4$

EXPLORATION

The Number *e*

You've solved problems exploring the amount of interest earned in a savings account when interest is compounded yearly, monthly, or daily. But what if interest is compounded continuously? That means that at every instant, your interest is redeposited in your account, and the new interest is calculated on it. Few if any banks calculate interest this way, but it turns out that this type of continuous growth is closely modeled by frequent compounding. Furthermore, this type of continuous growth is related to a number ***e***. It has a value of approximately 2.71, and, like π, it is a *transcendental number*—a number that has infinitely many digits in its decimal expansion with no pattern. The number e is irrational and cannot be written as a fraction of whole numbers. The logarithmic function with base e, $\log_e x$, is also written as $\ln x$ and is called the *natural logarithm function.*

Activity

The Swiss mathematician Jacob Bernoulli (1654–1705) explored the following problem in 1683.

Suppose you put \$1 in an account that earns 100% interest per year. If the interest is compounded only once, at the end of the year you will have earned \$1 in interest, and your balance will be \$2. What if interest were compounded more frequently? Follow the steps below to analyze this situation.

Step 1 If interest is compounded ten times annually, how much money will be in your account at the end of one year? Remember that if 100% interest is compounded ten times, you'd earn 10% each time. Check your answer with another group to be sure you did this correctly.

Step 2 Predict what will happen if your money earns interest compounded continuously.

Step 3 What will the balance of your account be at the end of one year if interest is compounded 100 times? 1,000 times? 10,000 times? 1,000,000 times? How do your answers compare with your predictions in Step 2?

Step 4 Write an equation that tells you the balance if the interest were compounded x times annually, and graph it on your calculator. Does this equation seem to be approaching one particular long-term value, or limit? If so, what is it? If not, how is the graph behaving?

Step 5 Look for e on your calculator, and find the value of it to six decimal places. What is the relationship between this number and your answer to Step 4?

Step 6 When interest is compounded continuously, the formula $y = P(1 + r/n)^{nt}$ becomes $y = Pe^{rt}$, where P is the principal (the initial value of your investment), e is the natural exponential base, r is the interest rate, and t is the length of time the investment is left in the account. Use this formula to calculate the value of $1 deposited in an account earning 100% annual interest for one year. Calculate the value if it is left in the same account for ten years.

Questions

1. $y = Pe^{rt}$ is often a more accurate representation of growth or decay problems because things usually grow or decay continuously, not at specific time intervals. For example, a bacteria population may double every four hours, but it is increasing throughout those four hours, not suddenly doubling in value when exactly four hours have passed. Return to any problem involving growth or decay from this chapter, and use this new formula to solve the problem. How does your answer compare to your answer using regular exponential functions?

2. The intensity, I, of a beam of light after passing through t cm of liquid is given by $I = Ae^{-kt}$, where A is the intensity of the light when it enters the liquid, e is the natural exponential base, and k is a constant for a particular liquid. On a field trip, a marine biology class took light readings at Deep Lake. At a depth of 50 cm, the light intensity was 80% of the light intensity at the surface.
   a. Find the value of k for Deep Lake.
   b. If the light intensity is measured as 1% of the intensity at the surface, at what depth was the reading taken?

ScienceConnection *Phytoplankton* is the generic name of a great variety of micro-organisms, including algae, that live in lakes and oceans and constitute the lowest step of the food chain in some ecosystems. Because phytoplankton require light for photosynthesis, light levels determine the maximum depth at which these organisms can grow. Limnologists, scientists who study inland waters, estimate this depth to be the point at which the amount of light available is reduced to 0.5%–1% of the amount of light available at the lake surface.

PROJECT

ALL ABOUT *e*

The number you found in this Exploration is called the *natural exponential base, e.* Mathematicians have been exploring e and calculating more digits since the 1600s. There are a number of equations that can be used to calculate digits of e. Do some research in math books, in the library, or on the Internet, and find at least two. Write a report explaining the formulas you find and how they are used to calculate e. Include any interesting historical facts you find on e as well.

EXPLORATION
with Microsoft® Excel

Exponential and Power Functions

Download the Chapter 5 Excel Exploration from the website at www.keycollege.com/online or from your Student CD.

In this Exploration you will work with both power functions and exponential functions, and you will study how the two concepts are related. In particular, you will compare the growth of power functions and exponential functions.

On the spreadsheet you will find complete instructions and questions to consider as you explore.

Activity
Logarithms Using Bases 2, 3, 4, 5, 6, 7, 8, and 9

Excel skills required

- Data fill
- Absolute cell reference
- Name a cell

In this Activity you will build a table that allows you to take a deeper look at powers of numbers by using powers that are not themselves whole numbers.

You can find instructions for the required Excel skills in the Excel Notes at www.keycollege.com/online .

Launch Excel.

Step 1 Prepare the labels for this sheet as shown.

	A	B	C	D	E	F	G	H	I
1									
2									
3									
4	x	2	3	4	5	6	7	8	9

Step 2 In Cell A5 enter `=1/12` (Note that 1/12 is a formula as far as Excel is concerned, so using the "=" here is required, not optional).

Step 3 In Cell A6 enter `=A5+(1/12)`.

Step 4 In Cell B5 enter `=$B$4^A5`. The dollar signs make this an absolute cell reference. This means that when it is copied, the cell reference will be copied in a data fill without change.

Step 5 In Cell C5 enter `=$C$4^A5`.

Step 6 In Cell D5 enter `=$D$4^A5`.

Step 7 Continue in the same manner for Column E to Column I.

At this point you have

	A	B	C	D	E	F	G	H	I
1									
2									
3									
4	x	2	3	4	5	6	7	8	9
5	0.083333	1.059463	1.095873	1.122462	1.14353	1.161037	1.176047	1.189207	1.200937

Study Row 5 first. Cell B5 represents the numerical fact $2^{1/12} \approx 1.059463$. That is, 1/12 is the power to which you must raise 2 to get 1.059463, approximately. This is written as

$$\log_2 1.059463 = 1/12.$$

You have written a logarithmic equation based on Cell B5. Use Cells C5 to I5 to practice writing similar logarithmic equations.

Step 8 Highlight cells B5 to I5, and data-fill to Row 6.

Step 9 Now, highlight Cells A6 to I6, and data-fill down to Row 100.

Questions

1. Look for the number 3 in your table. Express what you find as an equation with exponents. Then, write a logarithmic equation related to the exponential equation.
2. Look for the number 2 in your table. Express what you find as an equation with exponents. Then, write a logarithmic equation related to the exponential equation.
3. Using your table, what is the $\log_5 25$?
4. Find these numbers in the table.
   - **a.** $\log_9 3$
   - **b.** $\log_4 2$
   - **c.** $\log_2 8$
   - **d.** $\log_8 2$
5. Find $\log_2 6$. Find $\log_2 8$. Find $\frac{\log_2 8}{\log_2 6}$. Compare this with $\log_6 8$. What property of logarithms does this illustrate?

CHAPTER 5 REVIEW

Exponential functions provide an explicit and continuous equation with which to model geometric sequences. They can be used to model the growth of populations, the decay of radioactive substances, and other phenomena. The general form of an exponential function is $y = ab^x$, where a is the initial amount and b is the base that represents the rate of growth or decay. Because the exponent can take on all real number values, including negative numbers and fractions, it is important to understand the meaning of these exponents.

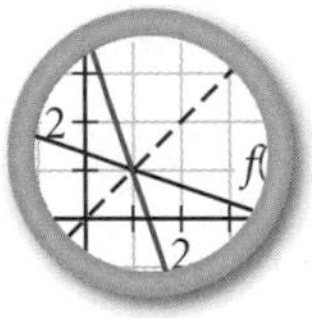

Until you read this chapter, you had no way to solve an exponential equation other than guess-and-check or graphing. Once you defined the **inverse** of the exponential function—the **logarithmic function**—you were able to solve exponential functions symbolically. The inverse of a function is the relation you get when all values of the independent variable are exchanged with the dependent variable. The graphs of a function and its inverse are reflected across the line $y = x$. The basic definition of a **logarithm** is that given $y = b^x$, then $\log_b y = x$. You learned that the properties of logarithms parallel those of exponents: the logarithm of a product is the sum of the logarithms, the logarithm of a quotient is the difference of the logarithms, and the logarithm of a number raised to a power is the product of the logarithm and that number. By looking at the logarithms of the x-value, the y-value, or both values in a set of data, you can determine what type of equation will best model the data by finding which of these creates the most linear graph.

EXERCISES

1. Evaluate each expression without using a calculator. Then check your work with a calculator.

a. 4^{-2} **b.** $(-3)^{-1}$ **c.** $\left(\frac{1}{5}\right)^{-3}$

d. $49^{1/2}$ **e.** $64^{-1/3}$ **f.** $\left(\frac{9}{16}\right)^{3/2}$

g. -7^0 **h.** $(3)(2)^2$ **i.** $(0.6^{-2})^{-1/2}$

2. Rewrite each expression in another form.

a. $\log x + \log y$ **b.** $\log\frac{z}{v}$ **c.** $(7x^{2.1})(0.3x^{4.7})$

d. $\log w^k$ **e.** $\sqrt[5]{x}$ **f.** $\log_5 t$

3. Solve each equation by using the properties of exponents and logarithms to evaluate the expressions. Confirm your answers by solving a graphing approximation.

a. $4.7^x = 28$

b. $4.7x^2 = 2209$

c. $\log_x 2.9 = 1.25$

d. $\log_{3.1} x = 47$

e. $7x^{2.4} = 101$

f. $9000 = 500(1.065)^x$

g. $\log x = 3.771$

h. $\sqrt[5]{x^3} = 47$

4. Solve for x. Round answers to the nearest 0.001.

a. $\sqrt[8]{2432} = 2x + 1$

b. $4x^{2.7} = 456$

c. $734 = 11.2(1.56)^x$

d. $f(f^{-1}(x)) = 20.2$

e. $147 = 12.1(1 + x)^{2.3}$

f. $2\sqrt{x - 3} + 4.5 = 16$

5. Once a certain medicine is in the bloodstream, its half-life is 16 hours. How long (to the nearest 0.1 hr) will it be before an initial 45 cc of the medicine has been reduced to 8 cc?

6. Given $f(x) = (4x - 2)^{1/3} - 1$, find:

a. $f(2.5)$

b. $f^{-1}(x)$

c. $f^{-1}(-1)$

d. $f(f^{-1}(12))$

7. Find the equation of an exponential curve through the points (1, 5) and (7, 32).

8. Draw the inverse of $f(x)$, shown in the graph at right.

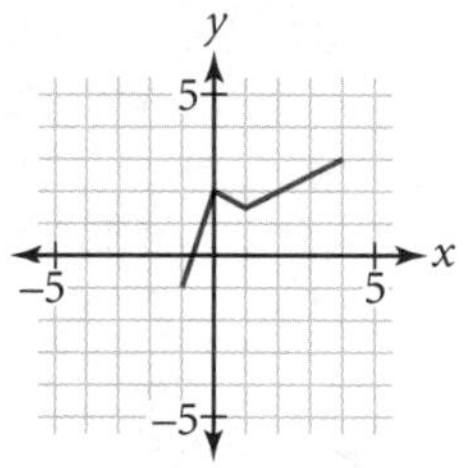

9. Your head gets larger as you grow. Most of the growth occurs in the first few years of life, and there is very little additional growth after you reach adolescence. The percent of adult size for your head is $y = 100 - 80(0.75)^x$, where x is your age in years, and y is the percent of the average adult size.

a. Graph this function.

b. What are the reasonable domain and range of this function?

c. Describe the transformations on the graph of $y = (0.75)^x$ that produces the graph in 9a.

d. A 2-year-old child's head is what percent of the adult size?

e. About how old would a person be if his or her head circumference were 75% that of an average adult?

CareerConnection Pediatricians routinely use a variety of growth charts to track the growth of infants, children, and adolescents. These charts are tools that contribute to an overall impression of the child being measured. If a child's measurements deviate greatly from the chart, the doctor can look for causes and recommend treatments. Charts are available for boys and for girls and include length vs. age, weight vs. age, weight vs. length or height, body mass index vs. age and head circumference vs. age, which is primarily used with infants from birth to 3 years old.

10. A new incentive plan for the Talk Alot long-distance phone company varies the cost of a call according to the formula $cost = a + b \log(time)$. For long-distance calls, the cost of the first minute is \$0.10. The charge for 15 min is \$0.70.

a. Find the value of a in the model.

b. Find the value of b in the model.

c. What is the x-intercept of the graph of this model? What is the real-world meaning of the x-intercept?

d. Use your model to predict the cost of a 30-min call.

e. If you decide you can afford to make only a \$0.40 call, how long can you talk?

11. A "learning curve" describes the rate at which a task can be learned. Suppose the equation

$$t = -144 \log\left(1 - \frac{N}{90}\right)$$

predicts t, the number of short daily sessions it will take to achieve a goal of typing N words per minute on a word processor.

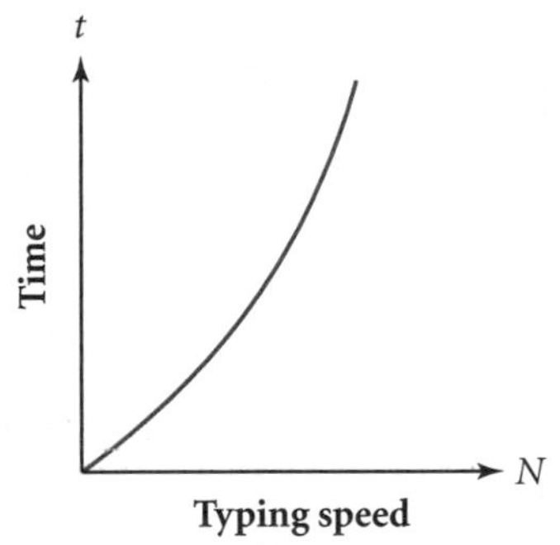

a. Using this equation, how long should it take someone to learn to type 40 words per minute?

b. If the typical person had 47 lessons, what speed would you expect them to have obtained?

c. Interpret the shape of the graph as it relates to learning time. What domain is realistic for this problem?

12. Every human starts as a single cell. This cell splits into two cells, then each of those two cells splits into two cells, and so on.

a. Write a recursive formula for cell division, starting with a single cell.

b. Write an explicit formula for cell division.

c. Sketch a graph to model the formulas in parts 12a and 12b.

d. Describe some of the features of the graph.

e. After how many divisions were there more than 1 million cells?

f. If there are about 1 billion cells after 30 divisions, then after how many divisions were there about 500 million cells?

Science Connection Embryology is the branch of biology that deals with the formation, early growth, and development of living organisms. In humans the growth of a fetus inside a mother takes about nine months. During this time a single cell will grow into many different cell types with different shapes and functions in the body.

A similar process will occur in the embryo of any animal. Historically, chicken embryos were among the first embryos studied. A chicken embryo develops and hatches in 20 to 21 days. Cutting a window in the eggshell allows direct observation for the study of embryonic growth.

Assessing What You've Learned

- Can you write equations modeling exponential data? Can you write an exponential equation modeling growth or decay?
- Do you know the properties of exponents? Can you give an example of each of them?
- What is a power function?
- How do you solve equations involving variables raised to a power? How do you solve equations involving exponential expressions?
- How do you find the inverse of a function algebraically? How do the graph of a function and its inverse compare?
- What are logarithms? How are logarithmic functions and exponential functions related?
- Give examples of three properties of logarithmic functions.

JOURNAL Look back at the journal entries you have written about Chapter 5. Then, to help you organize your thoughts, here are some additional questions to write about.

- What do you think were the main ideas of the chapter?
- Were there any exercises that you particularly struggled with? If so, are they clear now? If some ideas are still foggy, what plans do you have to clarify them?
- What type of function, or relation is the inverse of a logarithmic function? The inverse of an exponential function? The inverse of a linear function? The inverse of a square root function? You should include sketches of graphs of the different types of functions and their inverses.

CHAPTER

6 Quadratic and Other Polynomial Functions

American artist Sarah Sze (b 1969) creates flowing sculptures, such as this one created for an exhibit at the San Francisco Museum of Modern Art. This piece features a fractured sport utility vehicle whose pieces have been replaced with disposable household items, including foam packing peanuts. Some of the curves in this artwork's cascade resemble the graphs of polynomial functions.

Things Fall Apart: 2001, mixed media installation with vehicle; variable dimensions/San Francisco Museum of Modern Art, Accessions Committee Fund purchase © Sarah Sze

OBJECTIVES

In this chapter you will

- find polynomial functions that fit a set of data
- study quadratic functions in general form, vertex form, and factored form
- find roots of a quadratic equation from a graph by factoring and by the quadratic formula
- define complex numbers and operations involving them
- identify important features of the graph of a polynomial function
- use division and other strategies to find the roots of higher-degree polynomials

LESSON

6.1

Polynomial Degree and Finite Differences

Differences challenge assumptions.

ANNE WILSON SCHAEF

In Chapter 2 you studied arithmetic sequences, which have a common difference between consecutive terms. You can draw a line through the points of an arithmetic sequence, and this line has a constant slope. In other words, if you choose x-values along the line that form an arithmetic sequence, the corresponding y-values will also form an arithmetic sequence.

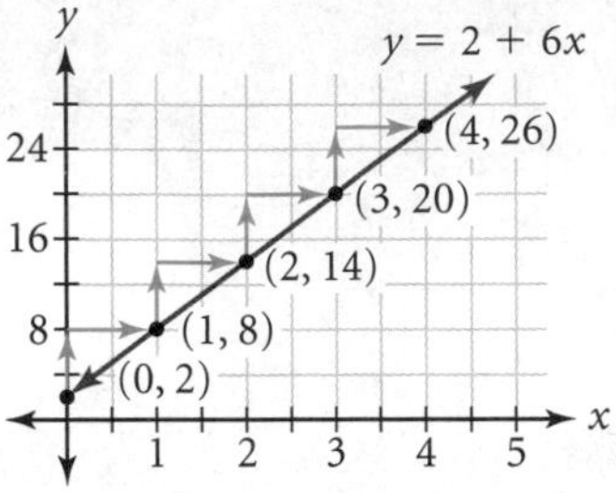

You have also studied many kinds of nonlinear sequences and functions, which do not have a common difference or a constant slope. However, in this lesson you will discover that even nonlinear functions sometimes have a special pattern in their differences.

A **polynomial** is a sum of terms containing the same variable raised to different powers. When a polynomial is set equal to a second variable, such as y or $f(x)$, it is a *polynomial function.*

Polynomial

A **polynomial** in one variable is any expression in the form

$$a_n x^n + a_{n-1}x^{n-1} + \cdots + a_1 x^1 + a_0,$$

where x is a variable, the exponents are nonnegative integers, and the coefficients are real numbers.

The **degree** of a polynomial or polynomial function is the power of the term that has the greatest exponent. Linear functions are 1st-degree polynomial functions because the largest power of x is 1. The polynomial function below has degree 3. If the degrees of the terms of a polynomial are arranged in decreasing order from left to right, the polynomial is in **general form.**

Coefficients (if a term shows no coefficient, the coefficient is 1)

Polynomial Function: $y = 1x^3 + 9x^2 + 26x + 24$

Highest-degree term — Constant term

Polynomial

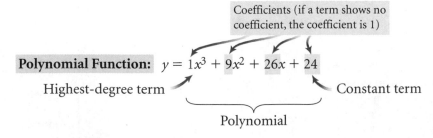

A polynomial with only one term is called a *monomial.* A polynomial with two terms is a *binomial,* and a polynomial with three terms is a *trinomial.* Polynomials with more than three terms are usually just called "polynomials."

In modeling linear functions you have already discovered that for x-values that are uniformly spaced, the differences between the corresponding y-values must be the same. With 2nd- and 3rd-degree polynomial functions, the differences between the corresponding y-values are not the same. However, finding the differences between those differences produces an interesting pattern.

1st degree
$y = 3x + 4$

x	y	D_1
2	10	3
3	13	3
4	16	3
5	19	3
6	22	3
7	25	

2nd degree
$y = 2x^2 - 5x - 7$

x	y	D_1	D_2
3.7	1.88	1	0.04
3.8	2.88	1.04	0.04
3.9	3.92	1.08	0.04
4.0	5.00	1.12	0.04
4.1	6.12	1.14	
4.2	7.28		

3rd degree
$y = 0.1x^3 - x^2 + 3x - 5$

x	y	D_1	D_2	D_3
−5	−57.5	52.5	−50	75
0	−5	2.5	25	75
5	−2.5	27.5	100	75
10	25	127.5	275	
15	152.5	302.5		
20	455			

Note that in each case the x-values are spaced equally. You find the first set of differences, D_1, by subtracting each y-value from the one after it. You find the second set of differences, D_2, by finding the differences of consecutive D_1 values in the same way. Notice that for the 2nd-degree polynomial function, the D_2 values are constant, and for the 3rd-degree polynomial function, the D_3 values are constant. What do you think will happen with a 4th- or 5th-degree polynomial function?

This way of analyzing differences is called the **finite differences method.** You can use this method to find the degree of the polynomial function that models a certain set of data.

CalculusConnection In calculus you study rates of change and shapes of graphs by taking derivatives of functions. If you take enough derivatives of derivatives of polynomials, you will eventually get a constant. As with finite differences, the number of repetitions before reaching the constant is the polynomial's degree.

EXAMPLE Find a polynomial function that models the relationship between the number of sides and the number of diagonals of a polygon. Use the function to find the number of diagonals of a dodecagon.

▶ Solution First, you need to collect data. Sketch a few polygons and all of their diagonals.

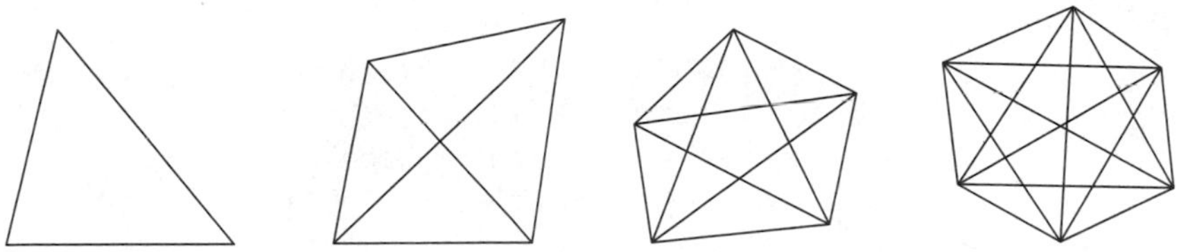

Record the number of sides and diagonals in a table. You may notice a pattern in the number of diagonals that will help you extend your table beyond the sketches you have made. Calculate the finite differences to determine the degree of the polynomial function.

Number of sides x	Number of diagonals y
3	0
4	2
5	5
6	9
7	14
8	20

D_1	D_2
2	
3	1
4	1
5	1
6	1

Stop the finite differences method when the values of a set of differences are constant. Because the values of D_2 are constant, the data can be modeled by a 2nd-degree polynomial function like $y = ax^2 + bx + c$.

To find the equation of a line, you need two points to find the parameters a and b in the equation $y = a + bx$. Similarly, to find the values of a, b, and c in $y = ax^2 + bx + c$, you need three points. Choose any three of the points, say (4, 2), (6, 9), and (8, 20), and find an equation of the form $y = ax^2 + bx + c$ modeling this data. [▶ Follow the instructions in **Calculator Note 6A.**◀]

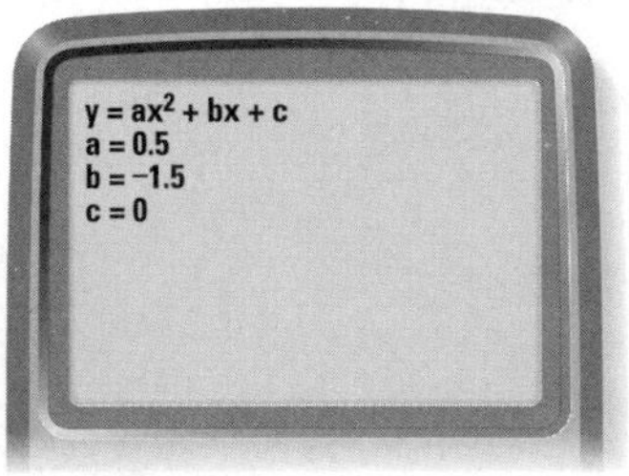

The equation $y = 0.5x^2 - 1.5x$ gives the number of diagonals of any polygon as a function of the number of sides.

Substitute 12 for x to find that a dodecagon has 54 diagonals.

$$y = 0.5x^2 - 1.5x$$
$$y = 0.5(12)^2 - 1.5(12)$$
$$y = 54$$

With exact data you can expect the differences to be equal when you find the right degree. But with experimental or statistical data, as in the Investigation, you may have to settle for differences that are nearly constant and do not show an increasing or decreasing pattern when graphed.

HistoryConnection Galileo Galilei (1564–1642), Italian mathematician, physicist, and astronomer, performed experiments with free-falling objects and discovered that the speed of a falling object at any moment is proportional to the length of time it has been falling. In other words, the longer an object falls, the faster it falls. To learn more about Galileo's experiments and discoveries, see the links at **www.keycollege.com/online** .

Investigation
Free Fall

You will need

- a motion sensor
- a pillow or other soft object

Procedure Note

1. Set up the sensor to read every 0.05 s for 2 s. [▶ See **Calculator Note 6B** to learn how to set up your calculator. ◀]
2. From a height of about 2 m, drop a small pillow toward the sensor resting on the floor.
3. Start the sensor, and then drop the pillow.

What function models the position of an object falling due to the force of gravity? Use a motion sensor to collect data, and analyze the data to find a function to model them.

Step 1 Follow the Procedure Note to collect data for a falling object. Let x represent time in seconds, and let y represent height in meters. Select about ten points during the free-fall portion of your graph, with x-values forming an arithmetic sequence. Record this information in a table. Round all table values to thousandths.

Step 2 Use the finite differences method to find the degree of the polynomial function that models your data. Stop when the differences are nearly constant.

Step 3 Enter your time values, x, into list L_1 on your calculator. Enter your height values, y, into list L_2. For your first differences enter your time values without the first value into list L_3, and enter the first differences, D_1, in list L_4. For your second differences enter the time values without the first two values into L_5, and enter the second differences into L_6. Continue this process for any other differences you have calculated. Then make scatter plots of (L_1, L_2), (L_3, L_4), (L_5, L_6), etc. [▶ See **Calculator Note 6C** to learn how to calculate and graph finite differences. ◀]

Step 4 In writing, describe each graph from Step 3 and state what these graphs tell you about the data.

Step 5 Based on your results from using finite differences, what is the degree of the polynomial function that models free fall? Write the general form of this polynomial function.

Step 6 Find an equation to model the position of a free-falling object dropped from a height of 2 m. You can choose three points that are relatively far apart to represent your data, or you can use all your data when you find a quadratic regression.

When using experimental data, you must choose your points carefully. When the data are measured, as they are in the Investigation, your equation will most likely not fit all of the data points exactly, due to some errors in measurement and rounding. To minimize the effects of these errors, choose representative points that are not close together, just as you did when fitting a line to data.

HistoryConnection The method of finite differences was used by the Chinese astronomer Li Shun-Fêng in the 7th century to find a quadratic equation to express the Sun's apparent motion across the sky as a function of time. The Iranian astronomer Jamshid Masud al-Khashi, who worked at the Samarkand Observatory in the 15th century, also used the finite differences method when calculating the celestial longitudes of planets.

The finite differences method was further developed in 17th- and 18th-century Europe. It was used to eliminate calculations involving multiplication and division in the construction of tables of polynomial values. It was not uncommon for a late–18th-century European scientist to have more than 125 volumes of various kinds of tables. In the 19th century, early automatic calculating machines were programmed to calculate differences and were called "difference engines."

Note that some functions, such as logarithmic functions, cannot be expressed as polynomials. The finite differences method will not produce a set of constant difference for functions other than polynomial functions.

EXERCISES

Practice Your Skills

1. Identify the degree of each polynomial.

a. $x^3 + 9x^2 + 26x + 24$

b. $7x^2 - 5x$

c. $x^7 + 3x^6 - 5x^5 + 24x^4 + 17x^3 - 6x^2 + 2x + 40$

d. $16 - 5x^2 + 9x^5 + 36x^3 + 44x$

2. Determine which of these expressions are polynomials. State the degree of each polynomial, and write it in general form. If it is not a polynomial, explain why.

a. $-3 + 4x - 3.5x^2 + \frac{5}{9}x^3$

b. $5p^4 + 3.5p - \frac{4}{p^2} + 16$

c. $4\sqrt{x^3} + 12$

d. $\sqrt{15}x^2 - x - 4^{-2}$

3. For each data set decide whether the last column shows constant values. If not, calculate the next set of finite differences.

a.

x	y
2	4.4
3	6.6
4	9.2
5	11
6	10.8
7	7.4

b.

x	y	D_1
3.7	−8.449	
		−0.257
3.8	−8.706	
		−0.251
3.9	−8.956	
		−0.244
4.0	−9.200	
		−0.236
4.1	−9.436	
		−0.227
4.2	−9.662	

c.

x	y	D_1	D_2
−5	−101		
		95	
0	−6		−100
		−5	
5	−11		50
		45	
10	34		200
		245	
15	279		350
		595	
20	874		

4. Find the degree of the polynomial function that can model these data.

x	0	2	4	6	8	10	12
y	12	−4	−164	−612	−1492	−2948	−5124

5. Consider these data.

a. Fill in the missing numbers to represent a quadratic function with 2nd differences of 5.

x	−1	0	1	2	3	4
y			3	11		

b. Find an equation that fits the data in the table you have created.

6. Consider these data.

a. Fill in the missing numbers to represent a cubic function with 3rd differences of 12.

x	−1	0	1	2	3	4
y		−4	−2	12		

b. Find an equation that fits the data in the table you have created. [▶ Use a procedure similar to that described in **Calculator Note 6A.** Which of the regression choices should you use in place of quadratic regression? ◀]

Reason and Apply

7. Consider these data.

n	1	2	3	4	5	6
s	1	3	6	10	15	21

a. Calculate finite differences to find the degree of a polynomial that models these data.

b. Describe how the degree of the polynomial is related to the finite differences you have calculated.

c. What is the minimum number of data points required to determine the degree of this polynomial? Why?

d. Find the polynomial function that models this table of values, and use it to find s when n is 12.

e. The values in the s row are called *triangular numbers.* Why do you think they are called triangular? (Hint: Find some pennies, and try arranging them into a triangle. Which numbers of pennies can form a triangle?)

8. The data in these tables represent the heights of two objects at different times during free fall.

i.

Time (s) t	0	1	2	3	4	5	6
Height (m) h	80	95.1	100.4	95.9	81.6	57.5	23.6

ii.

Time (s) t	0	1	2	3	4	5	6
Height (m) h	4	63.1	112.4	151.9	181.6	201.5	211.6

a. Calculate the finite differences for each table.

b. What is the degree of the polynomial function that you would use to model each data set?

c. Write a polynomial function to model each set of data. Check your answer by substituting another data point in your function.

9. You can use blocks to build pyramids such as these. All of the pyramids are solid.

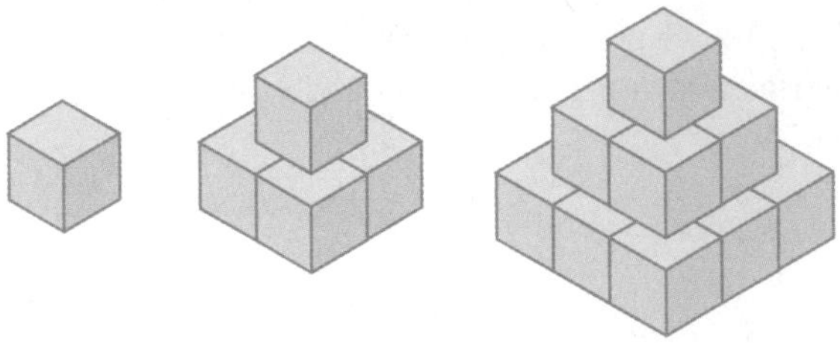

a. Create a table to record the number of layers, x, in each pyramid and the total number of blocks, y, needed to build it. You may need to build or sketch a few more pyramids or look for patterns in the table.

b. Use finite differences to find a polynomial function that models these data.

c. Find the number of blocks needed to build a pyramid with eight layers.

d. Find the number of layers in a pyramid built with 650 blocks. (You can graph and trace your polynomial function.)

10. Andy has measured his son's height every three months since his son was $9\frac{1}{2}$ years old. At right are his measurements in meters.

a. Find the first differences for Andy's son's heights, and make a scatter plot of points in the form (*age*, D_1). Remember to shorten the list of ages to match D_1. Describe the pattern you see.

b. Repeat 10a until the differences are nearly constant.

c. What type of model do you think will fit this data? Why?

d. Define variables and find a model. Show that the model approximates Andy's son's height for at least three data points.

Age (yr)	Height (m)
9.5	1.14
9.75	1.21
10	1.27
10.25	1.31
10.5	1.33
10.75	1.34
11	1.35
11.25	1.35
11.5	1.35
11.75	1.35
12	1.36
12.25	1.37
12.5	1.39
12.75	1.42
13	1.47
13.25	1.54

11. **APPLICATION** In an atom electrons spin rapidly around a nucleus. An electron can occupy only specific energy levels, and each energy level can only hold a certain number of electrons. The greatest number of electrons possible in any one level is given in this table.

Energy level	1	2	3	4	5	6	7
Maximum number of electrons	2	8	18	32	50	72	98

Is it possible to find a polynomial function that expresses the relationship between the energy level and the maximum number of electrons? If so, find the function. If not, explain why not.

ScienceConnection The electrons in an atom exist at various energy levels. When an electron moves from a lower energy level to a higher energy level, the atom absorbs energy. When an electron moves from a higher to a lower energy level, energy is released, often as light. This is the principle underlying neon lights. The electricity running through a tube of neon gas makes the electrons in the neon atoms jump to higher energy levels. When they drop down to their original levels, they emit light.

Review

12. (Lessons 4.4 and 4.5) Sketch a graph of each function without using your calculator.

a. $y = (x - 2)^2$ **b.** $y = x^2 - 4$ **c.** $y = (x + 4)^2 + 1$

13. (Algebra and Lesson 5. 7) Solve.

a. $12x - 17 = 13$ **b.** $2(x - 1)^2 + 3 = 11$ **c.** $3(5^x) = 48$

14. (Lesson 1.4) The mean salary of seven employees at the Raven Travel Agency is $40,000. The median salary is only $25,000. Give possible salaries for the employees at Raven.

15. (Algebra) Find the product $(x + 3)(x + 4)(x + 2)$.

LESSON

6.2

Equivalent Quadratic Forms

I'm very well acquainted, too,
with matters mathematical.
I understand equations both
the simple and quadratical.

"THE MAJOR GENERAL'S SONG," FROM *The Pirates of Penzance* BY WILLIAM S. GILBERT AND SIR ARTHUR SULLIVAN

In Lesson 6.1 you were introduced to many polynomial functions, including the special cases of 2nd-degree polynomial functions, or *quadratic functions.* The general form of a quadratic function is $y = ax^2 + bx + c$. In this lesson you will work with two additional, equivalent forms of quadratic functions.

This fountain near the center of Pompidou in Paris, France, contains 16 animated surreal sculptures inspired by the music of Russian American composer Igor Stravinsky (1882–1971). It was designed by artists Jean Tinguely (1925–1991) and Niki de Saint-Phalle (b 1930). The arc formed by spouting water can be described with a quadratic equation.

Recall from Chapter 4 that every quadratic function can be considered as a transformation of the graph of the parent function $y = x^2$. A quadratic function in the form $\frac{y-k}{a} = \left(\frac{y-h}{b}\right)^2$ or $y = a\left(\frac{y-h}{b}\right)^2 + k$ identifies the location of the vertex after a translation, (h, k), and the vertical and horizontal scale factors, a and b. For example, consider this parabola with vertex $(4, -2)$. If you consider the point $(7, 4)$ to be the image of the point $(1, 1)$ on the graph of $y = x^2$, then the horizontal scale factor is 3, and the vertical scale factor is 6. So the quadratic function is $\frac{y+2}{6} = \left(\frac{x-4}{3}\right)^2$ or $y = 6\left(\frac{x-4}{3}\right)^2 - 2$. Choosing a point other than $(7, 4)$ would still give the same equation.

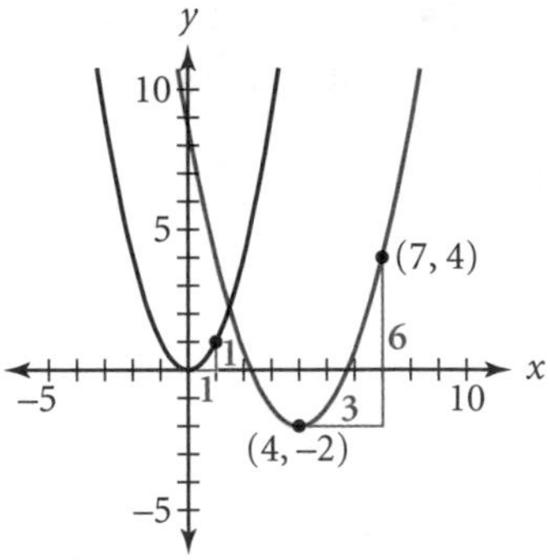

If you square the denominator, the quadratic function can also be written as $y = \frac{6}{9}(x-4)^2 - 2$ or $y = \frac{2}{3}(x-4)^2 - 2$. Notice that the horizontal and vertical scale factors are now combined into one vertical scale factor of $\frac{2}{3}$. In fact, changing a quadratic function from the form $y = c\left(\frac{x-h}{d}\right)^2 + k$ to $y = \frac{c}{d^2}(x - h) + k$ is always possible. The coefficient $\frac{c}{d^2}$ combines the horizontal and vertical scale factors into one vertical scale factor, which can be thought of as a single coefficient, say, a. This new form, $y = a(x - h)^2 + k$, is called the **vertex form** of a quadratic function because it identifies the vertex, (h, k), and a vertical scale factor, a. If you know the vertex of a parabola and one other point, then you can write the quadratic function in vertex form.

Now consider these parabolas. The x-intercepts have been marked. It should be no surprise that the y-coordinate is zero at each x-intercept,

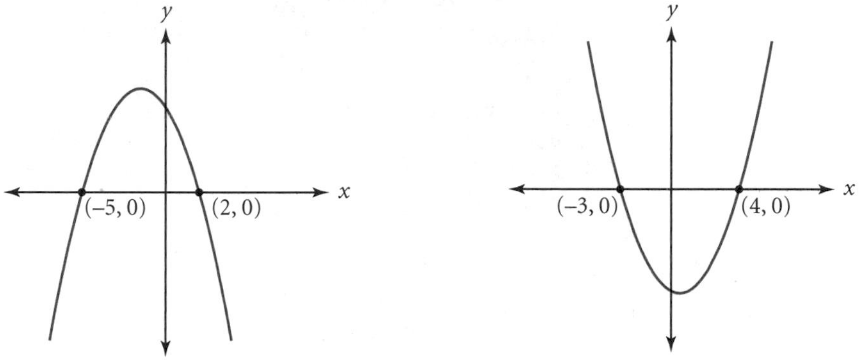

because every point on the x-axis has a y-coordinate of zero. For this reason the x-intercepts of the graph of a function are called the *zeros* of the function. You will use this information and the **zero-product property** to find the zeros of a function without graphing.

The Zero-Product Property

For all real numbers a and b, if $ab = 0$, then $a = 0$, $b = 0$, or $a = 0$ and $b = 0$.

To understand the zero-product property, think of numbers that, when multiplied, equal zero. Whatever numbers you think of will have this characteristic: *At least one of the factors must be zero.* Before moving on, think about numbers that satisfy each of these equations:

$\underline{\ ?\ } \cdot 16.2 = 0$

$3(\underline{\ ?\ } - 4)(\underline{\ ?\ } - 9) = 0$

EXAMPLE A Given the function $y = -1.4(x - 5.6)(x + 3.1)$, find the zeros.

▶ Solution The zeros will be the x-values that make $y = 0$. First set the function equal to 0.

$$0 = -1.4(x - 5.6)(x + 3.1)$$

As the product of three factors equals 0, the zero-product property tells you that at least one of the factors must equal 0.

$-1.4 = 0$	or	$x - 5.6 = 0$	or	$x + 3.1 = 0$
not possible		$x = 5.6$		$x = -3.1$

The solutions, or **roots,** of the equation $0 = -1.4(x - 5.6)(x + 3.1)$ are $x = 5.6$ or $x = -3.1$. That means the zeros of the function $y = -1.4(x - 5.6)(x + 3.1)$ are $x = 5.6$ and $x = -3.1$.

Use your graphing calculator to check your work. You should find that the x-intercepts of the graph of $y = -1.4(x - 5.6)(x + 3.1)$ are 5.6 and -3.1.

$[-10, 10, 1, -40, 40, 10]$

If you know the x-intercepts of a parabola, then you can write the quadratic function in **factored form,** $y = a(x - r_1)(x - r_2)$. This form identifies the locations of the x-intercepts, r_1 and r_2, and a vertical scale factor, a.

EXAMPLE B

Consider this parabola.

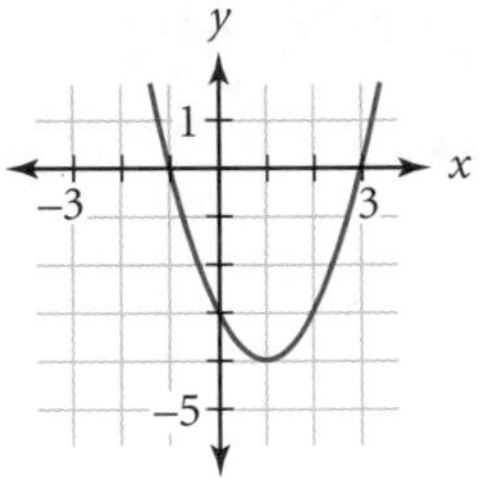

a. Write an equation of the parabola in vertex form.

b. Write an equation of the parabola in factored form.

c. Show that the equations are equivalent by converting to general form.

▸ Solution

a. The vertex is $(1, -4)$. If you consider the point $(2, -3)$ to be the image of point $(1, 1)$ on the graph of $y = x^2$, then the vertical and horizontal scale factors are both 1. As a vertical scale factor, $a = \frac{1}{1^2} = 1$. The vertex form is

$$y = (x - 1)^2 - 4$$

b. The x-intercepts are -1 and 3. From the vertex form you know that the scale factor, a, is 1. The factored form is

$$y = (x + 1)(x - 3)$$

c. To convert to general form, expand by multiplying the binomial expressions, and then combine like terms.

$$\begin{aligned} y &= (x - 1)^2 - 4 \\ y &= (x - 1)(x - 1) - 4 \\ y &= (x^2 - x - x + 1) - 4 \\ y &= x^2 - 2x - 3 \end{aligned} \qquad \begin{aligned} y &= (x + 1)(x - 3) \\ y &= x^2 + x - 3x - 3 \\ y &= x^2 - 2x - 3 \end{aligned}$$

The vertex form and the factored form are equivalent because they are both equivalent to the same general form.

You now know three different forms of a quadratic function.

Three Forms of a Quadratic Function

General form	$y = ax^2 + bx + c$
Vertex form	$y = a(x - h)^2 + k$
Factored form	$y = a(x - r_1)(x - r_2)$

The Investigation will give you practice in using the three forms with real-world data. You'll find that the form you use dictates which features of the data you focus on. Conversely, if you know only a few features of the data, you may need to focus on a particular form of the function.

Investigation
Rolling Along

You will need

- a motion sensor
- an empty coffee can
- a long table

Procedure Note

Prop up one end of the table slightly. Place the motion sensor at the low end of the table, and aim it toward the high end. With tape or chalk, mark a starting line 0.5 m from the sensor on the table.

Step 1 Practice rolling the can up the table directly in front of the motion sensor. Start the can behind the starting line. Give the can a gentle push so that it rolls up the table on its own momentum, stops near the end of the table, and then rolls back. Stop the can after it crosses the line and before it hits the motion sensor.

Step 2 Set up your calculator to collect data for 6 seconds. [▶ See **Calculator Note 6D.**◀] When the sensor begins, roll the can up the table.

Step 3 The data collected by the sensor will have the form (*time*, *distance*). Adjust for the position of the starting line by subtracting 0.5 from all values in the distance list.

Step 4 Let x represent time in seconds, and let y represent distance from the line in meters. Draw an accurate graph of your data. What shape is the graph of the data points? What type of function would model the data? Use finite differences to justify your answer.

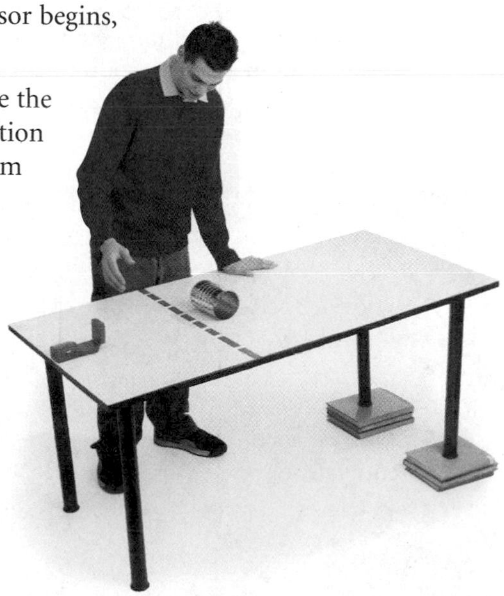

Step 5 Mark the vertex and another point on your graph. Approximate the coordinates of these points, and use them to write the equation of a quadratic model in vertex form.

Step 6 From your data, find the general form of a quadratic model in general form.

Step 7 Mark the x-intercepts on your graph. Approximate the values of these x-intercepts. Use the zeros and the value of a from Step 5 to find a quadratic model in factored form.

Step 8 Verify by graphing that the three equations in Steps 5, 6, and 7 are equivalent or nearly so. Write a few sentences explaining in what instances you would use each of the three forms to find a quadratic model to fit parabolic data.

The ablilty to create a model of data in different ways allows you to find equations according to the type of data you have. The ability to convert from one form to another allows you to find facts about the curve and to compare equations written in different forms. In the exercises in this lesson, you will convert from both the vertex form and the factored form to the general form. In later lessons you will learn the other conversions.

EXERCISES

Practice Your Skills

For the exercises in this lesson, you may find a background grid on your calculator helpful.

1. Identify the form of each quadratic function as general, vertex, factored, or none of these.

a. $y = -3.2(x + 4.5)^2$ **b.** $y = 2.5(x + 1.25)(x - 1.25) + 4$

c. $y = 2x(3 + x)$ **d.** $y = 2x^2 - 4.2x - 10$

2. Each quadratic function below is written in vertex form. What are the coordinates of each vertex? Graph each function to check your answers.

a. $y = (x - 2)^2 + 2$ **b.** $y = 0.5(x + 4)^2 - 2$ **c.** $y = 4 - 2(x - 5)^2$

3. Each quadratic function below is written in factored form. What are the zeros of each function? Graph each function to check your answers.

a. $y = (x + 1)(x - 2)$ **b.** $y = 0.5(x - 2)(x + 3)$ **c.** $y = -2(x - 2)(x - 5)$

4. Convert each function to general form. Graph both forms to check that the equations are equivalent.

a. $y = (x - 2)^2 + 3$ **b.** $y = 0.5(x + 4)^2 - 2$ **c.** $y = 4 - 2(x - 5)^2$

5. Convert each function to general form. Graph both forms to check that the equations are equivalent.

a. $y = (x + 1)(x - 2)$ **b.** $y = 0.5(x - 2)(x + 3)$ **c.** $y = -2(x - 2)(x - 5)$

Reason and Apply

6. As you learned in Chapter 4, the graphs of all quadratic functions have a line of symmetry that contains the vertex and divides the parabola into mirror-image halves. Consider this table of values generated by a quadratic function.

x	y
1.5	−8
2.5	7
3.5	16
4.5	19
5.5	16
6.5	7
7.5	−8

a. What is the line of symmetry for the graph of this quadratic function?

b. The vertex of a parabola represents either the maximum or the minimum value of the quadratic function. Name the vertex of this function, and determine whether it is a maximum or minimum.

c. Use the table of values to write the quadratic function in vertex form.

7. Here is the graph of the quadratic function that passes through $(-2.4, 0)$, $(0.8, 0)$, and $(1.2, -2.592)$.

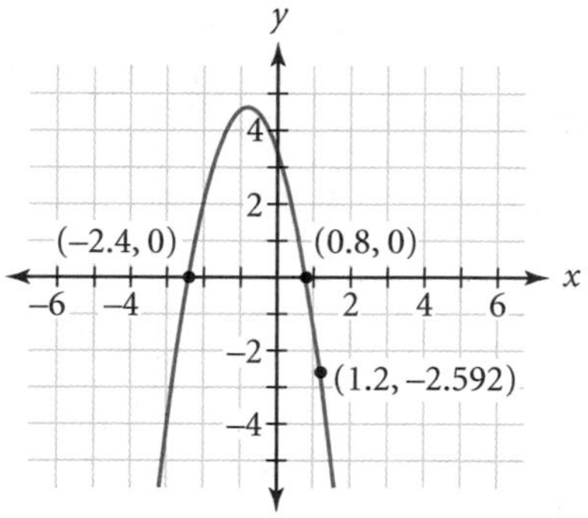

a. Use the x-intercepts to write the quadratic function in factored form. For now, leave the vertical scale factor as a.

b. Substitute the coordinates $(1.2, 2.592)$ into your function from 7a, and solve for a. Write the complete quadratic function in factored form.

c. The line of symmetry of the graph of this quadratic function passes through the vertex and the midpoint of the two x-intercepts. What is the x-coordinate of the vertex? What is the y-coordinate?

d. Write this quadratic function in vertex form.

8. Write the factored form of each polynomial function. (Hint: Substitute the coordinates of the y-intercept to solve for the scale form factor a.)

a.

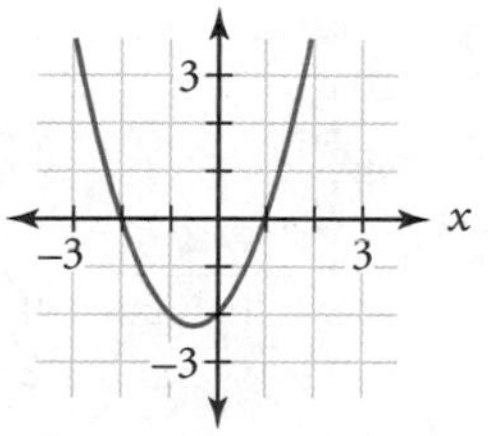

b.

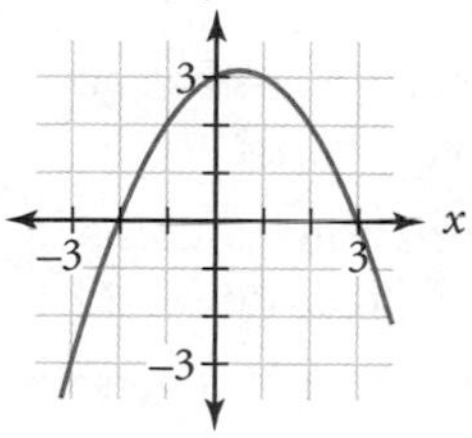

c.

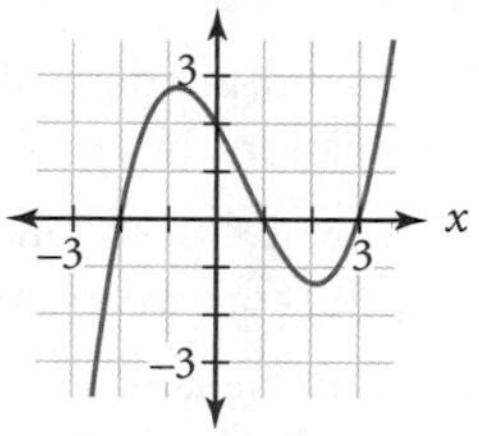

9. Write each function in general form.

a. $y = 4 - 0.5(x + h)^2$

b. $y = a(x - 4)^2$

c. $y = a(x - h)^2 + k$

d. $y = -0.5(x + r)(x + 4)$

e. $y = a(x - 4)(x + 2)$

f. $y = a(x - r)(x - s)$

10. **APPLICATION** The local outlet store charges \$2.00 for a pack of four AA batteries. On an average day, 200 packs are sold. A survey indicates that the sales will decrease by 5 packs per day for each \$0.10 increase in price.

a. Complete this table according to the results of the survey.

Selling price (\$)	2.00	2.10	2.20	2.30	2.40
Number sold	200	?	?	?	?
Revenue (\$)	400	?	?	?	?

b. Calculate the 1st and 2nd differences for the revenue.

c. Let x represent the selling price in dollars, and let y represent the revenue in dollars. Write a function that describes the relationship between the revenue and the selling price.

d. Graph your function, and find the maximum revenue. What selling price provides maximum revenue?

11. **APPLICATION** Delores has 80 m of fence with which to protect her vegetable garden. She wants to enclose the largest possible rectangular area.

Width (m)	5	10	15	20	25
Length (m)	?	?	?	?	?
Area (m^2)	?	?	?	?	?

a. Complete this table.

b. Let x represent the width in meters, and let y represent the area in square meters. Write a function that describes the relationship between the area and the width of the garden.

c. Which width provides the largest possible area? What is that area?

d. Which widths result in an area of 0 m^2?

12. **APPLICATION** Photosynthesis is the process in which plants use energy from the Sun, together with carbon dioxide and water, to make their own food and produce oxygen. Various factors affect the rate of photosynthesis, such as light intensity, light wavelength, CO_2 concentration, and temperature. Below is a graph of how temperature relates to the rate of photosynthesis for a particular plant. (All other factors are assumed to be held constant.)

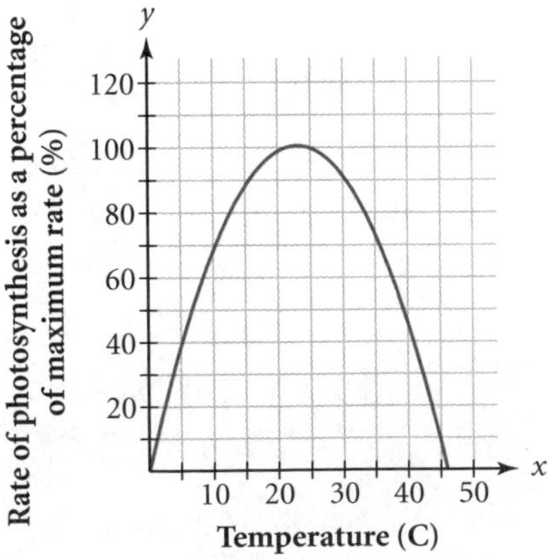

a. Describe the general shape of the graph. What does the shape of the graph mean in the context of photosynthesis?

b. Approximate the optimum temperature for photosynthesis in this plant and the corresponding rate of CO_2 consumed.

c. Temperature must be kept within a certain range. If it gets too hot, the enzymes in the chlorophyll are killed, and photosynthesis stops altogether. If it gets too cold, temperature becomes a limiting factor, and the enzymes stop working. At approximately what temperatures is the rate of photosynthesis equal to 0?

d. Write a function in at least two forms that will produce this graph.

Review

13. (Lessons 2.1 and 2.3) Find a recursive formula that models each sequence. What are the sequences' long-run values?

a. $-8, 2, -0.5, 0.125, \ldots$

b. $8, 5, 3.5, 2.75, \ldots$

(Hint for 13b: You will divide *and* add as part of your recursive rule).

14. (Lesson 3.5) You can solve systems of equations to find quadratic equations instead of using the procedure outlined in Lesson 6.1, Example A. For instance, if a parabola has the equation $y = ax^2 + bx + c$ and passes through (0, 4), (1, 5), and (2, 7), by substituting the given points into the general form of the equation, we can get a system of equations:

$$\begin{aligned} 4 &= a(0)^2 + b(0) + c \\ 5 &= a(1)^2 + b(1) + c \\ 7 &= a(2)^2 + b(2) + c \end{aligned} \qquad \text{or} \qquad \begin{aligned} 4 &= c \\ 5 &= a + b + c \\ 7 &= 4a + 2b + c \end{aligned}$$

a. Solve this system of equations. Write the quadratic equation that passes through (0, 4), (1, 5), and (2, 7).

b. Look back at Example A in Lesson 6.1. Write a system of equations of a quadratic equation passing through (4, 2), (6, 9), and (8, 20).

In Chapter 7 you will learn to solve systems like those given in 14b.

15. (Lesson 4.2) Use the function $f(x) = 3x^3 - 5x^2 + x - 6$ to find these values.

a. $f(2)$ **b.** $f(-1)$ **c.** $f(0)$ **d.** $f\left(\frac{1}{2}\right)$ **e.** $f\left(-\frac{4}{3}\right)$

LESSON

6.3 Completing the Square

This graph of $y = -5.33(x - 0.86)^2 + 0.6$ models one bounce of a ball, where x is time in seconds and y is height in meters. The maximum height of this ball occurs at the vertex (0.86, 0.6), which means that after 0.86 s the ball reaches its maximum height of 0.6 m. To answer questions about data, it is often necessary to find the **maximum** or **minimum** value of a quadratic function. Finding the vertex is simple when you are given an equation in vertex form and sometimes when you are using a graph. However, often you have to estimate values on a graph. In this lesson you will learn a procedure called *completing the square,* which converts a quadratic equation from general form to vertex form accurately.

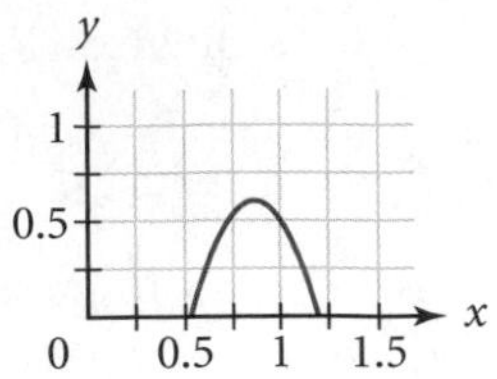

Problems that involve objects rising or falling under the influence of gravity are modeled using quadratic functions. The distance measurements for these projectile motion problems are typically in either meters or feet. The leading coefficient of the polynomial used to model projectile motion is based on the acceleration due to gravity, g, which on Earth has a numerical value of 9.8 m/s^2 when height is measured in meters, and 32 ft/s^2 when height is measured in feet. The leading coefficient of the **projectile motion function** is $-\frac{1}{2}g$.

The Projectile Motion Function

The height of an object rising or falling under the influence of gravity is modeled by the function

$$y = ax^2 + v_0x + s_0$$

where x represents time in seconds, y represents the object's height in meters or feet, a is the acceleration due to gravity (-4.9 m/s^2 or -16 ft/s^2), v_0 is the initial velocity of the object in meters per second or feet per second, and s_0 is the initial height of the object in meters or feet.

EXAMPLE A

A motion sensor records that when Julani jumps in the air, his feet leave the ground at 0.25 s and touch the ground again at 0.83 s. How high was his jump in feet?

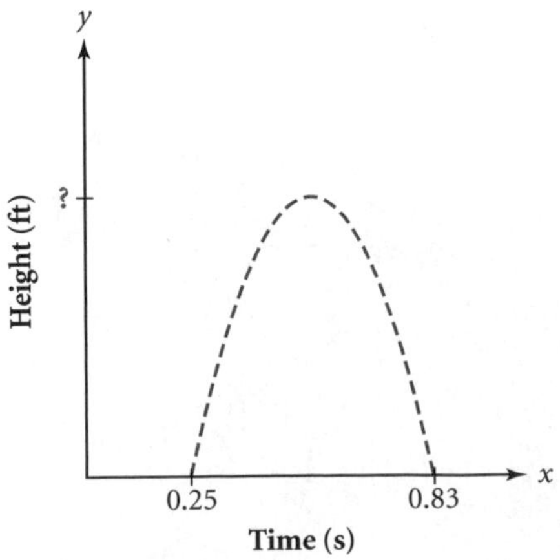

▶ Solution

You don't know the initial velocity, so you can't yet use the projectile motion function, but you do know that height is modeled by a quadratic function and that the leading coefficient needs to be -16. Use this information, along with 0.25 and 0.83 as the x-intercepts (the times when Julani leaves and returns to the floor), to write the function

$$y = -16(x - 0.25)(x - 0.83)$$

The vertex of the graph of $y = -16(x - 0.25)(x - 0.83)$ represents Julani's maximum height. The x-coordinate of the vertex will be midway between the two x-intercepts 0.25 and 0.83. The mean of 0.25 and 0.83 is $\frac{0.25 + 0.83}{2}$, or 0.54.

$y = -16(x - 0.25)(x - 0.83)$	The original function.
$y = -16(0.54 - 0.25)(0.54 - 0.8)$	Substitute the x-coordinate of the vertex.
$y = 1.3456$	Evaluate for the y-coordinate.

Julani jumped approximately 1.3 ft.

Example A showed how to find the vertex of a quadratic function in factored form. If the equation were in general form instead, you might be able to put it in factored form and then follow the same steps, but there's another way. First, note that each rectangle diagram below represents a **perfect square** because both factors are the same.

$$\begin{aligned}(x + 5)^2 &= (x + 5)(x + 5)\\ &= x^2 + 5x + 5x + 25\\ &= x^2 + 10x + 25\end{aligned}$$

The square of the first term of the binomial.

Twice the product of the first and second terms of the binomial.

The square of the second term of the binomial.

	x	5
x	x^2	$5x$
5	$5x$	25

$$\begin{aligned}(a + b)^2 &= (a + b)(a + b)\\ &= a^2 + ab + ab + b^2\\ &= a^2 + 2ab + b^2\end{aligned}$$

The square of the first term of the binomial.

Twice the product of the first and second terms of the binomial.

The square of the second term of the binomial.

	a	b
a	a^2	ab
b	ab	b^2

You'll notice a pattern. For a perfect square, the first and last terms of the trinomial are squares, and the middle term is twice a product. This knowledge will be useful in the Investigation, which helps you convert a quadratic function from general form to vertex form.

Investigation

Square Me: The Details

You can use rectangle diagrams to help convert from general form to vertex form.

Step 1 Consider the expression $x^2 + 6x$.

	x	3
x	x^2	$3x$
3	$3x$	?

a. What could you add to the expression to make it a perfect square? That is, what must be added to complete this rectangle diagram?

b. If you add a number to an expression, you must also subtract the same amount in order to preserve the value of the original expression. Fill in the blanks to rewrite $x^2 + 6x$ as the difference between a perfect square and a number.

$$x^2 + 6x = x^2 + 6x + \underline{\ ?\ } - \underline{\ ?\ } = (x+3)^2 - \underline{\ ?\ }.$$

c. Use a graph or table to verify that your expression in the form $(x - h)^2 + k$ is the equivalent of the original expression $x^2 + 6x$.

Step 2 Consider the expression $x^2 + 6x - 4$.

	x	3
x	x^2	$3x$
3	$3x$	?

a. Focus on the 2nd- and 1st-degree terms of the expression, $x^2 + 6x$. What must be added and subtracted to these terms to complete a perfect square yet preserve the value of the expression?

b. Rewrite the expression $x^2 + 6x - 4$ in the form $(x - h)^2 + k$.

c. Use a graph or table to verify that your expression is equivalent to the original expression $x^2 + 6x - 4$.

Step 3 Rewrite each expression in the form $(x - h)^2 + k$. If you use a rectangle diagram, focus on the 2nd- and 1st-degree terms first. Use a graph or table to verify that your expression is equivalent to the original expression.

a. $x^2 - 14x + 3$

b. $x^2 + bx + 10$

When the 2nd-degree term has a coefficient, you can first factor it out of the 2nd- and 1st-degree terms. For example, $3x^2 + 24x + 5$ becomes $3(x^2 + 8x) + 5$. Completing a diagram for $x^2 + 8x$ can help you rewrite the expression in the form $a(x - h)^2 + k$.

	x	4
x	x^2	$4x$
4	$4x$	16

$3x^2 + 24x + 5$	The original expression.
$3(x^2 + 8x) + 5$	Factor the 2nd- and 1st-degree terms.
$3(x^2 + 8x + 16) - 3(16) + 5$	Complete the square. You add $3 \cdot 16$, so you must subtract $3 \cdot 16$.
$3(x + 4)^2 - 43$	An equivalent expression in the form $a(x - h)^2 + k$.

Step 4 | Rewrite each expression in the form $a(x-h)^2+k$. Use a graph or table to verify that your expression is equivalent to the original expression.

a. $2x^2-6x+1$

b. $ax^2+10x+7$

In the Investigation you saw how to convert a quadratic expression from the form ax^2+bx+c to the form $a(x-h)^2+k$. This process is called **completing the square.** You can also complete the square to convert a quadratic function from general form, $y=ax^2+bx+c$, to vertex form, $y=a(x-h)^2+k$.

EXAMPLE B

Convert each quadratic function to vertex form. Identify the vertex.

a. $y=x^2-18x+100$

b. $y=3x^2+21x-35$

▶ *Solution*

a. To complete the square, separate the constant from the first two terms.

$y=(x^2-18x)+100$ — Original equation.

$y=(x^2-18x+81)+100-81$ — Add 81 to complete the square, and subtract 81 to keep the equation equivalent.

To find what number you must add to complete the square, use a rectangle diagram.

$x^2-18x+100 \rightarrow (x^2-18x+?)+100-? \rightarrow$
$(x^2-18x+81)+100-81 \rightarrow (x-9)^2+19$

When you add 81 to complete the square, you must also subtract 81 to keep the expression equivalent.

	x	-9
x	x^2	$-9x$
-9	$-9x$	81

$y=(x-9)^2+19$ — Factor into a perfect square; add the constant terms.

The vertex form of this quadratic function is $y=(x-9)^2+19$, and the vertex is (9, 19).

b. $y=3x^2+21x-35$ — Original equation.

$y=3(x^2+7x)-35$ — Factor the lead coefficient.

$y=3(x^2+7x+12.25)-36.75-35$ — Add and subtract 3(12.25), or 36.75.

$y=3(x+3.5)^2-71.75$ — Factor into a perfect square; add the constant terms.

The vertex is $(-3.5, -71.75)$, and the vertex form is $y=3(x+3.5)^2-1.75$.

EXAMPLE C

Nora hits a baseball straight up at a speed of 120 ft/s. If her bat contacts the ball at a height of 3 ft above the ground, how high does the ball travel?

▶ Solution

Using the projectile motion function, you know that the height of the object at time x is represented by the equation $y = ax^2 + v_0x + s_0$. The initial velocity, v_0, is 120 ft/s, and the initial height, s_0, is 3 ft. Because the distance is measured in feet, the leading coefficient is -16. Thus, the function is $y = -16x^2 + 120x + 3$. To find the maximum height, you must locate the vertex.

$y = -16x^2 + 120x + 3$	Original equation.
$y = -16(x^2 - 7.5x) + 3$	Factor the lead coefficient.
$y = -16(x^2 - 7.5x + 14.0625) + 225 + 3$	Add and subtract $-16(14.0625)$, or 225.
$y = (x - 3.75)^2 + 228$	Factor into a perfect square; add the constant terms.

The baseball reaches a maximum height of 228 ft at 3.75 s.

Examples B and C show that completing the square gives you a method for finding the vertex of a quadratic function. You can then use the coordinates, (h, k), of the vertex to rewrite an equation in general form as an equation in vertex form.

EXERCISES

Practice Your Skills

1. Factor each quadratic expression.

a. $x^2 + 5x + \frac{25}{4}$ **b.** $4x^2 - 12x + 9$ **c.** $x^2 - 2xy + y^2$

2. What value is required to complete the square?

a. $x^2 + 20x +$ __?__ **b.** $x^2 - 7x +$ __?__ **c.** $6x^2 - 24x +$ __?__

3. Convert each quadratic function to vertex form.

a. $y = x^2 + 20x + 94$ **b.** $y = x^2 - 7x + 16$

c. $y = 6x^2 - 24x + 147$ **d.** $y = 5x^2 + 8x$

4. Rewrite each expression in the form $ax^2 + bx + c$, and then identify the coefficients, a, b, and c.

a. $3x^2 + 2x - 5$ **b.** $14 + 2x^2$ **c.** $-3 + 4x^2 - 2x + 8x$

Reason and Apply

5. What is the vertex of the graph of quadratic function $y = -2x^2 - 16x - 20$?

6. Convert the function $y = 7.51x^2 - 47.32x + 129.47$ to vertex form. Use a graph or table to verify that the functions are equivalent.

7. Imagine that an arrow is shot from the bottom of a well. It passes ground level at 1.1 s and lands on the ground at 4.7 s.

a. Define variables, and write a quadratic function that describes the height of the arrow. (Use $-4.9\ \text{m/s}^2$ as the acceleration due to gravity.)

b. What was the initial velocity of the arrow in meters per second?

c. How deep was the well in meters?

8. A rock is thrown with an initial velocity of 17.2 m/s from the edge of a 50-m cliff overlooking Lake Superior. Define variables, and write an equation that models the height of the rock.

9. **APPLICATION** A store in Yosemite National Park charges $6.60 for a flashlight. Approximately 200 of them are sold each month. A survey indicates that the sales will decrease by 10 flashlights per month for each $0.50 increase in price.

a. Write a function that describes the monthly revenue in dollars, y, as a function of selling price in dollars, x. (A table might help.)

b. What selling price provides maximum monthly revenue? What is the maximum revenue?

10. **APPLICATION** You are enclosing a rectangular area to create a rabbit cage. You have 80 ft of fence and want to build a pen with the largest possible area for your rabbit, so you build the cage using an existing building as one side.

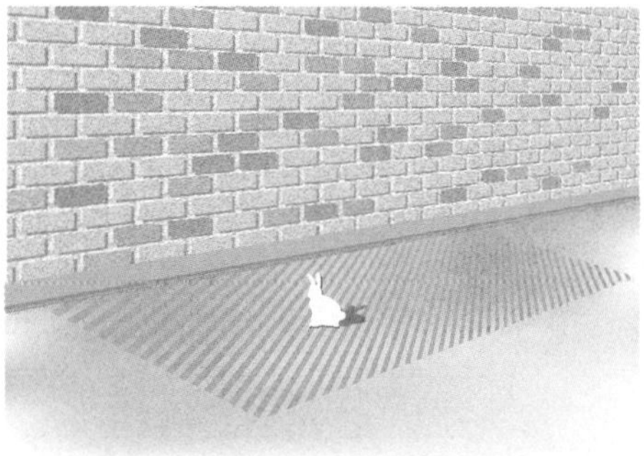

a. Make a table showing the areas for some selected values of x, and write a function that gives the area, y, as a function of the width, x.

b. What width maximizes the area? What is the maximum area?

11. An object is projected upward, and the following data are collected.

Time (s) t	1	2	3	4	5	6
Height (m) h	120.1	205.4	280.9	346.6	402.4	448.4

a. Write a function that relates time and height for this object.

b. What was the initial height? The initial velocity?

c. At what time does the object reach its maximum height? What is the maximum height?

12. **APPLICATION** The members of the Student Service Club decide to sell T-shirts in their school colors for a fund-raiser. In a marketing survey the members ask students whether or not they would buy a T-shirt for a specific price. In analyzing the data they find that at a price of $20 they would sell 60 T-shirts. For each $5 increase in price, they would sell 10 fewer T-shirts.

a. Find a linear function that relates the price in dollars, p, and the number of shirts sold, n.

b. Write a function that gives revenue as a function of price. (Use your function in 12a as a substitute for the number of shirts sold.)

c. Convert the revenue function to vertex form. What is the real-world meaning of the vertex?

d. If the club wanted to receive at least $1,050 in revenue, what price should they charge for the T-shirts?

Review

13. (Algebra) Multiply.

a. $(x - 3)(2x + 4)$

b. $(x^2 + 1)(x + 2)$

14. (Algebra) Solve.

$(x - 2)(x + 3)(2x - 1) = 0$

15. (Lesson 4.7) Consider a graph of the unit circle, $x^2 + y^2 = 1$. Stretch it vertically by a factor of 3, and translate it left 5 units and up 7 units.

a. Write the equation of this new shape. What is it called?

b. Sketch a graph of this shape. Label the center and at least four points.

16. **APPLICATION** (Lesson 3.3) This table shows the number of endangered species in the United States from 1980 to 2000.

Endangered Species

Year	1980	1985	1990	1995	1996	1997	1998	1999	2000
Number of endangered species	224	300	442	756	837	896	924	939	961

(*2002 The New York Times Almanac*)

a. Define variables, and create a scatter plot of these data.

b. Find a linear model for these data.

c. Predict the number of endangered species in 2008 and in 2050 using your model.

LESSON

6.4 The Quadratic Formula

Although you can always graph a function to approximate the x-intercepts, often you are not able to find exact solutions. This lesson develops a procedure to find the exact roots of equations by first converting them to vertex form.

EXAMPLE A

Nora hits a baseball straight up at a speed of 120 ft/sec. Her bat contacts the ball at a height of 3 ft above the ground. Recall that the equation relating height in meters, y, and time in seconds, x, is $y = -16x^2 + 120x + 3$. How long will it be until the ball hits the ground?

▸ Solution

The height will be zero when the ball hits the ground, so you want to find the solutions to the equation $-16x^2 + 120x + 3 = 0$. You can approximate the x-intercepts by graphing, but you may not be able to find the exact x-intercept.

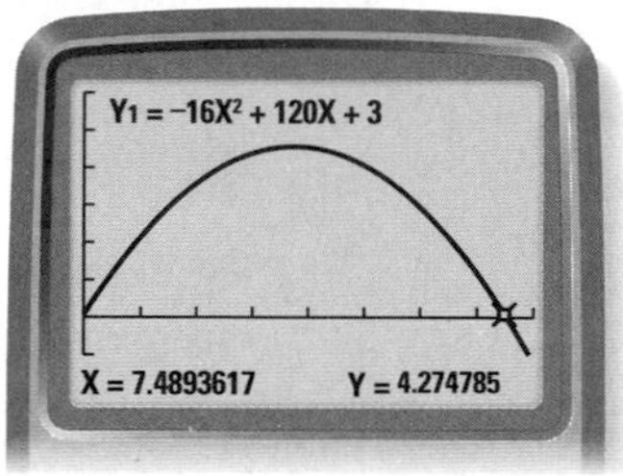

[0, 8, 1, −50, 300, 50]

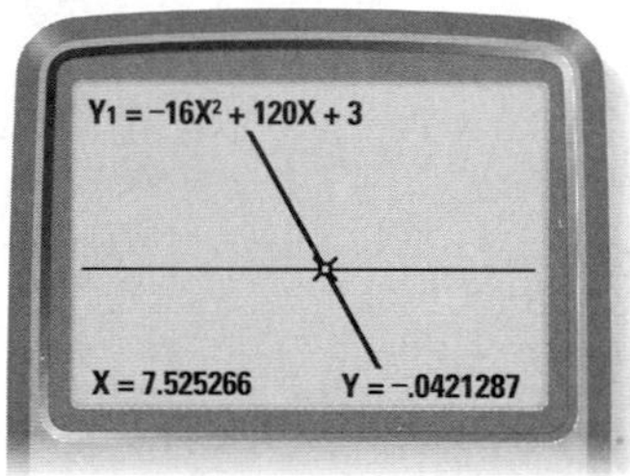

[7.46, 7.58, 1, −2.28, 3.18, 50]

As with most quadratic equations, you will not be able to factor this equation by inspection, so you can't use the zero-product property. Instead, to solve this equation symbolically, you will first write the equation in the form $0 = a(x - h)^2 + k$.

$-16x^2 + 120x + 3 = 0$	Original equation.
$-16(x^2 + 7.5x) + 3$	
$-16(x^2 + 7.5x + 14.0625) + 225 + 3 = 0$	Complete the square.
$-16(x - 3.75)^2 + 228 = 0$	
$-16(x - 3.75)^2 = -228$	Subtract 228 from both sides.
$(x - 3.75)^2 = 14.25$	Divide both sides by -16.
$x - 3.75 = \pm\sqrt{14.25}$	Take the square root of both sides.
$x = 3.75 \pm \sqrt{14.25}$	Add 3.75 to both sides.
$x = 3.75 + \sqrt{14.25}$ or $x = 3.75 - \sqrt{14.25}$	Write the two exact solutions to the equation.
$x \approx 7.525$ or $x \approx -0.025$	Approximate the values of x.

The zeros of the function $y = -16x^2 + 120x + 3$ are $x \approx 7.525$ and $x \approx -0.025$. The negative time, -0.025 s, does not make sense in this situation, so the ball hits the ground after approximately 7.525 seconds.

If you follow these same steps with a general quadratic equation, you can develop the **quadratic formula,** which provides solutions to $ax^2 + bx + c = 0$ in terms of a, b, and c.

$ax^2 + bx + c = 0$	Original equation.
$a\left(x^2 + \frac{b}{a}x\right) + c = 0$	
$a\left(x^2 + \frac{b}{a}x + \frac{b^2}{4a^2}\right) - \frac{b^2}{4a} + c = 0$	Complete the square.
$a\left(x + \frac{b}{2a}\right)^2 + c - \frac{b^2}{4a} = 0$	Rewrite the equation in the form $a(x - h)^2 + k = 0$.
$a\left(x + \frac{b}{2a}\right)^2 = \frac{b^2}{4a} - c$	Subtract c from both sides. Add $\frac{b^2}{4a}$ to both sides.
$a\left(x + \frac{b}{2a}\right)^2 = \frac{b^2}{4a} - \frac{4ac}{4a}$	Rewrite the right side with a common denominator.
$a\left(x + \frac{b}{2a}\right)^2 = \frac{b^2 - 4ac}{4a}$	Add terms with a common denominator.
$\left(x + \frac{b}{2a}\right)^2 = \frac{b^2 - 4ac}{4a^2}$	Divide both sides by a.
$x + \frac{b}{2a} = \pm\sqrt{\frac{b^2 - 4ac}{4a^2}}$	Take the square root of both sides.
$x + \frac{b}{2a} = \pm\frac{\sqrt{b^2 - 4ac}}{2a}$	Use the power of a quotient property to take the square roots of the numerator and denominator.
$x = -\frac{b}{2a} \pm \frac{\sqrt{b^2 - 4ac}}{2a}$	Subtract $\frac{b}{2a}$ from both sides.
$x = \frac{-b \pm \sqrt{b^2 - 4ac}}{2a}$	Add terms with a common denominator.

The Quadratic Formula

Given a quadratic equation written in the form $ax^2 + bx + c = 0$, the solutions are

$$x = \frac{-b \pm \sqrt{b^2 - 4ac}}{2a}$$

To use the quadratic formula on the equation in Example A, $-16x^2 + 120x + 3 = 0$, first identify the coefficients as $a = -16$, $b = 120$, and $c = 3$. The solutions would be:

$$x = \frac{-120 \pm \sqrt{120^2 - 4(-16)(3)}}{2(-16)}$$

$$x = \frac{-120 + \sqrt{14592}}{-32} \quad \text{or} \quad x = \frac{-120 - \sqrt{14592}}{-32}$$

$x \approx -0.025$ or $x \approx 7.525$

The quadratic formula gives you a way to find the roots of any equation in the form $ax^2 + bx + c = 0$. The Investigation will give you an opportunity to apply the quadratic formula in different situations.

Investigation
How High Can You Go?

Salvador hits a baseball at a height of 3 ft and with an initial upward velocity of 88 feet per second.

Step 1 Let x represent time in seconds after the ball is hit, and let y represent the height of the ball in feet. Write a function that gives the height as a function of time.

Step 2 Write an equation to find the times when the ball is 24 ft above the ground.

Step 3 Put your equation from Step 2 into the form $ax^2 + bx + c = 0$, then use the quadratic formula to solve. What is the real-world meaning of each of your solutions? Why are there two solutions?

Step 4 Confirm that the vertex of this parabola has a y-coordinate of 124.

Step 5 Write an equation to find the time when the ball reaches a height of 124 ft. Solve it using the quadratic formula. At what point does it become obvious that there is only one solution to this equation?

Step 6 Write an equation to find the time at which the ball reaches a height of 200 ft. What happens when you try to solve this impossible situation with the quadratic formula?

It's important to note that a quadratic equation must be in the general form $ax^2 + bx + c = 0$ before you use the quadratic formula. Otherwise, the signs of a, b, and c may be incorrect.

EXAMPLE B Solve $3x^2 = 5x + 8$.

► ***Solution***

To use the quadratic formula, first write the equation in the form $ax^2 + bx + c = 0$, and identify the coefficients.

$$3x^2 - 5x - 8 = 0$$

$$a = 3,\ b = -5,\ c = -8$$

Substitute a, b, and c in the quadratic formula to solve.

$$x = \frac{-b \pm \sqrt{b^2 - 4ac}}{2a}$$

$$x = \frac{-(-5) \pm \sqrt{(-5)^2 - 4(3)(-8)}}{2(3)}$$

$$x = \frac{5 \pm \sqrt{121}}{6}$$

$$x = \frac{5 \pm 11}{6}$$

$$x = \frac{5 + 11}{6} = \frac{8}{3} \text{ or } x = \frac{5 - 11}{6} = -1$$

The two roots are $x = \frac{8}{3}$ or $x = -1$.

To check your work, substitute these values in the original equation. Here's a way to use your calculator to check.

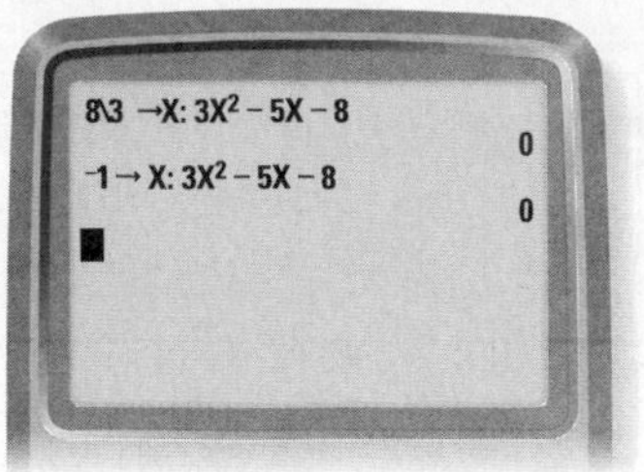

Remember, you can find exact solutions to some quadratic equations by factoring. However, most quadratics don't factor. The quadratic formula can be used to solve any quadratic equation.

EXERCISES

Practice Your Skills

1. Solve.

a. $(x - 2.3)^2 = 25$ **b.** $(x + 4.45)^2 = 12.25$ **c.** $\left(x - \frac{3}{4}\right)^2 = \frac{25}{16}$

2. Rewrite each equation in general form, $ax^2 + bx + c = 0$. Identify a, b, and c.

a. $3x^2 - 13x = 10$ **b.** $x^2 - 13 = 5x$ **c.** $3x^2 + 5x = -1$

d. $3x^2 - 2 - 3x = 0$ **e.** $14(x - 4) - (x + 2) = (x + 2)(x - 4)$

3. Using your calculator, evaluate each expression. Round your answers to the nearest thousandth.

a. $\dfrac{-30 + \sqrt{30^2 - 4(5)(3)}}{2(5)}$ **b.** $\dfrac{-30 - \sqrt{30^2 - 4(5)(3)}}{2(5)}$

c. $\dfrac{8 - \sqrt{(-8)^2 - 4(1)(-2)}}{2(1)}$ **d.** $\dfrac{8 + \sqrt{(-8)^2 - 4(1)(-2)}}{2(1)}$

4. Solve by any method.

a. $x^2 - 6x + 5 = 0$ **b.** $x^2 - 7x - 18 = 0$ **c.** $5x^2 + 12x + 7 = 0$

5. Use the roots of the equations in Exercise 4 to write each of these functions in factored form, $y = a(x - r_1)(x - r_2)$.

a. $y = x^2 - 6x + 5$ **b.** $y = x^2 - 7x - 18$ **c.** $y = 5x^2 + 12x + 7$

Reason and Apply

6. Beth uses the quadratic formula to solve an equation and gets $x = \frac{-9 \pm \sqrt{9^2 - 4(1)(10)}}{2(1)}$.

a. Write the quadratic equation with which Beth started.

b. Write the simplified forms of the exact answers.

c. Give an example of a quadratic function that would have the answers from 6b as zeros. Write your answer in standard form.

7. Write a quadratic function the graph of which has

a. x-intercepts at 3 and -3

b. x-intercepts at 4 and $\frac{-2}{5}$

c. x-intercepts at r_1 and r_2

8. Use the quadratic formula to find the zeros of $y = 2x^2 + 2x + 5$. Explain what happens. Graph $y = 2x^2 + 2x + 5$ to confirm what happens with the quadratic formula. Is there a way to recognize this situation before using the quadratic formula?

9. Write a quadratic function that has no x-intercepts.

10. Show that the mean of the two solutions provided by the quadratic formula is $-\frac{b}{2a}$. Explain what this tells you about the graph.

11. These data give the amount of water in a draining bathtub and the amount of time after the plug was pulled.

Time (min) x	1	1.5	2	2.5
Amount of water (L) y	38.4	30.0	19.6	7.2

a. Write a function that gives the amount of water as a function of time.

b. How much water was in the tub when the plug was pulled?

c. How long did it take the tub to empty?

12. A *golden rectangle* is a rectangle that can be divided into a square and another rectangle that is also a golden rectangle similar to the original. (Two rectangles are similar if the ratio of the longer side to the shorter side is the same in each rectangle). In the figure below, *ABCD* is a golden rectangle because it can be divided into square *ABFE* and golden rectangle *FCDE.*

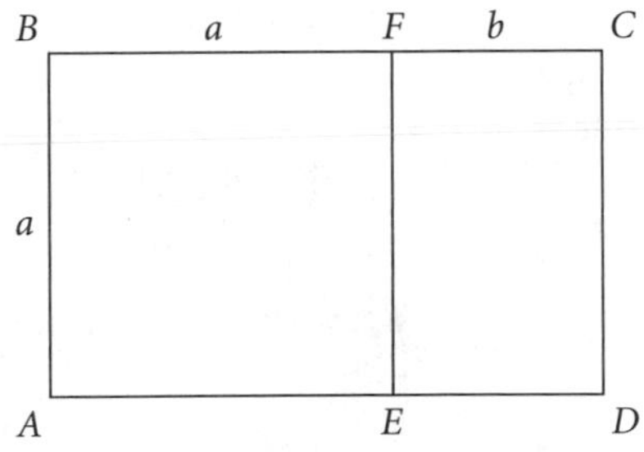

Setting up a proportion of the side lengths of the similar rectangles leads to $\frac{a}{a+b} = \frac{b}{a}$. Let $b = 1$, and solve this equation for a.

HistoryConnection Throughout history many people and cultures have held the golden rectangle to be one of the most visually pleasing geometric shapes. It was used in the architectural designs of the Cathedral of Notre Dame as well as in music and famous works of art. The ancient Egyptians used the ratio when building their pyramids, temples, and tombs and knew the value of the **golden ratio** to be $\frac{1+\sqrt{5}}{2}$.

Review

13. (Lesson 6.3) Complete each equation.

a. $x^2 + \underline{\ ?\ } + 49 = (x + \underline{\ ?\ })^2$

b. $x^2 - 10x + \underline{\ ?\ } = (\underline{\ ?\ })^2$

c. $x^2 + 3x + ? = (\underline{\ ?\ })^2$

d. $2x^2 + \underline{\ ?\ } + 8 = 2(x^2 + \underline{\ ?\ }) + \underline{\ ?\ } = (\underline{\ ?\ })^2 + \underline{\ ?\ }$

14. (Lessson 5.5) Find the inverse of each function. (The inverse does not need to be a function.)

a. $y = (x + 1)^2$ **b.** $y = (x + 1)^2 + 4$ **c.** $y = x^2 + 2x - 5$

15. (Lesson 6.1) Convert these quadratic functions to general form.

a. $y = (x - 3)(2x + 5)$ **b.** $y = -2(x - 1)^2 + 4$

16. (Lesson 2.2) Consider these two investments.

a. Write the equation of a sequence that gives the amount of money in a bank account after n years if \$6,000 is invested at a rate of 7%, compounded annually.

b. Write the equation of a sequence that gives the amount of money in a bank account after n years if \$7,000 is invested at a rate of 6%, compounded annually.

c. After how many years will the \$6,000 invested at 7% be worth more than the \$7,000 invested at 6%?

PROJECT

QUADRATIC FORMULA PROGRAM

Write a calculator program that uses the quadratic formula to solve equations. The program should prompt the user to input values for a, b, and c for a quadratic equation in the form $ax^2 + bx + c = 0$, and it should calculate and display the two solutions. Your program may be quite elaborate or very simple.

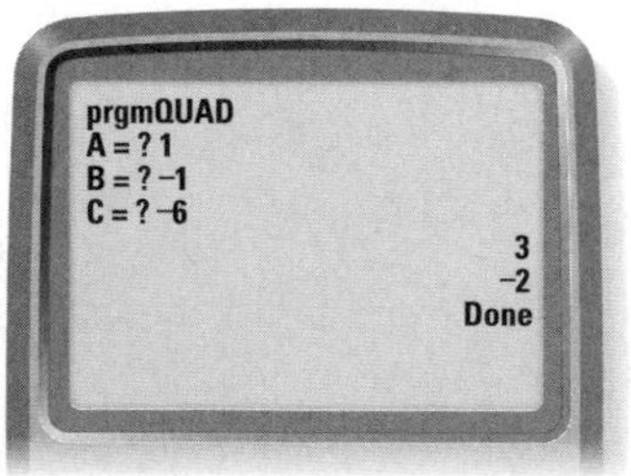

Your project should include

- a written record of the steps your program uses
- an explanation of how the program works
- the results of solving at least two equations by hand and with your program, to verify that your program works

LESSON

6.5

Complex Numbers

Everything should be made as simple as possible, but not simpler.

ALBERT EINSTEIN

You have explored several ways to solve quadratic equations. You can find the x-intercepts on a graph, you can solve by completing the square, or you can use the quadratic formula. What happens when you try to use the quadratic formula on an equation the graph of which has no x-intercepts?

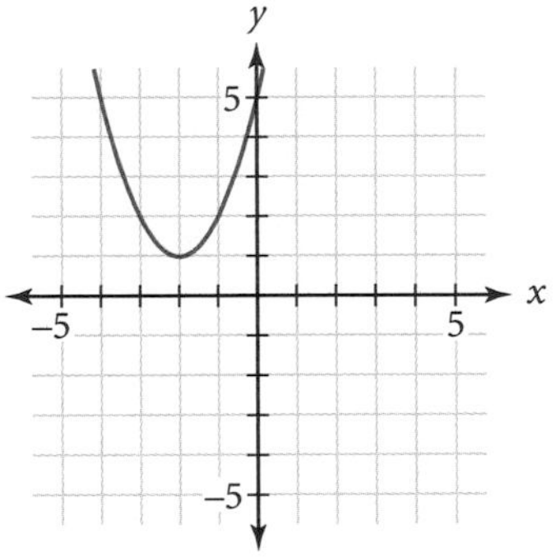

The graph of $y = x^2 + 4x + 5$ shows that this function has no x-intercepts. Using the quadratic formula to try to find x-intercepts, you will get

$$x = \frac{-4 \pm \sqrt{16 - 4(1)(5)}}{2(1)} = \frac{-4 \pm \sqrt{-4}}{2}$$

How do you take the square root of a negative number? The two numbers $\frac{-4 + \sqrt{-4}}{2}$ and $\frac{-4 - \sqrt{-4}}{2}$ are unlike any of the numbers you have worked with in this course—they are nonreal, but they are still numbers. For the same reasons we have fractions and not just whole numbers, negative numbers and not just positive numbers, and irrational numbers and not just fractions, we also have square roots of negative numbers and not just square roots of positive numbers. The set of numbers that include the square roots of negatives are called **complex numbers.**

HistoryConnection Since the 1500s the square root of a negative number has been called an **imaginary number.** In the late 1700s the Swiss mathematician Leonhard Euler (1707–1783) introduced the symbol i to represent $\sqrt{-1}$. He wrote:

> It is evident that we cannot rank the square root of a negative number amongst possible numbers, and we must therefore say that it is an impossible quantity . . . But notwithstanding this these numbers present themselves to the mind; they exist in our imagination, and we still have a sufficient idea of them; since we know that by $\sqrt{-4}$ is meant a number which, multiplied by itself, produces -4; for this reason also, nothing prevents us from making use of these imaginary numbers, and employing them in calculation.

Defining imaginary numbers gave a name to the solutions of problems once thought unsolvable.

To express the square root of a negative number, we use an imaginary unit called i, defined by $i^2 = -1$ or $i = \sqrt{-1}$. You can rewrite $\sqrt{-4}$ as $\sqrt{4}\sqrt{-1}$, or $2i$. Therefore, the two solutions of the quadratic equation above can be written as the complex numbers $\frac{-4 + 2i}{2}$ and $\frac{-4 - 2i}{2}$, or $-2 + i$ and $-2 - i$. Note that these two solutions are a **conjugate pair**. That is, one is $a + bi$, and the other is $a - bi$. Can you explain why nonreal solutions to the quadratic formula will always give answers that are a conjugate pair? A complex number in the form $a - bi$ is also known as the *complex conjugate* of $a + bi$.

Roots of polynomials can be real or complex, or there may be some of each. However, as long as the polynomial has real coefficients, any complex roots will come in conjugate pairs, such as $2i$ and $-2i$ or $3 + 4i$ and $3 - 4i$.

Complex Numbers

A **complex number** is a number of the form $a + bi$, where a and b are real numbers and $i = \sqrt{-1}$.

For any complex number in the form $a + bi$, a is the real part, and bi is the imaginary part. This means the set of complex numbers contains all real numbers and all imaginary numbers, as shown in this tree diagram.

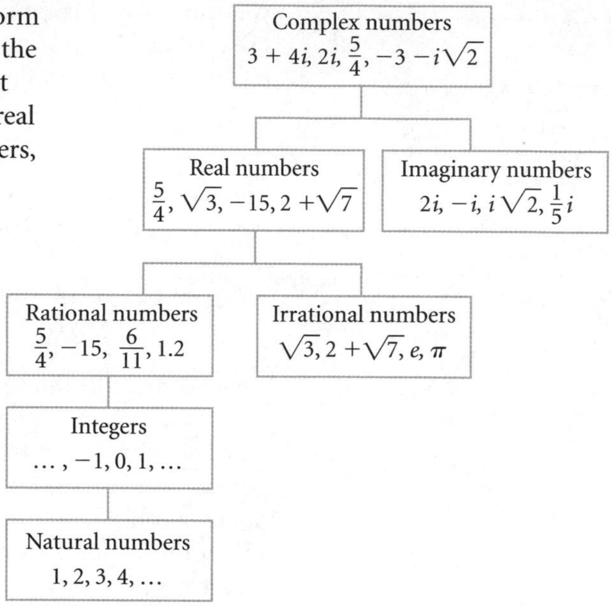

EXAMPLE A | Solve $x^2 + 3 = 0$.

▶ Solution

One method is to isolate the x^2 and take the square root of both sides.

$$\begin{aligned} x^2 + 3 &= 0 \\ x^2 &= -3 \\ x &= \pm\sqrt{-3} \\ x &= \pm\sqrt{3} \cdot \sqrt{-1} \\ x &= \pm\sqrt{3} \cdot i \\ x &= \pm i\sqrt{3} \end{aligned}$$

To check the two solutions, substitute them in the original equation.

$$\begin{aligned} x^2 + 3 &= 0 \\ (i\sqrt{3})^2 + 3 &\stackrel{?}{=} 0 \\ i^2 \cdot 3 + 3 &\stackrel{?}{=} 0 \\ -1 \cdot 3 + 3 &\stackrel{?}{=} 0 \\ -3 + 3 &\stackrel{?}{=} 0 \\ 0 &= 0 \end{aligned} \qquad \begin{aligned} x^2 + 3 &= 0 \\ (-i\sqrt{3})^2 + 3 &\stackrel{?}{=} 0 \\ i^2 \cdot 3 + 3 &\stackrel{?}{=} 0 \\ -1 \cdot 3 + 3 &\stackrel{?}{=} 0 \\ -3 + 3 &\stackrel{?}{=} 0 \\ 0 &= 0 \end{aligned}$$

The two imaginary numbers, $\pm i\sqrt{3}$, are solutions to the original equation, but as they are not real, the graph of $y = x^2 + 3$ shows no x-intercepts.

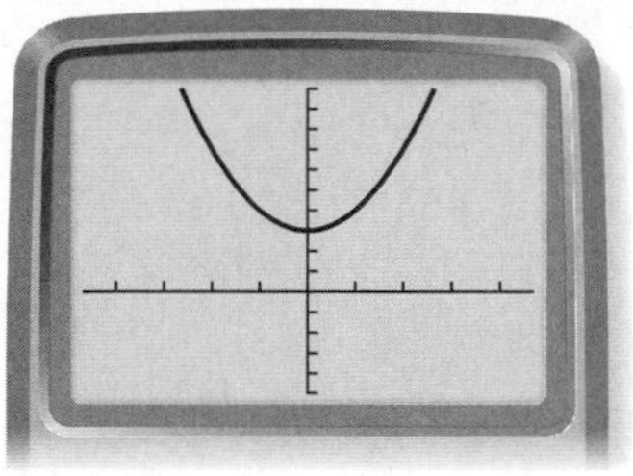

$[-4.7, 4.7, 1, -5, 10, 1]$

Complex numbers are used to model many applications, particularly in science and engineering. To measure the strength of an electromagnetic field, a real number denotes the amount of electricity, and an imaginary number represents the amount of magnetism. The state of a component in an electronic circuit is also measured by a complex number, where the voltage across it is a real number, and the current flowing through it is an imaginary number. The properties of calculations with complex numbers apply to these types of physical states more accurately than do calculations with real numbers. In this Investigation you'll explore patterns in arithmetic with complex numbers.

Investigation
Complex Arithmetic

Computation with complex numbers has conventional rules similar to those for working with real numbers. The purpose of this Investigation is to discover these rules. You may use your calculator to check your work or to explore further examples. [► See **Calculator Note 6F** to learn how to enter complex numbers on your calculator. ◄]

Part 1: Addition and Subtraction

Addition and subtraction of complex numbers is similar to combining like terms. Use your calculator to add these complex numbers. Make a conjecture about how to add complex numbers.

a. $(2 - 4i) + (3 + 5i)$ **b.** $(7 + 2i) + (-2 + i)$

c. $(2 - 4i) - (3 + 5i)$ **d.** $(4 - 4i) - (1 - 3i)$

Part 2: Multiplication

Use your knowledge of multiplying binomials to multiply these complex numbers. Your products should be in the form $a + bi$. Recall that $i^2 = -1$.

a. $(2 - 4i)(3 + 5i)$ **b.** $(7 + 2i)(-2 + i)$

c. $(2 - 4i)^2$ **d.** $(4 - 4i)(1 - 3i)$

Part 3: The Complex Conjugates

Recall that every complex number $a + bi$ has a complex conjugate, $a - bi$. Complex conjugates have some special properties and uses. Each example below shows either the sum or product of a complex number and its conjugate. Simplify these expressions by putting them into the form $a + bi$, and generalize what happens.

a. $(2 - 4i) + (2 + 4i)$

b. $(7 + 2i) + (7 - 2i)$

c. $(2 - 4i)(2 + 4i)$

d. $(-4 + 4i)(-4 - 4i)$

Part 4: Division

To divide two complex numbers in the form $a + bi$, you must eliminate the imaginary part of the denominator. First, use your work from Part 3 to decide how to change each denominator into a real number. Once you have a real number in the denominator, it should be easy to divide.

a. $\frac{7 + 2i}{1 - i}$

b. $\frac{2 - 5i}{3 + 4i}$

c. $\frac{2 - i}{8 - 6i}$

d. $\frac{2 - 4i}{2 + 4i}$

You cannot graph a complex number, such as $3 + 4i$, on a real number line, but you can graph it on a **complex number plane** where the horizontal axis is the **real axis**, and the vertical axis is the **imaginary axis**. In this graph, $3 + 4i$ is located at the point with coordinates $(3, 4)$. Any complex number $a + bi$ has (a, b) as its coordinates on the complex number plane.

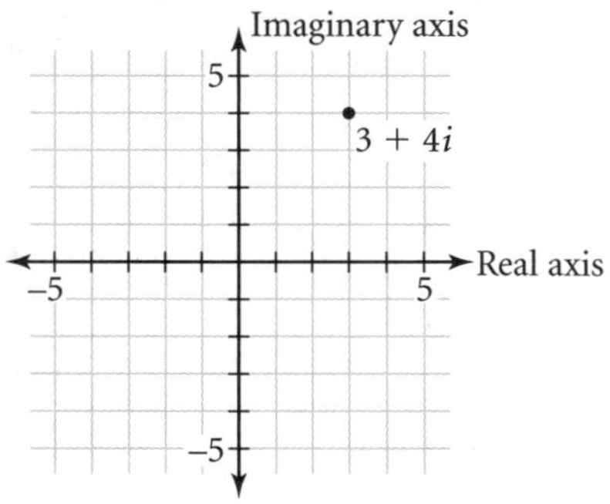

You'll observe some properties of the complex number plane in the exercises.

EXERCISES

Practice Your Skills

1. Add or subtract.

a. $(5 - 1i) + (3 + 5i)$

b. $(6 + 2i) - (-1 + 2i)$

c. $(2 + 3i) + (2 - 5i)$

d. $(2.35 + 2.71i) - (4.91 + 3.32i)$

2. Multiply.

a. $(5 - 1i)(3 + 5i)$

b. $6(-1 + 2i)$

c. $3i(2 - 5i)$

d. $(2.35 + 2.71i)(4.91 + 3.32i)$

3. Find the conjugate of each complex number.

a. $5 - i$ **b.** $-1 + 2i$ **c.** $2 + 3i$ **d.** $-2.35 - 2.71i$

4. Name the complex number associated with each point on the complex plane.

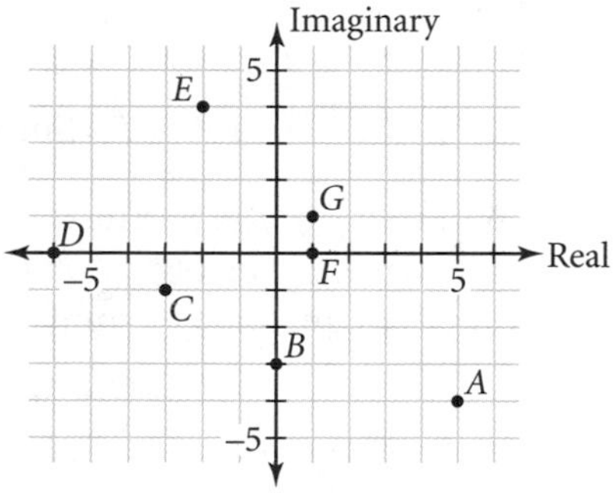

5. Draw Venn diagrams (see illustration) to show the relationships between these sets of numbers.

a. real numbers and complex numbers

b. rational numbers and irrational numbers

c. imaginary numbers and complex numbers

d. imaginary numbers and real numbers

e. complex numbers, real numbers, and imaginary numbers

HistoryConnection In his book *Symbolic Logic*, English logician John Venn (1834–1923) proposed the use of diagrams to represent logic relationships.

For example, this diagram shows that "If it's an eagle, then it's a bird," and "If it's a bird, then it's not a whale." It must also be true, then, that "If it's an eagle, then it's not a whale." A Venn diagram of this situation shows this conclusion clearly. Venn diagrams have become a popular tool for representing all kinds of relationships.

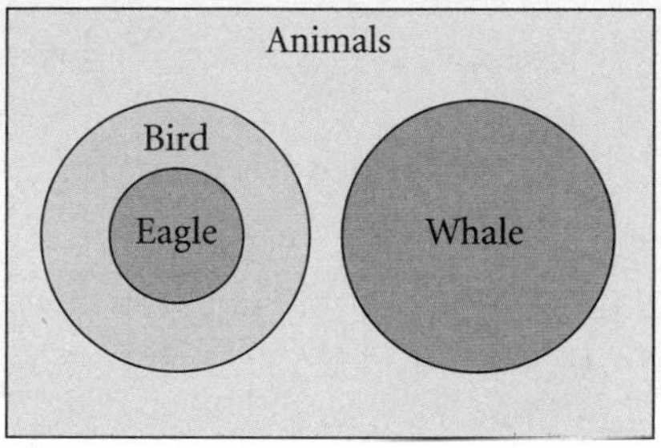

Reason and Apply

6. Rewrite the equation $(x - (2 + i))(x - (2 - i)) = 0$ in general form.

7. Use the definitions $i = \sqrt{-1}$ and $i^2 = -1$ to rewrite each power of i as 1, i, -1, or $-i$.

a. i^3 **b.** i^4

c. i^5 **d.** i^{10}

8. ***Mini-Investigation*** Plot the numbers $i, i^2, i^3, i^4, i^5, \ldots$ in the complex plane. What pattern do you see? What would you expect the value of i^{17} to be? Verify your conjecture.

9. Divide $\frac{2 + 3i}{2 - i}$, and express your answer in $a + bi$ form.

10. Solve each equation. Label each solution as real, imaginary, and/or complex.

a. $x^2 - 4x + 6 = 0$ **b.** $x^2 + 1 = 0$

c. $x^2 + x = -1$ **d.** $x^2 - 1 = 0$

11. Write a quadratic function in general form that has the given zeros.

a. $x = -3$ or $x = 5$ **b.** $x = -3.5$ (and only $x = -3.5$)

c. $x = 5i$ or $x = -5i$ **d.** $x = 2 + i$ or $x = 2 - i$

12. Write in general form a quadratic function that has a zero of $x = 4 + 3i$ and the graph of which has y-intercept 50.

13. Solve.

a. $x^2 - 10ix - 9i^2 = 0$

b. $x^2 - 3ix = 2$

c. Why don't the solutions to 13a and 13b come in conjugate pairs?

14. The quadratic formula, $x = \frac{-b \pm \sqrt{b^2 - 4ac}}{2a}$, provides solutions to $ax^2 + bx + c = 0$. Write some rules involving a, b, and c that determine each of these conditions without using the entire quadratic formula.

a. The solutions are complex.

b. The solutions are real.

c. There is only one real solution.

15. Use these recursive formulas to find the first six terms (z_0 to z_5) of each sequence. Describe what happens in the long run for each.

a. $\begin{cases} z_0 = 0 \\ z_n = z_{n-1}^2 + 0 \quad \text{where } n \geq 1 \end{cases}$

b. $\begin{cases} z_0 = 0 \\ z_n = z_{n-1}^2 + i \quad \text{where } n \geq 1 \end{cases}$

c. $\begin{cases} z_0 = 0 \\ z_n = z_{n-1}^2 + 1 - i \quad \text{where } n \geq 1 \end{cases}$

d. $\begin{cases} z_0 = 0 \\ z_n = z_{n-1}^2 + 0.2 + 0.2i \quad \text{where } n \geq 1 \end{cases}$

MathematicsConnection Fractals are incredibly complicated, yet they are very beautiful geometric shapes made up of similar shapes combined in different scales. Fractals can be generated by quite simple rules. Complex numbers, c, which converge (get close to a point in the complex plane) when following the recursive formula $z_0 = 0$ and $z_n = z_{n-1}^2 + c$ where $n \geq 1$ are members of the Mandelbrot set. A picture of the Mandelbrot set reveals a fractal. In the graph of the Mandelbrot set in the next Project, these convergent points are plotted as black points in a complex plane. In the late 1900s Polish mathematician Benoit Mandelbrot (b 1924) noticed that fractals weren't just a mathematical curiosity but rather the geometry of nature. Clouds, coastlines, and trees can be described using fractal geometry. Fractals are used in medicine to study the growth of cancer tissue, in art to date early paintings, and in computer programming to encode large sets of data.

Review

16. (Lesson 6.4) Consider the function $y = 2x^2 + 6x - 3$.

a. List the zeros in exact radical form and as approximations to the nearest hundredth.

b. Graph the function, and label the exact coordinates of the x-intercepts, the y-intercepts, and the vertex.

17. (Chapter 4) Consider the function $f(x) = |x|$.

a. What transformations of the graph of $f(x) = |x|$ are needed to create the graph of $g(x) = 2|x - 3| + 5$?

b. Sketch a graph of $g(x) = 2|x - 3| + 5$.

PROJECT

THE MANDELBROT SET

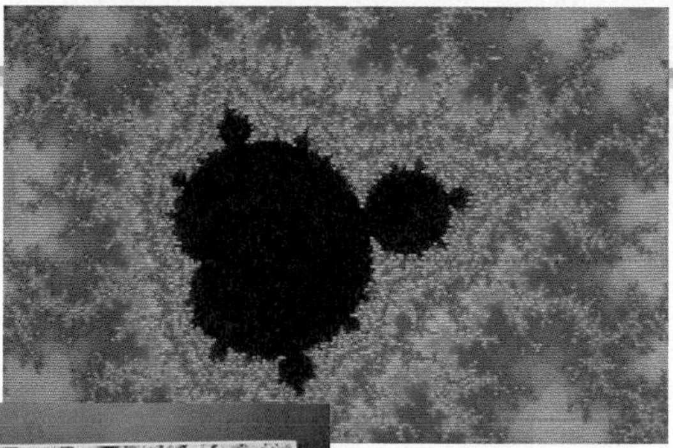

You have seen geometric fractals such as the Sierpiński triangle, and you may have seen other fractals that look much more complicated. The Mandelbrot set is a famous fractal that relies upon repeated calculations with complex numbers.

This Mandelbrot set shows how fractal geometry creates order out of what seem like irregular patterns. Points that are not in the Mandelbrot set are colored based on how quickly they diverge. Benoit Mandelbrot (b 1924), left, was the first person to study and name fractal geometry.

To create the Mandelbrot set, use the quadratic formula $z_0 = 0$ and $z_n = z_{n-1}{}^2 + c$, where $n \geq 0$. Depending upon the complex number you choose for the constant, c, one of two things will happen: either the magnitude of the values of z will get increasingly large, or it will not. You already explored a few values of c in Exercise 15. Try a few more. Which values of c make the magnitude of z increase? Which values of c make z converge to a single value or alternate between values?

Check these results using your calculator. What will happen if $z_1 = 0$ and $c = 0.25$? What will happen if c or z is a complex number, for example, if $z_1 = 0$ and $c = -0.4 + 0.5i$?

The Mandelbrot set is all the values of c that do not make the magnitude of z increase. If you plot these points on a complex plane, then you'll get a pattern that looks like this one. Your project is to choose a small region on the boundary of the black area of this graph and create a graph of that smaller region. [▶ **Calculator Note 6G** includes a program that will analyze every point in the window to determine if it should be in the Mandelbrot set. Look at this graph, select a window, and then run the program. This may take several hours. ◀]

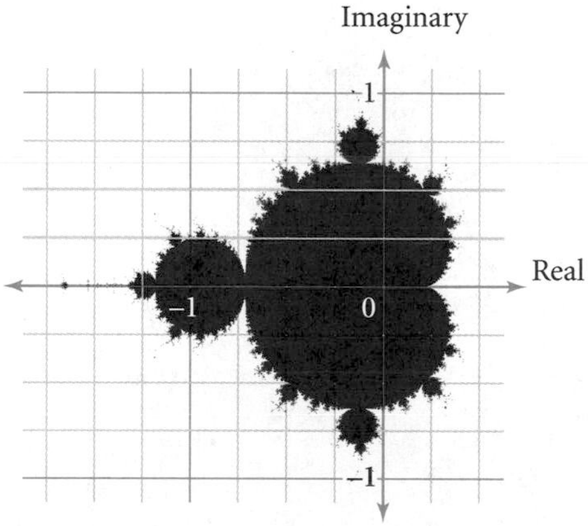

Your project should include

- a sketch of your graph
- a report that describes any similarities between your portion of the Mandelbrot set and the complete graph shown here
- any additional research you do on the Mandelbrot set or fractals in general

You can learn more about the Mandelbrot set and other fractals by using the Internet links at www.keycollege.com/online .

LESSON

6.6 Factoring Polynomials

Ideas are the factors that lift civilization.

JOHN H. VINCENT

Imagine a cube with any side length. Imagine increasing the height by 2 cm, the width by 3 cm, and the length by 4 cm.

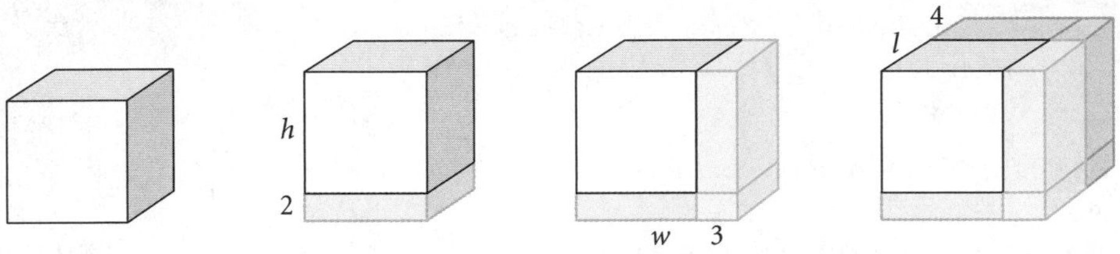

Imagine a cube. Increase its height 2 cm. Increase the width 3 cm. Increase the length 4 cm.

The starting figure is a cube, so you can let x be the length of all its sides. So $l = w = h = x$. The volume of the starting figure is x^3. To find the volume of the expanded box, you can see it as the sum of the volumes of eight different boxes. Find the volume of each piece by multiplying length by width by height.

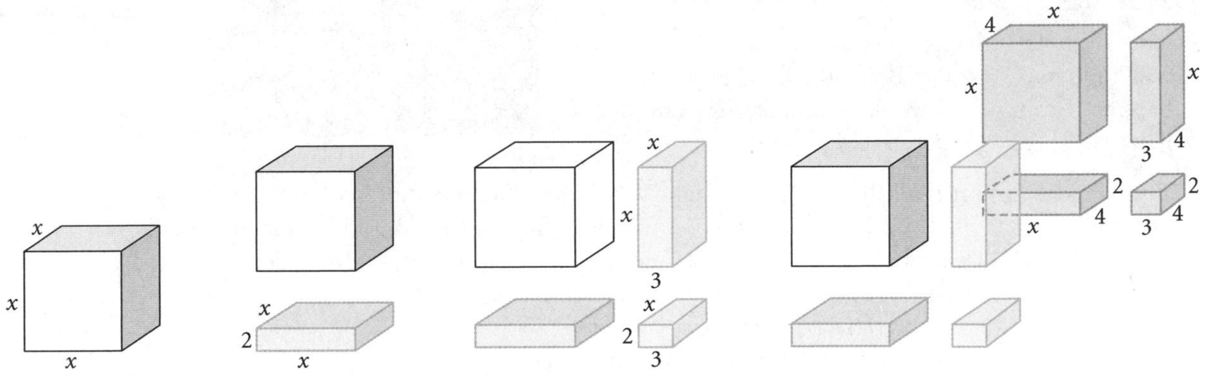

The total volume is this sum:

$$V = x^3 + 2x^2 + 3x^2 + 6x + 4x^2 + 8x + 12x + 24 = x^3 + 9x^2 + 26x + 24$$

You can also think of the expanded volume as the product of the new height, width, and length.

$$V = (x + 2)(x + 3)(x + 4)$$

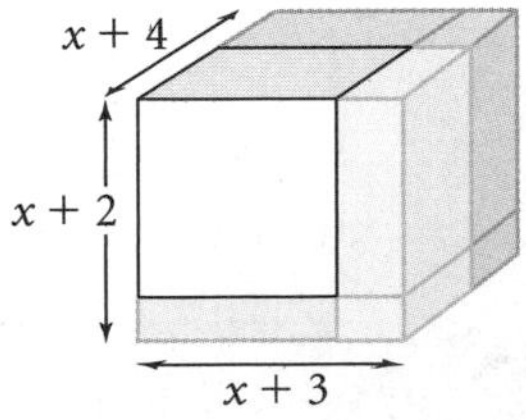

This equation is equivalent to the polynomial equation above. (Try graphing both on your calculator.) However, this second equation is in factored form.

You already know that there is a relationship between the factored form of a quadratic equation and the roots and x-intercepts of that quadratic equation. In this lesson you will learn how to write higher-degree polynomial equations in factored form when you know the roots of the equation. You'll also discover useful techniques for converting a polynomial in general form to the factored form.

Recall that a 3rd-degree polynomial function is called a *cubic function.* Let's examine the shape of the graph of a cubic function. The graphs of the two cubic functions below have the same x-intercepts: -2.5, 3.2, and 7.5. This means that both functions have the factored form $y = a(x + 2.5)(x - 7.5)(x - 3.2)$, with the vertical scale factor a yet to be determined.

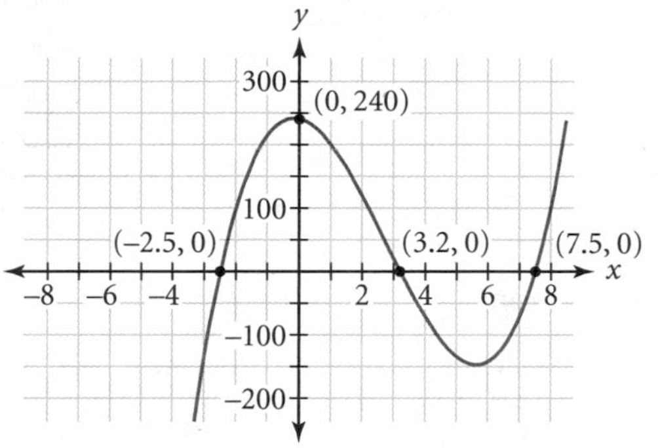

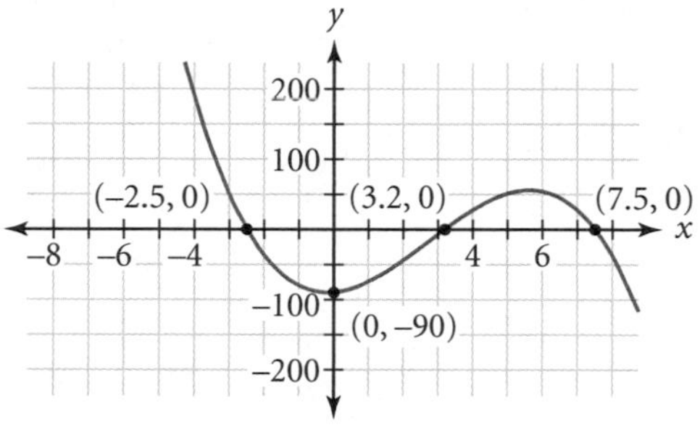

As you know by now, one way to find a is to substitute in the function the coordinates of one other point, such as the y-intercept.

The curve on the left has y-intercept (0, 240). Substituting this point in the equation gives $240 = a(2.5)(-7.5)(-3.2)$. Solving for a, you get $a = 4$. This means the equation of the cubic function on the left is

$$y = 4(x + 2.5)(x - 7.5)(x - 3.2)$$

The curve on the right has y-intercept $(0, -90)$. Substituting this point in the equation gives $-90 = a(2.5)(-7.5)(-3.2)$. So $a = -1.5$, and the equation of the cubic function on the right is

$$y = -1.5(x + 2.5)(x - 7.5)(x - 3.2)$$

The factored form of a polynomial function tells you the zeros of the function and the x-intercepts of the graph of the function. Recall that zeros are solutions to the equation $f(x) = 0$. Factoring, if a polynomial can be factored, is one strategy for finding the real roots of a polynomial equation. You will practice writing a higher-degree polynomial function in the next Investigation.

Investigation
The Box Factory

You will need

- graph paper
- scissors

What are the different ways to construct an open-top box from a 16-by-20-unit sheet of material? What is the maximum volume this kind of box can have? What is the minimum volume? Your group will investigate this problem by constructing open-top boxes using several possible integer values for x.

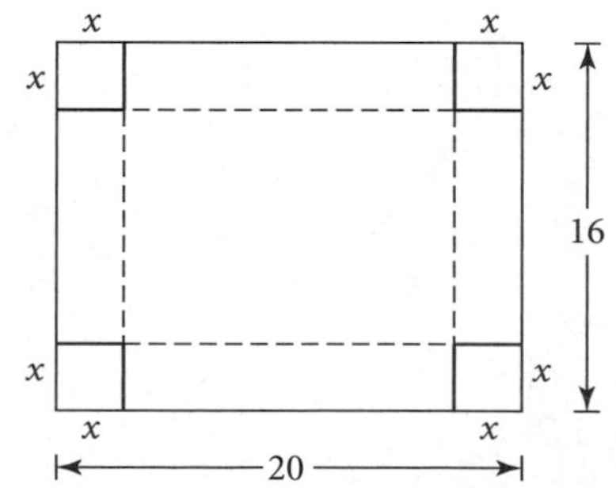

Step 1 Follow the Procedure Note to construct several different-sized boxes from 16-by-20 sheets of paper. Record the dimensions of each box, and calculate its volume. Make a table to record the x-values and volumes of the boxes.

Procedure Note

1. Cut several 16-by-20-unit rectangles out of graph paper.
2. Choose several different values for x.
3. For each value of x, construct a box by cutting squares with side length x from each corner and folding up the sides.

Step 2 What are the length, width, and height of each box in terms of x? Use these expressions to write a function that gives the volume of the box as a function of x.

Step 3 Graph your volume function from Step 2. Plot your data points on the same graph. How do the points relate to the function?

Step 4 What is the degree of this function? Give some reasons to support your answer.

Step 5 Locate the x-intercepts of your graph. (There should be three.) Call these three values r_1, r_2, and r_3. Use these values to write the function $y = (x - r_1)(x - r_2)(x - r_3)$.

Step 6 Graph the function from Step 5 as well as your function from Step 3. What are the similarities and differences between the graphs? How can you alter the function from Step 5 to make both functions equivalent?

Step 7 What happens when you try to make boxes by using the values r_1, r_2, and r_3 as x? What domain of x-values makes sense in this context? What x-value maximizes the volume of the box?

The connection between the roots of a polynomial equation and the x-intercepts of a polynomial function helps you factor any polynomial that can be factored.

EXAMPLE

Find the factored form of each function.

a. $y = x^2 - x - 2$

b. $y = 4x^3 + 8x^2 - 36x - 72$

▶ Solution

You can find the x-intercepts of each function by graphing. The x-intercepts tell you the real roots, which help you factor the function.

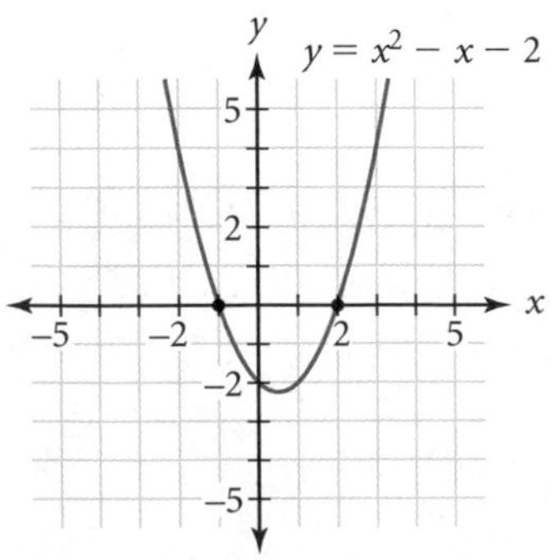

a. The x-intercepts are -1 and 2. Since the coefficient of the highest degree term, x^2, is 1, the vertical scale factor is 1. The factored form is

$$y = (x + 1)(x - 2)$$

You can verify that the expressions $x^2 - x - 2$ and $(x + 1)(x - 2)$ are equivalent by graphing $y = x^2 - x - 2$ and $y = (x + 1)(x - 2)$.

You can also check your work algebraically by finding the product $(x + 1)(x - 2)$. This rectangle diagram confirms that the product is $x^2 - x - 2$.

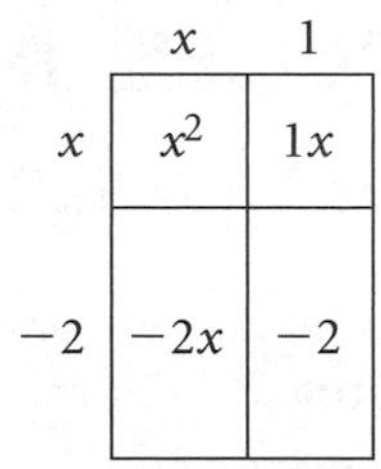

b. The x-intercepts are -3, -2, and 3. So you can write the function as

$$y = a(x + 3)(x + 2)(x - 3)$$

Because the leading coefficient needs to be 4, the vertical scale factor is also 4.

$$y = 4(x + 3)(x + 2)(x - 3)$$

To check your answer, you can compare graphs or algebraically find the product of the factors.

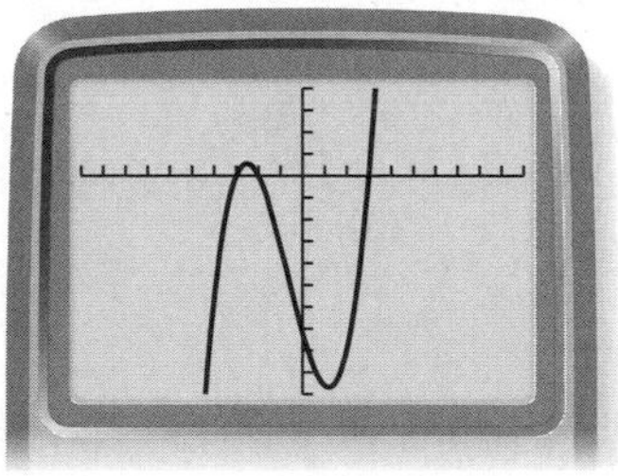

$[-10, 10, 1, -100, 40, 10]$

In Example A you converted a function from general form to factored form by using a graph and looking for the x-intercepts. This method works especially well when the zeros are integer values. However, polynomials can be separated into three types: polynomials that can't be factored with real numbers; polynomials that can be factored with real numbers but the roots of which are not "nice" integer or rational values; and polynomials that can be factored and that have integer or rational roots. For example, consider these cases of quadratic functions:

$y = x^2 - x + 1$

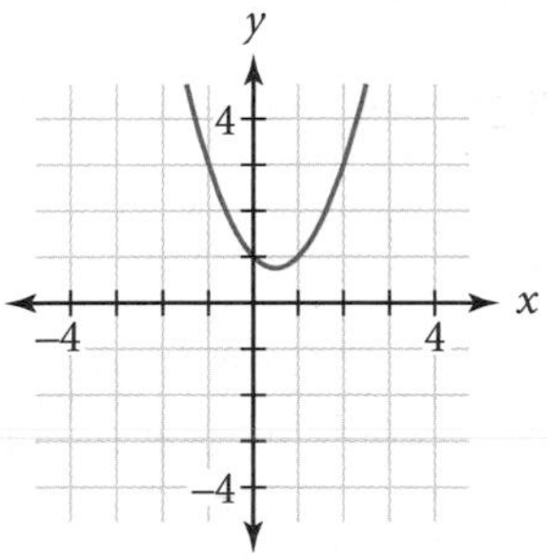

If the graph of a quadratic function does not cross the x-axis, you cannot factor it using real numbers. However, you can use the quadratic formula to find the complex zeros, which will be a conjugate pair.

$y = x^2 - x - 1$

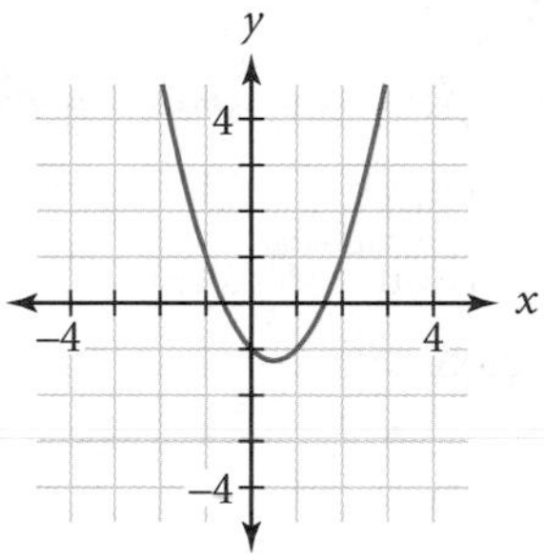

If the graph of a quadratic function crosses the x-axis but not at integer or rational values, then you cannot factor it. However, you can use the quadratic formula to find the real zeros.

$y = x^2 - x - 2$

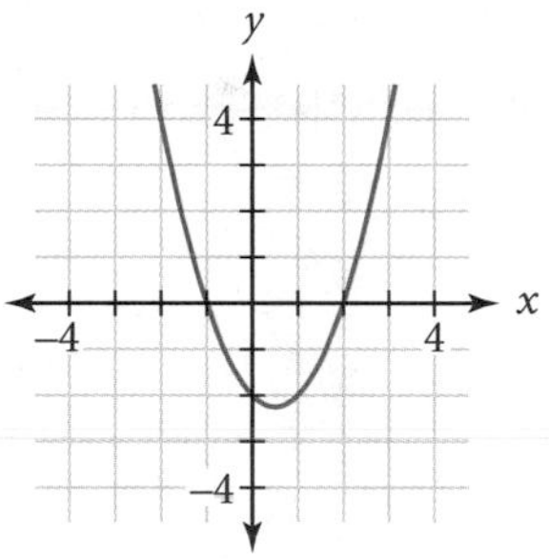

If the graph of a quadratic function crosses the x-axis at integer or rational values, then you can use the x-intercepts to factor it, which is usually quicker and easier than using the quadratic formula or a rectangle diagram.

What happens when the graph of a quadratic function has exactly one point of intersection with the x-axis?

EXERCISES

Practice Your Skills

1. Without actually graphing, find the x-intercepts and y-intercept of the graph of each equation. Check each answer by graphing.

a. $y = -0.25(x + 1.5)(x + 6)$

b. $y = 3(x - 4)(x - 4)$

c. $y = -2(x - 3)(x + 2)(x + 5)$

d. $y = 5(x + 3)(x + 3)(x - 3)$

2. Write the factored form of the quadratic function for each graph. Don't forget the vertical scale factor.

a.

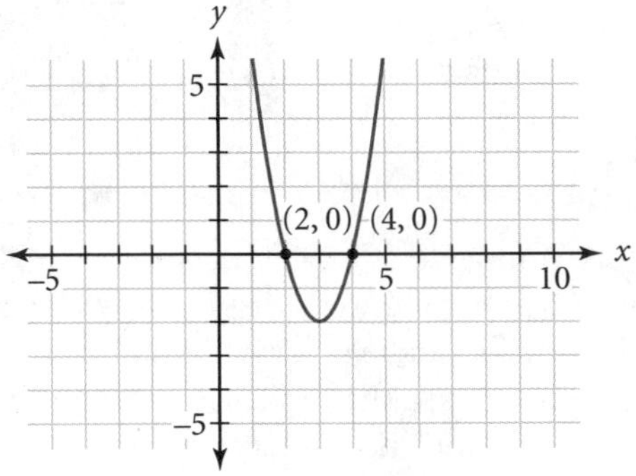

b.

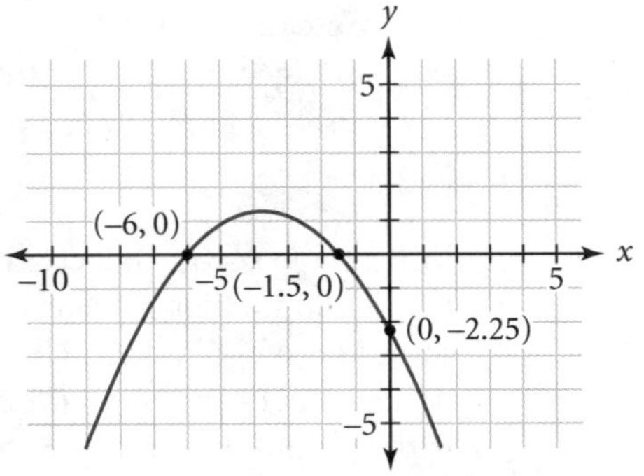

3. Convert each polynomial function to general form.

a. $y = (x - 4)(x - 6)$

b. $y = (x - 3)(x - 3)$

c. $y = x(x + 8)(x - 8)$

d. $y = 3(x + 2)(x - 2)(x + 5)$

4. Given the function $y = 2.5(x - 7.5)(x + 2.5)(x - 3.2)$,

a. Find the x-intercepts without graphing.

b. Find the y-intercept without graphing.

c. Write the function in general form.

d. Graph both the factored form and the general form of the function to check your work.

Reason and Apply

5. Use your work from the Investigation to answer these questions.

a. What x-value maximizes the volume of your box? What is the maximum volume?

b. What x-value(s) give a volume of 300 cubic units?

c. The portion of the graph with domain $x > 10$ shows positive volume. What does this mean in the context of the problem?

d. Explain the meaning of the parts of the graph showing negative volume.

6. Write each polynomial in factored form.

a. $4x^2 - 88x + 480$
b. $6x^2 - 7x - 5$
c. $x^3 + 5x^2 - 4x - 20$
d. $2x^3 + 16x^2 + 38x + 24$
e. $a^2 + 2ab + b^2$
f. $x^2 - 64$
g. $x^2 + 64$
h. $x^2 - 7$

7. If possible, sketch a graph of each situation.

a. a quadratic function with only one real zero
b. a quadratic function with no real zeros
c. a quadratic function with three real zeroes
d. a cubic function with only one real zero
e. a cubic function with two real zeros
f. a cubic function with no real zeros

8. Consider the function in this graph.

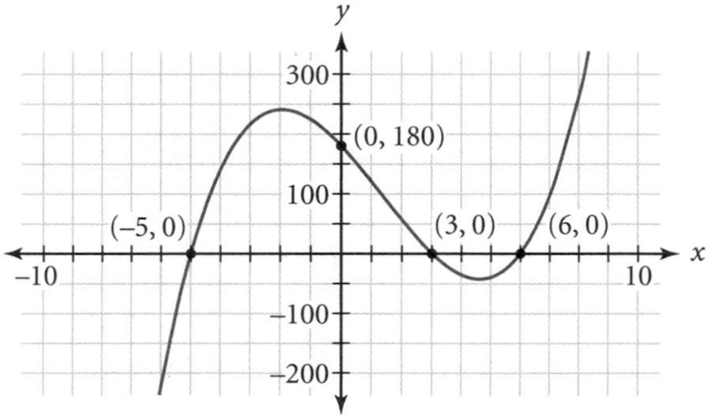

a. Write the equation of a polynomial function that contains the x-intercepts shown in the graph. Use a for the vertical scale factor.

b. Use the y-intercept to determine the vertical scale factor. Write the function with a in place.

c. Imagine that this graph is translated up 100 units. Write the function of the image.

d. Imagine that this graph is translated left 4 units. Write the function of the image.

Review

9. (Lesson 6.1) Is it possible to find a quadratic function that contains the points $(-4, -2)$, $(-1, 7)$, and $(2, 16)$? Explain why or why not.

10. (Lesson 6.1) Find the quadratic function whose graph has vertex $(-2, 3)$ and contains the point $(4, 12)$.

11. (Lesson 5. 3) Solve.

a. $x^2 = 50.41$
b. $x^4 = 169$
c. $(x - 2.4)^2 = 40.21$
d. $x^3 = -64$

12. (Lesson 5.5) Find the inverse of each function algebraically. Then choose a value of x, and check your answer.

a. $f(x) = \frac{2}{3}(x + 5)$
b. $g(x) = -6 + (x + 3)^{2/3}$
c. $h(x) = 7 - 2^x$

13. (Lesson 6.1) Use the method of finite differences to find the function that generates this table of values. Explain your reasoning.

x	2.2	2.6	3.0	3.4
f(x)	−4.5	−5.5	−6.5	−7.5

LESSON

6.7

Higher-Degree Polynomials

It is good to have an end to journey towards, but it is the journey that matters in the end.

URSULA K. LEGUIN

In this section we'll call polynomials of higher than degree 3 *higher-degree polynomials.* Frequently, 3rd-degree polynomials are associated with volume measures, as you saw in Lesson 6.6. If you create a box by removing small squares of side length x from each corner of a square piece of cardboard 30 inches on each side, the volume of the box in cubic inches is modeled with the function $y = x(30 - 2x)^2$, or $y = 4x^3 - 120x^2 + 900x$.

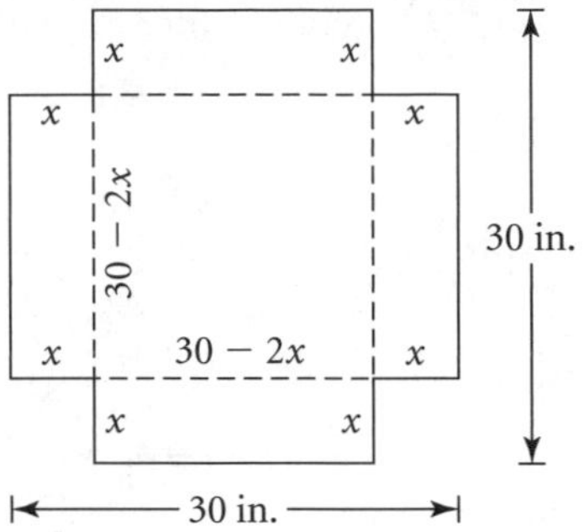

x
x
x
x
30 − 2x
30 − 2x
x
x
x
x
30 in.
30 in.

The zero-product property tells you that the zeros are $x = 0$ or $x = 15$; that is, 0 and 15 are the two values of x for which the volume is 0. The x-intercepts on the graph below confirm this.

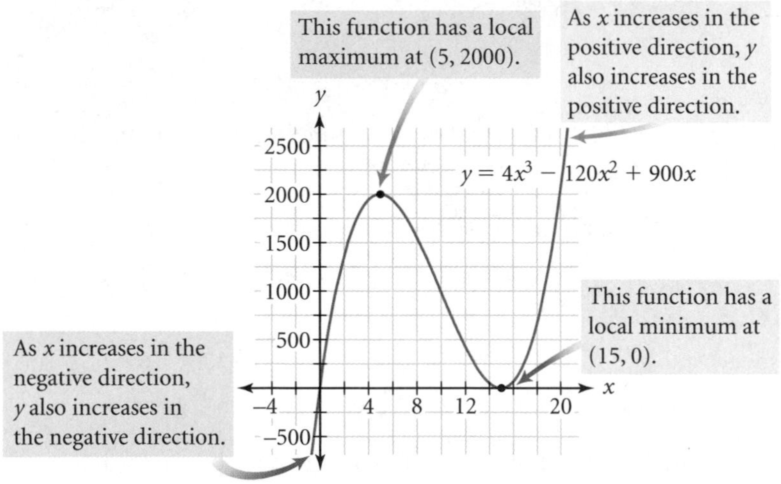

This function has a local maximum at (5, 2000).
As x increases in the positive direction, y also increases in the positive direction.
y = 4x³ − 120x² + 900x
This function has a local minimum at (15, 0).
As x increases in the negative direction, y also increases in the negative direction.
y
x
2500
2000
1500
1000
500
−500
−4
4
8
12
20

The shape of this graph is typical of the higher-degree polynomial graphs you will work with in this lesson. Note that it has one **local maximum** at (5, 2000) and one **local minimum** at (15, 0). These are the points that are higher or lower than all other points near them. You can also describe the *end behavior*—what happens to $f(x)$ as x takes on large positive and negative values of x. In the case of this 3rd-degree polynomial function, as x takes on larger positive values, y becomes increasingly large. As x takes on larger negative numbers, y becomes decreasingly more negative.

Graphs of polynomials with real coefficients have a y-intercept, possibly one or more x-intercepts, other features like local maximums or minimums, and end behavior. Maximums and minimums, collectively, are called *extreme values.*

Investigation
The Largest Triangle

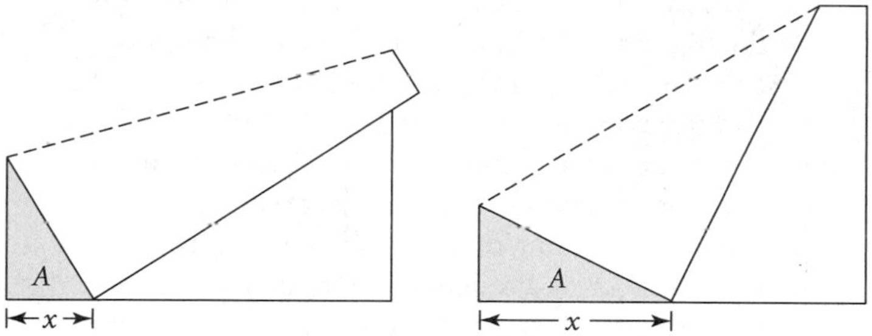

Take a sheet of notebook paper, and fold the upper left corner so that it touches some point on the bottom edge. Find the area, A, of the triangle formed in the lower left corner of the paper. What distance, x, along the bottom of the paper produces the triangle with the greatest area?

Work with your group to find a solution. You may want to use strategies you've learned in several lessons in this chapter. Write a report that explains your solution and your group's strategy for finding the largest triangle. Include any diagrams, tables, or graphs that you used.

In this lesson you will discover the connection between a polynomial function and its graph, which will allow you to predict when certain features will occur in the graph.

EXAMPLE A Find a polynomial function whose graph has x-intercepts 3, 5, and -4 and y-intercept 180. Describe the features of its graph.

▶ Solution A polynomial function with three x-intercepts has too many x-intercepts to be a quadratic function. It could be a 3rd-, 4th-, 5th-, or higher-degree polynomial function. Consider a 3rd-degree polynomial function, because that is the lowest degree that has three x-intercepts. Therefore, using the x-intercepts, $y = a(x - 3)(x - 5)(x + 4)$ works as the function when $a \neq 0$.

Substitute the coordinates of the y-intercept, (0, 180), into this function to find the vertical scale factor.

$$180 = a(0 - 3)(0 - 5)(0 + 4)$$
$$180 = a(60)$$
$$a = 3$$

The polynomial function of the lowest degree through the given intercepts is

$$y = 3(x - 3)(x - 5)(x + 4)$$

Graph this function to confirm your answer, and look for features.

This graph shows a local minimum at about $(4, -25)$ because it is the lowest point in its immediate neighborhood of x-values. There is also a local maximum at about $(-1.5, 225)$ because it is the highest point in its immediate neighborhood of x-values. This small domain already suggests the end behavior. As x takes on larger positive values, y becomes increasingly large. As x takes on larger negative numbers, y becomes decreasingly more negative. If you increase the domain of this graph to include more x-values at the right and left extremes of the x-axis, you'll discover that the graph continues this end behavior.

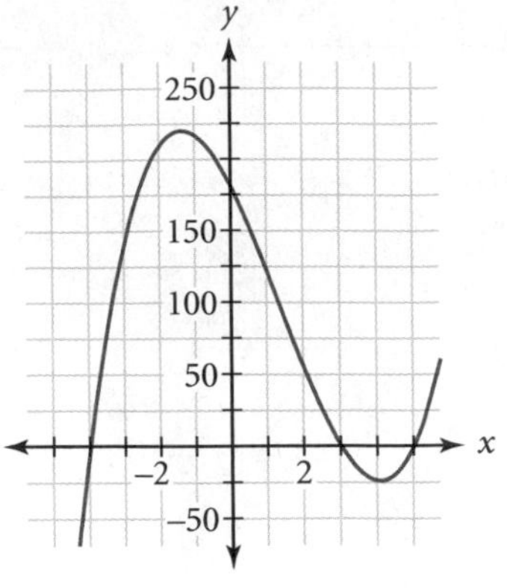

You can identify the degree of many polynomial functions by looking at the shapes of their graphs. Every 3rd-degree polynomial function has essentially one of the shapes shown below. Graph A shows the graph of $y = x^3$. It can be translated, stretched, or reflected. Graph B shows one possible transformation.

Graph A

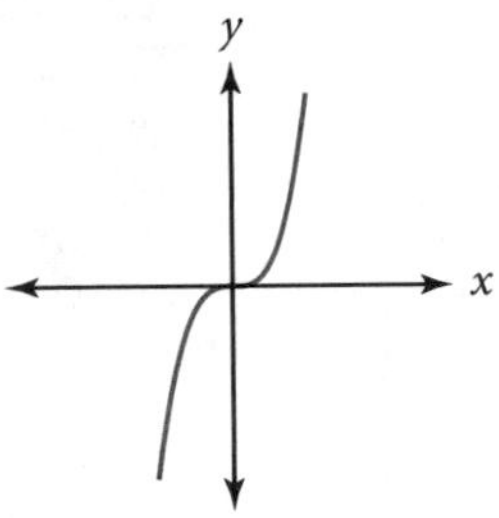

Graph B

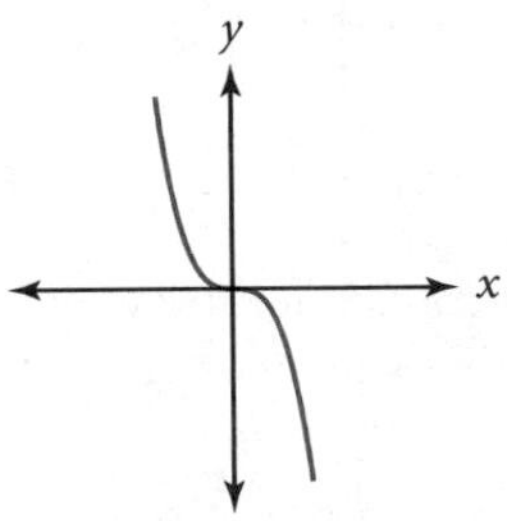

Graph C

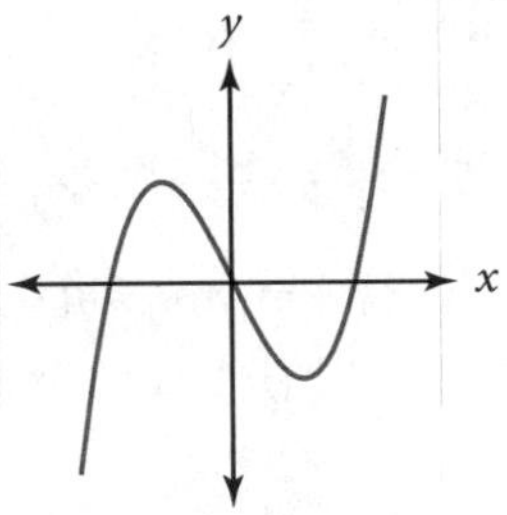

Graph D

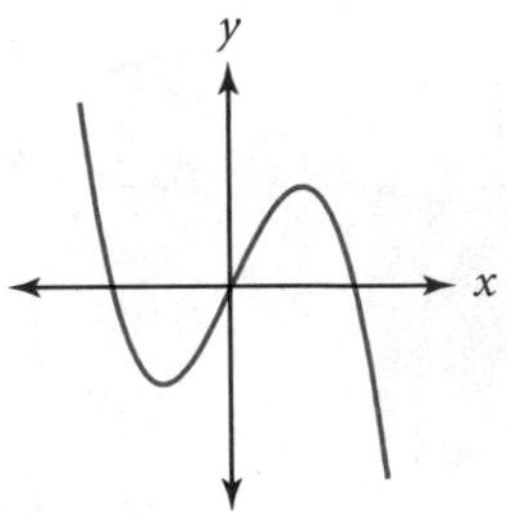

Graphs C and D show the graphs of general cubic functions in the form $y = ax^3 + bx^2 + cx + d$. In Graph C the leading coefficient is positive, and in Graph D it is negative.

You'll explore the general shapes and characteristics of other higher-degree polynomials in the exercises.

EXAMPLE B

Write a polynomial function with real coefficients and zeros $x = 2$, $x = -5$, and $x = 3 + 4i$.

▶ *Solution*

For a polynomial function with real coefficients, complex zeros occur in conjugate pairs, so $x = 3 - 4i$ must also be a zero. In factored form the polynomial function of the lowest degree is

$$y = (x - 2)(x + 5)(x - (3 + 4i))(x - (3 - 4i))$$

Multiplying the last two factors to eliminate complex numbers gives $y = (x - 2)(x + 5)(x^2 - 6x + 25)$. Multiplying all factors gives the polynomial equation in general form.

$$y = x^4 - 3x^3 - 3x^2 + 135x - 250$$

Graph this function to check your solution. You can't see the complex zeros, but you can see x-intercepts 2 and -5.

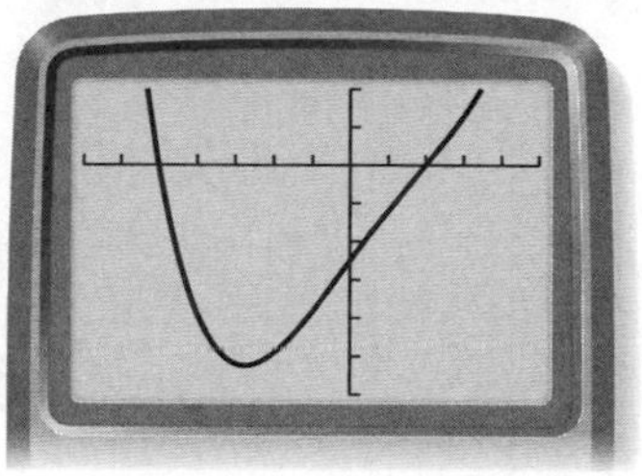

$[-7, 5, 1, -600, 200, 100]$

Note that in Example B the solution was a 4th-degree polynomial function. It had four complex zeros, but the graph had only two x-intercepts, corresponding to the two real zeros. Any polynomial function of degree n always has n complex zeros (including repeated zeros) and at most n x-intercepts. Remember that complex zeros always come in conjugate pairs.

EXERCISES

Practice Your Skills

Exercises 1–4 use these four graphs.

a.

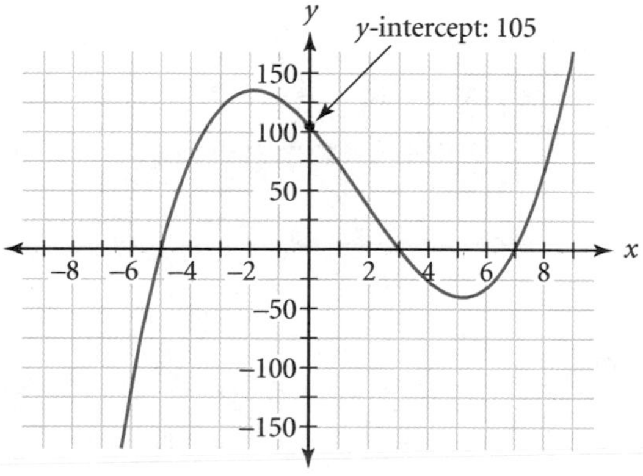

b.

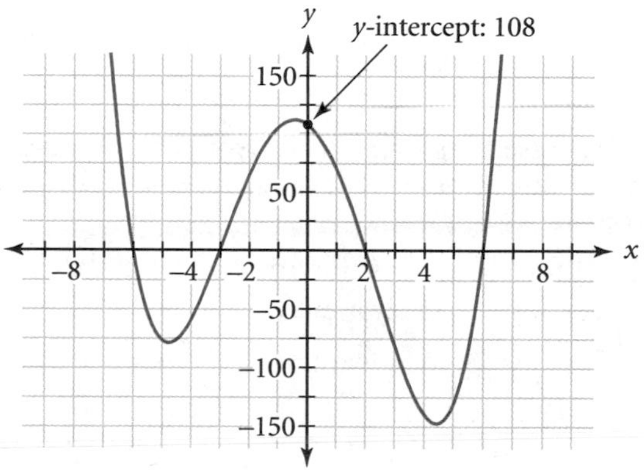

c.

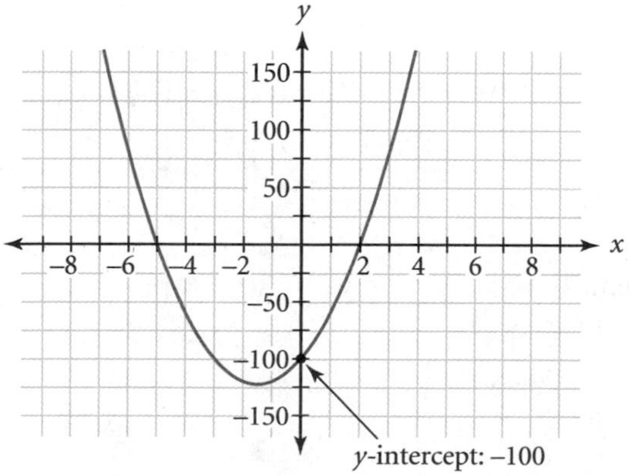

d.

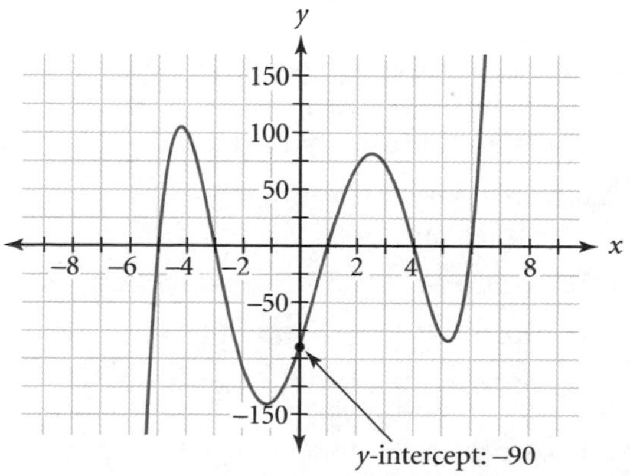

1. Identify the zeros of each polynomial function.

2. Estimate the y-intercept of each graph.

3. Identify the lowest possible degree of each polynomial function.

4. Write the factored form for each polynomial function. Check your work by graphing on your calculator.

Reason and Apply

5. Write polynomial functions with these features.

a. a linear function the graph of which has x-intercept 4

b. a quadratic function the graph of which has only one x-intercept, 4

c. a cubic function the graph of which has only one x-intercept, 4

6. The graph of $y = 2(x - 3)(x - 5)(x + 4)^2$ has x-intercepts 3, 5, and -4 because they are the only possible x-values that make $y = 0$. This is a 4th-degree polynomial, but it has only 3 x-intercepts. Make a *complete graph*—one that displays all of the relevant features, including local extreme values—of each of the functions in parts 6a–6f.

a. $y = 2(x - 3)(x - 5)(x + 4)^2$

b. $y = 2(x - 3)^2(x - 5)(x + 4)$

c. $y = 2(x - 3)(x - 5)^2(x + 4)$

d. $y = 2(x - 3)^2(x - 5)(x + 4)^2$

e. $y = 2(x - 3)(x - 5)(x + 4)^3$

f. $y = 2(x - 3)(x - 5)^2(x + 4)^3$

g. Based on your graphs from 6a–6f, describe a connection between the power of a factor and what happens at that x-intercept.

7. This graph is a complete graph of a polynomial function.

a. How many x-intercepts are there?

b. What is the lowest possible degree of this polynomial function?

c. Write the function of this graph such that it includes the points $(0, 0)$, $(-5, 0)$, $(4, 0)$, $(-1, 0)$, and $(1, 216)$.

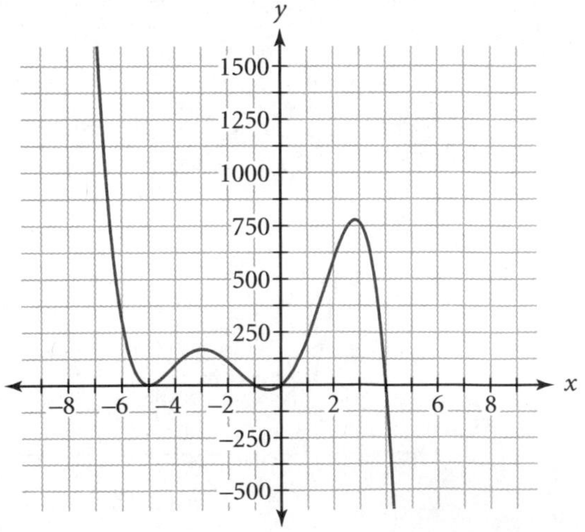

8. Write the lowest-degree polynomial function with real coefficients that has the given set of zeros and whose graph has the given y-intercept.

a. zeros: $x = -4$, $x = 5$, $x = -2$, $x = -2$; y-intercept: -80

b. zeros: $x = -4$, $x = 5$, $x = -2$, $x = -2$; y-intercept: 160

c. zeros: $x = \frac{1}{3}$, $x = \frac{-2}{5}$, $x = 0$; y-intercept: 0

d. zeros: $x = -5i$, $x = -1$, $x = -1$, $x = -1$, $x = 4$; y-intercept: 100

9. Look back at Exercises 1–4. Find the product of the zeros in Exercise 1. How does the value of the leading coefficient, a, relate to the y-intercept, the product of the roots, and the degree of the function?

10. A 4th-degree polynomial function has the general form $y = ax^4 + bx^3 + cx^2 + dx + e$ for real values of a, b, c, d, and e. Graph several 4th-degree polynomial functions by trying different values for each coefficient. Be sure to include positive, negative, and zero values. Make a sketch of each different type of curve you get. Concentrate on the shape of the curve. You do not need to include axes in your sketches. Compare your graphs with the graphs of your classmates and come up with six or more different shapes that describe all 4th-degree polynomial functions.

11. Each of these graphs is the graph of a polynomial function with leading coefficient $a = 1$ or $a = -1$.

i.

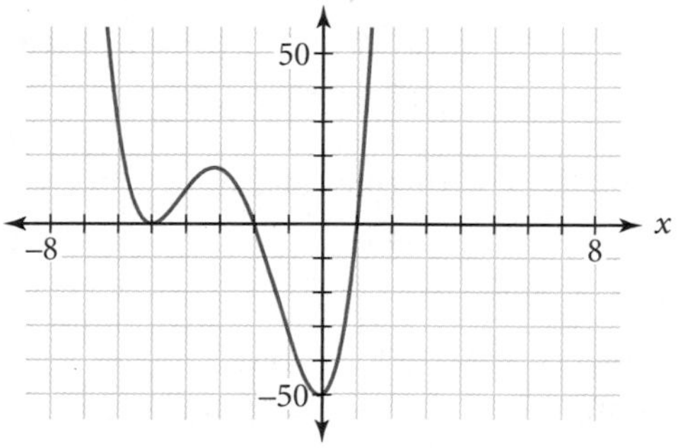

ii.

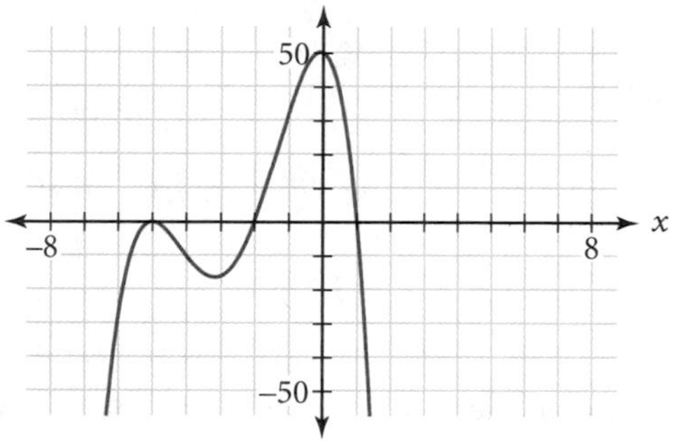

iii.

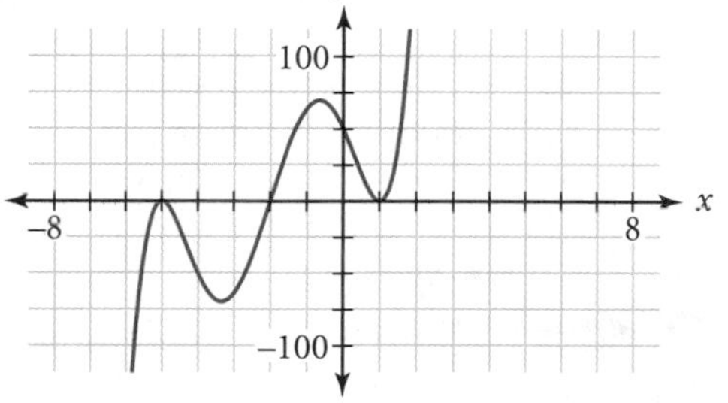

iv.

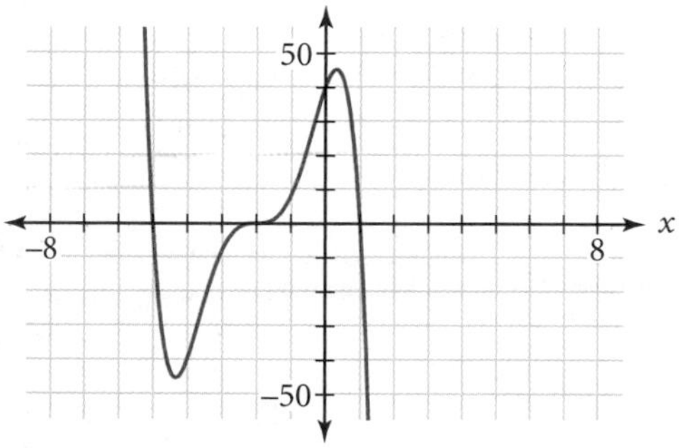

a. Write functions in factored form that will produce each graph.

b. Name the zeros of each polynomial function in 11i–11iv. If a factor is raised to the power of n, list the zero n times.

12. Consider the polynomial functions in Exercise 11.

a. What is the degree of each polynomial function?

b. How many extreme values does each graph have?

c. What is the relationship between the degree of the polynomial function and the number of extreme values?

d. Complete these statements:

i. The graph of a polynomial curve of degree n has at most __?__ x-intercepts.

ii. A polynomial function of degree n has at most __?__ real zeros.

iii. A polynomial function of degree n has at most __?__ complex zeros.

iv. The graph of a polynomial function of degree n has at most __?__ extreme values.

13. In Lesson 6.7 you saw various possible appearances of the graph of a 3rd-degree polynomial function, and in Exercise 10 you explored possible appearances of the graph of a 4th-degree polynomial function. In Exercises 11 and 12 you found a relationship between the degree of a polynomial function and the number of zeros and extreme values. Use all the patterns you have noticed in these problems to sketch one possible graph of each of these functions:

a. a 5th-degree function

b. a 6th-degree function

c. a 7th-degree function

Review

14. (Lesson 6.6) Find the roots of these quadratic equations. Express them as fractions, not decimals.

a. $0 = 3x^2 - 13x - 10$

b. $0 = 6x^2 - 8x + 3$

c. $0 = 8x^3 - 2x^2 - 3x$

d. $0 = 15x^2 + 14x - 8$

e. List all the factors of the constant term and the leading coefficient for 14a through 14d. What do you notice about the roots of the functions and the relationship between the factors of a and c?

15. (Lesson 6.4) If $3 + 5\sqrt{2}$ is a solution of a quadratic equation with rational coefficients, then what other number also must be a solution? Write an equation in general form that has these solutions.

16. (Lesson 4.2) Given the function $Q(x) = x^2 + 2x + 10$, find these values.

a. $Q(-3)$

b. $Q(-\frac{1}{5})$

c. $Q(2 - 3\sqrt{2})$

d. $Q(-1 + 3i)$

LESSON 6.8

More about Finding Solutions

It isn't that they can't see the solution. It is that they can't see the problem.

G. K. CHESTERTON

You can find zeros of a quadratic function by factoring or by using the quadratic formula. How can you find the zeros of a higher-degree polynomial? Sometimes a graph will show you zeros in the form of x-intercepts, but only if they are real, and often this method is accurate only if the zeros have integer values. Fortunately, there is a method for finding exact zeros of higher-degree polynomial functions; it is based on long division. First, find one or more zeros, and then divide them out of the polynomial function until you have a simpler polynomial for which you can find the zeros by factoring or using the quadratic formula. Let's start with an example in which we already know several zeros. Then you'll learn a technique for finding some zeros when they're not so obvious.

EXAMPLE A What are the zeros of $P(x) = x^5 - 6x^4 + 20x^3 - 60x^2 + 99x - 54$?

▶ Solution

The graph appears to have x-intercepts at 1, 2, and 3. You can confirm that these values are zeros of the function by substituting these values in P.

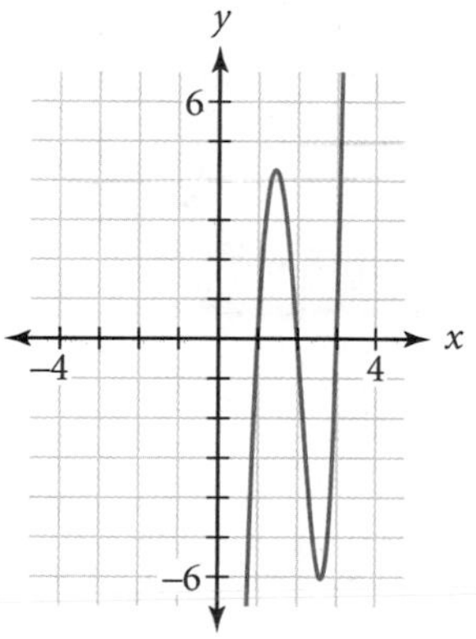

$$P(1) = (1)^5 - 6(1)^4 + 20(1)^3 - 60(1)^2 + 99(1) - 54 = 0$$
$$P(2) = (2)^5 - 6(2)^4 + 20(2)^3 - 60(2)^2 + 99(2) - 54 = 0$$
$$P(3) = (3)^5 - 6(3)^4 + 20(3)^3 - 60(3)^2 + 99(3) - 54 = 0$$

For all three values you get $P(x) = 0$, which shows that $x = 1$, $x = 2$, and $x = 3$ are zeros. This also means that $(x - 1)$, $(x - 2)$, and $(x - 3)$ are factors of $x^5 - 6x^4 + 20x^3 - 60x^2 + 99x - 54$. None of the x-intercepts have the appearance of a repeated root, and you know that a 5th-degree polynomial function has five zeros, so it appears that this function has two additional complex zeros.

You know that $(x - 1)$, $(x - 2)$, and $(x - 3)$ are all factors, so the product of these three factors, $x^3 - 6x^2 + 11x - 6$, must be a factor also. Your task is to find another factor such that

$$(x^3 - 6x^2 + 11x - 6)(\textit{factor}) = x^5 - 6x^4 + 20x^3 - 60x^2 + 99x - 54$$

You can find this factor by using division.

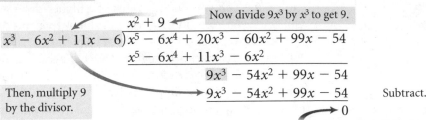

First, divide x^5 by x^3 to get x^2.

$$\begin{array}{r}
x^2 \qquad\qquad\qquad\qquad\qquad\qquad\quad \\
x^3 - 6x^2 + 11x - 6\overline{)\,x^5 - 6x^4 + 20x^3 - 60x^2 + 99x - 54} \\
\underline{x^5 - 6x^4 + 11x^3 - 6x^2 \qquad\qquad\quad} \\
9x^3 - 54x^2 + 99x - 54
\end{array}$$

Then, multiply x^2 by the divisor.

Subtract.

Now divide $9x^3$ by x^3 to get 9.

$$\begin{array}{r}
x^2 + 9 \qquad\qquad\qquad\qquad\qquad\quad \\
x^3 - 6x^2 + 11x - 6\overline{)\,x^5 - 6x^4 + 20x^3 - 60x^2 + 99x - 54} \\
\underline{x^5 - 6x^4 + 11x^3 - 6x^2 \qquad\qquad\quad} \\
9x^3 - 54x^2 + 99x - 54 \\
\underline{9x^3 - 54x^2 + 99x - 54} \\
0
\end{array}$$

Then, multiply 9 by the divisor.

Subtract.

The remainder is zero, so the division is finished, resulting in two factors.

Now you can rewrite the original polynomial as a product of factors.

$$\begin{aligned}
x^5 - 6x^4 + 20x^3 - 60x^2 + 99x - 54 &= (x^3 - 6x^2 + 11x - 6)(x^2 + 9) \\
&= (x - 1)(x - 2)(x - 3)(x^2 + 9)
\end{aligned}$$

Now that the polynomial is in factored form, you can easily find the zeros. You knew three of them from the graph. Because this is a 5th-degree polynomial, there are two more zeros to find. They are contained in $x^2 + 9$. What values of x make $x^2 + 9$ equal zero?

$$\begin{aligned}
x^2 + 9 &= 0 \\
x^2 &= -9 \\
x &= \pm\sqrt{-9} = \pm 3i
\end{aligned}$$

Therefore, the five zeros are $x = 1$, $x = 2$, $x = 3$, $x = 3i$, or $x = -3i$.

To confirm that 1, 2, and 3 were zeros in the example, you used the **factor theorem.**

The Factor Theorem

$(x - r)$ is a factor of the polynomial that defines the function $P(x)$ if and only if $P(r) = 0$.

By this theorem $P(3i)$ and $P(-3i)$ will also equal zero.

Division of polynomials is similar to long division, which you may have learned in elementary school. Both the original polynomial and the divisor are written in descending order of the powers of x. If any degree is missing, insert a term with coefficient 0 as a placeholder. For example, you can write the polynomial $x^4 + 3x^2 - 5x + 8$ as

$$x^4 + 0x^3 + 3x^2 - 5x + 8$$

Insert a zero placeholder because the polynomial did not have a 3rd-degree term.

Often, you won't be able to find any zeros for certain by looking at a graph. However, there is a pattern to rational numbers that might be zeros.

The Rational Root Theorem

If the polynomial equation $P(x) = 0$ has rational roots, they are of the form $\frac{p}{q}$, where p is a factor of the constant term and q is a factor of the leading coefficient.

The **rational root theorem** helps you narrow down the values that might be zeros of a polynomial function. Notice that this theorem will identify only possible *rational* roots. It won't find roots that are irrational or contain imaginary numbers.

EXAMPLE B

Find the roots of this polynomial equation.

$$3x^3 + 5x^2 - 15x - 25 = 0$$

▶ Solution

First, graph the function $y = 3x^2 + 5x^2 - 15x - 25$ to see if there are any identifiable integer x-intercepts.

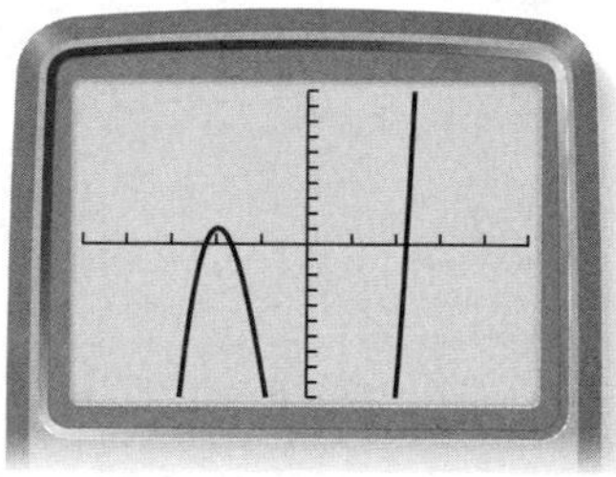

$[-5, 5, 1, -10, 10, 1]$

There are no integer x-intercepts, but the graph shows x-intercepts between -3 and -2, -2 and -1, and 2 and 3. Any rational root of this polynomial will be a factor of 25, the constant, divided by a factor of 3, the leading coefficient. The factors of 25 are ±1, ±5, and ±25, and the factors of 3 are ±1 and ±3, so the possible rational roots are ±1, ±5, ±25, $\pm\frac{1}{3}$, $\pm\frac{5}{3}$, or $\pm\frac{25}{3}$. The only one of these that looks like a possibility on the graph is $-\frac{5}{3}$. Try substituting $-\frac{5}{3}$ in the original polynomial.

$$3\left(-\frac{5}{3}\right)^3 + 5\left(-\frac{5}{3}\right)^2 - 15\left(-\frac{5}{3}\right) - 25 = 0$$

Because the result is 0, you know that $-\frac{5}{3}$ is a root of the equation.

If $-\frac{5}{3}$ is a root of the equation, then $\left(x + \frac{5}{3}\right)$ is a factor. Use long division to divide out this factor.

$$\begin{array}{r l}
 & 3x^2 - 15 \\
x + \frac{5}{3} \big) & 3x^3 + 5x^2 - 15x - 25 \\
 & \underline{3x^3 + 5x^2} \\
 & 0 - 15x - 25 \\
 & \underline{- 15x - 25} \\
 & 0
\end{array}$$

Now, you know that $3x^3 + 5x^2 - 15x - 25 = 0$ is equivalent to the equation $\left(x + \frac{5}{3}\right)(3x^2 - 15) = 0$. You already knew $-\frac{5}{3}$ was a root. Now, solve $3x^2 - 15 = 0$.

$$3x^2 = 15$$
$$x^2 = 5$$
$$x = \pm\sqrt{5}$$

As a decimal approximation, $-\frac{5}{3}$ is about -1.7, $\sqrt{5}$ is about 2.2, and $-\sqrt{5}$ is about -2.2. These values appear to be correct according to the graph. The three roots are $x = -\frac{5}{3}$, $x = \sqrt{5}$, and $x = -\sqrt{5}$.

When you divide a polynomial by a linear factor, such as $\left(x + \frac{5}{3}\right)$, you can use a shortcut method called *synthetic division.* Synthetic division is simply an abbreviated form of long division.

Consider this division of a cubic polynomial by a linear factor.

$$\frac{6x^3 + 11x^2 - 17x - 30}{x + 2}$$

Here are the procedures for long division and synthetic division.

Long Division

$$\begin{array}{r} 6x^2 - 1x - 15 \\ x + 2\overline{)6x^3 + 11x^2 - 17x - 30} \\ (-)\ \underline{6x^3 + 12x^2 \qquad\qquad\quad} \\ -1x^2 - 17x \qquad \\ (-)\ \underline{-1x^2 - \ \ 2x} \qquad \\ -15x - 30 \\ (-)\ \underline{-15x - 30} \\ 0 \end{array}$$

Synthetic Division

$$\begin{array}{r|rrrr} -2 & 6 & 11 & -17 & -30 \\ & & -12 & 2 & 30 \\ \hline & 6 & -1 & -15 & \boxed{0} \end{array}$$

$$6x^2 - 1x - 15$$

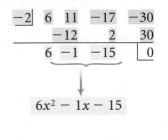

Synthetic division certainly looks faster. The corresponding numbers in each process are shaded. Notice that synthetic division contains all of the same information, but in a condensed form. Here's how to perform synthetic division:

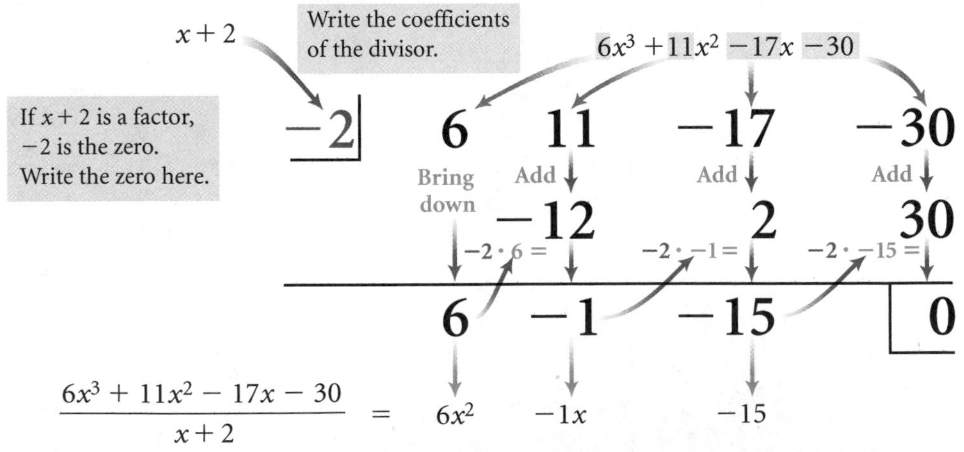

The number farthest to the right in the last row of a synthetic division is the remainder, which in this case is 0. When the remainder in a division problem is zero, you know that the divisor is a factor. This means -2 is a root, and the polynomial $6x^3 + 11x^2 - 17x - 30$ factors into the product of the divisor and the quotient, or $(x + 2)(6x^2 - 1x - 15)$. You could now use any of the methods you've learned—simple factoring, quadratic formula, synthetic division, or perhaps graphing—to factor the quotient even further.

EXERCISES 

Practice Your Skills

1. Find the missing polynomial in each long-division problem.

a.
$$
\begin{array}{rl}
 & \boxed{a} \\ \cline{2-2}
x+5) & 3x^3 + 22x^2 + 38x + 15 \\
 & -3x^3 + 15x^2 \\ \cline{2-2}
 & 7x^2 + 38x + 15 \\
 & -7x^2 - 35x \\ \cline{2-2}
 & 3x + 15 \\
 & 3x - 15 \\ \cline{2-2}
 & 0
\end{array}
$$

b.
$$
\begin{array}{rl}
 & 2x^2 + 5x - 3 \\ \cline{2-2}
3x-2) & 6x^3 + 11x^2 - 19x + 6 \\
 & \boxed{b} \\ \cline{2-2}
 & 15x^2 - 19x + 6 \\
 & -15x^2 + 10x \\ \cline{2-2}
 & -9x + 6 \\
 & +9x - 6 \\ \cline{2-2}
 & 0
\end{array}
$$

c.
$$
\begin{array}{rl}
 & 4x^2 + 4x - 2 \\ \cline{2-2}
2x+1) & 8x^3 + 12x^2 - 2 \\
 & -8x^3 - 4x^2 \\ \cline{2-2}
 & \boxed{c} \\
 & -8x^2 - 4x \\ \cline{2-2}
 & -4x - 2 \\
 & 4x + 2 \\ \cline{2-2}
 & 0
\end{array}
$$

2. Use the dividend, divisor, and quotient to rewrite each long-division problem in Exercise 1 as a factored product in the form $P(x) = D(x) \cdot Q(x)$. For example, $x^3 + 2x^2 + 3x - 6 = (x - 1)(x^2 + 3x + 6)$.

3. Find the missing value in each synthetic-division problem.

a.
$$
\begin{array}{r|rrrr}
4 & 3 & -11 & 7 & -44 \\
 & & \boxed{a} & 4 & 44 \\ \hline
 & 3 & 1 & 11 & 0
\end{array}
$$

b.
$$
\begin{array}{r|rrrr}
-3 & 1 & 5 & -1 & -21 \\
 & & -3 & -6 & 21 \\ \hline
 & 1 & \boxed{b} & -7 & 0
\end{array}
$$

c.
$$
\begin{array}{r|rrrr}
1.5 & 4 & -8 & \boxed{c} & -6 \\
 & & 6 & -3 & 6 \\ \hline
 & 4 & -2 & 4 & 0
\end{array}
$$

d.
$$
\begin{array}{r|rrrr}
\boxed{d} & 1 & 7 & 11 & -4 \\
 & & -4 & -12 & 4 \\ \hline
 & 1 & 3 & -1 & 0
\end{array}
$$

4. Use the dividend, divisor, and quotient to rewrite each synthetic-division problem in Exercise 3 as a factored product in the form $P(x) = D(x) \cdot Q(x)$.

5. Sometimes, graphing can simplify the task of finding rational roots of polynomials.

a. Make a list of the possible rational roots of $0 = 2x^3 + 3x^2 - 32x + 15$.

b. Graph $f(x) = 2x^3 + 3x^2 - 32x + 15$. Which of the possible rational roots from 5a still appear to be possible roots?

Reason and Apply

6. Division often results in a remainder. In each of these problems, use the polynomial that defines P as the dividend and the polynomial that defines Q as the quotient. Write the result of the division in the form $P(x) = D(x) \cdot Q(x) + R$, where R is an integer remainder. For example, $x^3 + 2x^2 + 3x - 4 = (x - 1)(x^2 + 3x + 6) + 2$.

a. $P(x) = 47, D(x) = 11$

b. $P(x) = 6x^4 - 5x^3 + 7x^2 - 12x + 15, D(x) = x - 1$

c. $P(x) = x^3 - x^2 - 10x + 16 , D(x) = x - 2$

7. Consider the function $p(x) = 2x^3 - x^2 + 18x - 9$.

a. Verify that $3i$ is a zero.

b. Find the remaining zeros of the function p.

8. Consider the function $y = x^4 + 3x^3 - 11x^2 - 3x + 10$.

a. How many zeros does this function have?

b. Name the zeros.

c. Write the polynomial function in factored form.

9. Use the rational roots from Exercise 5 to write this function in factored form.
$y = 2x^3 + 3x^2 - 32x + 15$

10. When you trace the graph of a function on your calculator to find the value of an x-intercept, you often see the y-value jump from positive to negative when you pass over the zero. By using smaller windows, you can find more and more accurate approximations for x. This process can be automated by your calculator. The automation uses successive midpoints of each region above and below zero and is called the *bisection method.* Approximate the x-intercepts of each equation by using the program BISECTN, then use synthetic or long division to find any nonreal zeros. [► See **Calculator Note 6H** for the instructions for the BISECTN program. ◄]

a. $y = x^5 - x^4 - 16x + 16$

b. $y = 2x^3 + 15x^2 + 6x - 6$

c. $y = 0.2(x - 12)^5 - 6(x - 12)^3 - (x - 12)^2 + 1$

d. $y = 2x^4 + 2x^3 - 14x^2 - 9x - 12$

Technology Connection Many computer programs employ search methods that find a particular data item in a large collection of such items. It would be inefficient to search for the item one piece of data at a time, so a binary search algorithm is used instead. First, the data items are sorted, and the computer checks the middle entry. If the entry is too low, the computer will move halfway up toward the highest entry to check; if it's too high, the computer will check halfway toward the lowest entry. Each time, it will cut the item list in half (going higher or lower) until it reaches the item for which it is searching. In a list of n items the maximum number of times the list must be cut in half before finding the target is $\log_2 n + 1$. Try this technique in a number guessing game. Have a friend think of a whole number from 1 to 100. When you guess a number, your friend will tell you only to guess higher or lower until you guess the number. Can you do it in seven guesses at most?

Review

11. (Chapter 5) The relationship between the height and the diameter of a tree is approximately determined by the equation $f(x) = kx^{2/3}$, where x is the height in feet, $f(x)$ is the diameter in inches, and k is a constant that depends on the object you are measuring.

a. A British Columbian pine 221 ft high is about 21 in. in diameter. Find the value of k, and use it to express diameter as a function of height.

b. Give the inverse function.

c. What would be the diameter of a 300-ft British Columbian pine?

d. How high would be a similar pine that is 15 in. in diameter?

12. (Lesson 6.1) Find a polynomial function of lowest possible degree whose graph passes through the points $(-2, -10.5)$, $(-1, 4.2)$, $(0, 3.5)$, $(1, 0)$, $(2, 6.3)$, and $(3, 3.5)$.

13. (Lesson 6.2) Write each quadratic function in general form and in factored form. Identify the vertex, y-intercept, and x-intercepts of each parabola.

a. $y = (x - 2)^2 - 12$

b. $y = 3(x + 1)^2 - 27$

c. $y = -\frac{1}{2}(x - 5)^2 + \frac{49}{2}$

d. $y = 2(x - 3)^2 + 3$

14. (Lessons 6.3 and 6.4) Solve.

a. $6x + x^2 + 5 = -4 + 4(x + 3)$

b. $7 = x(x + 3)$

c. $2x^2 - 3x + 1 = x^2 - x - 4$

EXPLORATION
with Microsoft® Excel

Polynomials and Tables of Differences

Download the Chapter 6 Excel Exploration from the website at www.keycollege.com/online or from your Student CD.

In this Exploration you will experiment with polynomials of different degree. You will study tables of differences and search for patterns relating those differences to coefficients.

On the spreadsheet you will find complete instructions and questions to consider as you explore.

Activity

Using a Data Table to Solve Quadratic Equations

Excel skills required

- Data fill
- Name a cell
- Create chart with one or more series graphed

In this Activity you will study quadratic polynomials, solving quadratic equations to the nearest hundredth.

You can find instructions for the required Excel skills in the Excel Notes at www.keycollege.com/online .

Launch Excel.

	A	B	C	D	E
1		Coefficients	a	b	c
2					
3	x	P(x)			

Step 1 Prepare the labels to describe your columns as shown.

Cells C2, D2, and E2 will store the coefficients for the polynomial to be studied. The P(x) column will store the values of the quadratic function to be studied.

Step 2 Name the cells.

Name Cell C2 `COEFFA`.

Name Cell D2 `COEFFB`.

Name Cell E2 `COEFFC`.

Step 3 In Cell A4 enter `=-2`.

Step 4 In Cell B4 enter `=coeffa*A4^2+coeffb*A4+coeffc`.

You are ready to deal with quadratic equations of the form $ax^2 + bx + c = k$. You can build your table to solve such equations accurately to the hundredths place.

Step 5 In Cell A5 enter `=A4+.01`.

Step 6 Use data fill to fill the formula from Cell B4 down to Cell B5. You can use this completed row to generate as many rows as you like to help you solve problems.

Step 7 Select Columns A and B. Then, from the Format menu, choose Cells. Select Number Format with 3 decimal places.

Step 8 Highlight Cells A5 and B5, and data-fill to Row 700. You can data-fill more cells later if necessary. Note the zeros in the P(x) column. As you have not yet entered any values for the coefficients, Excel treats empty cells as 0.

Questions

1. In Cells C2, D2, and E2, enter the numbers 1, -4, and 1, respectively. Suppose you need to solve the equation $x^2 - 4x + 1 = -2$. Scan down Column B until you find the number -2. Record the value of x that will give you that answer. This will be a solution of the equation, with x expressed to three decimal places. Is there more than one answer? Explain why or why not.

2. Study Column B for numeric patterning. As you move down from row to row, do the numbers always increase? Do they always decrease? Is there a point at which the answers to these questions change? Record that point. How many solutions do you find in the table for the equation $x^2 - 4x + 1 = -2$? What are these solutions, to the nearest hundredth?

3. Change the equation to $x^2 - 2x + 1 = 4$, and answer Question 2 again.

4. Change the equation to $x^2 - 2x + 1 = 2$ and answer Question 2 once more.

5. Change the coefficients to 1, 1, and -1 for a, b, and c respectively. How many solutions can you find for the equation $x^2 + x - 1 = 0$? What are they? A positive root of this equation is called the Golden Mean. What is that number, accurate to two decimal places?

6. Change the coefficients to 1, 1, and 1. Study the table as it represents the values for $x^2 + x + 1$. Can you find values of x for which $x^2 + x + 1 = 0$? Use the table to explain your answer.

Extension

7. Change the lead coefficient, a, to a number other than 1. Answer the queries in Question 2. Find solutions as you study the table.

8. Change the lead coefficient to a negative number. Answer the queries in Question 2. Find solutions as you study the table.

9. How would you change the spreadsheet to allow you to study a cubic equation like $x^3 - 3x^2 + 2x + 1 = 1$?

CHAPTER 6 REVIEW

Polynomials can be used to represent the motions of projectiles, the areas of regions, and the volumes of boxes. When examining a set of data whose x-values form an arithmetic sequence, you can use the **finite differences method** to calculate the degree of a polynomial that will fit the data. When you know the **degree** of the polynomial, you can use regression to find an equation for the polynomial. Polynomial equations can be written in several forms. The form $a_nx^n + a_{n-1}x^{n-1} + \cdots + a_1x^1 + a_0$ is called the **general form.** Quadratic equations, which are 2nd-degree polynomial equations, can also be written in **vertex form** or **factored form,** with each factor corresponding to a **root** of the equation. In a quadratic equation the roots can be found by using the **quadratic formula.**

There are the same number of roots as the degree of the polynomial. In some cases these roots may include **imaginary** or **complex numbers.** If the coefficients of a polynomial are real, then any nonreal complex roots will come in **conjugate pairs.** The degree of a polynomial function determines the shape of its graph. The graphs may have hills and valleys where you will find **local minimums** and **maximums.** By varying the coefficients you can change the vertical scale factor and the relative sizes of these hills and valleys.

EXERCISES

1. Factor each expression completely.

   a. $2x^2 - 10x + 12$　　b. $2x^2 + 7x + 3$　　c. $x^3 - 10x^2 - 24x$

2. Solve each equation by setting equal to zero and factoring.

   a. $x^2 - 8x = 9$　　b. $x^4 + 2x^3 = 15x^2$

3. Using three noncollinear points as vertices, how many different triangles can you draw? Given a choice of four points no three of which are collinear, how many different triangles can you draw? Given a choice of five points? n points? Hint: Make a table with two columns—number of points and number of triangles. Use the finite differences method.

4. Is each equation written in general form, vertex form, or factored form? Write each equation in the other two forms if possible.

   a. $y = 2(x - 2)^2 - 16$　　b. $y = -3(x - 5)(x + 1)$

   c. $y = x^2 + 3x + 2$　　d. $y = (x + 1)(x - 3)(x + 4)$

   e. $y = 2x^2 + 5x - 6$　　f. $y = -2 - (x + 7)^2$

5. Sketch a graph of each function, and label the coordinates of all zeros and of local maxima and minima. (Each coordinate should be accurate to the nearest hundredth.)

a. $y = 2(x - 2)^2 - 16$

b. $y = -3(x - 5)(x + 1)$

c. $y = x^2 - 3x + 2$

d. $y = (x + 1)(x - 3)(x + 4)$

e. $y = x^3 + 2x^2 - 19x + 20$

f. $y = 5x^5 + 38x^4 + 79x^3 - 8x^2 - 102x + 36$

6. Write an equation of each graph.

a.

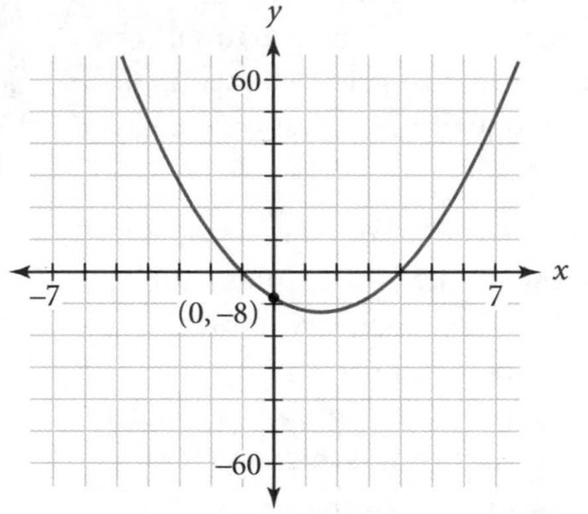

b.

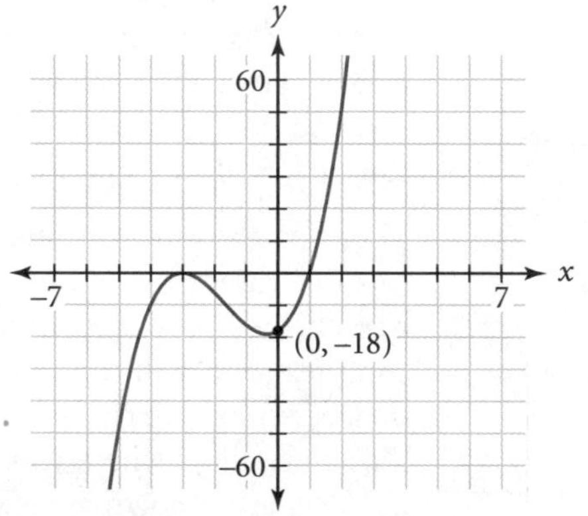

c.

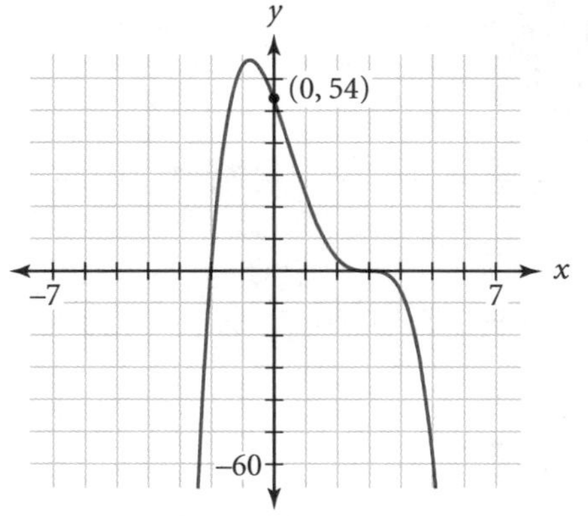

d.

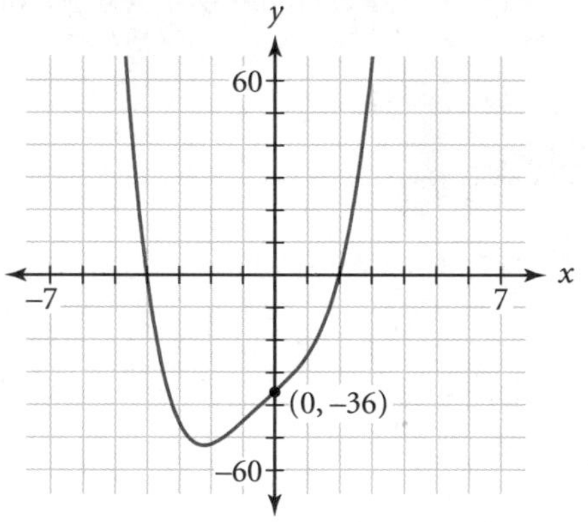

(*Hint:* One of the zeros occurs at $x = 3i$.)

7. **APPLICATION** According to postal regulations, a rectangular package must have a maximum combined girth and length of 108 in. The length is the longest dimension, and the girth is the perimeter of the cross section. Find the dimensions of the package with maximum volume that can be sent through the mail. (Assume the cross section is always a square with side length x.) Making a table might be helpful.

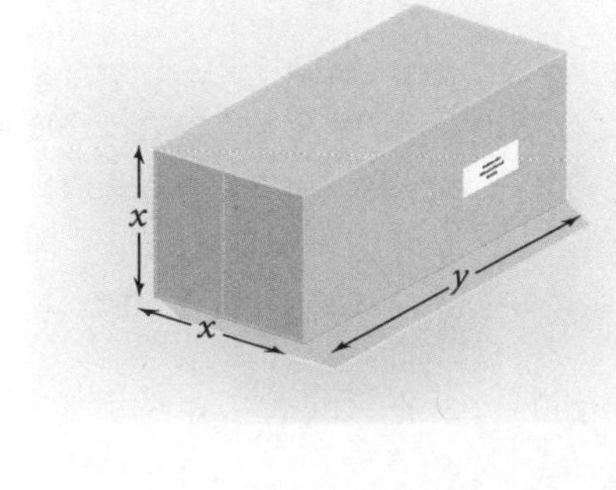

8. An object is dropped from the top of a building into a pool of water at ground level. There is a splash 6.8 sec after the object is dropped. How high is the building in meters? In feet?

9. Consider this puzzle.

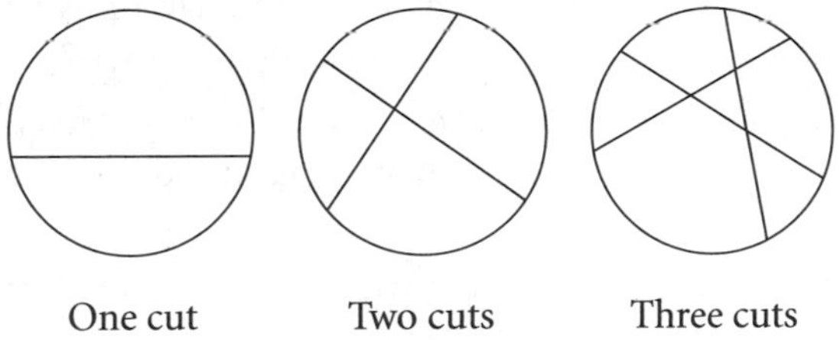

a. Write a formula relating the greatest number of pieces, y, that you can obtain from x cuts.

b. Use the formula to find the maximum number of pieces with five cuts and with ten cuts.

10. This 26-by-21-cm. rectangle has been divided into two regions. The width of the unshaded region is x cm.

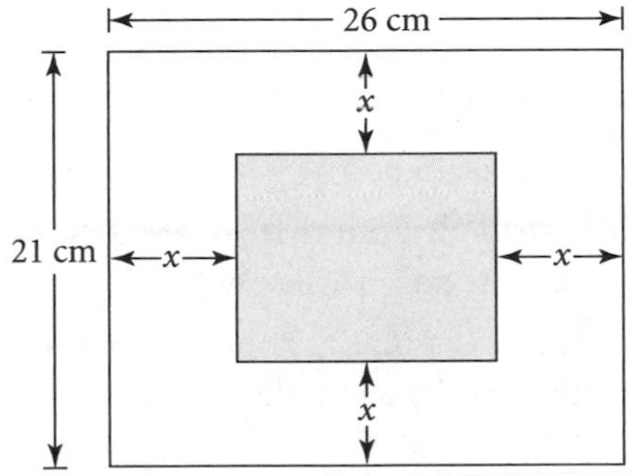

a. Express the area of the shaded part as a function of x, and graph it.

b. Find the domain and range of this function.

c. Find the x-value that makes the two regions (shaded and unshaded) equal in area.

11. Consider the polynomial function $y = 3x^4 - 20x^3 + 68x^2 - 92x - 39$.

a. List all possible rational roots.

b. Find the four zeros of the equation.

12. Simplify these expressions.

a. $(4 - 2i)(-3 + 6i)$ **b.** $(-3 + 4i) - (3 + 13i)$ **c.** $\dfrac{2 - i}{3 - 4i}$

13. Divide $\dfrac{6x^3 + 8x^2 + x - 6}{3x - 2}$.

Assessing What You've Learned

- How do you use the finite difference method to determine the degree of a polynomial? How can you find the equation of a polynomial?
- How do the general form, the vertex form, and the factored form of a quadratic equation differ? Do you know how to convert each form to the others?
- How do you complete the square in a quadratic equation? Why do you complete the square? Many students say that completing the square is harder if the lead coefficient of an expression, or function, is not 1. Why?
- What must you do to a quadratic equation before using the quadratic formula? Do you know the quadratic formula?
- What is the number i? Can you graph complex numbers? Can you add, subtract, multiply, and divide complex numbers? If $3 + 7i$ is a solution to a polynomial equation, and the polynomial has real number coefficients, what other number must be a solution? Why?
- Do you know how to perform polynomial long division? Do you know how to perform synthetic division? How do long division and polynomial division help you find zeros of higher-order polynomial equations? Be specific.

JOURNAL Look back at the journal entries you have written about Chapter 6. Then, to help you organize your thoughts, here are some additional questions to write about.

- What do you think were the main ideas of the chapter?
- Were there any exercises you particularly struggled with? If so, are they clear now? If some ideas are still foggy, what plans do you have to clarify them?
- If a polynomial function has six zeros, what can you say about the degree of the polynomial function? If a polynomial function has degree 6, then what can you say about the number of zeros it has?
- If a polynomial function changes from increasing to decreasing or from decreasing to increasing, we say it *turns*. For instance, all quadratic functions have one turn. Cubic functions can have no turns, or two turns. Use your calculator to graph $y = x^3$ and $y = x^3 - 9x$ to see this. In general, how are the number of turns in a polynomial's graph and its degree related?

CHAPTER

7 Matrices and Linear Systems

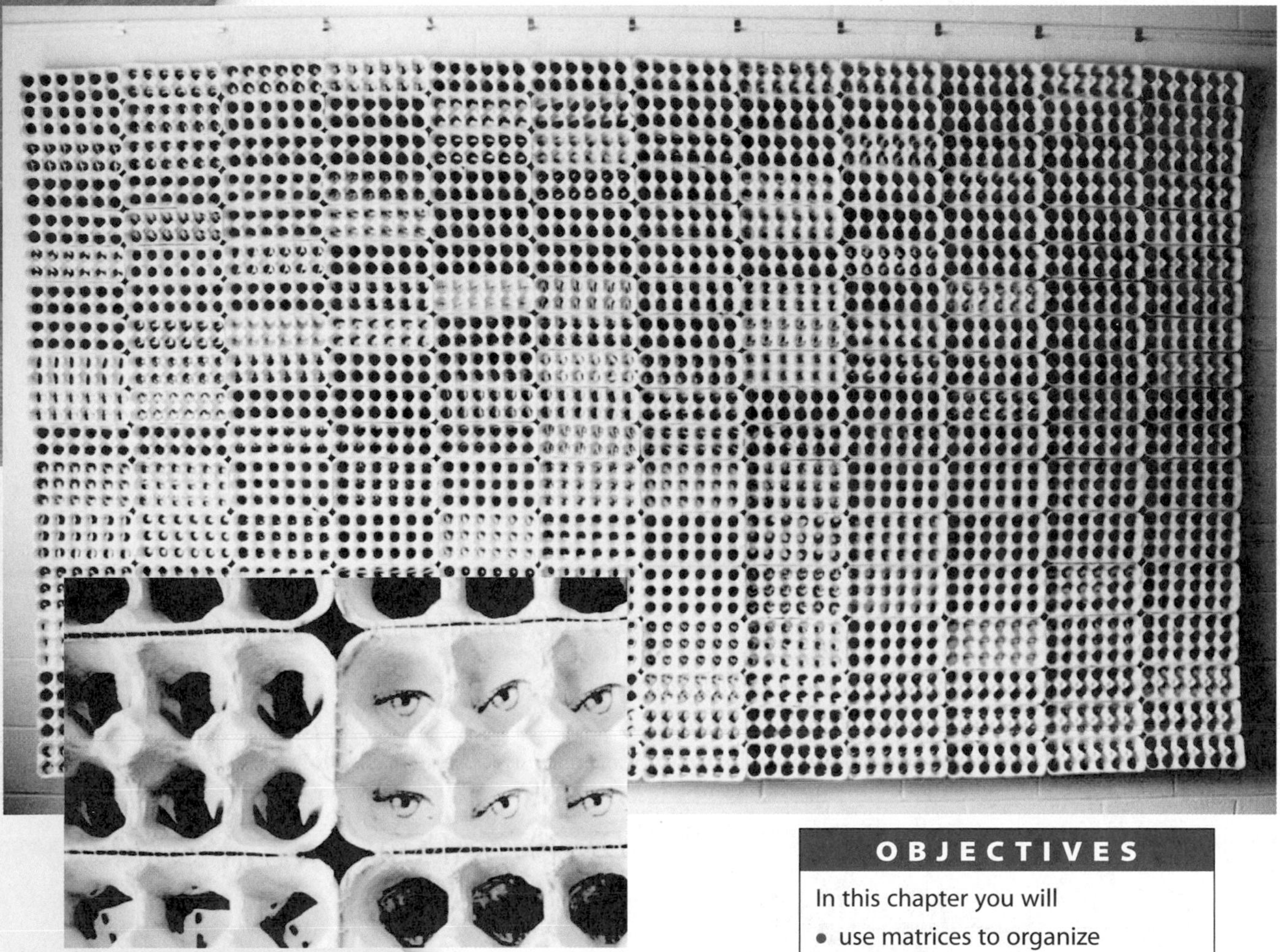

American installation artist Amy Stacey Curtis (b 1970) created this sculpture. The rectangular arrangement of egg cartons is used to organize an even larger arrangement of photocopied images. The egg cartons and their compartments divide the piece into rows and columns, while the small images—some darker than others, some lighter—help certain elements of the piece to stand out.

Fragile and detail of *Fragile* by Amy Stacey Curtis
Egg cartons, acrylic, dye, thread, beads, photocopies

OBJECTIVES

In this chapter you will

- use matrices to organize information
- add, subtract, and multiply matrices
- solve systems of linear equations with matrices
- graph two-variable inequalities on the coordinate plane and solve systems of inequalities
- identify and write inequalities that represent conditions that must be met simultaneously

LESSON

7.1 Matrix Representations

All dimensions are critical dimensions, otherwise why are they there?

RUSS ZANDBERGEN

On Saturday, Karina surveyed visitors to Snow Mountain who had weekend passes and found that 75% of skiers planned to ski again the next day, whereas 25% planned to snowboard. Of the snowboarders, 95% planned to snowboard the next day, and 5% planned to ski. To display this information, she made a diagram like the one shown.

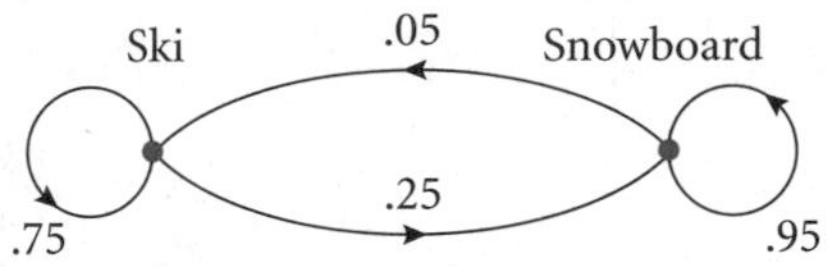

The arrows and labels show the patterns of the visitors' next-day activities. For instance, the circular arrow labeled .75 indicates that 75% of the visitors skiing one day plan to ski again the next day. The arrow labeled .25 indicates that 25% of the visitors who ski one day plan to snowboard the next day.

Diagrams like these are called **transition diagrams** because they show how something changes from one time to the next. The same information is sometimes represented in a **transition matrix.** A **matrix** (plural: *matrices*) is a rectangular arrangement of numbers. For the Snow Mountain information the transition matrix looks like this:

Tomorrow

Today	Ski	Snowboard
Ski	.75	.25
Snowboard	.05	.95

$$\begin{bmatrix} .75 & .25 \\ .05 & .95 \end{bmatrix}$$

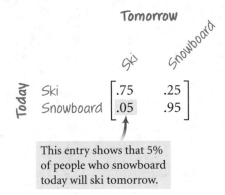

This entry shows that 5% of people who snowboard today will ski tomorrow.

In the Investigation you will create a transition diagram and matrix for another situation. You will also use the information to determine how the numbers of people in two different categories change from one period of time to the next.

Investigation
Chilly Choices

Big Ben's House of Beans sells two types of frozen mocha drink: dark chocolate and white chocolate. During the 1st week 220 customers choose dark chocolate, whereas only 20 choose white chocolate. During each of the following weeks, 10% of the white chocolate drinkers switch to dark chocolate, and 5% of those who preferred dark chocolate switch to white chocolate.

Step 1 | Complete a transition diagram that displays this information.

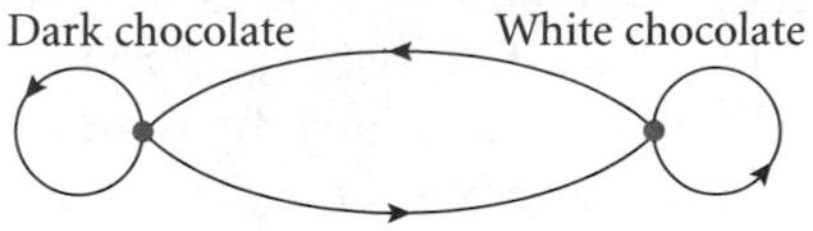

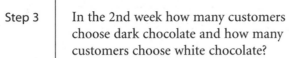

Step 2 | Complete a transition matrix that represents this information. The rows should indicate the present condition, and the columns should indicate the next condition after the transition.

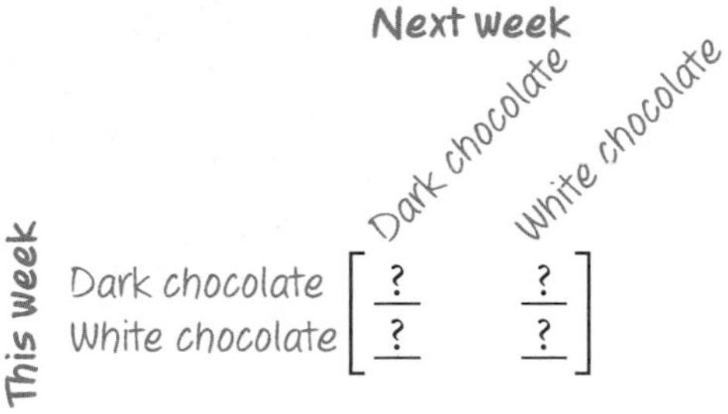

Step 3 | In the 2nd week how many customers choose dark chocolate and how many customers choose white chocolate?

Step 4 | How many will choose each option in the 3rd week?

Step 5 | Write a recursive routine to take any week's values and give the next week's values.

Step 6 | What do you think will happen to the long-run values of the number of customers who choose dark chocolate and the number who choose white chocolate?

Matrices can be used to organize many kinds of information. For example, the matrix below can be used to represent the number of math, science, and history texts sold this week at the main and branch campus bookstores. The rows represent math, science, and history, from top to bottom, and the columns represent the main and branch bookstores, from left to right.

$$[A] = \begin{bmatrix} 83 & 33 \\ 65 & 20 \\ 98 & 50 \end{bmatrix}$$

This entry is the number of history books sold at the main book store.

The **dimensions** of the matrix are the number of rows and columns, in this case, 3×2 (say "three by two"). Each number in the matrix is called an **entry,** or *element,* and is identified as $a_{i,j}$, where i and j are the row number and column number respectively. Entries also may be designated by enclosing the row and column numbers in brackets, for example, $A[i,j]$. In matrix $[A]$ above, $a_{2,1} = 65$ because 65 is the entry in row 2, column 1. Note that in this text commas are used to separate the row and column numbers.

The next example shows how matrices can be used to represent coordinates of geometric figures. This kind of representation is the basis of many computer graphics applications, such as video games.

EXAMPLE A | Represent quadrilateral $ABCD$ as a matrix $[M]$.

▸ Solution

You can use a matrix to organize the coordinates of a geometric figure. Because each point has two coordinates, and there are four points, a 2 × 4 matrix with each column containing the x- and y-coordinates of a vertex makes sense. Row 1 contains consecutive x-coordinates, and row 2 contains the corresponding y-coordinates.

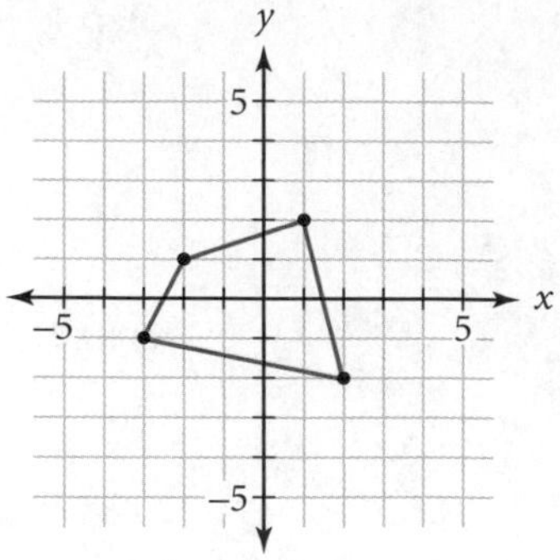

$$[M] = \begin{bmatrix} 1 & -2 & -3 & 2 \\ 2 & 1 & -1 & -2 \end{bmatrix}$$

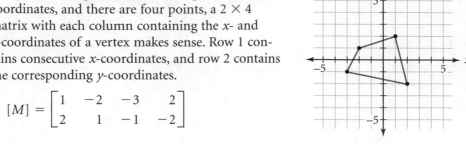

Example B shows how a transition matrix makes it easy to compute data. In Lesson 7.2 you'll learn more ways to operate with matrices.

EXAMPLE B

In the survey at the start of this lesson, Karina interviewed 260 skiers and 40 snowboarders. If her transition predictions are correct, how many people will engage in each activity the next day?

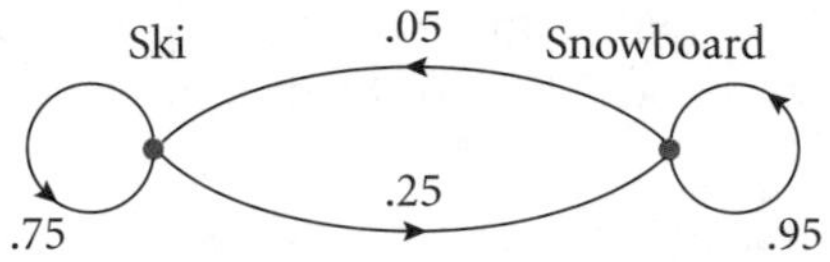

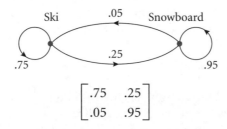

$$\begin{bmatrix} .75 & .25 \\ .05 & .95 \end{bmatrix}$$

▸ Solution

The next day 75% of the 260 skiers will ski again, and 5% of the 40 snowboarders will switch to skiing.

Skiers: $260 \cdot (0.75) + 40 \cdot (0.05) = 197$

So, 197 people will ski the next day.

Of the 260 skiers, 25% will switch to snowboarding, and 95% of the 40 snowboarders will snowboard again.

Snowboarders: $260 \cdot (0.25) + 40 \cdot (0.95) = 103$

So, 103 people will snowboard the next day.

You can organize the information for the second day in a matrix in the form

[*number of skiers*   *number of snowboarders*].

$$[197 \quad 103]$$

Transition diagrams and transition matrices can be used to show changes in a closed system. Although very informative for simple problems, the diagram is difficult to use when you have 5 or more starting conditions, as this would create 25 or more arrows, or paths. The transition matrix is relatively easy to read, even with a large number of starting conditions. It grows in size, but each entry shows what percent changes from one condition to another.

EXERCISES

Practice Your Skills

1. At Powder Hill Resort, Russell collected data similar to Karina's. He found that 86% of the skiers planned to ski the next day, and 92% of snowboarders planned to snowboard the next day.

a. Draw a transition diagram for Russell's information.

b. Write a transition matrix for the same information. Remember that rows indicate the present condition, and columns indicate the next condition. List skiers first and snowboarders second.

2. Complete this transition diagram:

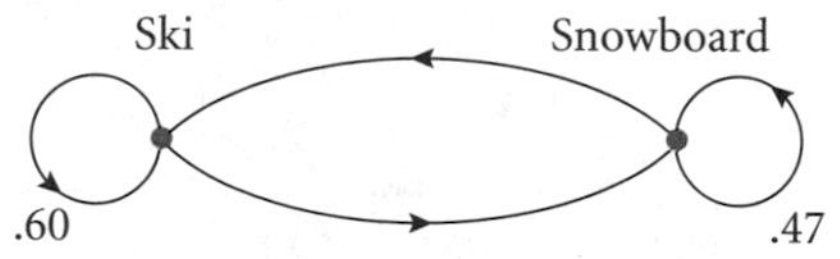

3. Write a transition matrix for the diagram in Exercise 2. Order your information as in Exercise 1b.

4. Matrix $[M]$ represents the vertices of triangle ABC.

$$[M] = \begin{bmatrix} -3 & 1 & 2 \\ 2 & 3 & -2 \end{bmatrix}$$

a. Name the coordinates of the vertices, and draw the triangle.

b. What matrix represents the image of ΔABC after a translation (slide) down 4 units?

c. What matrix represents the image of ΔABC after a translation (slide) right 4 units?

5. During a recent softball tournament, information about players' batting sides was recorded in a matrix. Row 1 represents women, and row 2 represents men. Column 1 represents left-handed batters, column 2 represents right-handed batters, and column 3 represents those who can bat with either hand.

$$[A] = \begin{bmatrix} 5 & 13 & 2 \\ 4 & 18 & 3 \end{bmatrix}$$

a. How many women and how many men participated in the tournament?

b. How many men were right handed?

c. What does the value of $a_{1,2}$ mean?

Reason and Apply

6. A mixture of 40 mL of NO (nitric oxide) and 200 mL of N_2O_2 (dinitrogen dioxide) is heated. At the new temperature 10% of the NO changes to N_2O_2 each second, and 5% of the N_2O_2 changes to NO each second.

a. Draw a transition diagram that displays this information.

b. Write a transition matrix that represents the same information. List NO first and N_2O_2 second.

c. Assume that the combined amount remains at 240 mL and the transition percentages stay the same throughout. Find the amounts in mL of NO and N_2O_2 after 1 s and after 2 s. Write your answers in the matrix form [NO N_2O_2].

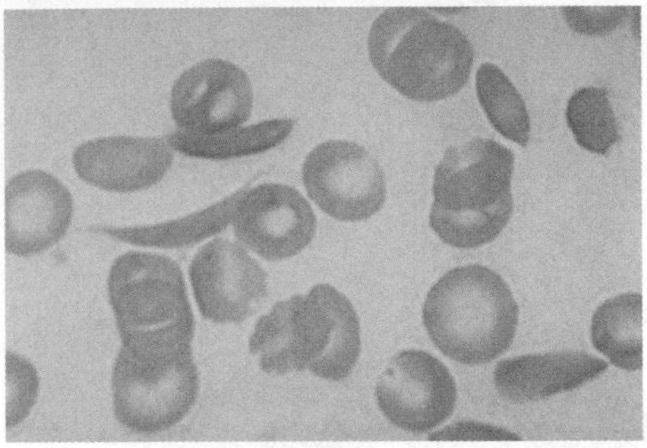

This photo shows red blood cells, some deformed by sickle-cell anemia. Researchers have found that nitric oxide (NO) counteracts the effects of sickle-cell anemia.

7. In many countries more people move into the cities than out of them. Suppose that in a certain country 10% of the rural population moves to the city each year, and 1% of the urban population moves to the country each year.

a. Draw a transition diagram that displays this information.

b. Write a transition matrix that represents this same information. List urban dwellers first and rural dwellers second.

c. Assume that 16 million of the country's 25 million people live in the city initially. Find the urban and rural populations in millions after 1 year and after 2 years. Write your answers in the matrix form [*urban* *rural*].

8. Matrix $[A]$ represents the number of math, science, and history texts sold at the main and branch campus bookstores this week.

$$[A] = \begin{bmatrix} 83 & 33 \\ 65 & 20 \\ 98 & 50 \end{bmatrix}$$

a. Explain the meaning of the value of $a_{3,2}$.

b. Explain the meaning of the value of $a_{2,2}$.

c. Matrix $[B]$ represents last week's sales. Compare this week's sales of math books with last week's sales.

$$[B] = \begin{bmatrix} 80 & 25 \\ 65 & 15 \\ 105 & 55 \end{bmatrix}$$

d. Write a matrix that represents the total sales during last week and this week.

9. The three most popular kinds of passenger vehicles are the sedan, the SUV, and the minivan. When it is time to purchase a new vehicle, buyers are influenced by many factors. Suppose that of the buyers in a particular community who presently own a minivan, 18% will change to an SUV, and 20% will change to a sedan. Of the buyers who presently own a sedan, 35% will change to a minivan, and 20% will change to a SUV; and of those who presently own an SUV, 12% will buy a minivan, and 32% will buy a sedan.

a. Draw a transition diagram that displays these changes.

b. Write a transition matrix that represents this scenario. List the rows and columns in the order minivan, sedan, SUV.

c. What is the sum of the entries in row 1? Row 2? Row 3? Why does this make sense?

10. APPLICATION Lisa Crawford is getting into the moving-truck rental business. She has the funds to buy about 100 trucks, and she will have offices in three nearby counties. Her studies show that in Bay County 20% of rentals go to Sage County, and 15% go to Thyme County. The rest start and end in Bay County. From Sage County 25% of rentals go to Bay, and 55% stay in Sage; the rest move to Thyme County. From Thyme County 40% of rentals end in Bay and 30% in Sage.

a. Draw a transition diagram that displays this information.

b. Write a transition matrix that represents this scenario. List your rows and columns in the order Bay, Sage, Thyme.

c. What is the sum of the entries in row 1? Row 2? Row 3? Why does this make sense?

d. If she starts with 45 trucks in Bay County, 30 trucks in Sage County, and 25 trucks in Thyme County, and all trucks are rented one Saturday, how many trucks will she expect to be at each office the next morning?

11. APPLICATION Fly-Right Airways operates routes out of five cities as shown in the route map here. Each segment connecting two cities represents a round-trip flight between them. Matrix $[M]$ displays the map information in matrix form with the cities A, B, C, D, and E listed in order in the rows and columns. The rows represent starting conditions (departure cities), and columns represent next conditions (arrival cities). This matrix is called an **adjacency matrix**. For instance, the value of the entry in row 1, column 5, shows that there are two round-trip flights between City A and City E.

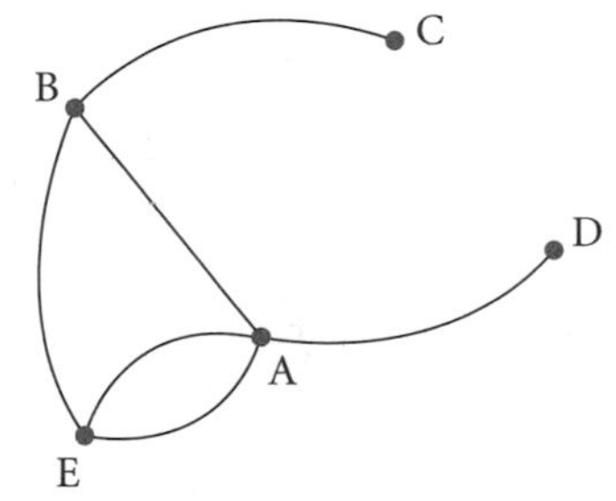

$$[M] = \begin{bmatrix} 0 & 1 & 0 & 1 & 2 \\ 1 & 0 & 1 & 0 & 1 \\ 0 & 1 & 0 & 0 & 0 \\ 1 & 0 & 0 & 0 & 0 \\ 2 & 1 & 0 & 0 & 0 \end{bmatrix}$$

a. What are the dimensions of this matrix?

b. What is the value of $m_{3,2}$? What does this entry represent?

c. Which city has the most flights? Explain how you can tell using the graph and using the matrix.

d. Matrix $[N]$ below represents Americana Airways' routes connecting four cities, J, K, L, and M. Sketch a possible route map.

$$[N] = \begin{bmatrix} 0 & 1 & 2 & 1 \\ 1 & 0 & 2 & 0 \\ 2 & 2 & 0 & 1 \\ 1 & 0 & 1 & 0 \end{bmatrix}$$

Mathematics Connection Graph theory is a branch of mathematics that deals with connections between items. The route diagram that shows the connections between locations in Exercise 11 is an example of a vertex-edge graph. A paragraph could have been written to describe the flight routes, but a vertex-edge graph allows you to show the routes quickly and clearly. Graphs could also be used to show a natural gas pipeline, the chemical structure of a molecule, a family tree, or a Program Evaluation and Review Technique (PERT) chart (often used in project management). Matrices are the tool with which the data in a graph can be manipulated mathematically and investigated further.

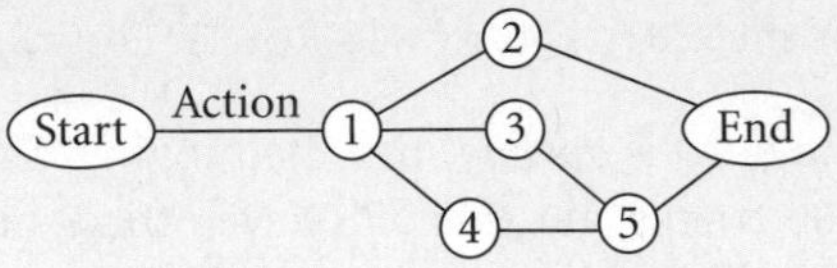

Review

12. (Lesson 3.5) Solve this system using either substitution or elimination.

$$5x - 4y = 25$$
$$x + y = 3$$

13. (Lesson 3.1) Solve the equation $2x + 3y = 12$ for y, and then graph it.

14. **APPLICATION** (Lesson 5.1) The following table shows the number of cellular telephone subscribers in the United States from 1985 to 2000.

Cellular Phone Subscribers

Year	Number of Subscribers	Year	Number of Subscribers
1985	340,000	1993	16,009,000
1986	682,000	1994	24,134,000
1987	1,231,000	1995	33,786,000
1988	2,069,000	1996	44,043,000
1989	3,509,000	1997	55,312,000
1990	5,283,000	1998	69,209,000
1991	7,557,000	1999	86,047,000
1992	11,033,000	2000	109,478,000

(*The World Almanac and Book of Facts 2002*)

a. Create a scatter plot of the data.

b. Find an exponential function to model the data.

c. Use your model to predict the number of subscribers in 2010. Do you think this is a realistic prediction? Do you think an exponential model is appropriate?

LESSON

7.2

Matrix Operations

It is not once nor twice but times without number that the same ideas make their appearance in the world.

ARISTOTLE

Like the table, the matrix is a compact way of organizing data. Representing data as a matrix instead of a table allows you to perform operations such as addition and multiplication with the data. In this lesson you will see how this is useful.

Consider this problem from Lesson 7.1. Matrix $[A]$ represents the number of math, science, and history books sold this week at the main and branch campus bookstores. Matrix $[B]$ contains the same information for last week. What are the total sales by category and location for both weeks?

$$[A] = \begin{bmatrix} 83 & 33 \\ 65 & 20 \\ 98 & 50 \end{bmatrix} \qquad [B] = \begin{bmatrix} 80 & 25 \\ 65 & 15 \\ 105 & 55 \end{bmatrix}$$

To solve this, add matrices $[A]$ and $[B]$.

$$83 + 80 = 163$$

$$\begin{bmatrix} 83 & 33 \\ 65 & 20 \\ 98 & 50 \end{bmatrix} + \begin{bmatrix} 80 & 25 \\ 65 & 15 \\ 105 & 55 \end{bmatrix} = \begin{bmatrix} 163 & 58 \\ 130 & 35 \\ 203 & 105 \end{bmatrix}$$

If 83 math books were sold at the main bookstore this week, and 80 math books were sold at the main bookstore last week, a total of 163 math books were sold at the main bookstore for both weeks.

To add two matrices, simply add corresponding entries. Adding (or subtracting) two matrices requires that they both have the same dimensions. The corresponding rows and columns should also have similar interpretations if the results are to make sense. [► See **Calculator Note 7A** to learn how to enter matrices into your calculator. **Calculator Note 7B** shows how to compute with matrices. ◄]

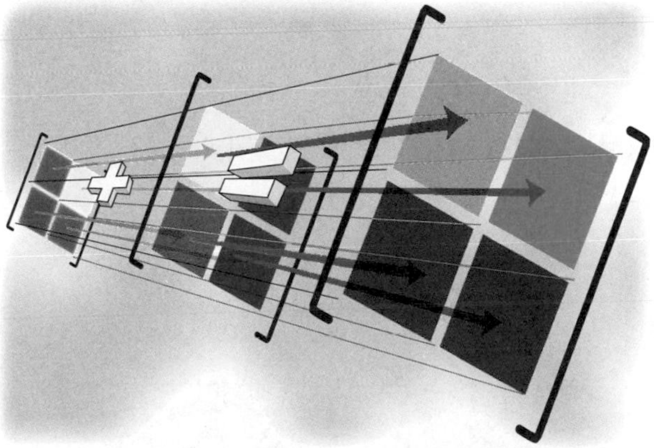

When you add matrices, you add corresponding entries. This illustration uses color to show how the addition carries through to the matrix representing the sum.

In Lesson 7.1 you used a matrix to record the coordinates of the vertices of a triangle. You can use matrix operations to transform a figure such as a triangle just as you transformed the graph of a function.

EXAMPLE A

This matrix represents a triangle.

$$\begin{bmatrix} -3 & 1 & 2 \\ 2 & 3 & -2 \end{bmatrix}$$

a. Graph the triangle and its image after a translation (slide) left 3 units. Write a matrix equation to represent the transformation.

b. Describe the transformation represented by this matrix expression.

$$\begin{bmatrix} -3 & 1 & 2 \\ 2 & 3 & -2 \end{bmatrix} + \begin{bmatrix} -4 & -4 & -4 \\ -3 & -3 & -3 \end{bmatrix}$$

c. Describe the transformation represented by this matrix expression.

$$2 \cdot \begin{bmatrix} -3 & 1 & 2 \\ 2 & 3 & -2 \end{bmatrix}$$

▶ *Solution*

The original matrix represents a triangle with vertices $(-3, 2)$, $(1, 3)$, and $(2, -2)$.

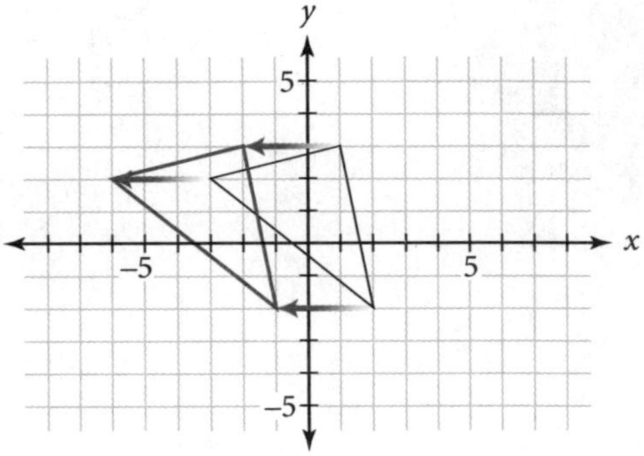

a. After a translation left 3 units, the x-coordinates of the image are reduced by 3. There is no change to the y-coordinates. You can represent this transformation as addition.

$$\begin{bmatrix} -3 & 1 & 2 \\ 2 & 3 & -2 \end{bmatrix} + \begin{bmatrix} -3 & -3 & -3 \\ 0 & 0 & -0 \end{bmatrix} = \begin{bmatrix} -6 & -2 & -1 \\ 2 & 3 & -2 \end{bmatrix}$$

b. $$\begin{bmatrix} -3 & 1 & 2 \\ 2 & 3 & -2 \end{bmatrix} + \begin{bmatrix} -4 & -4 & -4 \\ -3 & -3 & -3 \end{bmatrix} = \begin{bmatrix} -7 & -3 & -2 \\ -1 & 0 & -5 \end{bmatrix}$$

This matrix addition represents a translation left 4 units and down 3 units.

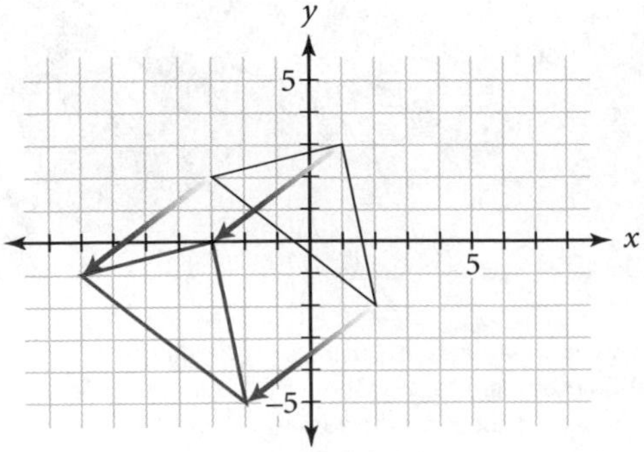

c. Just as 2×5 is the same as $5 + 5$, $2[A] = [A] + [A]$.

$$2 \cdot \begin{bmatrix} -3 & 1 & 2 \\ 2 & 3 & -2 \end{bmatrix} = \begin{bmatrix} -3 & 1 & 2 \\ 2 & 3 & -2 \end{bmatrix} + \begin{bmatrix} -3 & 1 & 2 \\ 2 & 3 & -2 \end{bmatrix} = \begin{bmatrix} -6 & 2 & 4 \\ 4 & 6 & -4 \end{bmatrix}$$

Multiplying a matrix by a number is called **scalar multiplication.** Each entry in the matrix is simply multiplied by the **scalar,** in this case 2.

The resulting matrix represents a stretch both horizontally and vertically by a factor of 2. A transformation that stretches or shrinks both horizontally and vertically by the same scale factor is called a **dilation.**

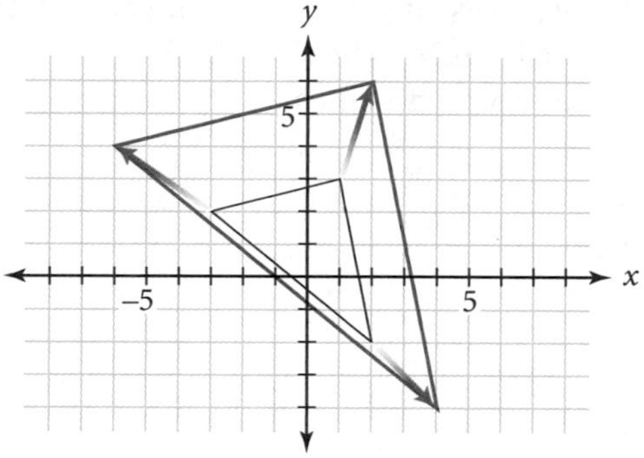

[▶ See **Calculator Note 7C** to learn how to graph polygons on your calculator with matrices. ◀]

Addition and scalar multiplication operate on one entry at a time. The multiplication of two matrices is more involved and uses several entries to find one entry of the answer matrix. The scenario in Example B is from the Investigation in Lesson 7.1.

EXAMPLE B

Big Ben's House of Beans sells two types of frozen mocha drink, dark chocolate and white chocolate. During the 1st week 220 customers choose dark chocolate, whereas only 20 choose white chocolate. During each of the following weeks, 10% of the white chocolate drinkers switch to dark chocolate, and 5% of those who preferred dark chocolate switch to white chocolate. How many customers will choose each drink in the 2nd week? In the 3rd week?

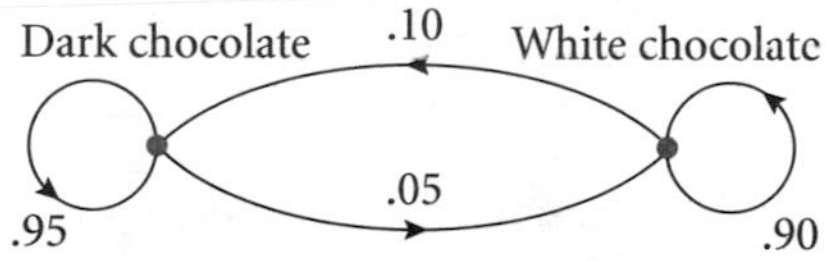

▶ Solution

You can use this matrix equation to find the answer for the 2nd week.

$$[220 \quad 20]\begin{bmatrix} .95 & .05 \\ .10 & .90 \end{bmatrix} = [\textit{dark chocolate} \quad \textit{white chocolate}]$$

The initial matrix, $[A] = [220 \quad 20]$, represents the original numbers of dark chocolate and white chocolate drinkers.

The top row of the transition matrix, $[B] = \begin{bmatrix} .95 & .05 \\ .10 & .90 \end{bmatrix}$, represents the transitions in the present number of dark chocolate drinkers, and the bottom row represents the transitions in the present number of white chocolate drinkers.

You can define matrix multiplication by looking at how you calculate the numbers for the 2nd week. The 2nd week's number of dark chocolate drinkers will be

$$220 \cdot (.95) + 20 \cdot (.10)$$

or 211 customers, because 95% of the 220 original dark chocolate mocha drinkers remain with the same preference, and 10% of the 20 original white chocolate customers switch to dark chocolate. In effect, you multiply the two entries in row 1 of $[A]$ by the two entries in column 1 of $[B]$ and add the products. The result, 211, is entry $c_{1,1}$ in the answer matrix $[C]$.

Initial matrix		**Transition matrix**		**Answer matrix**
A	$\cdot$	$[B]$	$=$	C

$$[220 \quad 20] \begin{bmatrix} .95 & .05 \\ .10 & .90 \end{bmatrix} = [211 \quad \textit{white chocolate}]$$

Likewise, the second week's number of white chocolate drinkers will be 220 (.05) + 20 · (.90), or 29 customers because 5% of the dark chocolate drinkers convert to white chocolate, and 90% of the white chocolate drinkers remain unchanged in their preference. This is the sum of the products of the entries in row 1 of $[A]$ and column 2 of $[B]$. The answer, 29, is entry $c_{1,2}$ in the answer matrix $[C]$.

$$[220 \quad 20] \cdot \begin{bmatrix} .95 & .05 \\ .10 & .90 \end{bmatrix} = [211 \quad 29[$$

To find the numbers for the 3rd week, multiply the result of your previous calculations by the transition matrix again.

$$\begin{bmatrix} 211 & 29 \end{bmatrix} \begin{bmatrix} .95 & .05 \\ .10 & .90 \end{bmatrix} = \begin{bmatrix} 203.35 & 36.65 \end{bmatrix}$$

Multiply row 1 by column 1.

$$211\ (.95) + 29\ (.10) = 203.35$$

Multiply row 1 by column 2.

$$211\ (.05) + 29\ (.90) = 36.65$$

In the next week approximately 203 customers will choose dark chocolate, and 37 will choose white chocolate.

You can continue multiplying to find the numbers for the 4th week, the 5th week, and so on. [▶ Revisit **Calculator Note 7B** to learn how to multiply matrices on your calculator.◀]

Investigation
Find Your Place

In this Investigation you and your classmates will simulate weekly transitions of rental cars between cities. After completing the classroom simulation, you will analyze the results.

Procedure Note

Rental Car Simulation

1. Your instructor will designate three areas of your class room as City A, City B, and City C. You and your classmates will each be assigned to one of these three locations. Each person represents a rental car. Record the starting quantities at each city.
2. Use your calculator to generate a random number, x. [▶ See **Calculator Note 1H** to learn how to generate random numbers.◀] Determine your location for next week as follows:
   - **a.** If you are at City A, move to City B if $x \leq .2$, move to City C if $.2 < x \leq .7$, or stay at City A if $x > .7$.
   - **b.** If you are at City B, move to City A if $x \leq .5$, or stay at City B if $x > .5$.
   - **c.** If you are at City C, move to City B if $x \leq .1$, move to City A if $.1 < x \leq .3$, or stay at City C if $x > .3$.
3. Record the quantities at each city after Week 1.
4. This simulation represents a situation in which proportions of the car fleet are moved to various destinations. Use the proportions from the simulation, along with your calculator, to do Step 2 a–f below.

Step 1 Follow the Procedure Note, and perform the simulation with your class.

Step 2 Work with your group to perform these tasks.

a. Draw a transition diagram that displays the rules of the simulation.

b. Write a transition matrix that represents the same information.

c. Write an initial condition matrix for the starting quantities at each city.

d. Show how to multiply the initial condition matrix and the transition matrix for the first transition. How do these theoretical results for Week 1 compare to the experimental data from your simulation?

e. Use your calculator to multiply matrices repeatedly to find the theoretical number of cars in each city for the next four weeks.

f. Use your calculator to find the theoretical long-run values of the number of cars in each city.

Just as only some matrices can be added (those with the same dimensions), only some matrices can be multiplied. Example C and Exercise 3 will help you explore what kinds of matrices can be multiplied.

EXAMPLE C

Determine the dimensions of the answer to this product, and describe how to calculate entries in the answer.

$$\begin{bmatrix} -1 & 0 \\ 0 & 1 \end{bmatrix} \cdot \begin{bmatrix} -3 & 1 & 2 \\ 2 & 3 & -2 \end{bmatrix}$$

▸ Solution

a. To multiply two matrices, multiply each entry in a row of the first matrix by each entry in a column of the second matrix.

You can multiply a 2×2 matrix by a 2×3 matrix because the **inside dimensions** are the same—the two row entries match up with the two column entries.

2×2 , 2×3

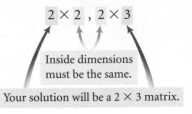

Inside dimensions must be the same.

Your solution will be a 2×3 matrix.

The **outside dimensions** tell you the dimensions of your answer.

The answer to this product has dimensions 2×3.

$$\begin{bmatrix} -1 & 0 \\ 0 & 1 \end{bmatrix} \cdot \begin{bmatrix} -3 & 1 & 2 \\ 2 & 3 & -2 \end{bmatrix} = \begin{bmatrix} c_{1,1} & c_{1,2} & c_{1,3} \\ c_{2,1} & c_{2,2} & c_{2,3} \end{bmatrix}$$

b. To find the values of entries in the first row of your solution matrix, sum the products of the entries in the first row of the first matrix and the entries in the columns of the second matrix.

$-1(-3) + 0(2) = 3$

$-1(1) + 0(3) = -1$

$-1(2) + 0(-2) = -2$

$$\begin{bmatrix} -1 & 0 \\ 0 & 1 \end{bmatrix} \begin{bmatrix} -3 & 1 & 2 \\ 2 & 3 & -2 \end{bmatrix} = \begin{bmatrix} 3 & -1 & -2 \\ c_{2,1} & c_{2,2} & c_{2,3} \end{bmatrix}$$

To find the values of entries in the second row of your solution matrix, sum the products of the entries in the second row of the first matrix and the entries in the columns of the second matrix.

$$\begin{bmatrix} -1 & 0 \\ 0 & 1 \end{bmatrix} \begin{bmatrix} -3 & 1 & 2 \\ 2 & 3 & -2 \end{bmatrix} = \begin{bmatrix} 3 & -1 & -2 \\ 2 & 3 & -2 \end{bmatrix}$$

$0(-3) + 1(2) = 2$

$0(1) + 1(3) = 3$

$0(2) + 1(-2) = -2$

The product is

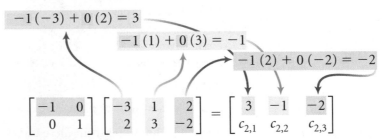

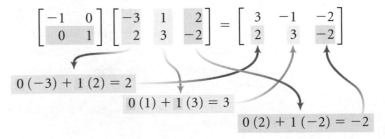

$$\begin{bmatrix} 3 & -1 & -2 \\ 2 & 3 & -2 \end{bmatrix}$$

The following definitions review the matrix operations you've learned in this lesson.

Matrix Operations

Matrix Addition

To add matrices, add corresponding entries.

$$\begin{bmatrix} 2 & 3 \\ -1 & 0 \end{bmatrix} + \begin{bmatrix} -1 & 2 \\ 1 & 4 \end{bmatrix} = \begin{bmatrix} 1 & 5 \\ 0 & 4 \end{bmatrix}$$

$$-1 + 1 = 0$$

Only matrices that have the same dimension can be added.

Scalar Multiplication

To multiply a scalar by a matrix, multiply the scalar by each value in the matrix.

$$3 \cdot \begin{bmatrix} 2 & -1 \\ 1 & 3 \\ 0 & -2 \end{bmatrix} = \begin{bmatrix} 6 & -3 \\ 3 & 9 \\ 0 & -6 \end{bmatrix}$$

$$3(-2) = -6$$

Matrix Multiplication

To multiply two matrices $[A]$ and $[B]$, multiply each entry in a row of $[A]$ by the corresponding entry in a column of $[B]$.

$$\begin{bmatrix} 0 & -1 \\ 2 & 1 \end{bmatrix} \begin{bmatrix} 3 & -2 & 4 \\ 0 & -3 & 5 \end{bmatrix} = \begin{bmatrix} 0 & 3 & -5 \\ 6 & -7 & 13 \end{bmatrix}$$

$$2(-2) + 1(-3) = -7$$

An entry $c_{i,j}$ in the answer matrix represents the sum of the products of each entry in row i of the first matrix and the entry in the corresponding position in column j of the second matrix. The number of entries in a row of $[A]$ must equal the number of entries in a column of $[B]$, that is, the inside dimensions must be equal. The answer matrix will have the same number of rows as $[A]$ and the same number of columns as $[B]$, or the outside dimensions.

EXERCISES

Practice Your Skills

1. Look back at the calculations in Example B. Multiply these matrices to calculate how many customers will choose each drink in the 4th week.

$$[203.35 \quad 35.65] \begin{bmatrix} .95 & .05 \\ .10 & .90 \end{bmatrix} = [\textit{dark chocolate} \quad \textit{white chocolate}]$$

2. Find the missing values.

a. $[13 \quad 23] + [-6 \quad 31] = [x \quad y]$

b. $\begin{bmatrix} .90 & .10 \\ .05 & .95 \end{bmatrix} \cdot \begin{bmatrix} .90 & .10 \\ .05 & .95 \end{bmatrix} = \begin{bmatrix} c_{1,1} & c_{1,2} \\ c_{2,1} & c_{2,2} \end{bmatrix}$

c. $\begin{bmatrix} 18 & -23 \\ 5.4 & 32.2 \end{bmatrix} + \begin{bmatrix} -2.4 & 12.2 \\ 5.3 & 10 \end{bmatrix} = \begin{bmatrix} a & b \\ c & d \end{bmatrix}$

d. $10 \cdot \begin{bmatrix} 18 & -23 \\ 5.4 & 32.2 \end{bmatrix} = \begin{bmatrix} m_{1,1} & m_{1,2} \\ m_{2,1} & m_{2,2} \end{bmatrix}$

e. $\begin{bmatrix} 7 & -4 \\ 18 & 28 \end{bmatrix} + 5\begin{bmatrix} -2.4 & 12.2 \\ 5.3 & 10 \end{bmatrix} = \begin{bmatrix} a & b \\ c & d \end{bmatrix}$

3. Perform the matrix arithmetic problems in 3a–f. If a problem is impossible, explain why.

a. $\begin{bmatrix} 1 & 2 \\ 3 & -2 \\ 0 & 1 \end{bmatrix} \begin{bmatrix} -3 & -1 & 2 \\ 5 & 2 & -1 \end{bmatrix}$

b. $\begin{bmatrix} 1 & -2 \\ 6 & 3 \end{bmatrix} + \begin{bmatrix} -3 & 7 \\ 2 & 4 \end{bmatrix}$

c. $[5 \quad -2 \quad 7] \begin{bmatrix} -2 & 3 \\ -1 & 0 \\ 3 & 2 \end{bmatrix}$

d. $\begin{bmatrix} 3 & -8 & 10 & 2 \\ -1 & 2 & 3 & 4 \end{bmatrix} \begin{bmatrix} 2 & -5 & 3 & 12 \\ 8 & -4 & 0 & 2 \end{bmatrix}$

e. $\begin{bmatrix} 3 & 6 \\ -4 & 1 \end{bmatrix} - \begin{bmatrix} -1 & 7 \\ -8 & 3 \end{bmatrix}$

f. $\begin{bmatrix} 4 & 11 \\ 7 & 3 \\ 4 & 2 \end{bmatrix} + \begin{bmatrix} 3 & -2 & 7 \\ 5 & 0 & 2 \end{bmatrix}$

4. Find matrix $[B]$ such that

$$\begin{bmatrix} 8 & -5 & 4.5 \\ -6 & 9.5 & 5 \end{bmatrix} - [B] = \begin{bmatrix} 5 & -1 & 2 \\ -4 & 3.5 & 1 \end{bmatrix}$$

Reason and Apply

5. This matrix represents a triangle.

$$\begin{bmatrix} -3 & 1 & 2 \\ 2 & 3 & -2 \end{bmatrix}$$

a. Graph the triangle.

b. Find the answer to this matrix multiplication.

$$\begin{bmatrix} -1 & 0 \\ 0 & 1 \end{bmatrix} \cdot \begin{bmatrix} -3 & 1 & 2 \\ 2 & 3 & -2 \end{bmatrix}$$

c. Graph the image represented by the matrix in part 5b.

d. Describe how the original triangle changed position as a result of the matrix multiplication in 5b.

6. Find matrix $[A]$ and matrix $[C]$ such that the triangle represented by $\begin{bmatrix} -3 & 1 & 2 \\ 2 & 3 & -2 \end{bmatrix}$ is reflected (flipped) across the x-axis.

$$\begin{bmatrix} a_{1,1} & a_{1,2} \\ a_{2,1} & a_{2,2} \end{bmatrix} \cdot \begin{bmatrix} -3 & 1 & 2 \\ 2 & 3 & -2 \end{bmatrix} = \begin{bmatrix} c_{1,1} & c_{1,2} & c_{1,3} \\ c_{2,1} & c_{2,2} & c_{2,3} \end{bmatrix}$$

7. Of two-car families in a small city, 88% remain two-car families in the following year, and 12% become one-car families in the following year. Of one-car families, 72% remain one-car families, and 28% become two-car families. Suppose these trends continue for a few years. Presently 4,800 families have one car, and 4,200 have two cars.

a. Draw a transition diagram that displays this information.

b. What matrix represents the present situation? Let $a_{1,1}$ represent one-car families.

c. Write a transition matrix that represents the same information as your transition diagram.

d. Write a matrix equation to find the numbers of one-car and two-car families one year from now.

e. Find the numbers of one-car and two-car families two years from now.

8. ***Mini-Investigation*** Enter these matrices into your calculator.

$$[A] = \begin{bmatrix} 2 & 3 \\ -1 & 1 \end{bmatrix} \quad [B] = \begin{bmatrix} 3 & 4 \\ 0 & -2 \end{bmatrix} \quad [C] = \begin{bmatrix} -2 & 3 & 0 \\ -1 & 5 & 4 \end{bmatrix} \quad [D] = \begin{bmatrix} 1 & 0 \\ 0 & 1 \end{bmatrix}$$

a. Find $[A] \cdot [B]$ and $[B] \cdot [A]$. Are they the same?

b. Find $[A] \cdot [C]$ and $[C] \cdot [A]$. Are they the same? What do you notice?

c. Find $[A] \cdot [D]$ and $[D] \cdot [A]$. Are they the same? Can you explain why?

d. Is matrix multiplication commutative? That is, does order matter?

9. Find the missing values.

a. $\begin{bmatrix} 2 & a \\ b & 1 \end{bmatrix} \cdot \begin{bmatrix} 5 \\ 3 \end{bmatrix} = [19 \quad 17]$ **b.** $\begin{bmatrix} a & -2 \\ 3 & 1 \end{bmatrix} \cdot \begin{bmatrix} -3 \\ b \end{bmatrix} = [-29 \quad -5]$

10. Recall the frozen mocha drink problem in Example B. Enter these matrices into your calculator, and use them to find the long-run values for the numbers of customers who choose dark chocolate or white chocolate. Explain why your answer makes sense.

$$[A] = [220 \quad 20] \quad [B] = \begin{bmatrix} .95 & .05 \\ .10 & -.95 \end{bmatrix}$$

11. In the study of genetics, the terms *dominant* and *recessive* are used to denote how children inherit traits, such as eye color or left-handedness, from their parents. Suppose that for a given trait, 95% of the population has the recessive characteristic, and 5% of the population has the dominant characteristic. In this population if an individual possesses the recessive characteristic, then 98% of her offspring can be expected to show the recessive trait, and 2% of her offspring will show the dominant trait. If an individual possesses the dominant trait, then 45% of her offspring will show the recessive trait, and 55% of her offspring will show the dominant trait.

a. Assume that the population size stays completely stable; each new generation has exactly the same population as the previous generation. Write a transition matrix to represent the genetic trait information given above.

b. Use matrix multiplication to determine what percentage of the next generation will possess the recessive trait.

c. Use repeated matrix multiplication to explore how the percentage of the population that possesses the recessive trait will change in subsequent generations.

d. Does it make sense in this situation to use repeated matrix multiplication to determine how the population's traits will change over time? Explain why or why not.

12. **APPLICATION** A researcher studies the birth weights of women and their daughters. The weights were each assigned to one of three categories: low (under 6 lb), average (between 6 and 8 lb), and high (over 8 lb). Sometimes, a very high infant birth weight can indicate a medical problem in the mother or child. However, in general, high birth weight is considered healthiest. This transition diagram shows how birth weights changed from mother to daughter.

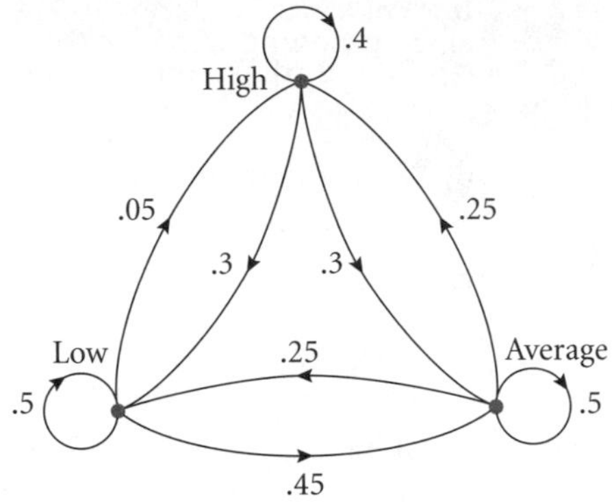

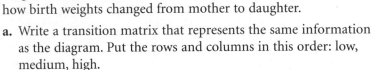

a. Write a transition matrix that represents the same information as the diagram. Put the rows and columns in this order: low, medium, high.

b. In the initial generation of women, 25% had birth weights in the low category, 60% in the average category, and 15% in the high category. Use repeated matrix multiplication on your calculator to determine the percentages after one generation, after two generations, after three generations, and in the long run.

13. A spider is in a building that has three rooms. Each room has two doors, each of the two doors connects to each of the other rooms. The spider moves from room to room by choosing a door at random. If the spider starts in Room 1, what is the probability that it will be in Room 1 again after four room changes? (Hint: As the spider starts in Room 1, use $[1 \quad 0 \quad 0]$ as the initial matrix.) What happens to the probabilities in the long run?

ConsumerConnection Universal Product Codes (UPCs) are used widely to identify products. You can find the UPC on almost every mass-produced product. The UPC consists of a sequence of vertical lines with a 12-digit number beneath it. When scanned, the lines are read by the scanning device as the series of numbers. The first 6 digits represent the manufacturer, and the next 5 represent the product. The last digit is a check digit, so that when the item is scanned, the computer can verify the correctness of the item number and search its database to get the price. To verify the code, each digit in an odd position is multiplied by 3. These products are then summed with the digits in the even positions. The sum should be a multiple of 10. The check digit is chosen such that this sum is indeed divisible by 10.

14. **APPLICATION** Read the Consumer Connection about UPCs and answer these questions.

a. Write a 12×1 matrix that can be multiplied by a UPC to find the sum of each digit in an even position and 3 times each digit in an odd position.

b. Use the matrix from 14a to check these four UPCs. Which one(s) are valid?

$$\begin{bmatrix} 0 & 3 & 6 & 2 & 0 & 0 & 0 & 0 & 4 & 0 & 0 & 5 \\ 0 & 7 & 6 & 1 & 0 & 7 & 0 & 2 & 2 & 3 & 3 & 6 \\ 0 & 7 & 4 & 2 & 2 & 0 & 0 & 0 & 2 & 9 & 1 & 8 \\ 0 & 8 & 5 & 3 & 9 & 1 & 7 & 8 & 6 & 2 & 2 & 1 \end{bmatrix}$$

c. For the invalid UPC(s), what should the check digit be to result in a valid code?

Review

15. ***Mini-Investigation*** (Lessons 3.4, 3.5) A system of equations that has no solutions is referred to as **inconsistent,** and a system with infinite solutions is referred to as **dependent.** Follow these steps to make some discoveries about inconsistent and dependent systems.

a. By graphing, determine which of the following systems of linear equations are inconsistent and which are dependent.

i. $\begin{cases} y = 0.7x + 8 \\ y = 1.1x - 7 \end{cases}$ **ii.** $\begin{cases} y = \frac{3}{4}x - 4 \\ y = 0.75x + 3 \end{cases}$ **iii.** $\begin{cases} 4x + 6y = 9 \\ 1.2x + 1.8y = 4.7 \end{cases}$

iv. $\begin{cases} \frac{3}{4}x - \frac{1}{2}y = 4 \\ 0.75x + 0.5y = 3 \end{cases}$ **v.** $\begin{cases} y = 1.2x + 3 \\ y = 1.2x - 1 \end{cases}$ **vi.** $\begin{cases} y = \frac{1}{4}(2x - 1) \\ y = 0.5x - 0.25 \end{cases}$

vii. $\begin{cases} 4x + 6y = 9 \\ 1.2x + 1.8y = 2.7 \end{cases}$ **viii.** $\begin{cases} \frac{3}{5}x + \frac{2}{5}y = 3 \\ 0.6x + 0.4y = 3 \end{cases}$

b. Describe the graphs of the equations of the inconsistent systems.

c. Try to solve each inconsistent system either by substitution or by elimination. Show your steps. Describe the outcomes of these solutions.

d. How can you recognize an inconsistent linear system without graphing it?

e. Describe the graphs of equations of dependent systems.

f. Solve each dependent system by substitution or elimination. Show your steps. Describe the outcome of each solution.

g. How can you recognize a dependent linear system without graphing it?

16. (Lesson 5.6) If $\log_p x = a$ and $\log_p y = b$, find

a. $\log_p xy$ **b.** $\log_p x^3$ **c.** $\log_p \frac{y^2}{x}$

d. $\log_{p^2} y$ **e.** $\log_p \sqrt{x}$ **f.** $\log_m xy$

17. (Lesson 3.6) Solve this system of equations for x, y, and z.

$$\begin{cases} x + 2y + z = 0 \\ 3x - 4y + 5z = -11 \\ -2x - 8y - 3z = 1 \end{cases}$$

LESSON

7.3

The Row Reduction Method

Rather than denying problems, focus inventively, intentionally on what solutions might look or feel like...

MARSHA SINETAR

In Chapter 3 you learned how to solve systems of equations using elimination. You added equations or multiples of equations to reduce an equation to a single variable. In this lesson you will learn how to use matrices to simplify this elimination solution to systems of equations.

Any system of equations in standard form can be written as a matrix equation. For example,

$$\begin{cases} 2x + y = 5 \\ 5x + 3y = 13 \end{cases} \qquad \text{The original system.}$$

$$\begin{bmatrix} 2x + y \\ 5x + 3y \end{bmatrix} = \begin{bmatrix} 5 \\ 13 \end{bmatrix} \qquad \text{Rewrite with matrices.}$$

$$\begin{bmatrix} 2 & 1 \\ 5 & 3 \end{bmatrix}\begin{bmatrix} x \\ y \end{bmatrix} = \begin{bmatrix} 5 \\ 13 \end{bmatrix} \qquad \text{The product } \begin{bmatrix} 2 & 1 \\ 5 & 3 \end{bmatrix}\begin{bmatrix} x \\ y \end{bmatrix} \text{ is equivalent to } \begin{bmatrix} 2x + y \\ 5x + 3y \end{bmatrix}.$$

You can also write the system as an **augmented matrix,** which is a single matrix that contains columns for each variable and a final column for the constant terms.

$$\begin{matrix} 2x + y = 5 \\ 5x + 3y = 13 \end{matrix} \Rightarrow \left[\begin{array}{cc|c} 2 & 1 & 5 \\ 5 & 3 & 13 \end{array}\right]$$

You can use the augmented matrix to carry out a process that is similar to elimination but does not require you to rewrite each entire equation at each step.

The **row reduction method** transforms an augmented matrix into a solution matrix. Instead of combining equations and multiples of equations until one variable remains, you add multiples of rows to other rows until you obtain the solution matrix. A solution matrix contains the solution to the system in the last column. The rest of the matrix consists of 1's along the main diagonal and 0's above and below it.

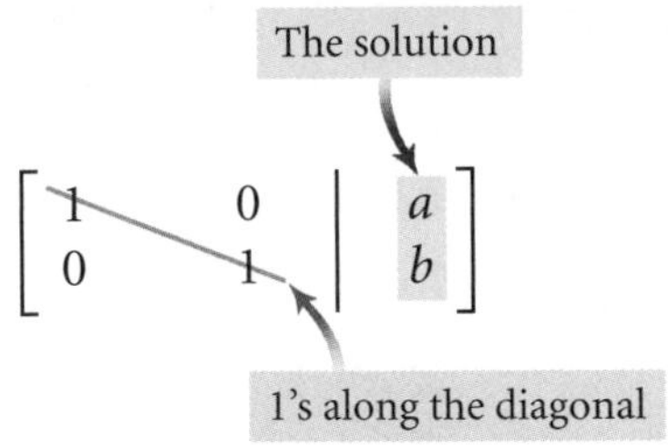

This matrix is in **reduced row-echelon form** because each row is reduced to a 1 and a solution, and the rest of the matrix entries are 0's. The 1's are in echelon, or step, formation. The ordered pair (a, b) is the solution to the system.

Because an augmented matrix represents a system of equations, the same rules apply to row operations in a matrix that apply to equations in a system of equations.

Row Operations in a Matrix

- ▶ You can multiply (or divide) all numbers in a row by a nonzero number.
- ▶ You can add a multiple of the numbers in one row to the corresponding numbers in another row.
- ▶ You can exchange two rows.

EXAMPLE A

Solve this system of equations.

$$\begin{cases} 2x + y = 5 \\ 5x + 3y = 13 \end{cases}$$

▶ Solution

You can solve the system using matrices or equations. Let's compare the row-reduction method using matrices with the elimination method using equations.

Because the equations are in standard form, you can copy the coefficients and constants from each equation into corresponding rows of the augmented matrix.

$$\begin{cases} 2x + y = 5 \\ 5x + 3y = 13 \end{cases} \Rightarrow \left[\begin{array}{cc|c} 2 & 1 & 5 \\ 5 & 3 & 13 \end{array}\right]$$

Let's call this augmented matrix $[M]$. Using only the elementary row operations, you can transform this matrix into the solution matrix. Both $m_{2,1}$ and $m_{1,2}$ must be 0, and both $m_{1,1}$ and $m_{2,2}$ must be 1.

Add -2.5 times row 1 to row 2 to get 0 for $m_{2,1}$.

$$\begin{bmatrix} 2 & 1 & 5 \\ 0 & 0.5 & 0.5 \end{bmatrix}$$

Multiply Equation 1 by -2.5 and add to row 2 to eliminate x.

$$\begin{array}{rcl} -5x - 2.5y &=& -2.5 \\ 5x + 3y &=& 13 \\ \hline 0.5y &=& 0.5 \end{array}$$

Multiply row 2 by 2 to change $m_{2,2}$ to 1.

$$\begin{bmatrix} 2 & 1 & 5 \\ 0 & 1 & 1 \end{bmatrix}$$

Multiply the equation by 2 to find y.

$$y = 1$$

Subtract row 2 from row 1 to get a 0 for $m_{1,2}$.

$$\begin{bmatrix} 2 & 0 & 4 \\ 0 & 1 & 1 \end{bmatrix}$$

Multiply -1 by new equation, and add it to the first equation to eliminate y.

$$\begin{array}{rcl} 2x + y &=& 5 \\ -y &=& -1 \\ \hline 2x &=& 4 \end{array}$$

Multiply row 1 by 0.5.

$$\begin{bmatrix} 1 & 0 & 2 \\ 0 & 1 & 1 \end{bmatrix}$$

Multiply the equation by 0.5 to find x.

$$x = 2$$

The last column of the solutin matrix indicates that the solution to the system is (2, 1).

You can represent row operations symbolically. For example, you can use R_1 and R_2 to represent the two rows of the matrix in Example A and show the steps this way:

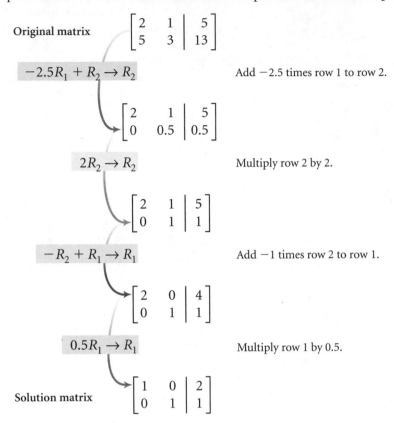

Original matrix $\left[\begin{array}{cc|c} 2 & 1 & 5 \\ 5 & 3 & 13 \end{array}\right]$

$-2.5R_1 + R_2 \rightarrow R_2$ Add -2.5 times row 1 to row 2.

$$\left[\begin{array}{cc|c} 2 & 1 & 5 \\ 0 & 0.5 & 0.5 \end{array}\right]$$

$2R_2 \rightarrow R_2$ Multiply row 2 by 2.

$$\left[\begin{array}{cc|c} 2 & 1 & 5 \\ 0 & 1 & 1 \end{array}\right]$$

$-R_2 + R_1 \rightarrow R_1$ Add -1 times row 2 to row 1.

$$\left[\begin{array}{cc|c} 2 & 0 & 4 \\ 0 & 1 & 1 \end{array}\right]$$

$0.5R_1 \rightarrow R_1$ Multiply row 1 by 0.5.

Solution matrix $\left[\begin{array}{cc|c} 1 & 0 & 2 \\ 0 & 1 & 1 \end{array}\right]$

Fortunately, technology can be a big help in performing row reduction operations. [▶ See **Calculator Note 7D** to learn how to use your calculator to find the reduced row-echelon form of a matrix.◀]

Investigation
A Martian Experiment

In the far distant future a team of astronauts visits Mars. They decide to perform a primitive experiment to determine how long it takes a dropped object to hit the Martian surface. The first astronaut drops a ball from various heights, using their space scooter. The second astronaut measures how many seconds it takes for the objects to hit the ground. Their data are shown in this table. In this Investigation you will find a function that describes the distance the ball will have fallen after a given number of seconds.

Number of seconds	0	1.8	2.2	5.2	8.7	9.3	10
Height (in meters)	0	6	9	50	140	160	185

Step 1 Make a scatter plot of these data. Let x represent the number of seconds, and let y represent the distance in meters. Describe the graph. Is it linear?

Step 2 Based on the shape of the graph, the equation looks like it may be quadratic. You can write a quadratic equation in the form $y = ax^2 + bx + c$. You can use each pair of values in the table to write an equation. Create three equations with the variables a, b, and c by substituting any three pairs of coordinates from the table. For example, the point (1.8, 6) creates the equation

$$a(1.8)^2 + b(1.8) + c = 6, \text{ or } 3.24a + 1.8b + c = 6$$

Step 3 You can now solve for the coefficients a, b, and c. Write a 3×4 augmented matrix for your system of three equations.

Step 4 Use your calculator's reduced row-echelon form function to create a matrix with 1's along the main diagonal. Your matrix should now be in the form

$$\begin{bmatrix} 1 & 0 & 0 & k_1 \\ 0 & 1 & 0 & k_2 \\ 0 & 0 & 1 & k_3 \end{bmatrix}.$$

What does this indicate about your equation $y = ax^2 + bx + c$?

Step 5 Use the other data points to verify that you have the correct values of a, b, and c.

Step 6 Complete the following table of values showing how far objects fall when dropped on Mars.

Seconds	0	1	5	10	50	100
Distance (meters)						

Here's another example of the solution of a large system of equations with the help of matrices.

EXAMPLE B

The Philosophy Club treasurer is totaling the sales and receipts from the last book sale. She has 50 receipts for sales of 3 different titles of books priced at \$14.00, \$18.50, and \$23.25. She has a total of \$909.00 and knows that 22 more of the \$18.50 books were sold than the \$23.25 books. How many copies of each book were sold?

▶ Solution

The number sold of the three different book titles is unknown, so you can assign three variables.

x = the number of \$14.00 books

y = the number of \$18.50 books

z = the number of \$23.25 books

Based on the information in the problem, write a system of three linear equations. The system can also be written as an augmented matrix.

$$\begin{aligned} x + y + z &= 50 \\ 14x + 18.50y + 23.25z &= 909 \\ y - z &= 22 \end{aligned} \quad \text{or} \quad \left[\begin{array}{ccc|c} 1 & 1 & 1 & 50 \\ 14 & 18.5 & 23.25 & 909 \\ 0 & 1 & -1 & 22 \end{array}\right].$$

The reduced row-echelon form function on your calculator [► See **Calculator Note 7E.** ◄] produces:

$$\begin{bmatrix} 1 & 0 & 0 & 12 \\ 0 & 1 & 0 & 30 \\ 0 & 0 & 1 & 8 \end{bmatrix}$$

Twelve \$14.00 books, 30 \$18.50 books, and 8 \$23.25 books were sold.

Some systems of equations have no solution or infinitely many solutions. Likewise, not all augmented matrices can be reduced to row-echelon form. An entire row of 0's means that one equation was a multiple of another, therefore not enough information was given, and infinitely many solutions are possible. If, on the other hand, an entire row reduces to 0's, except for a constant in the last entry, no solution exists, because a set of 0 coefficients cannot result in a constant on the right side of the equation.

EXERCISES

Practice Your Skills

1. Write a system of equations for each augmented matrix given.

a. $\left[\begin{array}{cc|c} 2 & 5 & 8 \\ 4 & -1 & 6 \end{array}\right]$ **b.** $\left[\begin{array}{ccc|c} 1 & -1 & 2 & 3 \\ 1 & 2 & -3 & 1 \\ 2 & 1 & -1 & 2 \end{array}\right]$

2. Write an augmented matrix for each system given.

a. $\begin{cases} x + 2y - z = 1 \\ 2x - y + 3z = 2 \\ 2x + y + z = -1 \end{cases}$

b. $\begin{cases} 2x + y - z = 12 \\ 2x + z = 4 \\ 2x - y + 3z = -4 \end{cases}$

3. Use this matrix to perform the row operations.

$$\left[\begin{array}{ccc|c} 1 & -1 & 2 & 3 \\ 1 & 2 & -3 & 1 \\ 2 & 1 & -1 & 2 \end{array}\right]$$

a. $-R_1 + R_2 \rightarrow R_2$

b. $-2R_1 + R_3 \rightarrow R_3$

4. Give the row operation or matrix missing from the table.

Description	Matrix
a. The original system. $\begin{cases} 2x + 5y = 8 \\ 4x + y = 6 \end{cases}$	$\left[\begin{array}{cc\|c} & & \\ & & \end{array}\right]$
b. $-2R_1 + R_2 \rightarrow R_2$	$\left[\begin{array}{cc\|c} & & \\ & & \end{array}\right]$
c.	$\left[\begin{array}{cc\|c} 2 & 5 & 8 \\ 0 & 1 & \frac{10}{11} \end{array}\right]$
d.	$\left[\begin{array}{cc\|c} 2 & 0 & \frac{38}{11} \\ 0 & 1 & \frac{10}{11} \end{array}\right]$
e.	$\left[\begin{array}{cc\|c} 1 & 0 & \frac{19}{11} \\ 0 & 1 & \frac{10}{11} \end{array}\right]$

Reason and Apply

5. Rewrite each system of equations as an augmented matrix. Transform the matrix into its reduced row-echelon form using row operations on your calculator.

a. $\begin{cases} x + 2y + 3z = 5 \\ 2x + 3y + 2z = 2 \\ -x - 2y - 4z = -1 \end{cases}$

b. $\begin{cases} -x + 3y + z = 4 \\ 2z = x + y \\ 2.2y + 2.2z = 2.2 \end{cases}$

c. $\begin{cases} 3x - y + z = 7 \\ x - 2y + 5z = 1 \\ 6x - 2y + 2z = 14 \end{cases}$

d. $\begin{cases} 3x - y + z = 5 \\ x - 2y + 5z = 1 \\ 6x - 2y + 2z = 14 \end{cases}$

6. A farmer raises only goats and chickens on his farm. Altogether he has 47 animals, and they have a total of 118 legs. Write a system of equations and an augmented matrix. How many of each animal does he have?

7. The largest angle of a triangle is 4° more than twice the smallest angle. The smallest angle is 24° less than the other angle. There are 180° in a triangle. What are the measures of the three angles?

8. Find a, b, and c such that the graph of $y = ax^2 + bx + c$ passes through the points $(1, 3)$, $(4, 24)$ and $(-2, 18)$.

9. The campus newspaper staff sells ads in three sizes. The full-page ads sell for $200, the half-page ads sell for $125, and the business–card-size ads sell for $20. This week the staff earned $1,715 from 22 ads. There were four times as many business–card-size ads sold as full-page ads. How many of each ad type did they sell?

10. **APPLICATION** The amount of merchandise that is available for sale is called supply. The amount of merchandise that consumers are prepared to buy is called demand. Supply and demand are in equilibrium, or balance, when a price is found at which supply and demand are equal.

Suppose the following data represent supply and demand for a bottled water.

Supply of Sparkling Rivers

Price (cents/gal.)	Quantity supplied (millions of gals.)
80	1304.4
90	2894
100	4483.6
110	6073.2
120	7662.8

Demand for Sparkling Rivers

Price (cents/gal.)	Quantity demanded (millions of gals.)
80	3268.47
90	2724.87
100	2181.27
110	1637.67
120	1094.07

a. Find linear models for the supply and demand.

b. Find the equilibrium point graphically.

c. Write the supply and demand equations from 10a as a system in an augmented matrix. Use row reduction to verify your answer to 10b.

EconomicsConnection Supply and demand are affected by many things. The supply may be affected by the price of the merchandise, the cost of making it, a competitive product, or unexpected events, like drought or hurricanes. Demand may be affected by the price, the income level of the consumer, and consumer tastes. An extra tax will cause a price increase and thereby lower consumer demand. An increased price may slow the purchase of the product and thus also increase supply. The stock market dramatically illustrates how prices are determined through the interaction of supply and demand in an auction-like environment.

Grocery store shelves empty as residents stock up in anticipation of a severe snowstorm. Events like this one can cause high demand and deplete supply.

Review

11. **APPLICATION** (Lesson 3.1) The Fancy Feat Dance Company has two choices of payment for their next series of performances. The first option is to receive \$12,500 for the series plus 5% of all ticket sales. The second option is \$6,800 for the series plus 15% of ticket sales. The company will perform three consecutive nights in a hall that seats 2,200 people. Tickets will cost \$12 a piece.

a. How much will the dance company receive under each plan if a total of 3,500 tickets are sold for all three performances?

b. Write an equation that gives the amount the company will receive under the first plan for any number of tickets sold.

c. Write an equation that gives the amount the company will receive under the second plan for any number of tickets sold.

d. How many tickets must be sold for the second plan to be the better choice?

e. Which plan should the company choose? Justify your choice.

12. (Lesson 3.4) Consider this graph of a system of two linear equations.

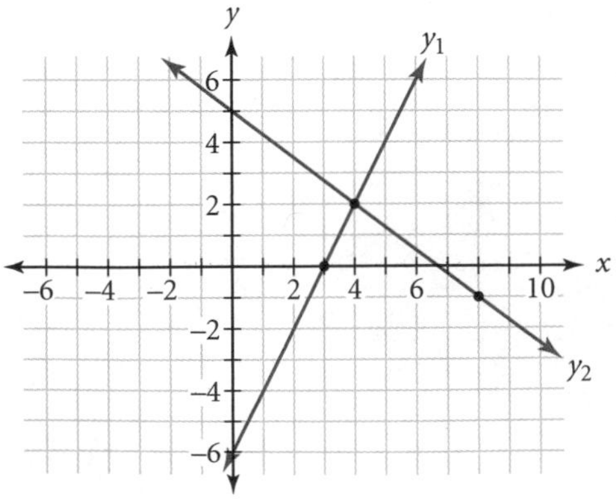

a. What is the solution to this system?

b. Write equations of the two lines.

13. (Lesson 3.1) For each segment shown in this pentagon, write an equation of the line that contains the segment. Check your equations by graphing them on your calculator.

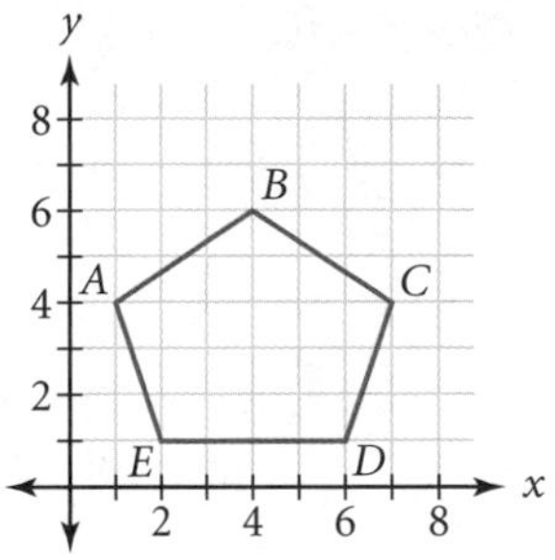

14. (Lesson 7.1) Consider the graph of ΔABC.

a. Represent ΔABC with a matrix $[M]$.

b. Find each product, and graph the image of the triangle represented by the result.

i. $\begin{bmatrix} 0 & 1 \\ 1 & 0 \end{bmatrix}[M]$ **ii.** $\begin{bmatrix} 0 & -1 \\ 1 & 0 \end{bmatrix}[M]$

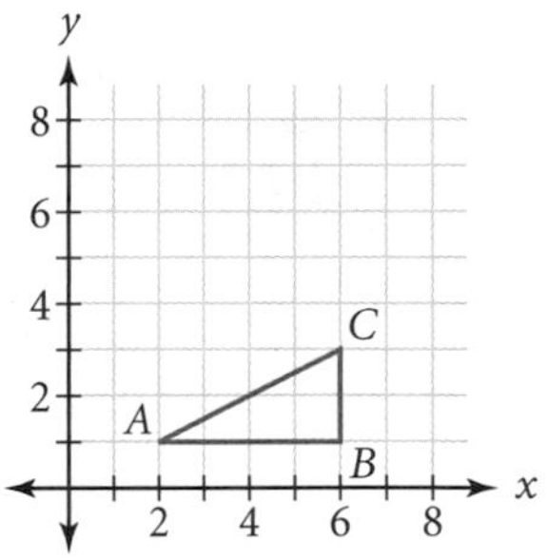

IMPROVING YOUR VISUAL THINKING SKILLS

Intersection of Planes

A system of two linear equations in two variables can be represented graphically by two lines. If the lines intersect, the point of intersection is the solution, and the system is referred to as *consistent.* If the lines are parallel, they never intersect, there is no solution, and the system is *inconsistent.* If the lines are the same, there are infinitely many solutions, and the system is referred to as *dependent.*

An equation such as $3x + 2y + 6z = 12$ is also called a *linear equation* because the highest power of any variable is 1. Because there are three variables, however, the graph is in three dimensions. There are three axes on the graph, representing x, y, and z. The graph of this equation is a plane.

A system of three linear equations in three variables can be represented graphically by three planes. Sketch all the possible outcomes for the graphs of three planes. Classify each outcome as consistent, inconsistent, or dependent.

LESSON

7.4

Solving Systems with Inverse Matrices

Things that oppose each other also complement each other.

CHINESE SAYING

Consider the equation $ax = b$. To solve for x, multiply both sides of the equation by $\frac{1}{a}$, the **multiplicative inverse** of a. The multiplicative inverse of a nonzero number such as 2.25 is the number that you can multiply by 2.25 to get 1. Also, the number 1 is the **multiplicative identity** because any number multiplied by 1 remains unchanged.

Similarly, to solve a system by using matrices you can use an **inverse matrix.** If an inverse matrix exists, then when you multiply it by your matrix you will get the matrix equivalent of 1, which is called the **identity matrix.** Any matrix multiplied by the identity matrix remains unchanged, just as any number multiplied by 1 remains unchanged. In Example A we will use this multiplicative identity to find a 2×2 identity matrix.

EXAMPLE A Find an identity matrix for $\begin{bmatrix} 2 & 1 \\ 4 & 3 \end{bmatrix}$.

▶ Solution

You want to find a matrix $\begin{bmatrix} a & b \\ c & d \end{bmatrix}$ that satisfies the definition of the identity matrix.

$$\begin{bmatrix} 2 & 1 \\ 4 & 3 \end{bmatrix}\begin{bmatrix} a & b \\ c & d \end{bmatrix} = \begin{bmatrix} 2 & 1 \\ 4 & 3 \end{bmatrix}$$ Multiplying by an identity matrix leaves the matrix unchanged.

$$\begin{bmatrix} 2a + c & 2b + d \\ 4a + 3c & 4b + 3d \end{bmatrix} = \begin{bmatrix} 2 & 1 \\ 4 & 3 \end{bmatrix}$$ Multiply the left side.

Because the two matrices are equal, their entries must be equal. Setting corresponding entries equal produces these equations.

$2a + c = 2$ $2b + d = 1$ $4a + 3c = 4$ $4b + 3d = 3$

You can treat these as two systems of equations. Use either substitution or elimination to solve each system.

$$\begin{cases} 2a + c = 2 \\ 4a + 3c = 4 \end{cases}$$ A system that can be solved for a and c.

$-6a - 3c = -6$ Multiply the first equation by -3.

$4a + 3c = 4$

$-2a = -2$ Add the equations to eliminate c.

$a = 1$ Solve for a.

$2(1) + c = 2$ Substitute 1 for a in the first equation to find c.

$c = 0$ Solve for c.

This system gives $a = 1$ and $c = 0$. You can use a similar procedure to find that $b = 0$ and $d = 1$.

The 2×2 identity matrix is

$$\begin{bmatrix} 1 & 0 \\ 0 & 1 \end{bmatrix}$$

Can you see why multiplying this matrix by any 2×2 matrix results in the same 2×2 matrix?

The identity matrix in Example A is the identity matrix for all 2×2 matrices. There are corresponding identity matrices for larger matrices. Will an identity matrix always be square? Why or why not?

The Identity Matrix

An *identity matrix*, symbolized by $[I]$, is a matrix that does not alter the entries of a matrix $[A]$ under multiplication.

$$[A]\,[I] = [A] \text{ and } [I]\,[A] = [A]$$

Matrix $[I]$ must have the same dimensions as matrix $[A]$ and has entries of 1's along the main diagonal (from top left to bottom right) and 0's in all other entries.

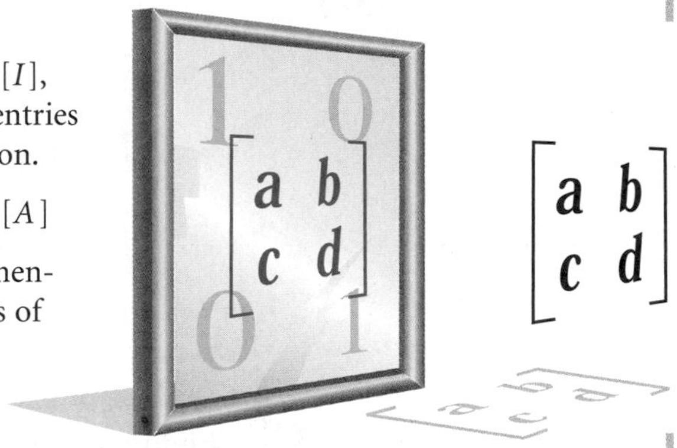

Now that you know the identity matrix for a 2×2 matrix, you can look for a way to find the inverse of a 2×2 matrix.

The Inverse Matrix

The *inverse matrix* of $[A]$, symbolized by $[A]^{-1}$, is the matrix that will produce an identity matrix when multiplied by $[A]$.

$$[A][A]^{-1} = [I] \text{ and } [A]^{-1}[A] = [I]$$

Investigation
The Inverse Matrix

In this Investigation you will learn ways to find the inverse of a 2×2 matrix, based on the definitions of the identity and inverse matrices. When we multiply a matrix by its inverse, the product is the identity matrix.

Step 1 Use the definition of inverse matrix to set up a matrix equation. Use these matrices and the 2×2 identity matrix for $[I]$.

$$[A] = \begin{bmatrix} 2 & 1 \\ 4 & 3 \end{bmatrix} \qquad [A]^{-1} = \begin{bmatrix} a & b \\ c & d \end{bmatrix}$$

Step 2 Use matrix multiplication to find the product of $[A]\,[A]^{-1}$. Set that product equal to matrix $[I]$.

Step 3 Use the matrix equation from Step 2 to write equations that you can solve to find values for a, b, c, and d. Solve systems to find the values in the inverse matrix.

Step 4 Use your calculator to find $[A]^{-1}$. If this answer does not match your answer to Step 3, check your work for mistakes. [▶ See **Calculator Note 7E** to learn how to find the inverse on your calculator.◀]

Step 5 Find the products $[A]\,[A]^{-1}$ and $[A]^{-1}\,[A]$. Do they both give you $\begin{bmatrix} 1 & 0 \\ 0 & 1 \end{bmatrix}$? Is matrix multiplication always commutative?

Step 6 Not every square matrix has an inverse. Find the inverse of each of these matrices.

a. $\begin{bmatrix} 2 & 1 \\ 4 & 2 \end{bmatrix}$ **b.** $\begin{bmatrix} 50 & -75 \\ 10 & -15 \end{bmatrix}$ **c.** $\begin{bmatrix} 10.5 & 1 \\ 31.5 & 3 \end{bmatrix}$ **d.** $\begin{bmatrix} 2 & 1 \\ 2 & 1 \end{bmatrix}$

Step 7 Can a nonsquare matrix have an inverse? Why or why not?

You now know how to find the inverse of a square matrix both by hand and on your calculator. You can use an inverse matrix to solve a system of equations quickly.

Solving a System Using the Inverse Matrix

A system of equations in standard form can be written in matrix form as $[A]\,[X] = [B]$, where $[A]$ is the coefficient matrix, $[X]$ is the variable matrix, and $[B]$ is the constant matrix. Multiplying both sides by the inverse matrix $[A]^{-1}$ gives the values of variables in matrix $[X]$, which is the solution to the system.

$[A]\,[X] = [B]$	The system in matrix form.
$[A]^{-1}[A]\,[X] = [A]^{-1}[B]$	Multiply both sides by the inverse.
$[I]\,[X] = [A]^{-1}[B]$	By the definition of inverse, $[A]^{-1}\,[A] = [I]$.
$[X] = [A]^{-1}[B]$	By the definition of identity, $[I][X] = [X]$.

EXAMPLE B Solve this system using an inverse matrix.

$$\begin{cases} 2x + 3y = 7 \\ x = 6 - 4y \end{cases}$$

▶ ***Solution*** First, rewrite the second equation in standard form.

$$\begin{cases} 2x + 3y = 7 \\ x + 4y = 6 \end{cases}$$

The matrix equation for this system is

$$\begin{bmatrix} 2 & 3 \\ 1 & 4 \end{bmatrix}\begin{bmatrix} x \\ y \end{bmatrix} = \begin{bmatrix} 7 \\ 6 \end{bmatrix}$$

The variables are x and y, so the variable matrix $[X]$ is $\begin{bmatrix} x \\ y \end{bmatrix}$. The coefficient matrix $[A]$ is $\begin{bmatrix} 2 & 3 \\ 1 & 4 \end{bmatrix}$, and the constant matrix $[B]$ is $\begin{bmatrix} 7 \\ 6 \end{bmatrix}$.

Use your calculator to find the inverse of $[A]$.

$$[A]^{-1} = \begin{bmatrix} 0.8 & -0.6 \\ -0.2 & 0.4 \end{bmatrix}$$

Multiply both sides of the equation by this inverse to find the solution to the system of equations.

$\begin{bmatrix} 2 & 3 \\ 1 & 4 \end{bmatrix}\begin{bmatrix} x \\ y \end{bmatrix} = \begin{bmatrix} 7 \\ 6 \end{bmatrix}$	$[A][X] = [B]$	The system in matrix form.
$\begin{bmatrix} 0.8 & -0.6 \\ -0.2 & 0.4 \end{bmatrix}\begin{bmatrix} 2 & 3 \\ 1 & 4 \end{bmatrix}\begin{bmatrix} x \\ y \end{bmatrix} = \begin{bmatrix} 0.8 & -0.6 \\ -0.2 & 0.4 \end{bmatrix}\begin{bmatrix} 7 \\ 6 \end{bmatrix}$	$[A]^{-1}[A][X] = [A]^{-1}[B]$	Multiply both sides by the inverse.
$\begin{bmatrix} 1 & 0 \\ 0 & 1 \end{bmatrix}\begin{bmatrix} x \\ y \end{bmatrix} = \begin{bmatrix} 0.8 & -0.6 \\ -0.2 & 0.4 \end{bmatrix}\begin{bmatrix} 7 \\ 6 \end{bmatrix}$	$[I][X] = [A]^{-1}[B]$	By the definition of inverse, $[A]^{-1}[A] = [I]$.
$\begin{bmatrix} x \\ y \end{bmatrix} = \begin{bmatrix} 0.8 & -0.6 \\ -0.2 & 0.4 \end{bmatrix}\begin{bmatrix} 7 \\ 6 \end{bmatrix}$	$[X] = [A]^{-1}[B]$	By the definition of identity, $[I][X] = [X]$
$\begin{bmatrix} x \\ y \end{bmatrix} = \begin{bmatrix} 2 \\ 1 \end{bmatrix}$		Matrix multiplication.

The solution to the system is (2, 1). Substitute the values in the original equations to check the solution.

$2x + 3y = 7$	$x = 6 - 4y$
$2(2) + 3(1) = 7$	$2 = 6 - 4(1)$
$4 + 3 = 7$	$2 = 6 - 4$
$7 = 7$	$2 = 2$

You can also solve larger systems of equations using an inverse matrix. First, decide what quantities are unknown, and write equations using the information given in the problem. Then, rewrite the system of equations as a matrix equation, and use either row reduction or an inverse matrix to solve the system. Even several equations with many unknowns can be solved quickly this way.

EXAMPLE C

On a recent trip to the grocery store, Duane, Marsha, and Parker purchased three food items. Duane bought two cans of soup, one frozen dinner, and two jars of peanut butter for a total of \$11.85. Marsha spent \$9.00 on a can of soup, two frozen dinners, and a jar of peanut butter. Parker spent \$12.35 on two frozen dinners and three jars of peanut butter but no soup. What was the price of each item?

▶ Solution

The prices of the items are the unknowns. Let c represent the price of a can of soup in dollars, let d represent the price of a frozen dinner in dollars, and let p represent the price of a jar of peanut butter in dollars. This system represents the three friends' purchases.

$$\begin{cases} 2c + 1d + 2p = 11.85 \\ 1c + 2d + 1p = 9.00 \\ 0c + 2d + 3p = 12.35 \end{cases}$$

Translate these equations into a matrix equation in the form $[A]\,[X] = [B]$.

$$\begin{bmatrix} 2 & 1 & 2 \\ 1 & 2 & 1 \\ 0 & 2 & 3 \end{bmatrix} \begin{bmatrix} c \\ d \\ p \end{bmatrix} = \begin{bmatrix} 11.85 \\ 9.00 \\ 12.35 \end{bmatrix}$$

The solution to the system is simply the product $[A]^{-1}[B]$.

$$[X] = [A]^{-1}[B] = \begin{bmatrix} 2.15 \\ 2.05 \\ 2.75 \end{bmatrix}$$

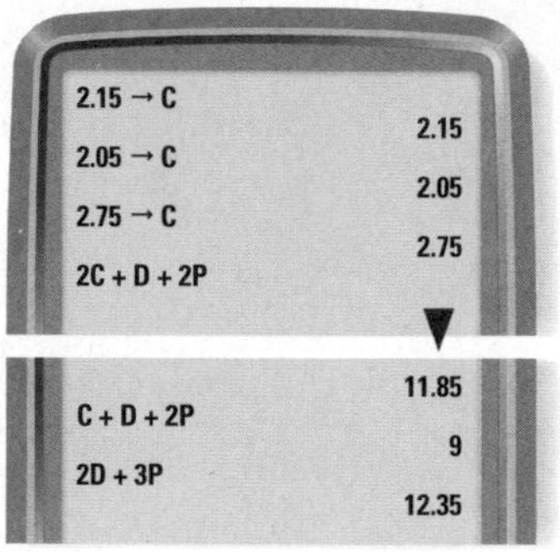

A can of soup costs \$2.15, a frozen dinner costs \$2.05, and a jar of peanut butter costs \$2.75.

Substituting these answers in the original system shows that they are correct. You can use your calculator to evaluate the expressions quickly and accurately.

You may have noticed that when you solve systems of equations with two variables, you must have two equations. When you solve a system of equations with three variables, you must have three equations. In general, you must have as many equations as variables; otherwise there will not be enough information to solve the problem. In terms of matrix equations, this means that the coefficient matrix must be square.

If there are more equations than variables, often one equation is a multiple of another equation and therefore just repeats the same information; or the extra information contradicts the other equations, and therefore there is no solution that will satisfy all the equations.

EXERCISES

Practice Your Skills

1. Rewrite each system of equations in matrix form.

a. $\begin{cases} 3x + 4y = 11 \\ 2x - 5y = -8 \end{cases}$

b. $\begin{cases} x + 2y + z = 0 \\ 3x - 4y + 5z = -11 \\ -2x - 8y - 3z = 1 \end{cases}$

c. $\begin{cases} 5.2x + 3.6y = 7 \\ -5.2x + 2y = 8.2 \end{cases}$

d. $\begin{cases} \frac{1}{4}x - \frac{2}{5}y = 3 \\ \frac{3}{8}x + \frac{2}{5}y = 2 \end{cases}$

2. Multiply each pair of matrices. If multiplication is not possible, explain why.

a. $\begin{bmatrix} 5 & 2 \\ 7 & 3 \end{bmatrix}\begin{bmatrix} 1 & -3 \\ 5 & -2 \end{bmatrix}$ **b.** $\begin{bmatrix} 4 & -1 \\ 3 & 6 \\ 2 & -3 \end{bmatrix}\begin{bmatrix} 2 & -5 & 0 \\ 1 & -2 & 7 \end{bmatrix}$ **c.** $\begin{bmatrix} 9 & -3 \end{bmatrix}\begin{bmatrix} 4 & -6 \\ 0 & -2 \\ -1 & 3 \end{bmatrix}$

3. Use matrix multiplication to expand each system. Then solve for each variable by using substitution or elimination.

a. $\begin{bmatrix} 1 & 5 \\ 6 & 2 \end{bmatrix}\begin{bmatrix} a & b \\ c & d \end{bmatrix} = \begin{bmatrix} -7 & 33 \\ 14 & -26 \end{bmatrix}$ **b.** $\begin{bmatrix} 1 & 5 \\ 6 & 2 \end{bmatrix}\begin{bmatrix} a & b \\ c & d \end{bmatrix} = \begin{bmatrix} 1 & 0 \\ 0 & 1 \end{bmatrix}$

4. Multiply each pair of matrices. Are the matrices inverses of each other?

a. $\begin{bmatrix} 5 & 2 \\ 7 & 3 \end{bmatrix}\begin{bmatrix} 3 & -2 \\ -7 & 5 \end{bmatrix}$ **b.** $\begin{bmatrix} 1 & 5 & 4 \\ 6 & 2 & -2 \\ 0 & 3 & 1 \end{bmatrix}\begin{bmatrix} 0.16 & 0.14 & -0.36 \\ -0.12 & 0.02 & 0.52 \\ 0.36 & -0.06 & -0.56 \end{bmatrix}$

5. Find the inverse of each matrix by solving the matrix equation $[A]\,[A]^{-1} = [I]$. To check your answer, find the inverse matrix on your calculator.

a. $\begin{bmatrix} 4 & 3 \\ 5 & 4 \end{bmatrix}$ **b.** $\begin{bmatrix} -3 & 8 \\ 5 & 6 \end{bmatrix}$ **c.** $\begin{bmatrix} 5 & 3 \\ 10 & 7 \end{bmatrix}$ **d.** $\begin{bmatrix} 1 & 2 \\ 2 & 4 \end{bmatrix}$

Reason and Apply

6. Rewrite each system in matrix form, and solve by using matrix multiplication. Check your solutions.

a. $\begin{cases} 8x + 3y = 41 \\ 6x + 5y = 39 \end{cases}$ **b.** $\begin{cases} 11x - 5y = -38 \\ 9x + 2y = -25 \end{cases}$

c. $\begin{cases} 2x + y - 2z = 1 \\ 6x + 2y - 4z = 3 \\ 4x - y + 3z = 5 \end{cases}$ **d.** $\begin{cases} 4w + x + 2y - 3z = -16 \\ -3w + 3x - y + 4z = 20 \\ 5w + 4x + 3y - z = -10 \\ -w + 2x + 5y + z = -4 \end{cases}$

7. Write a system of equations for each situation, and then solve for the values of the variables.

a. The perimeter of a rectangle is 44 cm. Its length is 2 cm more than twice its width.

b. The perimeter of an isosceles triangle is 40 cm. The length of the base is 2 cm less than the length of either leg.

c. The Fahrenheit reading on a dual thermometer is 0.4 less than three times the Celsius reading. ($F = \frac{9}{5}C + 32$)

8. At the High Flying Amusement Park there are three kinds of rides: Jolly rides, Adventure rides, and Thrill rides. Admission is free when you buy a book of tickets, which includes ten tickets for each type of ride. Or you can pay $5.00 for admission and then buy tickets for each of the rides individually. Noah, Rita, and Carey decide to pay the admission price and buy individual tickets.

Noah rides 7 Jolly rides, 3 Adventure rides, and 9 Thrill rides for a cost of \$19.55 for the rides only. Rita rides 9 Jolly rides, 10 Adventure rides, and no Thrill rides for a cost of \$13.00 for the rides. Carey pays \$24.95 for 8 Jolly rides, 7 Adventure rides, and 10 Thrill rides.

a. How much does each type of ride cost?

b. What is the total cost of a 30-ride book of tickets?

c. Would Noah, Rita, or Carey have been better off purchasing a ticket book?

9. A family invested a portion of \$5,000 in an account yielding 6% simple interest and the rest in an account yielding 7.5% simple interest. The total interest earned in the first year was \$340.50. How much was invested in each account?

10. One angle of a triangle is 30° greater than the smallest angle. The largest angle is 10° more than twice the midsized angle. What are the measures of the three angles?

11. The ability to solve a system of equations is definitely not "new" mathematics. The famous Chinese mathematician Liu Hui posed this problem in the 3rd century C.E. See if you can solve it.

A certain number of people are purchasing some chickens jointly. If each person contributes 9 wen, there is a surplus of 11 wen, and if each person contributes 6 wen, there is a deficiency of 16 wen. Find the number of people and the price of the chickens. (Burton, David M., *Burton's History of Mathematics: An Introduction*, Third Edition. Dubuque, IA: 1995).

HistoryConnection The row reduction method of solving linear systems is often called Gaussian elimination, or sometimes Gauss-Jordan elimination. These names honor Carl Friedrich Gauss, a famous German mathematician born in 1777, and German engineer Wilhelm Jordan, born in 1842. However, the method of row reduction was first documented by Chinese mathematician Liu Hui, who lived nearly 2,000 years ago.

12. **APPLICATION** The circuit shown here is made of two batteries (6-volt and 9-volt) and three resistors (47-ohm, 470-ohm, and 280-ohm). The batteries create an electric current in the circuit. Let x, y, and z represent the current in amps flowing through each resistor.

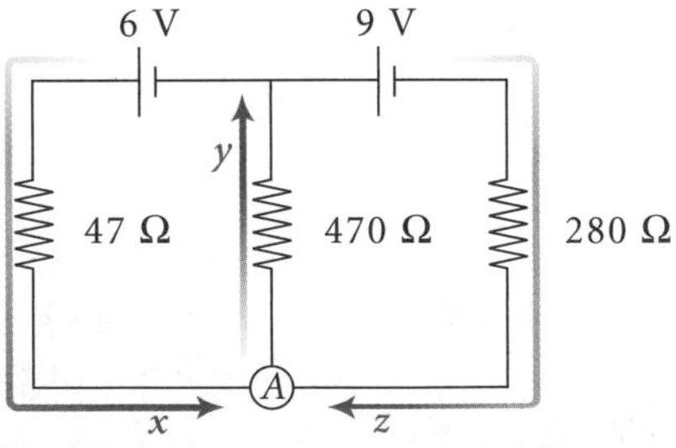

The formula for the voltage across each resistor is current times resistance ($V = IR$). This gives two equations for the two loops of the circuit:

$$47x + 470y = 6 \qquad 280z + 470y = 9$$

The electric current flowing into any point along the circuit must flow out. So, for instance, at junction A, $x + z - y = 0$. Find the current flowing through each resistor.

13. When you use your calculator to find the inverse of the coefficient matrix for this system, you get an error message. What does this indicate about the system?

$$\begin{cases} 3.2x + 2.4y = 9.6 \\ 2x + 1.5y = 6 \end{cases}$$

14. **APPLICATION** An important application in economics is the study of the relationship between industrial production and consumer demand. In creating an economic model, Russian American economist Wassily Leontief (1906–1999) noted that the total output less the internal consumption equals consumer demand. Mathematically, his input-output model looks like $[B] - [A][B] = [D]$, where $[B]$ is the total output matrix, $[A]$ is the input-output matrix, and D is the matrix, representing consumer demand.

Input \ Output	Agriculture	Manufacturing	Service
Agriculture	0.2	0.2	0.1
Manufacturing	0.2	0.4	0.1
Service	0.1	0.2	0.3

Here's an input-output matrix, $[A]$, for a simple three-sector economy.

For instance, the first column tells the economist that to produce (output) 1 unit of agricultural products requires the consumption (input) of 0.2 unit of agricultural products, 0.2 unit of manufacturing products and 0.1 unit of service products.

The demand matrix, $[D]$, represents millions of dollars. Use the equation $[X] - [A][X] = [D]$ to find the output matrix, $[X]$.

$$[D] \begin{bmatrix} 100 \\ 80 \\ 50 \end{bmatrix}$$

History Connection After the United States entered World War II in 1941, Wassily Leontief's method became a critical part of national wartime production planning. As a consultant to the U.S. Labor Department, he developed an input-output table for more than 90 economic sectors. During the early 1960s Leontief and Marvin Hoffenberg of Johns Hopkins University used input-output analysis to forecast the economic effects of disarmament. In 1973 Liontief was awarded a Nobel Prize in economics for his contributions to the field.

Wassily Leontief

15. One way to find an inverse of a matrix $\begin{bmatrix} a & b \\ c & d \end{bmatrix}$, if it exists, is to perform row operations on the augmented matrix $\begin{bmatrix} a & b & 1 & 0 \\ c & d & 0 & 1 \end{bmatrix}$ to change it to the form $\begin{bmatrix} 1 & 0 & e & f \\ 0 & 1 & g & h \end{bmatrix}$. The matrix $\begin{bmatrix} e & f \\ g & h \end{bmatrix}$ is the required inverse. Use this strategy to find the inverse of each matrix.

a. $\begin{bmatrix} 4 & 3 \\ 5 & 4 \end{bmatrix}$

b. $\begin{bmatrix} 6 & 4 & -2 \\ 3 & 1 & -1 \\ 0 & 7 & 3 \end{bmatrix}$

Review

16. (Lesson 3.5) For each equation write a 2nd linear equation that would create a dependent system.

a. $y = 2x + 4$

b. $y = \frac{-1}{3}x - 3$

c. $2x + 5y = 10$

d. $x - 2y = -6$

17. (Lesson 3.5) For each equation write a 2nd linear equation that would create an inconsistent system.

a. $y = 2x + 4$

b. $y = \frac{-1}{3}x - 3$

c. $2x + 5y = 10$

d. $x - 2y = -6$

18. (Lesson 7.1) Four towns—Lenox, Murray, Davis, and Terre—are connected by a series of roads.

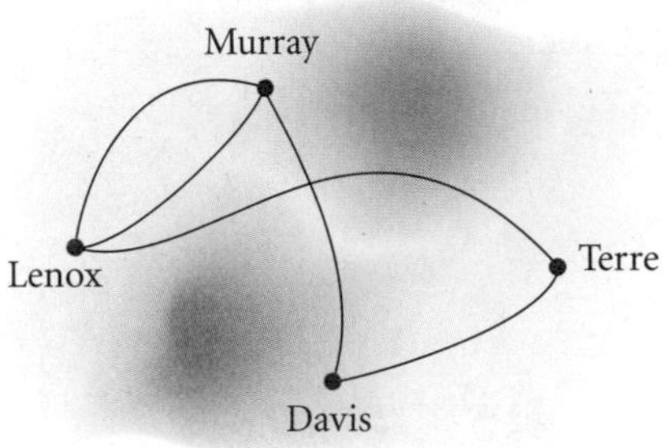

a. Represent the number of direct road connections between the towns in a matrix, $[A]$. List the towns in the order Lenox, Murray, Davis, Terre.

b. What does the value of $a_{2,2}$ mean?

c. Describe the symmetry of your matrix.

d. How many roads are there? What is the sum of the entries in the matrix? What is the relationship between these two answers?

e. Assume any one of the roads is a one-way road. How does this change your matrix in 18a?

19. (Lesson 2.1) The 3rd term of an arithmetic sequence is 28. The 7th term is 80. What is the 1st term?

HistoryConnection In World War II the Navajo language was used by the U.S. Marines to encode messages because it is a complex, unwritten language that is unintelligible to anyone who has not had extensive training. Navajos could encode, transmit, and decode three-line English messages in 20 seconds, whereas machines of the time required 30 minutes to perform the same job. Navajo recruits developed the code, including a dictionary and numerous words for military terms. To learn more about Navajo code-talkers, see the links at **www.keycollege.com/online** .

IMPROVING YOUR REASONING SKILLS

Coding and Decoding

One of the uses of matrices is in the mathematical field of cryptography, the science of enciphering and deciphering encoded messages. Here's one way that a matrix can be used to make a secret code.

Imagine encoding the word *CODE*. First, convert each letter to a numerical value based on its location in the alphabet. For example, $C = 3$ because it is the third letter in the alphabet.

So $CODE = 3, 15, 4, 5$. Arrange these numbers in a 2×2 matrix: $\begin{bmatrix} 3 & 15 \\ 4 & 5 \end{bmatrix}$.

Now, multiply by an encoding matrix.

Let's use $[E] = \begin{bmatrix} 1 & -1 \\ -1 & 2 \end{bmatrix}$, so $\begin{bmatrix} 3 & 15 \\ 4 & 5 \end{bmatrix}[E] = \begin{bmatrix} -12 & 27 \\ -1 & 6 \end{bmatrix}$.

Now, convert back to letters. Notice that three of these numbers are outside the series of numbers 1 to 26 that represent A to Z. To convert other numbers, simply add or subtract 26 to make them fall within the range of 1 to 26, like this:

$$\begin{bmatrix} -12 & 27 \\ -1 & 6 \end{bmatrix} \rightarrow \begin{bmatrix} 14 & 1 \\ 25 & 6 \end{bmatrix} \rightarrow \begin{bmatrix} N & A \\ Y & F \end{bmatrix}.$$

To *decode* this encoded message, convert back to numbers and multiply by the inverse of the coding matrix, and then convert each number back to the corresponding letter in the alphabet:

$$\begin{bmatrix} 14 & 1 \\ 25 & 6 \end{bmatrix}[E]^{-1} = \begin{bmatrix} 29 & 15 \\ 56 & 31 \end{bmatrix} = \begin{bmatrix} C & O \\ D & E \end{bmatrix}.$$

See if you can decode this Navajo saying. Begin by taking out the spaces and breaking the letters into groups of four letters each.

BD FSLG GTQN YP OPIIY UCB DKIC BYF BEQQ WW URQLPRE.

LESSON

7.5 Systems of Linear Inequalities

Real-world situations often involve a range of possible values. Algebraic representations of such situations are called **inequalities**.

Situation	Inequality
Write an essay between two and five pages in length.	$2 \le E \le 5$
Practice more than an hour each day.	$P > 1$
The post office is open from nine o'clock until noon.	$9 \le H \le 12$
Do not spend more than $10 on candy and popcorn.	$c + p \le 10$
A college fund has $40,000 to invest in stocks and bonds.	$s + b \le 40{,}000$

Recall that you can perform operations on inequalities much as you do with equations. You can add or subtract the same quantity to both sides, multiply by the same number or expression on both sides, and so on. The one exception to remember is that when you multiply or divide by a negative quantity or expression, the inequality symbol is reversed.

In this lesson, you will learn how to graph solutions of inequalities with two variables, such as the last two statements in the foregoing table.

Investigation
Paying for College

A sum of $40,000 has been donated to a college scholarship fund. The administrators of the fund are considering how much to invest in stocks and bonds. Stocks usually pay more but are a riskier investment, whereas bonds pay less but are safer.

Step 1 Let x represent the amount invested in stocks in dollars, and let y represent the amount invested in bonds in dollars. Graph the equation $x + y = \$40{,}000$.

Step 2 Name at least five pairs of x- and y-values that fit the inequality $x + y < 40{,}000$, and plot them on your graph. Why can $x + y$ be less than $40,000?

Step 3 Describe the location of all possible solutions to the inequality $x + y < 40{,}000$. Shade this region on your graph.

Step 4 Describe some points that fit the condition $x + y < 40{,}000$ but do not make sense for the situation.

Assume that either option, stocks or bonds, requires a minimum investment of $5,000 and that the fund administrators definitely want to purchase some stocks and some bonds. Based on the advice of their financial advisor, they decide that the amount invested in bonds should be at least twice the amount invested in stocks.

Step 5 — Translate all of the limitations, or **constraints**, into a system of inequalities. A table might help you to organize this information.

Step 6 — Graph all of the inequalities, and determine the region of your graph that will satisfy all the constraints. Find each corner, or **vertex**, of this region.

When there are one or two variables in an inequality, you can represent the solution as a set of ordered pairs by shading the coordinate plane.

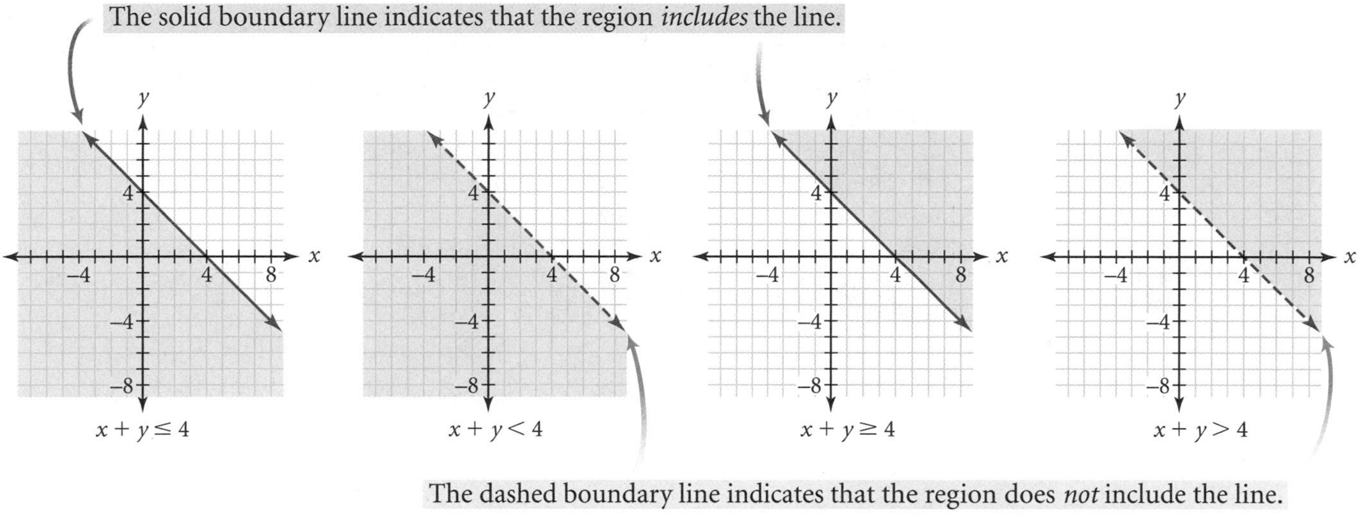

The solid boundary line indicates that the region *includes* the line.
$x + y \leq 4$
$x + y < 4$
$x + y \geq 4$
$x + y > 4$
The dashed boundary line indicates that the region does *not* include the line.

When you have several inequalities that must be satisfied simultaneously, you have a system. The solution to a system of inequalities with two variables will be a set of points rather than a single point. This set of points is called a **feasible region.** The feasible region can be shown graphically or described as a geometric shape with its vertices given.

EXAMPLE

Ellie has three hours to work on her homework tonight. She wants to spend more time working on mathematics than on chemistry, and she must spend at least half an hour working on chemistry. State the constraints of this system algebraically, with x representing mathematics time in hours and y representing chemistry time in hours. Graph your inequalities, shade the feasible region, and label its vertices.

▸ *Solution*

Convert each constraint into an algebraic inequality.

$x + y \leq 3$	Ellie has 3 h to do homework.
$x > y$	She wants to spend more time on mathematics than on chemistry.
$y \geq 0.5$	She must spend at least half an hour on chemistry.

To graph the line $x + y \leq 3$, or $y \leq -x + 3$, first graph $y = -x + 3$. Use a solid line because the inequality is *less than or equal to* 3.

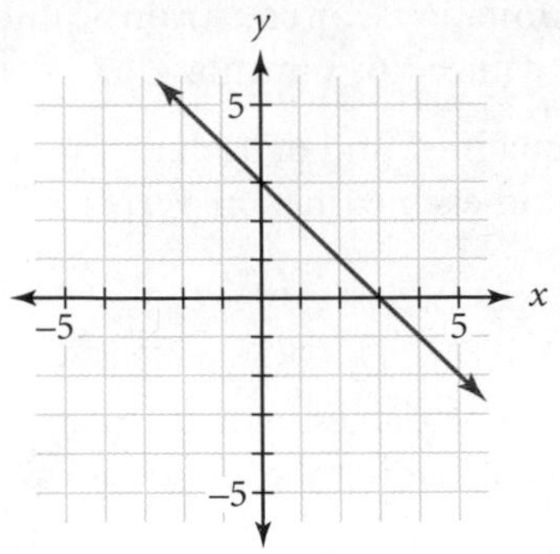

To determine which side to shade, choose a sample point on either side of the line. Test the coordinates of this point in the inequality to see whether it makes a true statement. If it does, shade on the side of the line where the point lies. If it doesn't, shade on the other side of the line. For example, if you choose (0, 0), $0 \leq -0 + 3$, or $0 \leq 3$, the result is a true statement, so shade the side of the line that contains (0, 0). [▶ See **Calculator Note 7F** to learn how to graph inequalities using your calculator.◀]

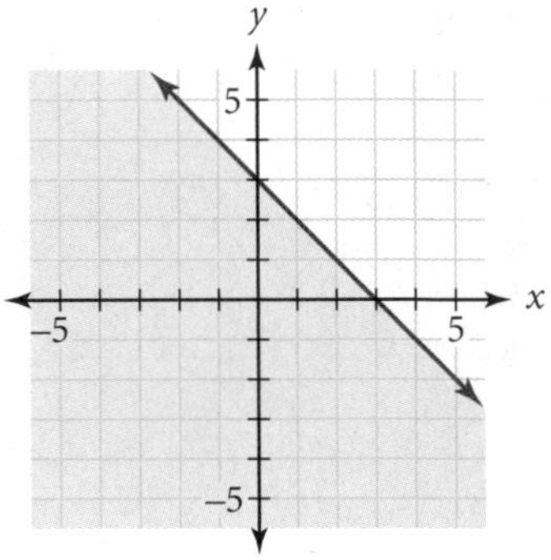

Follow the same steps to graph $x > y$ and $y \geq 0.5$ on the same axes. The graph of $x > y$ will have a dashed line.

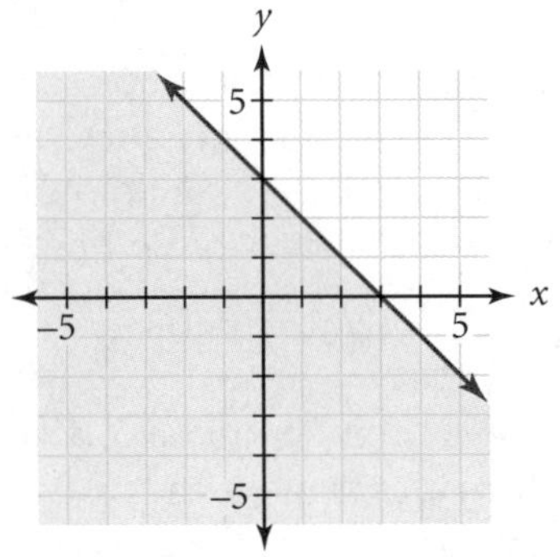

The solution of the system is the area that represents the overlap of all the shaded regions—the feasible region. Every point in this region is a possible solution of the system. [▶ See **Calculator Note 7F** to learn how to graph systems of inequalities with your calculator.◀]

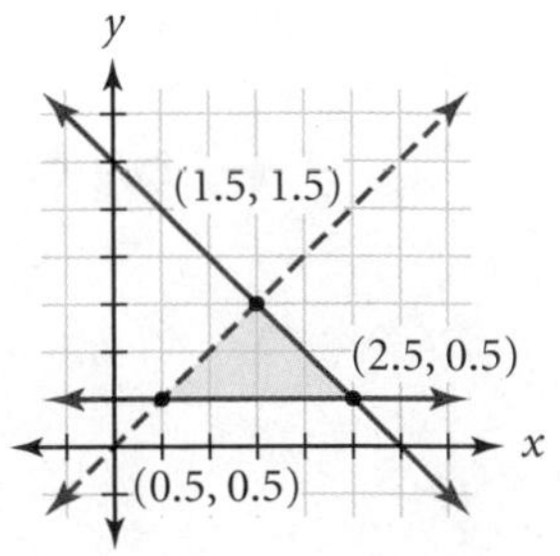

To clarify the area that represents the solutions, name the vertices of the region. You can find the vertices by finding the intersection of each pair of equations using substitution, elimination, or matrices.

Equations	Intersections
$x + y = 3$ and $x = y$	$(1.5, 1.5)$
$x + y = 3$ and $y = 0.5$	$(2.5, 0.5)$
$x = y$ and $y = 0.5$	$(0.5, 0.5)$

The solution to this system is a triangle with vertices $(1.5, 1.5)$, $(2.5, 0.5)$, and $(0.5, 0.5)$. Any point within this region represents a way that Ellie could divide her homework time. For example, the point $(1.5, 1)$ means she could spend 1.5 h on mathematics and 1 h on chemistry and still meet all her constraints.

When solving a system of equations based on real-world constraints, it is important to note that sometimes there are restrictions that are not specifically stated in the problem. In Example A, negative values of x and y would not make sense, because you can't study for a negative number of hours. You could have added the commonsense constraints $x \geq 0$ and $y \geq 0$, although in that example it would not affect the feasible region.

EXERCISES

Practice Your Skills

1. Solve each inequality for y.

a. $2x - 5y > 10$

b. $4(2 - 3y) + 2x > 14$

2. Graph each linear inequality.

a. $y \leq -2x + 5$

b. $2y + 2x > 5$

c. $x > 5$

3. Write the equation of each graph.

a.

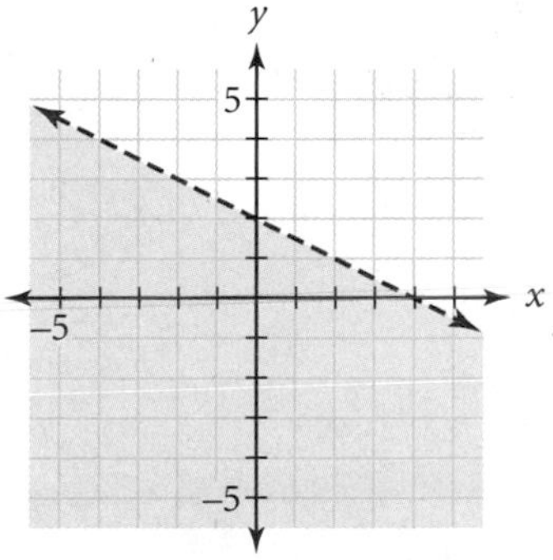

b.

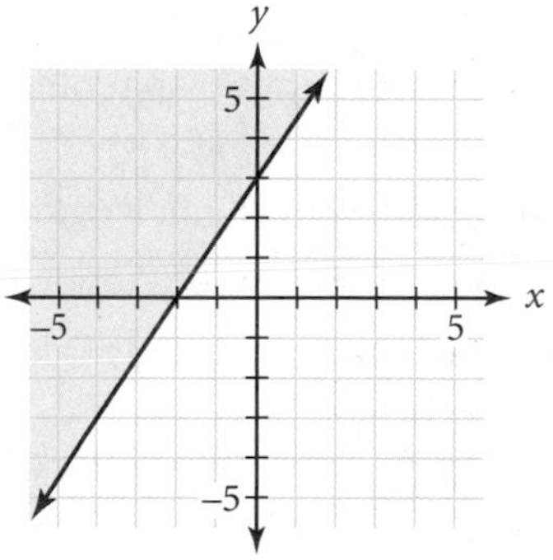

c.

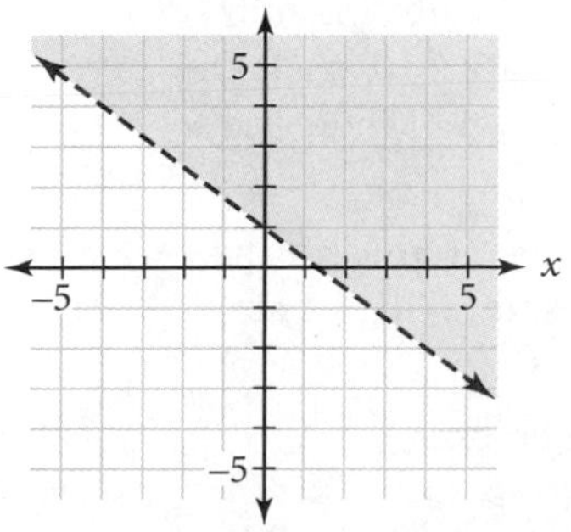

d.

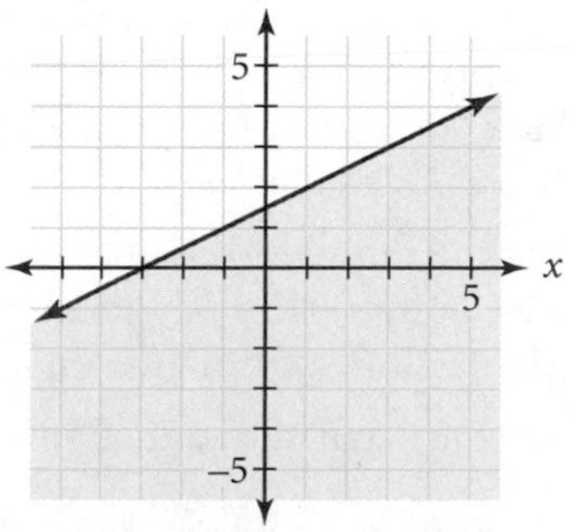

4. The graphs of $y = 2.4x + 2$ and $y = -x^2 - 2x + 6.4$ serve as the boundaries of this feasible region. What two inequalities identify this region?

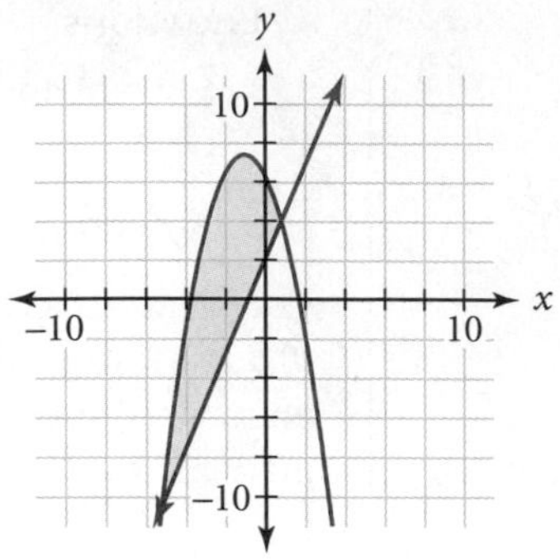

Reason and Apply

For Exercises 5–8, sketch the feasible region of each system of inequalities. Find the coordinates of each vertex.

5. $\begin{cases} y \le -0.51x + 5 \\ y \le -1.6x + 8 \\ y \ge 0.1x + 2 \\ y \ge 0 \\ x \ge 0 \end{cases}$

6. $\begin{cases} y \ge 1.6x - 3 \\ y \le -(x-2)^2 + 4 \\ y \ge 1 - x \\ y \ge 0 \\ x \ge 0 \end{cases}$

7. $\begin{cases} 4x + 3y \le 12 \\ 1.6x + 2y \le 8 \\ 2x + y \ge 2 \\ y \ge 0 \\ x \ge 0 \end{cases}$

8. $\begin{cases} y \ge |x - 1| \\ y \le \sqrt{9 - x^2} \\ y \le 2.5 \\ y \ge 0 \end{cases}$

9. In the Lux Art Gallery rectangular paintings must have an area between 200 and 300 in.^{2} and a perimeter between 66 in. and 80 in.

a. Write four inequalities involving length and width that represent these constraints.

b. Graph this system of inequalities to identify the feasible region.

c. Will the gallery accept a painting that measures

i. 12.4 in. by 16.3 in.?

ii. 16 in. by 17.5 in.?

iii. 14.3 in. by 17.5 in.?

10. Al is monitoring his intake of junk food calories. His two favorite snacks are cinnamon graham crackers and ice cream bars. One cinnamon graham cracker has 65 calories. One ice cream bar has 310 calories. He is trying to keep his total consumption of these snack items under 2,000 calories a week.

a. Let x represent the number of graham crackers Al might eat, and let y represent the number of ice cream bars he might eat. Write the inequality that shows that he ate fewer than 2,000 calories from graham crackers and ice cream bars.

b. Write the inequality that shows that he ate fewer than 10 graham crackers.

c. Write the inequality that shows that the total number of snack items he ate was no more than 20.

d. Graph the solution to the system of inequalities, including any commonsense constraints.

e. Name all of the vertices of the feasible region.

11. **APPLICATION** As the altitude of a spacecraft increases, an astronaut's weight decreases until a state of weightlessness is achieved. The weight of a 125-lb astronaut, W, at a given altitude in kilometers above Earth's sea level, x, is given by the formula

$$w = 57 \cdot \frac{6400^2}{(6400 + x)^2}$$

a. At what altitude will the astronaut weigh less than 5 lb?

b. At an altitude of 400 km, how much would the astronaut weigh?

c. Will the astronaut ever be truly weightless? Explain your answer.

ScienceConnection A typical space shuttle orbits at an altitude of about 400 km. At this height an astronaut still weighs about 88.8% of his or her weight on Earth. You have probably seen pictures in which astronauts in orbit on a space shuttle or space station appear to be weightless; this is not due to the absence of gravity but rather to an effect called *microgravity.* In orbit, astronauts and their craft are being pulled toward the Earth by gravity, but their speed is such that they are in free fall around Earth, rather than toward Earth. Because the astronauts and their spacecraft are falling through space at the same rate, astronauts appear to be floating inside the craft. This is similar to the fact that a car's passengers can appear to be sitting still, when actually they are traveling at a speed of 60 mi/h.

Astronauts floating in space during the 1994 testing of rescue system hardware appear to be weightless. By using a small control unit, one astronaut floats without being tethered to the spacecraft.

Review

12. (Lesson 6.4) A parabola with an equation in the form $y = ax^2 + bx + c$ passes through the points $(-2, -32)$, $(1, 7)$, and $(3, 63)$.

a. Use matrices and systems to find the values of a, b, and c for this parabola.

b. Write the equation of this parabola.

c. Describe how to verify your answer.

13. (Lesson 5.1) These data were collected from a bouncing ball experiment. Recall that the height in centimeters, y, is exponentially related to the number of the bounce, x. Find the values of a and b for an exponential model in the form $y = ab^x$.

Bounce number	Height (cm)
3	34.3
7	8.2

14. (Lesson 7.3) Complete the reduction of this augmented matrix to row-echelon form. Give each row operation, and find each missing matrix entry.

$$\left[\begin{array}{cc|c} 3 & -1 & 5 \\ -4 & 2 & 1 \end{array}\right] \xrightarrow{\underline{\ \ ?\ \ }} \left[\begin{array}{cc|c} 1 & \underline{\ \ ?\ \ } & \underline{\ \ ?\ \ } \\ -4 & 2 & 1 \end{array}\right] \xrightarrow{\underline{\ \ ?\ \ }} \left[\begin{array}{cc|c} 1 & \underline{\ \ ?\ \ } & \underline{\ \ ?\ \ } \\ 0 & \underline{\ \ ?\ \ } & \underline{\ \ ?\ \ } \end{array}\right] \xrightarrow{\underline{\ \ ?\ \ }} \left[\begin{array}{cc|c} 1 & \underline{\ \ ?\ \ } & \underline{\ \ ?\ \ } \\ 0 & 1 & \underline{\ \ ?\ \ } \end{array}\right] \xrightarrow{\underline{\ \ ?\ \ }} \left[\begin{array}{cc|c} 1 & 0 & \underline{\ \ ?\ \ } \\ 0 & 1 & \underline{\ \ ?\ \ } \end{array}\right]$$

15. **APPLICATION** (Lesson 5.1) The growth of a water fungus is modeled by the function $y = f(x) = 2.58\,(3.84)^x$, where x represents the number of hours elapsed and y represents the number of fungi spores.

a. How many spores are there initially?

b. How many spores are there after 10 h?

c. Find an inverse function that uses y as the dependent variable and x as the independent variable.

d. Use your answer to 15c to determine when the number of spores will exceed one billion.

LESSON

7.6

Linear Programming

Love the moment, and the energy of that moment will spread beyond all boundaries.

CORITA KENT

Industrial managers often investigate more economical ways of doing business. They must consider physical limitations, standards of quality, customer demand, availability of materials, and manufacturing expenses as restrictions, or constraints, that determine how much of an item they could produce. Then they will determine the optimum, or best, production level—usually to minimize production costs or maximize profit. The process of finding a feasible region and determining the point that gives the maximum or minimum value to a specific expression is called **linear programming**.

Problems that can be modeled with linear programming may involve two variables or hundreds of variables. Computerized modeling programs that analyze up to 200 constraints and 400 variables are regularly used to help businesses choose the best plan of action. In this lesson you will look at problems that involve two variables because you are relying on the visual assistance of a two-dimensional graph to help you find the feasible region.

In this Investigation you'll explore a linear programming problem and reach conclusions about how to find the optimum value in the most efficient way.

Investigation
Maximizing Profit

The Elite Pottery Shoppe makes two kinds of birdbaths: a fancy, glazed one and a simple, unglazed one. The unglazed birdbath requires 0.5 h to make using a pottery wheel and 3 h in the kiln. The glazed birdbath takes 1 h on the wheel and 18 h in the kiln. The company's one pottery wheel is available for 8 h/d at most. The three kilns are available 60 h/d at most. The company has a standing order for 6 unglazed birdbaths per day, so they must produce at least that many. The company's profit on each unglazed birdbath is $10, and the profit on each glazed birdbath is $40. How many of each kind of birdbath should be produced each day in order to maximize profit?

Step 1 Organize the information into a chart like this one.

	Number of unglazed birdbaths x	Number of glazed birdbaths y	Constraining value
Wheel hours			
Kiln hours			
Profit			Maximize

Step 2 Use your chart to help you write inequalities that reflect the constraints given, and be sure to include any commonsense constraints. Graph the feasible region to show the combinations of glazed and unglazed birdbaths the shop could produce, and label the coordinates of the vertices. (Note: Profit is not a constraint; it is what you are trying to optimize.)

Step 3 It makes sense to produce only whole numbers of birdbaths. List the coordinates of all points within the feasible region (there should be 22). Remember that the feasible region includes points on the boundary lines.

Step 4 Write the equation that will determine profit based on the number of unglazed and glazed birdbaths produced. Calculate the profit that would be earned at each of the feasible points you found in Step 3. You may want to divide this task among the members of your group.

Step 5 What number of each kind of birdbath should the Elite Pottery Shoppe produce to maximize profit? What is the maximum profit possible? Plot this point on your graph from Step 2. What do you notice about this point?

Step 6 Suppose you want profit to be exactly \$100. What equation would express this? Carefully graph this line on your graph from Step 2.

Step 7 Suppose you want profit to be exactly \$140. What equation would express this? Carefully add this line to your graph.

Step 8 Suppose you want profit to be exactly \$170. What equation would express this? Carefully add this line to your graph.

Step 9 How do your results from Steps 6–8 show you that (14, 1) must be the point that maximizes profit? Generalize your observations to describe a method you can use with other problems to find the optimum value. What do you think you might do if this vertex point did not have integer coordinates? What if you wanted to *minimize* profit?

Linear programming is one of the most beneficial real-world applications of systems inequalities. Its value is not limited to business settings, as the next example shows.

EXAMPLE

Marco coaches track. He is planning a snack of graham crackers and blueberry yogurt for the team. Because he is concerned about health and nutrition, he wants to make sure that the snack contains no more than 700 calories and 20 grams of fat. He also wants the snack to contain at least 17 grams of protein and at least 30% of the daily recommended value of iron. The nutritional content of each food is listed in the table below. Each serving of yogurt costs \$0.30, and each graham cracker costs \$0.06. What combination of servings of graham crackers and blueberry yogurt should Marco provide to minimize cost?

	Serving	Calories	Fat	Protein	Iron (percent of daily recommended value)
Graham crackers	1 cracker	60	2 g	2 g	6%
Blueberry yogurt	4.5 oz	130	2 g	5 g	1%

▶ Solution

First, organize the constraint information into a chart; then, write inequalities that reflect the constraints. Be sure to include any commonsense constraints. Let x represent the number of servings of graham crackers, and let y represent the number of servings of yogurt.

	Amount per graham cracker	Amount per serving of yogurt	Limiting Value
Calories	60	130	≤ 700
Fat	2 g	2 g	≤ 20 g
Protein	2 g	5 g	≥ 17 g
Iron	6%	1%	$\geq 30\%$
Cost	$0.06	$0.30	Minimize

$$\begin{cases} 60x + 130y \leq 700 & \text{Calories} \\ 2x + 2y \leq 20 & \text{Fat} \\ 2x + 5y \geq 17 & \text{Protein} \\ 6x + 1y \geq 30 & \text{Iron} \\ x \geq 0 & \text{Common sense} \\ y \geq 0 & \text{Common sense} \end{cases}$$

Now, graph the feasible region, and find the vertices.

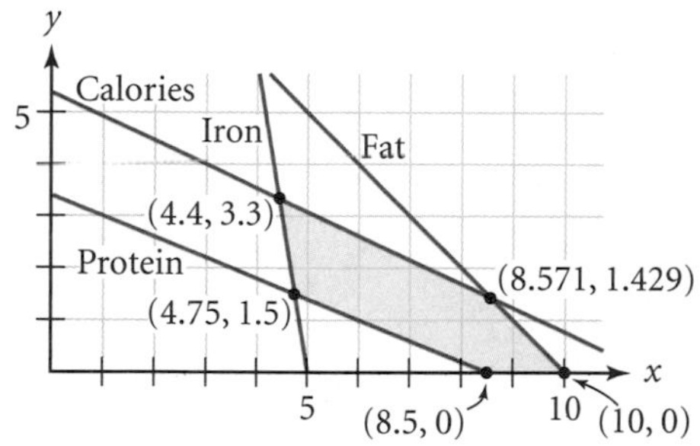

Next, determine an equation that will determine the cost of a snack based on the number of servings of graham crackers and yogurt.

$Cost = 0.06x + 0.30y$

You could try any possible combination of graham crackers and yogurt that is in the feasible region, but recall that in the Investigation it appeared that optimum values will occur at vertices. Calculate the cost at each of the vertices to see which one is a minimum.

The least expensive combination would be 8.5 crackers and no yogurt. However, Marco wants to serve only whole numbers of servings. The points (8, 1) and (9, 0) are the integer points within the feasible region closest to (8.5, 0), so test which one has a lower cost. The point (8, 1) gives a cost of $0.78, and (9, 0) will cost $0.54. Therefore, Marco should serve 9 graham crackers and no yogurt.

x	y	Cost
4.444	3.333	$1.27
4.75	1.5	$0.74
8.571	4.129	$1.75
10	0	$0.60
8.5	0	$0.51

The following box summarizes the steps of solving a linear programming problem. Refer to these steps as you do the exercises.

Solving a Linear Programming Problem

1. Define your variables, and write constraints using the information given in the problem. Don't forget commonsense constraints.
2. Graph the feasible region, and find the coordinates of all vertices.
3. Write the equation of the function you want to optimize, and decide whether you need to maximize or minimize it.
4. Evaluate your optimization function at each of the vertices of your feasible region, and decide which vertex provides the optimum value.
5. If your possible solutions need to be limited to integer values, and your optimum vertex does not contain integers, test the integer values within the feasible region that are closest to this vertex.

EXERCISES

Practice Your Skills

1. Carefully graph this system of inequalities, and label the vertices.

$$\begin{cases} x + y \leq 10 \\ 5x + 2y \geq 20 \\ -x + 2y \geq 0 \end{cases}$$

2. For the system in Exercise 1, find which vertex optimizes these expressions.

   a. maximize: $5x + 2y$

   b. minimize: $x + 3y$

   c. maximize: $x + 4y$

   d. minimize: $5x + y$

   e. Can you make any generalizations about which vertex provides a maximum or minimum value?

PROJECT

NUTRITIONAL ELEMENTS

Choose two food items from your home or a grocery store. Record information about calories, fat, protein, iron, or other appropriate nutrients for each of the two food items as given on its nutrition label. Be sure to record the price per serving as well. Use Marco's constraints from the example or create some of your own to write and solve a problem similar to the cracker-and-yogurt-snack combination problem in this lesson.

3. Graph this system of inequalities, and name the integer coordinates that maximize the equation $P = 0.08x + 0.10y$. What is this maximum value of P?

$$\begin{cases} x \geq 5500 \\ y \geq 5000 \\ y \leq 3x \\ x + y \leq 40{,}000 \end{cases}$$

4. **APPLICATION** During nesting season two different bird species inhabit a region with area 180,000 m^2. Dr. Chamberlin estimates that this ecological region can provide 72,000 kg of food during the season. Each nesting pair of species X needs 39.6 kg of food during a specified time period and 120 m^2 of land. Each nesting pair of species Y needs 69.6 kg of food and 90 m^2 of land. Let x represent the number of pairs of species X, and let y represent the number of pairs of species Y.

a. What is the meaning of the constraints $x \geq 0$ and $y \geq 0$?

b. What is the meaning of the constraint $120x + 90y \leq 180{,}000$?

c. What is the meaning of the constraint $39.6x + 69.6y \leq 72{,}000$?

d. Graph the system of inequalities, and identify each vertex of the feasible region.

e. Maximize the total number of nesting pairs, N, by considering the function $N = x + y$.

Reason and Apply

5. Use a combination of the four lines shown on the graph along with the axes to create a system of inequalities the graph of which satisfies each description.

a. The feasible region is a triangle.

b. The feasible region is a quadrilateral with one side on the y-axis.

c. The feasible region is a pentagon with sides on both the x-axis and the y-axis.

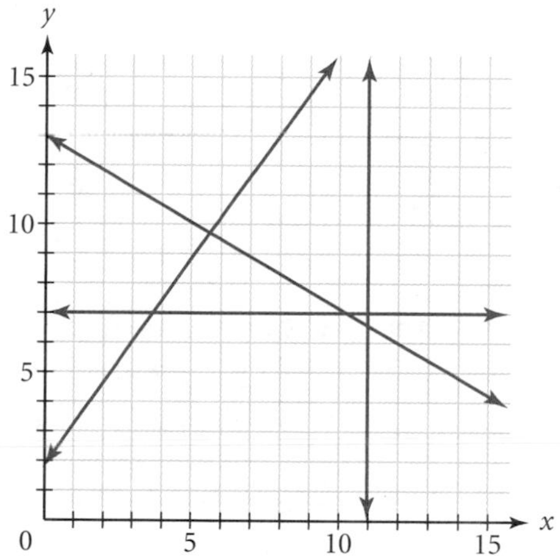

6. **APPLICATION** The International Canine Academy raises and trains Siberian sled dogs and dancing French poodles. Breeders supply the academy with at most 20 poodles and 15 huskies each year. Each poodle eats 2 lb/d of food, and each sled dog eats 6 lb/d. Food supplies are restricted to at most 100 lb/d. A poodle requires 1,000 h/yr of training, whereas a sled dog requires 250 h/yr. The academy cannot provide more than 15,000 h/yr of training time. If each poodle sells for a profit of \$200, and each sled dog sells for a profit of \$80, how many of each kind of dog should the academy raise in order to maximize profits?

7. **APPLICATION** The Elite Pottery Shoppe budgets a maximum of \$1,000 per month for newspaper and radio advertising. The newspaper charges \$50 per ad and requires at least four ads per month. The radio station charges \$100 per minute and requires a minimum of 5 minutes of advertising per month. It is estimated that each newspaper ad reaches 8,000 people and that each minute of radio advertising reaches 15,000 people. What combination of newspaper and radio advertising should the business use in order to reach the maximum number of people? What assumptions did you make in solving this problem? How realistic do you think they are?

8. APPLICATION A fair-trade farmers' cooperative in Chiapas, Mexico, is deciding how much coffee and cocoa to recommend that their members plant. Their 1,000 member families have a total of 7,500 acres to farm. Because of the geography of the region, 2,450 acres are suitable only for growing coffee, and 1,230 acres are suitable only for growing cocoa. A coffee crop produces 30 lb/acre, and a crop of cocoa produces 40 lb/acre. The cooperative has the resources to ship a total of 270,000 lb of product to the United States. Fair-trade organizations mandate a minimum price of \$1.26/lb for organic coffee and \$0.98 /lb for organic cocoa (note that price is per pound, not per acre). How many acres of each crop should the cooperative recommend planting in order to maximize income?

ConsumerConnection Many small coffee and cocoa farmers receive prices for their crops that are less than the costs of production, causing them to live in poverty and debt. Fair-trade certification has been developed to show consumers which products are produced with the welfare of farming communities in mind. To become fair-trade certified, an importer must meet stringent international criteria that include paying a minimum price per pound, providing credit to farmers, and providing technical assistance in farming upgrades. Fair-trade prices allow farmers to make enough to provide their families with food, education, and health care. Additionally, fair-trade cooperatives tend to support farmers who farm without the use of pesticides and chemical fertilizers and who raise shade-grown crops that provide habitats for many species of migrating birds that are destroyed when forests are cleared for large-scale farming.

9. APPLICATION A small electric generating plant must decide how much low-sulfur (2%) and high-sulfur (6%) oil to buy. The final mixture must have a sulfur content of no more than 4%. At least 1,200 barrels of oil are needed. Low-sulfur oil costs \$18.50 per barrel, and high-sulfur oil costs \$14.70 per barrel. How much of each type of oil should be used to keep the cost at a minimum? What is the minimum cost?

10. APPLICATION Teo sells a set of videos at an online auction. His preferred postal service puts the following restrictions on the size of packages.

up to 150 pounds
up to 130 inches in length and girth combined
up to 108 inches in length

Length is defined as the longest side of a package or object. Girth is the distance all the way around the package or object at its widest point perpendicular to the length. Teo is not concerned with the weight, as the videotapes weigh only 25 lb.

a. Write a system of inequalities that represents the constraints on the package size.

b. Sketch a feasible region for the dimensions of the package.

c. Teo packages the videos in a box the dimensions of which are 20 in. by 14 in. by 8 in. Does this box satisfy the restrictions?

ConsumerConnection In the packaging industry two sets of dimensions are used. Inside dimensions are used to ensure proper fit around a product to prevent damage, and outside dimensions are used in shipping classifications and determination of pallet patterns. The type of packaging material and its strength are also important. Corrugated cardboard is a particularly strong yet economical packaging material.

Review

11. (Lessons 3.5 and 7.3) Solve these systems in at least two different ways.

a. $\begin{cases} 8x + 3y = 41 \\ 9x + 2y = 25 \end{cases}$

b. $\begin{cases} 2x + y - 2z = 5 \\ 6x + 2y - 4z = 3 \\ 4x - y + 3z = 5 \end{cases}$

12. (Lesson 7.5) Sketch a graph of the feasible region described by the following system of inequalities.

$\begin{cases} y \geq (x-3)^2 + 5 \\ y \leq -|x-2| + 10 \end{cases}$

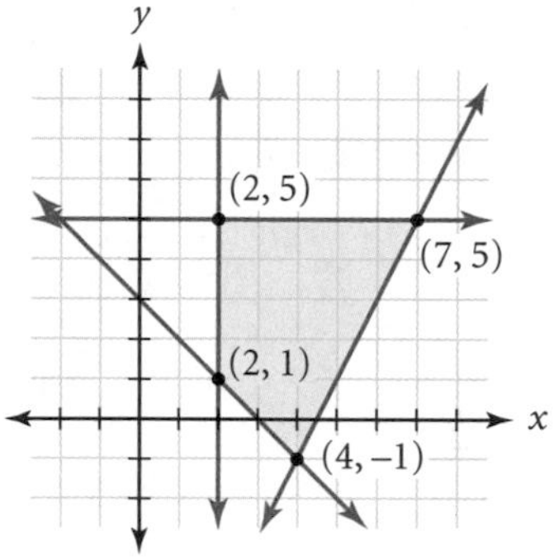

13. (Lesson 7.5) Give the system of inequalities the solution set of which is this polygon.

14. (Lesson 7.3) Consider this system of equations.

$\begin{cases} y = 2x - \dfrac{1}{2} \\ 5x - 2y + 5 = 0 \end{cases}$

a. Write the augmented matrix for this system.

b. Reduce the augmented matrix to row-echelon form. Write the solution as an ordered pair of coordinates.

c. Check your solution values for x and y by substituting them in the original equations.

15. (Lessons 3.5 and 6.2) Find the equation of this parabola.

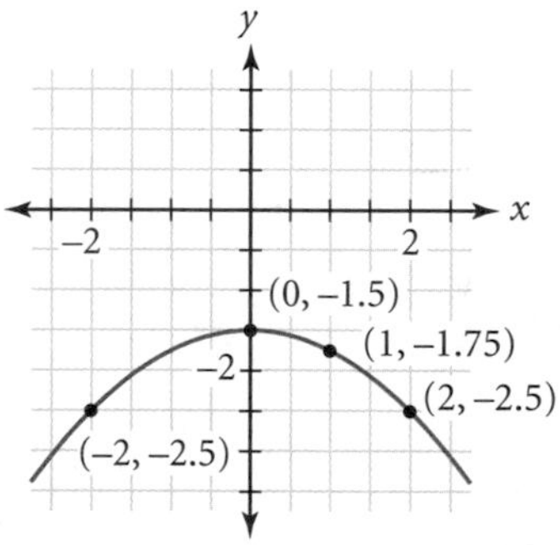

IMPROVING YOUR REASONING SKILLS

Secret Survey

Eric is doing a survey. He has a deck of cards and two questions written on a sheet of paper. He says, "Pick a card from the deck. Don't show it to me. If it's a red card, answer Question 1. If it's a black card, answer Question 2." You pick a card, look at the paper, and respond yes. Eric records your answer, shuffles the cards, and goes on to the next person. At the end of the survey, Eric has gathered 37 yeses and 23 noes. He calculates that 73% of those surveyed own a pet.

Question 1 (red card):
Does your phone number end in an even number?
Question 2 (black card):
Do you own a pet?

Explain how Eric was able to find this result without knowing which question each person was answering. Based on the number of responses, do you think that Eric's survey is valid?

EXPLORATION
with Microsoft® Excel

Analyzing Power Functions

Download the Chapter 7 Excel Exploration from the website at www.keycollege.com/online or from your Student CD.

Power functions always have consistent differences that tell you the degree of the function. For example, a power function of degree two (a quadratic function) has consistent second differences. A cubic function has consistent third differences, and so on. You will use technology to analyze any power function and determine its equation, using a table of differences.

On the spreadsheet you will find complete instructions and questions to consider as you explore.

Activity

Solving Systems of Equations using Matrix Inverses

Excel skills required

- Data-fill
- Name a cell or a cell range
- Multiply matrices

In this activity, you will investigate another way to solve systems of equations.

Launch Excel.

You can find instructions for the required Excel skills in the Excel Notes at www.keycollege.com/online or on your Student CD.

You will be working on the system of equations:

$$\begin{cases} 2x - 3y + 5z = -19 \\ 3x + 5y + z = 7 \\ 6x + 2y - 3z = 26 \end{cases}$$

Step 1 Make an augmented matrix in Excel that represents this system as follows:

	A	B	C	D
1	2	-3	5	-19
2	3	5	1	7
3	6	2	-3	26

Step 2 Name ranges as follows:

a. Highlight cells A1 to C3 and name this range of cells `CoeffMatrix`.

b. Highlight cells D1 to D3 and name this range of cells `ConstMatrix`.

Step 3 You will now find the inverse of the coefficient matrix and place it in the cell range F1 to H3. The process is explained in the Excel Notes but reviewed here since the details are important.

Start by highlighting cells F1 to H3. With cells F1 to H3 highlighted, press the **F2** key. Cell F1 will be prepared to receive a formula that will act on all of the cells marked. This is called an array formula. Whenever you are entering an array formula, you have to be sure that the range into which you are placing the result is the correct size. For example, if the result is supposed to be a 3 by 3 matrix, the range highlighted for the result should have 3 rows and 3 columns,

Step 4 In cell F1 type `=MINVERSE(CoeffMatrix)` being careful NOT to press the enter key. If you press the Enter key, the only calculated cell will be F1.

Step 5 To apply this formula to the entire array of marked cells, use the key combination CTRL+SHIFT+ENTER.

Step 6 Highlight the array F1 to H3 and name this range of cells `TheInverse`.

Step 7 Into cell A5 type `Determinant of the Coefficient Matrix` and into cell E5 type `=MDETERM(CoeffMatrix)`.

Step 8 Prepare labels for the answers in cells A7 to A9 as follows:

7	x =
8	y =
9	z =

Step 9 Highlight cells B7 to B9. These cells will receive the answer to the system of equations, if it exists.With cells B7 to B9 highlighted, press F2 and type `=MMULT(TheInverse,ConstMatrix)`. Remember to use CTRL+SHIFT+ENTER to apply the array formula properly.

Now you're ready to use the spreadsheet to solve linear systems of equations with three unknowns.

Questions

1. Record your answer from the above setup. If the original system is represented by the matrix equation $AX = B$, list A, A^{-1}, $A^{-1}B$. Record also the determinant of A.

2. Change the original system as follows:

	A	B	C	D
1	1	1	5	11
2	3	-3	1	23
3	2	5	4	1

a. What system of equations does this represent?

b. What is the determinant of the coefficient matrix?

c. What is the inverse of the coefficient matrix?

d. What is the answer to this system of equations?

3. Change the system as follows:

	A	B	C	D
1	1	2	11	2
2	-1	4	1	5
3	1	-2	3	-6

a. What system of equations does this represent?

b. What does the determinant of the coefficient matrix tell you?

c. What is the inverse of the coefficient matrix, if it exists?

d. Are there any answers to the original system? How can you tell?

4. Change the system as follows:

	A	B	C	D
1	1	2	11	-8
2	-1	4	1	5
3	1	-2	3	-6

a. What system of equations does this represent?

b. What does the determinant of the coefficient matrix tell you?

c. What is the inverse of the coefficient matrix, if it exists?

d. Are there any answers to the original system? How can you tell?

e. Compare this system to the system in Question 3 above.

5. At this point you have several methods at your disposal for solving systems of equations. You have the traditional manual methods using substitution and elimination. Here, you now have a method using matrix inverses. The companion Exploration deals with the method of row-reduction. What are the advantages and disadvantages of the different methods?

CHAPTER 7 REVIEW

Matrices are used in a variety of ways. They provide ways to organize data about such things as inventory or the coordinates of vertices of a polygon. A **transition matrix** represents repeated percent changes in a system over time. You can use matrix arithmetic to transform polygons on a coordinate graph. You can also use matrix multiplication to determine the quantities at various stages of a transition simulation.

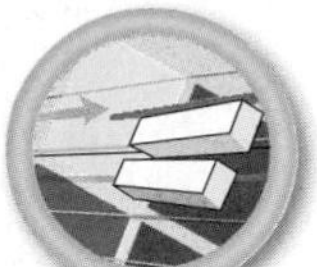

Another important application of matrices is to solve systems of equations. In Chapter 3 you solved systems of linear equations by looking for a point of intersection on a graph or by using substitution or elimination. With two linear equations, the system will have one solution (intersecting lines), will be **inconsistent** (parallel lines with no solution), or will be **dependent** (the same line with infinitely many solutions). You can use an **inverse matrix** and matrix multiplication to solve a system, or you can use an **augmented matrix** and the **row reduction** method. For systems that involve more than three equations and three variables, matrix methods are generally the simplest.

When a system is made up of inequalities, the solution usually consists of many points that can be represented by a region in the plane. One important use of systems of inequalities is in **linear programming.** In linear programming problems an equation of a quantity that is to be optimized (maximized or minimized) is evaluated at the vertices of the **feasible region.**

EXERCISES

1. (Lesson 7.2) Use these matrices to solve the following arithmetic problems. If a problem is impossible to solve, explain why.

$$[A] = \begin{bmatrix} 1 & -2 \end{bmatrix} \quad [B] = \begin{bmatrix} -3 & 7 \\ 6 & 4 \end{bmatrix} \quad [C] = \begin{bmatrix} 1 & 0 \\ 5 & 2 \end{bmatrix} \quad [D] = \begin{bmatrix} -3 & 1 & 2 \\ 2 & 3 & -2 \end{bmatrix}$$

a. $[A] + [B]$ **b.** $[B] - [C]$ **c.** $4 \cdot [D]$

d. $[C] \cdot [D]$ **e.** $[D] \cdot [C]$ **f.** $[A] \cdot [D]$

2. (Lesson 7.4) Find the inverse, if it exists, of each matrix.

a. $\begin{bmatrix} 2 & -3 \\ 1 & -4 \end{bmatrix}$

b. $\begin{bmatrix} 5 & 2 & -2 \\ 6 & 1 & 0 \\ -2 & 5 & 3 \end{bmatrix}$

c. $\begin{bmatrix} -2 & 3 \\ 8 & -12 \end{bmatrix}$

d. $\begin{bmatrix} 5 & 2 & -3 \\ 4 & 3 & -1 \\ 7 & -2 & -1 \end{bmatrix}$

3. (Lesson 7.3) Solve each system by using row reduction.

a. $\begin{cases} 8x - 5y = 17 \\ 6x + 4y = 33 \end{cases}$

b. $\begin{cases} 5x + 3y - 7z = 3 \\ 10x - 4y + 6z = 5 \\ 15x + y - 8z = -2 \end{cases}$

4. (Lesson 7.4) Solve each system by using an inverse matrix.

a. $\begin{cases} 8x - 5y = 17 \\ 6x + 4y = 33 \end{cases}$

b. $\begin{cases} 5x + 3y - 7z = 3 \\ 10x - 4y + 6z = 5 \\ 15x + y - 8z = -2 \end{cases}$

5. (Lesson 7.4) Identify each system as consistent (having a solution), inconsistent (having no solution), or dependent (having infinitely many solutions).

a. $y = -1.5x + 7$

$y = -3x + 14$

b. $y = \frac{1}{4}(x - 8) + 5$

$y = 0.25x + 3$

c. $2x + 3y = 4$

$1.2x + 1.8y = 2.6$

d. $\frac{3}{5}x - \frac{2}{5}y = 3$

$0.6x - 0.4y = -3$

6. Graph the feasible region of each system of inequalities. Find the coordinates of each vertex. Then, identify the point that maximizes the final expression.

a. $\begin{cases} 2x + 3y \le 12 \\ 6x + y \le 18 \\ x + 2y \ge 2 \\ x \ge 0, y \ge 0 \end{cases}$

maximize: $1.65x + 5.2y$

b. $\begin{cases} x + y < 50 \\ 10x + 5y < 440 \\ 40x + 60y < 2400 \end{cases}$

maximize: $6x + 7y$

7. (Lesson 7.2) A particular color of paint requires five parts red, six parts yellow, and two parts black. Thomas does not have the pure colors available, but he finds three premixed colors that he can use. The first is two parts red and four parts yellow; the second is one part red and two parts black; the third is three parts red, one part yellow, and one part black.

a. Write an equation that gives the correct portion of red by using the three available premixed colors.

b. Write an equation that gives the correct portion of yellow and another equation that gives the correct portion of black.

c. Solve the system of equations in 8a and b.

d. Find an integer that can be used as a scalar multiplier for your solutions in 8c to provide integer solution values.

e. Explain your solution in real-world terms.

8. (Lesson 7.1) Springfield State University freshmen are housed in three dormitories: Lupton, Snowdon, and Claytor. Lupton is an all-female dorm, Snowdon is an all-male dorm, and Claytor is coed. In August there were 800 students in Lupton, 600 students in Snowdon, and 700 students in Claytor. This transition graph shows the average monthly movements during fall semester.

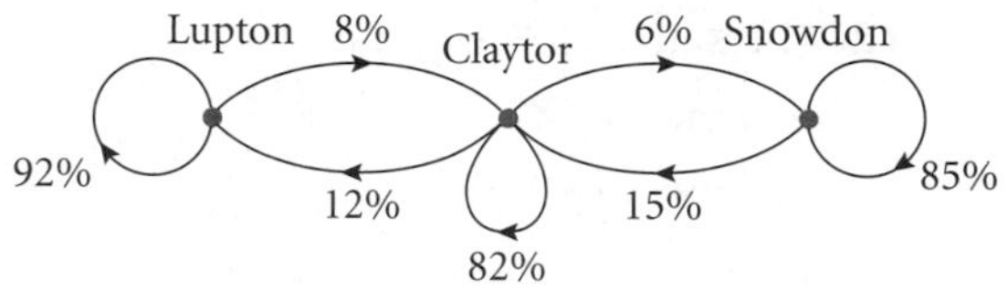

a. Write a transition matrix for this situation. List the dorms in the order Lupton, Claytor, Snowdon.

b. Based on the information provided, what were the populations of each of the three dormitories after four months of switches?

9. **APPLICATION** (Lesson 7.6) Matt, Yolanda, and Myriam have a small business producing handmade shawls and blankets. They spin the yarn, dye it, and weave it. A shawl requires 1 h of spinning, 1 h of dyeing, and 1 h of weaving. A blanket needs 2 h of spinning, 1 h of dyeing, and 4 h of weaving. They make a \$16 profit per shawl and a \$20 profit per blanket. Matt does the spinning on his day off, when he can spend at most 8 h spinning. Yolanda does the dyeing on her day off, when she can spend up to 6 h dyeing. Myriam does all the weaving on Friday and Saturday, when she has at most 14 h available. How many of each item should they make each week to maximize their profit?

Mixed Review

10. **APPLICATION** (Lesson 3.1) Heather's water heater needs repair. The repair service says it will cost \$300 to fix the unit, which currently costs \$75 per year to operate. Or Heather could buy a new water heater for \$500 including installation, and the new heater would save 60% on annual operating costs. How long would it take for the new unit to pay for itself?

11. (Lesson 2.1) Graphs 11a–c were all produced by a recursive formula in the form

$$u_0 = a$$
$$u_n = u_{n-1} \cdot (1 + p) + d \quad \text{where } n \geq 1$$

For each case tell if *a*, *p*, and *d* are greater than zero (positive), equal to zero, or less than zero (negative).

a.

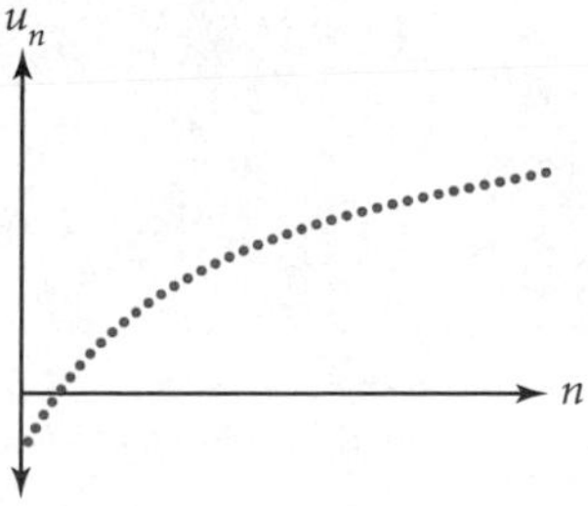

b.

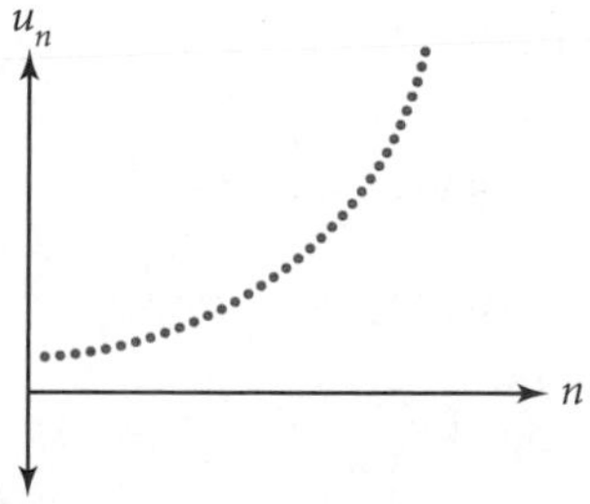

c.

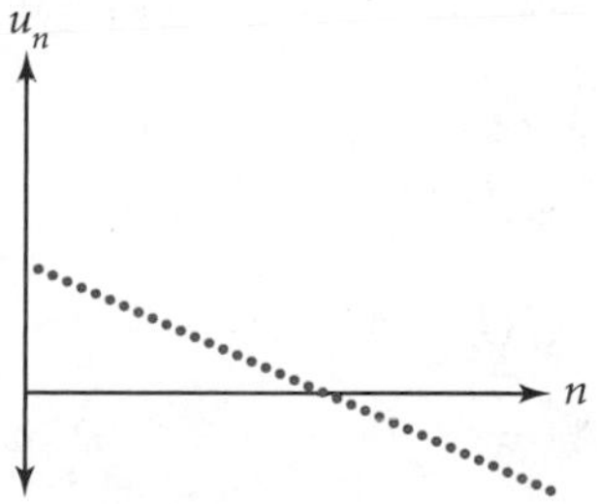

12. (Lesson 4.2) The graph of $y = f(x)$ is shown here.

a. Find $f(-3)$.

b. Find x such that $f(x) = 1$.

c. Is f a function? Explain why or why not.

d. What is the domain of f?

e. What is the range of f?

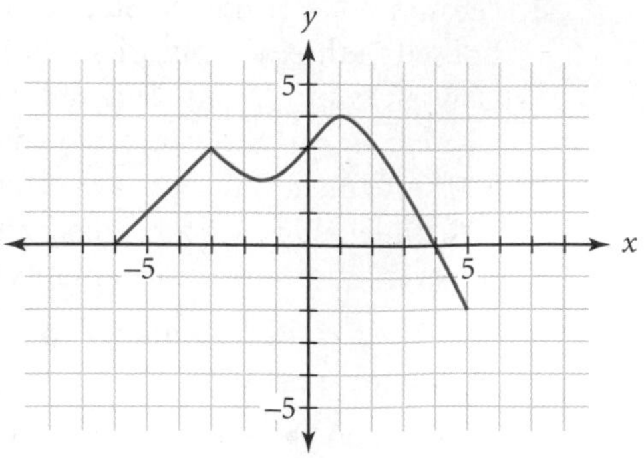

13. (Lesson 3.5) Last semester all of Dr. Nolte's students did projects. One-half of the students in her morning Math for Liberal Arts class investigated fractals, one-fourth of the students in that class did research projects, and the remaining students conducted surveys and analyzed their results. In her afternoon class one-third of the students investigated fractals, one-half of the students did research projects, and the remaining students conducted surveys and analyzed their results. In the evening class, one-fourth of the students investigated fractals, one-sixth of the students did research projects, and the remaining students conducted a survey and analyzed their results. Overall, 22 students investigated fractals, 18 students did research projects, and 22 students conducted surveys. How many students are in each of Dr. Nolte's classes?

14. APPLICATION (Lesson 1.3) Canada's oil production has increased over the last half-century. This table gives the per-day production of oil for various years.

Canada's Oil Production

Year	1960	1970	1980	1990	1995	1998	1999
Barrels per day (millions)	0.52	1.26	1.44	1.55	1.80	1.98	1.91

(*The New York Times Almanac, 2002*)

a. Define variables, and make a scatter plot of the data.

b. Use your scatter plot to predict oil production rates over the next 20 years.

c. According to the Canadian Association of Petroleum Producers (CAPP), by the year 2015 Canada will produce approximately 3.3 million barrels of oil per day. How does this prediction compare with what you might expect based on your scatter plot?

This oil-drilling ship was frozen in six feet of ice on the Beaufort Sea, Northwest Territories, Canada, in 1980.

15. (Lesson 5.7) Solve.

a. $\log 35 + \log 7 = \log x$

b. $\log 500 - \log 25 = \log x$

c. $\log\sqrt{\frac{1}{8}} = x\log 8$

d. $15(9.4)^x = 37{,}000$

e. $\sqrt[3]{(x+6)} + 18.6 = 21.6$

f. $\log_6 342 = 2x$

16. APPLICATION (Lesson 2.2) Suppose the signal strength in a fiber-optic cable diminishes by 15% every 10 mi.

a. What percent of the original signal is left after a stretch of 10 mi?

b. Create a table of the percentage of signal strength remaining in 10-mi intervals, and make a graph of the sequence.

c. If the phone company plans to boost the signal just before it falls to 1%, how far apart should the company place its booster stations?

TechnologyConnection Fiber-optic technology uses light pulses to transmit information from one transmitter to another down fiber lines made of silica (glass). Fiber-optic strands are used in telephone wires, cable television lines, power systems, and other communications. These strands operate on the principle of total internal reflection, which means that the light pulses cannot escape out of the glass tube and instead bounce information from transmitter to transmitter.

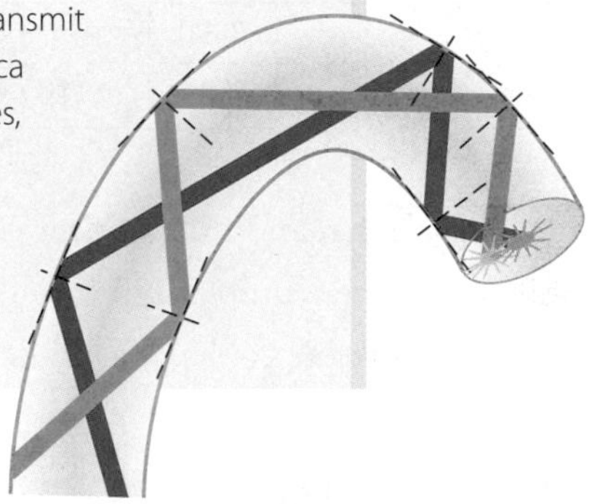

Light rays propagate by reflecting between walls of fiber-optic cable.

17. (Lesson 5.1) The graph of an exponential function passes through the points (4, 50) and (6, 25.92).

a. Find the equation of the exponential function.

b. What is the growth rate? Does the equation model growth or decay?

c. What is the y-intercept?

d. What is the long-run value?

18. (Lesson 1.4) This data set gives the weights in kilograms of the crew members participating in the 2002 boat race between Oxford University and Cambridge University. (*www.cubc.org.uk* and *www.ourcs.org*)

{83, 95, 91, 90, 93, 97.5, 97, 79, 55, 89, 89.5, 94, 89, 100, 90, 92, 96, 54}

a. Make a box plot of these data.

b. Are the data skewed left, skewed right, or symmetric?

c. Identify any outliers.

d. What is the percentile rank of the crew member who weighs 94 kg?

19. (Lesson 7.2) Consider the graph of $y = x^2$. Let matrix $[P]$, which organizes the coordinates of the points shown, represent the parabola.

$$[P] = \begin{bmatrix} -2 & -1 & 0 & 1 & 2 \\ 4 & 1 & 0 & 1 & 4 \end{bmatrix}$$

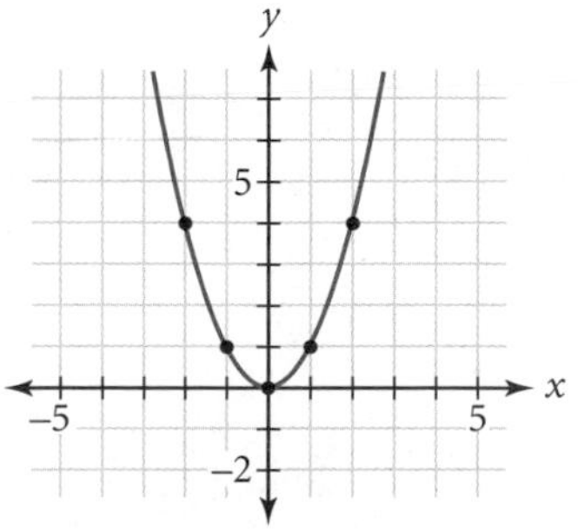

a. Describe the transformation(s) that would give the graph of $y = (x - 5)^2 - 2$. Sketch the graph.

b. Describe the transformation(s) that would give the graph of $y = -2x^2$. Sketch the graph.

c. Sketch the image of the graph of $y = x^2$ represented by the product $-1 \cdot [P]$. Describe the transformation(s).

d. Write a matrix equation that represents the image of the graph of $y = x^2$ after a translation left 2 units and up 3 units.

20. **APPLICATION** (Lesson 3.3) This table gives the mean population per U.S. household for several years.

Mean Household Population

Year	1890	1930	1940	1950	1960	1970	1980	1990	1998
Mean population	4.93	4.11	3.67	3.37	3.35	3.14	2.76	2.63	2.62

(*The New York Times Almanac, 2001*)

a. Define variables, and find a linear equation that fits these data.

b. Write and solve a problem that you could solve by interpolation with your line of fit.

c. Write and solve a problem that you could solve by extrapolation with your line of fit.

Assessing What You've Learned

- Can you name the elements of a matrix by row and column?
- Can you create a transition matrix to model a real-life problem?
- Can you perform matrix operations—addition, subtraction, and multiplication?
- Do you know what the *inverse matrix* and the *identity matrix* are?
- Can you use matrices to solve systems of equations and systems of inequalities?
- Do you know how to use linear programming as a tool in problem solving?

JOURNAL Choose one or more of the following questions to answer in your journal.

- What kinds of matrices can be added to or subtracted from one another? What kinds of matrices can you multiply? Is matrix multiplication commutative? That is, if you multiply $[A] \times [B]$, will you always get the same product as when you multiply $[B] \times [A]$? Why or why not? What are the identity matrix and inverse matrix, and what are they used for?
- You have learned five methods for finding a solution of a system of linear equations: graphing, substitution, elimination, matrix row reduction, and inverse matrix. Which method do you like best? Which one is the most challenging to you? What are the advantages and disadvantages of each method?
- You have now studied more than half of this book. What mathematical skills developed in the previous chapters were most crucial to your success with this chapter? Which concepts are your strengths and weaknesses?

CHAPTER

8 Parametric Equations and Trigonometry

Project Bandaloop is a group of contemporary dancers who combine dance, rock climbing, and rappelling. Their choreography relies on a blend of aerial, horizontal, and vertical movement. By performing in a variety of venues that range from skyscrapers to granite cliffs, the group hopes to stimulate their audiences' awareness of the natural environment.

OBJECTIVES

In this chapter you will

- use a third variable to write parametric equations that separately define x and y
- simulate objects in motion with parametric equations
- convert parametric equations to equations in x and y and vice versa
- review the trigonometric ratios: sine, cosine, and tangent
- use the trigonometric ratios and their inverses to solve applications

LESSON

8.1

Graphing Parametric Equations

Envisioning the end is enough to put the means in motion.

DOROTHEA BRANDE

In this chapter you will use a very powerful tool, the parametric equation, to model two variables interacting. Parametric equations allow you to use a graphing calculator to model simultaneously the vertical and horizontal paths of objects in motion. For example, we can use parametric equations to show the paths of boats crossing the ocean, including both north-south and east-west components. You will also extend your earlier function translation work, applying translations to parametric equations. Finally, you will use trigonometry to solve problems. In some cases you will use trigonometry to gather information about right triangles. In other cases you will use trigonometric laws to solve problems involving other kinds of triangles.

Sherlock Holmes followed footprints and clues to track down criminals. By following the clues, he discovered exactly where the person had been. The path could be drawn on a map, every location described by x- and y-coordinates. But how could you indicate *when* the criminal was at each place? Often, two variables are not enough to fully describe interesting graph situations. You can use **parametric equations** to describe separately the x- and y-coordinates of a point as functions of a third variable, t, called the **parameter**—which gives you more information and better control over what points you plot.

In this example the variable t represents time, and you will write parametric equations to simulate motion on your calculator screen. Observe how t controls the x- and y-values.

EXAMPLE A

Two tankers leave Corpus Christi, Texas, at the same time, traveling toward St. Petersburg, Florida, 900 mi east. Tanker A travels at a constant speed of 18 mi/h, and Tanker B travels at a constant speed of 22 mi/h. Write parametric equations and use your calculator to simulate the motion involved in this situation.

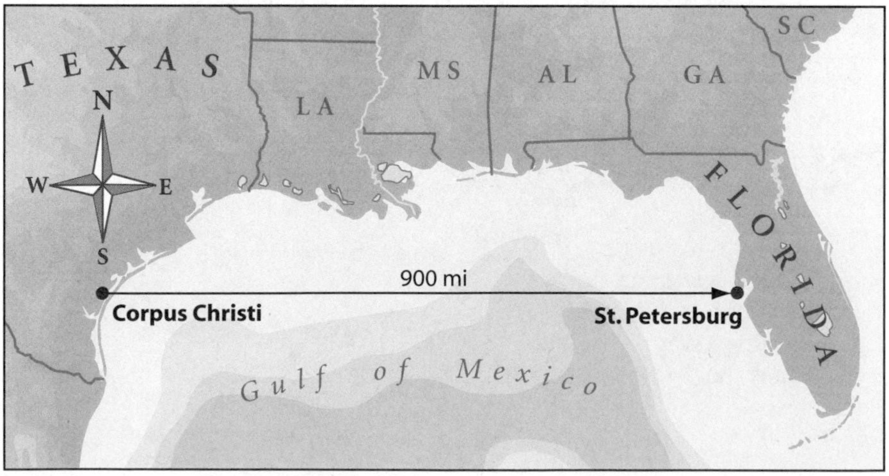

EnvironmentalConnection Many oil spills occur when overcrowded seaports and poor weather cause oil tankers to collide in the ocean. Whether tankers are traveling toward land or to an offshore refinery to deliver oil, these conditions threaten the safety of less sturdy, single-hulled ships. Since passage of the Oil Pollution Act of 1990, which placed federal safety regulations on the transport of oil, tanker spills have been at an all-time low.

▶ *Solution*

To graph the situation, you must first establish a coordinate system. Locate the origin, (0, 0), at Corpus Christi. Because St. Petersburg is directly east, its coordinates are (900, 0). The x-coordinate of each plotted point is the distance of a tanker from Corpus Christi, and the y-coordinate of each path will remain constant because the tankers travel directly east.

Set your calculator in *parametric* and *simultaneous* modes. [▶ See **Calculator Note 8A** to learn about the mode settings for parametric equations.◀] To determine your equations, think about the motion of the tankers. After 1 h traveling at 18 mi/h, Tanker A will be 18 mi from Corpus Christi. After 2 h it will be 36 mi away; after 10 h it will be 180 mi away; and after t h it will be $18t$ mi away. The horizontal distance traveled, then, is determined by $x = 18t$. This equation locates the horizontal position of Tanker A at any given time; the x-value is dependent on the time, or t-value. The tanker does not travel north or south at all, so the y-value must remain constant. If you choose a y-value of 0, the tanker will travel on the x-axis, and you will not be able to see it. To make the path more visible, indicate that this is the first tanker, Tanker A, by setting $y = 1$. The equations $x = 18t$ and $y = 1$ are a pair of parametric equations. You can model the motion of Tanker B with the parametric equations $x = 22t$ and $y = 2$. Why is its horizontal position determined by $22t$?

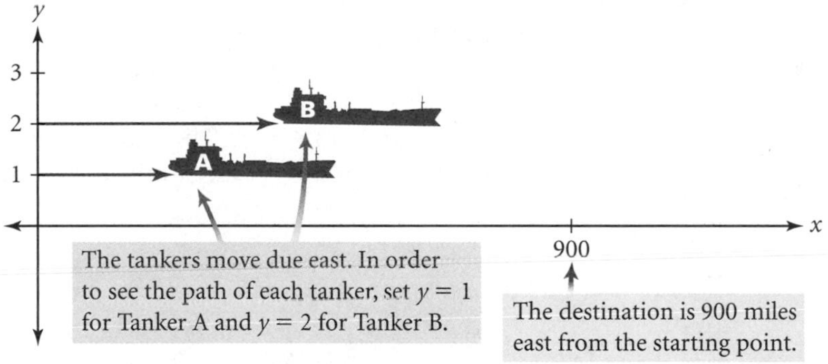
y
3
2
1
B
A
x
900
The tankers move due east. In order to see the path of each tanker, set y = 1 for Tanker A and y = 2 for Tanker B.
The destination is 900 miles east from the starting point.

To find a good graphing window, consider the situation. The tankers must travel 900 miles east, so the x-values should be in the domain $0 \leq x \leq 900$. You chose constant y-values of 1 and 2, so the range $-1 \leq y \leq 3$ is sufficient. When graphing parametric equations, you must also determine an interval for t. The slower tanker goes 18 miles per hour, so it takes 50 hours to go 900 miles; therefore, t-values must range from 0 to 50. Enter these values into your calculator as shown below. Note that you must also enter a t-step. A t-step of 0.1 means that a point is plotted every 0.1 h (or 6 min). Graph the equations on your calculator, and observe the motion of the tankers. [▶ See **Calculator Note 8B** to learn how to enter and graph parametric equations and to learn more about window settings.◀]

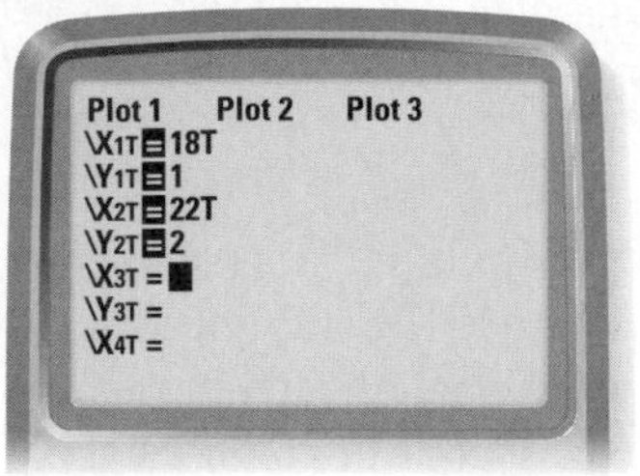

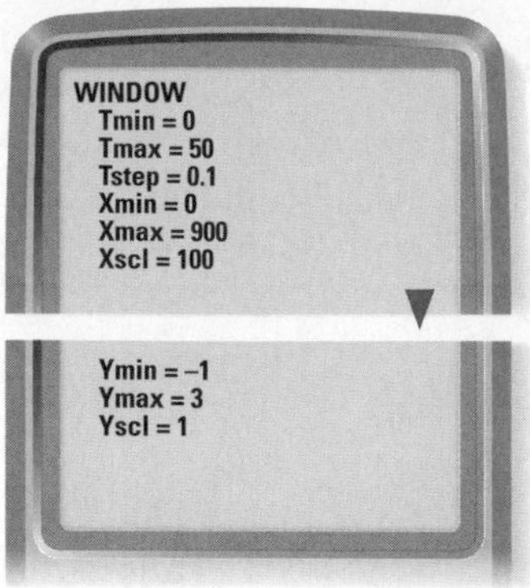

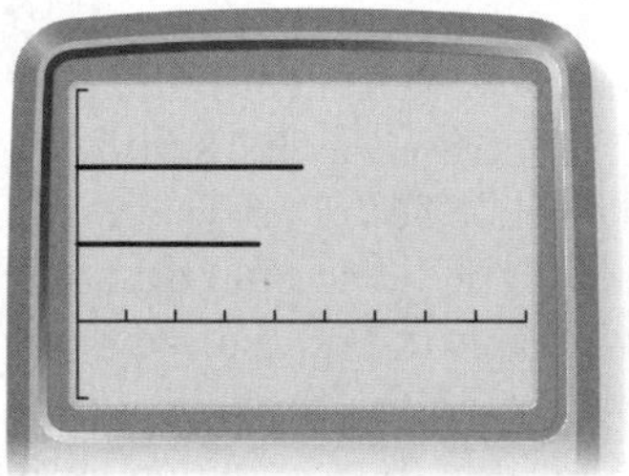

[0, 900, 100, −1, 3, 1]

The graph when $t = 20$.

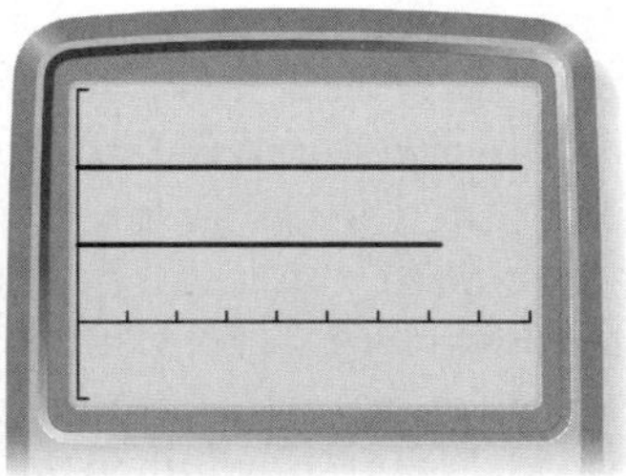

[0, 900, 100, −1, 3, 1]

The graph when $t = 40$.

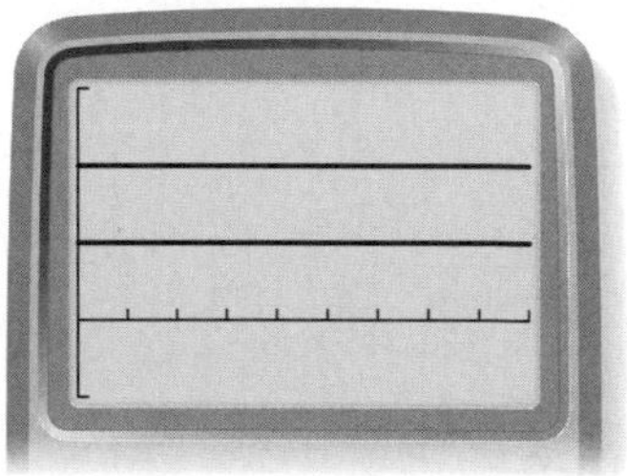

[0, 900, 100, −1, 3, 1]

The graph when $t = 50$.

When you model motion using parametric equations, you must consider the object's direction as well as its speed. Directed speed is called **velocity.** Velocity, unlike speed, can be either positive or negative. If you correctly determine whether the velocity is positive or negative, you will be able to see a realistic simulation of the motion on your calculator screen. This table shows how to relate the sign of the velocity to the direction of the motion.

Motion	Velocity
Up or north	Positive
Down or south	Negative
Left or west	Negative
Right or east	Positive

Investigation
Simulating Motion

In this Investigation you will explore the use of parametric equations and graphs to answer questions about motion.

Step 1 Before you begin this activity, enter the equations and graphing window from Example A into your calculator. Trace the path of the appropriate tanker, and write and solve equations to help you answer each question. [▶ See **Calculator Note 8C** to learn about tracing parametric equations. ◀]

a. How long does it take the faster tanker to reach St. Petersburg?

b. Where is the slower tanker when the faster tanker reaches its destination?

c. When is the faster tanker exactly 82 mi in front of the slower tanker?

d. During what part of the trip are the tankers less than 60 mi apart?

Step 2 Create another question involving these two tankers. Write an explanation of and an answer to your question on a separate sheet of paper. Exchange your question with another group, and try to answer each other's questions.

Parametric equations allow you to graph paths when the location of the points is dependent on time or some other parameter. A parametric representation lets you see the dynamic nature of the motion, and you can adjust the plotting speed by changing the *t*-step. The parameter, *t*, doesn't always have to represent time; it can be a unitless number. In the next example you will control which *x*- and *y*-values are plotted by limiting the range of the *t*-values, and you'll explore transformations of parametric functions.

EXAMPLE B Consider the parametric equations $x = t$ and $y = t^2$ for $-1 \le t \le 2$.

a. Graph the equations on graph paper and on your calculator.

b. Write equations to translate this parabola right 2 units and down 3 units.

▶ Solution

a. Use the equations to calculate *x*- and *y*-values that correspond to values of *t* in the range $-1 \le t \le 2$. If you use more *t*-values, you will have a smoother graph. Graph the points as *t* increases, connect each point to the previous one, and add arrows to indicate the direction. Verify this graph on your calculator.

t	*x*	*y*
−1	−1	1
0	0	0
1	1	1
2	2	4

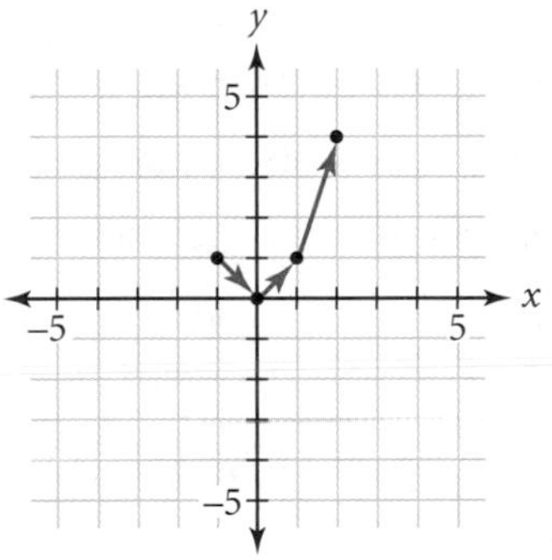

b. To translate the graph right 2 units, every *x*-coordinate should be increased by 2. To translate the graph down 3 units, every *y*-coordinate should be decreased by 3. The equations are

$$x = t + 2$$
$$y = t^2 - 3$$

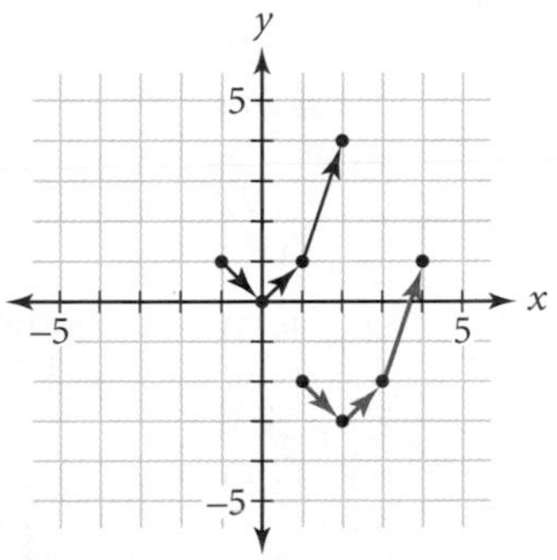

A graph defined parametrically is the set of points (x, y) defined by t. What was the effect of limiting the values of the variable t on the parabola in Example B? What do you think the effect will be if t-values are limited to $-2 \leq t \leq 1$? You will continue to explore translations and transformations of parametric functions in the Exercises.

EXERCISES

Practice Your Skills

1. Create a table for each equation with $t = \{-2, -1, 0, 1, 2\}$.

a. $x = 3t - 1$, $y = 2t + 1$

b. $x = t + 1$, $y = t^2$

c. $x = t^2$, $y = t + 3$

d. $x = t - 1$, $y = \sqrt{4 - t^2}$

2. Graph each pair of parametric equations, and use arrows to indicate the direction of increasing t-values along the graph. Limit your t-values as indicated; if an interval for t isn't listed, then find one that shows all of the graph that fits in a friendly window with a factor of 2.

a. $x = 3t - 1$, $y = 2t + 1$

b. $x = t + 1$, $y = t^2$

c. $x = t^2$, $y = t + 3$, $-2 \leq t \leq 1$

d. $x = t - 1$, $y = \sqrt{4 - t^2}$, $-2 \leq t \leq 2$

3. Explore these graphs, using a friendly window with a factor of 2 and $-10 \leq t \leq 10$.

a. Graph $x = t$ and $y = t^2$.

b. Graph $x = t + 2$ and $y = t^2$. How does this graph compare with the graph in 3a?

c. Graph $x = t$ and $y = t^2 - 3$. How does this graph compare with the graph in 3a?

d. Predict how the graph of $x = t + 5$ and $y = t^2 + 2$ will compare with the graph in 3a. Verify your conjecture.

e. Predict how the graph of $x = t + a$ and $y = t^2 + b$ will compare with the graph in 3a.

4. Explore the graphs, using a friendly window with a factor of 2 and $-10 \leq x \leq 10$.

a. Graph $x = t$ and $y = |t|$.

b. Graph $x = t - 1$ and $y = |t| + 2$. How does this graph compare with the graph in 4a?

c. Write a pair of parametric equations that will translate the graph in 4a left 4 units and down 3 units.

d. Graph $x = 2t$ and $y = |t|$. How does this graph compare with the graph in 4a?

e. Graph $x = t$ and $y = 3|t|$. How does this graph compare with the graph in 4a?

f. Describe how the numbers 2, 3, and 4 in the equations $x = t + 2$ and $y = 3|t| - 4$ change the graph in 4a. Verify your conjecture.

Reason and Apply

5. These two graphs display a simulation of a team's mascot walking toward the west-end goal on the college football field. Starting at a point 65 yards (yd) from the goal line and 50 ft from the sideline, the mascot moves toward the goal line; he waves at the crowd when he is 35 yd from the goal line. The parameter, t, represents time in seconds.

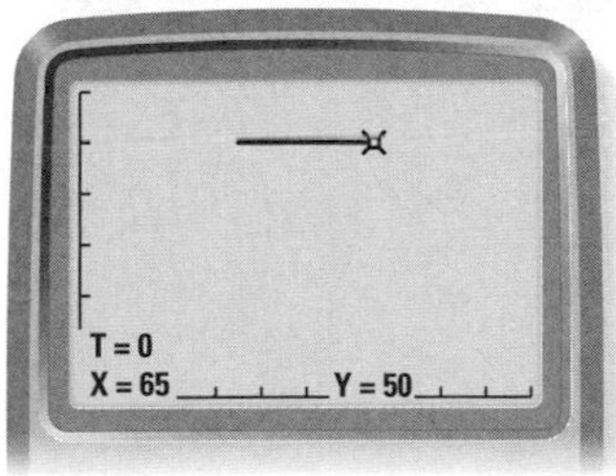

[0, 100, 10, 0, 60, 10]

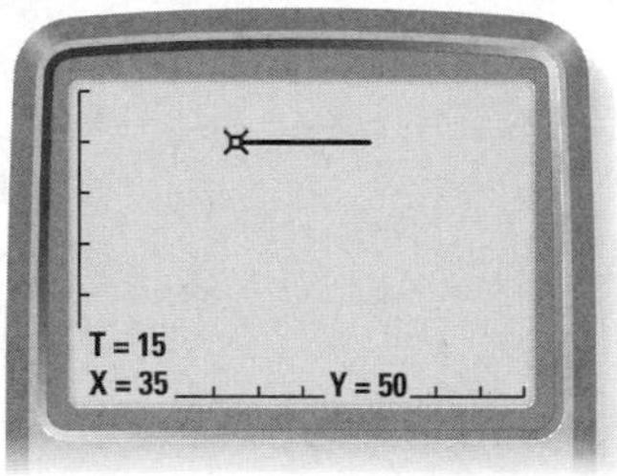

[0, 100, 10, 0, 60, 10]

a. According to the graphs, how much time has elapsed?

b. How far has the mascot traveled?

c. What is his average velocity?

d. Explain the meaning of each number in these parametric equations.

$x = 65 - 2t \qquad y = 50$

e. Using the equations from 5d, simulate the mascot's motion on your calculator screen in an appropriate window.

f. Increase the maximum t-value in your graphing window to determine when the mascot crosses the 10-yard line.

g. Write and solve an equation that will determine when the mascot crosses the 10-yard line.

6. Write parametric equations for each graph. (*Hint:* You can create parametric equations from a single equation in x and y by letting $x = t$ and changing y to a function of t by replacing x with t.)

a.

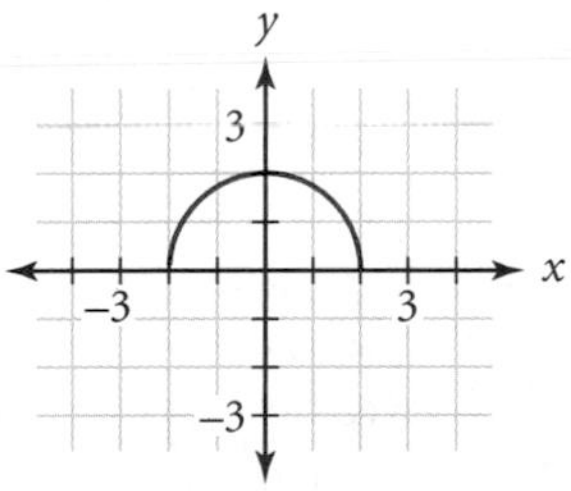

b.

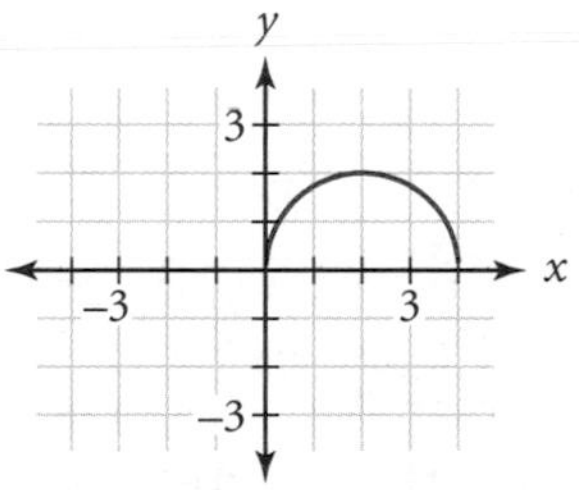

c.

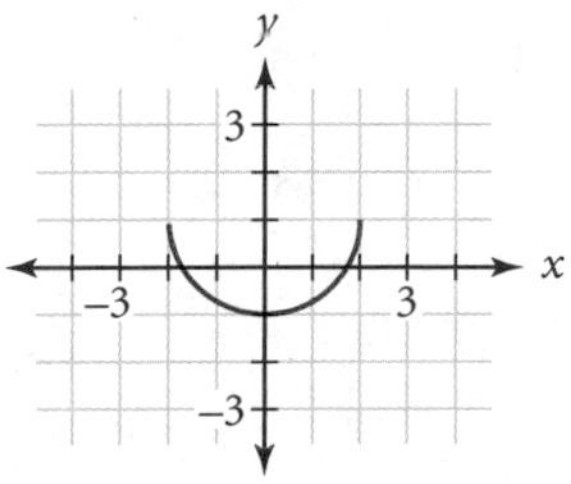

d.

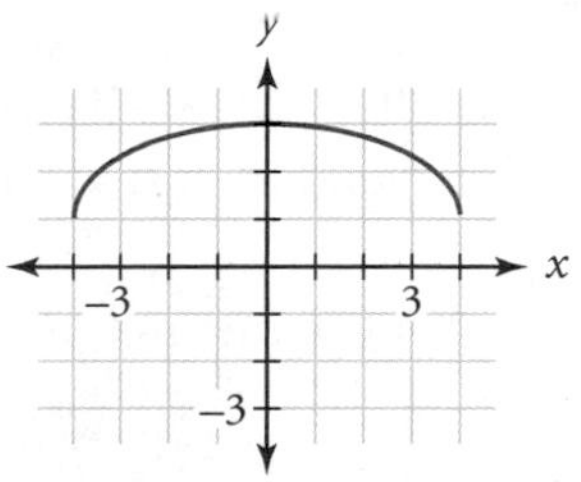

7. The graph of parametric equations $x = f(t)$ and $y = g(t)$ is pictured here.

a. Sketch a graph of $x = f(t)$ and $y = -g(t)$, and describe the transformation.

b. Sketch a graph of $x = -f(t)$ and $y = g(t)$, and describe the transformation.

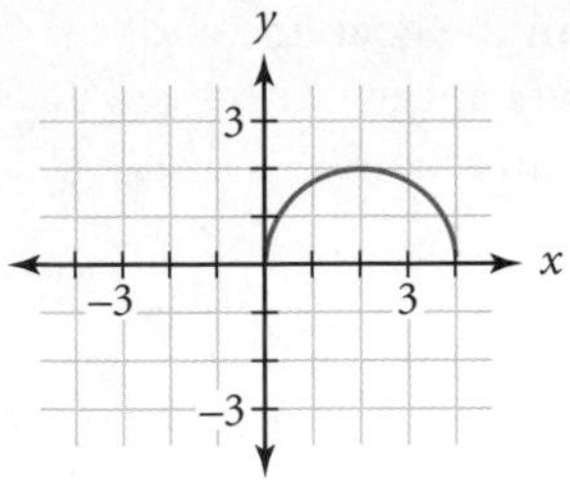

8. The graph of the parametric equations $x = f(t)$ and $y = g(t)$ is pictured here. Write the parametric equations of each graph 8a–d in terms of $f(t)$ and $g(t)$.

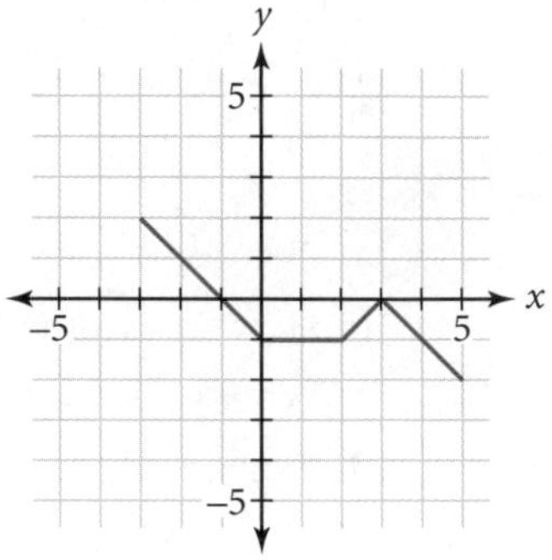

a.

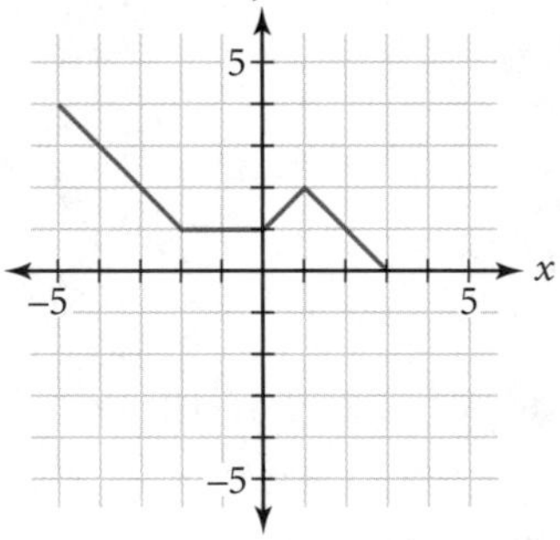

b.

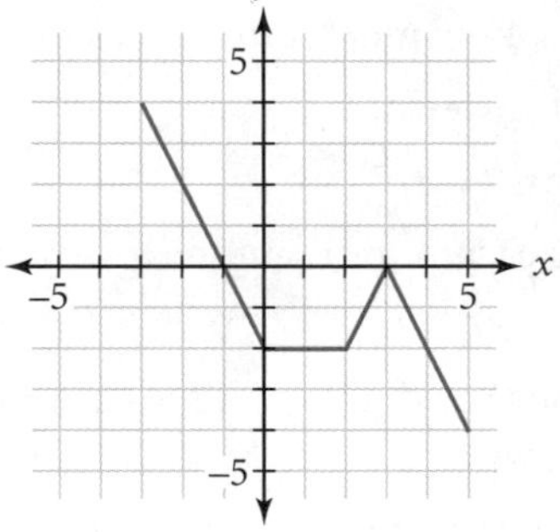

c.

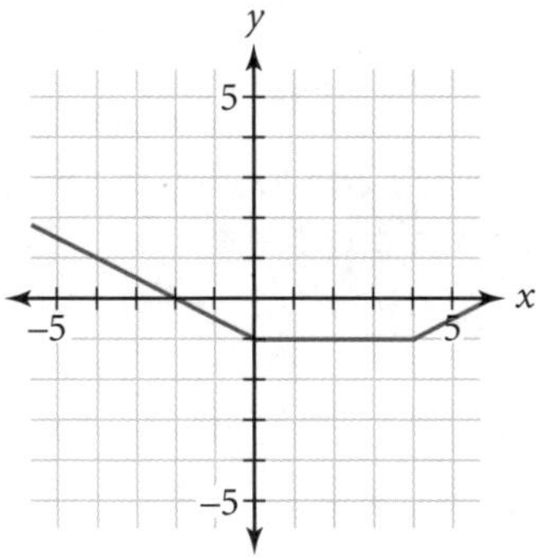

d.

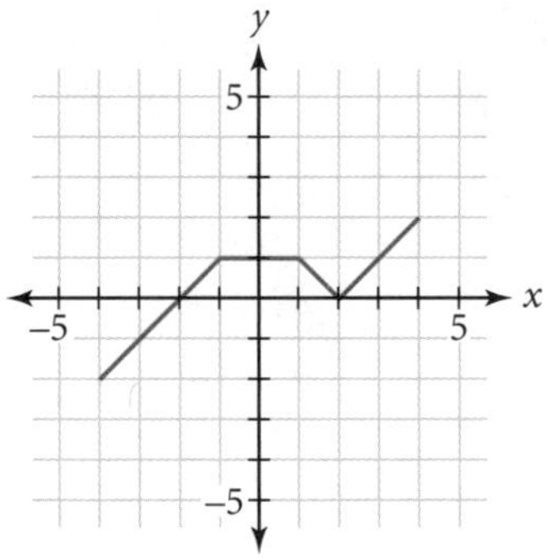

9. Each spring a hare challenges a tortoise to a 50-m race. The hare knows that he can run faster than the tortoise, and he boasts that he can still win the race even if he gives the tortoise a 100-s head start. The tortoise crawls at a rate of 0.4 m/s, while the hare's running speed is 1.8 m/s.

a. Write equations to simulate the motion of the tortoise.

b. Determine a graphing window and a range of t-values that will show the tortoise's motion.

c. Write equations to simulate the motion of the hare, who crosses the starting line at $t = 100$.

d. Who wins the race?

e. How long does it take each runner to finish the race?

10. These parametric equations simulate two walkers, with x and y measured in meters and t measured in seconds. Graph their motion for $0 \le t \le 5$.

$$x = 1.4t \quad \text{and} \quad x = 4.7$$
$$y = 3.1 \qquad\qquad y = 1.2t$$

a. Give real-world meanings of the values of 1.4, 3.1, 4.7, and 1.2 in the equations.

b. Where do the two paths intersect?

c. Do the walkers collide? How do you know this?

11. Los Angeles, California, is about 2500 km east of Honolulu, Hawaii. One plane flies from Los Angeles to Honolulu, and a second plane flies in the opposite direction.

a. Describe the meaning of each number in the x-equations. (The y-equations simply assign noncolliding flight paths, and positive values indicate distances traveled eastward.)

$$x = 450(t - 2) \quad \text{and} \quad x = 2500 - 525t$$
$$y = 1 \qquad\qquad y = 2$$

b. Use tables or a graph to find the approximate time and location of the planes relative to Los Angeles when the planes pass each other.

c. Write and solve an equation to find the time and location of the planes when they pass each other.

CareerConnection From airport control towers, air traffic controllers coordinate airplane takeoffs and landings. Controllers manage air traffic safety, track weather conditions, and communicate with pilots via radio. Sophisticated computer equipment and radar displays allow them to monitor airspace and prevent aircraft collisions. Air traffic control is considered one of the most demanding jobs in the country.

An airport control tower

12. **APPLICATION** The U.S. Navy is testing a weapon that hits a target that is floating in the ocean; the collision sends sound waves through the water as well as through the air. Sound can travel at approximately 340 m/s in air and 1500 m/s in water.

a. Write parametric equations that model the sound waves for 15 s, and graph them in an appropriate window.

b. How far does the sound wave in the water travel in 15 s? How long does it take the sound wave in the air to travel the same distance?

c. A ship monitoring the test hears the sound wave in the air 10 s after its sonar detects the sound wave in the water. How far is the ship from the target?

ScienceConnection The speed of a wave is the rate at which vibrations move through a medium. Sound moves faster in water than in air and faster in warm water than in cold water. The speed of a sound wave is slow in comparison to the speed of a light wave. Light travels through air at approximately 300,000,000 m/s, nearly 900,000 times faster than the speed of sound. This is why, during a thunderstorm, you hear the thunder several seconds after you see the lightning flash. If thunder is heard 5 s after lightning is seen, the storm is approximately 1700 m away.

Review

13. (Lesson 3.4) Solve this system.

$$\begin{cases} 3x + 5y = 6 \\ 4y = x - 19 \end{cases}$$

14. (Lesson 4.3) Write an equation of each line.

a. The line $y = 3x + 2$ translated right 5 units

b. The line $y = -1.4x$ translated down 3 units

c. The line $y = -x$ translated right 1.5 units and down 3 units.

15. (Lesson 2.1) Consider this sequence.

$-6, -4, 3, 15, 32, \ldots$

a. Using finite differences, find the nth term.

b. Find the 20th term.

16. (Lesson 3.1) Find the equation of the line through $(-3, 10)$ and $(6, -5)$.

(Lesson 6.2) Find the equation of the parabola that passes through the points $(-2, -20)$, $(2, 0)$, and $(4, -14)$.

LESSON

8.2

Converting Parametric to Nonparametric Equations

The future is not some place we are going to, but one we are creating.

JOHN SCHAAR

In the previous lesson you saw parametric equivalents of functions you have studied earlier, such as parabolas and absolute value graphs. You can create these by letting $x = t$ and making y a function of t instead of a function of x. For example, you can graph $y = x^2$ in parametric form using $x = t$ and $y = t^2$. The focus of this lesson is to do the reverse—to start with parametric equations and eliminate the parameter t to get a single equation using only x and y.

In this Investigation you will use parametric equations to model motion by measuring changes in both the x- and y-directions. You will discover the relationship between the graphs of the parametric equations, and the corresponding graph of the relation between x and y.

Investigation
Parametric Walk

You will need

- two motion sensors
- masking tape

This Investigation involves four participants: the walker, recorder X, recorder Y, and a director.

Procedure Note

1. The walker starts at one end of the segment and walks slowly for 5 s to reach the other end.
2. Recorder X points at the walker a motion sensor set for about 5 s and moves along the y-axis, keeping even with the walker, thus measuring the x-coordinate of the walker's path as a function of time.
3. Simultaneously, recorder Y points at the walker a motion sensor set for the same amount of time and moves along the x-axis, keeping even with the walker, measuring the y-coordinate of the walker's path as a function of time.
4. The director starts all three participants at the same moment and counts out the 5 s.

Step 1 Mark a quadrant of a coordinate graph on the floor with tape, identifying the x- and y-axes and a segment, as in the diagram.

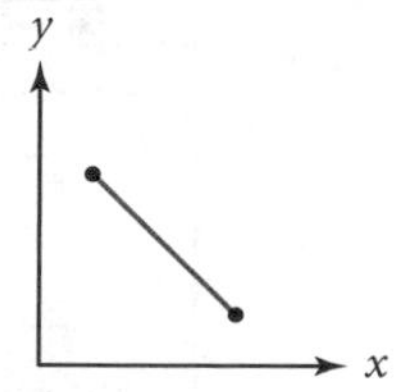

Step 2 Perform the activity as described in the Procedure Note. [▶ Revisit **Calculator Note 4D** for additional instructions for setting up your motion sensors. ◀]

Step 3 Download the data from recorder X's motion sensor into a calculator, and find a function in the form $x = f(t)$ that fits the data.

Step 4 Download the data from recorder Y's motion sensor into a different calculator, and find a function in the form $y = g(t)$ that fits the data.

Step 5 Transfer the distance values from recorder X's motion sensor into List L1 and the distance values from recorder Y's motion sensor into List L2. Plot these data as ordered pairs (x, y) along with the parametric functions $x = f(t)$ and $y = g(t)$. How do they compare?

Step 6 Solve $x = f(t)$ for t, and substitute this expression for t in $y = g(t)$.

Step 7 Change to function mode on your calculator, and graph your solution to Step 6 with the (x, y) data.

Step 8 Based on this Investigation, explain what eliminating the parameter does to parametric equations.

During this Investigation two motion sensors captured data that you used to create a parametric model of the walk. You then merged these two graphs into a single graphic representation of the relation between x and y, and you merged the two parametric equations into a single equation using only x and y. Similarly, in the next example you will merge a pair of parametric equations into a single equation by eliminating the parameter t.

EXAMPLE A Graph the curve described by the parametric equations $x = t^2 - 4$ and $y = \frac{t}{2}$. Then eliminate t from the equations, and graph the result.

▶ **Solution** Make a table of values, and plot the points, connecting them as t increases. Verify this graph on your calculator. Notice that the graph shows that y is not a function of x, even though both x and y are functions of t.

t	x	y
−3	5	−1.5
−2	0	−1
−1	−3	−0.5
0	−4	0
1	−3	0.5
2	0	1
3	5	1.5

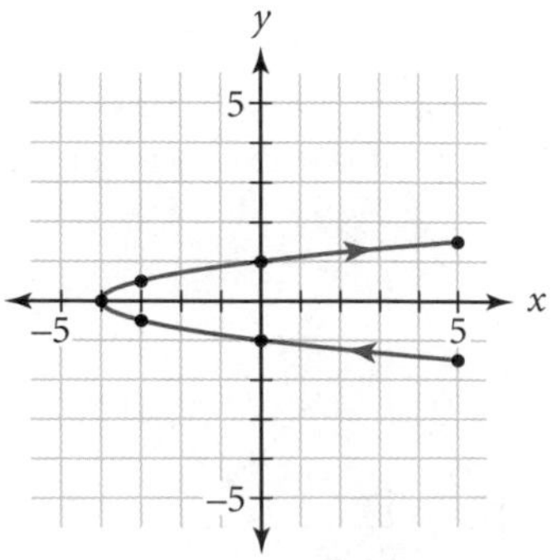

To eliminate the parameter, solve one of the parametric equations for t and substitute it in the other parametric equation.

$x = t^2 - 4$ and $y = \frac{t}{2}$	The parametric equations for x and y.
$t^2 = x + 4$	Add 4 to both sides in the parametric equation for x.
$t = \pm\sqrt{x+4}$	Take the square root of both sides to solve for t.
$y = \frac{t}{2} = \frac{\pm\sqrt{x+4}}{2}$	Substitute $\pm\sqrt{x+4}$ for t in the parametric equation for y to get a single equation using only x and y.
$y = \pm\sqrt{\frac{x+4}{2}}$	The single equation for y in terms of x.

You could also solve for t in the original parametric equation for y and substitute in the parametric equation for x.

$x = t^2 - 4,\ y = \frac{t}{2}$	The given equations.
$t = 2y$	Multiply by 2 to solve for t in the parametric equation for y.
$x = (2y)2 - 4$	Substitute $2y$ for t in the parametric equation for x.
$x = 4y^2 - 4$	Expand.
$x + 4 = 4y^2$	Add 4 to both sides.
$\sqrt{\frac{x+4}{4}} = y^2$	Divide both sides by 4.
$y = \pm\sqrt{\frac{x+4}{4}}$	Take the square root of both sides.
$y = \pm\sqrt{\frac{x+4}{2}}$	Take the square root of 4.

Notice that both methods result in an equivalent equation.

You might recognize this as the equation of a horizontally oriented parabola. Check your result by graphing to show that the graphs of

$$y = \pm\frac{\sqrt{x+4}}{2} \quad \text{and} \quad \begin{aligned} x &= t^2 - 4 \\ y &= \frac{t}{2} \end{aligned}$$

are the same. [▶ Revisit **Calculator Note 8B** to learn how to simultaneously graph parametric equations and equations that use only x and y. ◀]

Although often the same graph can be created with parametric equations or with a single equation using only x and y, parametric equations show your position at particular times and allow you to graph directly relations that are not functions. The next example shows how two independent actions can combine to make a single path.

EXAMPLE B

Hanna's hot-air balloon is ascending at a rate of 15 ft/s. A wind is blowing continuously from west to east at 24 ft/s. Write parametric equations to model this situation, and decide whether the hot-air balloon will clear power lines that are 300 ft to the east and 95 ft tall. Find the time it takes for the balloon to reach the power lines.

▶ Solution

Create a table of time, ground distance, and height for a few seconds of flight. Set the origin as the initial launching location of the balloon. Let x represent the ground distance traveled to the east in feet, and let y represent the balloon's height above the ground in feet. The table shows these values for the first 4 s of flight.

Time (s) t	Ground distance (ft) x	Height (ft) y
0	0	0
1	24	15
2	48	30
3	72	45
4	96	60

The parametric equations that model the motion are $x = 24t$ and $y = 15t$. Graph this pair of equations on your calculator. You can picture the power lines by plotting the point (300, 95). When you trace the graph to a time of 1 s, you will see that the balloon is 24 ft to the east at a height of 15 ft. At 12.5 s it has traveled 300 ft to the east and has reached a height of 187.5 ft. Hanna's balloon will not touch the power lines.

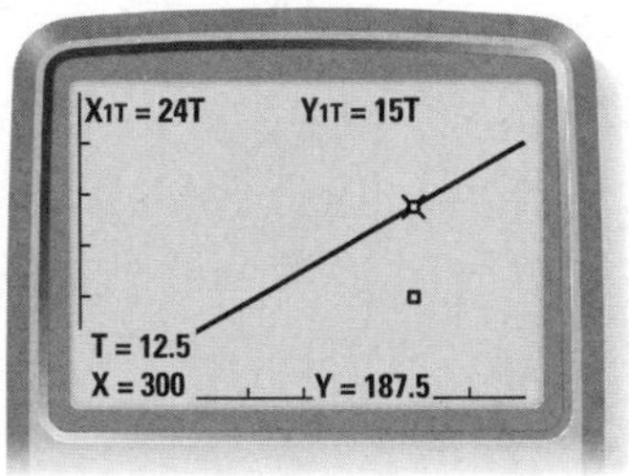

[0, 400, 100, 0, 300, 100]

Modeling motion with parametric equations is much like the graphing you have done in earlier chapters, but each of the directions is dealt with independently. Often, this will make difficult relations easier to model. Although some parametric equations cannot be written as a single equation using only x and y, most can. The ability to eliminate the parameter of these equations is an important skill because it gives you two different ways to study a relationship.

EXERCISES

Practice Your Skills

1. Solve each equation for t.

a. $x = t + 1$ **b.** $x = 3t - 1$ **c.** $x = t^2$ **d.** $x = t - 1$

2. Write each pair of parametric equations as a single equation using only x and y. Graph this new relation in a friendly graphing window. Verify that the graph of the new equation is the same as the graph of the pair of parametric equations.

a. $x = t + 1$
$y = t^2$

b. $x = 3t - 1$
$y = 2t + 1$

c. $x = t^2$
$y = t + 3$

d. $x = t - 1$
$y = \sqrt{4 - t^2}$

3. Write a single equation, using only x and y, that is equivalent to each pair of parametric equations.

a. $x = 2t - 3$
$y = t + 2$

b. $x = t^2$
$y = t + 1$

c. $x = \frac{1}{2}t + 1$
$y = \frac{t - 2}{3}$

d. $x = t - 3$
$y = 2(t - 1)^2$

4. The table gives x- and y-values for several values of t.

a. Write an equation of x in terms of t.

b. Plot the points in the table, and then sketch the graph of the equation of y in terms of t.

c. Write an equation of y in terms of t.

d. Eliminate the parameter and combine the equations in 4a and 4c. Verify that this equation fits the values of (x, y).

t	x	y
0	2	2
1	3	1
2	4	0
3	5	−1
4	6	0
5	7	1
6	8	2

5. Use the graphs of $x = f(t)$ and $y = g(t)$ to create a graph of y as a function of x.

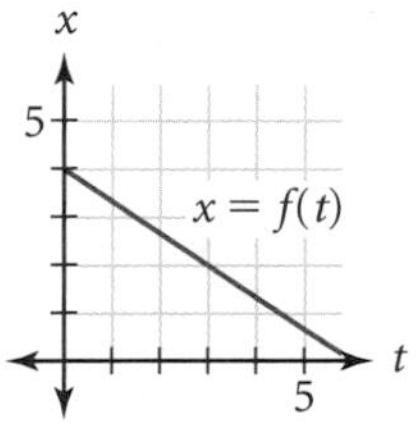

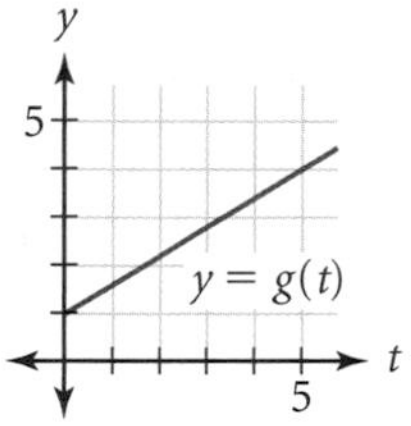

Reason and Apply

6. Write parametric equations for $x = f(t)$ and $y = g(t)$ and an equation for y as a function of x in Exercise 5. How do the slopes of these equations compare?

7. Find the smallest interval for t that produces a graph of the parametric equations $x = t + 2$ and $y = t^2$ that just fits in a window with $-5 \leq x \leq 5$ and $-6 \leq y \leq 6$.

8. Graph the parametric equations $x = f(t) = t + 2$ and $y = g(t) = \sqrt{1 - t^2}$.

a. Graph $x = f(t)$ and $y = -g(t)$, and identify the transformations of the original equations. Eliminate the parameter to write a single equation.

b. Graph $x = -f(t)$ and $y = g(t)$, and identify the transformations of the original equations. Eliminate the parameter to write a single equation.

c. Graph $x = -f(t)$ and $y = -g(t)$, and identify the transformations of the original equations. Eliminate the parameter to write a single equation.

9. A bug is crawling up a wall with locations given in the table. The variables x and y represent horizontal and vertical distances from the lower left corner of the wall, measured in inches, and t represents time measured in seconds.

a. Write parametric equations for x and y in terms of t that generate the information given in the table.

b. Graph your parametric equations from 9a with the data points (x, y). How do they relate?

c. Using only x and y, write the equation of the line that passes through the data points.

d. How can the parametric equations be used to find the slope of the line in 9c?

t	x	y
0	20	5
2	24	7
4	28	9
6	32	11
8	36	13
10	40	15

10. Write parametric equations of two perpendicular lines that intersect at the point (3, 2), with one line having a slope of -0.5.

11. The functions $d_1 = 1.5t$ and $d_2 = 12 - 2.5t$ represent Edna's and Maria's distances in miles from a trailhead, as functions of time in hours.

a. Write parametric equations to simulate Edna's hike north away from the trailhead. Let $x = 1$.

b. Write parametric equations to simulate Maria's hike south toward the trailhead. Use $x = 1.1$ so that Maria will come close to meeting Edna without actually bumping into her.

c. Name a graphing window and a range of t-values that show these hikes.

d. Use your graph to approximate when and where the two hikers meet.

e. What equation can you write to find when they meet? Solve this equation symbolically to verify your answer to 11d.

12. An egg is dropped from the roof of a 98-m building.

a. How long will it take the egg to reach the ground?

b. Write and graph parametric equations to model the motion.

c. When will the egg reach a height of 1.75 m?

d. A 1.75-m-high trampoline is rolled at a rate of 1.2 m/s toward the egg drop site. How far away should the trampoline start so that when the egg is dropped, it makes a direct hit?

e. Simulate both motions with parametric equations.

Review

13. (Lesson 8.1) Here are the graphs of $x = t$ and $y = t$, and $x = t$ and $y = t^2$. Write parametric equations of the parabola's reflection across the line $y = x$.

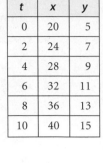

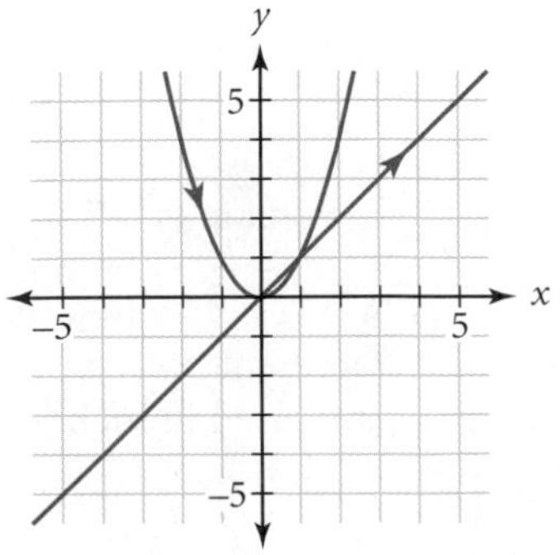

14. (Lesson 8.1) Tanker A moves at 18 mi/h, and Tanker B moves at 22 mi/h. Both are traveling from Corpus Christi, Texas, to St. Petersburg, Florida, 900 mi directly east. Simulate the tanker movements if Tanker A leaves Corpus Christi at noon and Tanker B leaves at 5 P.M.

a. Write parametric equations to simulate the motion.

b. Name a graphing window and a range of t-values that show this situation.

c. When and where does Tanker B pass Tanker A?

d. Simulate the tanker movements if both tankers leave at noon but Tanker A leaves from Corpus Christi and Tanker B leaves from St. Petersburg, each heading toward the other. Record your equations, and determine the time interval during which they are within 50 mi of each other.

15. (Lesson 4.3) What is the equation of the image of $y = \frac{2}{3}x - 2$ after a translation right 5 units and up 3 units?

16. (Lesson 4.2) Consider the function $f(x) = 3 + \sqrt[3]{(x - 1)^2}$.

a. Find $f(9)$, $f(1)$, $f(0)$, and $f(-7)$.

b. Find the equation(s) of the inverse of $f(x)$. Is the inverse a function?

c. Describe how you can use your calculations in 16a to check your inverse in 16b.

d. Use your calculator to graph $f(x)$ and its inverse on the same axes.

LESSON

8.3

Right Triangle Trigonometry

Happiness is not a destination. It is a method of life.

BURTON HILLIS

Panama City, Florida, is 750 mi from Corpus Christi, Texas, at a bearing of 73°. Model the movement of a tanker as it travels from Corpus Christi to Panama City.

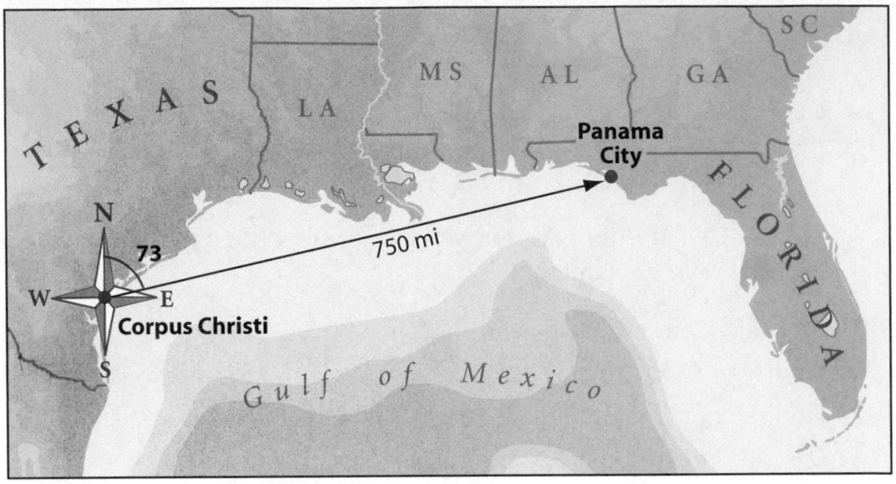

Trigonometry can be used with parametric equations to model motion that is at an angle to the horizontal, such as in the problem above. A **bearing** of 73° refers to a 73° angle measured clockwise from north, as shown here. In this case you must model a motion that isn't just horizontal or vertical. You can begin by separating any motion in two dimensions into its horizontal and vertical components.

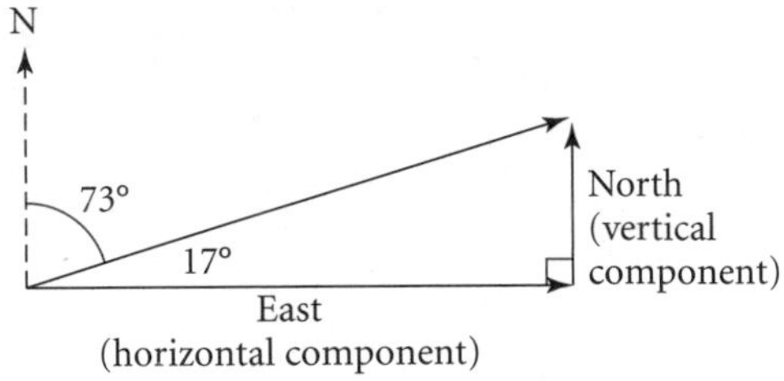

To solve a problem like this one, you will use the **trigonometric ratios.**

The word *trigonometry* comes from the Greek words for "triangle" and "measure." **Trigonometry** relates angle measures to ratios of sides in similar triangles. For example, in the similar right triangles shown here, all corresponding angles are congruent. The ratio of the length of the shorter leg to the length of the longer leg is always 0.75, and the ratios of the lengths of other pairs of corresponding sides are also equal. In a right triangle, each of these ratios has a special name.

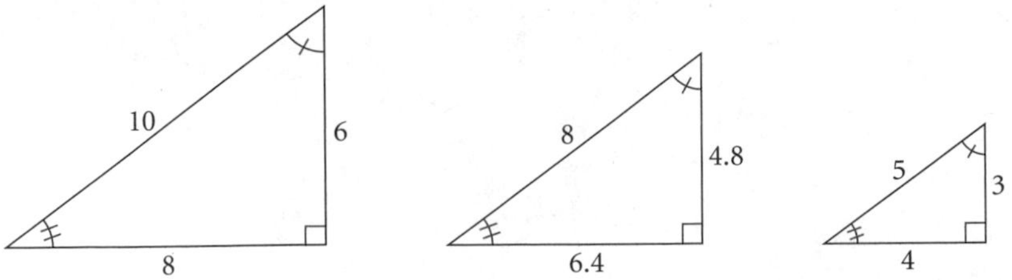

The Trigonometric Ratios

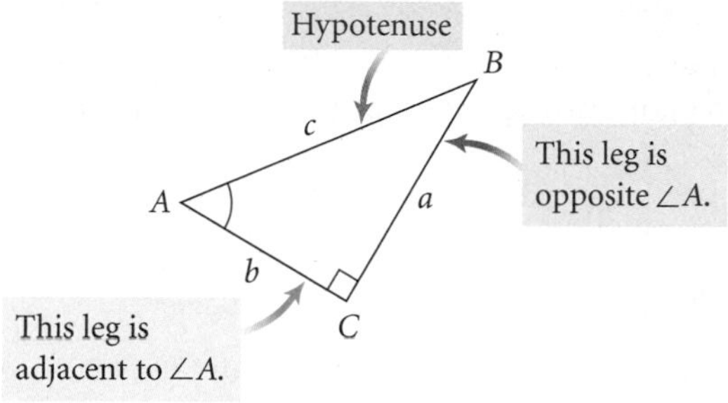

For any acute angle A in a right triangle, the **sine** of $\angle A$ is the ratio of the length of the leg opposite $\angle A$ to the length of the hypotenuse.

$$\sin A = \frac{\text{opposite leg}}{\text{hypotenuse}} = \frac{a}{c}$$

The **cosine** of $\angle A$ is the ratio of the length of the leg adjacent to $\angle A$ to the length of the hypotenuse.

$$\cos A = \frac{\text{adjacent leg}}{\text{hypotenuse}} = \frac{b}{c}$$

The **tangent** of $\angle A$ is the ratio of the length of the opposite leg to the length of the adjacent leg.

$$\tan A = \frac{\text{opposite leg}}{\text{adjacent leg}} = \frac{a}{b}$$

LanguageConnection The word *sine* has a curious history. The Sanskrit term was *jya-ardha* ("half-cord"), later abbreviated to *jya*. Islamic scholars, who learned about sine from India, called it *jiba*. In Segovia around 1140, Robert of Chester read *jiba* as *jaib* when he was translating al-Khwarizmi's book, *Kitab al-jabr wa'al-muqabalah,* from Arabic into Latin. One meaning of *jaib* is "indentation" or "gulf." So *jiba* was translated into Latin as *sinus*, meaning "fold" or "indentation," and from that we get the word *sine*.

You can use the trigonometric ratios to find unknown side lengths of a right triangle when you know the measure of one acute angle and the length of one of the sides. [▶ See **Calculator Note 8E** to learn about calculating the trigonometric ratios on your calculator. ◀]

EXAMPLE A

Find the unknown length, b.

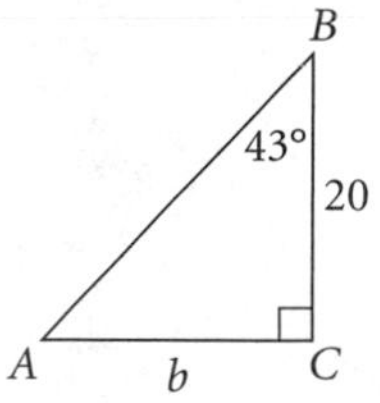

▶ Solution

In this problem you know the length of one leg, you know the measure of one acute angle, and you want to find the length of the other leg. The tangent ratio relates the lengths of the legs to the measure of the angle. Therefore, you can use the tangent ratio. (The sine or cosine would be harder to use because you would need to write an expression for the length of the hypotenuse.)

$\tan 43° = \dfrac{b}{20}$	Write the tangent ratio, and substitute the known values of the angle measure and length of the adjacent leg.
$b = 20 \tan 43°$	Multiply both sides by 20.
$b \approx 18.65$	Use a calculator to find the tangent of 43°, and multiply it by 20.

The length, b, of $\overline{AC}$ is approximately 18.65 units.

Note that the trigonometric ratios apply to both acute angles in a right triangle. In this triangle, b is the leg adjacent to A and the leg opposite B, whereas a is the leg opposite A and the leg adjacent to B. As always, c is the hypotenuse.

$\sin A = \frac{a}{c}, \sin B = \frac{b}{c}$

$\cos A = \frac{b}{c}, \cos B = \frac{a}{c}$

$\tan A = \frac{a}{b}, \tan B = \frac{b}{a}$

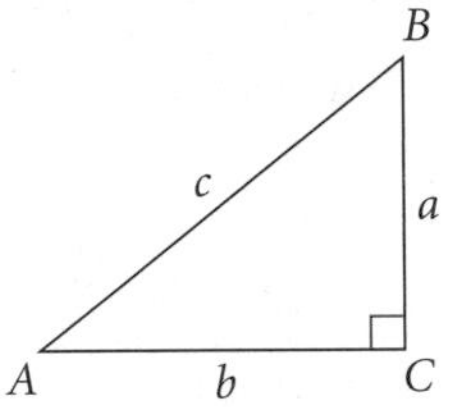

Investigation
Two Ships

In this Investigation you will model the motion of two cargo ships traveling from Corpus Christi. Ship A is traveling on a bearing of 73° toward Panama City, 750 mi away. Ship B is traveling on a bearing of 90° toward St. Petersburg, 900 mi away. Both ships are traveling at 23 mi/h.

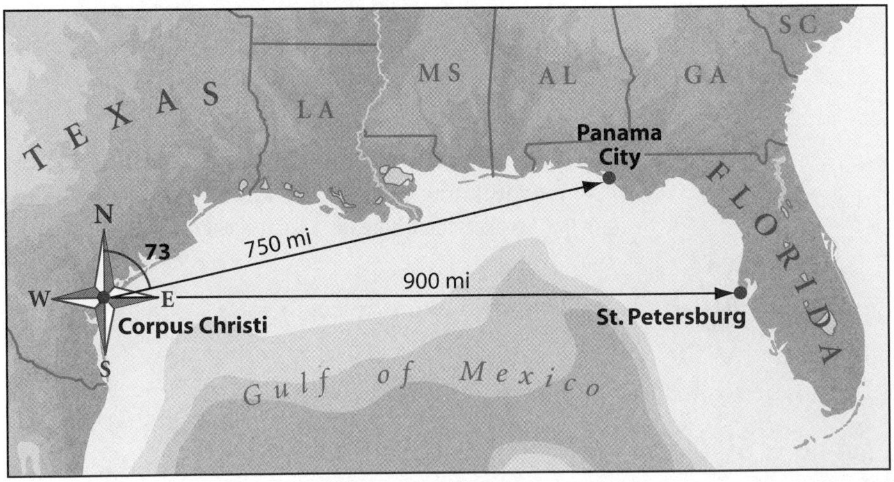

Step 1 For each ship write an equation that gives its distance from Corpus Christi as a function of time.

Step 2 | Solve each equation from Step 1 to find out when each ship will arrive at its destination.

As with other problems in this chapter, often you will need to choose a coordinate system or point of reference for application problems. Most of the time, it makes sense to locate the origin at the initial point of the problem.

Step 3 | How far east of Corpus Christi is each ship after 1 h? 2 h? How far north is each ship at those times? (*Hint:* You may need to use trigonometry to determine values for Ship A.)

Step 4 | Write a pair of parametric equations of each ship's location as a function of time. Explain where each number in each equation comes from. For numbers that are approximations, change them back with exact form (using trigonometric ratios).

Step 5 | How far north of Corpus Christi is Panama City? How far east?

Step 6 | Use the parametric equations you wrote in Step 4 to find out when each ship will arrive at its destination. Explain how your answers relate to your answers from Step 2.

Each trigonometric ratio is a function of the angle measure A because each angle measure has a unique ratio associated with it. The *inverse* of each trigonometric function gives the measure of the angle. For example, $\tan^{-1}\left(\frac{18}{20}\right) = 42°$.

EXAMPLE B

Two hikers leave their campsite. One walks east 2.85 km, and the other walks south 6.03 km.

a. After the hikers get to where they're going, what is the bearing from the southern hiker to the eastern hiker?

b. How far apart are they?

▶ Solution

Draw a diagram using the information given.

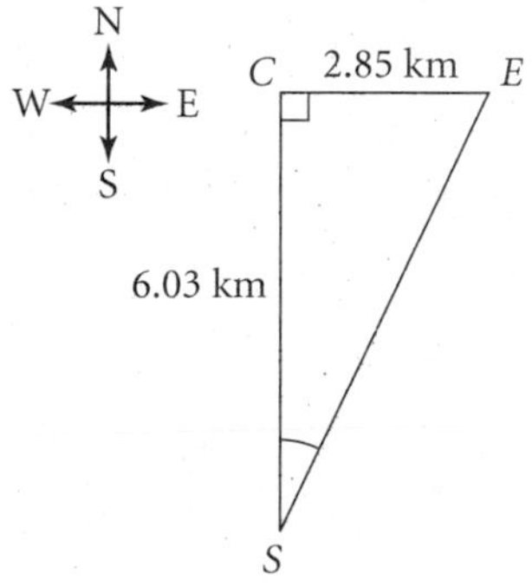

a. Angle S measures the bearing from the southern hiker to the eastern hiker. This is a right triangle, and you know the lengths of both legs, so you can use the tangent ratio.

$$\tan S = \frac{2.85}{6.03}$$

Take the inverse tangent of both sides to find the angle measure. Because you are composing tangent with its inverse, $\tan^{-1}(\tan S)$ is equivalent to S. [▶ Revisit **Calculator Note 8E** to learn about calculating the inverse trigonometric functions. ◀]

$$S = \tan^{-1}\left(\frac{2.85}{6.03}\right) \approx 25°$$

The eastern hiker is at a bearing of about 25° from the southern hiker.

b. You can use the Pythagorean Theorem to find the distance from S to E.

$$6.03^2 + 2.85^2 = (SE)^2$$
$$SE = \sqrt{6.03^2 + 2.85^2}$$
$$SE \approx 6.67$$

The distance between the two hikers is 6.67 km.

RecreationConnection A compass is a useful tool to have if you're hiking, camping, or sailing in an unfamiliar setting. The circular edge of a compass has 360 marks, which represent degrees of direction. North is 0° or 360°, east is 90°, south is 180°, and west is 270°. Each degree on a compass shows the direction of travel, or bearing. The needle inside a compass always points north. Magnetic poles on the Earth guide the direction of the needle and allow you to navigate along your desired course.

Your calculator will display each trigonometric ratio to many digits, so your final answer could be displayed to several decimal places. Often, it is appropriate to round your final answer to the nearest degree or 0.1 unit of length, as in the solution to part a in Example B. If the problem gives more precise measurements, you can use that same amount of precision in your answer. In the example, distances were given to the nearest 0.01 km, so the answer in part b also was rounded to the nearest 0.01.

EXERCISES

Practice Your Skills

1. For each of the following values, the approximate trigonometric function value of an angle of a right triangle is given. Find the angle measure, and round the value to one tenth of a degree.

a. $\sin\alpha = 0.824$ **b.** $\cos\alpha = 1.0$ **c.** $\cos\alpha = 0.891$ **d.** $\tan\alpha = 0.759$

2. Write all the trigonometric formulas (including inverses) relating the sides and angles in this triangle. There should be a total of 12.

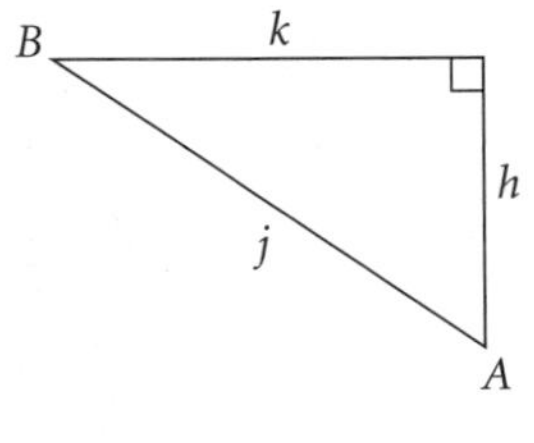

3. Draw a right triangle for each problem. Label the sides and angle, then solve to find the unknown measure.

a. $\sin 20° = \dfrac{a}{12}$ **b.** $\cos 80° = \dfrac{25}{b}$ **c.** $\tan 55° = \dfrac{c+4}{c}$ **d.** $\sin^{-1}\left(\dfrac{17}{30}\right) = D$

Reason and Apply

4. For each triangle find the length of the indicated side or the measure of the indicated angle.

a.

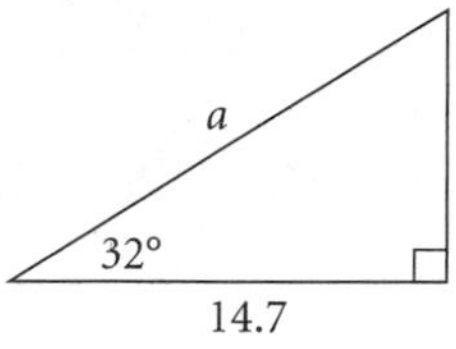

b.

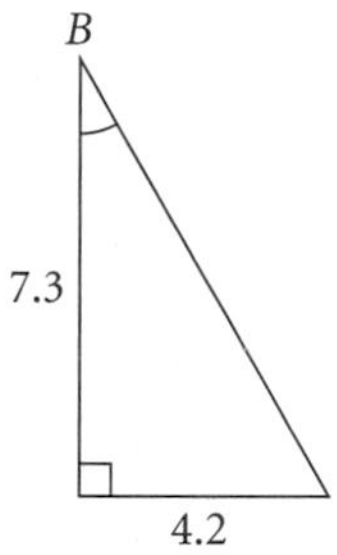

c.

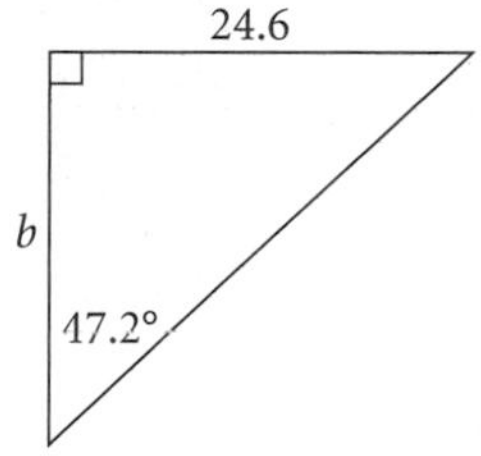

d.

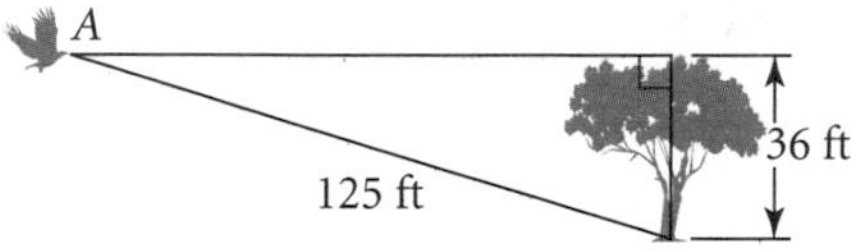

e.

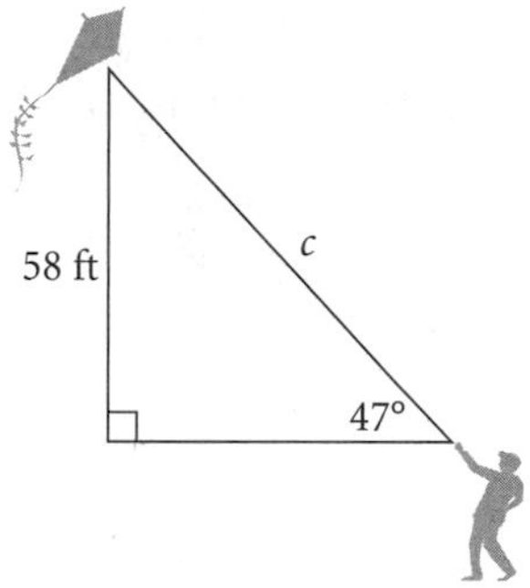

f.

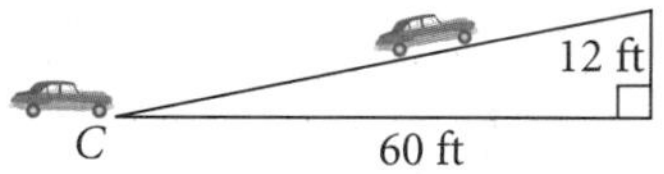

5. Draw a pair of intersecting horizontal and vertical lines, and label the four directions N, E, S, and W for north, east, south, and west. Sketch the path of a plane flying on a bearing of 30°.

a. What is the measure of the angle formed by the plane's path and the horizontal axis?

b. Choose any point along the plane's path. From this point draw a segment perpendicular to the horizontal axis to create a right triangle. Label the lengths of the horizontal leg x and the vertical leg y.

c. If the plane flies at 180 mi/h for 2 h, how far east and how far north will it travel?

6. Graph the parametric equations $x = t\cos 39°$ and $y = t\sin 39°$. Use a graphing window with $-1 \le x \le 10$, $-1 \le y \le 10$, and $0 \le t \le 10$.

a. Describe the graph. What happens when you change the minimum and maximum values of t?

b. Find the measure of the angle between this line and the x-axis. (*Hint:* Trace to a point on the line, and find the coordinates.)

7. Graph the parametric equations $x = 5t\cos 40°$ and $y = 5t\sin 40°$. Use a graphing window with $0 \le x \le 5$, $-2 \le y \le 5$, and $0 \le t \le 1$.

a. Describe the graph and the measure of the angle between the graph and the x-axis.

b. What is the relationship of this angle to the parametric equations?

c. What is the effect of the 5 in each equation? Change the 5 to a 1 in the first equation. What happens? What happens when you change 5 to 1 in both equations?

8. Write parametric equations of each graph.

a.

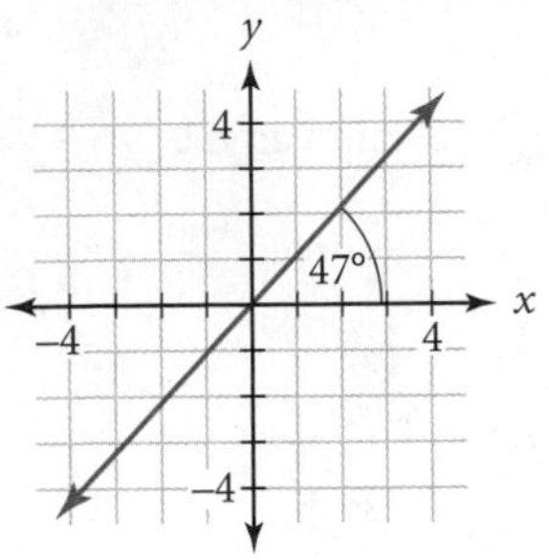

b.

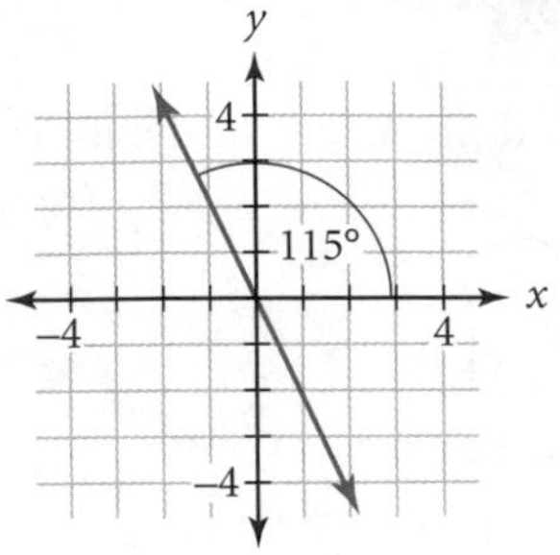

9. A plane is flying at 100 mi/h on a bearing of 60°.

a. Draw a diagram of the motion. Draw a segment perpendicular to the x-axis to create a right triangle. Write equations for x and y in terms of t to model the motion.

b. What range of t is required to display 500 mi of plane travel? (Assume t represents time in hours.)

c. Explain the real-world meaning of the numbers and variables you used in your equations.

10. Tanker A is moving at a speed of 18 mi/h from Corpus Christi, Texas, toward Panama City, Florida. Panama City is 750 mi from Corpus Christi at a bearing of 73°.

a. Sketch the tanker's motion, including coordinate axes.

b. How long does the tanker take to get to Panama City?

c. How far east and how far north is Panama City from Corpus Christi?

11. Tanker B is traveling at a speed of 22 mi/h from St. Petersburg, Florida, to New Orleans, Louisiana, on a bearing of 285°. The distance between the two ports is 510 mi.

a. Sketch the tanker's motion, including coordinate axes.

b. How long will it take the tanker to get to New Orleans?

c. How far west and how far north is New Orleans from St. Petersburg?

d. Suppose Tanker A in Exercise 10 leaves at the same time as Tanker B. Describe where the ships' paths intersect. Recall that St. Petersburg is 900 miles east of Corpus Christi. Will the ships collide? Explain your answer.

12. **APPLICATION** Civil engineers generally bank, or angle, a curve on a road so that a car going around the curve at the recommended speed does not skid off the road. Engineers use this formula to calculate the proper banking angle, θ, where v represents the velocity in meters per second, r represents the radius of the curve in meters, and g represents the gravitational constant, 9.8 m/s^2.

$$\tan\theta = \frac{v^2}{rg}$$

a. If the radius of an exit ramp is 60 m, and the recommended speed is 40 km/h, at what angle should the curve be banked?

b. A curve on a racetrack is banked at 36°. The radius of the curve is about 1.7 km. For what speed is this curve designed?

PhysicsConnection When a car rounds a curve, the driver must rely on the friction between the car's tires and the road surface to stay on the road. Unfortunately, this does not always work—especially if the road surface is wet!

In car racing, where cars travel at high speeds, tracks banked steeply allow cars to go faster, especially around the corners. Banking on NASCAR tracks ranges from 36° in the corners to a slight degree of banking in the straighter portions.

Review

13. (Lesson 3.2) The table below gives the position of a walker at several times.

Time (min) t	Horizontal distance (m) x	Vertical distance (m) y
0	6	5
2	14	11
4	22	17
6	30	23
8	38	29
10	46	35

a. Write a single equation for y in terms of x that fits the points (x, y), listed in the table.

b. Parametric equations modeling this table are $x = 6 + 4t$ and $y = 5 + 3t$. Eliminate the parameter, t, from these equations, and compare your final equation to the answer you found in 13a.

14. (Lesson 6.2) Graph the parabola $y = 35 - 4.9(x - 3.2)^2$.

a. What are the coordinates of the vertex?

b. What are the x-intercepts?

c. Where does the parabola intersect the line $y = 15$?

15. (Lesson 4.7) Write the equation of the circle with center at $(2.6, -4.5)$ and radius 3.6.

EXPLORATION

Parametric Equations of a Circle

By definition, a **circle** is the set of all the points in a plane at a given distance **(radius)** from a given point **center** in the plane. To graph a complete circle on your calculator, you use two separate functions: one for the top half of the circle and one for the bottom half. This task becomes much simpler with parametric equations.

Imagine a circle with radius r and center at the origin. For a right triangle in the first quadrant with central angle measuring t degrees, $\sin t = \frac{y}{r}$, and $\cos t = \frac{x}{r}$. Solving these equations for x and y, you get the **parametric equations of a circle:**

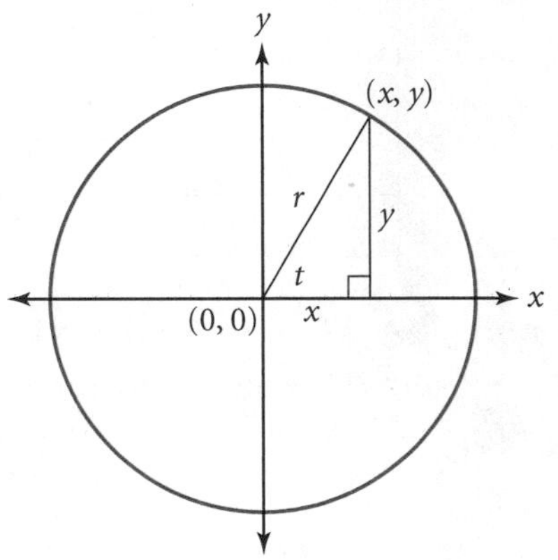

$x = r\cos t$

$y = r\sin t$

How do these equations work when the angles are greater than 90°? The definitions of the trigonometric ratios can still apply. However, when the angle is greater than 90°, you must use a right triangle situated in another quadrant. In each case, you form the triangle by connecting the point on the circle to the x-axis. For example, if the angle is 150°, you form a triangle where the angle with the x-axis is 30°. To find the values of the trigonometric ratios for 150°, you use the newly formed right triangle. Using a circle allows you to extend the definitions of sine, cosine, and tangent for any real angle value.

$\sin 150° = \frac{1}{2}$ $\quad\cos 150° = -\frac{\sqrt{3}}{2}$ $\quad\tan 150° = -\frac{1}{\sqrt{3}}$

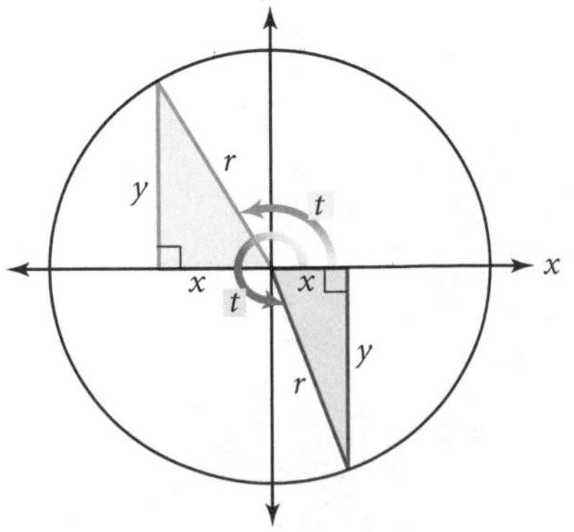

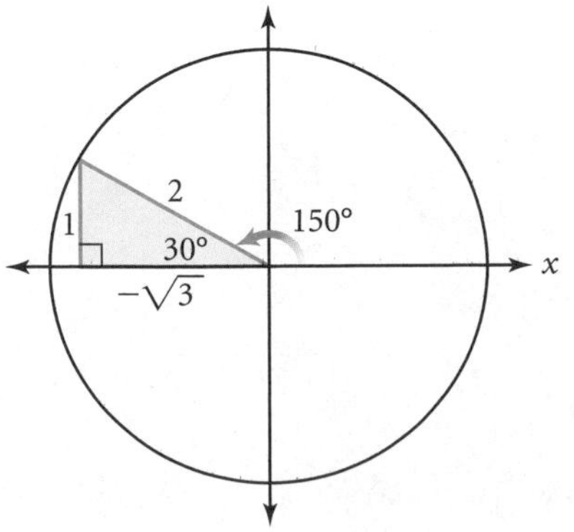

EXAMPLE Use parametric equations to graph a circle with radius 3 and center (0, 0).

▶ ***Solution*** The parametric equations of a circle with radius 3 and center (0, 0) are

$$x = 3\cos t$$
$$y = 3\sin t$$

The variable t represents the central angle of the circle. Use the range $0° \le t \le 360°$ with a t-step of 1° to graph the complete circle on your calculator.

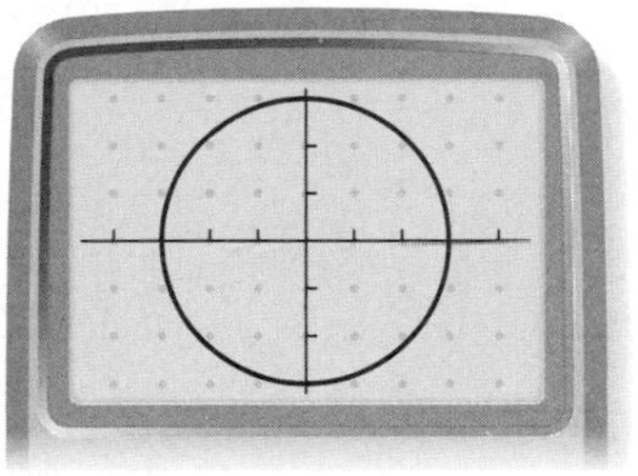

[−4.7, 4.7, 1, −3.1, 3.1, 1]

The t-step tells your calculator how often to plot a point. The calculator then connects these points with straight segments. When you make the t-step larger, the points are farther apart on the circle, and your calculator's graph stops looking like a circle. This gives you a way to draw some interesting geometric shapes on your calculator.

Activity
Variations on a Circle

In this Activity you'll explore what happens to the graph of a circle when you change the parameter t. Work individually on Step 1. Then work with a partner or group for the remainder of the activity.

Step 1 Start with the parametric equations $x = 3\cos t$ and $y = 3\sin t$. Experiment with the parameter t, changing the minimum and maximum values and the t-step. Use these questions to guide your exploration.

a. What effect does changing the range of t have on the graph?

b. What effect does changing the t-step have on the graph?

c. How can you make a triangle? A square? A hexagon? An octagon? Try to find more than one way to draw each figure.

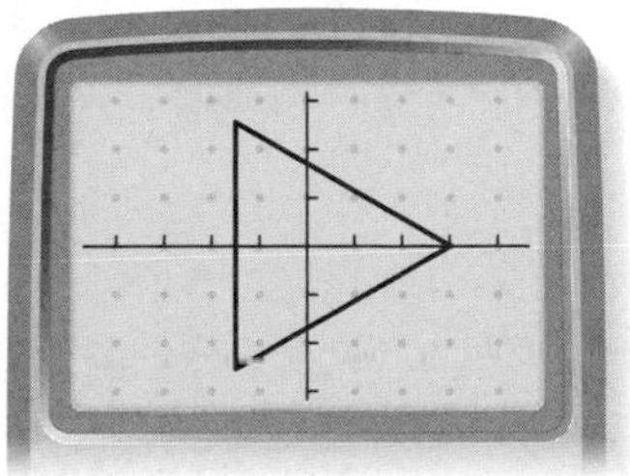

[−4.7, 4.7, 1, −3.1, 3.1, 1]

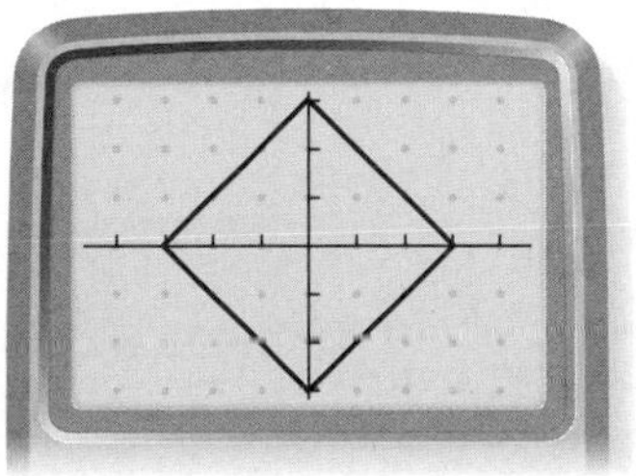

[−4.7, 4.7, 1, −3.1, 3.1, 1]

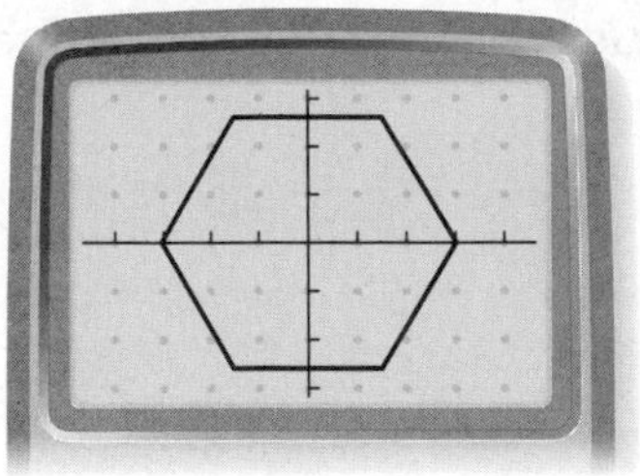

$[-4.7, 4.7, 1, -3.1, 3.1, 1]$

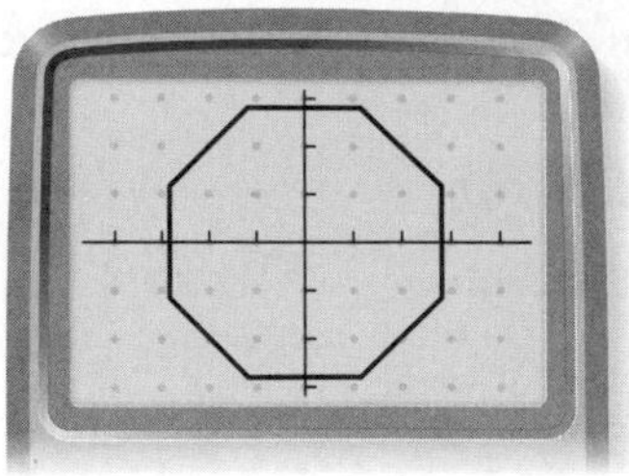

$[-4.7, 4.7, 1, -3.1, 3.1, 1]$

d. How can you make a square with sides parallel to the axes?

e. How can you rotate a polygon shape about the origin by altering the parametric equations?

Step 2 Share your results from Step 1 with a partner or your group. Work together to write a paragraph summarizing your discoveries.

Step 3 Find a way to translate the circle $x = 3 \cos t$ and $y = 3 \sin t$ so that it is centered at $(5, 2)$.

Step 4 Reflect the graph from Step 3 across the y-axis, across the x-axis, across the line $y = x$, and across the line $x = -1$. Describe the method you used for each of these reflections.

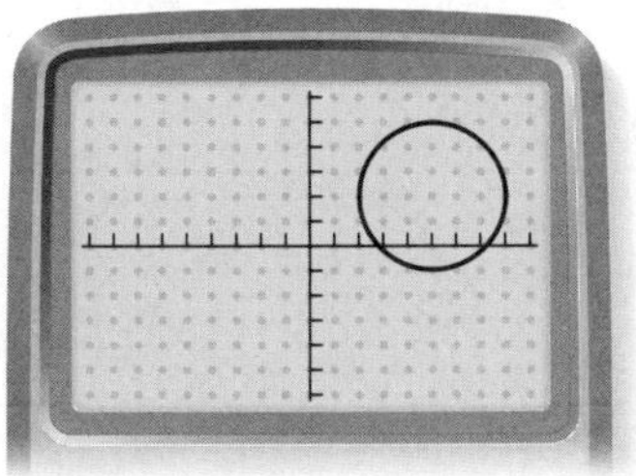

$[-9.4, 9.4, 1, -6.2, 6.2, 1]$

Step 5 Graph the equations $x = 3 \cos t$ and $y = 3 \sin t$, using the range $0° \leq t \leq 3600°$ with t-step $125°$. Then try other t-step values that are not factors of $360°$, such as $100°$, $150°$, and $185°$. (You may also need to increase the maximum value of t.) Write a paragraph explaining what happens in each case.

Questions

1. In the Investigation you used the parametric equations of a circle, $x = 3 \cos t$ and $y = 3 \sin t$, to graph other geometric shapes on your calculator. Mathematically, do you think it is correct to say that $x = 3 \cos t$ and $y = 3 \sin t$ are the parametric equations of a square or a hexagon? Explain your reasoning.

2. Find parametric equations of each translated circle.

a.

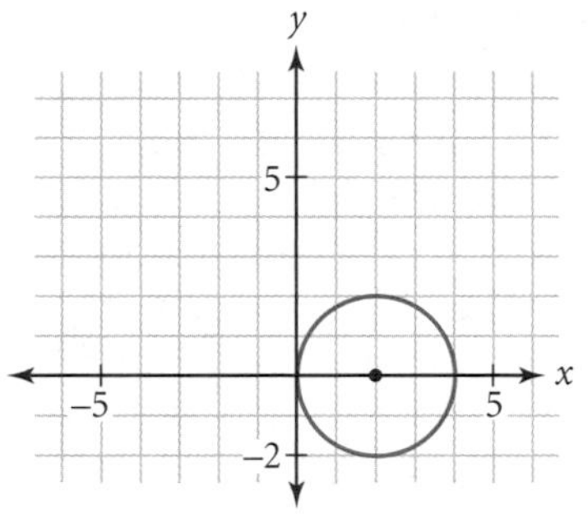

b.

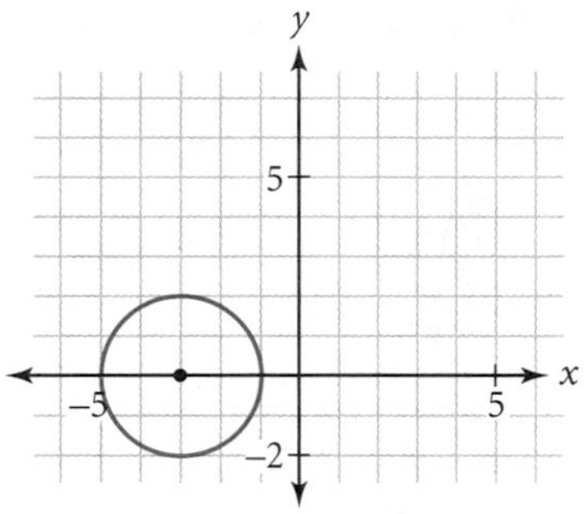

c.

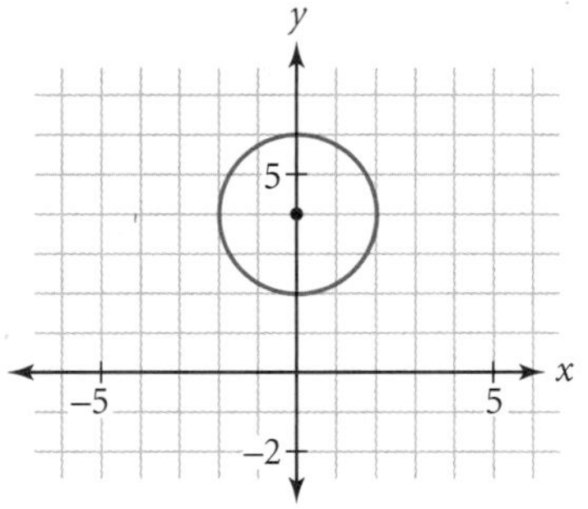

d.

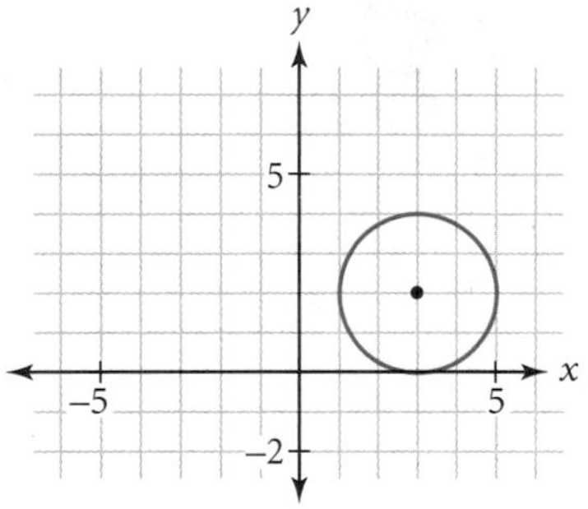

3. Use the parametric equations $x = 3 \cos t$ and $y = 3 \sin t$ to make each of these figures on your calculator. What range of t-values and what t-step are required for each?

a.

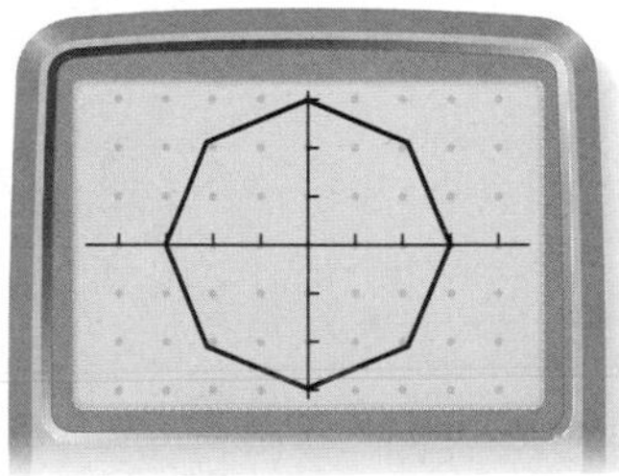

[−4.7, 4.7, 1, −3.1, 3.1, 1]

b.

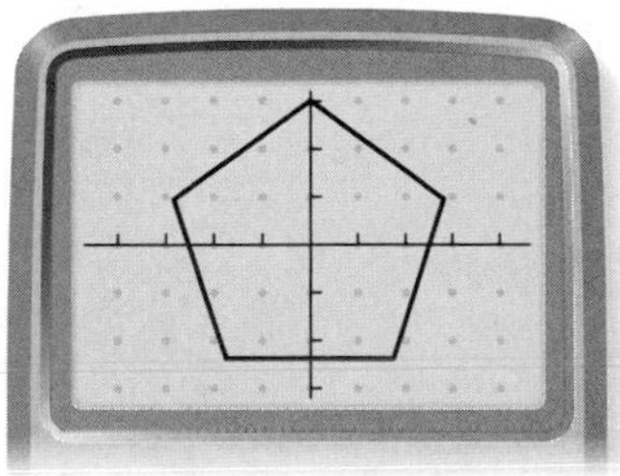

[−4.7, 4.7, 1, −3.1, 3.1, 1]

c.

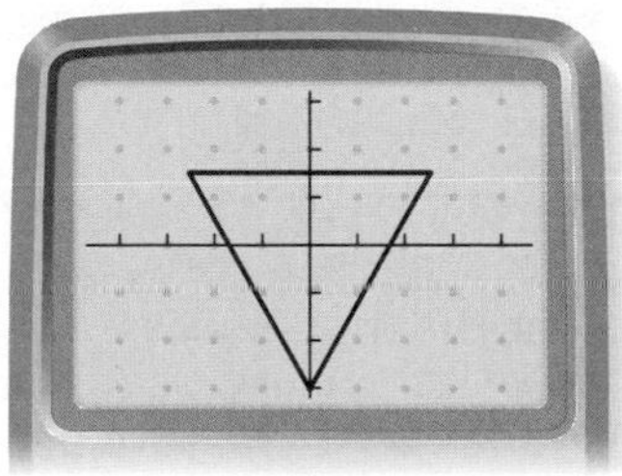

[−4.7, 4.7, 1, −3.1, 3.1, 1]

d.

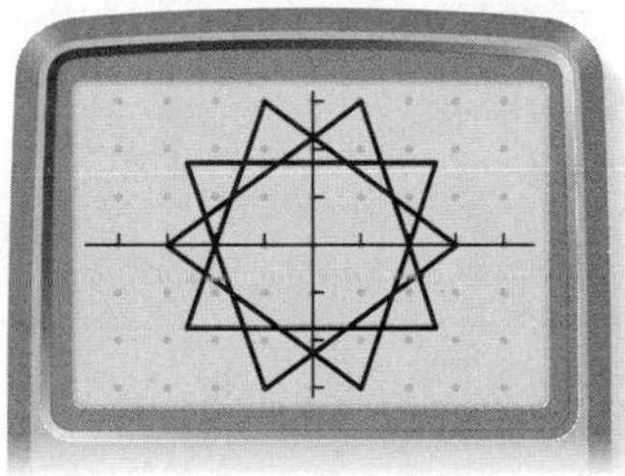

[−4.7, 4.7, 1, −3.1, 3.1, 1]

4. Consider a unit circle (a circle with radius 1).

a. What are the parametric equations of the unit circle?

b. From Chapter 4 you know that the Pythagorean Theorem yields the equation of the unit circle, $x^2 + y^2 = 1$. Substitute the parametric equations for x and y to get an equation in terms of sine and cosine.

c. Use your calculator to verify that your equation in 4b is true for $t = 47°$.

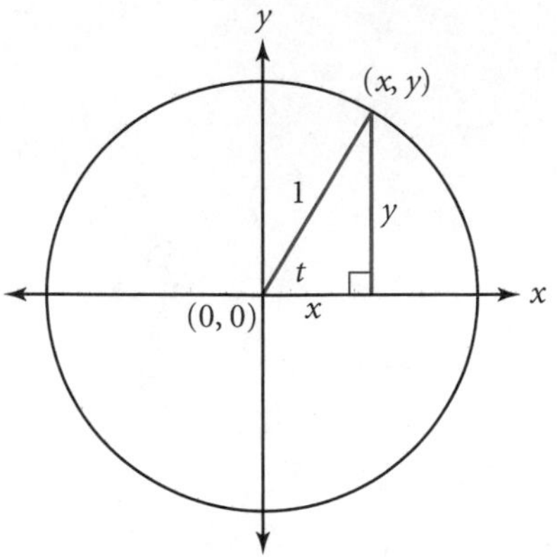

5. Experiment to find parametric equations of each ellipse.

a.

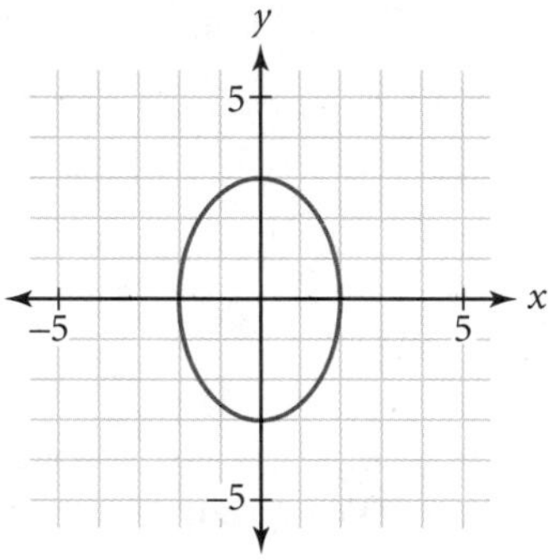

b.

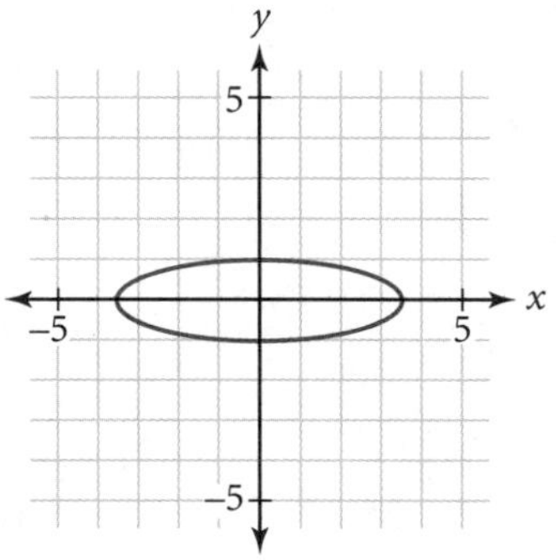

c.

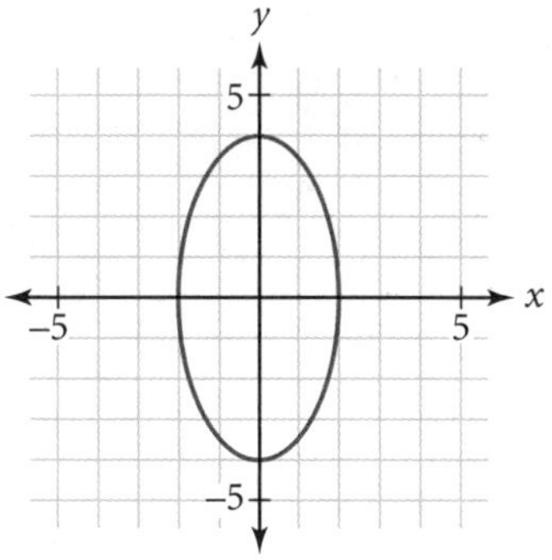

ConsumerConnection By about 2800 B.C.E., gemstone cutting and engraving were widely practiced throughout Egypt and Asia Minor (modern Turkey). Ancient gem cutters used the same geometric designs that today's gem cutters use. Working with multiples of 8, on each gem the cutter carves small symmetric planes, called facets, into an overall shape. Rubies, sapphires, and emeralds are often square or rectangular, whereas other gems are cut into triangular, diamond, or trapezoidal shapes. The oldest gemstone shape, however, is circular or rounded, which is how many opals and opaque gems are still cut today.

LESSON

8.4

Using Trigonometry to Set a Course

Not snow, no, nor rain, nor heat, nor night keeps them from accomplishing their appointed courses with all speed.

HERODOTUS

Winds and air currents affect the direction and speed a plane travels. A pilot flying an airplane must compensate for these effects. Similarly, if you swim across a river, a current can sweep you downstream, and when you reach the other side, you will not be directly across from the point where you began. You can use parametric equations to model and simulate motion affected by a combination of forces.

This Investigation simulates what happens when you try to swim or fly with, against, or across a strong water or air current.

Investigation
Motion in a Current

Asuka, Ben, and Chelsea are playing with remote-controlled boats in a pool. Each boat is moving 4 ft/s in the direction indicated. The boats all start in motion at the same time.

Step 1 Create a table like the one below showing the x- and y-positions of the three boats for 3 s.

Time	Boat A		Boat B		Boat C	
t	x	y	x	y	x	y
0						
1						
2						
3						

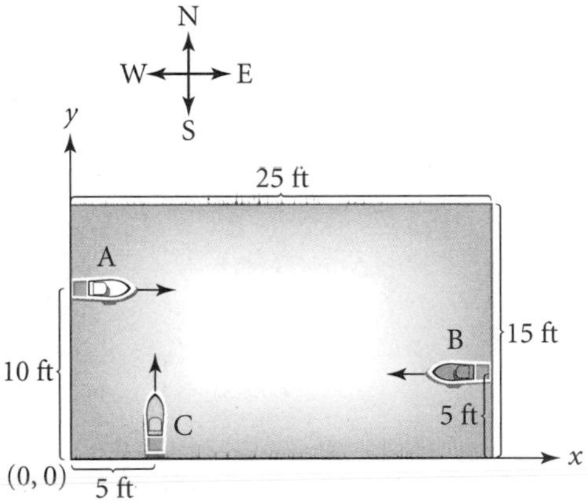

Step 2 Write and graph parametric equations to model the motion of the three boats. Identify the window and range of t-values you use to graph the simulation.

Step 3 Asuka, Ben, and Chelsea return their boats to the starting positions and repeat the motion, but a gust of wind begins blowing from west to east at a constant rate of 3 ft/s at the same time the boats start moving. Create a table, as you did in Step 1, of the positions of the boats over 3 s.

Step 4 Write and graph parametric equations to model the motion of the boats in Step 3. Show how to write each of these using the $4t$ from the boat's speed and the $3t$ from the wind's speed. Identify the window and range of t-values you use to graph the simulation.

Step 5 Use the Pythagorean Theorem to determine the distance Boat C travels in 3 s and its velocity.

Step 6 At what angle does Boat C travel?

Step 7 Use the velocity and angle you found in Steps 5 and 6 to write parametric equations in the form $x = x_1 + vt\cos A$ and $y = y_1 + vt\sin A$ to represent the motion of Boat C. The constants x_1 and y_1 represent the boats' starting coordinates. Verify that these equations model Boat C's motion accurately.

The Investigation showed you how motion can be affected by another factor, such as wind or current. In this example you will explore what happens when a pilot doesn't compensate for the effect of wind.

EXAMPLE A

A pilot heads a plane due west from Memphis, Tennessee, toward Albuquerque, New Mexico. The cities are 1,000 mi apart, and the pilot sets the plane's controls to fly at 250 mi/h. However, there is a constant 20 mi/h wind blowing from the north. Where does the plane end up?

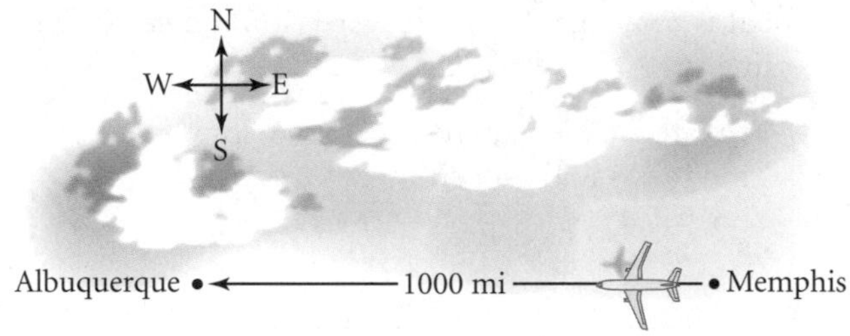

▶ Solution

Set up a coordinate system with Memphis at the origin. The plane's motion can be described by the equation $x = -250t$. Notice that velocity is negative because the plane is flying west. The effect of the wind on the plane's motion can be described by the equation $y = -20t$. Why is the wind's velocity negative?

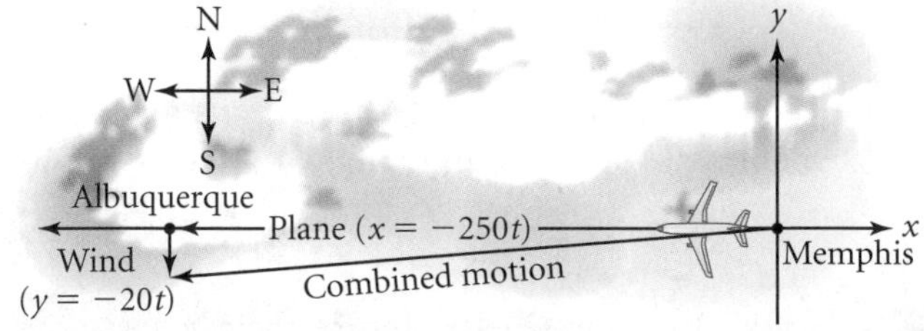

Graph $x = -250t$ and $y = -20t$ in an appropriate window with $0 \leq t \leq 4$.

Solving the equation $-1000 = -250t$, or tracing the graph, will show that after 4 h the plane has traveled the necessary 1000 mi west, but it is 80 mi south of Albuquerque, somewhere over the White Sands Missile Range!

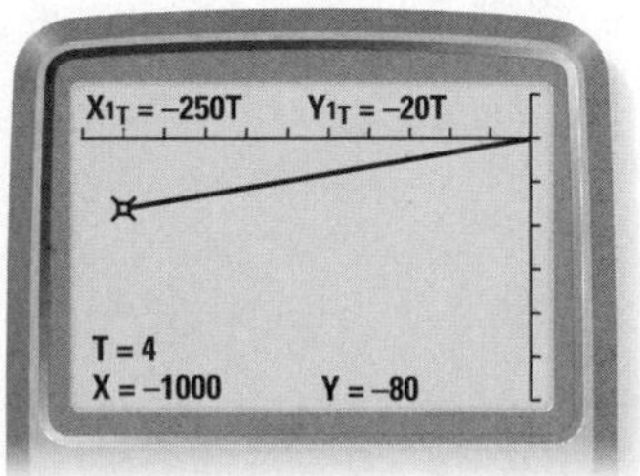

$[-1100, 10, 100, -300, 50, 50]$

The Pythagorean Theorem indicates that the plane has actually traveled $\sqrt{1000^2 + 80^2}$, or about 1,003 mi, in 4 h. Because speed is the distance traveled per unit of time, you can find

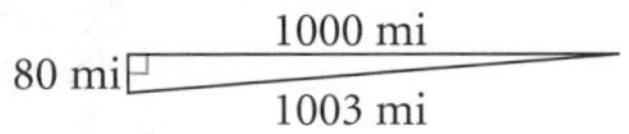

the plane's speed by dividing $\frac{1003 \text{ mi}}{4 \text{ h}}$ to get the rate of 250.8 mi/h. This tells you that, although the *air speed* of the plane was 250 mi/h, the *ground speed* was 250.8 mi/h. The air speed is the velocity at which the plane flies, and the ground speed is its rate of motion relative to a fixed point on the ground.

The actual angle of motion in Example A is $\tan^{-1}\left(\frac{80}{1000}\right)$, or about 4.57°. To determine the plane's bearing, look at the compass rose shown here. The bearing is always between 0° and 360° measured clockwise from north. This plane intended to travel west (270°) but actually traveled south of west. Its bearing was 270° − 4.6°, or 265.4°.

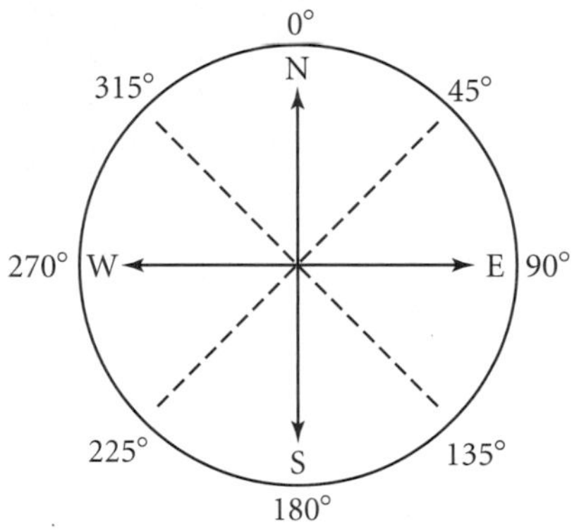

EXAMPLE B

Suppose the pilot from Example A had initially taken the wind into consideration in planning the trip. What angle and bearing, to the nearest hundredth of a degree, should the pilot have set?

▶ Solution

Set up a coordinate system with Memphis at the origin. Because the plane will be blown to the south, it actually should head in a direction slightly to the north. Sketch the plane's path. The plane flies 250 mi/h, so its distance along this path is $250t$. This distance is broken into two separate components. The east-west component is $x = -250t \cos A$, and the north-south component is $y = 250t \sin A$ where A is the angle toward the north that the pilot must set.

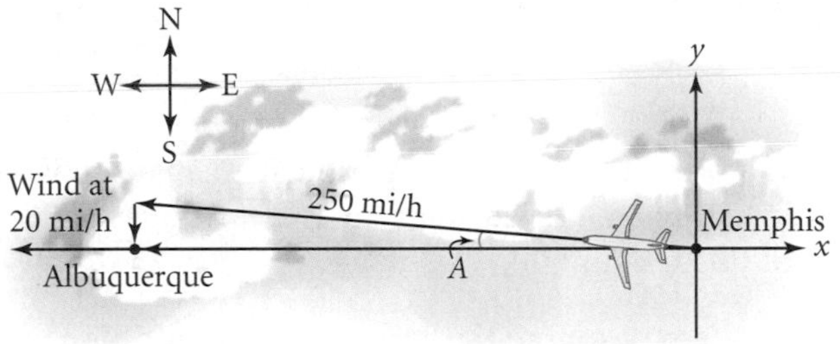

The effect of the wind on the plane's motion can be described by the equation $y = -20t$. The sum of the north-south component of the plane's course and the wind's velocity must be 0 if the pilot hopes to land in Albuquerque.

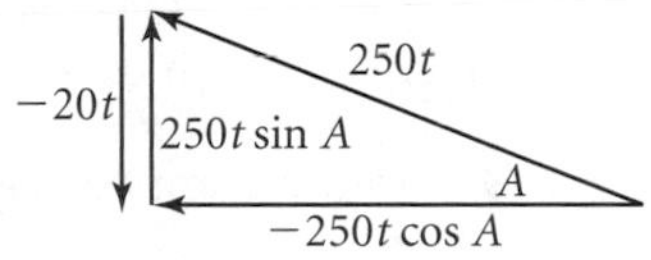

$250t \sin A + (-20t) = 0$	The sum of the north-south velocities of the wind and the plane is 0.
$250t \sin A = 20t$	Add $20t$ to both sides.
$250 \sin A = 20$	Divide by t. (Assume $t \neq 0$.)
$\sin A = 0.08$	Divide both sides by 250, and evaluate.
$A = \sin^{-1} 0.08$	Take the inverse sine of both sides.
$A \approx 4.59°$	Evaluate.

The plane must head west at 4.59° above the horizontal. This is equivalent to a bearing of 270° + 4.59°, or 274.59°.

Graph $x = -250t \cos 4.59°$ and $y = 250t \sin 4.59° - 20t$ to verify that the plane moves directly west. As you trace the graph, notice that the value of y increases slightly. By the time the plane has traveled east 996.80 mi, the plane is slightly north of Albuquerque by 0.02 mi. This means that the angle, 4.59°, is not accurate enough to result in motion directly west. Depending upon the sensitivity of the real-world situation, the pilot may need to increase accuracy by using more decimal places.

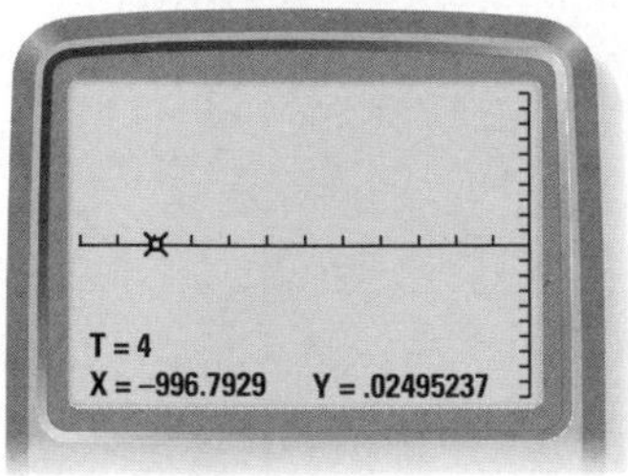

[−1200, 0, 100, −10, 10, 1]

Also notice that that the plane does not reach Albuquerque within 4 h. Because of the effect of the wind, the plane must travel slightly longer to complete the 1,000-mi trip.

The directed distances in this lesson are **vectors.** Vectors are used when you are working with things that have both a magnitude (size) and a direction, such as the velocity of the wind, the force of gravity, and the forces you feel when you ride a roller coaster. Working with vectors is a common practice in many fields, including physics, geology, and engineering. The Exercises will give you an introduction to this valuable tool.

EXERCISES

Practice Your Skills

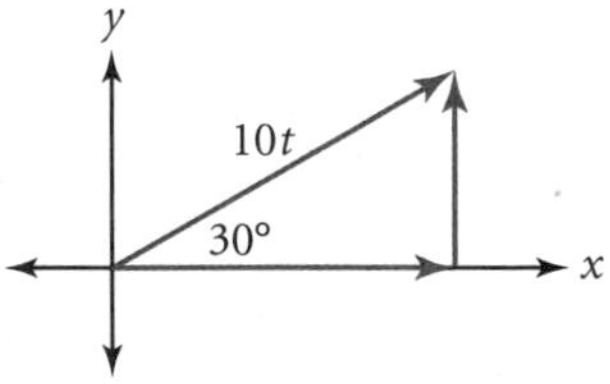

1. An object is moving at a speed of 10 units per second at an angle of 30° above the x-axis, as shown. What parametric equations provide the horizontal and vertical components of this motion?

2. Draw a compass rose and vector to find the bearing of each direction.

a. 14° south of east **b.** 14° east of south

c. 14° south of west **d.** 14° north of west

3. Draw a compass rose and vector for each bearing. Find the angle made with the x-axis.

a. 147° **b.** 204° **c.** 74° **d.** 314°

4. Give the sign of each component vector for the bearings in Exercise 3. For example, for a bearing of 290°, x is negative and y is positive.

Reason and Apply

5. **APPLICATION** A river is 0.3 km wide and flows south at a rate of 7 km/h. You start your trip on the river's west bank, 0.5 km north of the dock, as shown in the diagram.

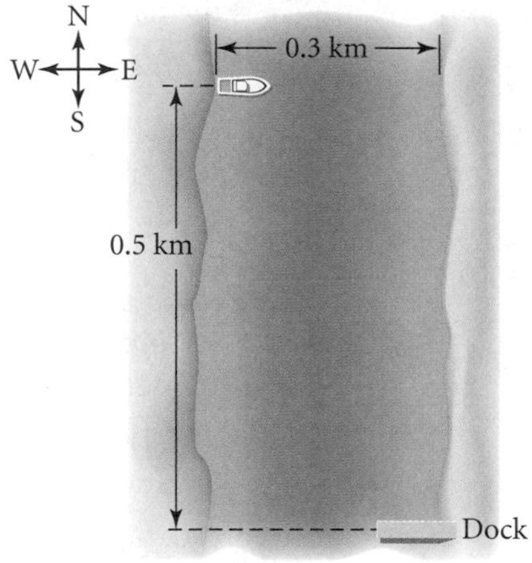

a. If the dock is at the origin, (0, 0), what are the coordinates of the boat's starting location?

b. Write an equation for x in terms of t that models the boat's horizontal position if you aim the boat directly east traveling at 4 km/h.

c. Write an equation for y in terms of t that models the boat's vertical position as a result of the flow of the river.

d. Enter the parametric equations from 5b and 5c into your calculator, determine a good viewing window and range of t-values, and make a graph to simulate this situation.

e. Determine when and where the boat meets the river's east bank. Does your boat arrive at the dock?

f. How far have you traveled?

6. **APPLICATION** A pilot wants to fly from Toledo, Ohio, to Chicago, Illinois, which lies 280 mi directly west. Her plane can fly at 120 mi/h. She ignores the wind and heads directly west. However, there is a 25 mi/h wind blowing from the south.

a. Write the equation that describes the effect of the wind.

b. Write the equation that describes the plane's contribution to the motion.

c. Graph these equations.

d. How far off course is the plane when it has traveled 280 mi west?

e. How far has the plane actually traveled?

f. What was the plane's ground speed?

7. Fred rows his small boat directly across a river, which is 4 mi wide. There is a 5-mi/h current. When he reaches the opposite shore, Fred finds that he has landed at a point 2 mi downstream.

a. Write the equation that describes the effect of the river current.

b. If Fred's boat can go s mi/h, what equation will describe his contribution to the motion?

c. Solve the system of equations from 7a and 7b for s so that Fred reaches the correct point 2 mi downstream on the opposite shore.

d. How far did Fred actually travel?

e. How long did it take him?

f. What was Fred's actual speed?

g. As the boat travels down the river, what angle does it form with the riverbank?

8. APPLICATION A plane takes off from Orlando, Florida, heading 975 mi due north toward Cleveland, Ohio. The plane flies at 250 mi/h, and a 25-mi/h wind is blowing from the west.

a. Where is the plane after it has traveled 975 mi north?

b. How far has the plane actually traveled?

c. How fast did the plane actually travel?

d. At what angle from the horizontal did the plane actually fly?

e. What was the bearing?

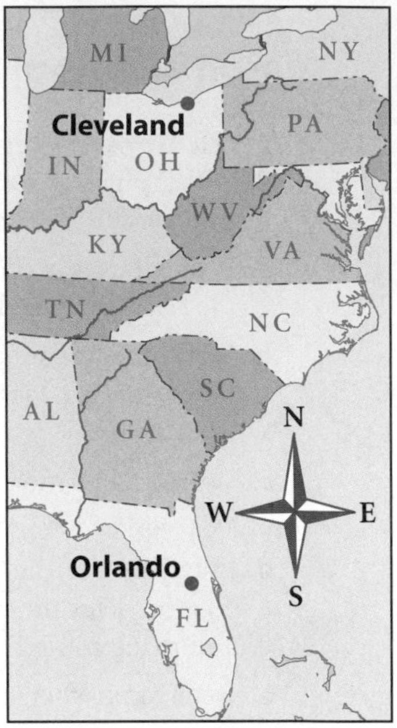

9. APPLICATION A plane is headed from Memphis, Tennessee, to Albuquerque, New Mexico, 1,000 mi due west. The plane flies at 250-mi/h, and the pilot encounters a 20-mi/h wind blowing from the northwest. (That means the direction of the wind makes a 45° angle with the x-axis.)

a. Write an equation modeling the southward component of the wind.

b. Write an equation modeling the eastward component of the wind.

c. If the pilot does not compensate for the wind, explain why the final equations for the flight are $x = -250t + 20t \cos 45°$ and $y = -20t \sin 45°$.

d. What graphing window and range of t-values can you use to depict this flight?

e. Solve the equation $-1000 = -250t + 20t \cos 45°$. What is the real-world meaning of your answer?

f. Use your answer from 9e to find how far south of Albuquerque the plane ends up.

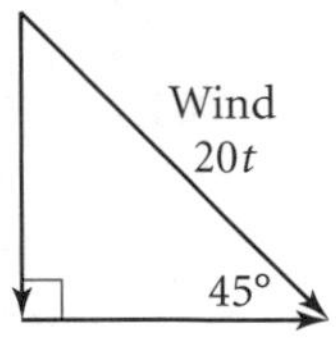

10. APPLICATION For the plane in Exercise 9 to land in Albuquerque, it must head a bit north.

a. For the plane to fly directly west, the northward component of the plane's motion and the wind's southward component must add up to zero. Write an equation that you can solve to find the angle at which the plane must fly to head directly to Albuquerque.

b. Solve your equation from 10a for A.

c. Using the value of A you found in 10b, write parametric equations to model the plane's flight, and graph them to verify that the plane travels directly west.

d. For what bearing should the pilot set his instruments?

11. APPLICATION A plane is flying on a bearing of 310° at a speed of 320 mi/h. The wind is blowing directly from the east at a speed of 32 mi/h.

a. Make a compass rose and vector to indicate the plane's motion.

b. Write equations that model the plane's motion without the wind.

c. Make a compass rose and vector to indicate the wind's motion.

d. Write equations that model the wind's motion.

e. What are the resulting equations that model the motion of the plane with the wind?

f. Where is the plane after 5 h?

12. Angelina wants to travel directly across the Wyde River, which is 2 mi wide in this stretch. Her boat can move at a speed of 4 mi/h. The river current flows south at 3 mi/h. At what angle upstream should she aim the boat so that she goes straight across?

13. APPLICATION Two forces are pulling on an object. Force A has a magnitude of 50 g and pulls at an angle of 40°, and Force B has a magnitude of 90 g and pulls at an angle of 140°, as shown. (Note that the lengths of the vectors show their magnitude.)

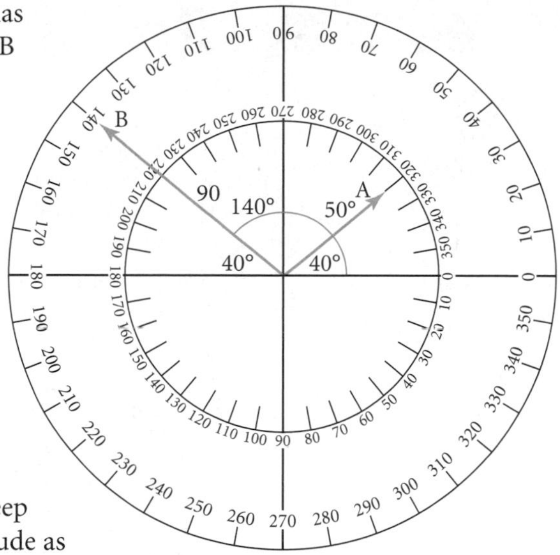

a. Find the x- and y-components of Force A.

b. Find the x- and y-components of Force B.

c. What is the sum of the x- and y-components of the two forces?

d. Using the Pythagorean Theorem, find the magnitude of the resulting force.

e. Find the angle of the resulting force. If your angle is negative, add 180° to get the direction.

f. What additional force will balance Forces A and B and keep the object in equilibrium? (It should be the same magnitude as the sum of Forces A and B but in the opposite direction.)

Science Connection Biomechanics, the application of the principles of mechanics to the study of human motion, provides an understanding of the internal and external forces acting on the human body during movement. Knowledge of the role muscles play in generating force and controlling movement is necessary to an understanding of the limitations of human motion. Information gained from biomechanics helps athletes prepare better for their sports and sporting goods manufacturers to produce better equipment. Research in biomechanics also contributes to better treatment and rehabilitation of injuries.

Review

14. (Lesson 6.4) Write a quadratic function whose graph has

a. x-intercepts at 1 and -1

b. x-intercepts at 10 and -10

15. (Lesson 6.5) Without graphing, determine that each quadratic equation has no real roots, one real root, or two real roots. If a root is real, indicate whether it is rational or irrational.

a. $y = 2x^2 - 5x - 3$

b. $y = x^2 + 4x - 1$

c. $y = 3x^2 - 3x + 4$

d. $y = 9x^2 - 12x + 4$

16. (Lesson 7.3) Consider this system of equations.

$$\begin{cases} 5x - 3y = -1 \\ 2x + 4y = 5 \end{cases}$$

a. Write the augmented matrix of the system of equations.

b. Use row reduction to write the augmented matrix in row-echelon form. Show each step, and indicate the operation you use.

c. Give the solution of the system. Check your answer.

LESSON

8.5

Projectile Motion

It is a mathematical fact that the casting of this pebble from my hand alters the centre of gravity of the universe.

THOMAS CARLYLE

In this lesson you will model motion affected by gravity, building on your earlier work with the quadratic equation $y = agt^2 + v_0t + s_0$. Recall that a is the acceleration due to gravity, so the equation can also be written as $y = -\frac{1}{2}gt^2 + v_0t + s_0$. This equation models the changing vertical height of an object in free fall. Recall that t represents time, v_0 represents the vertical component of the initial velocity, and x_0 represents the initial height. Some numerical values for the force of gravity, g, are given below. What are the corresponding values of a?

g	Earth	Moon	Mars
m/s^2	9.8	1.6	3.7
ft/s^2	32	5.3	12
cm/s^2	980	162	370
$in./s^2$	384	64	1446

In Lesson 8.3 you dealt only with the vertical component, the height, of objects in projectile motion. In this lesson you'll model the horizontal component as well, using parametric equations.

EXAMPLE A

With a horizontal velocity of 1.5 ft/s, a ball rolls off the end of a table. The table is 2.75 ft high.

a. What equation describes the x-direction position of the ball?

b. Write an equation to model the vertical position of the ball. Remember that gravity will have an effect.

c. Enter these two equations into your calculator, and graph them. Where and when does the ball appear to hit the floor?

d. Solve algebraically to find where and when the ball hits the floor.

▸ Solution

a. The horizontal component of the motion is not affected by the force of gravity, so it is described by $x = 1.5t$. The total distance traveled is the velocity multiplied by the time elapsed.

b. An equation of the height is $y = -\frac{1}{2} \cdot 32 \cdot t^2 + 0 \cdot t + 2.75$. Here, t is the time measured in seconds, 32 is the acceleration due to gravity in ft/s^2, the coefficient 0 implies that none of the initial velocity is directed upward or downward, and 2.75 is the initial height.

c. Graph the equations $x = 1.5t$ and $y = -16t^2 + 2.75$ with $0 \leq t \leq 1$ and a t-step of 0.01. The ball hits the floor when $y = 0$. Trace the graph to approximate when this happens. The graph shows that the ball hits the floor approximately 0.41 s after it leaves the edge of the table. At that time it is 0.62 ft from a point directly below the edge of the table.

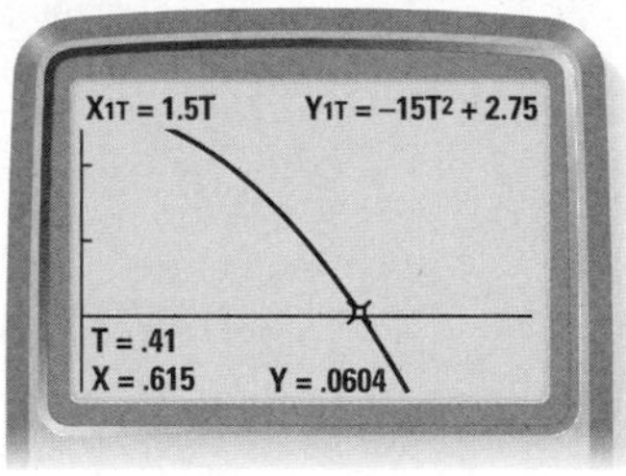

[0, 1, 1, −1, 3, 1]

d. The ball hits the floor at the point where $y = 0$. Solve the equation $0 = -16t^2 + 2.75$.

$0 = -16t^2 + 2.75$	Original equation.
$16t^2 = 2.75$	Add $16t^2$ to both sides.
$t^2 = \frac{2.75}{16}$	Divide both sides by 16.
$t = \pm\sqrt{\frac{2.75}{16}}$	Take the square root of both sides.
$t \approx \pm 0.41$	Evaluate.

Only the positive answer makes sense in this situation, so the ball hits the ground at approximately 0.41 s. To find where this occurs, substitute 0.41 in the parametric equation for x.

$$x = 1.5t$$
$$x = 1.5(0.41) \approx 0.62$$

The ball hits the ground at a horizontal distance of approximately 0.62 ft from the edge of the table.

In Example A the initial motion was only in a horizontal direction. In the previous chapter you explored motion that was only in a vertical direction, like the motion of a free-falling object. Using trigonometric functions and techniques similar to those in the previous lesson, you can parametrically model motion that begins at an angle.

Parametric Projectile Motion

You can model projectile motion parametrically with these equations, where x is a measure of horizontal position, y is a measure of vertical position, and t is a measure of time.

$$x = v_0 t \cos A + x_0$$

$$y = -\frac{1}{2}gt^2 + v_0 t \sin A + y_0$$

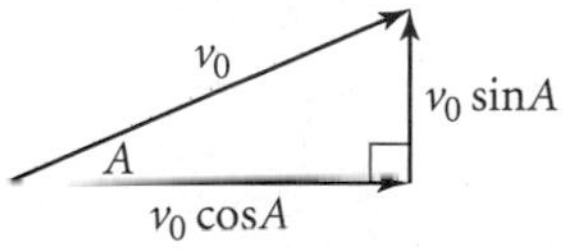

The point (x_0, y_0) is the initial position at time $t = 0$, v_0 is the velocity at time $t = 0$, A is the angle of motion from the horizontal at $t = 0$, and g is the acceleration due to gravity.

EXAMPLE B

Carolina hits a baseball so that it travels at an initial speed of 120 ft/s and at an angle of 30° above the ground. If her bat contacts the ball at a height of 3 ft above the ground, how far does the ball travel horizontally before it hits the ground?

▶ ***Solution***

Draw a picture and write equations for the *x*- and *y*-components of the motion.

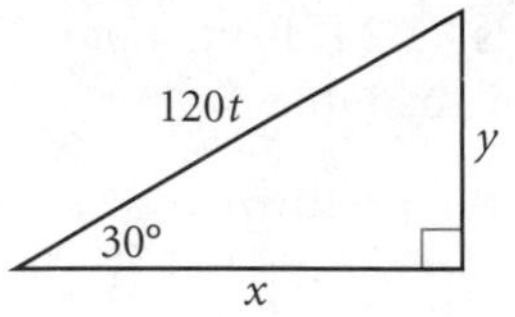

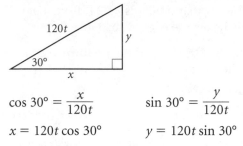

$\cos 30° = \frac{x}{120t}$ $\quad\quad$ $\sin 30° = \frac{y}{120t}$

$x = 120t\cos 30°$ $\quad\quad$ $y = 120t\sin 30°$

Because the horizontal motion is affected only by the initial speed and angle, the horizontal distance is modeled by $x = 120t\cos 30°$.

The vertical motion is also affected by the force of gravity pulling the ball down. In this case the equation that models the horizontal distance is $y = -16t^2 + 120t\sin 30° + 3$. Compare this equation with the one given in the box. Notice how the values from this problem appear in the equation.

To find out when the ball hits the ground, find when the *y*-value is 0.

$-16t^2 + 120t\sin 30° + 3 = 0$	Original equation.
$-16t^2 + 120t(0.5) + 3 = 0$	Evaluate sin 30°.
$-16t^2 + 60t + 3 = 0$	Multiply.
$t = \frac{-60 \pm \sqrt{60^2 - 4(-16)(3)}}{2(-16)}$	Use the quadratic formula. Substitute −16 for *a*, 60 for *b*, and 3 for *c*.
$t \approx -0.049$ or $t \approx 3.799$	Evaluate.

A negative value of *t* doesn't make sense, so use only the positive answer. The ball reaches the ground about 3.8 seconds after being hit.

To determine how far the ball has traveled, substitute this *t*-value in the parametric equation for *x*:

$x = 120(3.799)\cos 30°$

$x \approx 395$

The ball will travel about 395 ft horizontally.

In the Investigation you will explore simulating a basketball free throw. This is another instance where you must consider motion at an angle.

Investigation
Basketball Free Throw

(Based on an activity developed by Arne Engebretsen)

Basketball players decide on the angle and velocity of the ball necessary to make a basket from various positions on the court. In this Investigation you'll explore the relationship between the angle and velocity necessary for a successful free-throw shot.

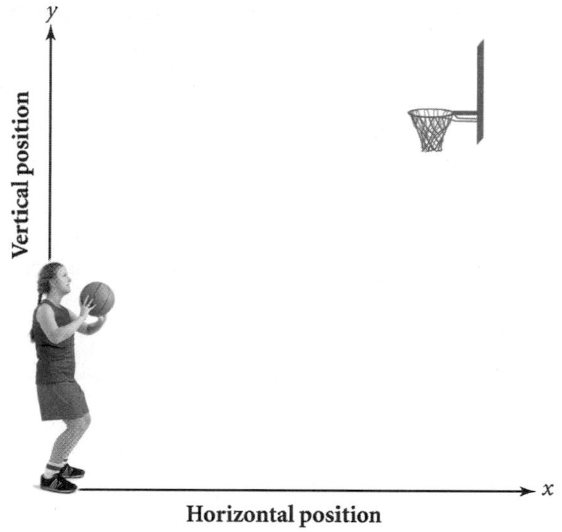

Step 1 Discuss with your class or get data from your instructor on free-throw measurements. You'll need the distance from the floor to the rim of the basket, the horizontal distance from the free-throw line to the backboard, the diameter of the basket, and the length of the bracket that fastens the basket to the backboard.

Step 2 Draw a diagram with the measurements from Step 1, and sketch the typical path of a successful free throw.

Step 3 Using the free-throw line as the origin, (0, 0), what are the coordinates of the front and back rims of the basket? Plot these two data points in a graphing window with $0 \leq x \leq 18.8$ and $0 \leq y \leq 12.4$.

Step 4 Create some reasonable data for a free-throw shooter. Include the height of the ball at release (y_0), the measure of the angle at release (A), and the initial velocity of the ball (v_0).

Step 5 Using your data from Step 4, write parametric equations, $x = v_0 t \cos A$ and $y = -16t^2 + v_0 t \sin A + y_0$, to model a free throw. Use your equations to simulate a free throw on your calculator.

Step 6 Experiment with different values of y_0, A, and v_0 until you can simulate a successful free throw on your graphing calculator.

Step 7 Is there only one combination of values that will produce a successful free throw? If so, why? If not, generalize patterns in the relationships between the variables.

EXERCISES

Practice Your Skills

1. A projectile's motion is described by the equations

$x = -50t\cos 30° + 400$

$y = -81t^2 + 50t\sin 30° + 700$

a. Is this projectile motion occurring on the Earth, the Moon, or Mars? What are the units used in the problem? (*Hint:* Look for the value of g.)

b. What is the initial position? Include units in your answer.

c. At time $t = 0$, what direction is the projectile moving (up, up–right, right, down–right, down, down–left, left, or up–left)?

d. What is the initial velocity? Include units.

2. Find the position at the time given of a projectile in motion described by these equations.

$x = -50t\cos 30° + 40$

$y = -81t^2 + 50t\sin 30° + 60$

a. 2 seconds **b.** 0 second **c.** 1 second **d.** 4 seconds

3. A ball rolls off the edge of a 12-m-tall cliff at a velocity of 2 m/s.

a. Write parametric equations to simulate this motion.

b. What equation can you solve to determine when the ball hits the ground?

c. When and where does the ball hit the ground?

d. Describe a graphing window that you can use to model this motion.

4. Consider the scenario in Exercise 3. When and where will the ball hit the ground if the motion occurs?

a. On the Moon?

b. On Mars?

Reason and Apply

5. These parametric equations model projectile motion of an object.

$x = 6t\cos 52°$

$y = -4.9t^2 + 6t\sin 52° + 2$

a. Name a graphing window and a range of t-values that allow you to simulate the motion.

b. Describe a scenario for this projectile motion. Include a description of every variable and number listed in the parametric equations.

6. A ball rolls off a 3-ft-high table and lands at a point 1.8 ft away from the table.

a. How long did it take the ball to hit the floor? Give your answer to the nearest hundredth of a second.

b. How fast was the ball traveling when it left the table? Give your answer to the nearest hundredth, and include units.

7. **APPLICATION** An archer aims at a target 70 m away with diameter 1.22 m. The bull's-eye is 1.3 m above the ground. She holds her bow level at a height of 1.2 m and shoots an arrow with an initial velocity of 83 m/s.

a. What equations model this motion?

b. Will she hit the target? If not, by how much will she miss it?

c. At what angles could the archer hold her bow to hit somewhere on the target?

d. What initial velocity is necessary in order to hit the target if she holds her bow level?

SportsConnection Archery is one of the oldest sports still practiced today. Modern archery became an Olympic event in 1972. Olympic archers aim at a target 1.22 m in diameter, 70 m away—the bull's-eye is just 12.2 cm in diameter. From where the archers stand, the target looks about the same size as the head of a thumbtack held at arm's length. Archery is a sport that easily can accommodate disabled athletes, with modified equipment and sometimes courses. Disabled archers compete in the Paralympics, and archery is used as a means of rehabilitation and recreation at physical rehabilitation centers worldwide.

8. With a horizontal velocity of 5 ft/s, a golf ball rolls off the top step of a flight of 14 stairs. The stairs are each 8 in. high and 8 in. wide. On which step does the ball first bounce?

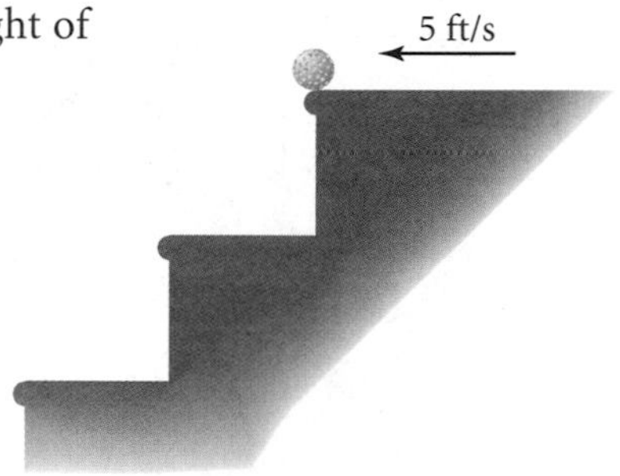

9. A golfer swings a 7-iron golf club with a loft of 38° and an initial velocity of 122 ft/s on level ground.

a. Write parametric equations to simulate this golf shot.

b. How far away does the ball first hit the ground?

c. How far away would the ball land if a golfer chose a 9-iron golf club with a loft of 46° and an initial velocity of 110 ft/s?

SportsConnection Golfers have a range of irons, or golf clubs, for hitting distance shots. Irons have faces with a variety of angles. An iron with a lesser angle, such as a 1-iron, is referred to as having less "loft"—the loft is the angle at which the ball should leave the ground. A low loft will cause the ball to leave the ground at a low angle and travel farther. An iron with a high loft, such as a 9-iron, will cause a ball to travel higher and not as far.

10. By how much does the ball from Example B clear a 10 ft fence that is 365 ft away if the wind is blowing directly from the fence toward Carolina at 8 mi/h?

11. Gonzo the Human Cannonball is fired out of a cannon 10 ft above the ground at a speed of 40 ft/s. The cannon is tilted at an angle of 60°. Gonzo's net hangs 5 ft above the floor. Where must his net be positioned so that he will land safely?

12. The tip of a metronome travels on a path modeled by the parametric equations $x = 0.7 \sin t$ and $y = \sqrt{1 - (0.7 \sin t)^2}$.

   a. Sketch the graph when $0° \leq t \leq 360°$. Describe the motion.

   b. Eliminate the parameter, t, and write a single equation using only x and y. Sketch this graph.

   c. Compare the two graphs.

   d. Rewrite the parametric equations such that the graph is reflected across the x-axis.

   e. Sketch the graph when $0° \leq t \leq 720°$. What does changing the 360° to 720° do?

ScienceConnection A trajectory is the path of an object in motion. The object might be a pendulum, a metronome, or a projectile, or it might be particles in the air or an oil spill at sea. By studying air movement trajectories, scientists can better understand the behavior and impact of air pollution. Trajectory models of hazardous materials spills in ocean and coastal areas have proven useful in locating the sites of plane crashes. Using ocean weather, current data, and debris fields, trajectory models assist in focusing on a particular search area.

Review

13. (Lesson 8.1) Two forces act simultaneously on a ball positioned at (4, 3). The first force imparts a velocity of 2.3 m/s to the east, and the second force imparts a velocity of 3.8 m/s to the north.

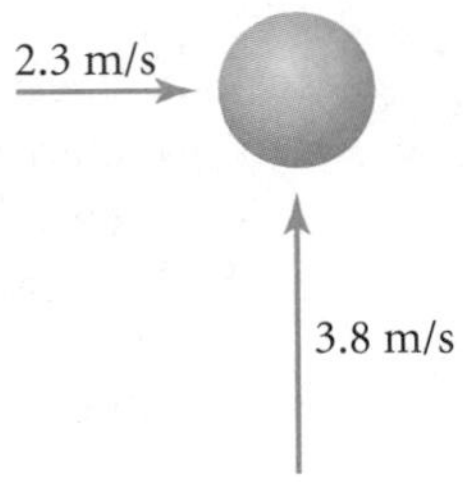

   a. Enter parametric equations into your calculator to simulate the resulting movement of the ball. Sketch your graph, and give your equations and graphing window.

   b. What is the ball's velocity? Give the magnitude and direction.

14. (Lesson 8.4) Find the height, width, and area specified.

   a. total height of the tree

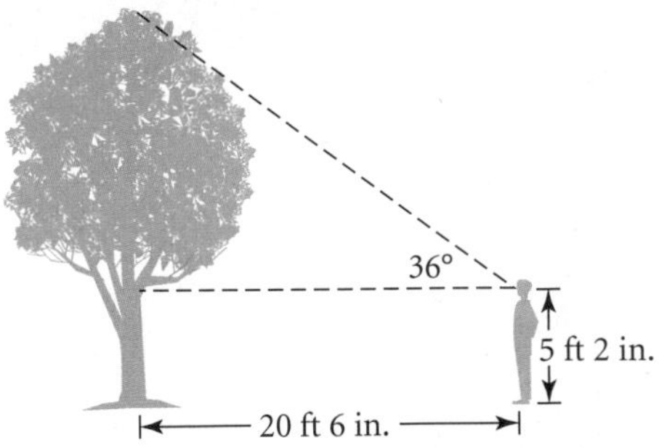

   b. width of the lake

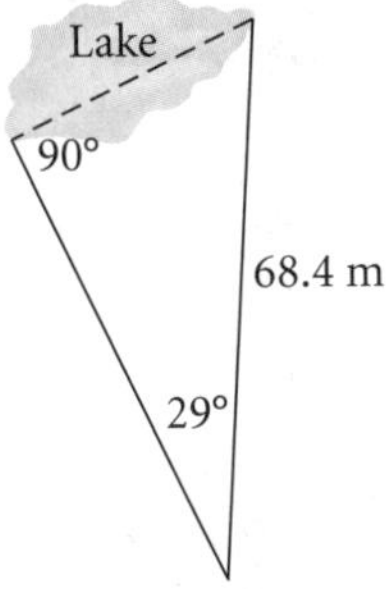

   c. area of the triangle

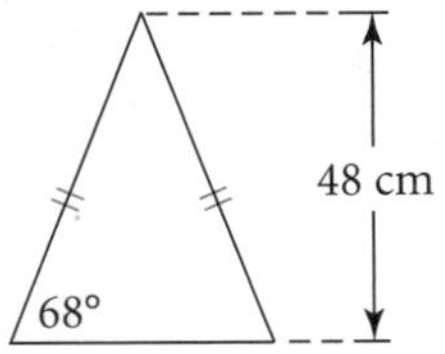

15. (Lesson 4.7) For this ellipse give parametric equations and a single equation using only x and y.

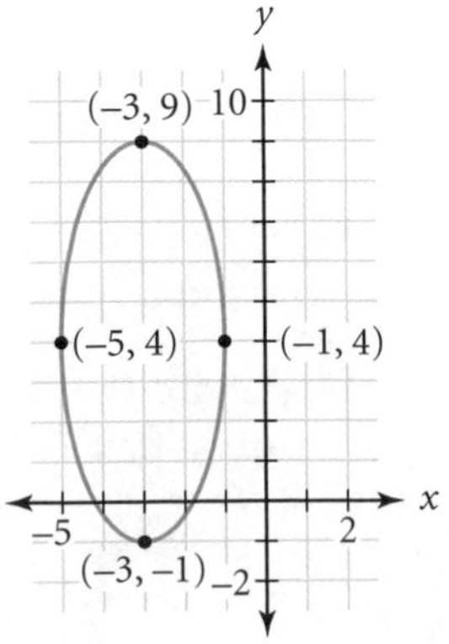

16. (Lesson 7.4) Identify each system as consistent (having a solution), inconsistent (having no solution), or dependent (having infinitely many solutions).

   a. $y = 3x - 7$
      $y = -x + 4$

   b. $y = 3x + 5$
      $y = 15x + 25$

   c. $x + y = 4$
      $x + y = 5$

LESSON

8.6

The Law of Sines

Two airplanes pass over Chicago at the same time. Plane A is cruising at 400 mi/h on a bearing of 105°, and Plane B is cruising at 450 mi/h on a bearing of 260°. How far apart will they be after 2 hours?

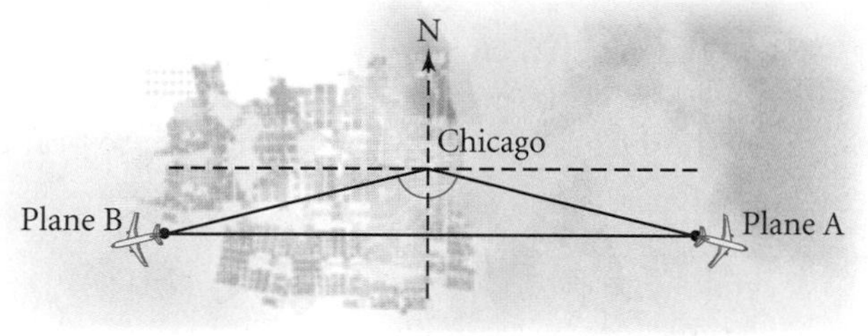

This problem is a familiar scenario, but the triangle formed by the planes' paths is not a right triangle. In the next two lessons you will discover useful relationships involving the sides and angles of nonright, or **oblique,** triangles and apply those relationships to situations much like the distance problem just presented.

Investigation
Oblique Triangles

You will need

- a ruler
- a protractor

In this Investigation you'll explore a special relationship between the sines of the angle measures of an oblique triangle and the lengths of the sides.

Step 1 Have each group member draw a different acute triangle ABC. Label the length of the side opposite $\angle A$ as a, the length of the side opposite $\angle B$ as b, and the length of the side opposite $\angle C$ as c. Draw the altitude from $\angle A$ to $\overline{BC}$. Label the height h.

Step 2 The altitude divides the original triangle into two right triangles, one containing $\angle B$ and the other containing $\angle C$. Use your knowledge of right triangle trigonometry to write an expression involving sin and B and h and an expression with sin C and h. Combine the two expressions by eliminating h. Write your new expression as a proportion in the form

$$\frac{\sin B}{?} = \frac{\sin C}{?}$$

Step 3 Now, draw the altitude from $\angle B$ to $\overline{AC}$, and label the height j. Repeat Step 2 using expressions involving j, sin C, and sin A. What proportion do you get when you eliminate j?

Step 4 Compare the proportions that you wrote in Steps 2 and 3. Use the transitive property of equality to combine them into an extended proportion:

$$\frac{?}{?} = \frac{?}{?} = \frac{?}{?}$$

Step 5 | Share your results with the members of your group. Did everyone get the same proportion in Step 4?

Sine, cosine, and tangent are defined for all real angle measures. Therefore, you can find the sines of obtuse angles as well as acute angles and right angles. Does your work from Steps 1–5 hold true for obtuse triangles as well?

Step 6 | Have each group member draw a different obtuse triangle. Measure the angles and the side of your triangle. Substitute the measurements, and evaluate to verify that the proportion from Step 4 holds for your obtuse triangles as well.

The relationships that you discovered in the Investigation allow you to solve many problems involving oblique triangles.

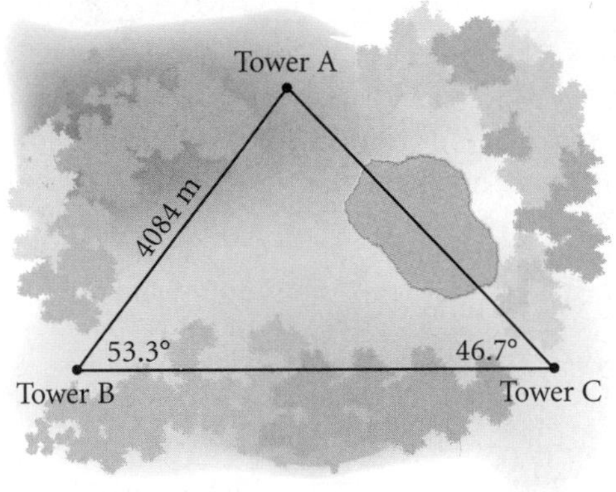

EXAMPLE A

Towers A, B, and C are located in a national forest. From Tower B the angle between Towers A and C is 53.3°, and from Tower C the angle between Towers B and C is 46.7°. The distance between Towers A and B is 4,084 m. A lake between Towers A and C makes it difficult to measure directly the distance between them. What is the distance between Towers A and C?

▸ Solution

First, convince yourself that *ABC* is not a right triangle. Because the sum of the measures of the angles in a triangle is 180°, $\angle A$ measures 80°.

On your paper, sketch and label a diagram. Include the altitude from $\angle A$ to $\overline{BC}$. As in the Investigation, two right triangles are formed. Set up sine ratios for each.

$$\sin 53.3° = \frac{h}{4084} \quad \text{or} \quad h = 4084 \sin 53.3°$$

and

$$\sin 46.7° = \frac{h}{b} \quad \text{or} \quad h = b \sin 46.7°$$

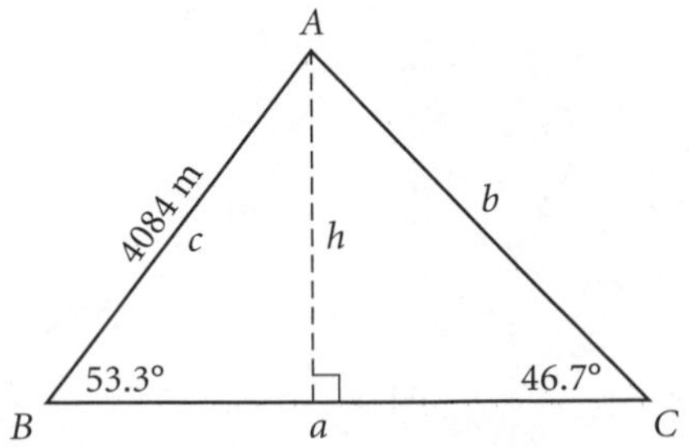

Substituting for h gives $4084 \sin 53.3 = b \sin 46.7$. Solving for b gives

$$b = \frac{4084 \sin 53.3°}{\sin 46.7°}$$

$$b \approx 4499$$

The distance between Towers A and C is approximately 4,499 m.

Notice that in Example A the height h was eliminated from the final calculations. You did not need a measurement for h! The equation $4084 \sin 53.3 = b \sin 46.7$ also can be written as $\frac{\sin 53.3°}{b} = \frac{\sin 46.7°}{4084}$, or $\frac{\sin B}{b} = \frac{\sin C}{c}$, which is the relationship you found in the Investigation.

The ratio of the sine of an angle to the length of the opposite side is constant throughout any triangle. This relationship is called the **law of sines.**

The Law of Sines

For any triangle with angles A, B, and C and sides of lengths a, b, and c (a is opposite $\angle A$, b is opposite $\angle B$, and c is opposite $\angle C$),

$$\frac{\sin A}{a} = \frac{\sin B}{b} = \frac{\sin C}{c}$$

Use the law of sines to find missing parts of triangles when you know the measures of two angles and the length of one side of a triangle.

EXAMPLE B

Find the length of $\overline{BC}$.

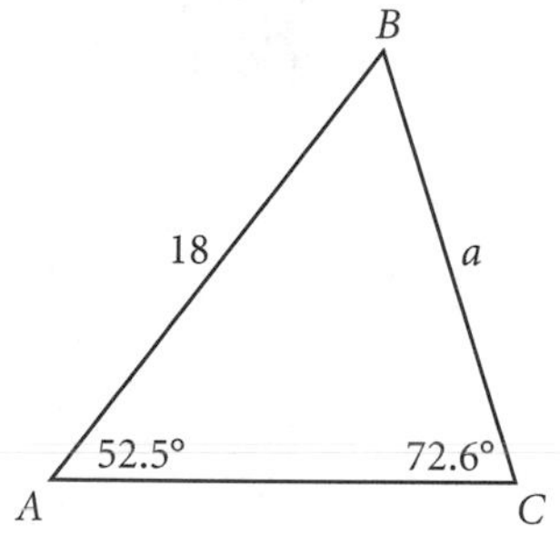

▶ Solution

Use the law of sines.

$$\frac{\sin A}{a} = \frac{\sin C}{c}$$ Select the proportion for the given angles and sides.

$$\frac{\sin 52.5°}{a} = \frac{\sin 72.6°}{18}$$ Substitute the angle measures and the known side length.

$$18 \sin 52.5° = a \sin 72.6°$$ Multiply both sides by $18a$ and reduce.

$$\frac{18 \sin 52.5°}{\sin 72.6°} = a$$ Divide both sides by sin 72.6°, and reduce the right side.

$$a \approx 15$$ Evaluate.

The length of $\overline{BC}$ is approximately 15 units.

You may also use the law of sines when you know two side lengths and the measure of the angle opposite one of the sides. However, in this case you may find more than one possible solution. This is because two different angles—one acute and one obtuse—may share the same value of sine. Look at this diagram to see how this works.

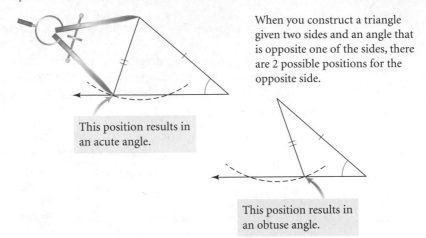

When you construct a triangle given two sides and an angle that is opposite one of the sides, there are 2 possible positions for the opposite side.

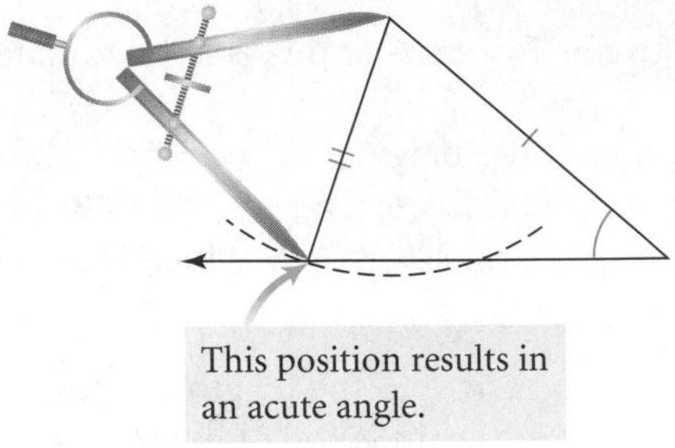
This position results in an acute angle.

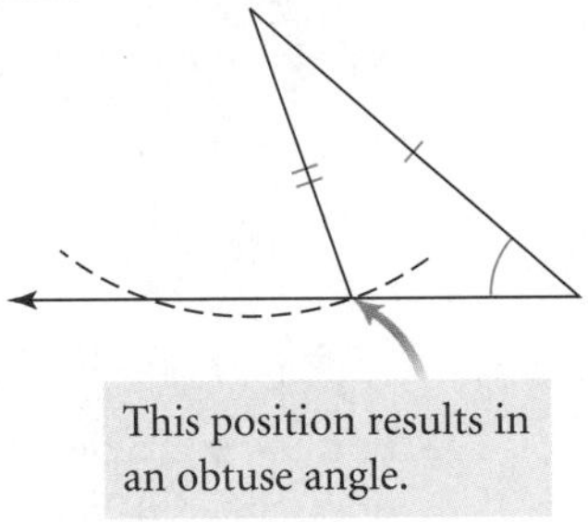
This position results in an obtuse angle.

EXAMPLE C

Tara and Yacin find a map that they think will lead to buried treasure. The map instructs them to start at the 47° fork in the river. They need to follow the line along the southern branch for 200 m, then walk to a point on the northern branch that is 170 m away. Where along the northern branch should they dig for the treasure?

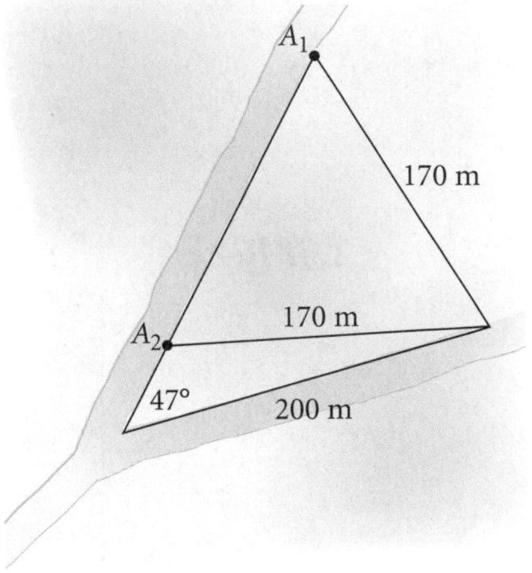
A_1
170 m
170 m
A_2
47°
200 m

▶ Solution

As the map shows, there are two possible places to dig. Consider the angles formed by the 170-m segment and the line along the northern branch as $\angle A$. Use the law of sines to find one possible measure of $\angle A$.

$$\frac{\sin 47°}{170} = \frac{\sin A}{200}$$ Set up a proportion using opposite sides and angles.

$$\sin A = \frac{200 \sin 47°}{170}$$ Solve for sin *A*.

$$A = \sin^{-1}\left(\frac{200 \sin 47°}{170}\right)$$ Take the inverse sine of both sides.

$$A \approx 59.4°$$ Evaluate.

If $\angle A$ is acute, it measures approximately 59.4°. The other possibility for $\angle A$ is the obtuse supplement of 59.4°, or 120.6°. You can verify this with geometry. Use your calculator to check that sin 59.4° is equivalent to sin 120.6° and that both angle measures satisfy the law of sines equation $\frac{\sin 47°}{170} = \frac{\sin A}{200}$.

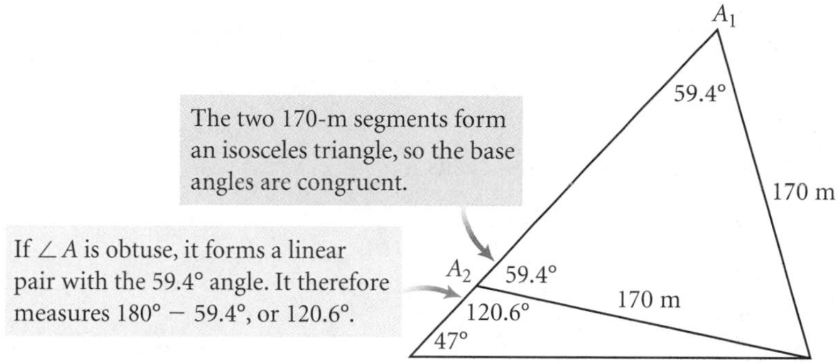

A_1
59.4°
The two 170-m segments form an isosceles triangle, so the base angles are congruent.
170 m
If $\angle A$ is obtuse, it forms a linear pair with the 59.4° angle. It therefore measures 180° − 59.4°, or 120.6°.
A_2
59.4°
170 m
120.6°
47°

To find the distance along the northern branch, you need the measure of the third angle in the triangle. Use the known angle measure, 47°, and the approximations for the measure of $\angle A$.

$$180° - (47° + 59.4°) \approx 73.6° \quad \text{or} \quad 180° - (47° + 120.6°) \approx 12.4°$$

Use the law of sines to find the distance along the northern branch, x.

$$\frac{\sin 73.6°}{x} = \frac{\sin 47°}{170} \quad \text{or} \quad \frac{\sin 12.4°}{x} = \frac{\sin 47°}{170}$$

$$x = \frac{170 \sin 73.6°}{\sin 47°} \quad \text{or} \quad x = \frac{170 \sin 12.4°}{\sin 47°}$$

$$x \approx 223 \quad \text{or} \quad x \approx 50$$

Tara and Yacin should dig two holes along the northern branch, one 50 m and the other 223 m from the fork of the river.

A situation like that in Example C is called an **ambiguous case,** that is, more than one possible solution exists. You can't tell which of the two possibilities is correct unless there is more information in the problem, such as whether the triangle is acute or obtuse. In general, in cases like this you should report both solutions.

EXERCISES

Practice Your Skills

1. Find the length of $\overline{AC}$.

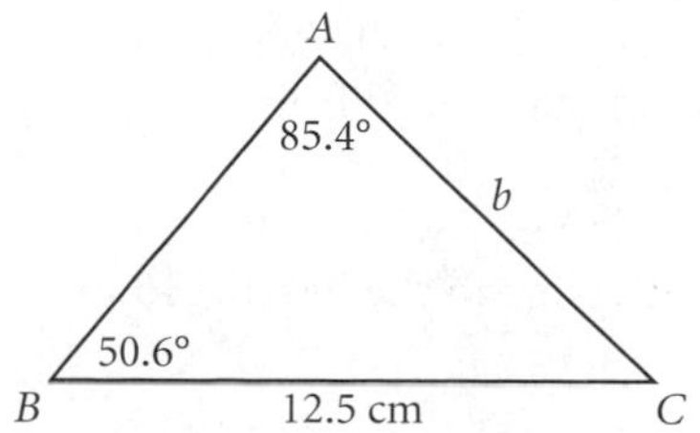

A
85.4°
b
50.6°
B
12.5 cm
C

2. Assume ΔPQR is an acute triangle. Find the measure of $\angle P$.

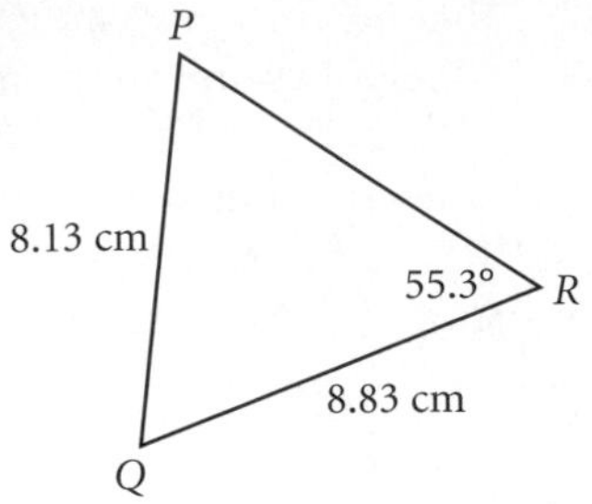

3. In ΔXYZ, $\angle Z$ is obtuse. Find the measures of $\angle X$ and $\angle Z$.

4. Find the length of $\overline{XY}$ in Exercise 3.

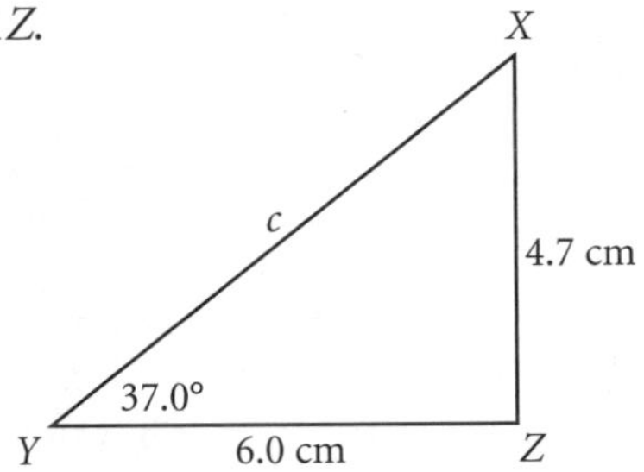

Reason and Apply

5. Find the unknown angle measures and side lengths.

a.

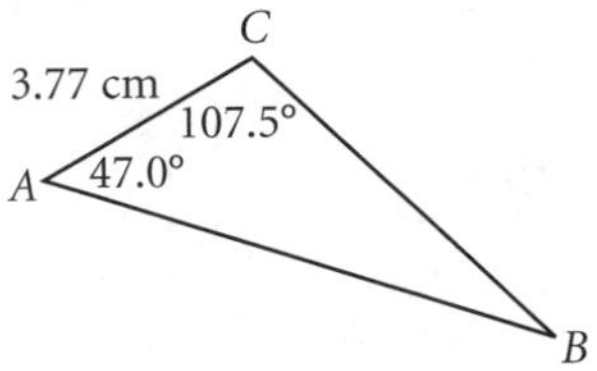

b.

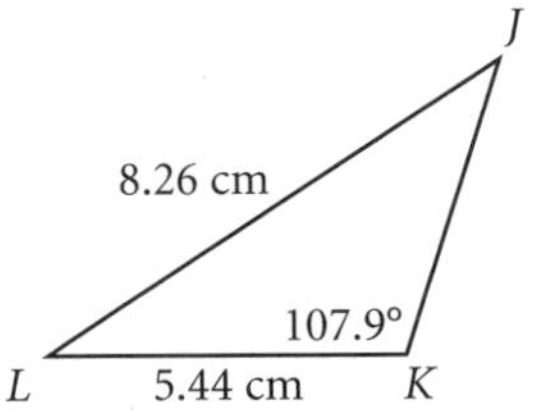

6. In an isosceles triangle one of the base angles measures 42°. The length of each leg is 8.2 cm.

a. Find the length of the base.

b. Even though you were given one angle and two sides not including the angle, this was not an ambiguous case. Why not?

7. **APPLICATION** Venus is 67 million miles from the Sun. Earth is 93 million miles from the Sun. Imagine a straight line that extends from the Sun to Earth and another line that extends from the Sun to Venus. The angle between the two lines is 14°. At this moment, how far is Venus from Earth? (*Hint:* There are two possible answers, depending on whether Venus is on the near side or the far side of the Sun.)

8. The Daredevil Cliffs rise vertically from the beach. The beach slopes gently down to the water at an angle of 3° from the horizontal. Scott lies at the water's edge, 50 ft from the base of the cliff, and determines that his line of vision to the top of the cliff makes a 70° angle with the line to the beach. How high is the cliff?

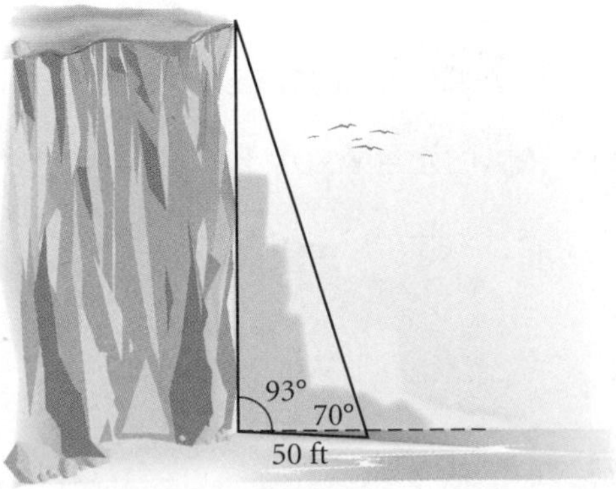

9. APPLICATION The SS *Minnow* is lost at sea in a deep fog. Moving on a bearing of 107°, the skipper sees a light at a bearing of 60°. The same light reappears through the fog after the skipper has sailed 1.5 km on his initial course. The second sighting of the light is at a bearing of 34°. How far is the boat from the source of the light at the time of the second sighting?

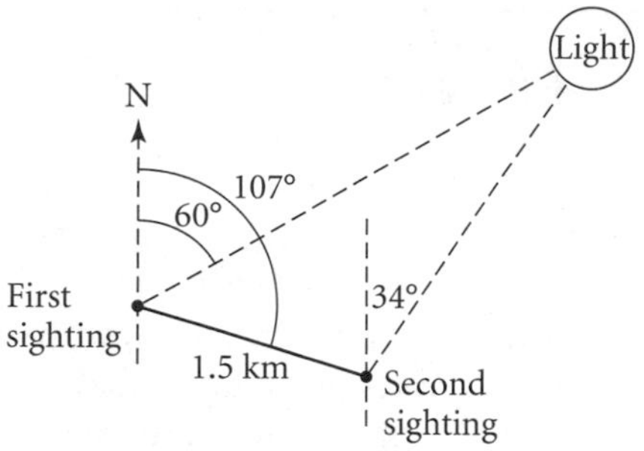

HistoryConnection Lighthouses project light through darkness and poor weather to help guide ships to shore. The first lighthouses were actually bonfires—one was even mentioned in the *Iliad,* a Greek epic poem written by Homer sometime before 700 B.C.E. Today, most lighthouses hold powerful electric lights that flash automatically. Every lighthouse has a distinct sequence of flashes that allow ship captains to identify the nearby harbor.

10. APPLICATION One way to calculate the distances between Earth and a nearby star is to measure the angle between the star and the ecliptic (the plane of Earth's orbit) at 6-month intervals. A star is measured at a 42.13204° angle. Six months later the angle is 42.13226°. The diameter of Earth's orbit is $3.13 \cdot 10^{-5}$ light-years. What is the distance to the star? Use this diagram to help you solve the problem.

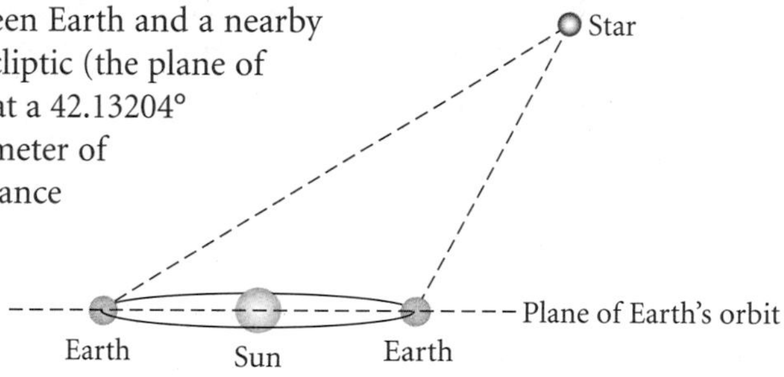

ScienceConnection A light-year is the distance that light travels in one year. Light moves at a velocity of about 300,000 km/s, so one light-year is equal to about 9,500,000,000,000 km. Distances in space are so large that it's difficult to express them with relatively small units like kilometers. For example, the distance to the next nearest big galaxy is 21 quintillion km! Another unit of distance used by astronomers to measure distances within our solar system is the astronomical unit (AU). One AU is the average distance between Earth and the Sun, about 150 million km. Pluto averages about 40 AU from the Sun.

11. APPLICATION When light travels from one transparent medium into another, the rays bend, or refract. Snell's law of refraction states that $\frac{\sin\theta_1}{n_2} = \frac{\sin\theta_2}{n_1}$, where θ_1 is the angle of incidence, θ_2 is the angle of refraction, and n_1 and n_2 are the indexes of refraction for the two mediums, as shown in the diagram.

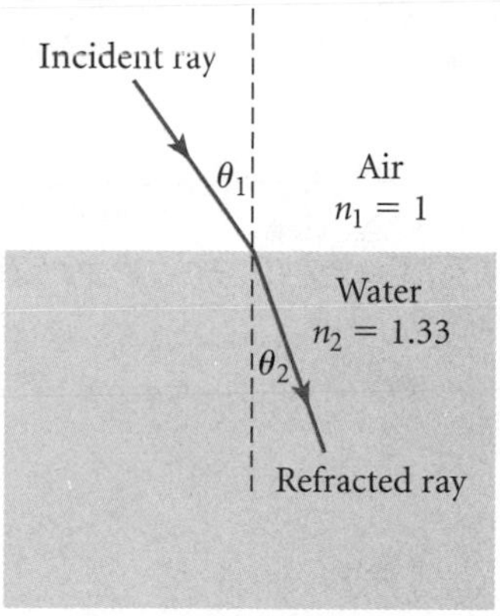

a. Find the angle of refraction in water if the angle of incidence from air is 60°.

b. If the angle of refraction from air to water is 45°, at what angle did the ray enter the water?

c. If the angle of incidence is 0°, what is the angle of refraction?

Review

12. (Lesson 6.4) Use the quadratic formula to solve each equation.

a. $2x^2 - 8x + 5 = 0$

b. $3x^2 + 4x - 2 = 7$

13. (Lesson 8.4) Draw a compass rose, and show each vector. Find the x- and y-components of each, and show the angle with the x-axis used to find the components. Indicate the direction of the components with the proper sign.

a. 12 units on a bearing of 168°

b. 16 units on a bearing of 221°

14. (Lesson 8.4) Find each vector. Give its magnitude and bearing.

a. x-component: -9.1 units
y-component: 4.1 units

b. x-component: 16.6 units
y-component: 14.4 units

15. (Lesson 2.5) The value of a building depreciates at a rate of 6% per year. When new, the building is worth \$36,500.

a. How much is the building worth after 5 years 3 months?

b. To the nearest month, when will the building be worth less than \$10,000?

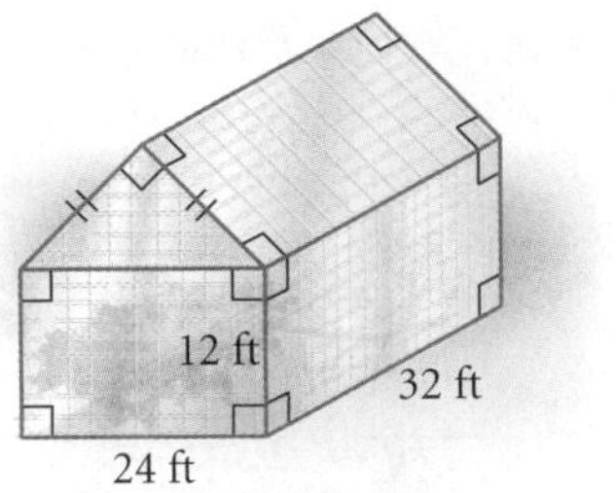

IMPROVING YOUR GEOMETRY SKILLS

A New Area Formula

You are given the lengths of two sides of a triangle, a and b, and the measure of the angle between them, C. How can you find the area of the triangle using only these three measurements?

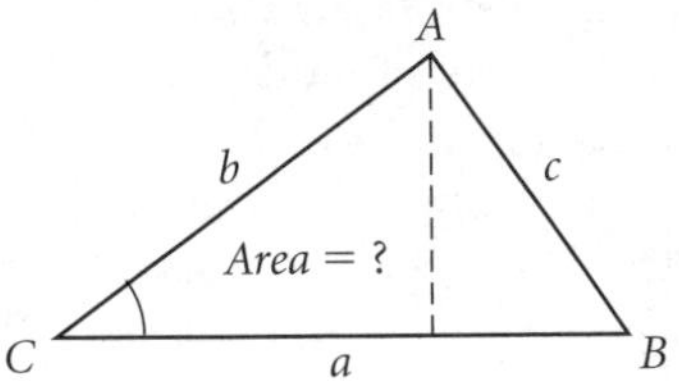

LESSON

8.7

The Law of Cosines

The law of sines enabled you to find the lengths of sides of a triangle or the measures of the angles in certain situations. To use the law of sines, you needed to know the measures of two angles and the length of any side, or the lengths of two sides and the measure of the angle opposite one of the sides. What if you know a different combination of sides and angles?

EXAMPLE A

Two hot-air balloons approach a landing field. One is 12 mi from the landing point, and the other is 17 mi from the landing point. The angle between the balloons is 70°. How far apart are the two balloons?

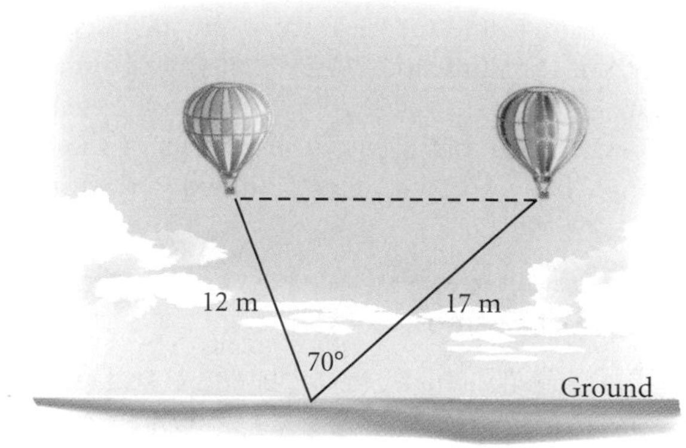

▶ Solution

In this case the law of sines does not help. You know only the included angle, or the angle between the two sides. If you try to set up an equation using the law of sines, you will always have more than one variable. So you must try something else.

Sketch one altitude to form two right triangles such that one of the right triangles contains the 70° angle.

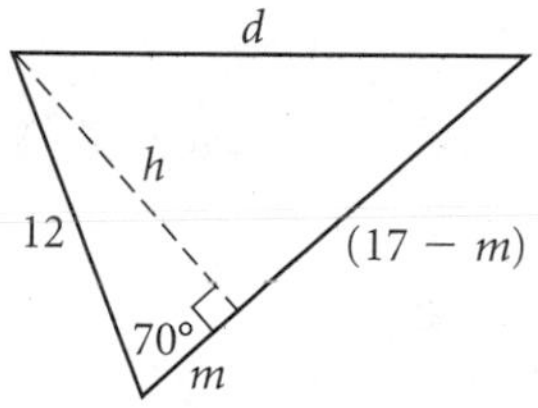

If you use the altitude from the balloon on the left, the 17-mi side is split into two parts. Label one part m and the other part $17 - m$. Label the height h, and label the distance between the balloons d. You can now write two equations using the Pythagorean Theorem.

$m^2 + h^2 = 12^2$ and $(17 - m)^2 + h^2 = d^2$ — Substitute the lengths of the sides of the two right triangles in the Pythagorean Theorem.

$m^2 + h^2 = 144$ and $289 - 34m + m^2 + h^2 = d^2$ — Multiply and expand.

$h^2 = 144 - m^2$ — Solve $m^2 + h^2 = 144$ for h^2.

$289 - 34m + m^2 + (144 - m^2) = d^2$ — Substitute $144 - m^2$ for h^2 into the second equation.

$289 + 144 - 34m = d^2$ — Combine like terms.

Use the right triangle that contains the 70° angle to write

$$\cos 70° = \frac{m}{12}$$

Solve this equation for m, and substitute in the equation for d^2.

$$289 + 144 - 34\,(12 \cos 70°) = d^2$$ Substitute $12 \cos 70°$ for m.

$$\sqrt{289 + 144 - 34(12\cos 70°)} = d$$ Take the square root of both sides. Because d represents a distance, you need only the positive root.

$$d \approx 17.1$$ Evaluate.

The distance between the two balloons is approximately 17.1 mi.

You can repeat the procedure used in Example A any time you know the lengths of two sides of a triangle and the measure of the included angle, and when you need to find the length of the third side. Notice that you could also write the equation for d^2 as $d^2 = 17^2 + 12^2 - 2(17)(12) \cos 70°$, which looks similar to the Pythagorean Theorem with an extra term that is twice the product of the length of the sides and the cosine of the angle between them. This modified Pythagorean relationship is called the **law of cosines.**

The Law of Cosines

For any triangle with angles A, B, and C and sides of lengths a, b, and c (a opposite A, b opposite B, and c opposite C),

$$a^2 = b^2 + c^2 - 2bc \cos A$$

$$b^2 = a^2 + c^2 - 2ac \cos B$$

$$c^2 = a^2 + b^2 - 2ab \cos C$$

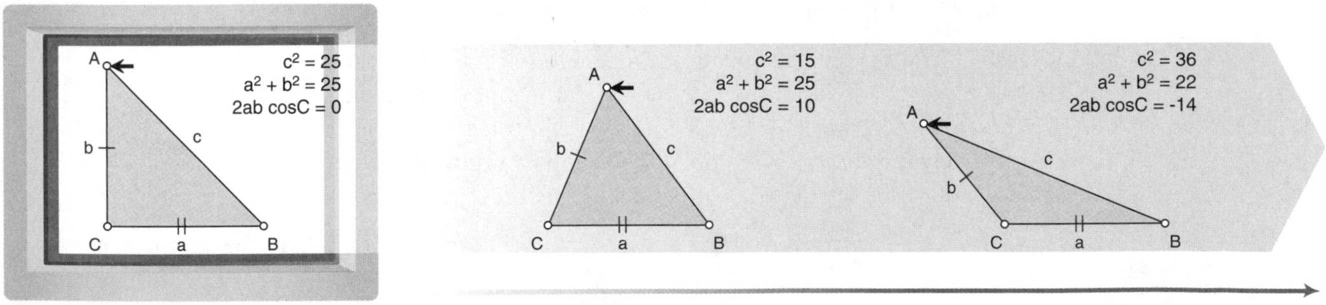

In this right triangle, c^2 is equivalent to $a^2 + b^2$.
$c^2 = a^2 + b^2$

In this acute triangle, c^2 is less than $a^2 + b^2$. The difference is $2ab \cos C$.
$c^2 = a^2 + b^2 - 2ab \cos C$

In this obtuse triangle, c^2 is more than $a^2 + b^2$. Again, the difference is $2ab \cos C$.
$c^2 = a^2 + b^2 - 2ab \cos C$

The Investigation gives you the opportunity to apply the law of cosines to a real-world situation.

Investigation
Around the Corner

You will need

- meter sticks or yardsticks
- a protractor

Position the metersticks or yardsticks so that the ends touch and form a 50° angle. Then put three chips of different colors along each stick. Use trigonometry to find the distance from each chip along the first stick to each chip along the second stick. Altogether there will be nine distances to be calculated. After you have used trigonometry to calculate the distances, measure to check your work.

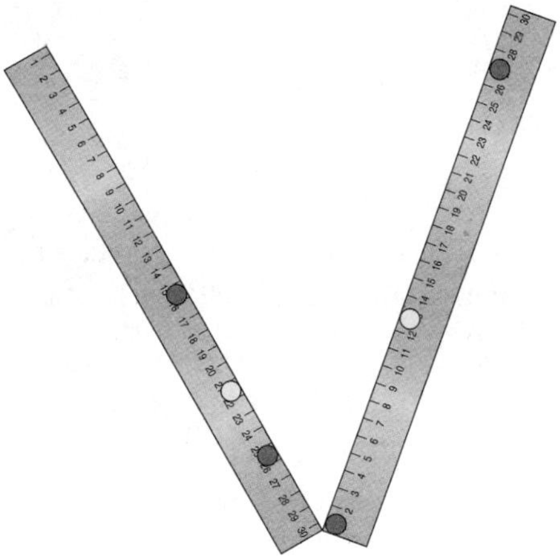

You can use the law of cosines alone or in combination with other triangle properties.

EXAMPLE B

Find the unknown angle measures and side lengths.

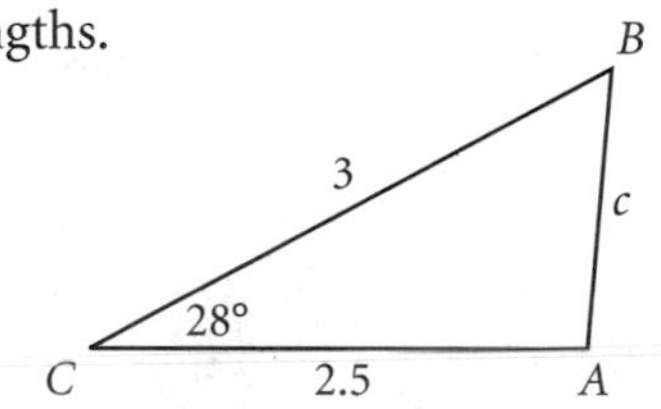

▶ *Solution*

First, use the law of cosines to find the length of $\overline{AB}$.

$c^2 = a^2 + b^2 - 2ab\cos C$	The law of cosines for finding c when a, b, and C are known.
$c^2 = 3^2 + 2.5^2 - 2(3)(2.5)\cos 28°$	Substitute 3 for a, 2.5 for b, and 28° for C.
$c^2 = 9 + 6.25 - 15\cos 28°$	Multiply.
$c = \sqrt{9 + 6.25 - 15\cos 28°}$	Take the positive square root of both sides.
$c \approx 1.42$	Evaluate.

The length of $\overline{AB}$ is approximately 1.42 units.

Now use the law of cosines to find the measure of $\angle A$.

$a^2 = b^2 + c^2 - 2bc\cos A$ — The law of cosines for finding A when a, b, and c are known.

$3^2 \approx 2.5^2 + 1.42^2 - 2(2.5)(1.42)\cos A$ — Substitute values for a, b, and c.

$9 \approx 8.2664 - 7.1\cos A$ — Multiply.

$0.7336 \approx -7.1\cos A$ — Subtract 8.2664 from both sides.

$\cos A \approx \dfrac{0.7336}{-7.1}$ — Divide by -7.1.

$A \approx \cos^{-1}\left(\dfrac{0.7336}{-7.1}\right)$ — Take the inverse cosine of both sides.

$A \approx 96°$ — Evaluate.

Angle A measures approximately 96°.

To find the measure of the last angle, use the fact that the measures of the three angles in any triangle sum to 180°. The measure of $\angle B$ is approximately $180° - 28° - 96°$, or 56°.

During a calculation it is best to use the entire previous answer for the next calculation. Rounding before the last step can jeopardize the accuracy of your answer. In Example A you could find the measure of $\angle A$ with more accuracy by using $\sqrt{9 + 6.25 - 15\cos 28°}$ for c instead of the approximation of 1.42. In all cases, you need to verify that the answers you get make sense in the context of the problem or in a sketch of the triangle.

In deciding whether to use the law of sines or the law of cosines, consider the triangle parts whose measurements you know and their relationships to each other.

The Law of Sines	**The Law of Cosines**
Angle-angle-side	Side-angle-side
Side-side-angle (ambiguous case)	Side-side-side

EXERCISES

Practice Your Skills

1. Find the length of $\overline{AC}$.

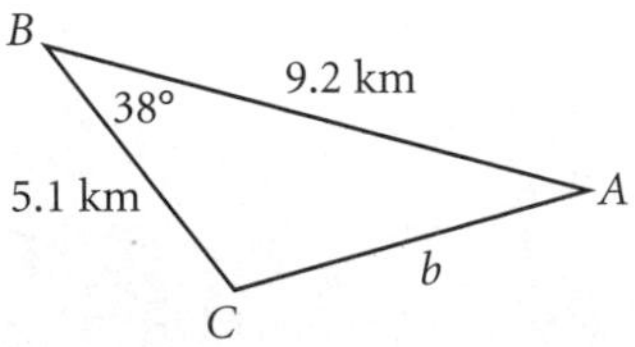

2. Find the measure of $\angle T$.

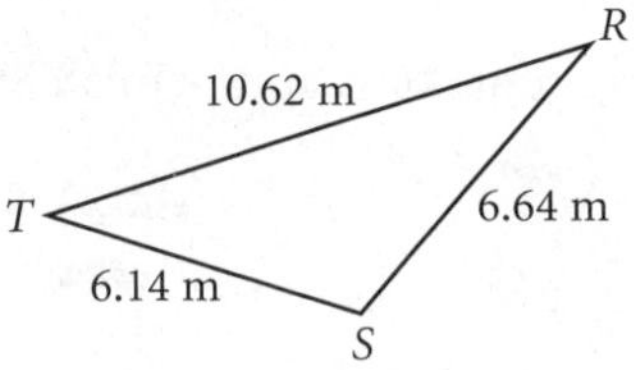

3. Solve for A and b. (Assume that b is positive.)

a. $16 = 25 + 36 - 2(5)(6)\cos A$

b. $49 = b^2 + 9 - 2(3)(b)\cos 60°$

4. Find the unknown angle measures and side lengths.

a.

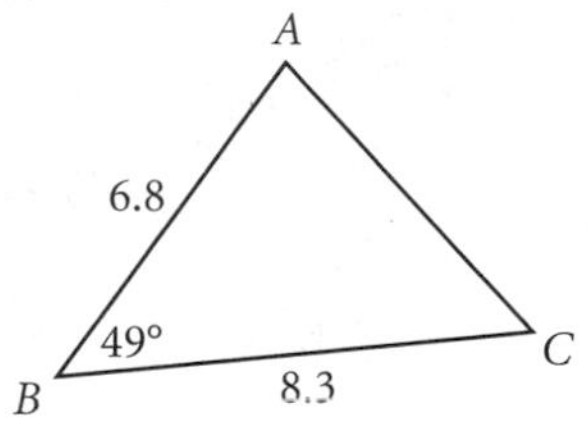

b.

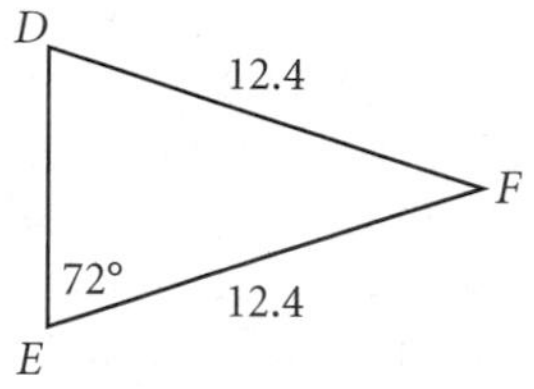

c.

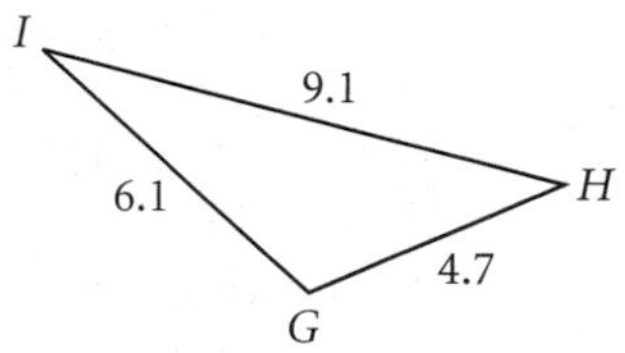

Reason and Apply

5. Two airplanes pass over Chicago, Illinois, at the same time. One is cruising at 400 mi/h on a bearing of 105°, and the other is cruising at 450 mi/h on a bearing of 260°. How far apart will they be after 2 h?

6. Find the measure of $\angle S$.

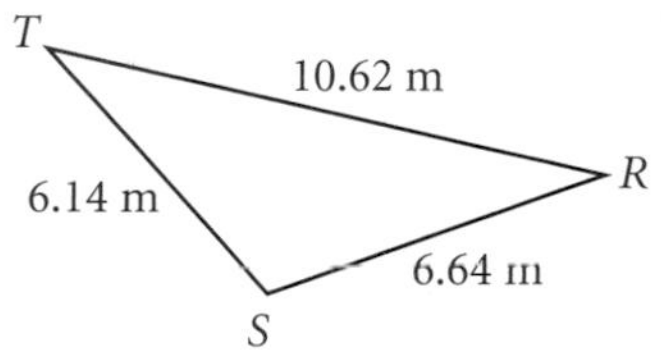

7. **APPLICATION** Seismic exploration identifies underground phenomena, such as caves, oil pockets, and rock layers, by transmitting sound into the ground and timing the echo of the vibration. From a sounding at point A, a thumper truck locates an underground chamber 7 km away. Moving to point B, 5 km from point A, the truck takes a second sounding and finds that the chamber is 3 km away from that point. Assume that the underground chamber lies in the same vertical plane as A and B. What more can you say about the location of the underground chamber?

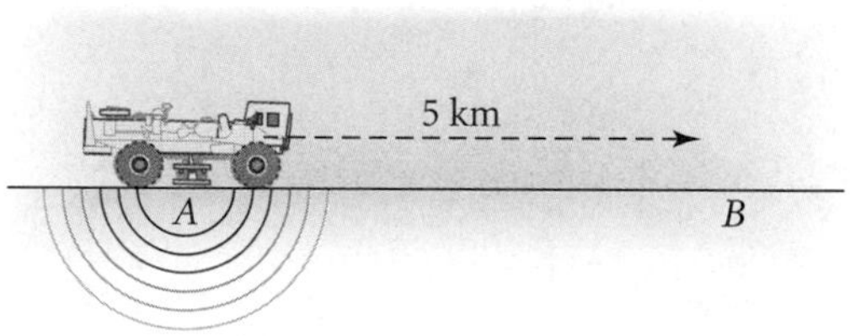

Geology Connection Locating oil deposits usually involves advanced geological research into the Earth's crust to find rock formations that typically produce petroleum. Once such an area is identified, "prospectors" build wells that periodically drill the ground for samples. More sophisticated methods for finding oil include measurement of the magnetic forces released by large oil deposits and transmitted to the Earth's surface, as well as the use of satellites to view ultraviolet images of the Earth that reveal oil-rich land.

8. A folding chair's legs meet to form a 50° angle. The rear leg is 55 cm long and attaches to the front leg at a point 75 cm from the front leg's foot. How far apart are the legs at the floor?

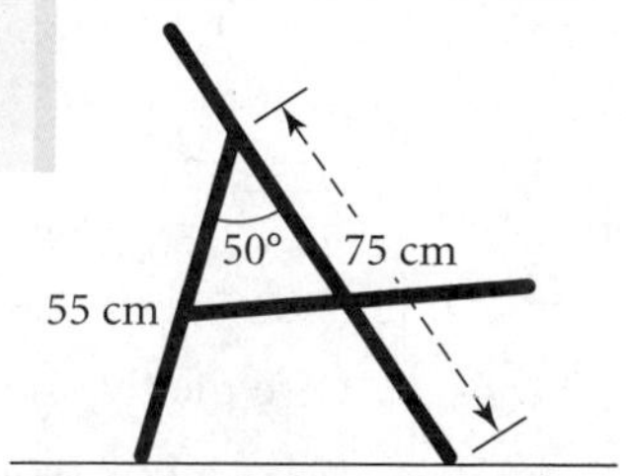

9. APPLICATION Triangulation is used to locate airplanes, boats, or vehicles that transmit radio signals. By measuring the strength of the signal at three fixed receiving locations, the distances to the vehicle are found and the directions calculated. Receiver B is 18 km from Receiver A at a bearing of 122°. Receiver C is 26 km from Receiver A at a bearing of 80°. The signal from a source vehicle to Receiver A indicates a distance of 15 km. The distance from the source vehicle to Receiver B is 8 km. The distance from the source vehicle to Receiver C is 25 km. What is the bearing from Receiver A to the source vehicle? (*Hint:* Try making a scale drawing of this situation, and use your compass to get a general idea of the location of the source before starting the calculations.)

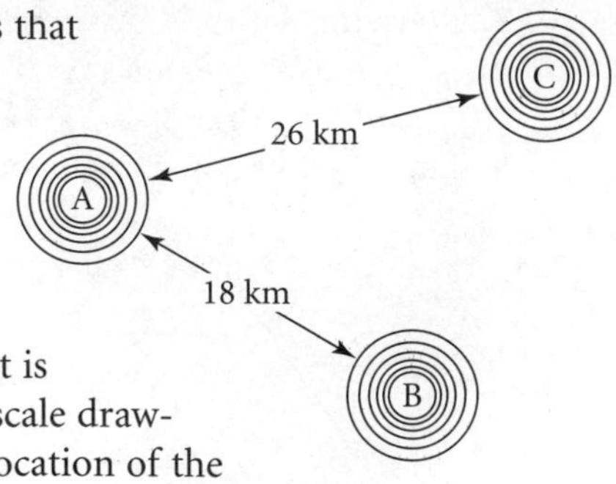

10. APPLICATION The Hear Me Now Phone Company plans to build a cell tower to serve the needs of Pleasant Beach and the beachfront. It decides to locate the cell tower such that Pleasant Beach is 1 mi away at a bearing of 60° from the tower. The range of the signal from the cell tower is 1.75 mi. The beachfront runs north to south. How far south of Pleasant Beach will customers be able to use their cell phones?

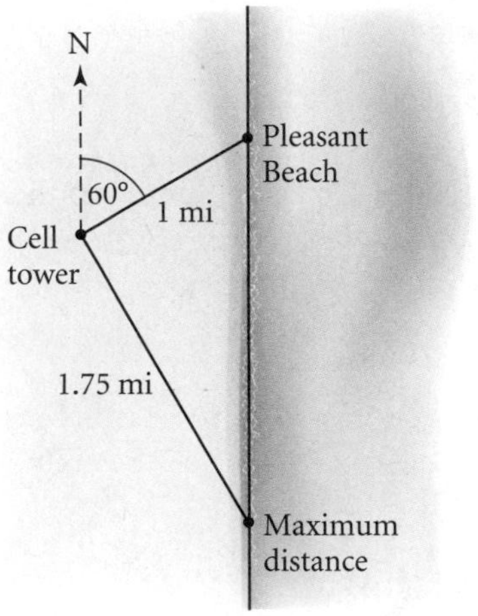

TechnologyConnection The placement of cell towers is crucial to a variety of cellular services. Mathematical models are used to analyze possible sites. Cell towers are located by individual companies based on the companies' business plans for the markets they serve. To place cell towers with the greatest cost efficiency, cellular companies consider, among other things, the physical design, environmental impact, engineering, and aesthetics of their towers, as well as the topography and population density of the sites.

Review

11. (Lesson 8.4) A sailor on a ship sees a lighthouse at a bearing of 105°. The ship sails 8.8 nautical miles on a bearing of 174°. The lighthouse is now at a bearing of 52°. How far is the ship from the lighthouse?

12. (Lesson 1.4) Here are the batting averages of the National League's Most Valuable Players from 1980 to 2000.

{.286, .316, .281, .302, .314, .353, .290, .287, .290, .291, .301, .319, .311, .336, .368, .319, .326, .366, .308, .319, .334}
(*The New York Times Almanac 2002*)

a. Give the five-number summary.
b. Find the range and interquartile range.
c. Make a box plot of these data.
d. Find the mean and standard deviation.

E X P L O R A T I O N
with Microsoft® Excel

Projectile Motion in the Plane

Download the Chapter 8 Excel Exploration from the website at www.keycollege.com/online or from your Student CD.

In this Exploration you will study projectile motion in the plane. You will study the impact of varying the angle of inclination and the initial speed. When you have finished, you should be able to answer the question, "For a given initial velocity, what is the best angle at which to hit a baseball if you want it to go as far as possible?"

On the spreadsheet you will find complete instructions and questions to consider as you explore.

Activity
Modeling a Trigonometric Identity and Studying Periodicity

Excel skills required

- Name a cell
- Data fill
- Create an XY (scatter) plot
- Modify the chart title and legend of a graph
- Fix the scale for the axes
- Resize a graph

In this Activity you will build tables and graphs to investigate the sine and cosine functions and the basic relation between these two functions. You will be able to model the most fundamental trigonometric identity by observing many specific examples in the data table. The trigonometric identity that you will study is $\sin^2 x + \cos^2 x = 1$.

You can find instructions for the required Excel skills in the Excel Notes at www.keycollege.com/online or on your Student CD.

Launch Excel.

Step 1 Create the data labels as shown.

	A	B	C	D	E	F
1			a			
2						
3	x	cos(ax)	sin(ax)	[cos(ax)]^2	[sin(ax)]^2	[cos(ax)]^2 + [sin(ax)]^2

Step 2 Name Cell D1 **a.**

Step 3 Enter these numbers and formulas.

a. In Cell D1 enter the number 1

b. In Cell A4 enter the number 0.

c. In Cell B4 enter `=COS(a*A4)`.

d. In Cell C4 enter `=SIN(a*A4)`.

e. In Cell D4 enter `=B4^2`.

f. In Cell E4 enter `=C4^2`.

g. In Cell F4 enter `= D4+E4`.

h. In Cell A5 enter `=A4+pi()/180`.

Step 4 Complete the data fill as stated.

a. Data-fill Cells B4 to F4 to fill Cells B5 to F5.

b. Data-fill Cells A5 to F5 to Row 364.

The first column represents radian measures in increments equivalent to one degree. The other columns represent trigonometric functions of the angles with the given radian measure.

Step 5 Highlight Cells A4 to C364, and create an XY(scatter) plot connected with smooth lines and without markers.

Step 6 Modify the chart title to read Cosine and Sine Functions.

Step 7 Position and size the graph to fit, and make it easily viewed.

Step 8 Highlight Cells B4 to C364. With these cells highlighted, create an XY(scatter) plot connected with smooth lines and without markers.

Step 9 On this graph specify the scale of the axes in both directions so that the minimum is -1 and the maximum is 1. By doing so, you ensure that Excel will not keep changing the scale of the graph. Then, resize the graph so that the box in which it appears is as close to square as possible.

Questions

1. What does the graph generated by Step 8 tell you about the relation between the sine and cosine functions?
2. Notice that Column F is constant, although the formulas change as you go down. What fundamental trigonometric identity is modeled by Column F? Do you think this relationship holds for all values of x, even those not showing in the table?
3. Change Cell D1 to the number 2. Describe what happens to the graph generated in Step 5. For which value of x does the behavior of the functions start to repeat?

4. Change Cell D1 first to the number 3, then to 4. Describe the patterns that you see in the sine and cosine graphs. In each case, for which value of x does the behavior of the functions start to repeat? This repetition is called *periodicity*.

5. In the graph generated in Step 8, a typical point can be described as having coordinates $(\cos(x), \sin(x))$. What does Column F tell you about the relationship of the two coordinates for any value of x?

6. The graph generated in Step 8 is the unit circle. What is the typical equation of the unit circle? Compare this with your result in Question 5.

7. Change the value of a, and inspect the changes in the graph of the circle. Do you see any changes as a increases, say, from $a = 0$ to $a = 6$? Double-click on a point on the graph, and in the Patterns tab set the marker at automatic. How does this point change as a changes? Explain this.

Extension

8. Add a new column for sin(2*x) and, yes, another column for 2*sin(x)*cos(x). Compare the results, and express the relation as an identity.

9. Model the trigonometric identity $\cos(2x) = \cos^2 x - \sin^2 x$.

CHAPTER 8 REVIEW

Parametric equations describe the locations of points by using a third variable, t, called a **parameter**. In many cases this third variable represents time. In other cases it can be interpreted as an angle or simply a number. By controlling the range of t-values, you can graph part of a function. To convert from parametric form to regular form, solve either the x- or the y-equation for t. Then substitute this expression in the other equation. The result is an equation involving only x and y.

One common use of parametric equations is to simulate motion. This often involves the use of right triangle **trigonometry**. The **trigonometric ratios**—sine, cosine, and tangent—relate the side lengths of a triangle to the measure of an angle. The **sine** of an angle is the ratio of the opposite leg to the hypotenuse in a right triangle. The **cosine** of an angle is the ratio of the adjacent leg to the hypotenuse, and the **tangent** of an angle is the ratio of the opposite leg to the adjacent leg. You can use these ratios to find missing side lengths and angles in right triangles. You can use the **law of sines** and the **law of cosines** to find missing side lengths and angles in triangles that do not contain right angles.

Modeling motion at an angle involves the use of the trigonometric ratios. If an object, such as a plane or a boat, is traveling in a direction that is not due north, south, east, or west, you must break the motion into east-west and north-south components. The east-west component is $x = vt \cos A$, and the north-south component is $y = vt \sin A$, where A is the angle the object makes with the x-axis, and v is the **velocity** of the object. If the motion is directed below the x-axis or to the left of the y-axis, then the velocity is negative in either the horizontal or the vertical direction or both. Sometimes the wind or a current is present and influences the motion. If this is the case, the motion of the wind or current also must be broken into its x- and y-components and then added to the equations of the object's motion. Parametric equations can also be used to model projectile motion with the equations $x = v_0 t \cos A + x_0$ and $y = -16t^2 + v_0 t \sin A + y_0$.

EXERCISES

1. A raft is being moved by a wind blowing east at 20 m/s and a current flowing south at 30 m/s.
   a. After 8 s what is the raft's position relative to its start?
   b. What equations simulate this motion?

2. Use the parametric equations $x = -3t + 1$ and $y = \frac{2}{t+1}$ to answer each question.

a. Find the x- and y-coordinates of the points that correspond to the values of $t = 3$, $t = 0$, and $t = -3$.

b. Find the y-value that corresponds to an x-value of -7.

c. Find the x-value that corresponds to a y-value of 4.

d. Sketch the curve for $-3 \leq t \leq 3$, showing the direction of movement. Trace the graph and explain what happens when $t = -1$.

3. Graph each pair of parametric equations in a friendly window with factor 2, then eliminate the parameter to get a single equation using only x and y. Graph the resulting equation, and describe how it compares with the original graph.

a. $x = 2t - 5, y = t + 1$

b. $x = t^2 + 1, y = t - 2, -2 \leq t \leq 6$

c. $x = \dfrac{t+1}{2}, y = t^2, -4 \leq t \leq 3$

d. $x = \sqrt{t+2}, y = t - 3$

4. Write parametric equations that will result in each transformation below for the equations $x = 2t - 5$ and $y = t + 1$.

a. Reflect the curve across the y-axis.

b. Reflect the curve across the x-axis.

c. Translate the curve up 3 units.

d. Translate the curve left 4 units and down 2 units.

5. Solve to find the measure of the angle or the length of the side, as indicated in each triangle.

a.

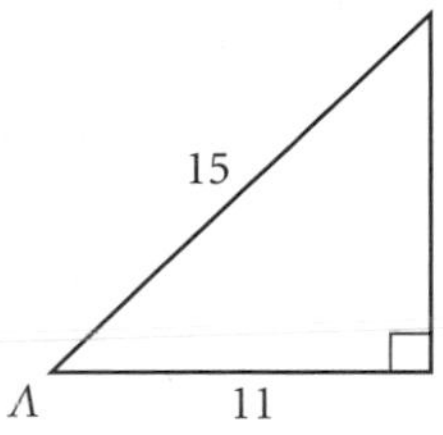

b.

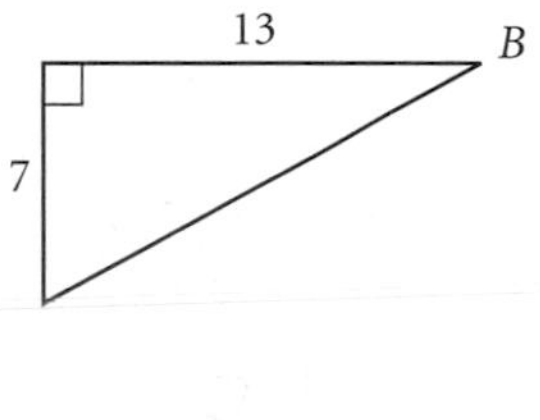

c.

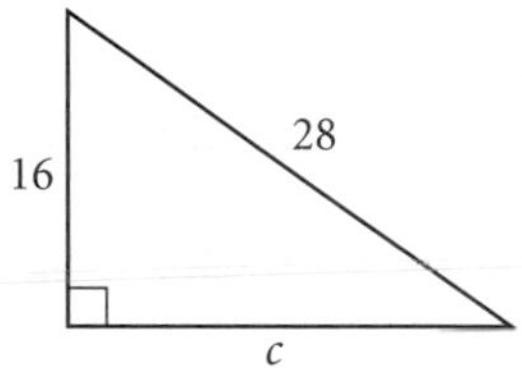

d.

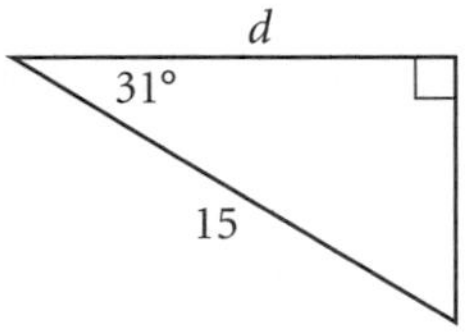

e.

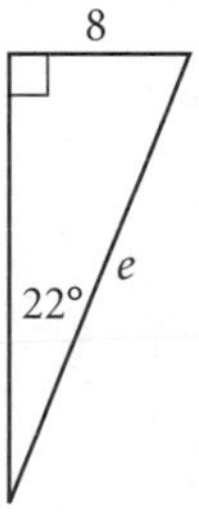

f.

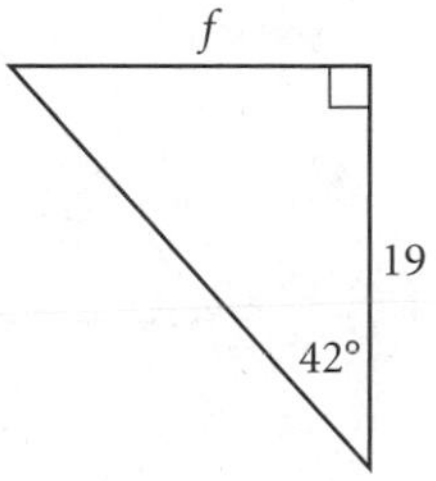

6. Sketch a graph of $x = t \cos 28°, y = t \sin 28°$. What is the angle between the graph and the x-axis?

7. A diver runs off a 10-m platform with an initial horizontal velocity of 4 m/s. The edge of the platform is directly above a point 1.5 m from the pool's edge. Simulate her motion on your graphing calculator. How far from the edge of the pool will she hit the water?

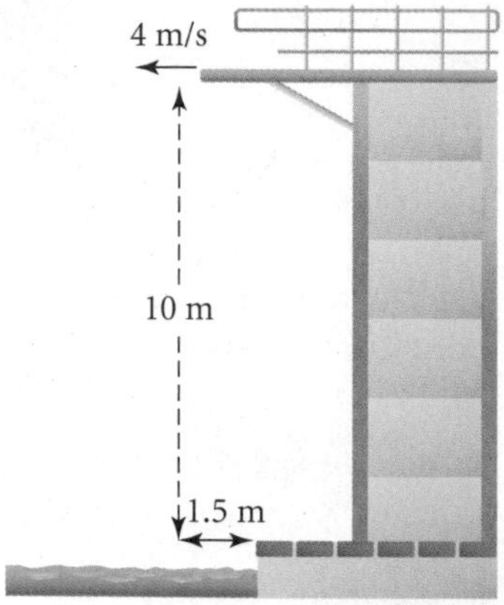

SportsConnection In competitive diving a running dive must be at least four steps long. The takeoff phase, just before flight, determines the diver's path through the air and away from the platform. Once in the air the diver has less than 2 seconds to finish the dive, which should end with the diver's body almost perpendicular to the water's surface.

8. A toy boat is placed into a 47-ft-wide river. The boat travels at a rate of 2.4 ft/s, aiming directly for the opposite bank of the river. When it reaches the other bank, the boat is 28 ft downstream from the point where it started. What is the speed of the current?

9. **APPLICATION** A pilot wishes to fly her plane to a destination 700 mi away at a bearing of 105°. The cruising speed of the plane is 500 mi/h, and the wind is blowing between 20 mi/h and 30 mi/h at a bearing of 30°. At what bearing should she aim the plane to compensate for the wind? With the variation in the wind, what is the greatest distance by which she could miss the airport?

10. Use the law of sines and the law of cosines to find the missing side lengths and angles.

a.

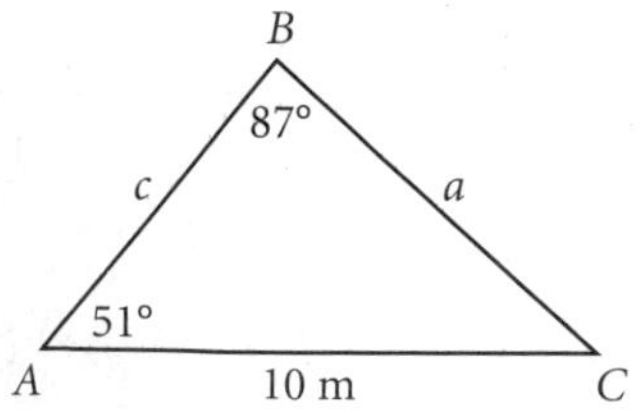

b.

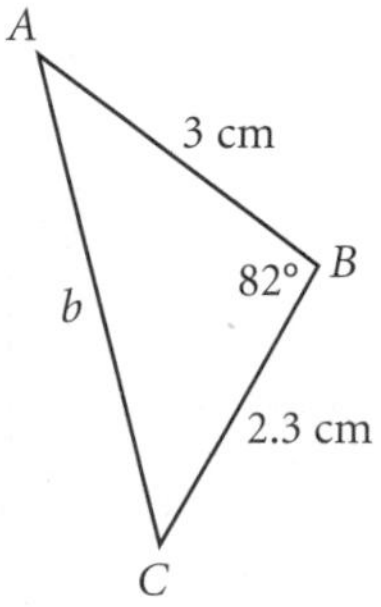

Assessing What You've Learned

- What are parametric equations?
- Can you use parametric equations to model the paths of objects in motion?
- Starting with a parametric equation, can you create equations using only the variables x and y?
- Can you use the the trigonometric ratios among sine, cosine, and tangent to solve problems involving right triangles?
- Do you know what $\sin^{-1}(x)$, $\cos^{-1}(x)$, and $\tan^{-1}(x)$ mean?
- Do you know how to use your calculator to work with trigonometry? For example, how do you set your calculator to use angles rather than radians?
- Can you use trigonometry and vectors to solve problems involving forces such as wind or water currents?
- Can you use the law of sines and law of cosines to solve problems involving oblique triangles?

JOURNAL Look back at the journal entries you have written about Chapter 8. Then, to help you organize your thoughts, here are some additional questions to write about.

- What do you think were the main ideas of the chapter?
- Were there any Exercises that you particularly struggled with? If so, are they clear now? If some ideas are still foggy, what plans do you have to clarify them?
- Trigonometry will be very important in more advanced work in mathematics. Now is a good time to organize your notes on trigonometric properties. How are sine, cosine, and tangent related to each other? What is the difference between $\frac{1}{\sin x}$ and $\sin^{-1}x$? How is the Pythagorean Theorem related to trigonometry? If you are struggling with these ideas, make a plan to get help now.

CHAPTER 9

Conic Sections and Rational Functions

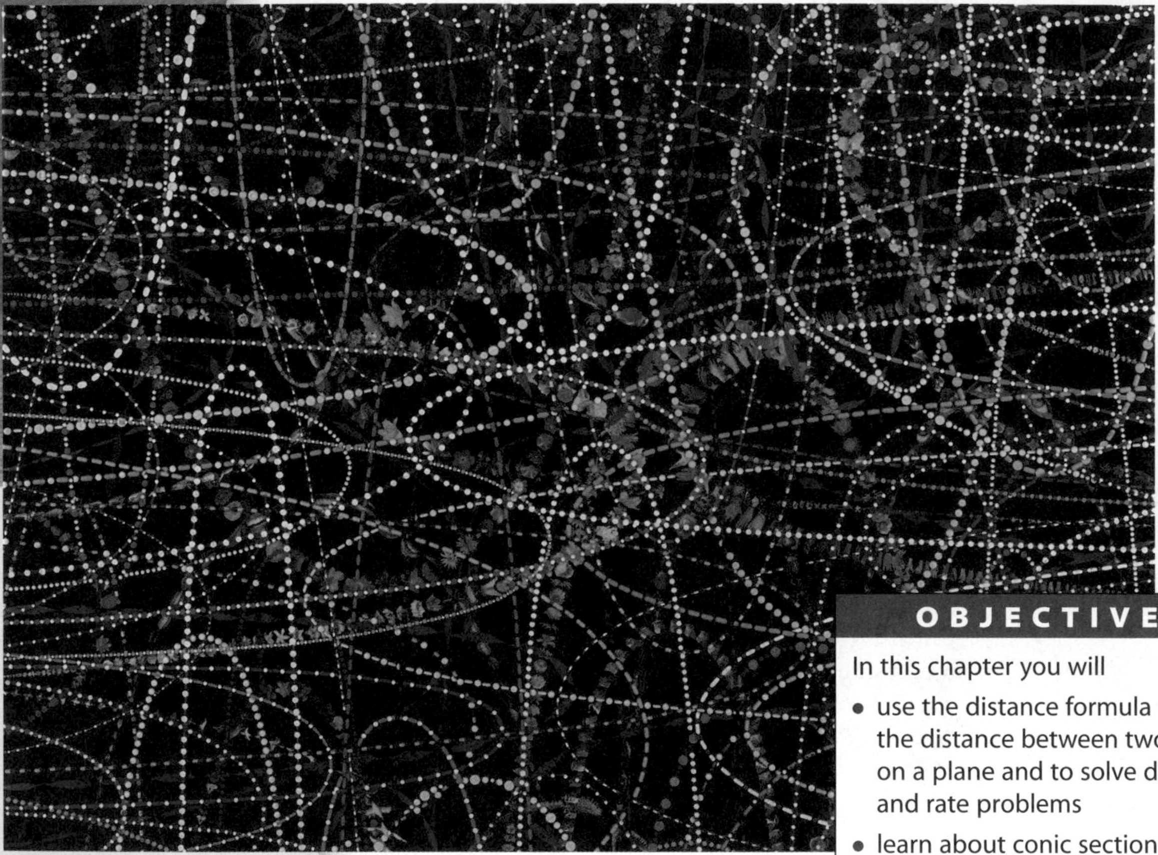

In some of artist Fred Tomaselli's (b 1956) work he uses nonstandard media to form geometric patterns. In *Gravity in Four Directions* (2001) Tomaselli uses leaves, pills, photocopies, and acrylic to form shapes suggesting motion along a parabola. Another Tomaselli work *Echo, Wow and Flutter* (2000) uses the same media to represent elliptical motion.

Gravity in Four Directions, 2001. Leaves, pills, photocopies, acrylic, resin on wood, 72 x 72 inches. Image courtesy of James Cohan Gallery, New York

OBJECTIVES

In this chapter you will

- use the distance formula to find the distance between two points on a plane and to solve distance and rate problems
- learn about conic sections—circles, ellipses, parabolas, and hyperbolas—which are created by the intersection of a plane and a cone
- investigate the properties of conic sections
- write the equations of conic sections
- solve nonlinear systems of equations
- study rational functions and learn special properties of their graphs
- add, subtract, multiply, and divide rational expressions

LESSON

9.1

Using the Distance Formula

The most dangerous thing in the world is to leap a chasm in two jumps.

DAVID LLOYD GEORGE

Imagine a race in which you carry an empty bucket from the starting line to the edge of a pool, fill the bucket with water, and then carry the bucket to the finish line. Luck, physical fitness, common sense, a calm attitude, and a little mathematics will make a difference in your performance. Finding the shortest path will help minimize the effort, distance, and time involved. In the Investigation you will mathematically analyze this situation.

Investigation

Bucket Race

You will need

- centimeter graph paper
- a ruler

The starting line of a bucket race is 5 m from one end of a pool, the pool is 20 m long, and the finish line is 7 m from the opposite end of the pool. In this Investigation you will find the shortest distance from point A to point C on the edge of the pool and then to point B. That is, you will find the value of x, the distance in meters from the end of the pool to point C, such that $AC + CB$ is the shortest path possible.

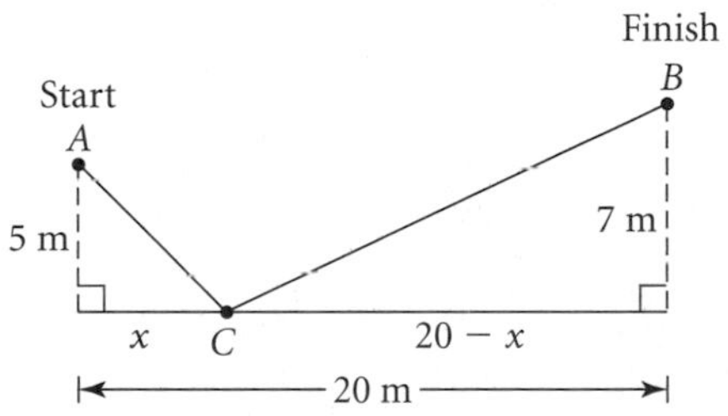

Step 1 Make a scale drawing of the situation on graph paper.

Step 2 Plot several different locations for point C. For each, measure the distance x, and find the total length $AC + CB$. Record your data.

Step 3 What is the best location for C such that the length $AC + BC$ is minimized? Are there more than one best location? Describe at least two different methods for finding the best location for C.

Step 4 Make a scale drawing of your solution.

Imagine that the amount of water you empty out at point B is an important factor in winning the race. You must move carefully so as not to spill water, and you'll be able to move faster with the empty bucket than with the bucket full of water. Assume that you can carry an empty bucket at a rate of 1.2 m/s and that you can carry a full bucket, without spilling, at a rate of 0.4 m/s.

Step 5 Go back to the data collected in Step 2, and find the time needed for each x-value.

Step 6 Now, find the best location for point C such that you minimize the time from point A to the pool edge and then to point B. What is your minimum time? Describe your solution process.

Many of the equations you will study in this chapter are based on finding the distance between two points. Consider the distance between two points with coordinates $(-5, -3)$ and $(9, 6)$. Based on the x- and y-coordinates, the horizontal distance between the points is 14 units, and the vertical distance between them is 9 units. The horizontal and vertical components of the distance create a right triangle. By the Pythagorean Theorem, the distance between the two points is $\sqrt{14^2 + 9^2}$, which is $\sqrt{227}$, or approximately 16.64 units. In general, if two points have coordinates (x_1, y_1) and (x_2, y_2), then the Pythagorean Theorem gives

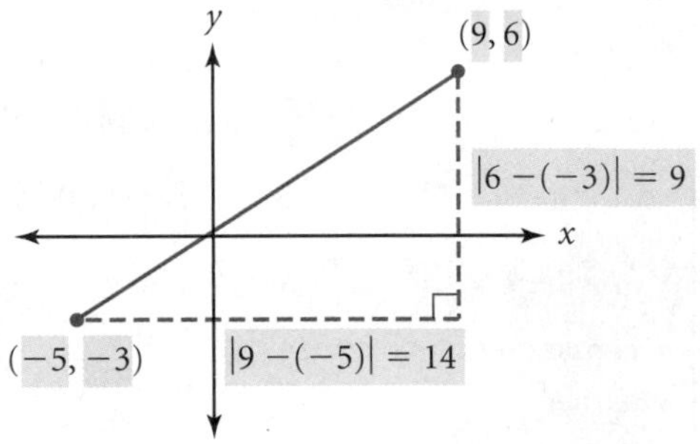

$$(x_2 - x_1)^2 + (y_2 - y_1)^2 = (\textit{distance between the two points})^2$$

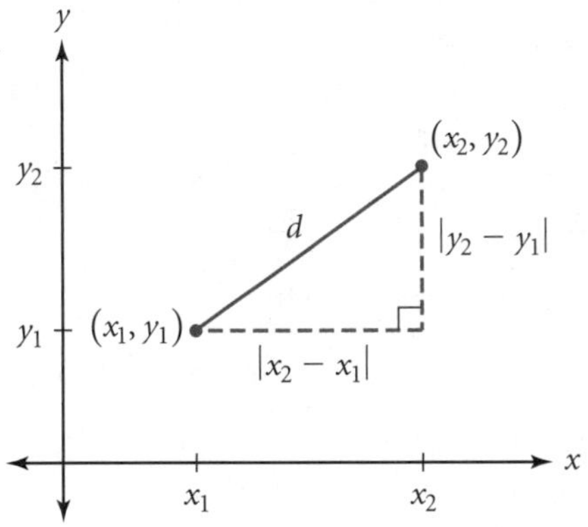

Taking the square root of both sides gives you a formula for distance on a coordinate plane. Because quantities are squared in the formula, absolute value signs are no longer necessary.

The Distance Formula

The distance, d, between two points on a coordinate plane, $P_1(x_1, y_1)$ and $P_2(x_2, y_2)$, is given by the formula

$$d = \sqrt{(x_2 - x_1)^2 + (y_2 - y_1)^2}$$

In Chapter 4 you used the Pythagorean Theorem to find the equation of a circle centered at the origin. You can also use the distance formula. If (x, y) is any point located on the circumference of a circle, its distance from the center, (0, 0), is $\sqrt{(x-0)^2+(y-0)^2}$. Because the distance from the center of the circle is defined as the radius, r, you get $r=\sqrt{x^2+y^2}$, or $r^2=x^2+y^2$.

The distance formula also enables you to write equations that represent other distance situations. One such equation describes a set of points all of which meet a certain condition. A set of points that fit a given condition is called a **locus** (plural: *loci*). For example, the locus of points that are 1 unit from the point (0, 0) is the circle with the equation $x^2+y^2=1$. In this chapter you will explore equations describing a variety of different loci.

EXAMPLE A

Find the equation of the locus of points that are equidistant from the points (1, 3) and (5, 6).

▶ Solution

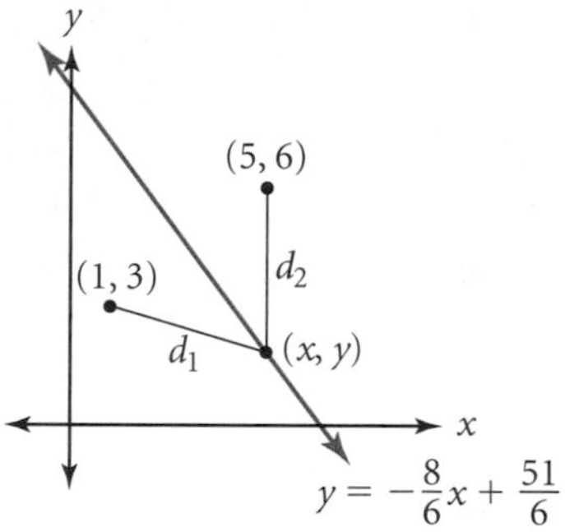

Let d_1 represent the distance between (1, 3) and any point, (x, y), on the locus. By the distance formula,

$$d_1=\sqrt{(x-1)^2+(y-3)^2}$$

Let d_2 represent the distance between (5, 6) and the same point on the locus; so,

$$d_2=\sqrt{(x-5)^2+(y-6)^2}$$

The locus of points contains all points the coordinates of which satisfy the equation

$$d_1=d_2, \text{ or } \sqrt{(x-1)^2+(y-3)^2}=\sqrt{(x-5)^2+(y-6)^2}$$

Use algebra to transform this equation into something more familiar.

$\sqrt{(x-1)^2+(y-3)^2}=\sqrt{(x-5)^2+(y-6)^2}$	Original equation.
$(x-1)^2+(y-3)^2=(x-5)^2+(y-6)^2$	Square both sides.
$x^2+2x+1+y^2-6y+9=x^2-10x+25+y^2-12y+36$	Expand the binomials.
$-2x+1-6y+9=-10x+25-12y+36$	Subtract x^2 and y^2 from both sides.
$8x+6y=51$	Rewrite in standard form by moving the variables to the left side and the constants to the right.

The locus is the line with the equation $8x+6y=51$, or $y=-\frac{8}{6}x+\frac{51}{6}$.

Just as the Pythagorean Theorem and the distance formula are useful for finding the equation of a locus of points, they are also helpful in solving real-world problems.

EXAMPLE B

An injured worker must be rushed from an oil rig 15 mi offshore to a hospital in the nearest town 98 mi down the coast from the oil rig.

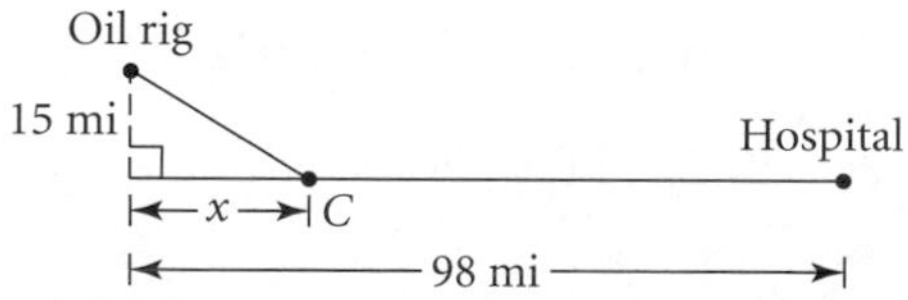

a. Let x represent the distance in miles from the point on the shore perpendicular to the oil rig and another point, C, on the shore. How far does the injured worker travel, in terms of x, if a boat takes him to C and then an ambulance takes him to the hospital?

b. Assume the boat moves 23 mi/h and the ambulance travels 70 mi/h. What value of x makes the trip 3 h?

HistoryConnection In 1790 the U.S. Coast Guard first began its duties for the government with ten small armed boats to prevent smuggling and maintain customs laws. Combined with the Life Saving Service in 1915, it now oversees rescue missions, environmental protection, navigation, safety during weather hazards, port security, boat safety, and oil tanker transfers. The U.S. Coast Guard has both military and volunteer divisions.

▶ ***Solution***

a. The boat must travel $\sqrt{15^2 + x^2}$, and the ambulance must travel $98 - x$. The total distance in miles is $\sqrt{15^2 + x^2} + 98 - x$.

b. Distance equals rate times time, or $d = rt$. Solving for time, $t = \frac{d}{r}$. The boat's time is $\frac{\sqrt{15^2 + x^2}}{23}$, and the ambulance's time is $\frac{98 - x}{70}$. The total time in hours, y, is represented by

$$y = \frac{\sqrt{15^2 + x^2}}{23} + \frac{98 - x}{70}$$

One way to find the value of x that gives a trip of 3 h is to graph the total time equation, graph $y = 3$, and find the intersection. The graphs intersect when x is approximately 51.6. For the trip to be 3 h, the boat and the ambulance should meet at the point on the shore 51.6 mi from the point closest to the oil rig.

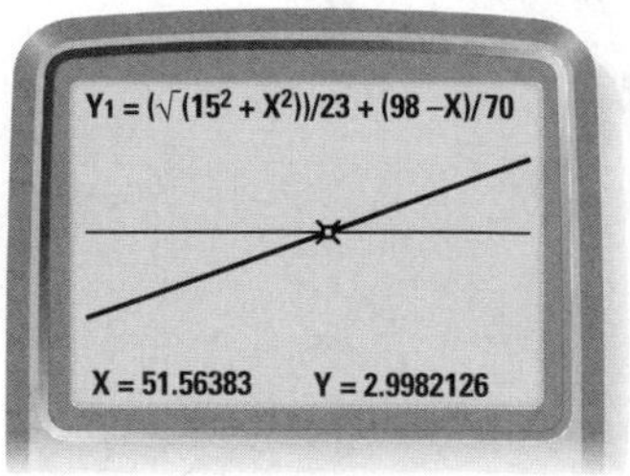

[38, 63, 10, 2.3, 3.6, 1]

EXERCISES

Practice Your Skills

1. Find the distance between each pair of points.

a. (2, 5) and (8, 13)

b. (0, 3) and (5, 10)

c. (−4, 6) and (−2, −3)

d. $(3, d)$ and $(-6, 3d)$

2. The distance between the points (2, 7) and $(5, y)$ is 5 units. Find the possible value(s) of y.

3. The distance between the points (−1, 5) and $(x, -2)$ is 47. Find the possible value(s) of x.

4. Which side of the triangle with vertices $A(1, 2)$, $B(3, -1)$, and $C(5, 3)$ is the longest?

5. Find the perimeter of the triangle with vertices $A(8, -2)$, $B(1, 5)$, and $C(4, -5)$.

Reason and Apply

6. Find the equation of the locus of points that are twice as far from the point (2, 0) as they are from (5, 0).

7. If you are too close to a radio tower, you will be unable to pick up its signal. Let the center of a town represent the origin of a coordinate plane. Suppose a radio tower is located 2 mi east and 3 mi north of the center of town, or at the point (2, 3). A highway runs north to south 2.5 mi east of the center of town along the line $x = 2.5$. Where on the highway will you be less than 1 mi from the tower and therefore unable to pick up the signal?

8. Josh is riding his mountain bike when he realizes that he needs to get home quickly to get ready for class. He is 2 mi from the road, and home is 3 mi down that road. He can ride 9 mi/h through the field separating him from the road, and he can ride 22 mi/h on the road.

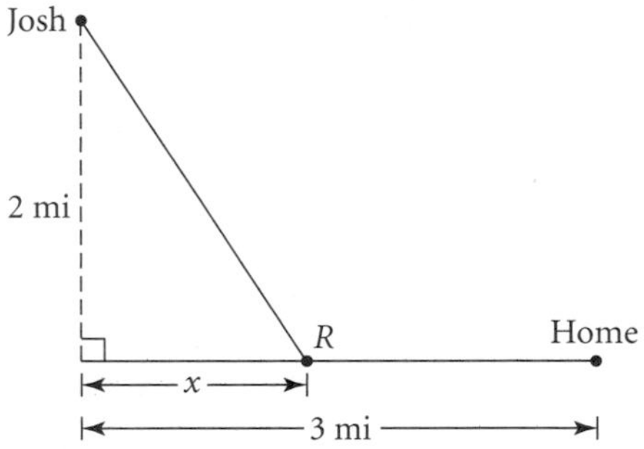

a. If Josh rides through the field to a point R on the road and then home along the road, how far will he ride through the field? How far on the road? Let x represent the distance in miles between point R and the point on the road that is closest to his current location.

b. How much time will Josh spend riding through the field? How much time on the road?

c. What value of x gets Josh home quickest? What is the minimum time?

9. A 10-m pole and a 13-m pole are 20 m apart at their bases. A wire connects the top of each pole with a point on the ground between them.

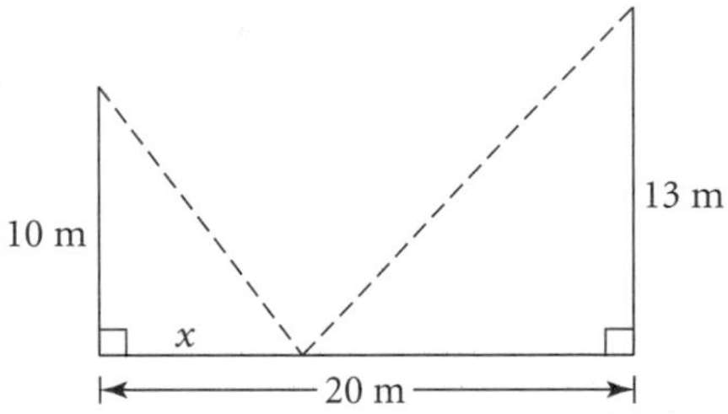

a. Let y represent the total length of the wire. Write an equation that relates x and y.

b. What domain and range make sense in this situation?

c. Where should the wire be fastened to the ground such that the length of wire is minimized? What is the minimum length?

This cable-stay bridge, the Fred Hartman Bridge, connects the Texas towns of Baytown and La Porte. Taut cables stretch from the tops of two towers to support the roadway.

10. A 24-ft ladder is placed upright against a wall. Then, the top of the ladder slides down the wall while the foot of the ladder is pulled outward along the floor at a steady rate of 2 ft/s.

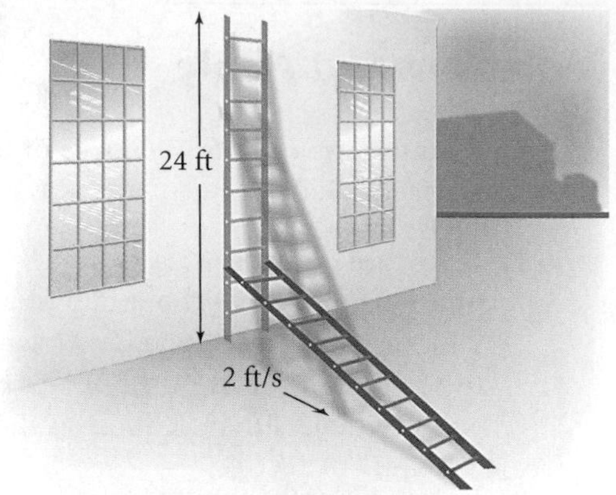

a. Find the heights reached by the ladder at 1 s intervals as it slides down the wall.

b. How long will it be before the entire ladder is lying on the ground?

c. Does the top of the ladder also slide down the wall at a steady rate of 2 ft/s? Explain your reasoning.

d. Write parametric equations that model the distance in feet of the foot of the ladder from the wall, x, and the height in feet that the top of the ladder reaches, y. Let t represent time in seconds.

e. Write a complete explanation of the rate at which the ladder slides down the wall.

11. Let d represent the distance between the point $(5, -3)$ and any point, (x, y), on the parabola $y = 0.5x^2 + 1$.

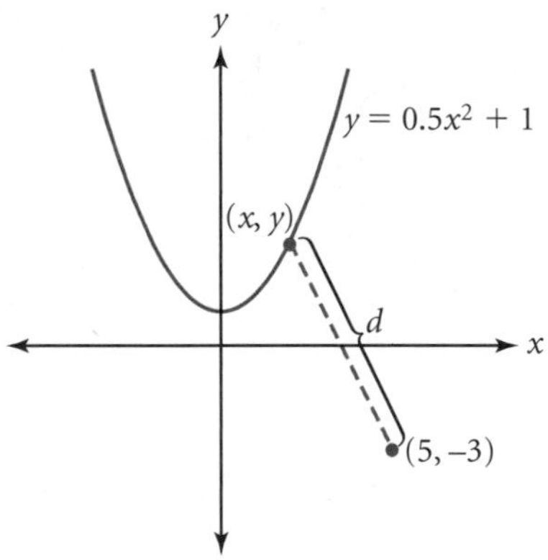

a. Write an equation for d in terms of x.

b. What is the minimum distance? What are the coordinates of the point on the parabola that is closest to the point $(5, -3)$?

12. **APPLICATION** The city councils of three neighboring towns—Ashton, Bradburg, and Carlville—decide to pool their resources and build a recreation center. To be fair, they decide to locate the recreation center equidistant from all three towns.

a. When a coordinate plane is placed on a map of the towns, Ashton is at $(0, 4)$, Bradburg is at $(3, 0)$, and Carlville is at $(12, 8)$. At what point on the map should the recreation center be located?

b. If the three towns were collinear (along a line), could the recreation center be located equidistant from all three towns? Explain your reasoning.

c. What other factors might the three city councils consider in deciding where to locate the recreation center?

Review

13. (Lesson 6.3) Complete the square in each equation such that the left side represents a perfect square or a sum of perfect squares.

a. $x^2 + 6x = 5$ **b.** $y^2 - 4y = -1$ **c.** $x^2 + 6x + y^2 - 4y = 4$

14. (Lesson 3.3) Triangle ABC has vertices $A(8, -2)$, $B(1, 5)$, and $C(4, -5)$.

a. Find the midpoint of each side.

b. Write the equations of the three medians of the triangle. (A median of a triangle is a segment connecting one vertex to the midpoint of the opposite side.)

c. Locate the point where the medians meet.

15. (Lessons 3.2 and 4.4) Give the domain and range of the function $f(x) = x^2 + 6x + 7$.

16. (Lesson 8.4) A ship leaves port and travels on a bearing of 205° for 2.5 h at 8 knots and then on a bearing of 150° for 3 h at 10 knots. How far is the ship from its port? (A knot is equivalent to 1 nautical mi/h.)

17. (Algebra) The sealed, 10-cm-tall cone shown here is filled to half its height with liquid. It is then turned upside down. What height, to the nearest tenth of a centimeter, does the liquid reach?

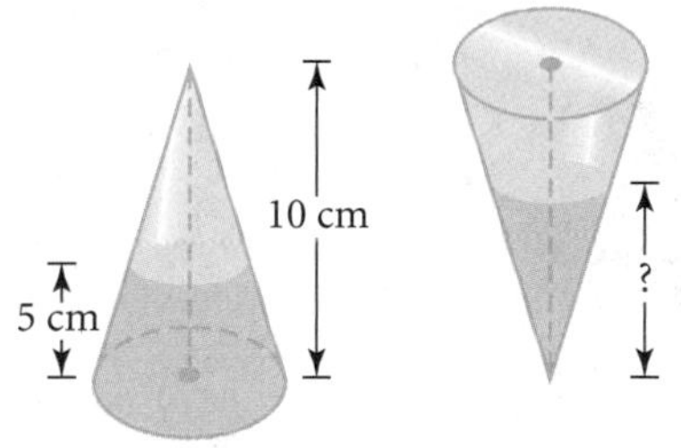

HistoryConnection Any cross section of Earth is close to a circle, so it can be divided into 360 degrees. Each degree can be divided into 60 minutes. The length of 1 minute of Earth's surface is called a nautical mile. However, Earth is not a perfect sphere, so the length of a nautical mile could vary depending on where you were. Since 1959 all countries have agreed on the definition of a nautical mile as 1.852 km. A *knot* is defined as a speed of 1 nautical mile per hour.

In the 1500s ships would trail a rope with a knot tied every 47 ft 3 in. The crew would measure the number of knots that were pulled into the water as 28 seconds passed. Counting the number of knots gave them their speed in nautical miles per hour.

LESSON

9.2

Circles and Ellipses

The best paths usually lead to the most remote places.

SUSAN ALLEN TOTH

In the next few lessons you will learn about circles, ellipses, parabolas, and hyperbolas. The orbital paths of the planets around the Sun are not exactly circular. These paths are examples of an important mathematical curve, the ellipse. The curved path of a stream of water from a water fountain, the path of a football kicked into the air, and the cables supporting the deck of the Golden Gate Bridge are all examples of parabolas. The designs of nuclear cooling towers, transmission gears, and the long-range navigational system known as LORAN all depend on hyperbolas. These curves all belong to a family of planar mathematical curves.

These interesting curves—the circle, the ellipse, the parabola, and the hyperbola—are called **conic sections** because each can be created by slicing a double cone.

Sonia Delaunay's (1885–1979) oil-on-canvas painting *Rhythm and Colors* (1939) contains many circles.

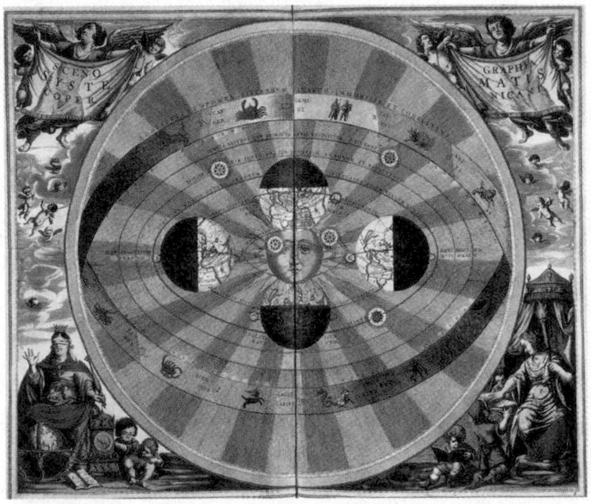

Dutch German mathematician and cosmographer Andreas Cellarius (ca 1595–1665) showed elliptical orbits in this celestial atlas entitled *Harmonia Macrocosmica* (1660).

The Swann Memorial Fountain at Logan Circle in Philadelphia, Pennsylvania, was designed by American sculptor Alexander Stirling Calder (1898–1976). The jets of water form parabolas.

Built in 1963, the McDonnell Planetarium in St. Louis, Missouri, is a **hyperboloid,** a three-dimensional shape created by revolving a hyperbola.

HistoryConnection Students of mathematics have studied conic sections since the time of the early Greeks. Menaechmus, a Greek mathematician who lived in the 4th century B.C.E., thought of conic sections as coming from different kinds of cones. Parabolas came from right-angled cones, ellipses from acute-angled cones, and hyperbolas from obtuse-angled cones. Three centuries later, Apollonius of Perga (ca 262–190 B.C.E.) showed how all three kinds of curves could be obtained from the same cone. Known as the "Great Geometer," Apollonius wrote the eight-volume *Treatise on Conic Sections* and named the ellipse, hyperbola, and parabola. His achievements still influence engineering and mathematics today.

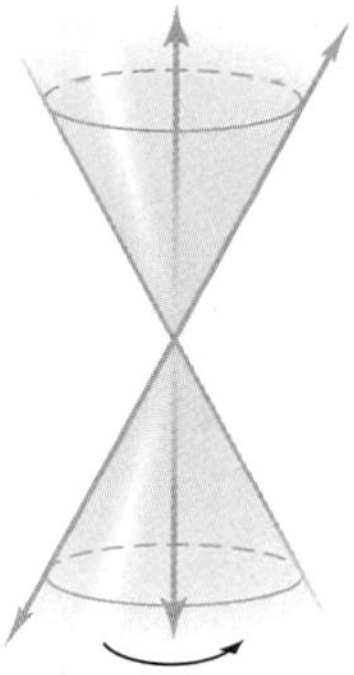

When two lines meet at an acute angle, rotating one of the lines about the other creates a double cone. These cones do not have bases. They continue infinitely, like the original lines.

Slicing these cones with a plane at different angles produces different conic sections.

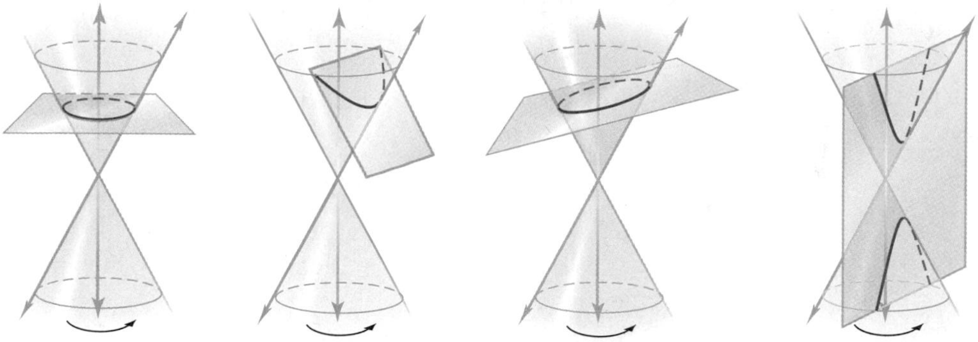

Circle Parabola Ellipse Hyperbola

Conic sections have some interesting properties. Each of the shapes can also be defined as a locus of points. For example, all the points on a circle are the same distance from the center. So, you can describe a circle as a locus of points that are a fixed distance from a fixed point.

The Definition of a Circle

A **circle** is a locus of points in a plane that are located a constant distance from a fixed point. In the diagram the fixed point, or **center,** is labeled *C*. The constant distance, or **radius,** is *r*.

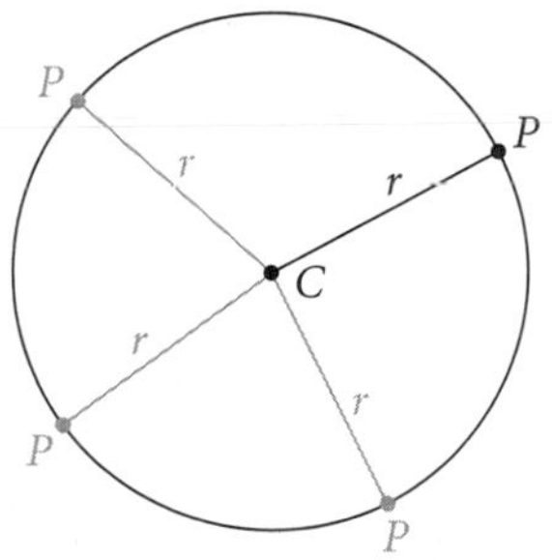

On a coordinate plane, you can also write an equation to describe all the points on a circle.

EXAMPLE A Write the equation for the locus of points (x, y) that are 4 units from the point $(0, 0)$.

▸ ***Solution***

The locus describes a circle with radius 4 and center (0, 0). The distance from each point of the circle, (x, y), to the center of the circle, (0, 0), is 4. Using the distance formula, you can write

$$\sqrt{(x-0)^2+(y-0)^2}=4$$

Squaring both sides gives the equation

$$x^2+y^2=16$$

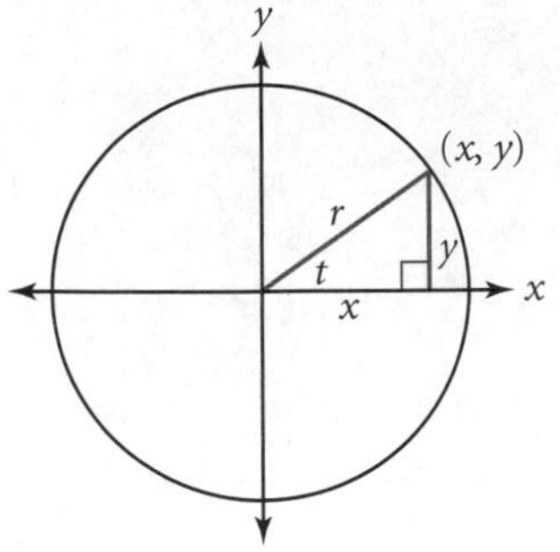

In general, the equation of a circle with center (0, 0) is $x^2 + y^2 = r^2$.

If the circle is translated horizontally and/or vertically, you can modify the equations by replacing x with $(x - h)$ and replacing y with $(y - k)$.

The Equation of a Circle

The **standard form** of the equation of a circle with center (h, k) and radius r is

$$(x-h)^2+(y-k)^2=r^2$$

EXAMPLE B

A circle has center $(3, -2)$ and is tangent to the line $y = 2x + 1$. Write the equation of the circle.

▸ ***Solution***

To write the equation of a circle, you need to know the center and the radius. You know the center, $(3, -2)$, but you need to find the radius.

A line tangent to a circle intersects the circle at only one point and is perpendicular to a diameter of the circle at the point of tangency. The tangent line has slope 2. A line perpendicular to this line will have slope $-\frac{1}{2}$. So, a line containing this diameter will have slope $-\frac{1}{2}$, and it will pass through the center of the circle, $(3, -2)$.

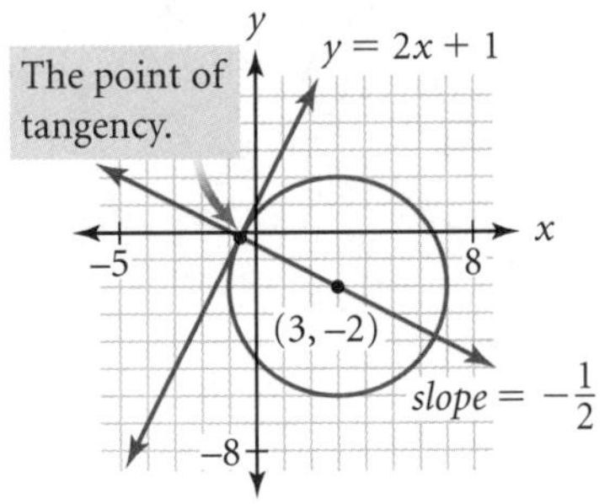

Use this information to write the equation of the line that contains the diameter.

$$y=-\frac{1}{2}-\frac{1}{2}x$$

Now, find the point of intersection of this line with the tangent line by solving the system of equations. The point of intersection is $(-0.6, -0.2)$. You can now find the radius, which is the distance from the point of tangency to the center.

$$\sqrt{(3+0.6)^2+(-2+0.2)^2}=\sqrt{16.2}\approx 4.025$$

The radius of the circle is about 4.025 units. Therefore, the equation of this circle is

$$(x-3)^2+(y+2)^2=16.2$$

In Chapter 4 you stretched a circle horizontally and vertically by different amounts to create ellipses. You can think of the equation of an ellipse as the equation of a unit circle that has been translated and stretched.

The Equation of an Ellipse

The standard form of the equation of an ellipse with center (h, k), horizontal scale factor of a, and vertical scale factor of b is

$$\left(\frac{x-h}{a}\right)^2 + \left(\frac{y-k}{b}\right)^2 = 1$$

An ellipse is like a circle except that it involves two points called **foci** (the plural of *focus*) instead of just one point at the center. You can construct an ellipse by tying a string around two pins and tracing a set of points.

The sum of the distances, $d_1 + d_2$, is the same for any point on the ellipse.

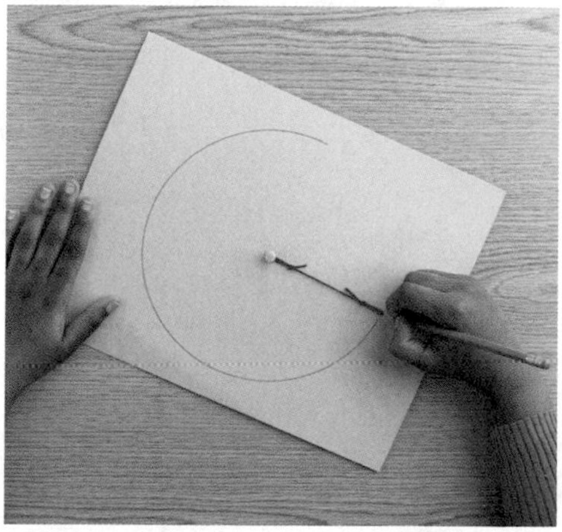

A piece of string attached to one pin helps you draw a circle. This is the same concept as using a compass to construct a circle.

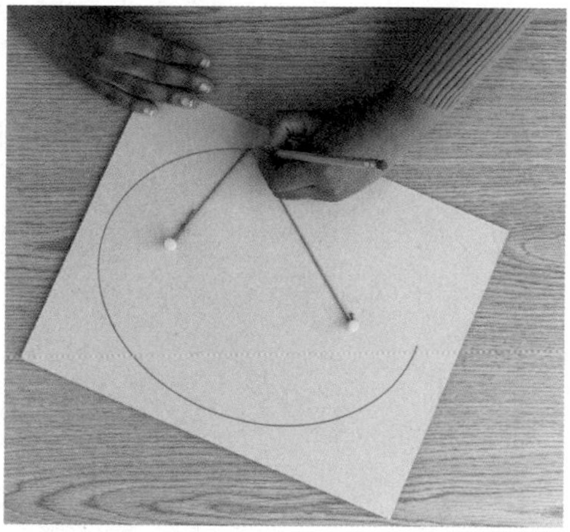

A piece of string attached to two pins helps you draw an ellipse.

The Definition of an Ellipse

An **ellipse** is a locus of points in a plane the sum of whose distances from two fixed points is always a constant. In the diagram the two fixed points, or *foci* (the plural of *focus*), are labeled F_1 and F_2. For each point on the ellipse, the distances d_1 and d_2 sum to the same value.

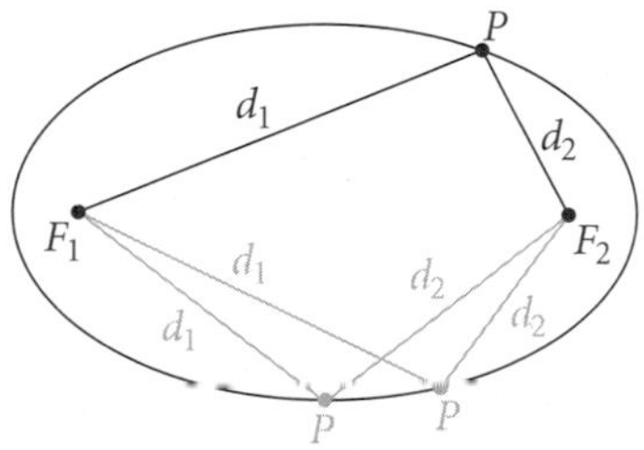

The longer dimension of an ellipse is called the **major axis.** The shorter dimension is called the **minor axis.** The foci are always located on the major axis. When the major axis is horizontal, the length of the major axis, $2a$, is equal to $d_1 + d_2$. When the major axis is horizontal, the length of the major axis, $2b$, is equal to $d_1 + d_2$.

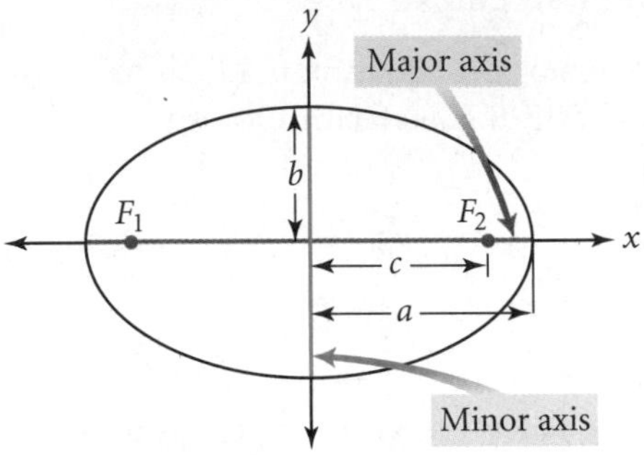

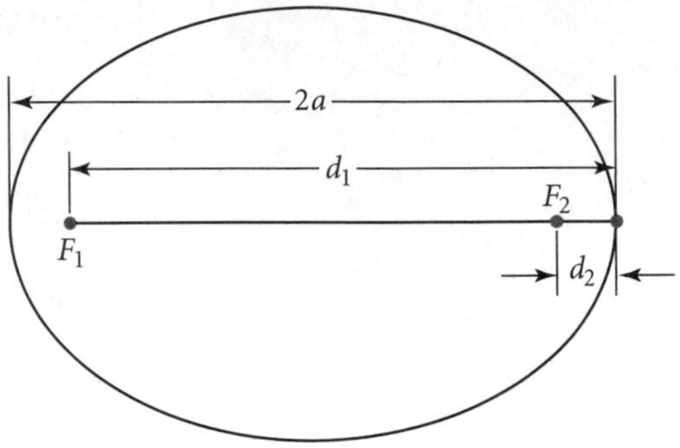

This means that connecting an endpoint of the minor axis to the foci forms two right triangles. Looking at the graph on the left below, let's call the lengths of the legs of each right triangle b and c. The length of the hypotenuse is a. Using the Pythagorean Theorem, you find $b^2 + c^2 = a^2$.

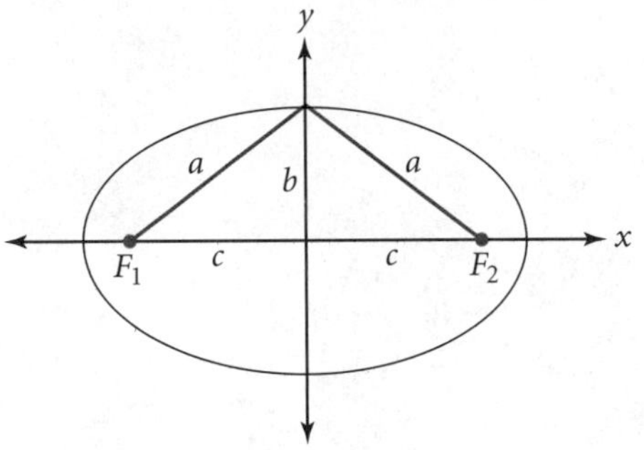

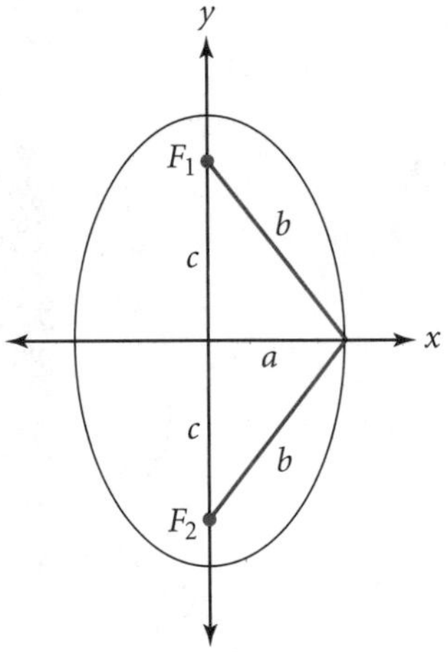

As shown on the graph on the right above, when the major axis is vertical, the relationship between a, b, and c is $a^2 + c^2 = b^2$.

EXAMPLE C

Graph an ellipse that is centered at the origin, with a vertical major axis of 6 units and a minor axis of 4 units. Where are the foci?

▸ Solution

Start with a unit circle, $x^2 + y^2 = 1$. The radius is 1 unit, and the diameter is 2 units. You can stretch this circle vertically by a factor of 3 to make it 6 units tall. To make it 4 units wide, you must stretch it horizontally by a factor of 2.

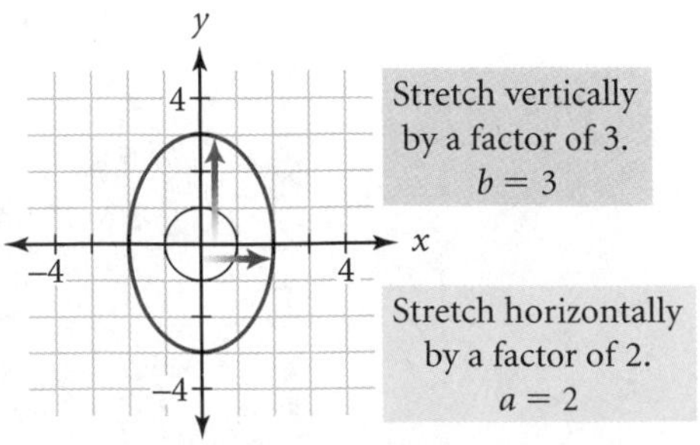

Replace y with $\frac{y}{3}$, and replace x with $\frac{x}{2}$. The equation is

$$\left(\frac{x}{2}\right)^2 + \left(\frac{y}{3}\right)^2 = 1, \qquad \text{or} \qquad \frac{x^2}{4} + \frac{y^2}{9} = 1$$

To sketch $\left(\frac{x}{2}\right)^2 + \left(\frac{y}{3}\right)^2 = 1$ by hand, plot the center, the endpoints of the major axis that are vertically 3 units from the center, and the endpoints of the minor axis that are horizontally 2 units from the center. Then connect the endpoints with a smooth curve.

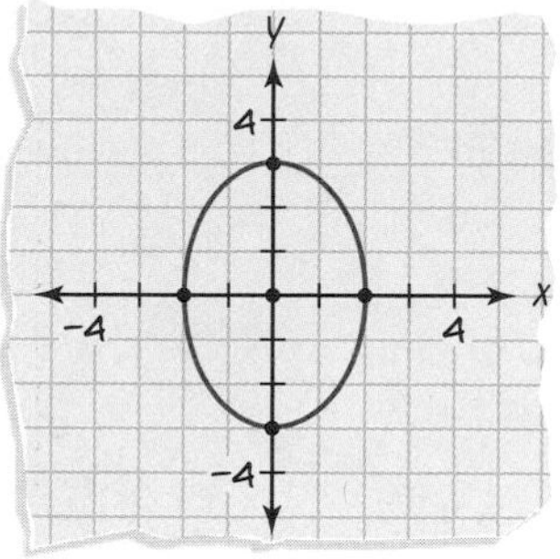

It is easy to sketch $\left(\frac{x}{2}\right)^2 + \left(\frac{y}{3}\right)^2 = 1$ by hand. However, to graph the ellipse on your calculator, you will need to solve for y. It then takes two equations to graph the entire shape.

$$\left(\frac{x}{2}\right)^2 + \left(\frac{y}{3}\right)^2 = 1$$

The equation in standard form.

$$\left(\frac{y}{3}\right)^2 = 1 - \left(\frac{x}{2}\right)^2$$

Subtract $\left(\frac{x}{2}\right)^2$ from both sides.

$$\frac{y}{3} = \pm\sqrt{1 - \left(\frac{x}{2}\right)^2}$$

Take the square root of both sides.

$$y = \pm 3\sqrt{1 - \left(\frac{x}{2}\right)^2}$$

Multiply both sides by 3.

Use your calculator to check the graph. Be sure to use a square setting such as ZOOM DECIMAL to accurately see which dimension is larger.

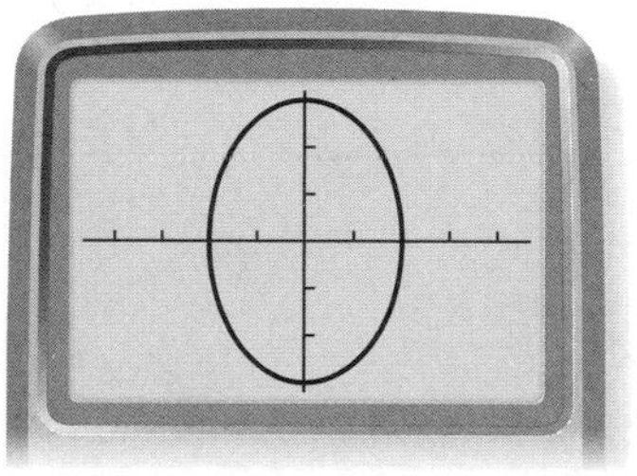

[–4.7, 4.7, 1, –3.1, 3.1, 1]

To locate the foci, recall the relationship $a^2 + c^2 = b^2$ for an ellipse with a vertical major axis:

$$a^2 + c^2 = b^2$$
$$2^2 + c^2 = 3^2$$
$$4 + c^2 = 9$$
$$c^2 = 5$$
$$c = \pm\sqrt{5}$$

So, the foci are $\left(0, \sqrt{5}\right)$ and $\left(0, -\sqrt{5}\right)$, or approximately $(0, 2.24)$ and $(0, -2.24)$.

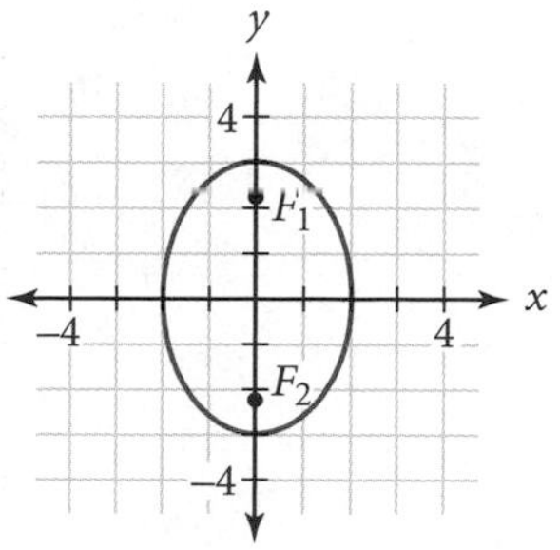

In the Investigation you will find the equation of an ellipse that you create yourself.

Investigation
A Slice of Light

You will need

- graph paper
- a flashlight
- a relatively dark classroom

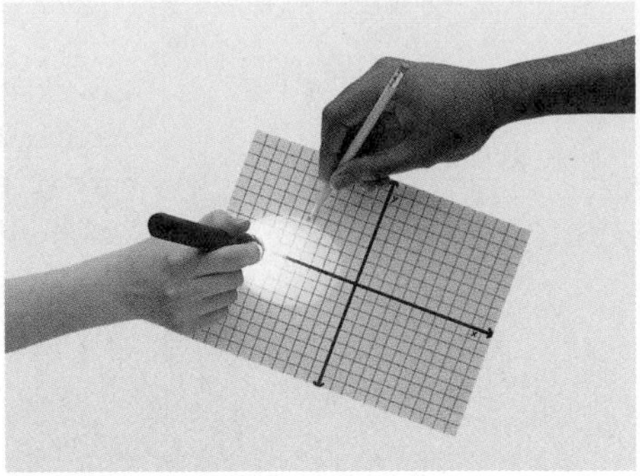

The beam of a flashlight is close to the shape of a cone. A sheet of paper in front of the flashlight shows different slices, or sections, of the cone of light.

Work with a partner, then share results with your group.

Procedure Note

1. Shine a flashlight on the graph paper at an angle.
2. Align the major axis of the ellipse formed by the beam with one axis of the paper. You might start by placing four points on the paper to help the person holding the flashlight stay on target.
3. Carefully trace the edge of the beam as your partner holds the light steady.

Step 1 Draw a pair of coordinate axes at the center of your graph paper. Follow the Procedure Note, and trace an ellipse.

Step 2 Write an equation that fits the data as closely as possible. Name the lengths of both the major and minor axes. Use the values in your equation to locate the foci. Finally, verify your equation by selecting any two points on the ellipse and checking that the sum of the distances to the foci is constant.

Eccentricity is a measure of the elongation of an ellipse. Eccentricity is defined as the ratio $\frac{c}{a}$ for an ellipse with a horizontal major axis, or $\frac{c}{b}$ for an ellipse with a vertical major axis. If the eccentricity is close to 0, then the ellipse looks almost like a circle. The higher the ratio, the more elongated the ellipse.

Step 3 Use your flashlight to make ellipses with different eccentricities. Trace three different ellipses. Calculate the eccentricity of each one, and label it on your paper. What is the range of possible values for the eccentricity of an ellipse?

Step 4 Continue to tilt your flashlight until the eccentricity becomes too large and you no longer have an ellipse. What shape can you trace now?

Exercises

Practice Your Skills

1. Sketch each circle on your paper, and label the center and the radius. For 1a–d, rewrite the equation as two functions. Use your calculator to check your work.

a. $x^2 + y^2 = 4$

b. $(x - 3)^2 + y^2 = 1$

c. $(x + 1)^2 + (y - 2)^2 = 9$

d. $x^2 + (y - 1.5)^2 = 0.25$

2. Graph the following equations using your graphing calculator.

a. $x = 2\cos t + 1$
$y = 2\sin t + 2$

b. $x = 4\cos t - 3$
$y = 4\sin t$

How can you write the equation of a circle using parametric equations?

3. Write an equation in standard form for each graph.

a.

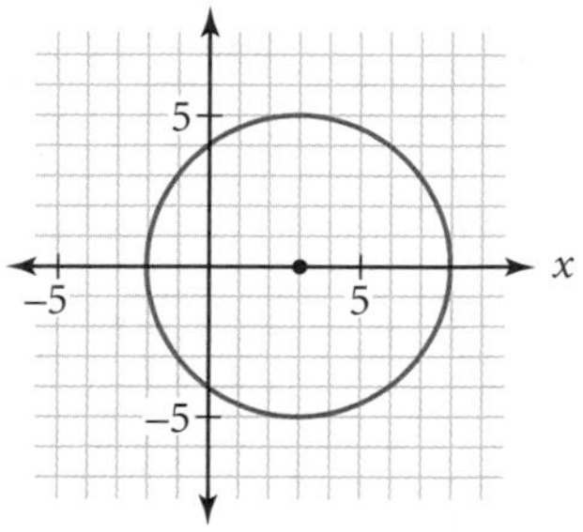

b.

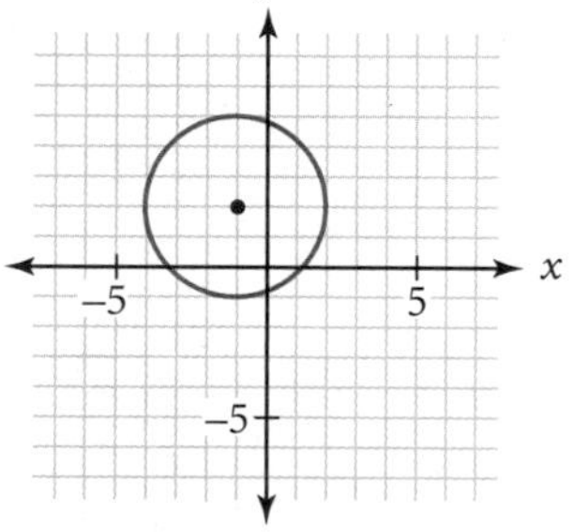

c.

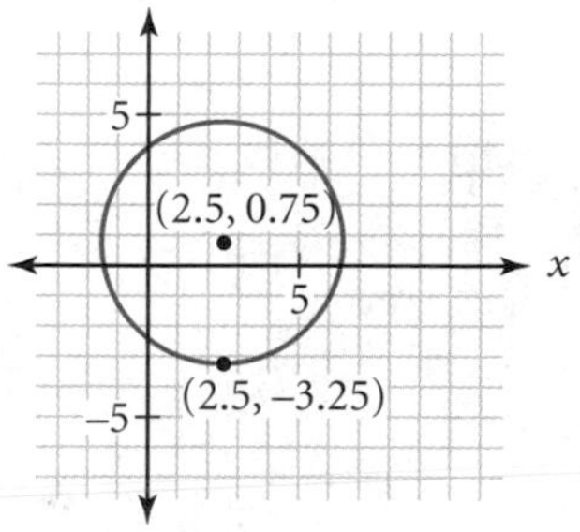

d.

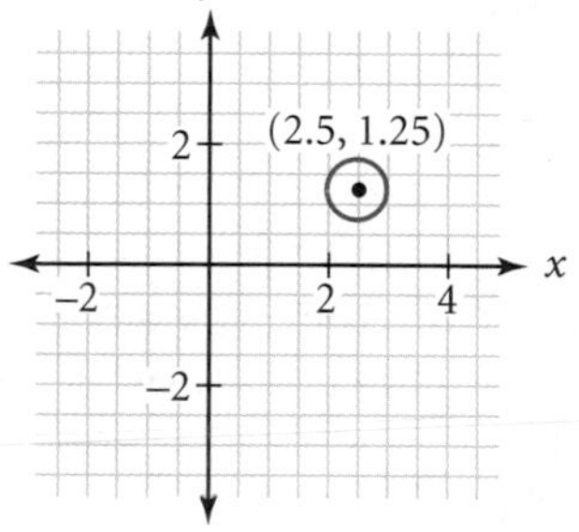

4. Sketch a graph of each equation. Label the coordinates of the endpoints of the major and minor axes.

a. $\left(\frac{x}{2}\right)^2 + \left(\frac{y}{4}\right)^2 = 1$

b. $\left(\frac{x-2}{3}\right)^2 + \left(\frac{y+2}{1}\right)^2 = 1$

c. $\left(\frac{x-4}{3}\right)^2 + \left(\frac{y-1}{3}\right)^2 = 1$

d. $y = \pm\sqrt{1 - \left(\frac{x+2}{3}\right)^2} - 1$

5. Graph each equation using your graphing calculator.

a. $x = 4\cos t - 1,\ y = 2\sin t + 3$

b. $x = 3\cos t + 3,\ y = 5\sin t$

How can you write the equation of an ellipse using parametric equations?

6. Write an equation in standard form for each graph.

a.

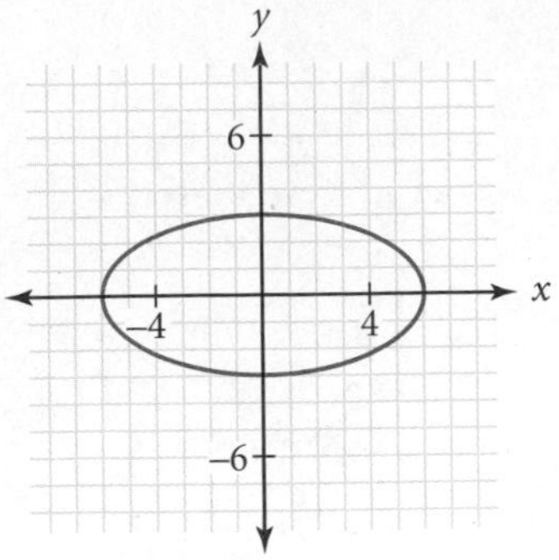

b.

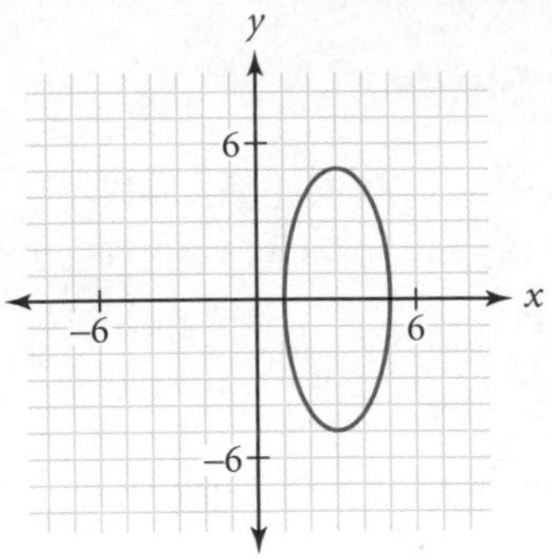

c.

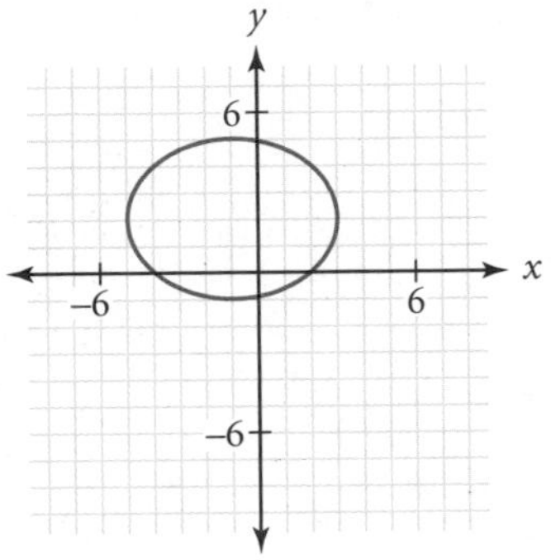

d.

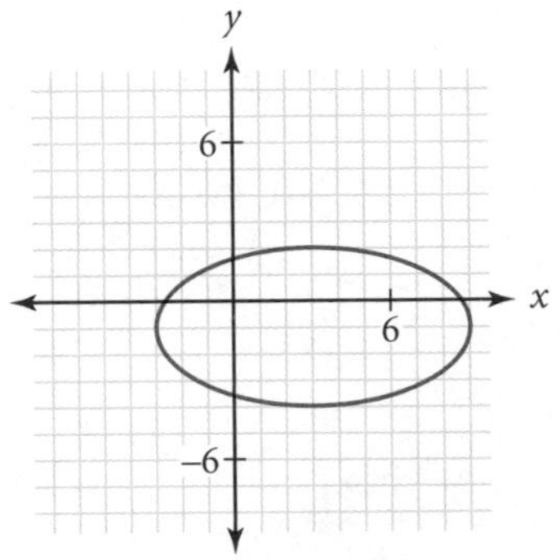

7. Find the exact coordinates of the foci for each ellipse in Exercise 6.

Reason and Apply

8. Suppose you placed a grid on the plane of a comet's orbit with the origin at the Sun and the *x*-axis running through the longer axis of the orbit, as shown in the diagram. The table gives the approximate coordinates of the comet as it orbits the Sun. Both *x* and *y* are measured in astronomical units (AU).

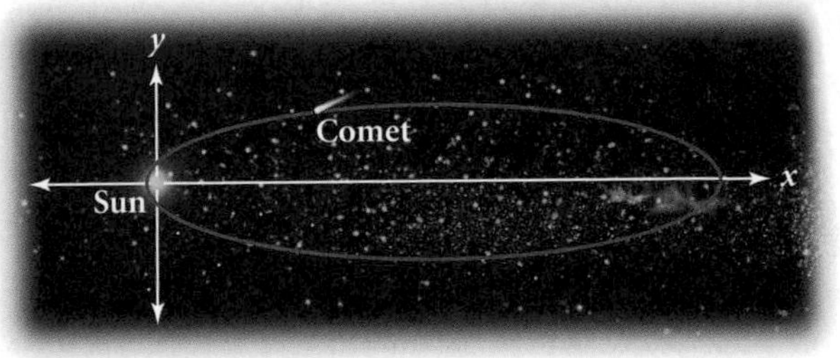

x	−2.1	12.9	62.6	244.5	579.3	778.1	900.1	982.4	923.4	663.0	450.0	141.6
y	5.5	16.3	31.5	54.6	62.0	51.6	36.1	10.9	−31.5	−59.2	−62.8	−44.5

a. Find an equation to fit these data.

b. Find the *y*-coordinate when the *x*-coordinate is 493.0 AU.

c. What is the greatest distance of the comet from the Sun?

d. What are the coordinates of the foci?

ScienceConnection Planets close to the Sun have nearly circular orbits, whereas those farther away have extremely elliptical ones. The orbits of comets can be long ellipses, parabolas, or hyperbolas. The gravitational pull of the Sun keeps planets and smaller bodies in their orbits.

German astronomer Johannes Kepler (1571–1630) was the first European to realize that planets move in elliptical rather than circular orbits with the Sun at one focus. Kepler worked as an assistant to the Danish astronomer Tycho Brahe (1546–1601), who had spent years collecting what was considered to be the finest pre-telescope data. Brahe hid most of his data from Kepler. After Brahe's death, Kepler obtained Brahe's data and was able to verify that Mars did indeed move in an elliptical orbit.

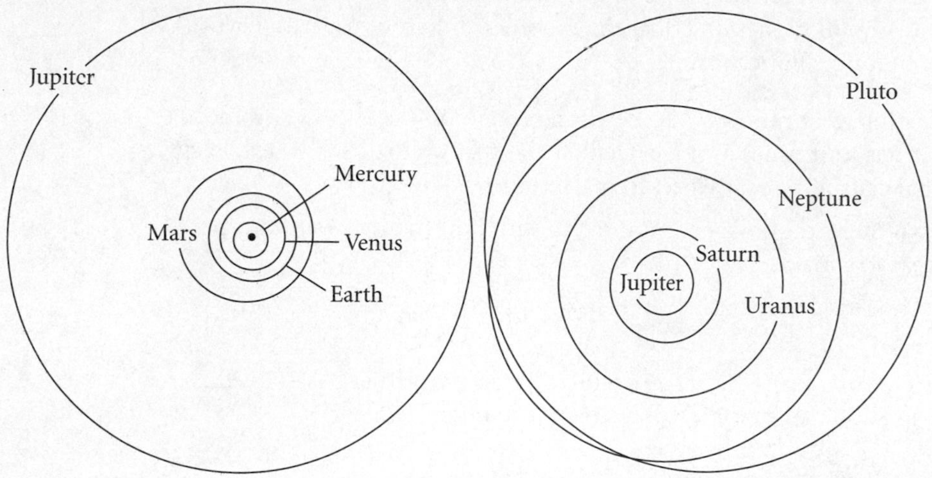

9. **APPLICATION** The top of a doorway is designed to be half an ellipse. The doorway is 1.6 m wide, and the height of the half-ellipse is designed to be 62.4 cm high. The crew have nails and string available. They want to trace the half-ellipse with a pencil before they cut the plywood to go over the doorway.

a. How far apart should they place the nails?

b. How long should the string be?

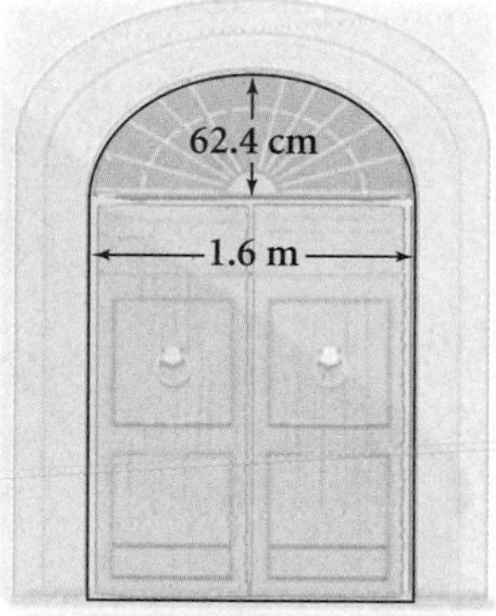

10. Read the Science Connection about the reflective property of an ellipse. If a room is constructed in the shape of an ellipse, and if you stand at one focus and speak softly, a person standing at the other focus will hear you clearly. Such rooms are often called "whispering chambers." Consider a whispering chamber that is 12 m long and 6 m wide.

a. Where should two whisperers stand to talk to each other?

b. How far does the sound travel from one person to the other, bouncing off the wall in between?

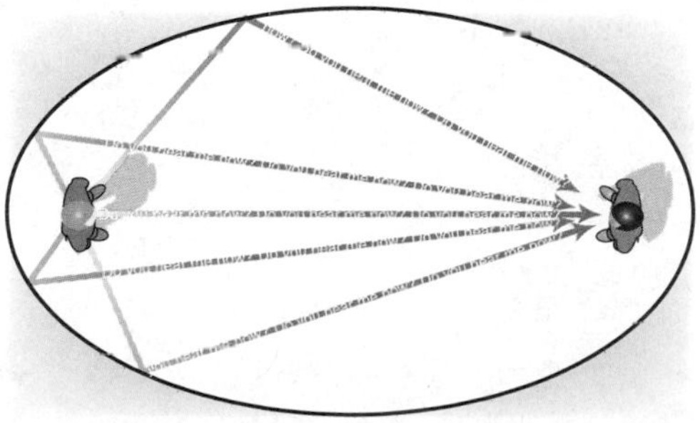

Science Connection Ellipses have an interesting property. A signal from one focus will always bounce off the ellipse in such a way that it will travel to the other focus.

This is called the **reflective property of an ellipse.**

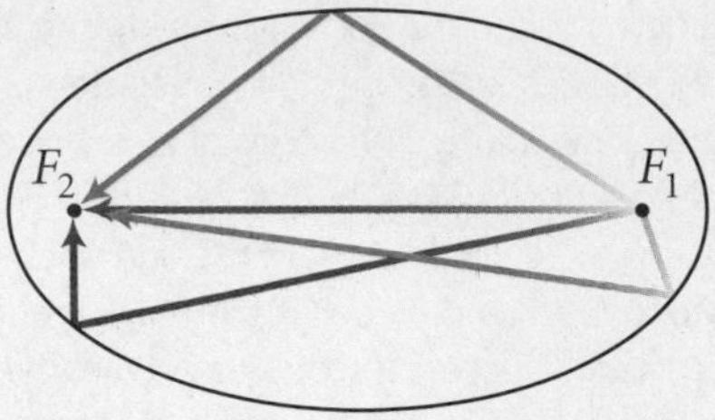

11. The moon's greatest distance from Earth is 252,710 mi, and its smallest distance is 221,643 mi. Write an equation that describes the moon's orbit around Earth. Earth is at one focus of the moon's elliptical orbit.

12. APPLICATION One possible gear ratio on Matthew's mountain bike is 4 to 1. This means that the pedaling gear has four times as many teeth as the gear on the back wheel. Each revolution of the pedal causes the rear wheel to make four revolutions.

a. If Matthew pedals at 60 revolutions per minute (r/min), how many revolutions per minute does the rear tire make?

b. If the diameter of the rear tire is 26 in., what speed in miles per hour will Matthew attain?

c. Matt downshifts to a front gear that has 22 teeth and a rear gear that has 30 teeth. If he keeps pedaling at 60 r/min, what will his new speed be?

Review

13. (Lesson 4. 4) Write the equation of a parabola congruent to $y = x^2$ that has been reflected across the x-axis and translated left 3 units and up 2 units.

14. (Algebra) Solve for y.

$$\frac{y-4}{0.5} = \left(\frac{x+2}{3}\right)^2$$

15. (Lesson 6.1) Find the quadratic equation, $y = ax^2 + bx + c$, that fits these data points.

x	y
0	117
1	95
2	77
3	63
4	53
5	47

16. (Lesson 9.1) Find the perimeter of this triangle.

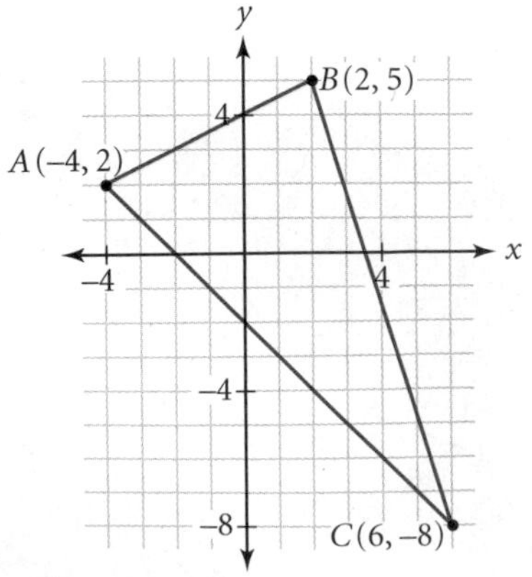

IMPROVING YOUR REASONING SKILLS

Elliptical Pool

You could use the reflective property of an ellipse to design an unusual pool table. On an elliptical pool table, if you start with a ball at one focus and hit it in any direction, it will always rebound off the side and roll toward the other focus. Suppose a pool table is designed in the shape of an ellipse with a pocket at one focus. Describe how you will hit Ball 1 to land in the pocket even though Ball 2 is in the way.

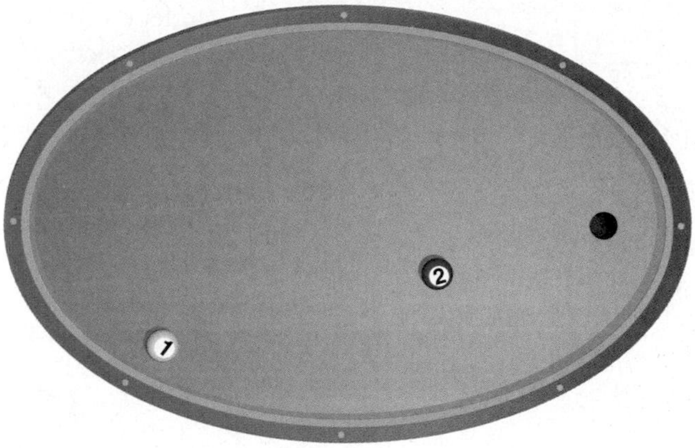

LESSON 9.3

Parabolas

When the telescope ends,
the microscope begins.
Who is to say of the two,
which has the grander view.
VICTOR HUGO

You have studied parabolas in several different lessons, and you have used parabolic equations to model a variety of situations. In this lesson you will study the parabola from a different perspective. In previous chapters you worked with parabolas as variations on the equation $y = x^2$ or the parametric equation $x = t$ and $y = t^2$. However, there is also a locus definition of a parabola.

TechnologyConnection A reflecting telescope is a type of optical telescope that uses a curved, mirrored lens to magnify objects. The most powerful reflecting telescope uses a parabolic or hyperbolic mirror and can bring the faintest light rays into clear view. The larger the mirror, the more distant the objects the telescope can detect. To avoid the expense and weight of producing a single massive lens, today's most sophisticated telescopes have a tile-like combination of hexagonal mirrors that produces the same effect as one concave mirror.

The designs of telescope lenses, spotlights, satellite dishes, and other parabolic reflecting surfaces are based on a remarkable property of parabolas: A ray that travels parallel to the axis of symmetry will strike the surface of the parabola or paraboloid and reflect toward the focus. Likewise, when a ray from the focus strikes the curve, it will reflect in a ray that is parallel to the axis of symmetry. A **paraboloid** is the three-dimensional shape formed when a parabola is rotated about its line of symmetry.

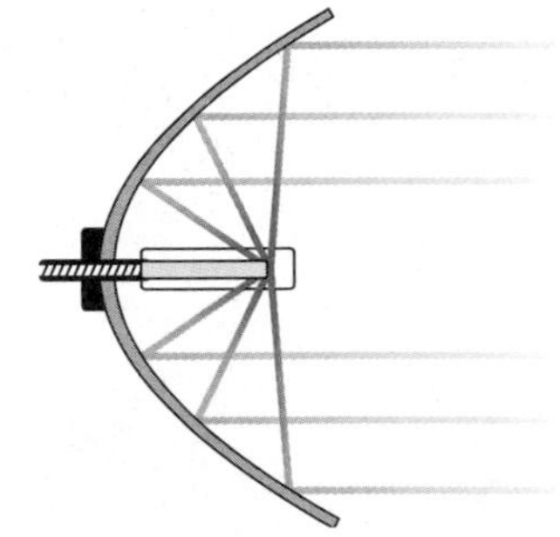

ScienceConnection Satellite dishes, which are used for television, radio, and other communications, are always parabolic. A satellite dish is set up to aim directly at a satellite. As the satellite transmits signals to the dish, the signals are reflected off the dish surface and toward the receiver, which is always located at the focus of the paraboloid dish. In this way, every signal that hits a parabolic dish can be directed into the receiver.

This reflective property of parabolas can be proved using the locus definition of a parabola. You'll see how in the exercises. Compare this locus definition of a parabola to that of an ellipse.

The Definition of a Parabola

A **parabola** is a locus of points in a plane whose distance from a fixed point called the **focus** is the same as the distance from a fixed line called the **directrix.** In the diagram F is the focus, and ℓ is the directrix.

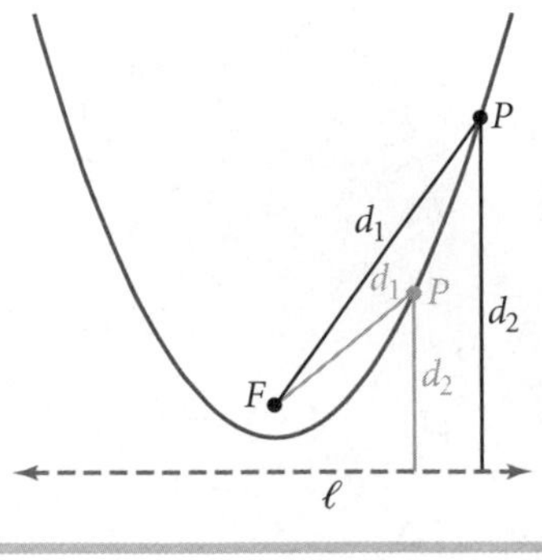

A parabola is the set of points for which the distances d_1 and d_2 are equal. If the directrix is a horizontal line, the parabola is vertically oriented, like the one in the definition box. If the directrix is a vertical line, the parabola is horizontally oriented, like the one shown here. The directrix can also be neither horizontal nor vertical, creating a parabola that is rotated at an angle.

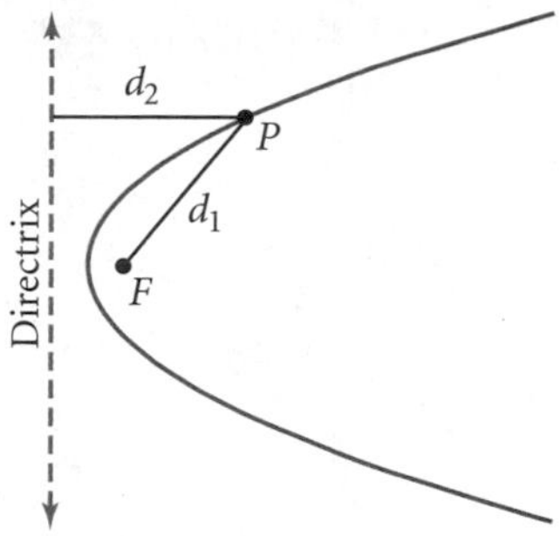

Directrix
d_2
P
d_1
F

How can you locate the focus of a given parabola? Suppose the parabola is horizontally oriented with vertex (0, 0). It has a focus inside the curve at a point, $(f, 0)$, as shown in the first diagram below. The vertex is on the curve and will be the same distance from the focus as it is from the directrix, as shown in the second diagram. This means the equation of the directrix is $x = -f$. You can use this information and the distance formula to find the value of f when the vertex is the origin, as shown in the third diagram.

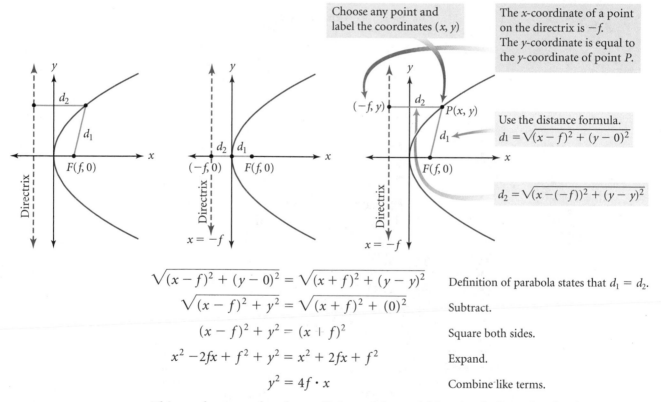

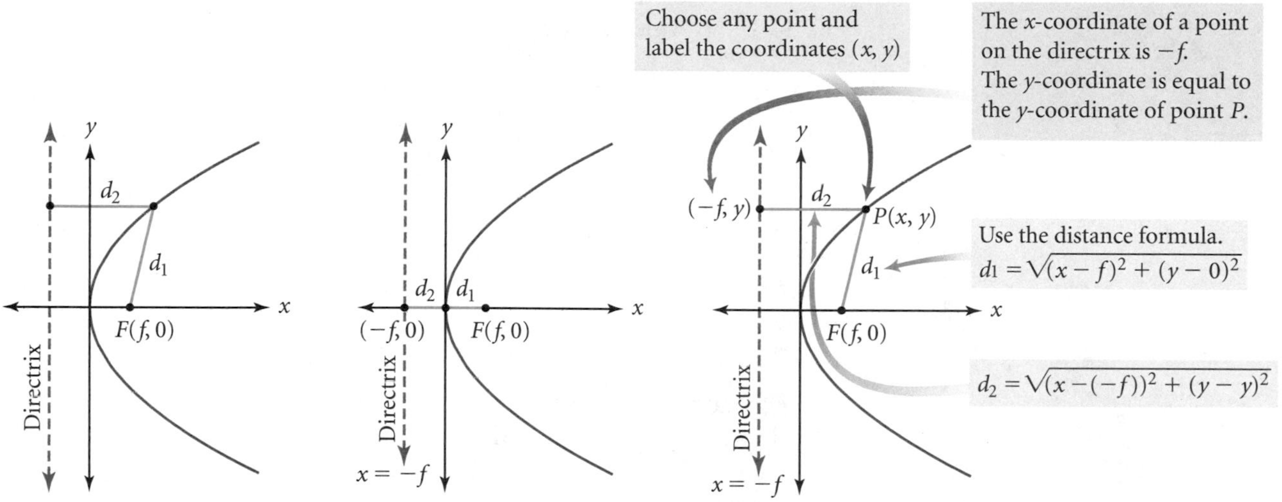

Choose any point and label the coordinates (x, y)
The x-coordinate of a point on the directrix is $-f$. The y-coordinate is equal to the y-coordinate of point P.
y
d_2
d_1
x
$F(f, 0)$
Directrix
$(-f, 0)$
$x = -f$
$(-f, y)$
$P(x, y)$
Use the distance formula. $d_1 = \sqrt{(x - f)^2 + (y - 0)^2}$
$d_2 = \sqrt{(x - (-f))^2 + (y - y)^2}$

$\sqrt{(x - f)^2 + (y - 0)^2} = \sqrt{(x + f)^2 + (y - y)^2}$	Definition of parabola states that $d_1 = d_2$.
$\sqrt{(x - f)^2 + y^2} = \sqrt{(x + f)^2 + (0)^2}$	Subtract.
$(x - f)^2 + y^2 = (x + f)^2$	Square both sides.
$x^2 - 2fx + f^2 + y^2 = x^2 + 2fx + f^2$	Expand.
$y^2 = 4f \cdot x$	Combine like terms.

This result means that the coefficient of the variable x is $4f$, where f is the distance from the vertex to the focus. What do you think it means when f is negative?

If the parabola is vertically oriented, the x- and y-coordinates are exchanged, giving a final equation of $x^2 = 4f \cdot y$, or $y = \frac{1}{4f}x^2$.

EXAMPLE A

Consider the parent equation, $y^2 = x$, of a horizontally oriented parabola.

a. Write the equation of the image of this graph after the following transformations have been performed in order: a vertical stretch by a factor of 3, a translation right 2 units, and then a translation down 1 unit. Graph the new equation.

b. Where is the focus of $y^2 = x$? Where is the directrix?

c. Where is the focus of the transformed parabola? Where is its directrix?

► ***Solution***

a. Begin with the parent equation, and perform the specified transformations.

$$y^2 = x \qquad \text{Original equation.}$$

$$\left(\frac{y}{3}\right)^2 = x \qquad \text{Stretch vertically by a factor of 3.}$$

$$\left(\frac{y+1}{3}\right)^2 = x - 2 \qquad \text{Translate right 2 units and down 1 unit.}$$

Graph the transformed parabola.

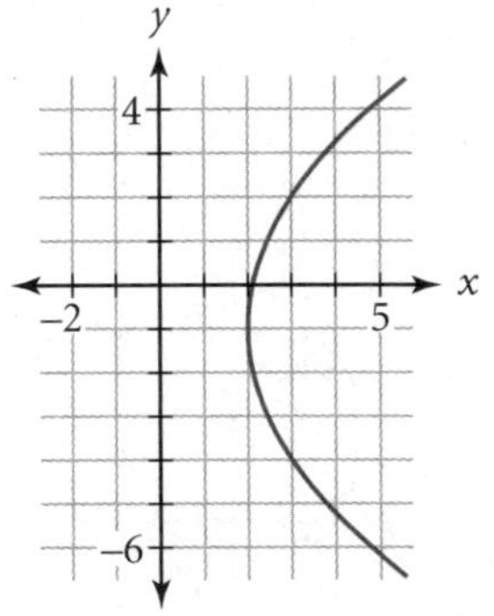

b. Use the general form $y^2 = 4f \cdot x$ to locate the focus and the directrix of the graph of the equation $y^2 = x$. The coefficient of x is $4f$ in the general form and 1 in the equation $y^2 = x$. So, $4f = 1$, or $f = \frac{1}{4}$. Recall that f is the distance from the vertex to the focus, so the focus is $\left(\frac{1}{4}, 0\right)$. Because the distance from the vertex to the focus is equal to the distance from the vertex to the directrix, the directrix has the equation $x = -\frac{1}{4}$.

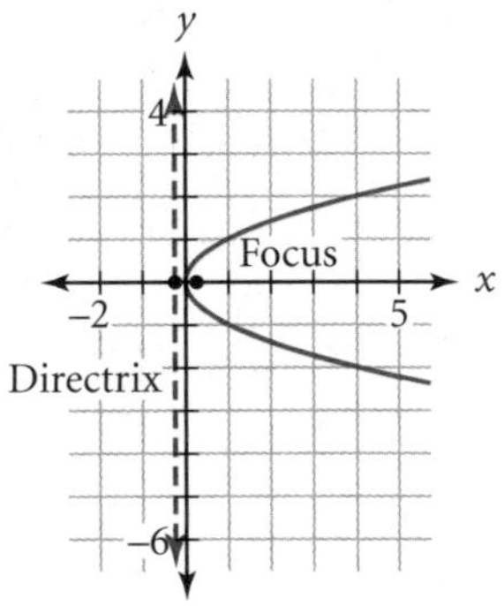

c. To locate the focus and the directrix of $\left(\frac{y+1}{3}\right)^2 = x - 2$, first rewrite the equation as $(y + 1)^2 = 9(x - 2)$. The coefficient of x in this equation is 9, so $4f = 9$, or $f = 2.25$. Both the focus and the directrix will be 2.25 units from the vertex, which is $(2, -1)$, in the horizontal direction. Therefore, the focus is $(4.25, -1)$ and the directrix is the line $x = -0.25$.

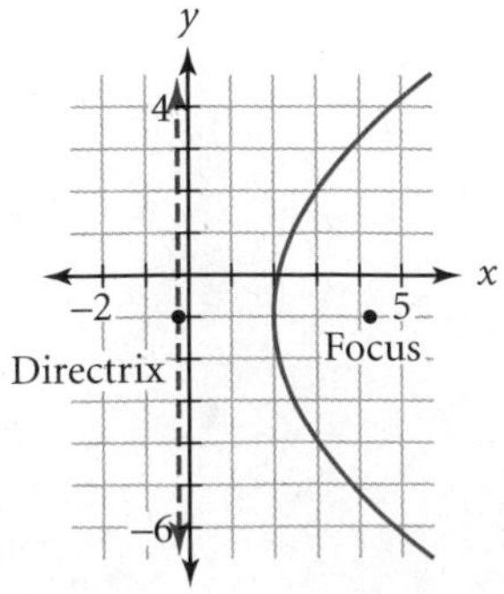

The Equation of a Parabola

The standard form of the equation of a vertically oriented parabola with vertex (h, k), horizontal scale factor of a, and vertical scale factor of b is

$$\left(\frac{y-k}{b}\right) = \left(\frac{x-h}{a}\right)^2$$

Because the focus of a vertically oriented parabola is $(h, k+f)$, where $\frac{a^2}{b} = 4f$, and the directrix is the line $y = k - f$, we can also write

$$4f(y-k) = (x-h)^2$$

The standard form of the equation of a horizontally oriented parabola with vertex (h, k), horizontal scale factor of a, and vertical scale factor b is

$$\left(\frac{x-h}{a}\right) = \left(\frac{y-k}{b}\right)^2$$

Because the focus of a horizontally oriented parabola is $(h+f, k)$, where $\frac{b^2}{a} = 4f$, and the directrix is the line $x = h - f$, we can also write $4f(x-h) = (y-k)^2$.

EXAMPLE B Write an equation of the parabola graphed below.

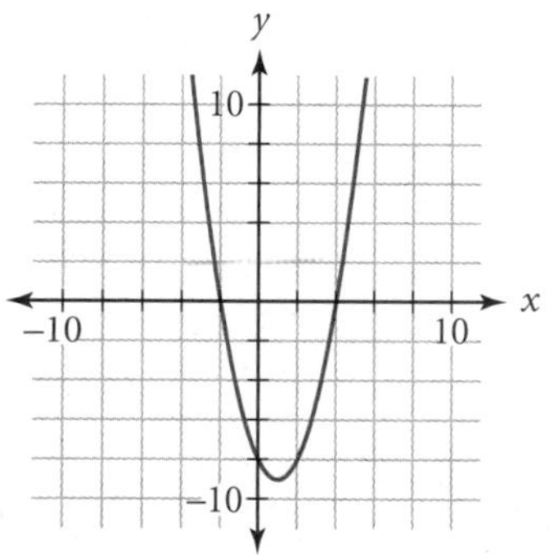

▶ Solution The vertex is $(1, -9)$ so $h = 1$, and $k = -9$. Because the parabola opens upward, use the equation $4f(y-k) = (x-h)^2$ to find that $4f(y+9) = (x-1)^2$. To find the value of f, substitute any other point on the parabola. The point $(4, 0)$ is on the parabola, so:

$$4f(0+9) = (4-1)^2$$
$$36f = 9$$
$$f = \frac{1}{4}$$

The equation of this parabola is $4\left(\frac{1}{4}\right)(y+9) = (x-1)^2$ or $y = (x-1)^2 - 9$.

EXAMPLE C Find the equation of the parabola with focus $(0, 3)$ and vertex $(0, 1)$.

▶ Solution

Because the vertex is $(0, 1)$, $h = 0$ and $k = 1$. We also see that $f = 3 - 1 = 2$.

The directrix is at $y = 1 - 2$ or $y = -1$.

Plotting all this, we see that the parabola opens upward, so using the equation $4f(y - k) = (x - h)^2$, we find that $4(2)(y - 1) = (x - 0)^2$, or $8y - 8 = x^2$.

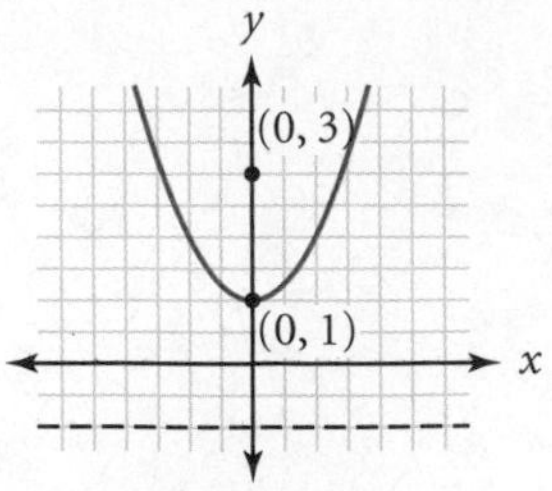

In the Investigation you will construct a parabola. As you create your model, think about how your process relates to the locus definition of a parabola.

Investigation
Fold a Parabola

You will need

- patty paper
- graph paper

Fold the patty paper parallel to one edge to form the directrix of a parabola. Mark a point on the larger portion of the paper to serve as the focus of your parabola. Fold the paper so that the focus lies on the diretrix. Unfold, and then fold again, so that the focus is at another point on the directrix. Repeat this many times. The creases from these folds should create a parabola. Lay the patty paper on top of a sheet of graph paper. Identify the coordinates of the focus and the equation of the directrix, and write an equation of your parabola.

EXERCISES

Practice Your Skills

1. For each parabola described, use the information given to find the location of the missing feature. It may be helpful to sketch the information.

a. If the focus is $(1, 4)$ and the directrix is $y = -3$, where is the vertex?

b. If the vertex is $(-2, 2)$ and the focus is $(-2, -4)$, what is the equation of the directrix?

c. If the directrix is $x = 3$ and the vertex is $(6, 2)$, where is the focus?

2. Sketch each parabola on your paper, and label the vertex and line of symmetry.

a. $\left(\frac{x}{2}\right)^2 + 5 = y$

b. $(y + 2)^2 - 2 = x$

c. $-(x + 3)^2 + 1 = 2y$

d. $2y^2 = -x + 4$

3. Locate the focus and directrix for each graph in Exercise 2.

4. Write an equation in standard form for each parabola.

a. 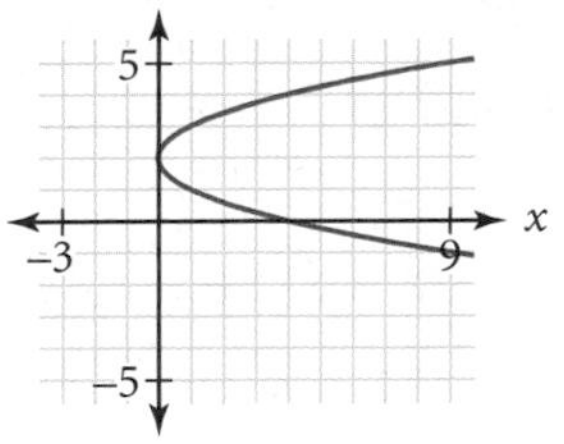

b. 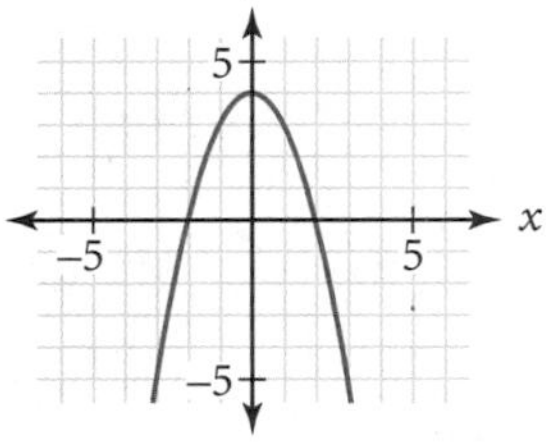

c. 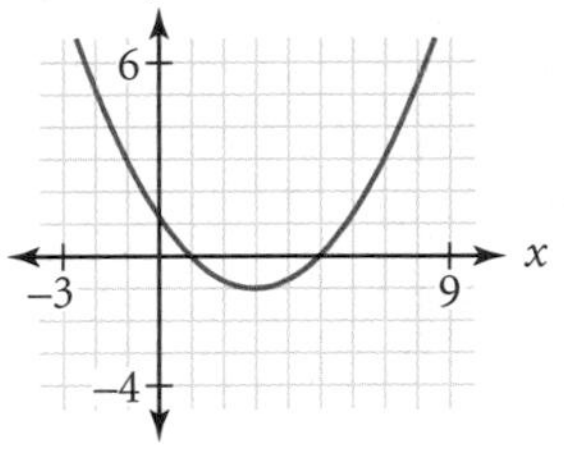

d. 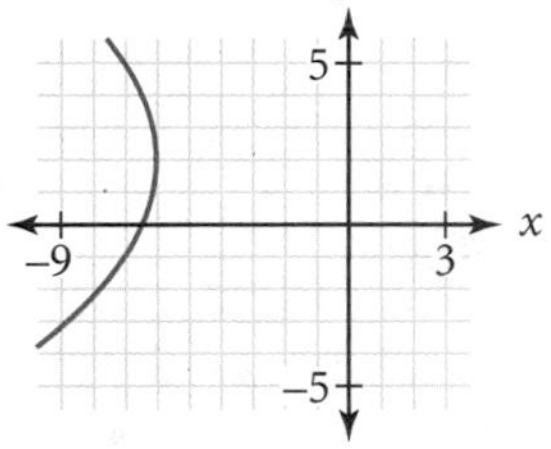

Reason and Apply

5. Find an equation of the parabola with directrix $x = 3$ and vertex $(0, 0)$.

6. The pilot of a small boat charts a course such that the boat will always be equidistant from an upcoming rock and the shoreline. Describe the path of the boat. If the rock is two miles offshore, write an equation of the path of the boat if the rock is at the origin.

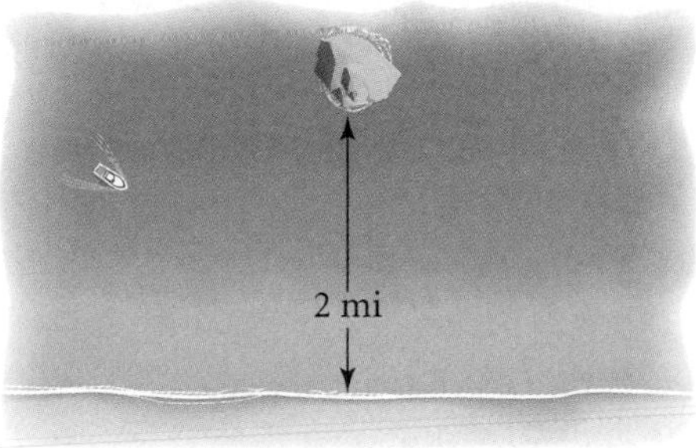

7. Consider this graph.

a. Because $d_1 = d_2$, you can write the equation

$$\sqrt{(x-0)^2 + (y-3)^2} = \sqrt{(x-x)^2 + (y+1)^2}.$$

Rewrite this equation by solving for y.

b. Describe the graph represented by your equation from 7a.

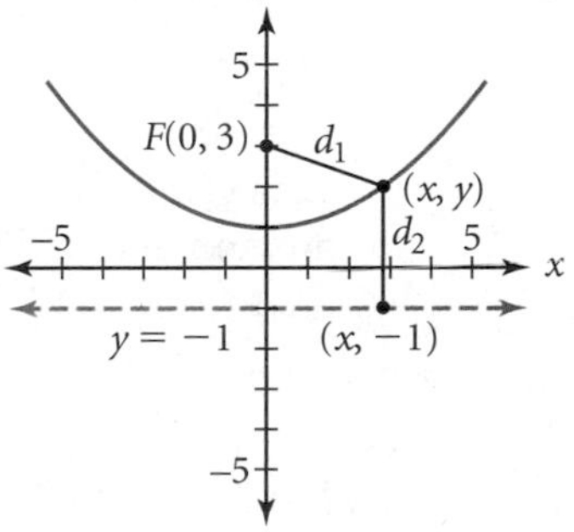

8. Write the equation of the parabola with focus $(1, 3)$ and directrix $y = -1$.

9. Sunny is designing a parabolic dish to use for cooking on a camping trip. She plans to make the dish 40 cm wide and 20 cm deep.

a. Sketch a cross section of this grill with the vertex located at (0, 0).

b. What are the coordinates of the points farthest from the vertex on your cross section?

c. Find the focus of the parabolic dish. Where should Sunny locate the cooking grill so that all of the light that enters the parabolic dish will be reflected to the food and will cook it?

Environmental Connection Solar cookers focus the heat of the Sun on a single spot to boil water or cook food. A well-designed cooker can create heat of up to 400°C. Solar cookers can be created with minimal materials and can help conserve natural resources. About one-third of the world's population currently depends on collecting firewood for cooking and heating. Every year the cutting of firewood results in the loss of about 25,000 km^2 of tropical forests. In developing countries, the environmental pollution caused by cooking fires is a significant public health problem. Inexpensive solar cookers are now being designed and distributed for use in underdeveloped countries. For more information on solar cookers and air quality in developing countries, see the links at **www.keycollege.com/online** .

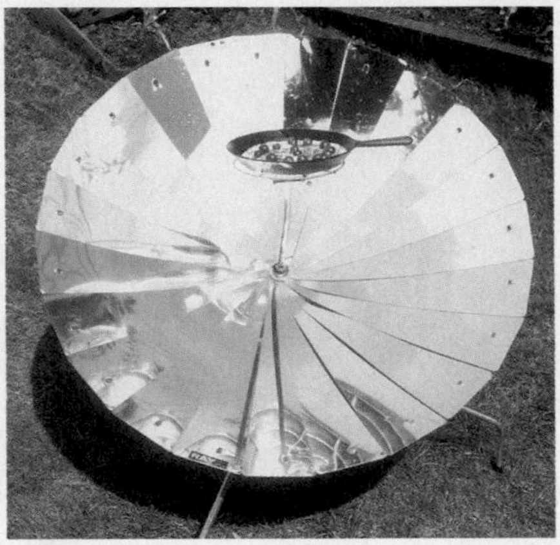
A solar cooker fries an omelette.

10. This diagram shows the reflection of a ray of light in a parabolic reflector. The angles A and B are equal. Follow these steps to verify this reflective property of parabolas.

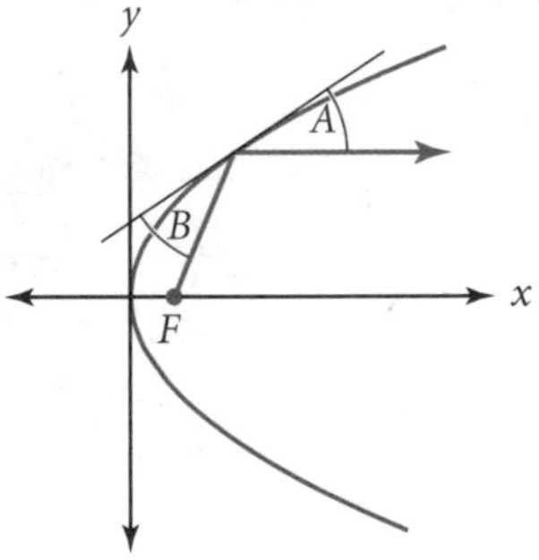

a. Sketch the parabola $y^2 = 8x$.

b. What are the coordinates of the focus of this parabola?

c. On the same graph sketch the line $y = 2x + 1$ tangent to the parabola. Find the
coordinates of the point of tangency. What is the slope of this line?

d. Sketch a ray through the point of tangency parallel to the axis of symmetry. What is the slope of this line?

e. Draw the segment from the point of intersection to the focus. What is the slope of this segment?

f. The formula $\tan A = \frac{m_2 - m_1}{1 + m_2 m_1}$ applies when $\angle A$ is the angle between two lines with slopes m_1 and m_2. Use this formula to find the angle between the tangent line and the horizontal line. Then find the angle between the tangent line and the segment joining the focus and the point of tangency. What do you notice about the angles?

TechnologyConnection The reflective backing in a flashlight is in the shape of a paraboloid. The light source is at the focus, and the light rays will reflect off the backing and travel outward parallel to the axis of symmetry. But not all light backings are paraboloidal. Next time you're sitting in the dentist's chair, see if you can determine the shape of the reflective surface of the dentist's lamp. It may be a parabola, but it could also be an ellipse. When the light source is put at one focus of an elliptical reflector, light will travel to the other focus, illuminating an area smaller than a parabola's.

Review

11. (Lesson 6.1) Find the equation that describes a parabola containing the points (3.6, 0.764), (5, 1.436), and (5.8, −2.404).

12. (Lesson 9.1) Find the minimum distance from the origin to the parabola $y = x^2 + 1$. What point or points on the parabola are closest to the origin?

13. (Lesson 9.2) Find the equation of the ellipse with foci (−6, 1) and (10, 1) that passes through the point (10, 13).

14. (Lesson 6. 7) Consider the polynomial function $f(x) = 2x^3 - 5x^2 + 22x - 10$.

a. What are the possible rational roots of $f(x)$?

b. Find all rational roots.

c. Write the equation in factored form.

15. (Lesson 7.3) On a three-dimensional coordinate system with variables x, y, and z, the general equation of a plane is in the form $ax + by + cz = d$. Find the intersection of the three planes described by $3x + y + 2z = -11$, $-4x + 3y + 3z = -2$, and $x - 2y - z = -3$.

LESSON 9.4

The Hyperbola

The 4th and final conic section is the hyperbola. Comets travel in orbits that are parabolic, elliptical, or hyperbolic. A comet that comes close to another object but never returns is on a hyperbolic path. The two light shadows on a wall next to a cylindrical lampshade form two branches of a hyperbola. The sonic-boom shock wave formed along the ground by a plane traveling faster than sound is one branch of a hyperbola.

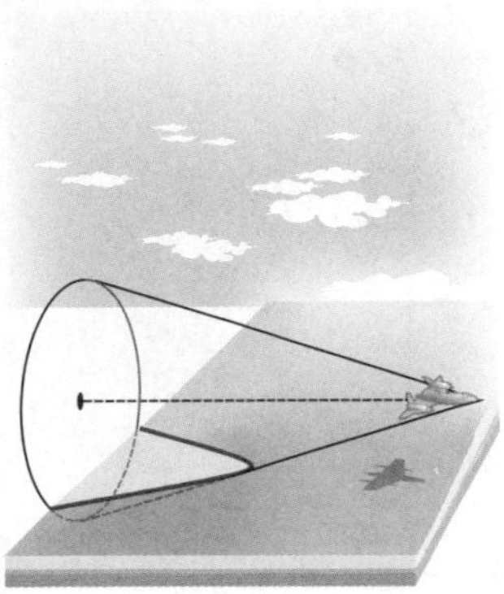

ScienceConnection When an aircraft reaches the speed of its own sound, the airflow around the craft changes significantly, causing a disturbance in the air particles but not an actual sound barrier. When an aircraft catches up with its own noise, which travels ahead of it at a limited speed, the sound then compresses the air and piles up at the nose of the aircraft. This air causes a shock wave against the craft, which may cause it to vibrate or lock its controls. If the aircraft exceeds the speed of sound, the shock waves fall behind the vehicle and cause a sonic boom, which can be heard from the ground. Because sonic booms cause noise pollution and are powerful enough to damage property on land, most supersonic planes fly over the ocean.

The Definition of a Hyperbola

A **hyperbola** is a locus of points in a plane the difference of whose distances from two fixed points, called *foci*, is always a constant. In the diagram F_1 and F_2 are the foci. The points at which the two branches of a hyperbola are closest to each other are called **vertices.** The center of a hyperbola is the midpoint of the segment connecting the vertices.

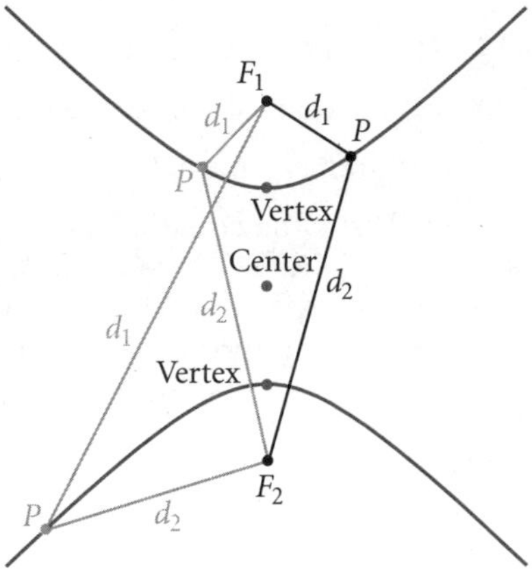

Regardless of a point's location on a hyperbola, the difference in the distances from the point to the two foci is constant. Notice that this constant is equal to the distance between the two vertices of the hyperbola. The hyperbola shown is oriented vertically. In Example A you will see a hyperbola that is oriented horizontally.

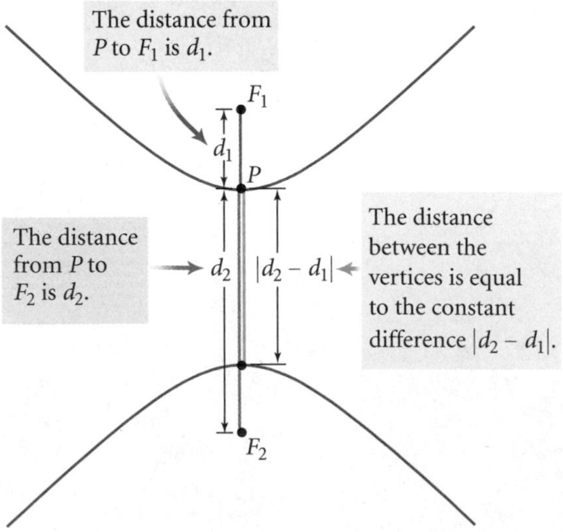

EXAMPLE A

Just as the parent equation of any circle is a unit circle, $x^2 + y^2 = 1$, the parent equation of a hyperbola is called the **unit hyperbola.** The horizontally oriented unit hyperbola has vertices $(1, 0)$ and $(-1, 0)$ and foci $(\sqrt{2}, 0)$ and $(-\sqrt{2}, 0)$. The distance between the vertices is 2, so the difference in the distances from any point on the hyperbola to the two foci is 2. Find the equation of a unit hyperbola.

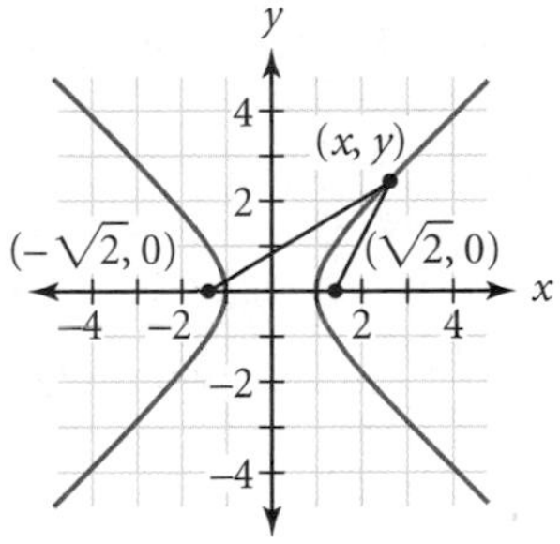

▸ Solution

Label a point on the hyperbola (x, y). The definition of a hyperbola gives this equation.

$\sqrt{(x+\sqrt{2})^2+y^2}-\sqrt{(x-\sqrt{2})^2+y^2}=c$	Definition of hyperbola states that $d_2 - d_1$ is constant.
$\sqrt{(x+\sqrt{2})^2+y^2}-\sqrt{(x-\sqrt{2})^2+y^2}=c$	$d_2 - d_1 = 2$
$\sqrt{(x+\sqrt{2})^2+y^2}=2+\sqrt{(x-\sqrt{2})^2+y^2}$	Add $\sqrt{(x-\sqrt{2})^2+y^2}$ to both sides of the equation.
$(x+\sqrt{2})^2+y^2=(x-\sqrt{2})^2+y^2+4\sqrt{(x-\sqrt{2})^2+y^2}+4$	Square both sides.
$x^2+2x\sqrt{2}+2+y^2=x^2-2x\sqrt{2}+2+y^2+4\sqrt{(x-\sqrt{2})^2+y^2}+4$	Expand.
$2x\sqrt{2}=-2x\sqrt{2}+4\sqrt{(x-\sqrt{2})^2+y^2}+4$	Combine like terms.
$4x\sqrt{2}-4=4\sqrt{(x-\sqrt{2})^2+y^2}$	Isolate the radical.
$x\sqrt{2}-1=\sqrt{(x-\sqrt{2})^2+y^2}$	Divide by 4.
$2x^2-2x\sqrt{2}+1=x^2-2x\sqrt{2}+2+y^2$	Square both sides and expand.
$x^2-y^2=1$	Combine like terms, and collect variables on one side of the equation.

Check your answer by graphing on a calculator. First, you must solve for y.

$$x^2 - y^2 = 1$$
$$-y^2 = 1 - x^2$$
$$y^2 = x^2 - 1$$
$$y = \pm\sqrt{x^2 - 1}$$

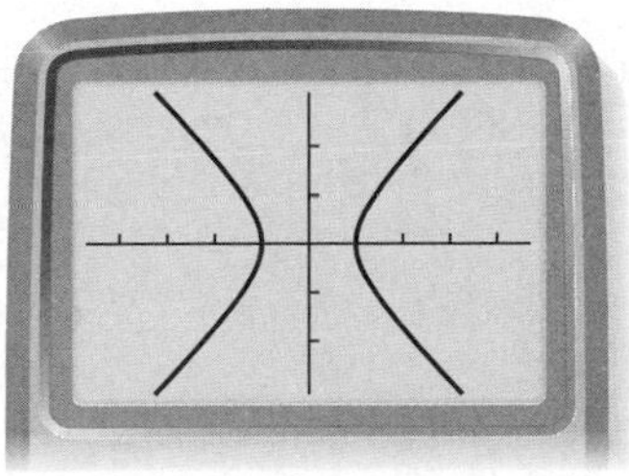

[−4.7, 4.7, 1, −3.1, 3.1, 1]

A special feature of the hyperbola is that both branches approach two lines called **asymptotes.** These are lines that the graph approaches when the curve is extended indefinitely. If you include the graphs of $y = x$ and $y = -x$ on the same coordinate axes, you will notice that they pass through the vertices of a square with corners at $(1, 1)$, $(1, -1)$, $(-1, -1)$, and $(-1, 1)$. As you zoom out, the hyperbola approaches the two lines. The asymptotes are not a part of the hyperbola, but sometimes they are shown to help you see the behavior of the curve. Sketching asymptotes will help you graph hyperbolas more accurately.

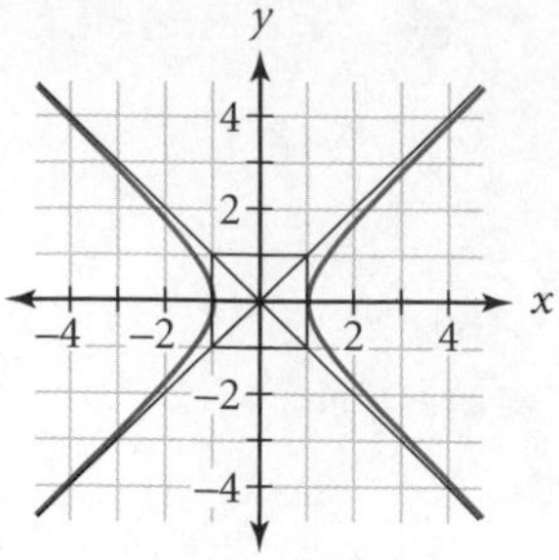

The equation $y^2 - x^2 = 1$ also defines a hyperbola. Look at the similarities to and the differences from the graph of $x^2 - y^2 = 1$. The features are the same, but the hyperbola is oriented vertically instead of horizontally.

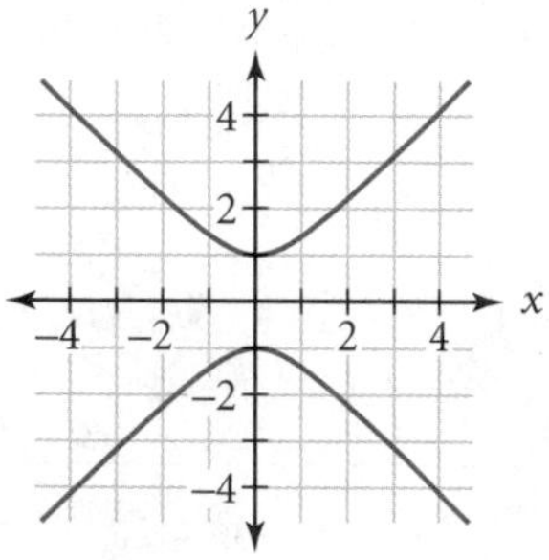

The equation of a hyperbola is similar to the equation of an ellipse except that the terms are subtracted rather than added. For example, the equation $\left(\frac{y}{4}\right)^2 - \left(\frac{x}{3}\right)^2 = 1$ describes a hyperbola, whereas $\left(\frac{y}{4}\right)^2 + \left(\frac{x}{3}\right)^2 = 1$ describes an ellipse.

The standard form of the equation of a hyperbola is

$$\left(\frac{x}{a}\right)^2 - \left(\frac{y}{b}\right)^2 = 1 \quad \text{or} \quad \left(\frac{y}{b}\right)^2 - \left(\frac{x}{a}\right)^2 = 1,$$

where a is the horizontal scale factor and b is the vertical scale factor.

EXAMPLE B | Graph $\left(\frac{y}{4}\right)^2 - \left(\frac{x}{3}\right)^2 = 1$.

▶ ***Solution***

From the equation you can see that this is a vertically oriented hyperbola with a vertical scale factor of 4 and a horizontal scale factor of 3. The hyperbola is not translated, so its center is at the origin. To graph this on your calculator, you must solve for y.

$$\left(\frac{y}{4}\right)^2 - \left(\frac{x}{3}\right)^2 = 1$$

$$\left(\frac{y}{4}\right)^2 = \left(\frac{x}{3}\right)^2 + 1$$

$$\frac{y}{4} = \pm\sqrt{\left(\frac{x}{3}\right)^2 + 1}$$

$$y = \pm 4\sqrt{\left(\frac{x}{3}\right)^2 + 1}$$

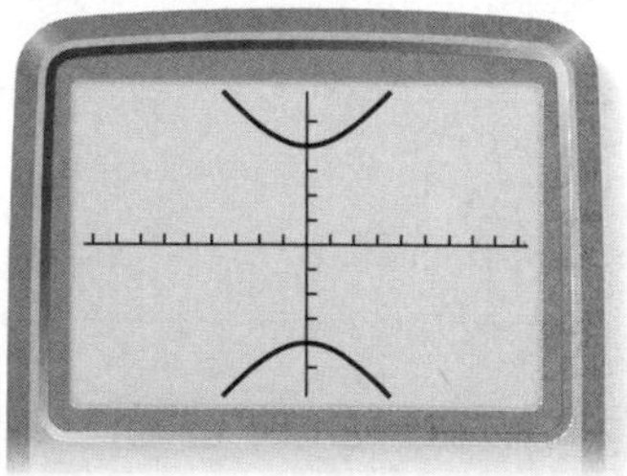

[–9.4, 9.4, 1, –6.2, 6.2, 1]

To sketch a hyperbola by hand, it is easiest to begin by sketching the asymptotes. Start by drawing a rectangle centered at the origin that measures $2a$, or 6, units vertically and $2b$, or 8, units horizontally. The unit hyperbola begins with a 2-by-2 rectangle. Because this hyperbola is stretched horizontally and vertically, it begins with a 6-by-8 rectangle.

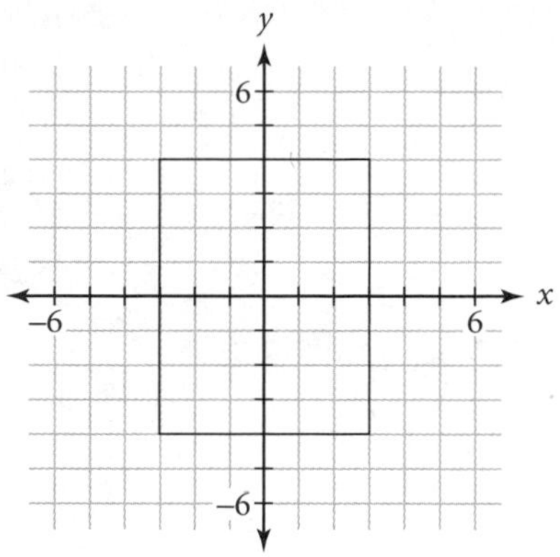

Draw the diagonals of this rectangle, and extend them outside the rectangle. These lines, with equations $y = \pm\frac{4}{3}x$, are the asymptotes of the hyperbola. In general, the slopes of the asymptotes are $\pm\frac{b}{a}$.

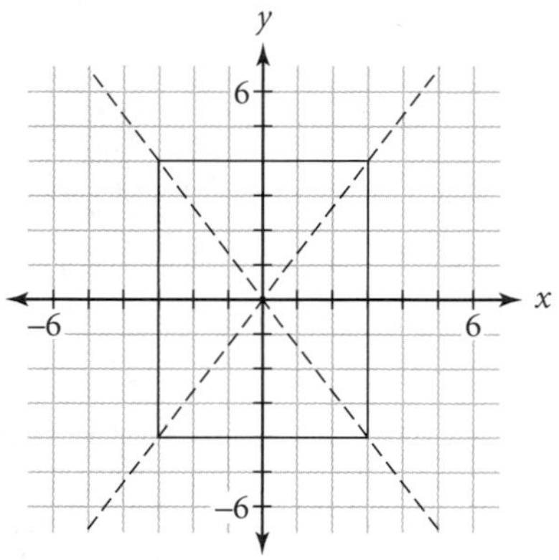

Because this is a vertically oriented hyperbola, the vertices will lie on the top and bottom sides of the rectangle at (0, 4) and (0, −4). Add two curves such that each one touches a vertex and extends outward approaching the asymptotes. You can graph the two asymptotes on your calculator to confirm that the hyperbola does approach them asymptotically.

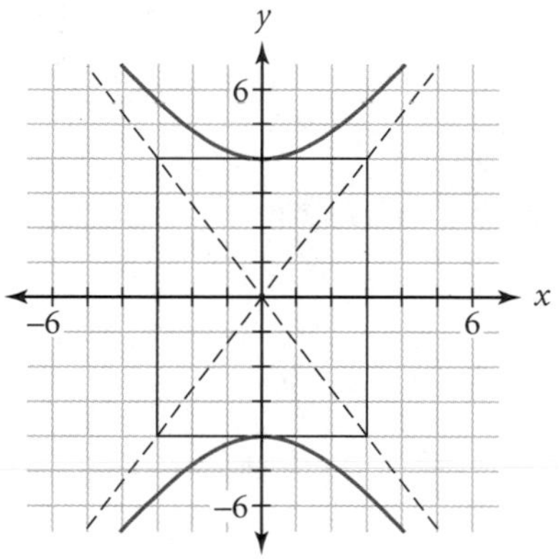

The location of foci in a hyperbola is related to a circle that can be drawn through the four corners of the asymptote rectangle. The distance from the center of the hyperbola to the foci is equal to the radius of the circle.

To locate the foci in a hyperbola, you can use the relationship $a^2 + b^2 = c^2$, where a and b are the horizontal and vertical stretches. In the hyperbola from Example B shown here, $3^2 + 4^2 = 5^2$, so the foci are 5 units above and below the center of the hyperbola.

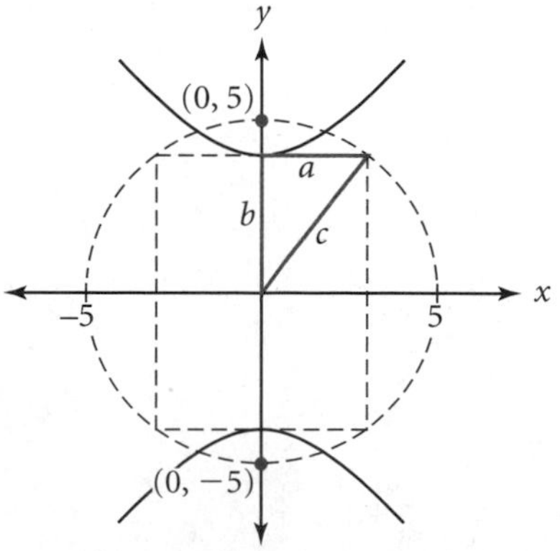

In the Investigation you will explore a situation that produces hyperbolic data and find a curve to fit your data.

Investigation
Passing By

You will need

- a motion sensor

Procedure Note

1. One member of your group will use the motion sensor to measure the distance to the walker for 10 s. The motion sensor must be kept pointed at the walker.
2. The walker should start about 5 m to the left of the sensor holder. He or she should walk at a steady pace, continuing past the sensor holder, and stop about 5 m to the right of the sensor holder. Be sure the walker stays at least 0.5 meter in front of the sensor holder.

Step 1 Collect data as described in the procedure note. Transfer these data from the motion sensor to each calculator in the group, and graph your data. The data should form one branch of a hyperbola.

Step 2 Assume the sensor was held at the center of the hyperbola, and find an equation to fit your data. It may be helpful to try to graph the asymptotes first.

Step 3 Transfer your graph to paper, and add the foci and the other branch of the hyperbola. To verify your equation, choose at least two points on the curve, and measure their distances from the foci. Calculate the differences between the distances from each focus. What do you notice? Why?

The Equation of a Hyperbola

The standard form of the equation of a horizontally oriented hyperbola with center (h, k), horizontal scale factor of a, and vertical scale factor of b is

$$\left(\frac{x-h}{a}\right)^2 - \left(\frac{y-k}{b}\right)^2 = 1$$

The equation of a vertically oriented hyperbola under the same conditions is

$$\left(\frac{y-k}{b}\right)^2 - \left(\frac{x-h}{a}\right)^2 = 1$$

The foci are located c units from the center, where $a^2 + b^2 = c^2$. The asymptotes pass through the center and have slope $\pm\frac{b}{a}$.

EXAMPLE C Write the equation of this hyperbola in standard form, and find the foci.

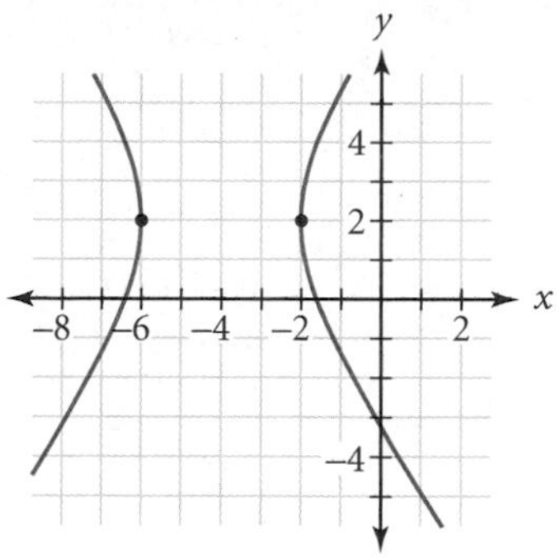

▶ Solution The center is halfway between the vertices, at the point $(-4, 2)$. The horizontal distance from the center to the vertex, a, is 2. If you knew the location of the asymptotes, you could find the value of b using the fact that the slopes of asymptotes of a hyperbola are $\pm\frac{b}{a}$. In this case, the value of b is not as easy to find. Write the equation, substituting the values you know.

$$\left(\frac{x+4}{2}\right)^2 - \left(\frac{y-2}{b}\right)^2 = 1$$

To solve for b, choose another point on the curve, and substitute. It appears that the point $(0, -3.2)$ is on the curve. Because this is an estimate, your value of b will be an approximation.

$$\left(\frac{0+4}{2}\right)^2 - \left(\frac{-3.2-2}{b}\right)^2 = 1$$ Substitute $(0, -3.2)$ for (x, y).

$$2^2 - \left(\frac{-5.2}{b}\right)^2 = 1$$ Add and divide.

$$4 - \frac{27.04}{b^2} = 1$$ Square.

$$-\frac{27.04}{b^2} = -3$$ Subtract 4 from both sides.

$$9.013 = b^2$$ Multiply by b^2 and divide by -3 on both sides.

$$3.002 \approx b$$ Take the square root of both sides.

The value of b is approximately 3, so the equation of the hyperbola is

$$\left(\frac{x+4}{2}\right)^2 - \left(\frac{y-2}{3}\right)^2 = 1$$

You can find the distance to the foci by using the equation $2^2 + 3^2 = c^2$. So, the foci are to the right and left of the center, or approximately $(-0.39, 2)$ and $(-7.6, 2)$.

In the Exercises you will continue to explore relationships between the equation and the graph of a hyperbola.

EXERCISES

Practice Your Skills

1. Sketch each hyperbola on your paper. Write the coordinates of each vertex and the equation of each asymptote.

a. $\left(\frac{x}{2}\right)^2 - \left(\frac{y}{4}\right)^2 = 1$

b. $\left(\frac{y+2}{1}\right)^2 - \left(\frac{x-2}{3}\right)^2 = 1$

c. $\left(\frac{x-4}{3}\right)^2 - \left(\frac{y-1}{3}\right)^2 = 1$

d. $y = \pm 2\sqrt{1 + \left(\frac{x+2}{3}\right)^2} - 1$

2. What are the coordinates of the foci of each hyperbola in Exercise 1?

3. Write an equation in standard form for each graph.

a.

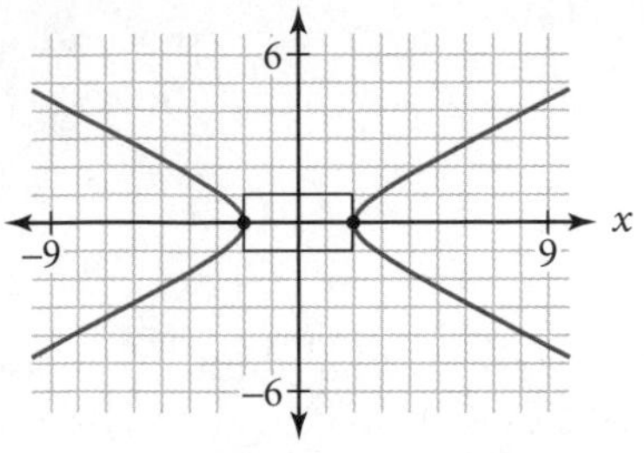

b.

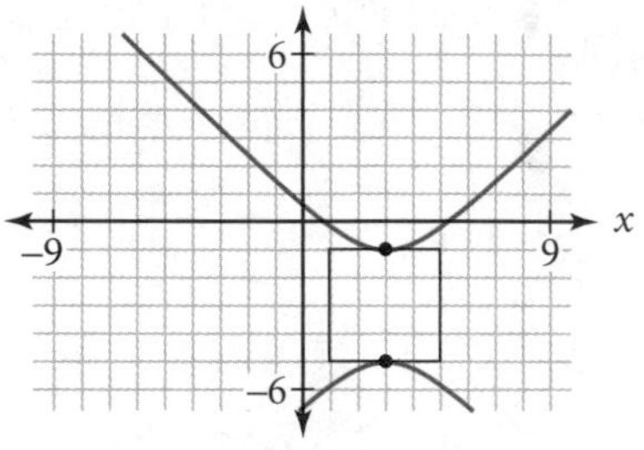

c.

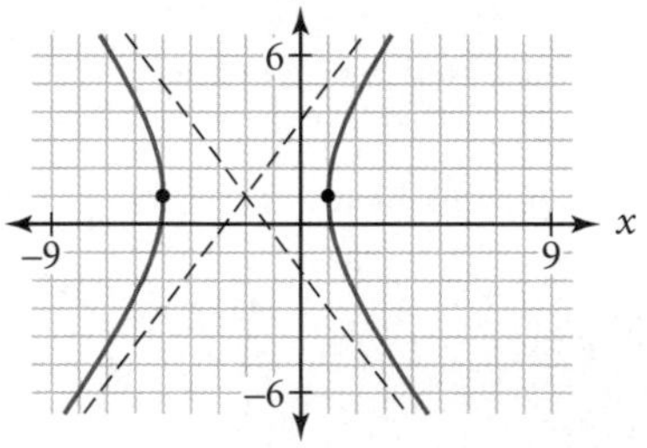

d.

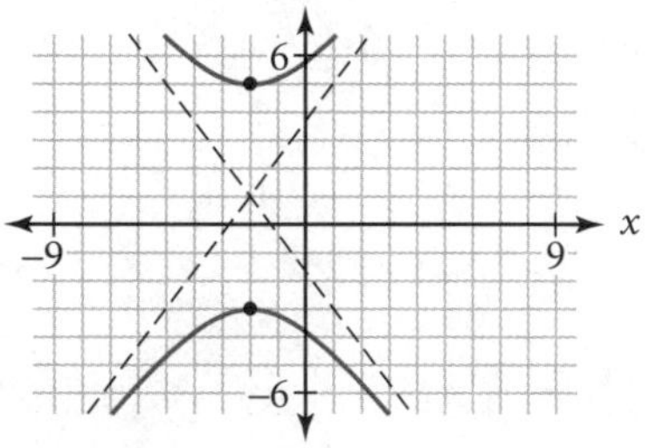

4. Write the equations of the asymptotes for each hyperbola in Exercise 3.

Reason and Apply

5. Another way to locate the foci of a hyperbola is by rotating the asymptote rectangle about its center so that opposite corners lie on the line of symmetry that contains the vertices of the hyperbola. From the diagram you can see that the distance from the origin to a focus is one-half the length of the diagonal of the rectangle.

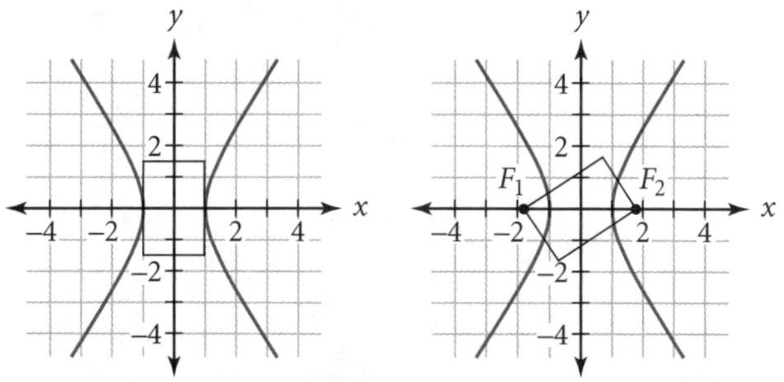

a. Show that this distance is $\sqrt{a^2 + b^2}$.

b. Find the coordinates of the foci for $\left(\frac{y+2}{1}\right)^2 - \left(\frac{x-2}{3}\right)^2 = 1$

6. A point moves in a plane so that the difference of its distances from $(-5, 1)$ and $(7, 1)$ is always 10 units. What is the equation of the path of this point?

7. Graph and write the equation of a hyperbola that has an upper vertex at $(-2.35, 1.46)$ and an asymptote of $y = 1.5x + 1.035$.

8. Approximate the equation of each hyperbola shown.

a.

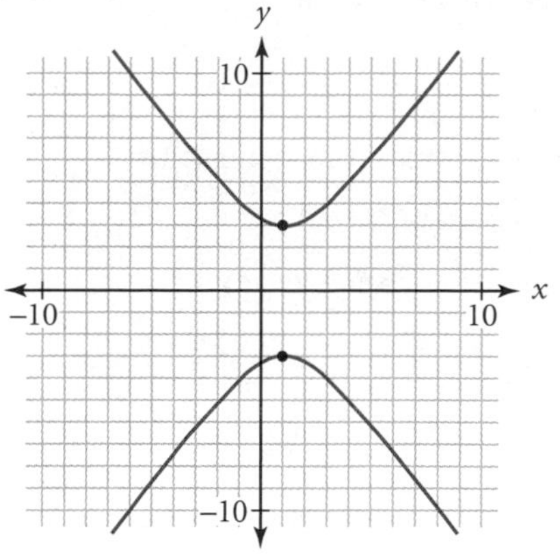

b.

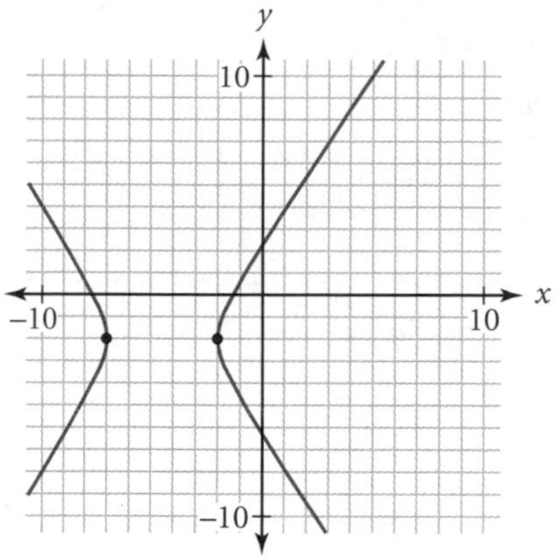

9. Find the vertical distance between a point on the hyperbola $\left(\frac{y+1}{2}\right)^2 - \left(\frac{x-2}{3}\right)^2 = 1$ and its nearest asymptote for each of the following x-values.

x-value	5	10	20	40
Distance				

10. APPLICATION A receiver can determine the distance to a homing transmitter by the transmitter's signal strength. These signal strengths were measured using a receiver in a car traveling due north.

Distance (mi)	0.0	2.0	4.0	6.0	8.0	10.0	12.0	14.0	16.0
Signal strength (w/m^2)	9.82	7.91	6.04	4.30	2.92	2.55	3.54	5.15	6.96

a. Find the equation of the hyperbola that best fits the data.

b. Name the center of this hyperbola.

c. What does this point tell you?

d. What are the possible locations of the homing transmitter?

11. Sketch the graphs of the conic sections in 11a–d:

a. $y = x^2$ **b.** $x^2 + y^2 = 9$ **c.** $\frac{x^2}{9} + \frac{y^2}{16} = 1$ **d.** $\frac{x^2}{9} - \frac{y^2}{16} = 1$

Review

12. (Lesson 6.3) Solve the quadratic equation $0 = -x^2 + 6x - 5$ by completing the square.

13. (Lesson 9.2) Mercury's orbit is an ellipse with the Sun at one focus and eccentricity 0.206. Mercury's major axis is about 1.16×10^8 km. If you consider Mercury's orbit with the Sun at the origin and the other focus on the positive x-axis, what equation models Mercury's orbit?

14. (Lesson 9.3) The setter on a volleyball team makes contact with the ball at a height of 5 ft. The parabolic path of the ball reaches a maximum height of 17.5 ft when the ball is 10 ft from the setter.

a. Find an equation that models the ball's path.

b. A hitter can smash the ball when it is 8.5 ft off the floor. How far from the setter is the hitter when she makes contact?

15. (Lesson 9.3) Sketch the graph of each parabola. Give the coordinates of each vertex and focus and the equation of each directrix.

a. $y = -(x + 1)^2 - 2$

b. $y = x^2 - 3x + 5$

16. (Lesson 5.1) The half-life of radium-226 is 1,620 yr.

a. Write a function that relates the amount s of a sample of radium-226 remaining after t years.

b. After 1,000 yr, how much of a 500-g sample of radium-226 will remain?

c. How long will it take for a 3-kg sample of radium-226 to decay so that only 10 g remain?

17. (Lesson 6.1) Use finite differences to write an equation for the nth term of the sequence $-20, -14, -6, 4, 16, \ldots$.

HistoryConnection In 1646 Dutch mathematician Frans van Schooten (1615–1660) wrote the book *Sive de Organica Conicarum Sectionum in Plano Descriptione, Tractus*, which is translated as *A Treatise on Devices for Drawing Conic Sections*. This book described several different ways to construct each of the conic sections. Some of the constructions used unique mechanical devices.

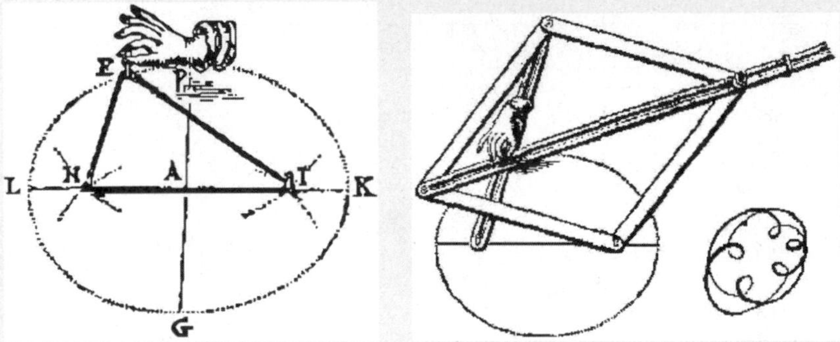

These illustrations from van Schooten's *Sive de Organica Conicarum Sectionum in Plano Descriptione, Tractatus* show two ways to construct an ellipse.

IMPROVING YOUR VISUAL THINKING SKILLS

Slicing a Cone

Describe how to slice a double cone to produce each of these geometric shapes: a circle, an ellipse, a parabola, a hyperbola, a point, one line, and two lines. Be sure to describe where and at what angle the plane must slice the cone. (*Hint:* Look back at the illustrations on page 475 [Lesson 9.2] to help.) Sketch a diagram of each slicing.

LESSON

9.5

Nonlinear Systems of Equations

A complex system that works is invariably found to have evolved from a simple system that works.

JOHN GAULE

In Chapter 3 you learned to solve linear systems of equations graphically and algebraically. Many practical systems of equations are nonlinear, however. For instance, Global Positioning System (GPS) equipment finds your location by solving three or four nonlinear equations in four variables.

ScienceConnection Twenty-four satellites orbit the Earth as part of the GPS system. Knowing the distance from your position to one of these satellites limits your possible location to a sphere. Knowing your distance to two of the satellites further limits your position to the points of a circle. Knowing your distance from a third satellite gives only two possible points where you could be. You can find your exact location by finding the distance from a fourth satellite or by noting that one of the two points of the three-satellite measurement is not on the surface of the Earth.

Investigation
Systems of Conic Equations

If you graph two conic sections on the same graph, in how many ways could they intersect? The five ways that an ellipse and a hyperbola could intersect are shown here.

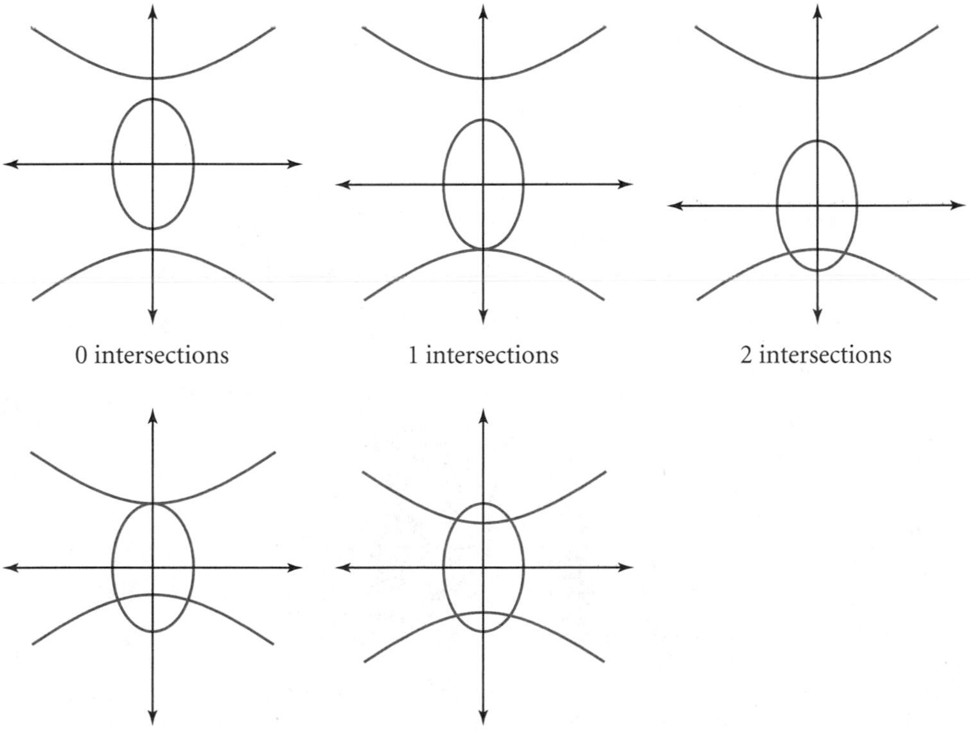

0 intersections

1 intersections

2 intersections

3 intersections

4 intersections

There are four conic sections: circles, ellipses, parabolas, and hyperbolas. Among the members of your group, investigate all ten possible pairs of shapes. For each pair, list the possible numbers of intersection points. Draw at least one sketch showing each possibility.

HistoryConnection A method of constructing an ellipse was discovered by Abū Ali Al-Hasan ibn al-Haytham (or Alhazen), who lived from about 965 to 1041 C.E. in Basra, Iraq, and Alexandria, Egypt. His book *Optics* explained light and vision using geometry. His main theory described how light reflected on a surface radiates light rays in every direction, even though only one perpendicular ray is visible to the eye.

EXAMPLE A Find the points of intersection of $\frac{(x-5)^2}{4} + y^2 = 1$ and $x = y^2 + 5$.

▸ Solution First, graph the curves, and estimate the points of intersection. Then you can solve the system algebraically.

You can graph this equation using your knowledge of transformations or by solving for y and graphing on your calculator. Graphing on the calculator enables you to trace the curve to find the point of intersection. First, solve for y in both equations.

$\frac{(x-5)^2}{4} + y^2 = 1$	The first equation.
$y^2 = 1 - \frac{(x-5)^2}{4}$	Solve for y^2.
$y = \sqrt{1 - \frac{(x-5)^2}{4}}$	Solve for y.
$x = y^2 + 5$	The second equation.
$y^2 = x - 5$	Solve for y^2.
$y = \pm\sqrt{x-5}$	Solve for y.

Graph the two equations, and trace to approximate the points of intersection.

There are two points of intersection, approximately (5.8, 0.9) and (5.8, −0.9). You can use algebraic methods to find the intersection points more accurately. To solve algebraically, you can use the two equations in terms of y and the substitution method.

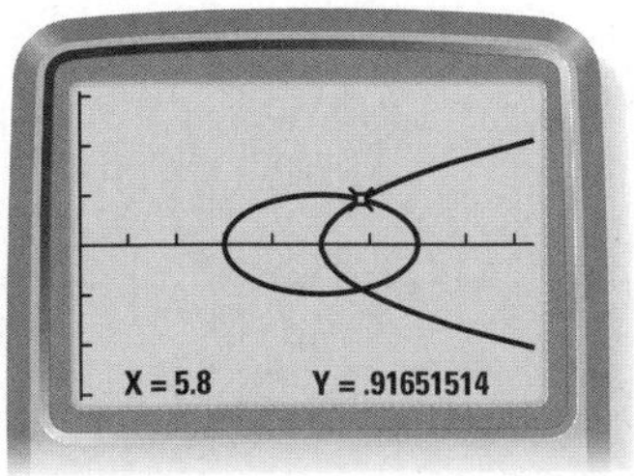

[0, 9.4, 1, −3.1, 3.1, 1]

Or, in this case you can use the two original equations and the substitution or elimination method. Notice that both equations have a y^2 term. Solve for y^2 in the second equation, and substitute.

$\frac{(x-5)^2}{4} + y^2 = 1$ and $x = y^2 + 5$	Original equations.
$y^2 = x - 5$	Solve the second equation for y^2.
$\frac{(x-5)^2}{4} + x - 5 = 1$	Substitute $(x - 5)$ for y^2 in the first equation.
$(x-5)^2 + 4x - 20 = 4$	Multiply both sides by 4 to eliminate the denominator.
$x^2 - 10x + 25 + 4x - 20 = 4$	Distribute the squared quantity.
$x^2 - 6x + 1 = 0$	Combine like terms.
$x = \frac{6 \pm \sqrt{36 - 4 \cdot 1 \cdot 1}}{2 \cdot 1} \approx 5.828$ and 0.172	Use the quadratic formula to solve for x.

Now, substitute these two values in one of the equations relating x and y, and solve for y.

$$y = \pm\sqrt{x-5} = \pm\sqrt{5.828-5} = \pm 0.910$$
$$y = \pm\sqrt{x-5} = \pm\sqrt{0.172-5} = \pm\sqrt{-4.828} = \pm 2.197i$$

The points of intersection are (5.828, 0.910) and (5.828, −0.910). When you substitute 0.172 for x, you find two imaginary values for y. These nonreal numbers are solutions, but they are not points of intersection. The two real solutions are close to the intersection points you estimated by graphing.

You can always use graphing to estimate real solutions. Graphing is valuable even when you are finding solutions algebraically, because a graph will tell you how many intersection points to look for. It can also help confirm that your algebraic solutions are correct.

EXAMPLE B

Solve

$$x^2 + 2y^2 = 10$$
$$x^2 - y^2 = 1$$

▶ Solution

It would be easy enough to graph the ellipse and hyperbola to estimate solutions. However, this system is easily solved using the elimination method from Chapter 3. You could begin by multiplying the 2nd equation by 2.

$x^2 + 2y^2 = 10$	
$2x^2 - 2y^2 = 2$	
$3x^2 = 12$	Add the equations.
$x^2 = 4$	
$x = \pm 2$	Solve for x.

Now, use the solutions you found for x to solve for y.

$$4 - y^2 = 1$$

$$y^2 = 3 \qquad x^2 \text{ is 4 whether } x = 2 \text{ or } x = -2.$$

$$y = \pm\sqrt{3}$$

There are four solutions: $(2, \sqrt{3})$, $(2, -\sqrt{3})$, $(-2, \sqrt{3})$ and $(-2, -\sqrt{3})$.

EXERCISES

Practice Your Skills

1. Using the substitution or elimination method, solve each system of equations algebraically.

a. $\begin{cases} y = x^2 \\ 2y + 3x = 2 \end{cases}$

b. $\begin{cases} y = x^2 + 4 \\ y = (x - 2)^2 + 3 \end{cases}$

c. $\begin{cases} x^2 - \dfrac{y^2}{4} = 1 \\ x^2 + (y + 4)^2 = 9 \end{cases}$

d. $\begin{cases} x^2 + y^2 = 8 \\ 4x^2 - 9y^2 = 19 \end{cases}$

2. Determine the number of solutions to the system by graphing.

a. $\begin{cases} x^2 + y^2 = 4 \\ y = x^2 - 2 \end{cases}$

b. $\begin{cases} y = x^2 - 4 \\ x = y^2 - 4 \end{cases}$

c. $\begin{cases} y = x^2 - 4 \\ 4x^2 + 9y^2 = 36 \end{cases}$

Reason and Apply

3. Point P is 5 units from $(0, 4)$ and 7 units from $(4, 0)$.

a. Given that P is 5 units from $(0, 4)$, what is the equation of one circle on which P must lie?

b. Given that P is 7 units from $(4, 0)$, what is the equation of a second circle on which P must lie?

c. Give possible locations of point P.

4. APPLICATION Two seismic monitoring stations recorded the vibrations of an earthquake. The second monitoring station is 50 mi due east of the first. The epicenter was determined to be 30 mi from the first station and 27 mi from the second station. Where could the epicenter of the earthquake be located?

5. Find the equation of the circle that passes through the four intersection points of the ellipses $\frac{x^2}{16} + \frac{y^2}{9} = 1$ and $\frac{x^2}{9} + \frac{y^2}{16} = 1$.

Review

6. (Lesson 9.3) Find the equation of two parabolas that pass through the points (2, 5), (0, 9), and (−6, 7). Sketch each parabola.

7. (Lesson 9.2) Find the coordinates of the foci of each ellipse.

a. $\left(\frac{x+2}{4}\right)^2 + \left(\frac{y-5}{6}\right)^2 = 1$

b. $(x-1)^2 + 4(y+2)^2 = 1$

8. (Lesson 9.4) Find the equations of the asymptotes of the hyperbola with vertices (5, 8.5) and (5, 3.5), and foci (5, 12.5) and (5, −0.5).

9. (Lesson 6.7) Write the polynomial expression $x^4 - 3x^3 + 4x^2 - 6x + 4$ in factored form.

LESSON

9.6 Introduction to Rational Functions

You probably know that a lighter tree climber can crawl farther out on a branch than a heavier climber can before the branch is in danger of breaking. What do you think the graph of (*length, mass*) will look like when mass is added to a length of pole until it breaks? Is the relationship linear, like line *A*, or does it resemble one of the curves, *B* or *C*?

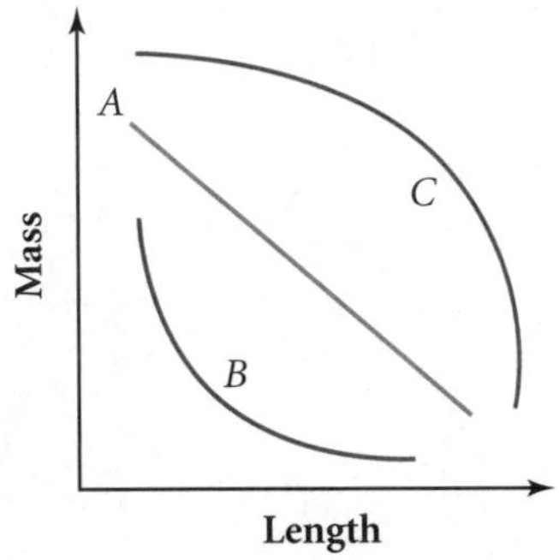

Engineers study problems like this because they need to know the weight that a beam can safely support. In the next Investigation you will collect data and experiment with this relationship.

Investigation
The Breaking Point

You will need

- several pieces of dry spaghetti
- a small film canister
- string
- some weights (pennies or other small, heavy objects)
- a ruler
- tape

Procedure Note

1. Lay a piece of spaghetti on a table so that its length is perpendicular to one side of the table and the end extends beyond the edge of the table.
2. Measure the length of the spaghetti that extends beyond the edge of the table. (See the photo.) Record this information in a table of (*length, mass*) data.
3. Tie the string to the film canister so that you can hang it from the end of the spaghetti. (You may need to use tape to hold the string in place.)
4. Place weights in the container one at a time until the spaghetti breaks. Record the maximum number of weights that the length of pasta was able to support.

Step 1 Work with a partner. Follow the Procedure Note to record at least five data points, and then compile your results along with those of other group members.

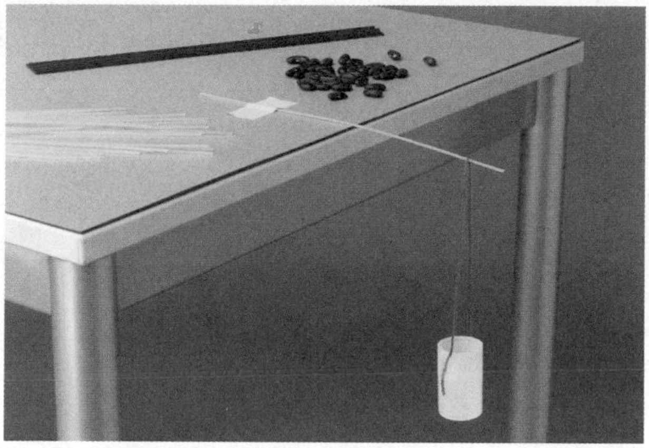

Step 2 Make a graph of your data with length as the independent variable, x, and mass as the dependent variable, y. Does the relationship appear to be linear? If not, describe the appearance of the graph.

Step 3 Write an equation that is a good fit with the plotted data.

The relationship between length and mass in the Investigation is an **inverse variation.** The parent function of an inverse variation curve, $f(x) = \frac{1}{x}$, is a **rational function.**

The Rational Function

A *rational function* is one that can be written as a quotient, $f(x) = \frac{p(x)}{q(x)}$, where $p(x)$ and $q(x)$ are both polynomial expressions.

This type of function can be transformed just like all of the other functions you have previously studied. The function you found in the Investigation was a transformation of $f(x) = \frac{1}{x}$.

Graph the function $f(x) = \frac{1}{x}$ on your calculator, and observe some of its special characteristics. Notice that the graph of $f(x) = \frac{1}{x}$ looks like a hyperbola. In fact, it's a rotation about the origin by 45° of the hyperbola $\frac{x^2}{2} - \frac{y^2}{2} = 1$, whose vertices have rotated to $(1, 1)$ and $(-1, -1)$. The graph is made up of two branches. One part occurs where x is negative and the other where x is positive. There is no value of this function when $x = 0$. What happens when you try to evaluate $f(0)$? As x gets closer to zero, the y-values become increasingly large in absolute value.

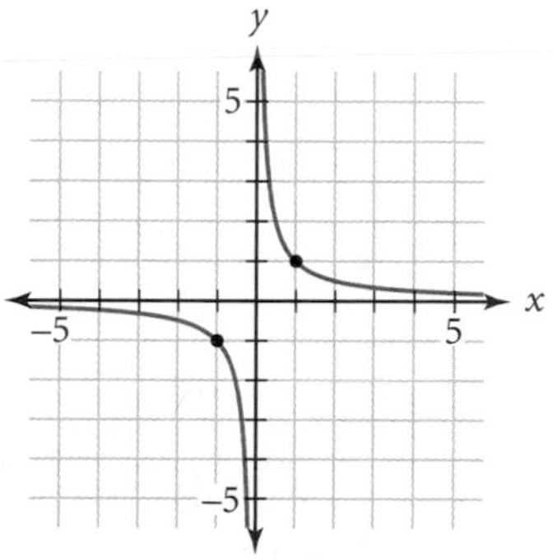

Consider these values of the function $f(x) = \frac{1}{x}$.

x	-1	-0.1	-0.01	-0.001	0	0.001	0.01	0.1	1
y	-1	-10	-100	-1000	undefined	1000	100	10	1

As x approaches zero from the negative side, the y-values have an increasingly large absolute value.

So $x = 0$ is a vertical asymptote.

As x approaches zero from the positive side, the y-values have an increasingly large absolute value.

The behavior of the y-values as x gets closer to zero shows that the y-axis is a vertical asymptote of this function.

x	-10000	-1000	-100	-10	0	10	100	1000	10000
y	-0.0001	-0.001	-0.01	-0.1	undefined	0.1	0.01	0.001	0.0001

As x takes on larger negative values, the y-values approach zero.

So $y = 0$ is a horizontal asymptote.

As x takes on larger positive values, the y-values approach zero.

As x approaches the extreme values at the left and right ends of the x-axis, the graph approaches the horizontal axis. The horizontal line $y = 0$, then, is a horizontal asymptote. This asymptote is an end-behavior model of the function. In general, the **end behavior** of a function is its behavior for x-values that are large in absolute value.

If you think of $y = \frac{1}{x}$ as a parent function, then $y = \frac{1}{x} + 1$, $y = \frac{1}{x-2}$, and $y = 3\left(\frac{1}{x}\right)$ are examples of transformed rational functions. What happens to a function when x is replaced with $(x - 2)$? The function $y = \frac{1}{x-2}$ is shown here.

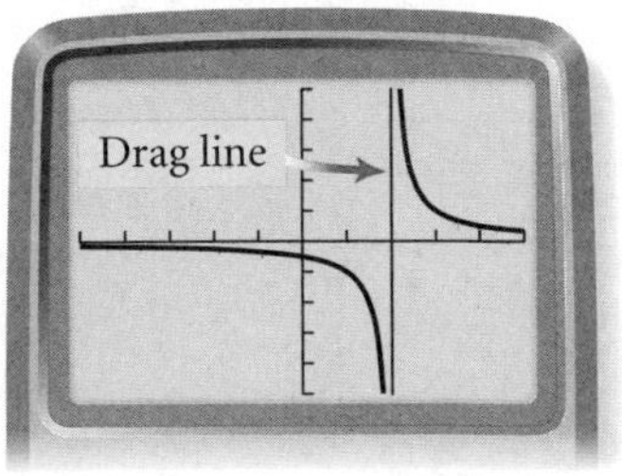

$[-5, 5, 1, -5, 5, 1]$

Rational function graphs on the calculator often include a nearly vertical drag line. The drag line is not part of the graph! However, it will look much like the graph of the vertical asymptote. [▶ See **Calculator Note 9A** to learn how to eliminate this line from your graph.◀]

EXAMPLE A

Describe how these functions have been transformed from the parent function $f(x) = \frac{1}{x}$.

$$g(x) = \frac{3x + 1}{x} \quad \text{and} \quad f(x) = \frac{2x - 5}{x - 1}$$

▶ **Solution**

Rewrite $g(x)$ as two fractions.

$$g(x) = \frac{3x}{x} + \frac{1}{x} \qquad g(x) = 3 + \frac{1}{x}$$

The first part of the expression reduces for nonzero values of x.

The parent function has been shifted up 3 units.

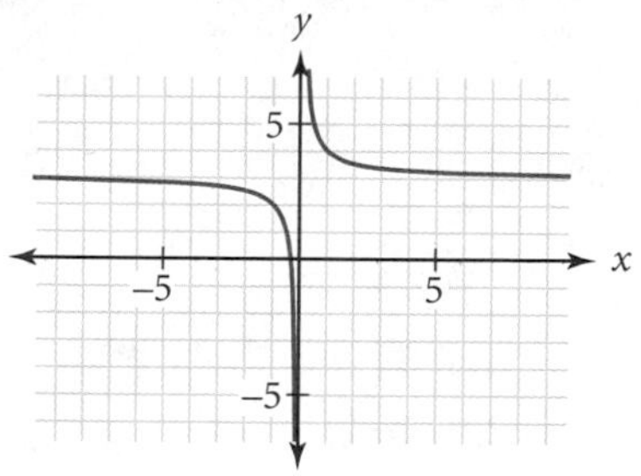

For $f(x)$ you can change the form of the equation so that the transformations are more obvious. Because the denominator is $(x - 1)$ rather than x, try to get the expression $(x - 1)$ in the numerator as well.

$f(x) = \frac{2x - 5}{x - 1}$ Original equation.

$f(x) = \frac{2(x - 1) - 3}{x - 1}$ Consider the numerator to be $2x - 2 - 3$, and then factor to get $2(x - 1) - 3$.

Now, look for a scale factor and an added term. Separate the rational expression into two fractions.

$f(x) = \frac{2(x - 1)}{x - 1} - \frac{3}{x - 1}$ Separate the numerator into two numerators over the same denominator.

$f(x) = 2 - \frac{3}{x - 1}$ Reduce $\frac{2(x - 1)}{x - 1}$ to 2.

Now you can see that the parent function has been stretched vertically by a factor of -3, then translated right 1 unit and up 2 units.

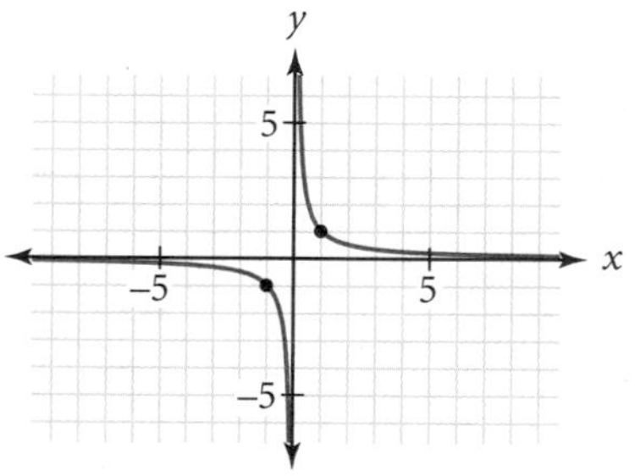

$y = \frac{1}{x}$

The parent rational function, $y = \frac{1}{x}$, has vertices $(1, 1)$ and $(-1, -1)$.

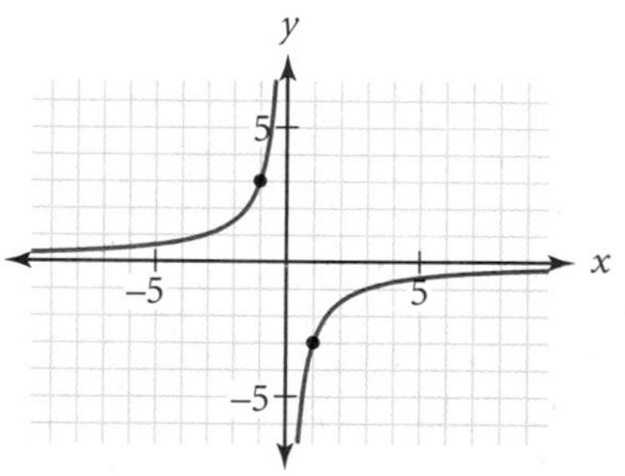

$y = -\frac{3}{x}$

A vertical stretch of –3 moves the vertices to $(1, -3)$ and $(-1, 3)$. The points $(3, 1)$ and $(-3, -1)$ are also on the curve. Notice that this graph looks more "spread out" than the graph of $y = \frac{1}{x}$.

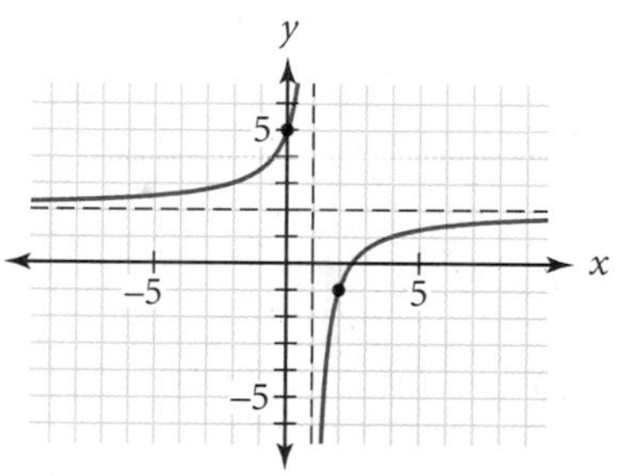

$y = 2 - \frac{3}{x - 1}$

A translation right 1 unit and up 2 units moves the vertices to $(2, -1)$ and $(0, 5)$. The asymptotes are also translated to $x = 1$ and $y = 2$.

Notice that the asymptotes too have been translated. How are the equations of the asymptotes related to your final equations above?

To identify an equation that will produce a given graph, use the procedure from Example A in reverse. You can identify translations simply by looking at the translations of the asymptotes. To identify stretch factors, pick a point, such as a vertex, the coordinates of which you would know after the translation. Then find a point on the stretched graph that has the same x-coordinate. The ratio of the y-coordinates of those two points is the vertical scale factor.

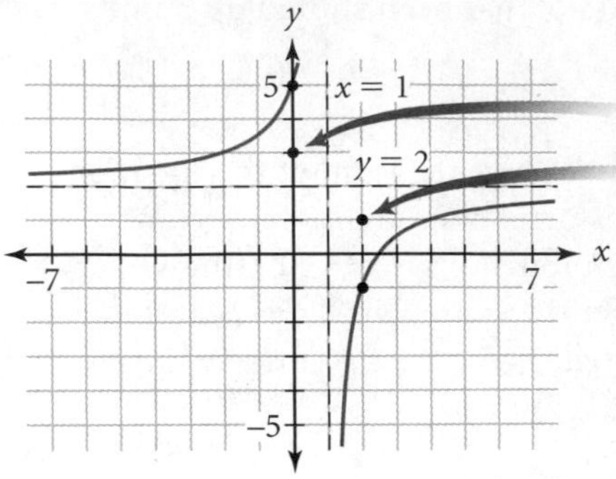

An unstretched (but reflected) inverse variation function with these translations would have vertices (2, 1) and (0, 3), 1 horizontal unit and 1 vertical unit away from the center. Because the distance is now 3 vertical units from the center, include a vertical stretch of 3 to get the equation

$$y = 2 - \frac{3}{x-1}$$

Rational expressions are very useful in chemistry. Scientists use them to model many situations, such as the concentration of a solution or mixture as it is diluted.

EXAMPLE B

Suppose you have 100 mL of a solution that is 30% acid and 70% water. How many mL of acid do you need to add to make a solution that is 60% acid? To make a 90% acid solution? Can the solution ever be 100% acid?

▸ Solution

Of the 100 mL of solution, 30%, or 30 mL, is acid. The percentage, P, can be written as $P = \frac{30}{100}$. If x mL of acid are added, there will be more acid, but also more solution. The concentration of acid will be

$$P = \frac{30 + x}{100 + x}$$

To find when the solution is 60% acid, substitute 0.6 for P, and solve the equation.

$0.6 = \frac{30 + x}{100 + x}$	Original equation.
$0.6(100 + x) = 30 + x$	Multiply both sides by $(100 + x)$.
$60 + 0.6x = 30 + x$	Distribute.
$30 = 0.4x$	Collect like terms.
$x = 75$	Divide by 0.4.

Adding 75 mL of acid will make a 60% acid solution.

To find the point at which the solution is 90% acid, solve the equation $0.9 = \frac{30 + x}{100 + x}$. You will find that 600 mL of acid must be added.

The graph of $P = \frac{30 + x}{100 + x}$ shows horizontal asymptote $y = 1$. No matter how many mL of acid you add, you will never have a mixture that is 100% acid. This is because the original 70 mL of water will remain, even though it is a smaller and smaller percentage of the entire solution as you continue to add acid.

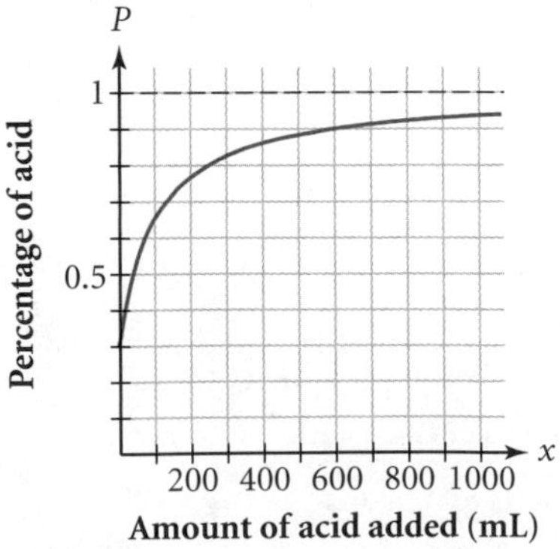

EXERCISES

Practice Your Skills

1. Write an equation and graph each transformation of the rational function $f(x) = \frac{1}{x}$.

a. Translate the graph up 2 units.

b. Translate the graph right 3 units.

c. Translate the graph down 1 unit and left 4 units.

d. Stretch the graph vertically by a scale factor of 2.

e. Stretch the graph horizontally by a factor of 3, and translate it up 1 unit.

2. What are the equations of the asymptotes of each hyperbola?

a. $y = \frac{2}{x} + 1$

b. $y = \frac{3}{x - 4}$

c. $y = \frac{4}{x + 2} - 1$

d. $y = \frac{-2}{x + 3} - 4$

3. Solve.

a. $12 = \frac{x - 8}{x + 3}$

b. $21 = \frac{3x + 8}{x - 5}$

c. $3 = \frac{2x + 5}{4x - 7}$

d. $-4 = \frac{-6x + 5}{2x + 3}$

4. As the rational function $y = \frac{1}{x}$ is translated, so too are its asymptotes. Write an equation for the translation of $y = \frac{1}{x}$ that has the asymptotes described.

a. horizontal asymptote $y = 2$ and vertical asymptote $x = 0$

b. horizontal asymptote $y = -4$ and vertical asymptote $x = 2$

c. horizontal asymptote $y = 3$ and vertical asymptote $x = -4$

5. If a basketball team's present record is 42 wins and 36 losses, how many consecutive games must it win to reach a winning record of 60%?

Reason and Apply

6. Write a rational equation to describe each graph. Some equations will need scale factors.

a.

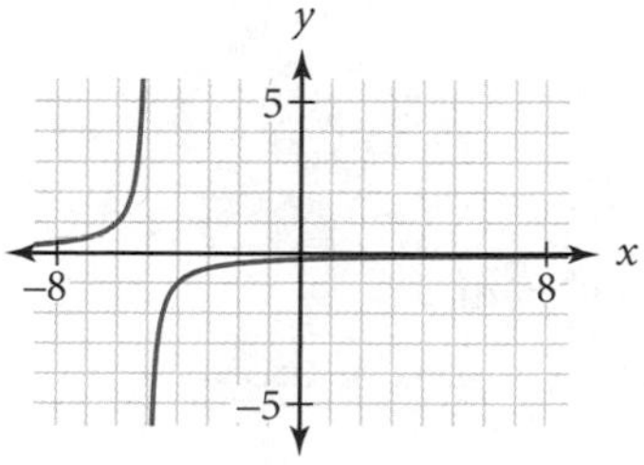

b.

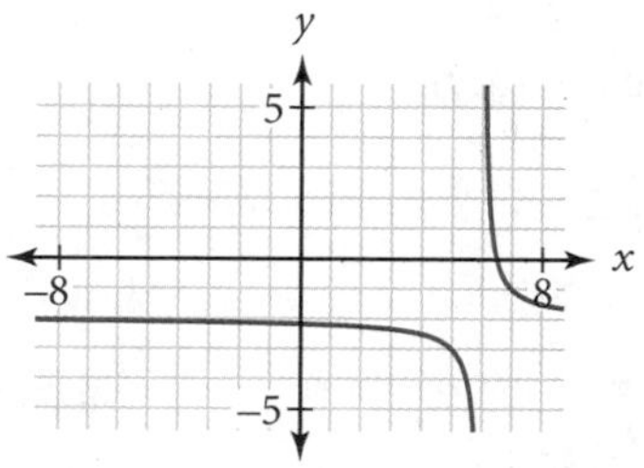

c.

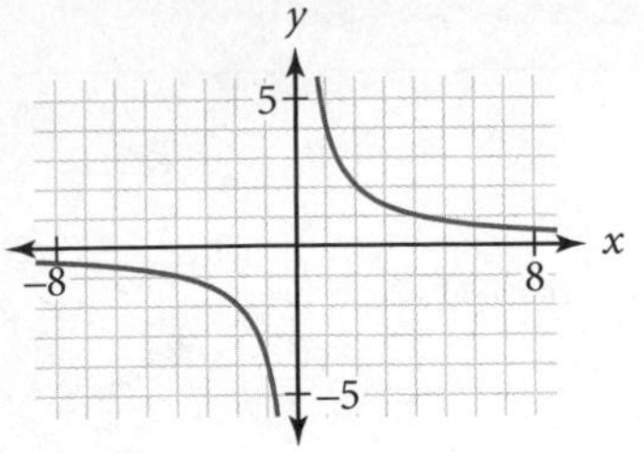

d.

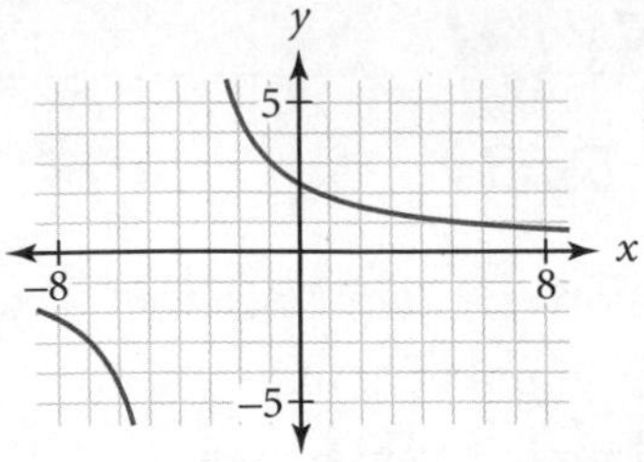

7. **APPLICATION** This graph depicts the concentration of acid in a solution as pure acid is added. The solution began as 55 mL of a 38% acid solution.

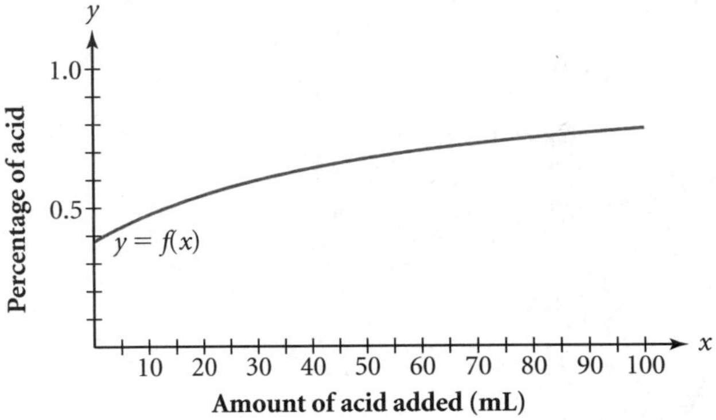

a. How many mL of pure acid were in the original solution?

b. Write an equation of $f(x)$.

c. Find the amount of pure acid that must be added to create a solution that is 64% acid.

d. Describe the end behavior of $f(x)$.

ScienceConnection Acids and bases are chemical compounds that have opposite properties. In food, acids taste sour, and bases taste bitter. When combined in water, acids and bases neutralize because acids release hydrogen ions, and bases accept them. This reaction often produces salt and water. The strength of an acid or a base in water is measured by a pH scale. The pH scale is logarithmic and indicates the level of an acid or base according to the pH level of pure water, which is 7.0.

8. **APPLICATION** In a container of 2% acne medication, 2% of the mixture is benzoyl peroxide. How much of the liquid in a 1-oz container of 2% acne medication would need to be emptied and replaced with pure benzoyl peroxide before the container could be labeled as maximum strength (3.25%) medication?

9. Consider these functions.

i. $y = \dfrac{2x - 13}{x - 5}$ **ii.** $y = \dfrac{3x + 11}{x + 3}$

a. Rewrite each rational function to show how it is a transformed version of $y = \frac{1}{x}$.

b. Describe the transformations of the graph of $y = \frac{1}{x}$ that will produce graphs of the equations in 9a.

c. Graph each equation on your calculator to confirm your answers to 9b.

10. **APPLICATION** Ohm's law states that $I = V/R$. This law can be used to determine the amount of current I, in amps, flowing in the circuit when a voltage V, in volts, is applied to a resistance R, in ohms.

a. If a hair dryer set on high is using a maximum of 8.33 amps on a 120-volt line, what is the resistance in the heating coils?

b. In the United Kingdom, power lines use 240 volts. If a traveler plugs in a hair dryer, and the resistance in the hair dryer is the same as in 10a, what is the flow of current?

c. The additional current flowing through the hair dryer would cause a meltdown of the coils and the motor wires. In order to reduce the current flow in 10b to the value in 10a, how much resistance would be needed?

ConsumerConnection Many travel appliances, such as hair dryers and shavers, are made with a 120V/240V switch. Moving the switch provides the necessary resistance for the appliance to work properly in different countries. A dimmer switch on a light fixture works in a similar way. When the dimmer switch is set low, there is higher resistance, causing less current to flow, and less illumination is produced. As the dimmer switch is turned up, there is less resistance, allowing more current to flow, and more illumination is produced. The volume control on a stereo system also works this way.

Review

11. (Lesson 6.6) Factor each expression completely.

a. $x^2 - 7x + 10$

b. $x^3 - 9x$

12. (Lesson 4.7) Write the equation of the circle with center $(2, -3)$ and radius 4.

13. (Lesson 9.1) A 2-m rod and a 5-m rod are mounted vertically 10 m apart on a long, straight iron rail. The ends of a 15-m wire are attached to the top of each rod. The wire is stretched taut and fastened to the rail. How far from the base of the 2-m rod is the wire fastened?

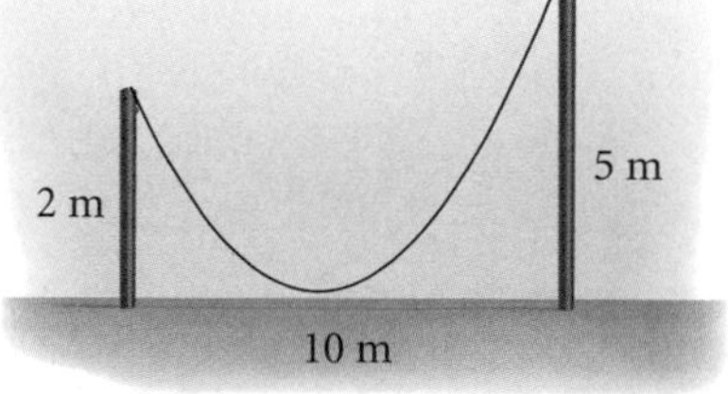

14. (Lesson 8.4) Sarah would like to row her boat directly across a river 500 m wide. The current flows at 3 km/h, and she is able to row at 5 km/h.

a. At what angle to the riverbank should she point her boat?

b. As she starts, how far upstream on the opposite bank should she head?

c. Write parametric equations to simulate Sarah's crossing.

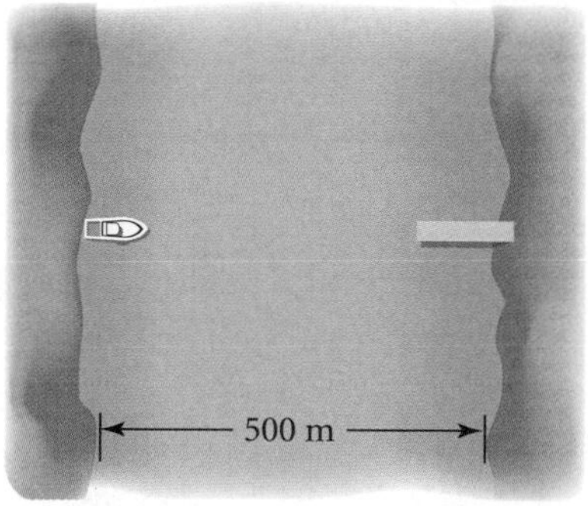

15. (Lesson 9.5) Solve the system $\begin{cases} y = \dfrac{1}{x} \\ x^2 - 5y^2 = 4 \end{cases}$

LESSON

9.7

Graphs of Rational Functions

Besides learning to see, there is another art to be learned—to see what is not.

MARIA MITCHELL

Some rational functions create very different kinds of graphs from those you have studied so far. The graphs of these functions are often in two or more parts. This is because the denominator, a polynomial function, may be equal to zero at some point, so the function will be undefined at that point. Sometimes it's difficult to see the different parts of the graph because they may be separated by only one missing point, called a **hole**. At other times you will see two parts that look very similar—one part may look like a reflection or rotation of the other part. Or you might get multiple parts that look totally different from each other. Look for these features in the graphs below.

$$y = \frac{x^3 + 2x^2 - 5x - 6}{x - 2}$$

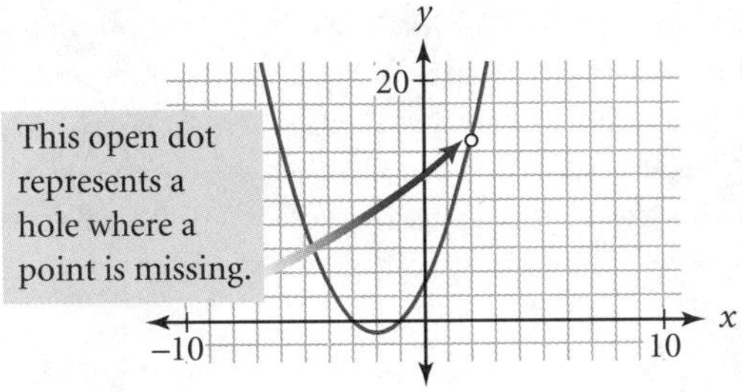

$$y = \frac{1}{x - 2}$$

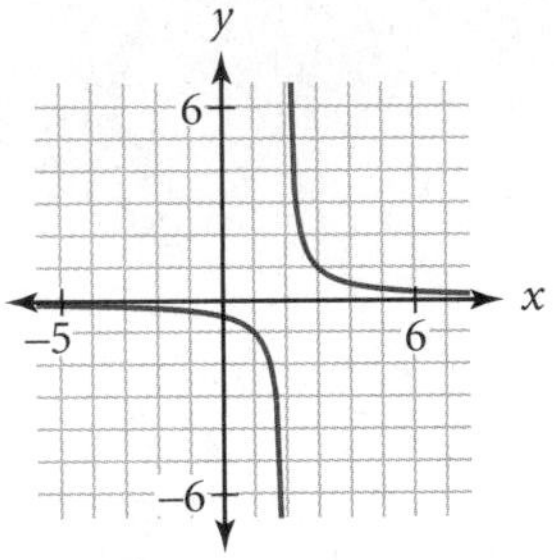

$$y = \frac{x^3 - x^2 - 8x + 12}{6x^2 + 6x - 12}$$

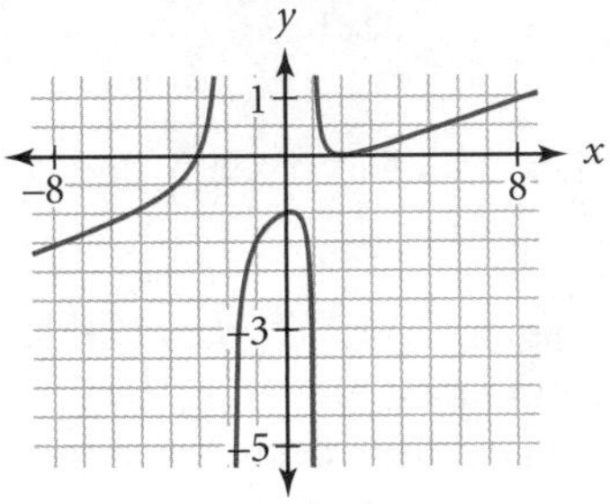

$$y = \frac{1}{x^2 + 1}$$

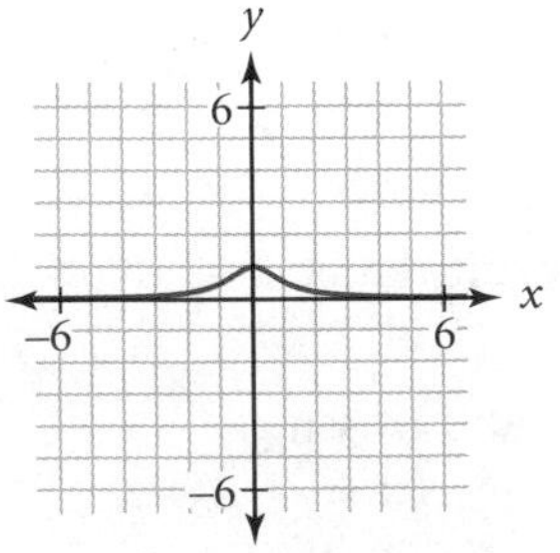

$$y = \frac{x - 1}{x^2 + 4}$$

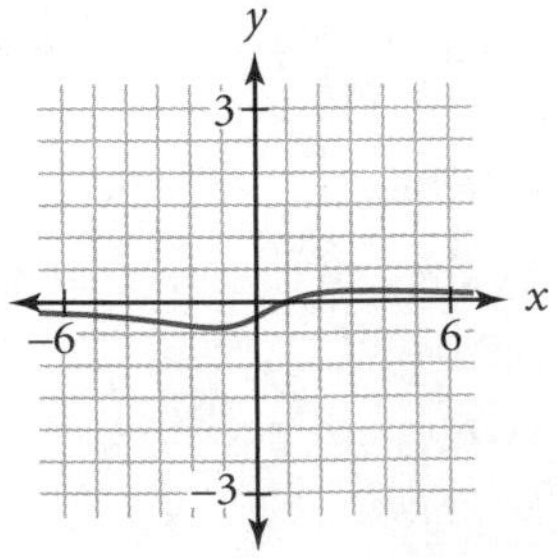

In this lesson you will explore local and end behavior of rational functions, and you will learn how to predict some of the features of a rational function's graph by studying its equation. When closely examining a rational function, you will often find it helpful to look at the equation in factored form.

Investigation
Predicting Asymptotes and Holes

In this Investigation you will consider the graphs of four rational functions and the local behavior of each at and near $x = 2$.

Step 1 Match each graph with a rational function. Use a friendly window as you graph, and trace the equations. Describe the unusual occurrences at and near $x = 2$, and try to explain what feature in the equation makes the graph look the way it does. (You will not actually see the hole pictured in Graph d unless you turn off the coordinate axes on your calculator.)

a.

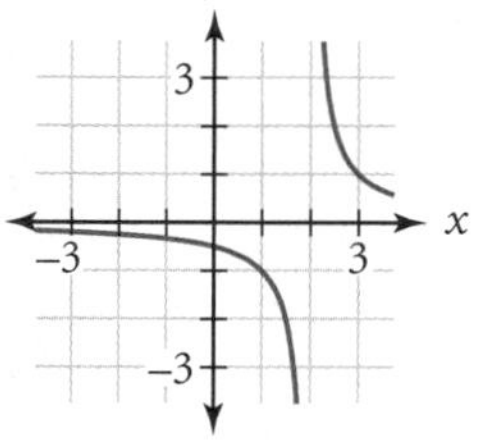

b.

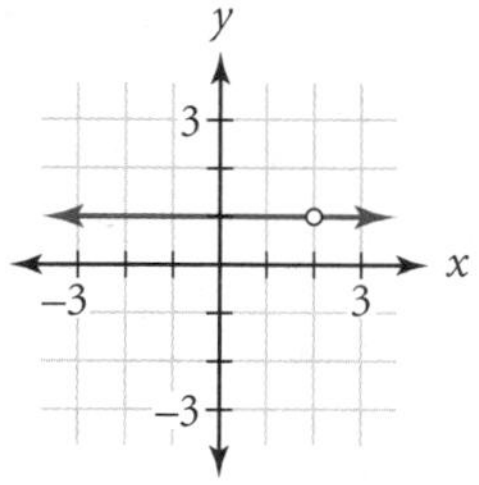

c.

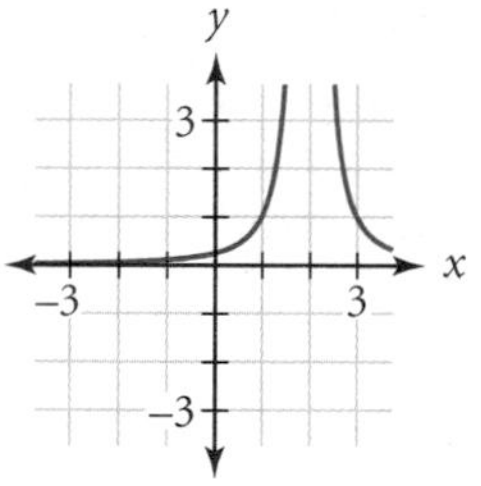

d.

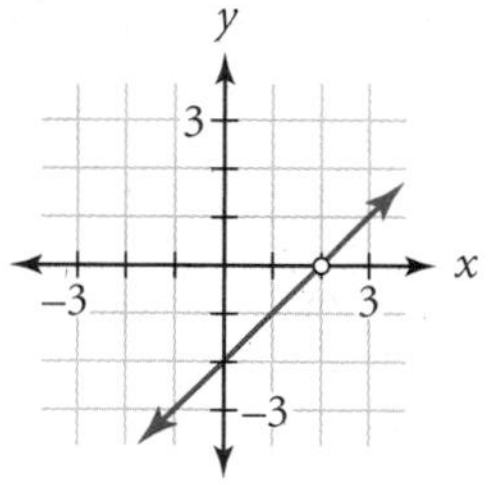

A. $y = \dfrac{1}{(x-2)^2}$ **B.** $y = \dfrac{1}{x-2}$ **C.** $y = \dfrac{(x-2)^2}{x-2}$ **D.** $y = \dfrac{x-2}{x-2}$

Step 2 Have each group member choose one of the graphs below. Find a rational function equation for your graph, and write a few sentences that explain the appearance of your graph.

a.

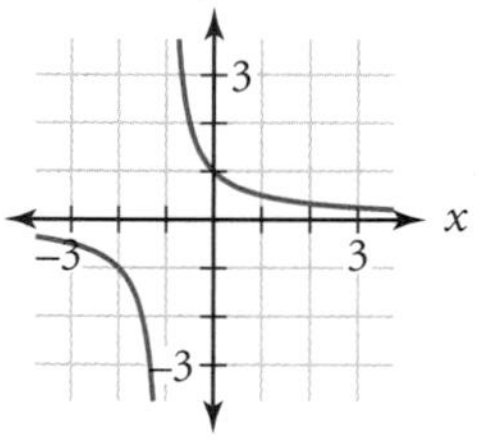

b.

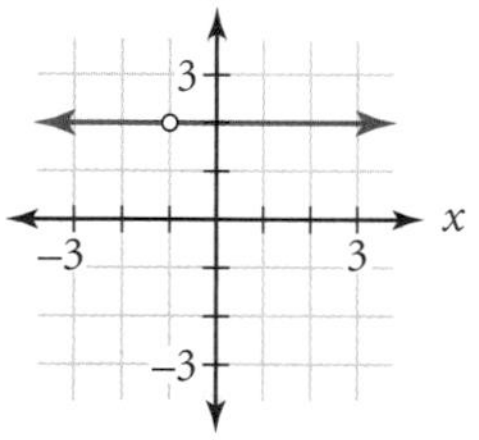

c.

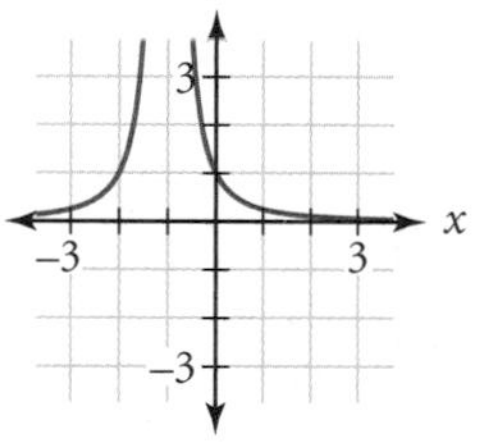

d.

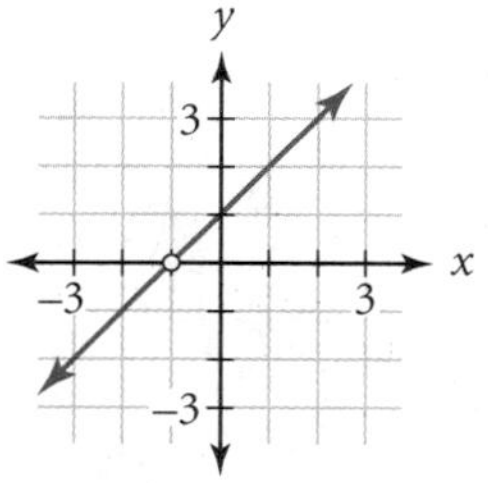

Step 3 Write a paragraph explaining how you can predict holes and asymptotes based on equations and how you can use these features in a graph to write an equation.

You have already studied the function $y = \frac{1}{x}$. You have seen transformations of this function and observed some of the peculiarities involving graphs of more complicated rational functions. You have seen that $y = \frac{1}{x}$ has both horizontal and vertical asymptotes. What do you think the graph would look like if you added x to $\frac{1}{x}$? Reflect on this question for a moment. Then, graph $y = x + \frac{1}{x}$ on your calculator.

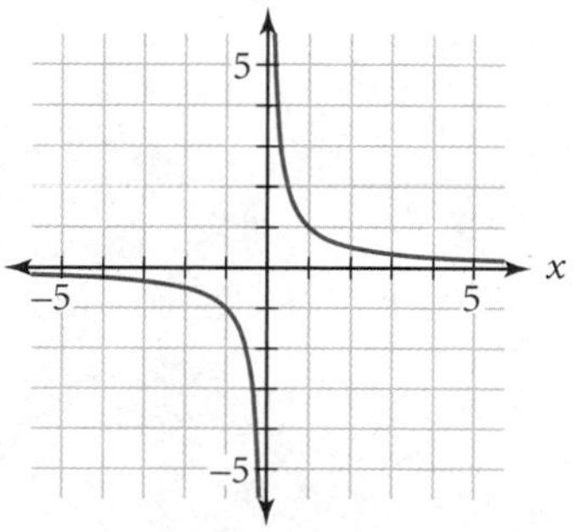

EXAMPLE A | Describe the graph of $y = x + \frac{1}{x}$.

▸ Solution

There is a vertical asymptote, or hole, at $x = 0$ because the function is undefined when $x = 0$. The graph shows that the feature at $x = 0$ is a vertical asymptote. The values of $\frac{1}{x}$ are added to the values of x. This means that as the absolute value of x increases, the absolute value of $y = x + \frac{1}{x}$ also increases. If you graph the curve and the line $y = x$, you see that the function approaches the line $y = x$ instead of the x-axis. The line $y = x$ is called a **slant asymptote** because it is a nonhorizontal, nonvertical line that the function approaches as x-values increase in the positive direction and as x-values decrease in the negative direction. This slant asymptote describes the end behavior of this function.

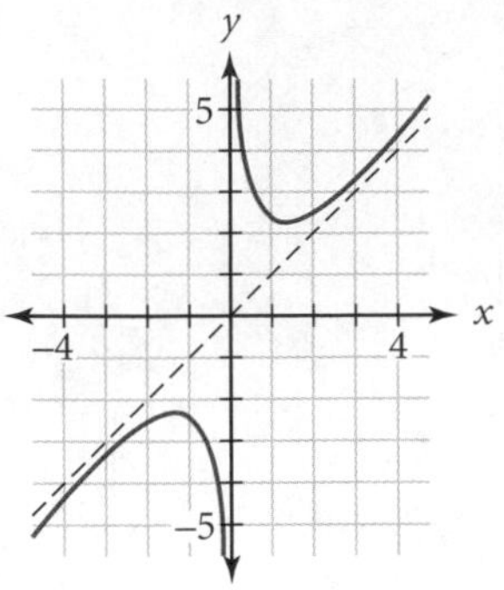

Often you can determine the features of a rational function graph before you actually graph it. Values that make the denominator or numerator equal to 0 give you important clues about the appearance of the graph.

EXAMPLE B | Describe the features of the graph of $y = \frac{x^2 + 2x - 3}{x^2 - 2x - 8}$.

▸ Solution

Features of rational functions are apparent when the numerator and denominator are factored.

$$y = \frac{x^2 + 2x - 3}{x^2 - 2x - 8} = \frac{(x + 3)(x - 1)}{(x - 4)(x + 2)}$$

No factors occur that are common to both the numerator and denominator, so there are no holes.

If $x = 4$ or $x = -2$, then the denominator is 0, and the function is undefined. There are vertical asymptotes at these values.

If $x = -3$ or $x = 1$, then the numerator is 0, so the x-intercepts are -3 and 1.

If $x = 0$, then $y = -0.375$. This is the y-intercept.

To find any horizontal asymptotes, consider what happens to the y-values as the x-values get very far from 0.

x	−100	−1000	−10000	100	1000	10000
y	0.9612441	0.9960129	0.9996001	1.0413603	1.0040131	1.0004001

A table shows that the y-values get closer and closer to 1 as x gets farther from 0. So, $y = 1$ is a horizontal asymptote.

A graph of the function confirms these features.

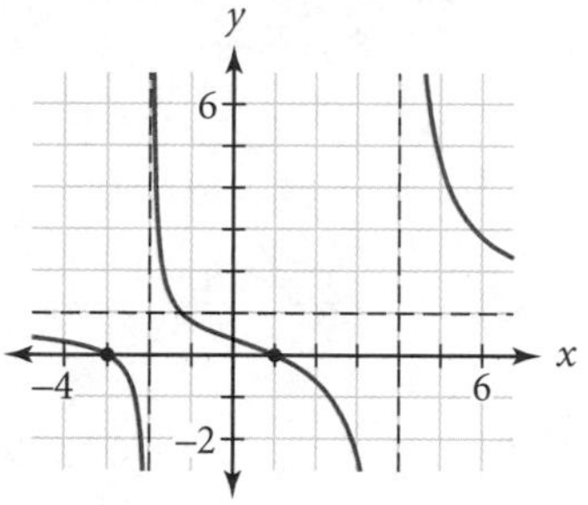

Rational functions can be written in different forms. The factored form is convenient for locating asymptotes and intercepts. And you saw in the previous lesson how rational functions can be written in a form that shows you clearly how the parent function has been transformed. Converting from one form to another can sometimes be challenging.

EXAMPLE C Rewrite $f(x) = \frac{3}{x-2} - 4$ in rational form.

▶ Solution

The original form shows that this function is related to the parent function, $f(x) = \frac{1}{x}$. It has been stretched vertically by a scale factor of 3 and translated right 2 units and down 4 units. To change to rational form, you must add the two parts to form a single fraction.

$$f(x) = \frac{3}{x-2} - 4$$ Original equation.

$$f(x) = \frac{3}{x-2} - \frac{4}{1} \cdot \frac{x-2}{x-2}$$ Create a common denominator of $(x - 2)$.

$$f(x) = \frac{3}{x-2} - \frac{4x-8}{x-2}$$ Rewrite second fraction.

$$f(x) = \frac{3 - 4x + 8}{x-2}$$ Combine the two fractions.

$$f(x) = \frac{11 - 4x}{x-2}$$ Add like terms.

No common factors occur in both the numerator and denominator, so there are no holes.

In this form you can see that $x = 2$ is a vertical asymptote because this value makes the denominator equal to zero. You can also see that the x-intercept is $\frac{11}{4}$, as this x-value makes the numerator equal to zero. Evaluating the function for large values shows that $y = -4$ is a horizontal asymptote.

x	−10000	−1000	−100	100	1000	10000
y	−4.0002999	−4.0029940	−4.0294118	−3.9693878	−3.9969940	−3.9996999

EXERCISES

Practice Your Skills

1. Rewrite each rational expression in factored form.

a. $\dfrac{x^2 + 7x + 12}{x^2 - 4}$

b. $\dfrac{x^3 - 5x^2 - 14x}{x^2 + 2x + 1}$

2. Identify the vertical asymptotes of each equation.

a. $y = \dfrac{x^2 + 7x + 12}{x^2 - 4}$

b. $y = \dfrac{x^3 - 5x^2 - 14x}{x^2 + 2x + 1}$

3. Rewrite each expression in rational form (as the quotient of two polynomials).

a. $3 + \dfrac{4x - 1}{x - 2}$

b. $\dfrac{3x + 7}{2x - 1} - 5$

4. Graph each equation on your calculator, and sketch the graph on your paper. Use a friendly graphing window. Indicate any holes on your sketches.

a. $y = \dfrac{5 - x}{x - 5}$

b. $y = \dfrac{3x + 6}{x + 2}$

c. $y = \dfrac{(x + 3)(x - 4)}{x - 4}$

d. What causes a hole to appear in the graph?

Reason and Apply

5. Write an equation of each graph.

a.

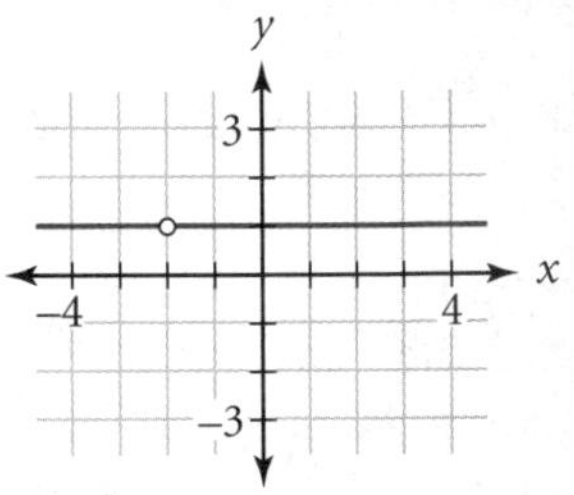

b.

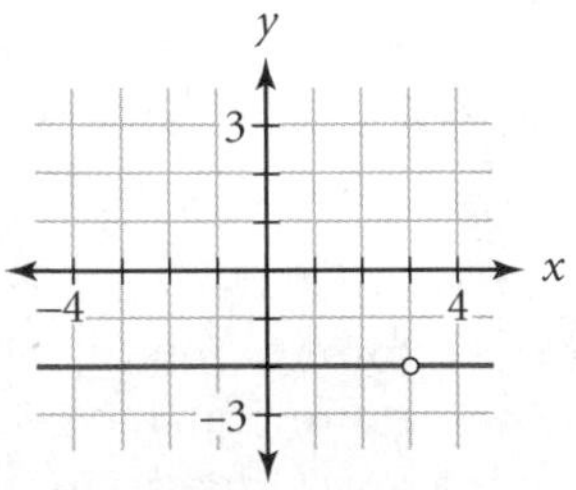

c.

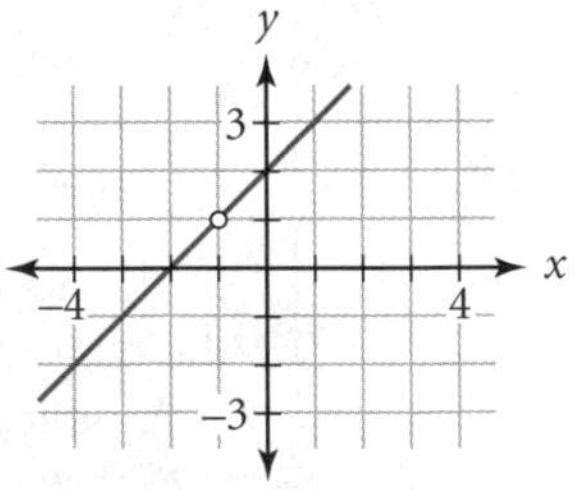

6. Graph $y = -x + \frac{4}{x - 3}$.

a. Describe the end behavior of the graph. (Use a large graphing window.)

b. Describe the behavior of the graph near $x = 3$.

c. Rewrite $y = -x + \frac{4}{x - 3}$ in rational function form.

d. Factor your answer from 6c. How do the factors show up in the graph?

7. Graph each function on your calculator. List all holes and asymptotes, including slant asymptotes.

a. $y = x - 2 + \frac{1}{x}$ **b.** $y = -2x + 3 + \frac{2}{x - 1}$ **c.** $y = 7 + \frac{8 - 4x}{x - 2}$

d. Convert each equation in 7a–c to rational form. Use the rational function form to help you verify the location of the holes and vertical asymptotes you found.

8. The two graphs below show the same function.

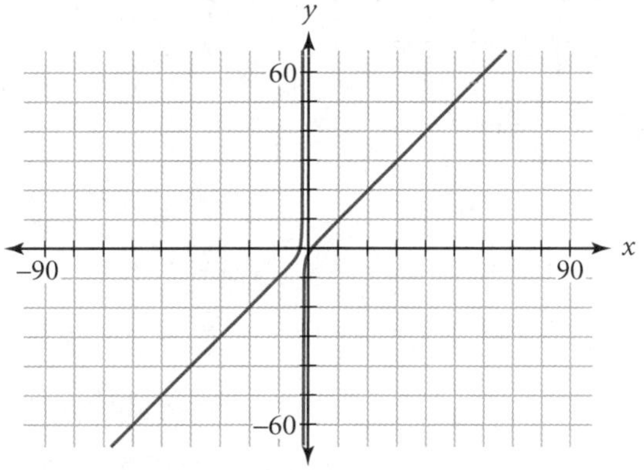

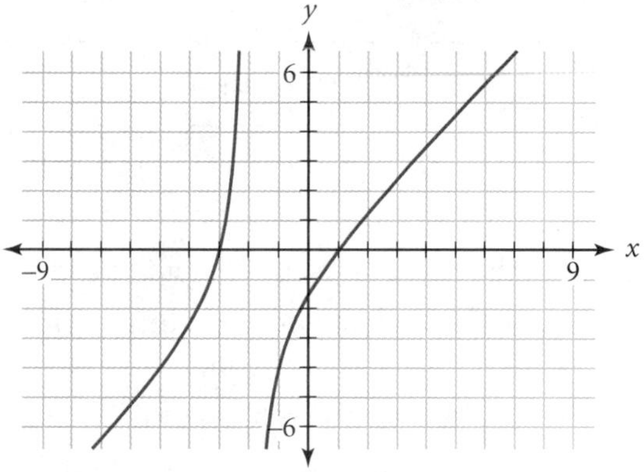

a. List as many important facts as you can about the graph.

b. Find the equation of the slant asymptote in the first graph.

c. Give an example of an equation with asymptote $x = -2$.

d. Name a polynomial with zeros at $x = -3$ and $x = 1$.

e. Write an equation of the function shown in the graphs. Graph the function to check your answer.

9. The two graphs below show the same function. Write an equation of this function.

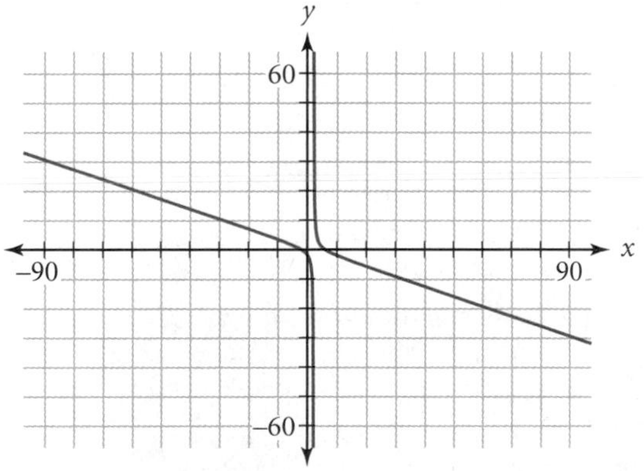

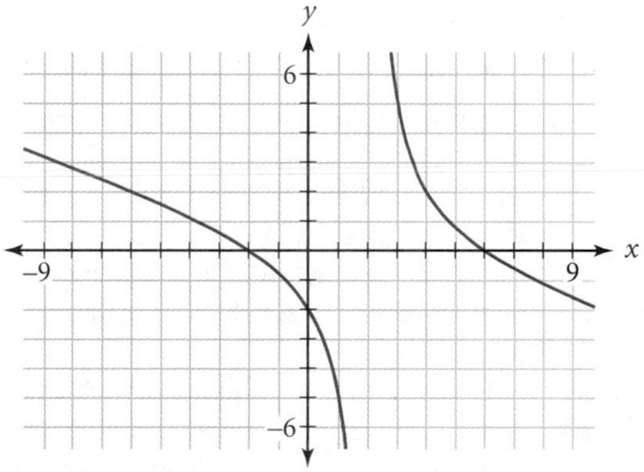

10. Consider the equation $y = \frac{(x - 1)(x + 4)}{(x - 2)(x + 3)}$.

a. Describe the features of the graph of this function.

b. Describe the end behavior of the graph.

c. Sketch the graph.

11. Solve. Give exact solutions. One strategy is to multiply both sides of each equation by $x - 1$ so no fractions remain. Solve the resulting equations. Check your answers by substituting in the original equation.

a. $\frac{2}{x-1} + x = 5$

b. $\frac{2}{x-1} + x = 2$

12. A machine drill removes a core from any cylinder. Suppose you want to hold constant the amount of material left after the core is removed. The table below compares the height and radius needed for the volume of the hollow cylinder to remain constant.

Radius x	2.5	3.0	3.5	4.0	4.5	5.0	5.5	6.0	6.5
Height h	56.6	25.5	15.4	10.6	7.8	6.1	4.9	4.0	3.3

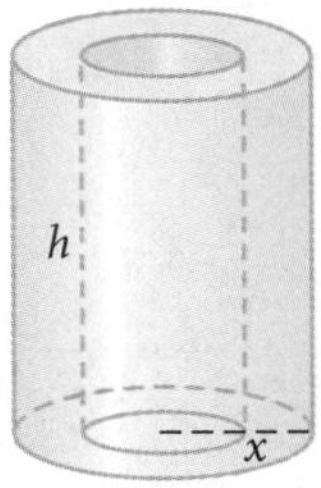

a. Plot the data points, (x, h), and draw a smooth curve through them.

b. Explain what happens to the height of the figure as the radius gets smaller. How small can x be?

c. Write a formula for volume of the hollow cylinder, V, in terms of x and h.

d. Solve the formula in 12c for h to get a function that describes the height as a function of the radius.

e. What is the constant volume?

13. **APPLICATION** The functional response curve given by the function $y = \frac{60x}{1 + .625x}$ models the number of moose attacked by wolves as the concentration of moose in an area increases. In this model, x represents the number of moose per 1,000 km^2, and y represents the number of moose attacked in 100-day periods.

a. How many moose are attacked each 100 days if a herd contains 260 moose in a land preserve with area 1,000 km^2?

b. Graph the function.

c. What are the asymptotes of this function?

d. Explain the significance of the asymptotes in this problem.

Environmental Connection Ecologists often look for a mathematical model to describe the interrelationship of organisms. In the late 1950s C. S. Hollings, a Canadian researcher, came up with an equation for what he called a Type II functional response curve. The equation describes the relationship between the number of prey attacked by a predator and the density of the prey population. For example, wolves will increase their number through reproduction as moose population density increases. Eventually, wolf populations stabilize at about 40 per 1,000 km^2, which is the optimum size of their range based on their ability to defend their territories. The functional response curve applies to all species of animals. It could be larvae-eating insects and mosquito larvae, fishermen and a particular species of fish, or pandas and the bamboo forest.

Review

14. (Lesson 9.5) Find the points of intersection, if any, of the circle with center (2, 1) and radius 5, and the line $x - 7y + 30 = 0$.

15. (Lesson 9.6) A 500-g jar of mixed nuts contains 30% cashews, 20% almonds, and 50% peanuts.

a. How many grams of cashews must you add to the mixture to increase the percentage of cashews to 40%?

b. After adding the cashews in 15a, what is the percentage of almonds and peanuts?

c. How many grams of almonds must you add to the original mixture to make the percentage of almonds the same as the percentage of cashews? Now what is the percentage of each type of nut?

16. (Lesson 6.4) Solve each quadratic equation.

a. $2x^2 - 5x - 3 = 0$ **b.** $x^2 + 4x - 4 = 0$ **c.** $x^2 + 4x + 1 = 0$

PROJECT

GOING DOWNHILL FAST

Design an investigation to determine the relationship between the angle of elevation of a long tube and the time it takes a ball to travel the length of the tube.

Your project should include

- A description of your investigation and the data you collect.
- A graph that shows the relationship between the tube's angle and the ball's time.
- The domain and range of the relationship. Include a description of what happens to the dependent variable as the independent variable approaches the extreme values of the domain.
- A description of the features of the graph of the relationship. Attach real-world meaning to each feature.

LESSON

9.8 Operations with Rational Expressions

Images
split the truth
in fractions.
DENISE LEVERTOV

In this lesson you will review adding, subtracting, multiplying, and dividing rational expressions. In the previous lesson you combined a rational expression with a single constant or variable by finding a common denominator. That process was much like adding a fraction to a whole number. Likewise, all the other arithmetic operations you will perform with rational expressions have their counterparts in working with fractions. Keeping the operations with fractions in mind will help you understand the procedures.

Recall that to add $\frac{7}{12} + \frac{3}{10}$, you need a common denominator. The smallest number that has both 12 and 10 as factors is 60. So, use 60 as the common denominator.

$\frac{7}{12} + \frac{3}{10}$	Original expression.
$\frac{7}{12} \cdot \frac{5}{5} + \frac{3}{10} \cdot \frac{6}{6}$	Multiply each fraction by the equivalent of 1 to get a denominator of 60.
$\frac{35}{60} + \frac{18}{60}$	Multiply.
$\frac{53}{60}$	Add.

You could use other numbers, such as 120, as a common denominator, but using the least common denominator keeps the numbers as small as possible and eliminates some of the reducing afterward. Recall that you can find the least common denominator by factoring the denominators to see what factors they share and what is unique to each one. In this example, 12 factors to $2 \cdot 2 \cdot 3$, and 10 factors to $2 \cdot 5$. The least common denominator must include factors that multiply to give each denominator, with no extras. So, in this case you need two 2's, a 3, and a 5. You can use this same process to add two rational expressions.

EXAMPLE A

Add rational expressions to rewrite the right side of this equation as a single rational expression in factored form.

$$y = \frac{x-3}{(x+1)(x-2)} + \frac{2x+1}{(x+2)(x-2)}$$

▶ Solution

First, identify the least common denominator. It must contain all of the factors of each denominator. The factors $(x + 1)$ and $(x - 2)$ are needed to create the first denominator, and an additional $(x + 2)$ is needed for the second denominator. So, use the common denominator $(x + 1)(x - 2)(x + 2)$.

$y = \frac{x-3}{(x+1)(x-2)} + \frac{2x+1}{(x+2)(x-2)}$	Original equation.
$y = \frac{x-3}{(x+1)(x-2)} \cdot \frac{(x+2)}{(x+2)} + \frac{2x+1}{(x+2)(x-2)} \cdot \frac{(x+1)}{(x+1)}$	Multiply each fraction by the equivalent of 1 to get a common denominator.
$y = \frac{x^2-x-6}{(x+1)(x-2)(x+2)} + \frac{2x^2+3x+1}{(x+2)(x-2)(x+1)}$	Multiply and expand the numerators.
$y = \frac{3x^2+2x-5}{(x+1)(x-2)(x+2)}$	Add like terms in the numerator.
$y = \frac{(3x+5)(x-1)}{(x+1)(x-2)(x+2)}$	If possible, factor the numerator.

In this case, the numerator factors. Often, however, it will not.

The advantage of expressing a rational function as a single rational expression in factored form is that you can easily identify some features of the graph. An x-value that makes the expression equal to zero will be an x-intercept. In this case, 1 and $-\frac{5}{3}$ are the x-intercepts. An x-value that leads to division by zero makes the expression undefined. This results in a hole (when the associated factor appears in the numerator and denominator) or a vertical asymptote (when the factor appears only in the denominator). The graph of the equation above has vertical asymptotes $x = -1$, $x = 2$, and $x = -2$, as shown.

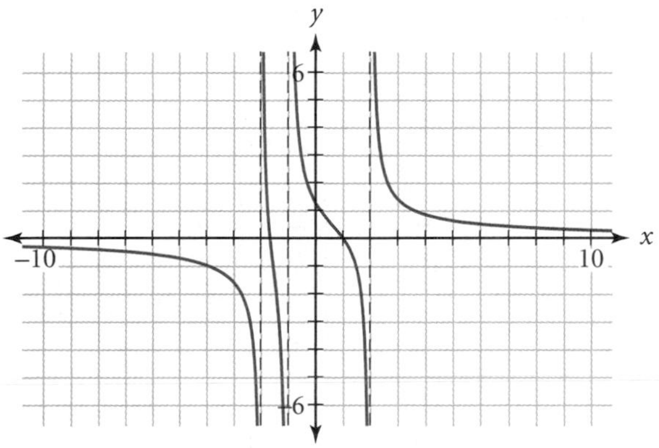

Subtraction of rational expressions is much like addition. You must begin by finding a common denominator.

EXAMPLE B Find any x-intercepts, vertical asymptotes, or holes in the graph of

$$y = \frac{x+2}{(x-3)(x+4)} - \frac{5}{x+1}$$

▶ Solution Begin by finding a common denominator so that you can write the expression on the right side as a single rational expression in factored form. The common denominator is $(x-3)(x+4)(x+1)$.

$$y = \frac{(x+2)}{(x-3)(x+4)} \cdot \frac{(x+1)}{(x+1)} - \frac{5}{(x+1)} \cdot \frac{(x-3)(x+4)}{(x-3)(x+4)}$$ Find a common denominator.

$$y = \frac{x^2+3x+2}{(x-3)(x+4)(x+1)} - \frac{5(x^2+x-12)}{(x+1)(x-3)(x+4)}$$ Expand the numerators.

$$y = \frac{x^2+3x+2-(5x^2+5x-60)}{(x-3)(x+4)(x+1)}$$ Write with a single denominator.

$$y = \frac{x^2+3x+2-5x^2-5x+60}{(x-3)(x+4)(x+1)}$$ Eliminate parentheses.

$$y = \frac{-4x^2-2x+62}{(x-3)(x+4)(x+1)}$$ Combine like terms.

Check to see whether the numerator factors. There is a common factor of -2 in the numerator, so you may rewrite it as $y = \frac{-2(2x^2+x-31)}{(x-3)(x+4)(x+1)}$. The equation is undefined when $x = 3$, $x = -4$, or $x = -1$, so these are the vertical asymptotes. Because the numerator cannot be factored further, the x-intercepts are not obvious. But you can use the quadratic formula to find the values that make the expression in the numerator equal to zero.

$$x = \frac{-1 \pm \sqrt{1-4(2)(-31)}}{2(2)}$$

$$x \approx 3.695 \quad \text{or} \quad x \approx -4.195$$

Check the graph to confirm the location of the vertical asymptotes and x-intercepts.

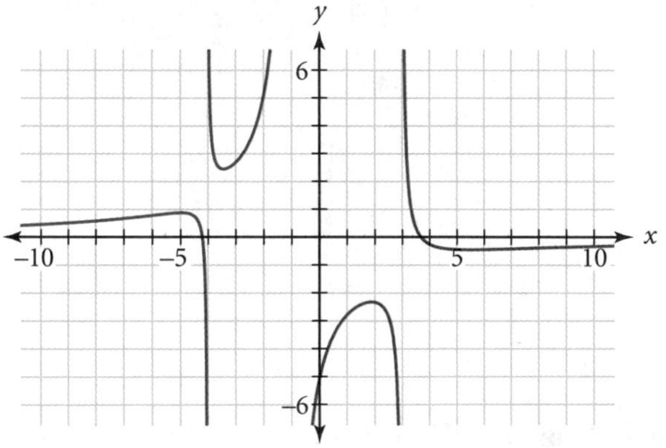

To multiply and divide rational expressions, you don't need to find a common denominator. Look at the following problems to remind yourself how multiplication and division work with fractions.

$$\frac{5}{12} \cdot \frac{3}{4} = \frac{5 \cdot 3}{12 \cdot 4} = \frac{5}{4 \cdot 4} = \frac{5}{16}$$ Multipy straight across. Reduce common factors if any appear.

$$\frac{\frac{2}{5}}{\frac{6}{7}} = \frac{2}{5} \cdot \frac{7}{6} = \frac{2 \cdot 7}{5 \cdot 6} = \frac{7}{5 \cdot 3} = \frac{7}{15}$$ To divide, multiply by the reciprocal of the denominator.

In multiplication and division problems with rational expressions, it is best to factor all expressions first. This will make it easy to reduce common factors and identify x-intercepts, holes, and vertical asymptotes.

EXAMPLE C Multiply $\frac{(x+2)}{(x-3)} \cdot \frac{(x+1)(x-3)}{(x-1)(x^2-4)}$.

▶ Solution

$$\frac{(x+2)}{(x-3)} \cdot \frac{(x+1)(x-3)}{(x-1)(x-2)(x+2)}$$ Factor any expressions that you can. The expression (x^2-4) factors to $(x-2)(x+2)$.

$$\frac{(x+2)(x+1)(x-3)}{(x-3)(x-1)(x-2)(x+2)}$$ Combine the two expressions.

$$\frac{\cancel{(x+2)}(x+1)\cancel{(x-3)}}{\cancel{(x-3)}(x-1)(x-2)\cancel{(x+2)}}$$ Reduce common factors.

$$\frac{(x+1)}{(x-1)(x-2)}$$ Rewrite.

Although $\frac{(x+2)}{(x-3)} \cdot \frac{(x+1)(x-3)}{(x-1)(x^2-4)}$ and $\frac{(x+1)}{(x-1)(x-2)}$ are equivalent expressions, the graph of $y = \frac{(x+2)}{(x-3)} \cdot \frac{(x+1)(x-3)}{(x-1)(x^2-4)}$ differs slightly from the graph of $y = \frac{(x+1)}{(x-1)(x-2)}$. The original multiplication expression had two factors that were eventually reduced, $(x-3)$ and $(x+2)$. When a factor can be reduced, it represents a hole rather than an asymptote. So, the graph of $y = \frac{(x+2)}{(x-3)} \cdot \frac{(x+1)(x-3)}{(x-1)(x^2-4)}$ has holes at $x = 3$ and $x = -2$, but the graph of $y = \frac{(x+1)}{(x-1)(x-2)}$ does not. Either graph has vertical asymptotes $x = 1$ and $x = 2$ and x-intercept -1.

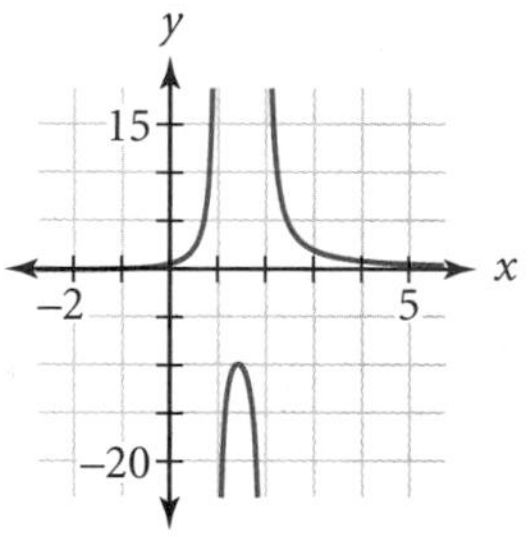

Multiplication and division with rational expressions will give you a good chance to practice your factoring skills.

EXAMPLE D Divide $\dfrac{\frac{x^2-1}{x^2+5x+6}}{\frac{x^2-3x+2}{x+3}}$.

▶ Solution

$$\frac{x^2-1}{x^2+5x+6} \cdot \frac{x+3}{x^2-3x+2}$$ Invert the second fraction and multiply.

$$\frac{(x+1)(x-1)}{(x+2)(x+3)} \cdot \frac{(x+3)}{(x-2)(x-1)}$$ Factor all expressions.

$$\frac{(x+1)(x-1)(x+3)}{(x+2)(x+3)(x-2)(x-1)}$$ Multiply.

$$\frac{(x+1)}{(x+2)(x-2)}$$ Reduce all common factors.

Rational expressions like those in Example D can look imposing. However, the rules are the same as for regular fraction arithmetic. Work carefully and stay organized. Check the graph of the original problem and any step along the way to see whether you have made an error. After you have reduced common factors, the graphs should be identical except for holes.

EXERCISES

Practice Your Skills

1. Factor each expression completely, and reduce common factors.

a. $\dfrac{x^2+2x}{x^2-4}$

b. $\dfrac{x^2-5x+4}{x^2-1}$

c. $\dfrac{3x^2-6x}{x^2-6x+8}$

d. $\dfrac{x^2+3x-10}{x^2-25}$

2. What is the least common denominator for each pair of rational expressions?

a. $\dfrac{x}{(x+3)(x-2)}, \dfrac{x-1}{(x-3)(x-2)}$

b. $\dfrac{x^2}{(2x+1)(x-4)}, \dfrac{x}{(x+1)(x-2)}$

c. $\dfrac{2}{x^2-4}, \dfrac{x}{(x+3)(x-2)}$

d. $\dfrac{x+1}{(x-3)(x+2)}, \dfrac{x-2}{x^2+5x+6}$

3. Add each pair of rational expressions.

a. $\dfrac{x}{(x+3)(x-2)} + \dfrac{x-1}{(x-3)(x-2)}$

b. $\dfrac{2}{x^2-4} + \dfrac{x}{(x+3)(x-2)}$

c. $\dfrac{x+1}{(x-3)(x+2)} + \dfrac{x-2}{x^2+5x+6}$

d. $\dfrac{2x}{(x+1)(x-2)} + \dfrac{-3}{x^2-1}$

4. Multiply or divide as indicated. Reduce any common factors to simplify.

a. $\dfrac{x+1}{(x+2)(x-3)} \cdot \dfrac{x^2-4}{x^2-x-2}$

b. $\dfrac{x^2-16}{x+5} \div \dfrac{x^2+8x+16}{x^2+3x-10}$

c. $\dfrac{x^2+7x+6}{x^2+5x-6} \cdot \dfrac{2x^2-2x}{x+1}$

d. $\dfrac{\dfrac{x+3}{x^2-8x+15}}{\dfrac{x^2-9}{x^2-4x-5}}$

Reason and Apply

5. Rewrite as a single rational expression by first adding or subtracting in the numerators and denominators, then following Example D.

a. $\dfrac{1-\dfrac{x}{x+2}}{\dfrac{x+1}{x^2-4}}$

b. $\dfrac{\dfrac{1}{x-1}+\dfrac{1}{x+1}}{\dfrac{x}{x-1}-\dfrac{x}{x+1}}$

6. Graph $y = \frac{x+1}{x^2 - 7x - 8} - \frac{x}{2(x-8)}$ on your calculator.

a. Based on your calculator's graph, list all asymptotes, holes, and intercepts.

b. Rewrite the right side of the equation as a single rational expression.

c. Use your answer from 6b to verify your observations in 6a. Explain.

7. Consider the equation $y = \frac{x-3}{x^2-4}$.

a. Without graphing, identify the zeros and asymptotes of the graph of the equation. Explain your methods.

b. Verify your answers by graphing the function.

8. Consider the equation $y = x + 1 + \frac{1}{x-1}$.

a. Without graphing, name the asymptotes of the function.

b. Rewrite the equation as a single rational function.

c. Sketch a graph of the function without using your calculator.

d. Confirm your work by graphing the function with your calculator.

9. Graph the equation $xy = 4$. Describe this graph as it relates to one of the conic sections.

10. **APPLICATION** How long should a traffic light stay yellow before turning red? One study suggests that for a car approaching a 70-ft-wide intersection under normal driving conditions, the length of time a light should stay yellow, in seconds, is given by the equation $y = 1 + \frac{v}{30} + \frac{90}{v}$, where v is the velocity of an approaching car in feet per second, and y is the suggested length of the yellow light in seconds.

a. Rewrite the equation in rational function form.

b. Enter into your calculator the original equation as Y1 and your simplified equation as Y2. Using the table feature, check that the values of both functions are the same. What does this tell you?

c. If the speed limit at a particular intersection is 45 mi/h, how long should the light stay yellow?

d. If cars typically travel at speeds ranging from 20 mi/h to 55 mi/h through that intersection, what is the possible range of lengths of time that a light should stay yellow?

CareerConnection Sometimes it may seem that the length of time a light stays yellow is too short for a driver to safely stop. That's when traffic engineers get to work. Traffic engineers time traffic signals to minimize the "dilemma zone." The dilemma zone occurs when a driver has to decide whether to brake hard to stop or accelerate to get through the intersection. Either decision can be risky. It is estimated that 22% of traffic accidents are caused by drivers running red lights. On the other hand, an abrupt stop can lead to a rear-end collision. Some towns and cities have installed red-light cameras at many of their intersections. When a car runs a red light, the camera takes a picture, and the owner of the car is mailed a ticket.

Review

11. (Lesson 9.6) The graph at right is the image of $y = \frac{1}{x}$ after a transformation.

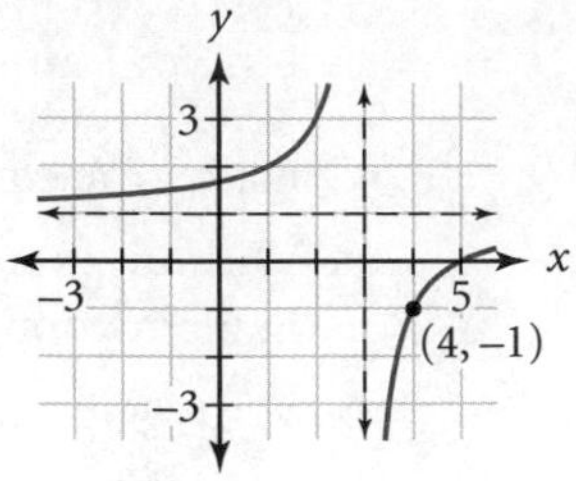

a. Write an equation of each asymptote.

b. What translations are involved in transforming $y = \frac{1}{x}$ to its image?

c. The point $(4, -1)$ is on the image. What is the vertical scale factor in the transformation?

d. Write an equation of the image.

e. Name the intercepts.

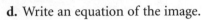

12. (Lesson 9.7) Consider the function $f(x) = \frac{2x^2 - 2}{x^2 - x - 12}$.

a. Write the function in factored form. Identify all intercepts and asymptotes.

b. Evaluate $f(30)$ and $f(-30)$.

c. Use the information from 12a and b to sketch a graph of $f(x)$.

d. Graph the function on your calculator to check your sketch.

13. (Lesson 9.7) A rational function, $h(x)$, has x-intercept 5 and y-intercept 5, vertical asymptotes $x = -5$ and $x = 1$, and horizontal asymptote $y = 0$.

a. Sketch a graph of $h(x)$ showing its general characteristics and the specific given information.

b. Write an equation for $h(x)$.

c. Graph your function from 13b on your calculator, and compare it to your sketch in 13a.

14. (Lesson 1.5) If you invest \$1,000 at 6.5% interest for five years, how much interest do you earn in each of these scenarios?

a. The interest is compounded annually.

b. The interest is compounded monthly.

c. The interest is compounded weekly.

d. The interest is compounded daily.

E X P L O R A T I O N
with Microsoft® Excel

Properties of the Ellipse

Download the Chapter 9 Excel Exploration from the website at www.keycollege.com/online or from your Student CD.

Whispering galleries such as the one in our nation's capital are built on the principles of the ellipse. You will explore this reflective property of the ellipse.

On the spreadsheet you will find complete instructions and questions to consider as you explore.

Activity
Mixture Problems

Excel skills required

- Data fill
- Name a cell
- Format a cell as percent

Pharmacists frequently must mix different concentrations of the same substance, and they need to know how much of each is needed to produce a required concentration. In this Activity you will explore mixing different concentrations. Assume that you have a fixed amount of Solution A. and explore the results of varying the amount of Solution B.

You can find instructions for the required Excel skills in the Excel Notes at www.keycollege.com/online or on your Student CD.

Launch Excel.

Step 1 Set up labels and some of the initial values as shown.

	A	B	C	D	E	F
1	Solution A		Solution B		Result	
2	Percent	Volume	Percent	Volume	Percent	Volume
3	0.3	100	1	50		

Step 2 Click on Cell E3. Use Format/Cells/Number/Percentage to set the number of decimal places at 2.

Step 3 In Cell E3 type `=($A$3*$B$3+C3*D3)/($B$3+D3)`. This formula gives you the concentration that results when A and B are mixed. Note: You want the values in Cells A3 and B3 to remain fixed when you generate the table later.

Step 4 In Cell F3 type `=$B$3+D3`.

This completes the initial values and formulas. Your screen should look like this one.

	A	B	C	D	E	F
1	Solution A		Solution B		Result	
2	Percent	Volume	Percent	Volume	Percent	Volume
3	30.00%	100	100.00%	50	53.33%	150

Leave Solution A alone. You will use Cell A3 to change the values in the other columns.

Step 5 In Cell C4 type `=C3`.

Step 6 In Cell D4 type `=D3+10`.

Step 7 Data-fill Cells E3 and F3 to fill Cells E4 and F4. Then data-fill Cells C4 to F4 to at least Row 73.

Questions

Answer these questions by using your spreadsheet to model mixing volumes from Solution A and Solution B in a new container.

1. If you mix 50 cc of Solution B with 100 cc of Solution A, what will be the percentage of concentrate in the mixture? What will be the volume of the resulting mixture?
2. If you mix 200 cc of Solution B with 100 cc of Solution A, what will be the percentage of concentrate in the mixture? What will be the volume of the resulting mixture?
3. How much of Solution B must you add to 100 cc of Solution A to get an 80% concentrate? What will be the resultant volume?
4. How much of Solution B must you add to 100 cc of Solution A to get an 85% concentrate? What will be the resultant volume?
5. How much of Mixture B must you add to 100 cc of Mixture A to get a 99.5% concentrate?
6. How much of Solution B must you add to 100 cc of Mixture A to get a 99.9% concentrate?
7. Change the percentage of Solution A to 50%. With 100 cc of Solution A at 50%, how much of Solution B is required to get a 75% concentrate?
8. Suppose that Solution A is 100 cc pure water. How much of Solution B is required to get a 50% concentrate? An 80% concentrate?

Extension

9. Create a graph with the volume of Solution B as the independent variable and the percent of the mixture as the dependent variable. What sort of function does the graph represent? What is its equation?
10. Repeat Questions 1–6 assuming a 90% concentration in Solution B. Explain any difficulties you have in solving these problems.

CHAPTER 9 REVIEW

In this chapter you saw some special relations described as a set of points, or **locus,** that satisfied some criteria. These relations are called **conic sections** because the shapes can be formed by slicing a double cone at various angles. The simplest of the conic sections is the **circle,** the set of points at a fixed distance from a fixed point called the **center.** Closely related to the circle is the **ellipse.** The ellipse can be defined as the set of points such that the sum of their distances from two fixed points, the **foci,** is a constant. The **parabola** is another conic section, which you studied in earlier chapters. It can be defined as the set of points that are equally distant from a fixed point called the **focus** and a fixed line called the **directrix.** The last of the conic sections is the **hyperbola.** The definition of a hyperbola is similar to the definition of an ellipse, except that the difference between the distances from the foci remains constant. Each of these conics can be either vertically oriented or horizontally oriented. You learned how to solve systems of nonlinear equations to find the intersections of two conic sections.

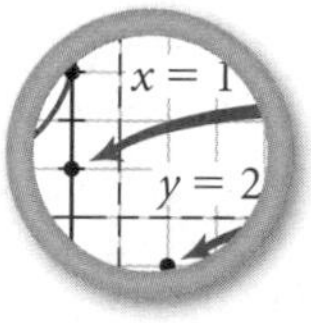

You were also introduced to **rational functions** in the form $y = \frac{p(x)}{q(x)}$, where $p(x)$ and $q(x)$ are polynomial expressions, and $q(x) \neq 0$. The graphs of rational functions can contain special features, such as vertical asymptotes and **holes** at values where the denominator is 0, when the denominator is undefined. Finally, you learned to do arithmetic with rational expressions.

EXERCISES

1. Sketch the graph of each equation. Label foci when appropriate.

   a. $\frac{x-3}{-2} = (y-2)^2$

   b. $x^2 + (y-2)^2 = 16$

   c. $\left(\frac{y+2.5}{3}\right)^2 - \left(\frac{x-1}{4}\right)^2 = 1$

   d. $\left(\frac{x}{5}\right)^2 + 3y^2 = 1$

2. Consider this ellipse.

   a. Write the equation of the graph in standard form.

   b. Write the parametric equations of this graph.

   c. Name the coordinates of the center and foci.

   d. Write the general quadratic form of the equation of this graph.

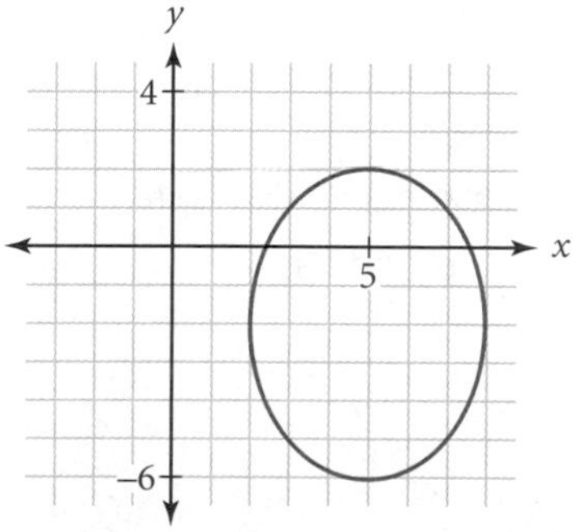

3. Consider this hyperbola.

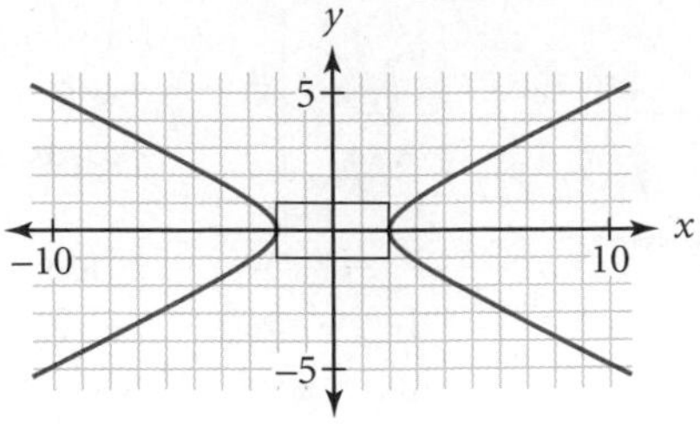

a. Write the equations of the asymptotes of this hyperbola.

b. Write the general quadratic equation of this hyperbola.

c. Write an equation that will give the vertical distance, d, between the asymptote with positive slope and a point on the upper portion of the right branch of the hyperbola as a function of this point's x-coordinate, x.

d. Use the function from 3c to fill in this table. What does this tell you about the relationship between the function and its asymptote?

x	2	10	20	100
d				

4. Solve the system $\begin{cases} x + y = 7 \\ x^2 + y^2 = 25 \end{cases}$

5. Solve the system $\begin{cases} x^2 + 4y^2 = 16 \\ x^2 - 2y^2 = 10 \end{cases}$

6. **APPLICATION** Pure gold is too soft to be used to make jewelry, so gold is always mixed with other metals. 18-karat gold is 75% gold and 25% other metals. How much pure gold must be mixed with 5 oz of 18-karat gold to make a 22-karat (91.7%) gold mixture?

7. Write an equation of each rational function described as a translation of the graph of $y = \frac{1}{x}$.

a. The rational function has asymptotes $x = -2$ and $y = 1$.

b. The rational function has asymptotes $x = 0$ and $y = -4$.

8. Graph $y = \frac{2x - 14}{x - 5}$. Write equations for the horizontal and vertical asymptotes.

9. How can you modify the equation $y = \frac{2x - 14}{x - 5}$ so that the graph of the new equation is the same as the original graph except for a hole at $x = -3$? Verify your new equation by graphing it on your calculator.

10. On her way to work, Ellen drives at a steady speed for the first 2 mi. After glancing at her watch, she drives 20 mi/h faster during the remaining 3.5 mi. How fast does she drive during the two portions of this trip if the total time of her trip is 10 min?

11. Rewrite each expression as a single rational expression in factored form.

a. $\frac{2x}{(x - 2)(x + 1)} + \frac{x + 3}{x^2 - 4}$ **b.** $\frac{x^2}{x + 1} \cdot \frac{3x - 6}{x^2 - 2x}$ **c.** $\frac{x^2 - 5x - 6}{x} \div \frac{x^2 - 8x + 12}{x^2 - 1}$

12. Solve this system of equations algebraically, then confirm your answer by graphing.

$$\begin{cases} x^2 + y^2 = 4 \\ (x + 1)^2 - \frac{y^2}{3} = 1 \end{cases}$$

Mixed Review

13. (Lesson 4.6) Write an equation of the image of the absolute-value function, $y = |x|$, after performing each of the following transformations in order. Sketch a graph of your final equation.

a. Stretch vertically by a factor of 2.

b. Then translate right 4 units.

c. Then translate down 3 units.

14. (Lesson 9.2) Earth's orbit is an ellipse with the Sun at one of the foci. *Aphelion* is the point at which Earth is farthest from the Sun, and *perihelion* is the point at which it is closest to the Sun. The distances from perihelion to the Sun and from the Sun to aphelion are in an approximate ratio of 59:61. (This drawing emphasizes that the Sun is at a focus and is not to scale). If the total distance from aphelion to perihelion along the major axis is about 186 million miles, approximate:

a. The distance from perihelion to the Sun.

b. The distance from aphelion to the Sun.

c. The distance from aphelion to the center.

d. The distance from the center to the Sun.

e. In this ellipse it can be shown that the distance from the Sun to point *P* equals the distance from aphelion to the center. Using this information, find the distance from the center to *P*.

f. Write an equation that models the orbit of the Earth around the Sun.

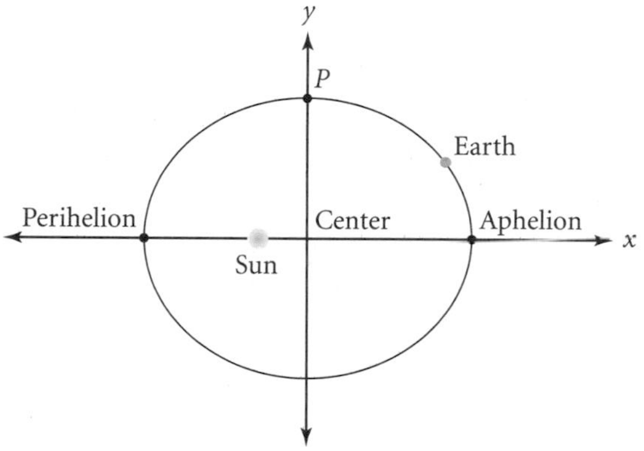

These four circular snow sculptures by British environmental sculptor Andy Goldsworthy (b 1956) were made at the Arctic Circle from bricks of packed snow.

15. (Lesson 7.2) Perform the following matrix computations. If a computation is not possible, explain why.

$$[A] = \begin{bmatrix} -2 & 0 \\ 1 & -3 \end{bmatrix} \qquad [B] = \begin{bmatrix} 0 & -2 \\ 5 & 0 \\ 3 & -1 \end{bmatrix} \qquad [C] = \begin{bmatrix} 1 & -1 \\ 0 & 2 \end{bmatrix}$$

a. $[A]\,[B]$ **b.** $[A] + [B]$ **c.** $[A] - [C]$ **d.** $[B]\,[A]$ **e.** $3[C] - [A]$

16. (Lesson 2.1) Identify each sequence as arithmetic, geometric, or neither. State the next three terms, and then write a recursive formula to generate each sequence.

a. 9, 12, 15, 18, . . . **b.** 1, 1, 2, 3, 5, 8, 13, . . . **c.** $-3, 6, -12, 24, \ldots$

17. (Lesson 8.4) Evan is standing at one corner of the football field when he sees his dog, Spot, start running diagonally across the field as shown. Evan knows that Spot can run to the opposite corner in 15 s. The dimensions of the football field are 100 yd by 52 yd.

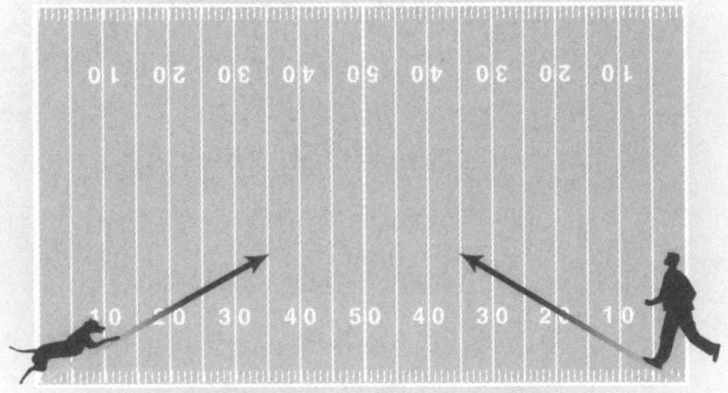

a. What is Spot's rate in yd/s?

b. What is Spot's angle with the horizontal axis?

c. Write equations that model the motion of Spot running from corner to corner.

d. Write equations that show Evan running at the same rate as Spot from one corner to an opposite corner as shown in the diagram.

e. If Spot and Evan both start at the same time, when and where do they meet?

18. (Lesson 1.5) D'Andre surveyed a randomly chosen group of 15 teachers at his school and asked them how many students were enrolled in their 3rd-period classes. Here is the data set he collected.

{27, 29, 18, 34, 42, 38, 34, 33, 25, 28, 45, 35, 32, 19, 36}

a. List the mean, median, and mode.

b. Make a box plot of the data.

c. Calculate the standard deviation. What does this tell you about the data? If the standard deviation were smaller, what would it tell you about the data?

19. (Lesson 9.3) The towers of a parabolic suspension bridge are 400 m apart and reach 50 m above the suspended roadway. The cable is 4 m above the roadway at the halfway point. Write an equation that models the shape of the cable. Assume the origin, (0, 0), is located at the halfway point of the roadway.

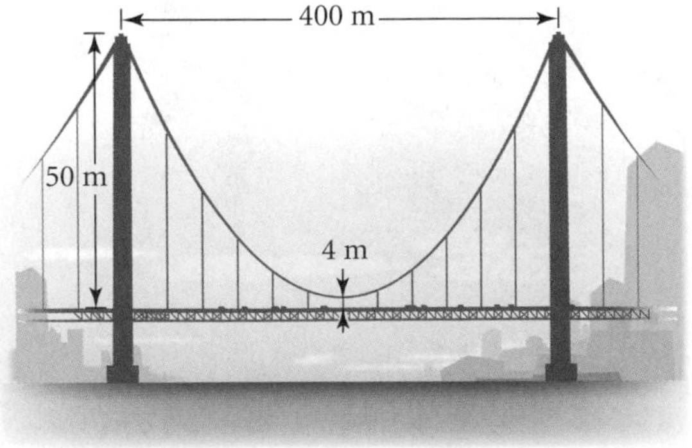

20. (Lesson 5.8) Solve algebraically. Round answers to the nearest hundredth.

a. $4 + 5^x = 18$

b. $12(0.5)^{2x} = 30$

c. $\log_3 15 = \dfrac{\log x}{\log 3}$

d. $\log_6 100 = x$

e. $2 \log x = 2.5$

f. $\log_5 5^3 = x$

g. $4 \log x = \log 16$

h. $\log(5 + x) - \log 5 = 2$

21. (Lesson 3.3) The chart below shows the average fuel efficiency of new U.S. passenger cars.

Year	1970	1975	1980	1985	1990	1995	1998	1999
Mi/gal	14.1	15.1	22.6	26.3	26.9	27.7	28.1	28.2

(*The New York Times Almanac, 2002*)

a. Find a regression line for the data.

b. Find the fuel efficiency predicted by your line for cars in 2010.

c. For how long do you think your line will accurately model fuel efficiency? Why?

22. (Lesson 8.1) The bases on a baseball diamond form a square that is 90 ft on each side. Deanna has a 12-ft lead and can run the remaining 78 ft from first to second base at 28 ft/s. The catcher releases the ball from home plate toward second base 1.5 s after Deanna starts to steal the base, and the ball travels 125 ft/s. Write parametric equations to simulate this situation, and determine whether Deanna is successful. Explain your solution.

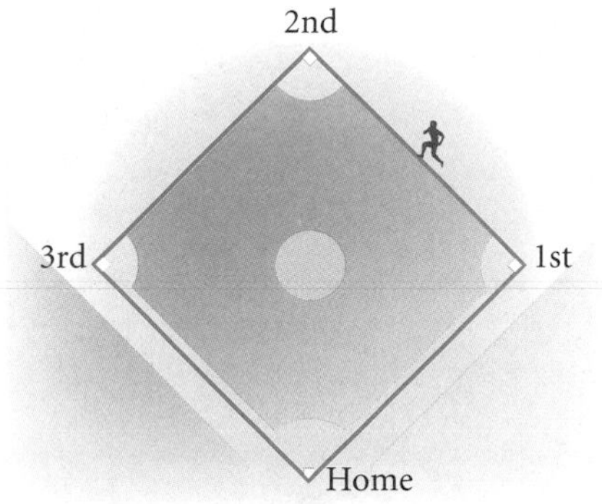

23. (Lesson 6.5) Use $P = 1 + 3i$, $Q = -2 + i$, and $R = 3 - 5i$ to evaluate each expression. Give answers in the form $a + bi$.

a. $P + Q - R$

b. PQ

c. Q^2

d. $P \div Q$

Assessing What You've Learned

- Do you know how to use the distance formula? On what famous theorem is the distance formula based?
- Can you describe the locus of points that make up an ellipse? Can you give an ellipse's equation given its graph? Its graph given its equation?
- Can you describe the locus of points that make up a parabola? Can you give a parabola's equation given its graph? Its graph given its equation?
- Can you describe the locus of points that make up a hyperbola? Can you give a hyperbola's equation given its graph? Its graph given its equation?
- What techniques can you use to solve a system of nonlinear equations? How many solutions can a system of nonlinear equations have?
- Can you sketch an accurate graph of $f(x) = \frac{1}{x}$? A translation of $f(x) = \frac{1}{x}$?
- How do you find horizontal, vertical, and slant asymptotes?
- How do you add, subtract, multiply, and divide rational expressions?

JOURNAL Look back at the journal entries you have written about Chapter 9. Then, to help you organize your thoughts, here are some additional questions to write about.

- What do you think were the main ideas of the chapter?
- Were there any exercises that you particularly struggled with? If so, are they clear now? If some ideas are still foggy, what plans do you have to clarify them?

CHAPTER

10 Series

Korean artist Do-Ho Suh (b 1962) is perhaps best known for his sculptures that use numerous miniature items that contribute to a greater mass. Shown here is a detail from his installation *Floor* (1997–2000), which is made of thousands of plastic figurines that support a 40 m^2 floor of glass.

OBJECTIVES

In this chapter you will

- learn about mathematical patterns called series, and distinguish between arithmetic and geometric series
- write recursive formulas and explicit formulas for series
- find the sum of a finite number of terms of an arithmetic or geometric series
- determine when an infinite geometric series has a sum, and find the sum if it exists

LESSON

10.1

Arithmetic Series

In this chapter you will build on the work you did in Chapter 2. You may recall that in chapter 2 you learned about sequences, such as the geometric sequence {1, 2, 4, 8, 16, 32, . . .}. A series looks like a sequence, except that successive terms are added, for example, in the series {1 + 2 + 4 + 8 + 16 + 32 + . . .}. Sometimes it is mathematically useful to compute the sum of only a limited number of terms of the series. For example, we may wish to know the sum of the first five terms of the series. On other occasions, it may be useful to know the sum of all the terms of the series. In this chapter you will learn some techniques for computing the sum of part of the series, as well as—in some situations—for computing the sum of the entire series even if the number of terms is infinite.

According to the U.S. Environmental Protection Agency, in 1960 each American produced an average of 2.68 lb/d of trash. This had increased to 3.66 lb/d by 1980. By 2000 trash production had risen to 4.51 lb/d. In previous chapters you learned how to write a sequence to describe the amount of trash production each year. If you added the terms in this sequence, what would the resulting number represent?

EnvironmentalConnection Mount Everest, part of the Himalayan mountain range of Southern Asia, rises about 29,035 ft and is the highest mountain in the world above sea level. It has been called "the world's highest junkyard" because decades' worth of litter left by explorers—for example, climbing gear and inorganic materials like plastic, glass, and metal—has piled up along Mount Everest's trails and camps—an estimated 50 tons of junk from the 1950s to the 1990s. Although environmental agencies such as the World Wildlife Fund have helped protect the area, the number of climbers and the amount of waste they leave continues to increase.

More than 50 tons of garbage have been left by climbers along the routes to the summit of Mount Everest since the first successful climb in 1953.

Series

The sums of consecutive terms of a sequence form a **series.** The sum of the first n terms of a sequence is represented by S_n. That is, you can calculate S_n by finding the sum $u_1 + u_2 + u_3 + \cdots + u_n$. A series consists of the sums S_n. The sums S_n are called the *partial sums of the series.*

Finding the value of a series is a problem that has intrigued mathematicians throughout history. The 13th-century Chinese mathematician Chu Shih-chieh called the sum $1 + 2 + 3 + \cdots + n$ a "pile of reeds" because it can be pictured like this illustration. The illustration shows S_9, the sum of the first nine terms of this sequence, $1 + 2 + 3 + \cdots + 9$. A sum of any *finite,* or limited, number of terms is called a **partial sum** of the series.

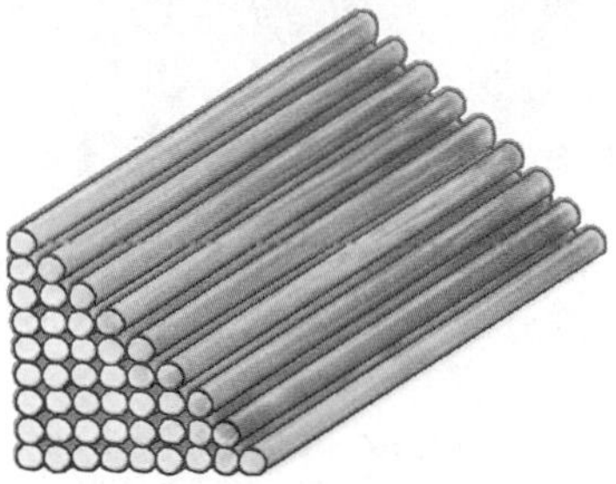

HistoryConnection Chu Shih-chieh (ca 1280–1303) (also known as Zhu Shijie) was a celebrated mathematician from Peking, China, whose theories on arithmetic series, geometric series, and finite differences were most notable. His two mathematical works, *Introduction to Mathematical Studies* and *Precious Mirror of the Four Elements,* went missing and weren't rediscovered until the 19th century.

The expressions S_9 and $\sum_{n=1}^{9} u_n$ are shorthand ways of writing $u_1 + u_2 + u_3 + \cdots + u_9$. The solution to the "pile of reeds" problem can be expressed with sigma notation, using the capital of the Greek letter sigma, as $\sum_{n=1}^{9} n$.

This expression tells you to substitute the integers 1 through 9 for n in the explicit formula $u_n = n$ and then sum the resulting nine values. You get $1 + 2 + 3 + \cdots + 9 = 45$.

How could you find the sum of the integers from 1 to 100? The most obvious method is to add the terms one by one. You can use a recursive formula and a calculator to do this quickly.

First, write the sequence recursively as

$u_1 = 1$

$u_n = u_{n-1} + 1$ where $n \geq 2$

A recursive formula for the series S_n can be written

$S_1 = 1$

$S_n = S_{n-1} + u_n$ where $n \geq 2$

This states that the sum of the first n terms is equal to the sum of the first $(n - 1)$ terms plus the nth term. From the recursive formula for the sequence, you know that u_n is equivalent to $u_{n-1} + 1$, so the recursive formula for the series becomes

$S_1 = 1$

$S_n = S_{n-1} + u_{n-1} + 1$ where $n \geq 2$

Enter the recursive formulas in your calculator as shown. A table shows each term in the sequence and the sequence of partial sums.

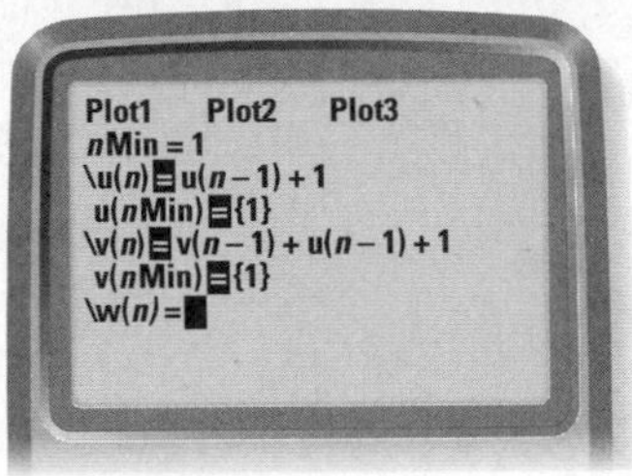

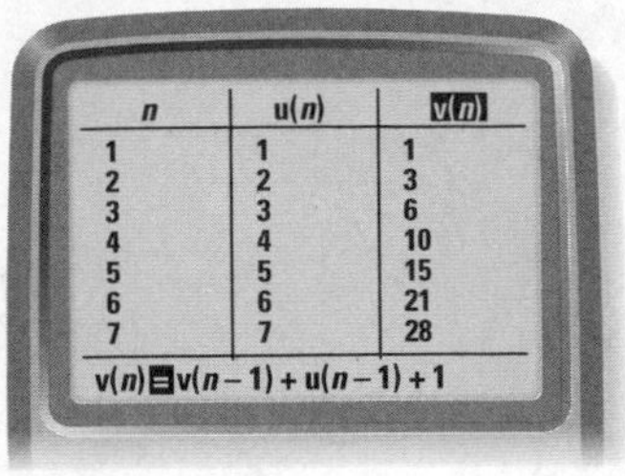

The graph of S_n appears to form a solid curve, but it is actually a set of 100 discrete points representing each partial sum from S_1 through S_{100}. Each point is (n, S_n) for integer values of n, $1 \leq n \leq 100$. The sum of the first 100 terms, S_{100}, is 5050. [▶ See **Calculator Note 10A** for more information on graphing and calculating partial sums.◀]

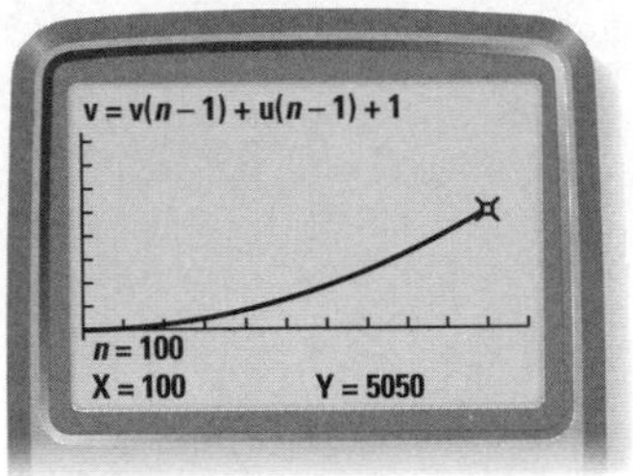

[0, 110, 10, –3000, 10000, 1000]

When you compute this sum recursively, you or the calculator must compute each individual term. You may notice, as you use the calculator, that this can take a lot of time and computing power. There is a way to calculate the partial sum without finding all 100 terms and adding. The Investigation will give you an opportunity to discover at least one explicit formula for calculating the partial sum of an **arithmetic series.**

Investigation
The Arithmetic Series Formula

Select three integers between 2 and 9 for your group to use. Each person will write his or her own arithmetic series using one of the three values for the first term and another for the common difference. Make sure that each person is using a different series.

Step 1 Find the first ten terms of your sequence. Then, find the first ten partial sums of the corresponding series. For example, using $u_1 = 7$ and $d = 8$, you would write

Sequence: $u_n = \{7, 15, 23, 31, \ldots\}$

Series: $S_n = \{7, 22, 45, 76, \ldots\}$

Step 2 Use finite differences to find the degree of the polynomial that would fit the series (n, S_n). Then find a polynomial equation to fit the data (n, S_n).

Step 3 Create a new series by exchanging either the first term or the common difference with another one of the three numbers your group selected. Repeat Steps 1 and 2.

Step 4 Combine your group members' results into a table like this one. In the partial sum column, enter the polynomials found in Step 2.

First term u_1	Common difference d	Partial sum S_n

Step 5 Look for a relationship between the coefficients of each polynomial and the values u_1 and d. Then, write an explicit formula for S_n in terms of u_1, d, and n.

In the Investigation you found a formula that can be used to find a partial sum of an arithmetic series. Use your formula to verify that when you sum $1 + 2 + 3 + \cdots + 100$, you get 5050.

EXAMPLE

Find the sum of the integers from 1 to 100 without using a calculator or formula.

HistoryConnection According to legend, when German mathematician Karl Friedrich Gauss (1777–1855) was 9 years old, his teacher asked the class to find the sum of the integers from 1 to 100. The teacher was hoping to keep the students busy, but Gauss immediately wrote the correct answer, 5050. The example shows Gauss's solution method.

▶ Solution

Karl Friedrich Gauss solved this problem by adding the terms in pairs. Consider the series written in ascending and descending order as shown.

$$\begin{array}{ccccccccccccc} 1 & + & 2 & + & 3 & + & \cdots & + & 98 & + & 99 & + & 100 = S_{100} \\ 100 & + & 99 & + & 98 & + & \cdots & + & 3 & + & 2 & + & 1 = S_{100} \\ \hline 101 & + & 101 & + & 101 & + & \cdots & + & 101 & + & 101 & + & 101 = 2S_{100} \end{array}$$

The sum of every column is 101, and there are 100 columns. Thus, the sum of the integers from 1 to 100 is

$$\frac{100(101)}{2}$$

You must divide by 2 because the series was added twice.

The method in the example can be extended to any arithmetic series. Before you continue, take a moment to consider why the sum of the reeds in the original pile from Chu Shih-chieh's problem can be calculated using the expression

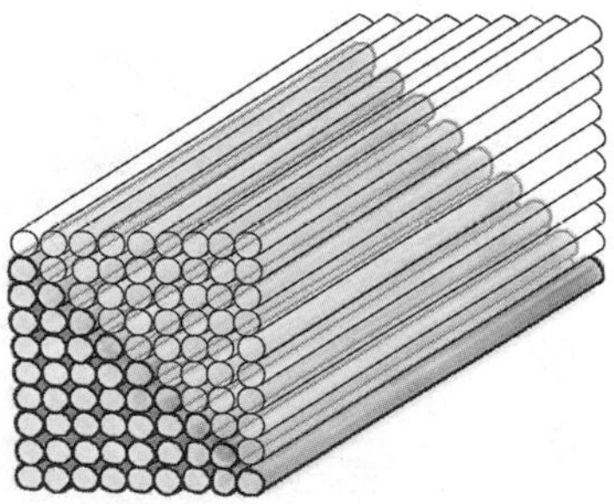

$$\frac{9(1 + 9)}{2}$$

What do the 9, 1, 9, and 2 represent in this context?

Partial Sums of an Arithmetic Series

The partial sums of an arithmetic series are given by the explicit formula

$$S_n = \frac{n(u_1 + u_n)}{2}$$

where n is the number of terms, u_1 is the first term, and u_n is the last term.

In the exercises you will use this formula to find the sum of consecutive terms of an arithmetic series.

EXERCISES

Practice Your Skills

1. List the first five terms of this sequence. Name the first term and the common difference.

$u_1 = -3$

$u_n = u_{n-1} + 15$ where $n \geq 2$

2. Find S_1, S_2, S_3, S_4, and S_5 for this sequence: 2, 6, 10, 14, 18.

3. Write each expression as a sum of terms, then calculate the sum.

a. $\sum_{n=1}^{4}(n+2)$ **b.** $\sum_{n=1}^{3}(n^2-3)$

4. Find the sum of the first 50 multiples of 6: $6 + 12 + 18 + \cdots + u_{50}$.

5. Find the sum of the first 75 even numbers, starting with 2.

Reason and Apply

6. Find the values described.

a. Find u_{75} if $u_n = 2n - 1$.

b. Find $\sum_{n=1}^{75}(2n-1)$.

c. Find $\sum_{n=20}^{75}(2n-1)$.

7. Consider this graph of an arithmetic sequence.

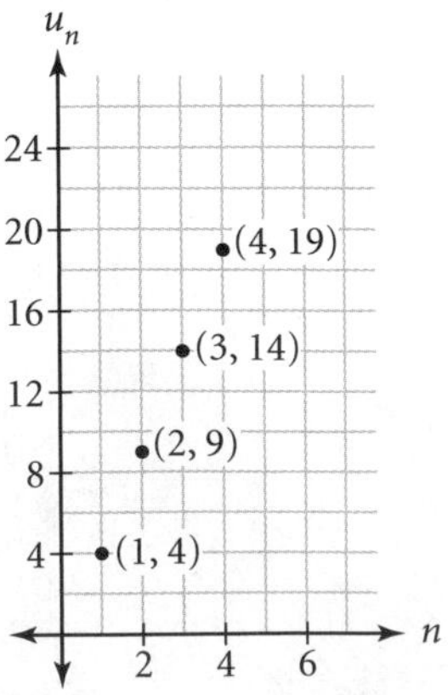

a. What is the 46th term?

b. Write a formula for u_n.

c. Find the sum of the heights from the horizontal axis of the first 46 points of the sequence's graph.

8. Suppose you begin an exercise program on the first day of the new semester. You run for 10 min on the first day of the semester and increase your running time by 5 min each day. How much total time will you have devoted to running

a. 15 days into the semester?

b. 35 days into the semester?

9. Mark arranges a display of soup cans as shown.

a. List the number of cans in the top row, in the 2nd row, in the 3rd row, and so on, down to the 10th row.

b. Write a recursive formula for the terms of the sequence in 9a.

c. If the cans are to be stacked 47 rows high, how many cans will it take to build the display?

d. If Mark uses six cases (288 cans), how tall can he make the display?

10. Find each value.

a. Find the sum of the first 1000 positive integers (the integers from 1 to 1000).

b. Find the sum of the second 1000 positive integers (the numbers from 1001 to 2000).

c. Make a guess at the sum of the third 1000 positive integers (the numbers from 2001 to 3000).

d. Now calculate the sum for 10c.

e. Describe a way to find the sum of the third 1000 positive integers if you know the sum of the first 1000 positive integers.

11. Suppose $y = 65 + 2(x - 1)$ is an explicit representation of an arithmetic sequence for integer values $x \geq 1$. Express the partial sum of the arithmetic series as a quadratic expression, with x representing the term number.

12. **APPLICATION** There are 650,000 people in a city. Every 15 minutes the local media broadcast a tornado warning. During each 15-minute time period 42% of the people who had not yet heard the news become aware of the approaching tornado. How many people have heard the news

a. After 1 hour?

b. After 2 hours?

13. It takes 5 toothpicks to build the top trapezoid shown. You need 9 toothpicks to build 2 joined trapezoids and 13 toothpicks for 3 trapezoids.

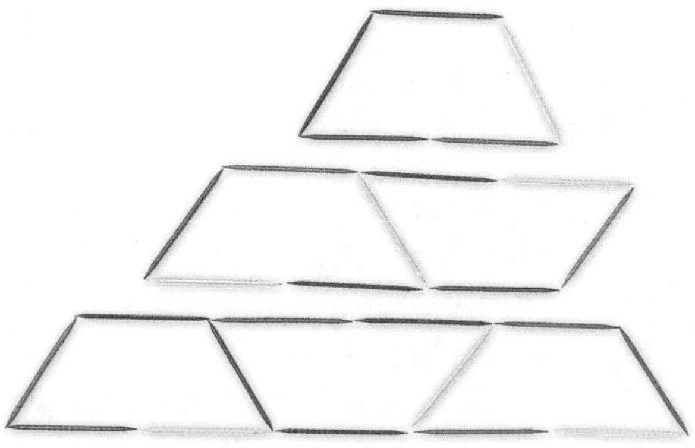

a. If 1,000 toothpicks are available, how many trapezoids will be in the last row?

b. How many rows will there be?

c. How many toothpicks will be used?

d. Use the numbers in this exercise to describe carefully the difference between a sequence and a series.

14. APPLICATION If an object falls from rest, then the distance fallen during the first second is about 4.9 m. In each subsequent second the object falls 9.8 m farther than in the preceding second.

a. Write a recursive formula to describe the distance fallen during each second of free fall.

b. Find an explicit formula for 14a.

c. How far will the object fall during the 10th second?

d. How far does the object fall during the first 10 seconds?

e. Find an explicit formula for the distance fallen by an object in n seconds.

f. Suppose a quarter is dropped from the Royal Gorge Bridge. How long would it take to reach the Arkansas River 331 m below?

ScienceConnection The acceleration due to gravity is 9.8 m/s^2. An object that is dropped will continue to fall with increasing velocity until a terminal velocity is reached. Terminal velocity is reached when a falling object encounters enough air resistance to balance the force of gravity. The object will continue to fall but with a constant velocity.

15. Consider these two geometric sequences.

i. 2, 4, 8, 16, 32, . . .

ii. $2, 1, \frac{1}{2}, \frac{1}{4}, \frac{1}{8}, \ldots$

a. What is the long-run value of each sequence?

b. What is the common ratio of each sequence?

c. What will happen if you try to sum all of the terms of each sequence?

Review

16. (Lesson 2.5) Suppose you invest $500 in a bank that pays 5.5% annual interest, compounded quarterly.

a. How much money will you have after five years?

b. Suppose you also deposit an additional $150 every three months? How much will you have after five years?

17. (Lesson 5.1) Consider the explicit formula $u_n = 81\left(\frac{1}{3}\right)^{n-1}$.

a. List the first six terms, u_1 to u_6.

b. Write a recursive formula for the sequence.

18. (Lesson 5.1) Consider this recursive formula.

$u_1 = 0.39$

$u_n = 0.01 \cdot u_{(n-1)}$ where $n \geq 2$

a. List the first six terms.

b. Write an explicit formula for the sequence.

LESSON

10.2

Infinite Geometric Series

Beauty itself is but the sensible image of the infinite.

GEORGE BANCROFT

In Lesson 10.1 you developed an explicit formula for a partial sum of an arithmetic series. This formula works when you have a finite, or limited, number of terms. You also saw that as the number of terms, n, increases, the partial sum, S_n, increases. If you had an *infinite* number of terms of an arithmetic series, then the sum would also be infinite.

But some geometric sequences have terms that get smaller and smaller. What happens to the sum of these sequences in the long run?

For example, the geometric sequence

0.4, 0.04, 0.004, . . .

is created with common ratio $\frac{1}{10}$ such that the terms get smaller. Adding the terms as a geometric series shows a pattern of repeating decimals.

$$S_3 = 0.4 + 0.04 + 0.004 = 0.444$$

$$S_4 = 0.4 + 0.04 + 0.004 + 0.0004 = 0.4444$$

$$S_5 = 0.4 + 0.04 + 0.004 + 0.0004 + 0.00004 = 0.44444$$

If you sum an infinite number of terms of this sequence, would the result be infinite?

An **infinite geometric series** is a geometric series with infinitely many terms. In this lesson you will look specifically at **convergent series,** for which the sequence of partial sums approaches a long-run value as the number of terms increases.

EXAMPLE A

Jack baked a pie. It was so delicious that he ate one-half the pie on the first day. Determined to make the pie last longer, he decided to eat each day only one-half of the pie that remained in the plate. (Assume that the pie does not spoil.)

a. Record the amount of pie eaten each day for the first seven days.

b. For each of the seven days, record the total amount of pie eaten since it was baked.

c. If Jack were to live forever, then how much of this pie will he have eaten?

▸ *Solution*

a. This is a geometric sequence. Use $\frac{1}{2}$ as the first term and $\frac{1}{2}$ as the common ratio.

$$\frac{1}{2}, \frac{1}{4}, \frac{1}{8}, \frac{1}{16}, \frac{1}{32}, \frac{1}{64}, \frac{1}{128}$$

b. Find the partial sums of the terms in part a.

$$\frac{1}{2}, \frac{3}{4}, \frac{7}{8}, \frac{15}{16}, \frac{31}{32}, \frac{63}{64}, \frac{127}{128}$$

c. It seems that eating pie "forever" would result in eating a lot of pie. However, if we look at the pattern of the partial sums, it looks as though for any finite number of days, Jack's total is slightly less than 1. This leads to the conclusion that Jack would eat exactly 1 pie in the long run. This is a convergent infinite geometric series with long-run value 1.

Recall from Chapter 5 that a geometric sequence can be represented with an explicit formula in the form $u_n = u_o \cdot r_n$ or $u_n = u_1 \cdot r^{n-1}$. In either form, r represents the common ratio between the terms.

This Investigation will now help you create an explicit formula for the sum of a convergent infinite geometric series.

Investigation
The Infinite Geometric Series Formula

Select three integers between 2 and 9 for your group to use. Each person will write his or her own geometric sequence using one of the three values for the first term and one-tenth of another for the common ratio. Make sure that each person is using a different sequence.

Step 1 Use your calculator to find the partial sum of the first 400 and the first 500 terms of your sequence. If your calculator rounds these partial sums to the same value, then use this number as the long-run value. If not, then continue summing terms until the sum does not change. [▶ Revisit **Calculator Note 10A** to learn how to calculate partial sums. ◀]

Step 2 Create a new series by exchanging either the first term or the common ratio with another one of the three numbers that your group selected. Remember that the common ratio is one-tenth of the number. Repeat Step 1.

Step 3 Combine your group members' results into a table like this one. Your long-run value is equivalent to the sum of infinitely many terms, or the **infinite sum.** The infinite sum is represented by S with no subscript.

First term u_1	Common ratio r	Infinite sum S

Step 4 Find an explicit formula for S in terms of u_1 and r. (*Hint:* Include and calculate another column for the ratio $\frac{u_1}{S}$, and look for relationships.) Will your formula work if the ratio is equal to 1? If the ratio is greater than 1? Justify your answers with examples.

In the Investigation you used partial sums with large values of n to determine the long-run value. A graph of the sequence of partial sums is another useful tool.

EXAMPLE B Consider an ideal (frictionless) bouncing ball. The ball is dropped from a height of 200 in. Each time the ball bounces straight up, it only goes eight-tenths as high as the previous time. Therefore the distances in inches that the ball falls after each bounce are 200, 200(0.8), $200(0.8)^2$, $200(0.8)^3$, and so on. Summing these distances creates a series. Find the total distance the ball falls during an infinite number of bounces.

Infinity by American sculptor José de Rivera (1904–1985)

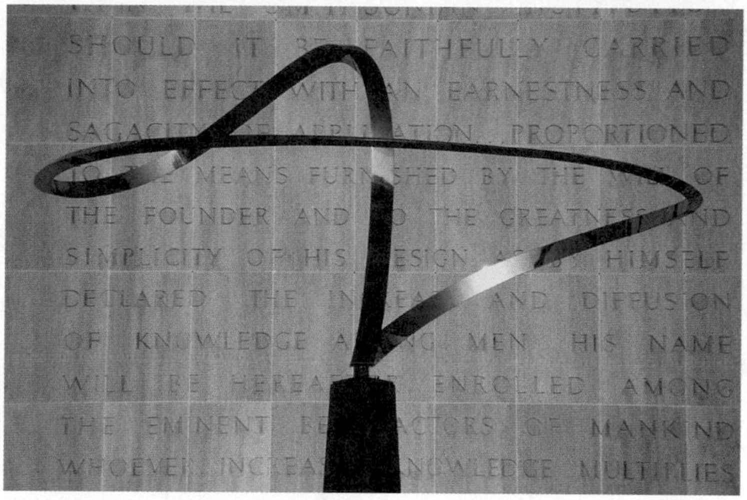

▶ *Solution*

The sequence of partial sums is represented by the recursive formula

$S_1 = 200$

$S_n = S_{n-1} + u_n$ where $n \geq 2$

The explicit formula for the sequence of terms is $u_n = 200(0.8)^{n-1}$. So the recursive formula for the series is equivalent to

$S_1 = 200$

$S_n = S_{n-1} + 200(0.8)^{n-1}$ where $n \geq 2$

The graph of S_n levels off as the number of bounces increases. This means the total sum of all the distances continues to grow but seems to approach a long-run value for larger values of n.

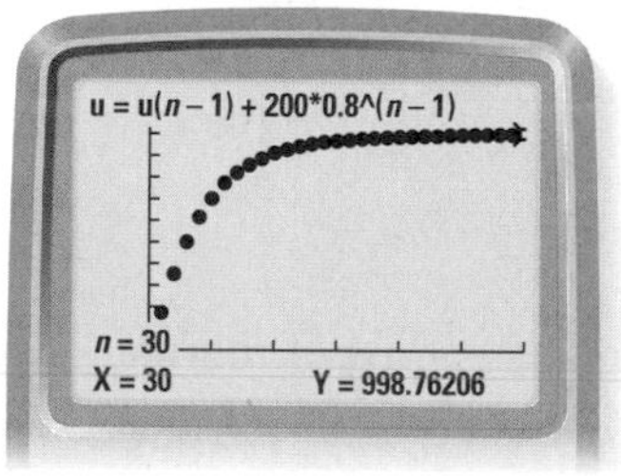

[–5, 30, 5, –200, 1200, 100]

By looking at larger and larger values of n, you'll find that the sum of this series approaches 1000 in. This is because the expression 0.8^{n-1} gets smaller and smaller and, in fact, approaches 0 as n gets larger. Therefore, the infinite sum of the distances the ball falls is 1000 in. too. Use the formula that you found in the Investigation to verify this answer.

You now have several ways to determine the long-run value of an infinite geometric series. If the series is convergent, the formula that you found in the Investigation gives you the infinite sum.

Convergent Infinite Geometric Series

An infinite geometric series is a convergent series if the absolute value of the common ratio is less than 1 ($|r| < 1$). The infinite sum, S, of a convergent infinite geometric series is given by the explicit formula

$$S = \frac{u_1}{1 - r}$$

where u_1 is the first term and r is the common ratio ($|r| < 1$).

In mathematics the symbol ∞ is used to represent infinity, or a number without bound. You can use ∞ and sigma notation to represent infinite series.

EXAMPLE C

Consider the infinite series

$$\sum_{n=1}^{\infty} 0.3(0.1)^{n-1}$$

a. Express this infinite sum as a decimal.

b. Identify the first term, u_1, and the common ratio, r.

c. Express the infinite sum as a fraction of integers.

▸ Solution

a. When you substitute $n = \{1, 2, 3, \ldots\}$ in the expression $0.3(0.1)^{n-1}$, you get $0.3 + 0.03 + 0.003 + \ldots$

The infinite sum is the repeating decimal $0.333\ldots$, or $0.\overline{3}$.

b. $u_1 = 0.3$ and $r = 0.1$

c. Use the formula for the infinite sum, and reduce to a fraction of integers.

$$S = \frac{0.3}{1 - 0.1} = \frac{0.3}{0.9} = \frac{1}{3}$$

Now that you have an understanding of infinite geometric series and infinite sums, try your hand at them in the exercises. In Lesson 10.3 you'll learn about partial sums of geometric series.

EXERCISES

Practice Your Skills

1. Consider the repeating decimal $0.444\ldots$, or $0.\overline{4}$.

a. Express it as the sum of terms of an infinite geometric series.

b. Identify the first term and the ratio.

c. Use the formula to express the infinite sum as a fraction of integers.

2. Repeat 1a–c with the repeating decimal 0.474747 . . . , or $0.\overline{47}$.

3. Repeat 1a–c with the repeating decimal 0.123123123 . . . , or $0.\overline{123}$.

4. A sequence has first term 20, and the infinite sum of the series is 400. What are the first five terms?

Reason and Apply

5. A geometric sequence contains the consecutive terms 128, 32, 8, and 2. The infinite sum of the series is 43690.6. What is the first term?

6. Consider the sequence $u_1 = 47$ and $u_n = 0.8u_{n-1}$ where $n \geq 2$. Find

a. $\sum_{n=1}^{10} u_n$ **b.** $\sum_{n=1}^{20} u_n$ **c.** $\sum_{n=1}^{30} u_n$ **d.** $\sum_{n=1}^{\infty} u_n$

7. Consider the sequence $u_n = 96(0.25)^{n-1}$.

a. List the first ten terms, u_1 to u_5.

b. Find the sum $\sum_{n=1}^{10} 96(0.25)^{n-1}$.

c. Make a graph of partial sums for $1 \leq n \leq 10$.

d. Find the infinite sum $\sum_{n=1}^{\infty} 96(0.25)^{n-1}$.

8. A ball is dropped from an initial height of 100 in. The rebound heights are 80, 64, 51.2, 40.96, and so on. What is the total distance the ball will travel, both up and down?

9. APPLICATION A sporting event is to be held at Ford Field in Detroit, Michigan, which holds about 65,000 people. Suppose 50,000 visitors arrive in the Detroit area and spend \$500 each. After the event 60% of the income from visitors is spent again in the local economy. The next month 60% of that income is spent again, and so on.

a. What is the initial amount spent by the visitors?

b. In the long run how much money is infused into the metropolitan Detroit economy by this sporting event?

c. The ratio of the long-run amount to the initial amount is called the *economic multiplier.* What is the economic multiplier in this example?

d. If the initial amount spent by visitors is \$10,000,000, and the economic multiplier is 1.8, what percent of the initial amount is spent again and again in the local economy?

ConsumerConnection Many cities actively pursue sponsorship of major sporting events. A city's local economy is given a major boost by the monies initially spent. Part of the money spent goes to workers, who then use their wages to purchase more goods and services within the community. Part may go to investors, who will then reinvest it in the community. With a high economic multiplier, economic growth is enhanced, thereby generating more economic activity and a stronger economy.

10. A flea jumps $\frac{1}{2}$ ft right, then $\frac{1}{4}$ ft left, then $\frac{1}{8}$ ft right, and so on. To what point is the flea zooming in?

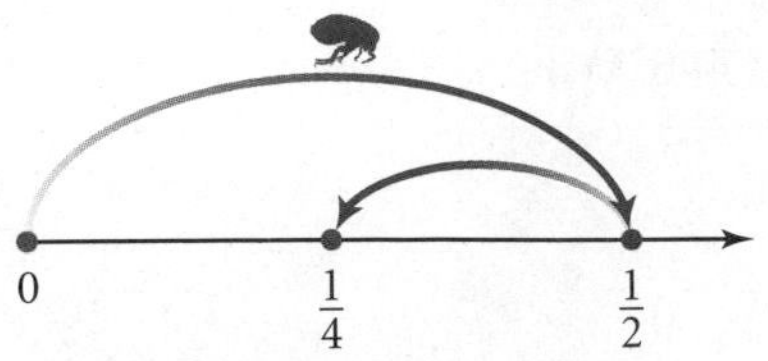

According to Guinness World Records, the longest recorded jump by a flea was in 1910 and measured 13 inches.

11. Suppose square $ABCD$ with side length 8 in. is cut out of paper. Another square, $EFGH$, is placed with its corners at the midpoints of $ABCD$. A third square is placed with its corners at midpoints of $EFGH$, and so on.

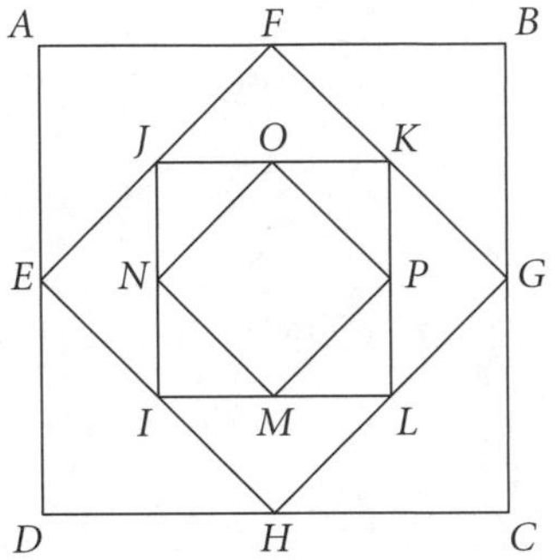

a. What is the perimeter of the 10th square?

b. What is the area of the 10th square?

c. If the pattern could be repeated forever, what would be the infinite sum of the perimeters of all of the nested squares?

d. If the pattern could be repeated forever, what would be the infinite sum of the areas of all of the nested squares?

12. A fractal is a geometric shape that has *self-similarity.* That is, if you were to zoom in on a small portion of the shape, you would see a replica of the entire shape. The fractal known as the Sierpiński triangle begins as an equilateral triangle with side length 1 unit and area $\frac{\sqrt{3}}{4}$ square units. The fractal is created recursively by replacing the triangle with three smaller, congruent equilateral triangles such that each smaller triangle shares a vertex with the larger triangle. This removes the area from the middle of the original triangle.

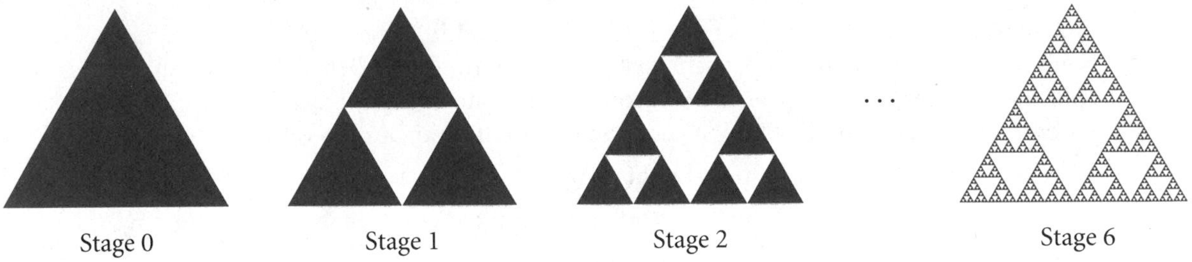

In the long run what happens to

a. The perimeter of each of the smaller triangles?

b. The area of each of the smaller triangles?

c. The sum of the perimeters of the smaller triangles? (*Hint:* You can't use the infinite sum formula from this lesson.)

d. The sum of the areas of the smaller triangles?

Review

13. (Lesson 3.2) A large barrel contains 12.4 gal of oil 18 min after a drain is opened. How many gallons of oil were in the barrel initially, if it drains at 4.2 gal/min?

14. (Lesson 4.6) Match each equation to a graph.

a. $y = 10(0.8)^x$

b. $y = 10 - 10(0.75)^x$

c. $y = 3 + 7(0.7)^x$

d. $y = 10 - 7(0.65)^x$

A.

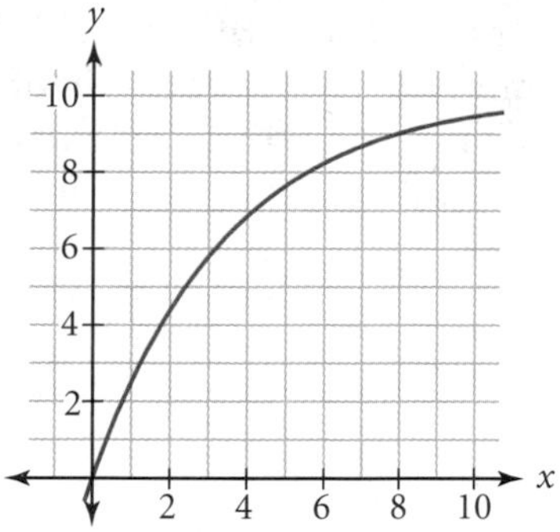

B.

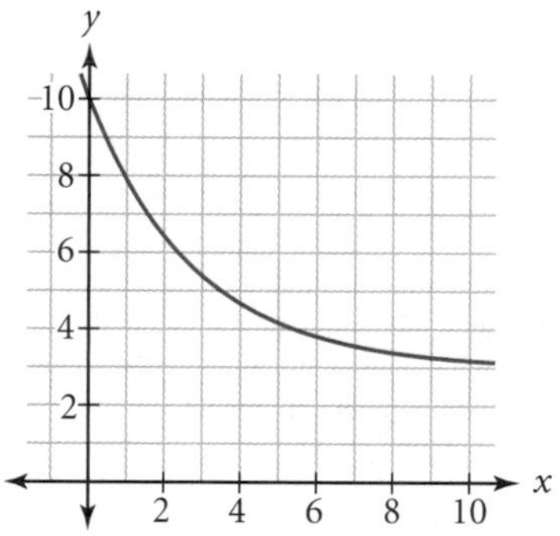

C.

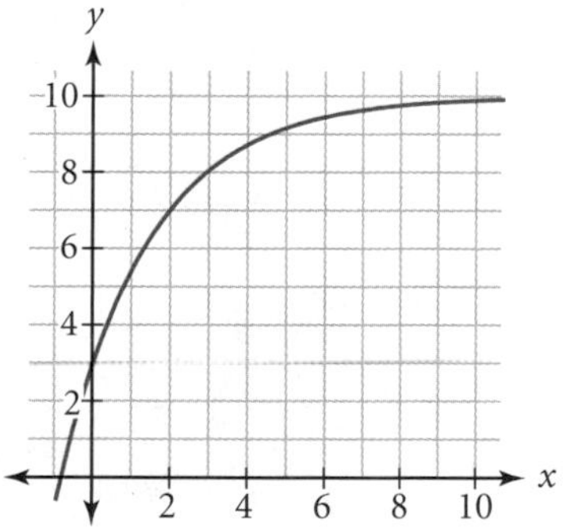

D.

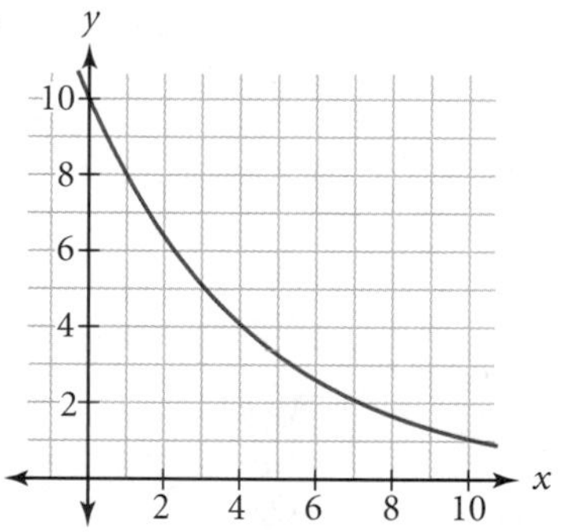

15. **APPLICATION** (Lesson 3.2) A computer software company decides to set aside $100,000 to develop a new video game. They estimate that development will cost $955 the first week and that expenses will increase by $65 each week.

a. After 25 weeks how much of the development budget is left?

b. How long can the company keep the development phase going before the budget will not support another week?

16. (Lesson 5.1) Hans sees a dog. The dog has four puppies with it. Four cats follow each puppy. Each cat has four kittens. Four mice follow each kitten. How many legs does Hans see? Express your answer using sigma notation.

LESSON

10.3

Partial Sums of Geometric Series

If a pair of calculators can be linked and a program transferred from one calculator to the other in 20 s, how long will it be before everyone in a lecture hall of 250 students has the program? During the first time period, the program is transferred to one calculator; during the second time period, to two calculators; during the third time period, to four more calculators; and so on. The number of students who have the program doubles every 20 s. To solve this problem, you must determine the maximum value of n before S_n exceeds 250. This problem is an example of a partial sum of a geometric series. It requires the sum of a finite number of terms of a geometric sequence.

EXAMPLE A

Consider the sequence 2, 6, 18, 54, ...

a. Find u_{15}.

b. Graph the partial sums S_1 through S_{15}, and find the partial sum S_{15}.

▶ Solution

The sequence is geometric with $u_1 = 2$ and $r = 3$.

a. A recursive formula for the sequence is $u_1 = 2$ and $u_n = 3u_{n-1}$ where $n \geq 2$. The sequence can also be defined explicitly as $u_n = 2(3)^{n-1}$. Substituting 15 for n in either equation gives $u_{15} = 9{,}565{,}938$.

b. Use your calculator to graph the partial sums. (You'll see the data points better if you turn the axes off.) Trace to find that S_{15} is 14,348,906.

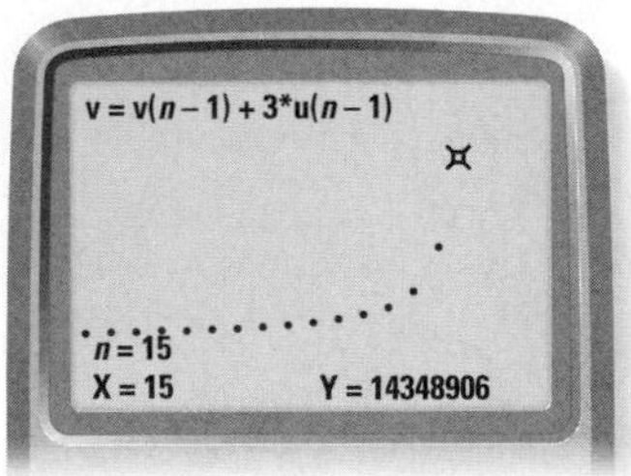

[0, 18, 1, –5000000, 20000000, 5000000]

In the example you used a recursive method to find the partial sum of a geometric series. For some partial sums, especially those involving a large number of terms, it can be faster and easier to use an explicit formula. This Investigation will help you develop an explicit formula.

Investigation
The Geometric Series Formula

Select three integers between 2 and 9 for your group to use. Each person will write his or her own geometric sequence using one of the three values for the first term and one-tenth of another for the common ratio. Make sure that each person is using a different sequence.

Step 1 Find the first ten terms of your sequence. Then, find the first ten partial sums of the corresponding series.

Step 2 Graph data points in the form (n, S_n), and find a translated exponential equation to fit the data. The equation will be in the form $S_n = L - a \cdot b^n$. (*Hint:* L is the long-run value of the partial sums. In Lesson 10.2 how did you find this value?)

Step 3 Rewrite your equation from Step 2 in terms of n, u_1, and r. Use algebraic techniques to write your explicit formula as a single rational expression.

Step 4 Create a new series as in Step 1, this time using a whole integer for the common ratio.

Step 5 Use your formula to find S_{10}, and compare the result to the partial sum from Step 4. Does your formula work when the ratio is greater than 1? If not, then what changes must be made?

In the Investigation you found an explicit formula for a partial sum of a geometric series that uses only three pieces of information—the first term, the common ratio, and the number of terms. Now, you do not need to write out the terms to find a sum.

EXAMPLE B Find S_{10} for the series $16 + 24 + 36 + \ldots$.

▶ Solution The first term, u_1, is 16; the common ratio, r, is 1.5; and the number of terms, n, is 10. Use the formula developed in the Investigation to calculate S_{10}.

$$S_{10} = \frac{16(1 - 1.5^{10})}{1 - 1.5} = 1813.28125$$

EXAMPLE C Each day, the imaginary Caterpillarsaurus eats 25% more leaves than it did the day before. If a 30-day-old Caterpillarsaurus has eaten 151,677 leaves in its brief lifetime, how many will it eat the next day?

▶ Solution To solve this problem, you must find u_{31}. The information in the problem tells you that r is $(1 + 0.25)$, or 1.25, and when n equals 30, S_n equals 151,677. Substitute these values in the formula for S_n, and solve for the unknown value, u_1.

$$\frac{u_1(1 - 1.25^{30})}{1 - 1.25} = 151677 \quad u_1(3227.174268) = 151677 \quad u_1 \approx 47$$

Now, you can write an explicit formula for the geometric sequence, $u_n = 47(1.25)^{n-1}$. Substitute 31 for n to find that on the 31st day the Caterpillarsaurus will consume 37,966 leaves.

The explicit formula for the sum of a geometric series can be written in several ways, but they are all equivalent. You may have discovered these two ways during the Investigation. Notice the relationship between the explicit formula for a partial sum and the explicit formula for the convergent infinite geometric series, given in Lesson 10.2.

Partial Sums of Geometric Series

The partial sum of a geometric series is given by the explicit formula

$$S_n = \left(\frac{u_1}{1-r}\right) - \left(\frac{u_1}{1-r}\right)r^n \text{ or } S_n = \frac{u_1(1-r^n)}{1-r}$$

where n is the number of terms, u_1 is the first term, and r is the common ratio ($r \neq 1$).

EXERCISES

Practice Your Skills

1. For each partial sum equation identify the first term, the ratio, and the number of terms.

a. $\frac{12}{1-0.4} - \frac{12}{1-0.4}0.4^8 \approx 19.9869$

b. $\frac{75(1-1.2^{15})}{1-1.2} \approx 5402.633$

c. $\frac{40 - 0.46117}{1-0.8} \approx 197.69$

d. $-40 + 40(2.5)^6 = 9725.625$

2. Consider this geometric sequence.

256, 192, 144, 108, . . .

a. What is the 8th term?

b. Which term is the first smaller than 20?

c. Find u_7.

d. Find S_7.

3. Find each partial sum for this sequence.

$u_1 = 40$

$u_n = 0.6u_{n-1}$ where $n \geq 2$

a. S_5

b. S_{15}

c. S_{25}

4. Identify the first term and the common ratio or common difference of each series. Then find the partial sum.

a. $3.2 + 4.25 + 5.3 + 6.35 + 7.4$

b. $3.2 + 4.8 + 7.2 + \cdots + 36.45$

c. $\sum_{n=1}^{27} 3.2 + 2.5n$

d. $\sum_{n=1}^{10} 3.2(4)^{n-1}$

Reason and Apply

5. Find the missing value in each set of numbers.

a. $u_1 = 3$, $r = 2$, $S_{10} = \underline{?}$

b. $u_1 = 4$, $r = 0.6$, $S_{\underline{?}} = 9.999868378$

c. $u_1 = \underline{?}$, $r = 1.4$, $S_{15} = 1081.976669$

d. $u_1 = 5.5$, $r = \underline{?}$, $S_{18} = 66.30642497$

6. Find the nearest integer value for n if $\frac{3.2(1 - 0.8^n)}{1 - 0.8}$ is approximately 15.

7. Consider the sequence $u_1 = 8$ and $u_n = 0.5u_{n-1}$ where $n \geq 2$. Find

a. $\sum_{n=1}^{10} u_n$ b. $\sum_{n=1}^{20} u_n$ c. $\sum_{n=1}^{30} u_n$

d. Explain what happens to these partial sums as you add more terms.

8. Suppose you begin a job with an annual salary of \$27,500. Each year you can expect a 3.2% raise.

a. What is your salary 10 years later?

b. What is the total amount you earn in ten years?

c. How long must you work at this job before your total earnings exceeds \$1 million?

9. As a contest winner you are given the choice of two prizes. The first choice awards \$1,000 the first hour, \$2,000 the second hour, \$3,000 the third hour, and so on. For one entire year you will be given \$1,000 more each hour than you were given during the previous hour. The second choice awards 1¢ the first week, 2¢ the second week, 4¢ the third week, and so on. For one entire year you will be given double the amount you received during the previous week. Which of the two prizes will be more profitable and by how much?

10. Consider this geometric series.

$5 + 10 + 20 + 40 + \ldots$

a. Find the first seven partial sums, $S_1, S_2, S_3, \ldots, S_7$.

b. Do the partial sums create a geometric sequence?

c. If u_1 is 5, find value(s) of r such that the partial sums also form a geometric sequence.

11. Consider this series.

$$\sum_{n=1}^{8} \frac{1}{n} = \frac{1}{1} + \frac{1}{2} + \frac{1}{3} + \frac{1}{4} + \cdots + \frac{1}{8}$$

a. Is this series arithmetic, geometric, or neither?

b. Find the sum of this series.

12. List terms to find

a. $\sum_{n=1}^{7} n^2$ b. $\sum_{n=3}^{7} n^2$

13. An Indian folktale recounted by al-Yaqubi in the 9th century begins, "It is related by the wise men of India that when Husiya, the daughter of Balhait, was queen . . ." and goes on to tell how the game of chess was invented. The queen was so delighted with the game that she told the inventor, "Ask what you will." The inventor asked for one grain of wheat on the first square of the chessboard, two grains on the second, four grains on the third, and so on, so that each square contained twice that of the one before. (There are 64 squares on a chessboard.)

a. How many grains are needed

   i. For the 8th square? **ii.** For the 64th square?

   iii. For the first row? **iv.** To fill the board?

b. Write the series using sigma notation.

14. The 32 members of the Lake City College Chess team decide to have a tournament. They are considering either a round-robin tournament or an elimination tournament. (Read the Recreation Connection.)

a. If the tournament is round-robin, how many games need to be scheduled?

b. If it is an elimination tournament, how many games need to be scheduled?

Recreation Connection In setting up tournaments, organizers must decide the type of play. One method is "round-robin," in which every player or team plays every other player or team. Scheduling is different for an odd number of teams and an even number of teams and can be tricky if there is to be a minimum number of rounds. Most intramural sports programs are set up in this format. Another method is the elimination tournament, in which teams or players are paired and only the winners progress to the next round. The NCAA Basketball Tournament is an example of an elimination tournament. Once again, scheduling can be tricky. Difficulties arise when the initial number of teams is not a power of 2.

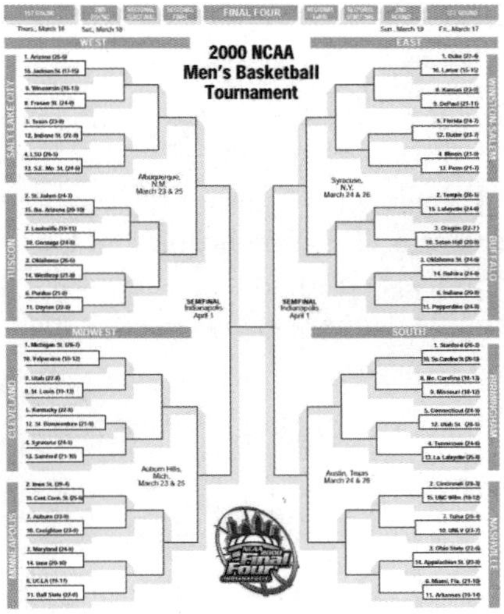

Review

15. (Lesson 2.5) What monthly payment is required to pay off an $80,000 mortgage at 8.9% interest in 30 years?

16. (Lesson 2.5) The enrollment at the local university is currently 5,678. From now on, each year the university will graduate 24% of its students and add 1250 new ones. What will the enrollment be during the 6th year? What will the enrollment be in the long run?

17. (Lesson 5.8) The Magic Garden Seed Catalog advertises a bean with unlimited growth. It guarantees that with proper watering the bean will grow 6 in. the first week and then will increase its height by three-quarters during each subsequent week. Pretty soon, the catalogue claims, "Your beanstalk will touch the clouds!" How long will it take for the beanstalk to be 1,000,000 inches high?

EXPLORATION
With Microsoft® Excel

Geometric Series

Download the Chapter 10 Excel Exploration from the website at www.keycollege.com/online or from your Student CD.

You will explore geometric series by looking at a bouncing ball problem. This problem illustrates a famous paradox: Suppose a ball is one meter above the ground, and on each bounce it rises to a position one-half its previous height. Will it ever come to rest? Will it travel a finite distance?

On the spreadsheet you will find complete instructions and questions to consider as you explore.

Activity
Jack's Magical Pie

Excel skills required

- Data fill
- Name a cell
- Create "clustered column" chart

In Lesson 10.2, Example A, Jack tried to make his pie last longer by eating only half of the pie each day. In this Activity, you will use Excel to make a bar graph of various "magic pie" scenarios.

You can find instructions for the required Excel skills in the Excel Notes at www.keycollege.com/online or on your Student CD.

In Lesson 10.2, Example A, Jack ate half of what was left each day. This Activity begins with Jack eating one-fourth of what is left each day.

Step 1 Prepare labels and starting values as shown.

	A	B	C	D
1	Percent	0.25		
2				
3	Day	Eaten	Remaining	Total Eaten
4	0	0	1	0

Label the starting situation Day 0.

Step 2 Give the name Percent to Cell B1, and enter the value 0.25 in Cell B1. This sets the percent for this part of the work at 25.

Step 3 Enter Day 1 as described.

a. In Cell A5 enter `=A4+1`.

b. In Cell B5 enter `=percent*C4`.

c. In Cell C5 enter `=C4-B5`.

d. In Cell D5 enter `=D4+B5`.

This sets up the formulas for Day 1 so that you can generate as many new days as you like. Your spreadsheet will now look like this:

	A	B	C	D
1	Percent	0.25		
2				
3	Day	Eaten	Remaining	Total Eaten
4	0	0	1	0
5	1	0.25	0.75	0.25

Step 4 Using Cells A5 to D5, data-fill down to Row 54. This will fill down to the 50th day.

Step 5 Highlight Cells B5 to B54. Leaving these highlighted, hold the CTRL button down as you highlight Cells D5 to D54. With these two sequences highlighted, use Insert/Chart/Standard Types, and select Clustered Columns. Then click on Finish. You may want to resize the chart to obtain a reasonable view of the data.

Your chart will look something like this, if you remove the legend:

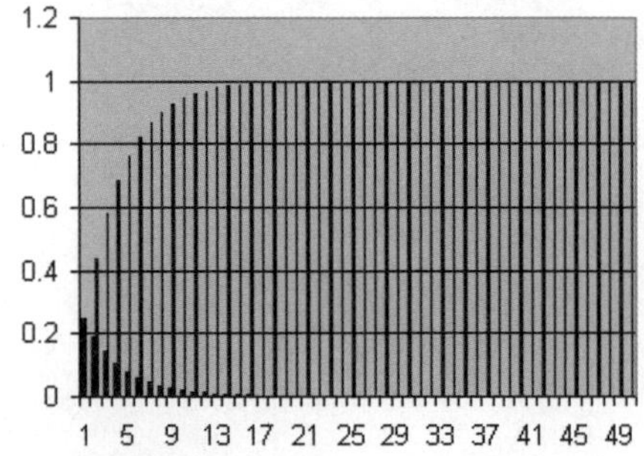

Questions

1. Using the chart, describe the relationship between the two sequences.
2. What geometric series is represented by Column D in the original model (eating one-fourth each day)? Write your answer in sigma notation.
3. Change the value in Cell B1 to view the results if Jack eats three-fourths of what was left. How do the bar charts change?
4. Change the value in Cell B1 to view the results if Jack eats one-tenth of what was left. How do the bar charts change?
5. According to the spreadsheet values, what is the infinite sum of each series that you graphed: one-fourth eaten each day, three-fourths, one-tenth? Use the explicit formula $S = \frac{u_1}{1-r}$ to confirm these infinite sums.

6. Repeat the Activity, but this time change the conditions. Each day Jack eats half of what is left. Then one-fourth of a pie magically appears next to whatever is left. In other words, on the first night he eats half a pie and then adds one-fourth of a pie to the remaining half, leaving three-fourths of a pie. The next day he eats half of what's left (half of three-fourths), then adds another one-fourth of a pie. What happens with this series model? How does the bar graph change as you change the formula?

7. Based on the graph you made for Question 6, estimate the amount of pie Jack will have eaten at the end of 365 days.

8. Write a formula to predict how much pie Jack will have eaten after n days under the conditions described in Question 6. For what values of n is the formula most accurate?

CHAPTER 10 REVIEW

A **series** is a sum of consecutive terms in a sequence. Series can be defined recursively with the rule $S_n = S_{n-1} + u_n$. Series can also be defined explicitly. With an explicit formula you can find any **partial sum** of a sequence without having to know the preceding term(s). The explicit formula for an **arithmetic series** is

$$S_n = \left(\frac{n(u_1 + u_n)}{2}\right)$$

Explicit formulas for a **geometric series** are

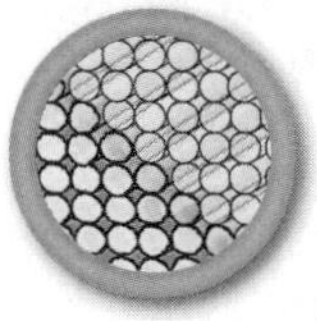

$$S_n = \left(\frac{u_1}{1-r}\right) - \left(\frac{u_1}{1-r}\right) r^n \text{ and } S_n = \frac{u_1(1-r^n)}{1-r}$$

Where $r \neq 1$. If a geometric series is **convergent,** then you can calculate the **infinite sum** with the formula $S = \frac{u_1}{1-r}$.

EXERCISES

1. Consider this arithmetic sequence.

   3, 7, 11, 15, . . .

   a. What is the 128th term?

   b. Which term has the value 159?

   c. Find u_{20}.

   d. Find S_{20}.

2. Consider this geometric sequence.

   100, 84, 70.56, . . .

   a. Which term is the first smaller than 20?

   b. Find the sum of all the terms greater than 20.

   c. Find the value of $\sum_{n=1}^{20} u_n$.

   d. What happens to S_n as n gets very large?

3. Given plenty of food and space, a particular bug species will reproduce geometrically, with each pair hatching 24 young every 5 days. (Assume that half the newborn bugs are male and half female, and all are ready to reproduce in 5 days.) Initially, there are 12 bugs, half of them male and half of them female.

   a. How many bugs are born from the original 12 bugs in 5 days? In 10 days? In 15 days? In 35 days?

   b. Write a recursive formula for the terms of the sequence in 3a.

   c. Write an explicit formula for the terms of the sequence from 3a.

   d. Find the *total* number of bugs after 60 days.

4. Consider this sequence.

$125.3 + 118.5 + 111.7 + 104.9 + \ldots$

a. Find S_{67}.

b. Write an equivalent expression for S_{67} using sigma notation.

5. Emma's golf ball lies 12 ft from the last hole on the golf course. She putts; unfortunately the ball rolls to the other side of the hole, two-thirds as far away as it was before. On her next putt, the same thing happens.

a. If this pattern continues, how far will her ball travel in seven putts?

b. How far will the ball travel in the long run?

6. A flea jumps $\frac{1}{2}$ ft, then $\frac{1}{4}$ ft, then $\frac{1}{8}$ ft, and so on. It always jumps to the right.

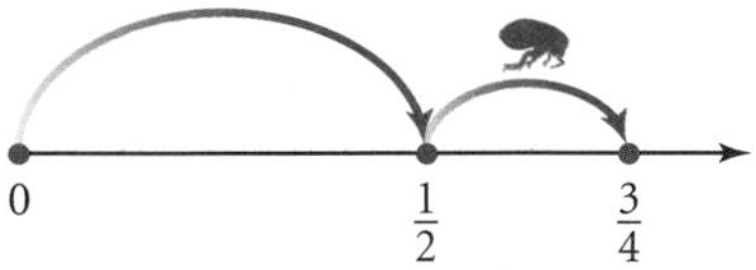

a. Do the distances of the jumps form an arithmetic sequence or a geometric sequence? What is the common difference or common ratio?

b. How long is the flea's 8th jump, and how far is the flea from its starting point?

c. How long is the flea's 20th jump? Where is it after 20 jumps?

d. Write explicit formulas for jump length and the flea's location for any jump.

e. To what point is the flea zooming in?

7. For 7a–c, use $u_1 = 4$. Round your answers to the nearest thousandth.

a. For a geometric series with $r = 0.7$, find S_{10} and S_{40}.

b. For $r = 1.3$ find S_{10} and S_{40}.

c. For $r = 1$ find S_{10} and S_{40}.

d. Make a graph of the partial sums for the series in 7a–c.

e. For which value of r (0.7, 1.3, or 1) do you have a convergent series?

8. Consider this series.

$0.8 + 0.08 + 0.008 + \ldots$

a. Find S_{10}.

b. Find S_{15}.

c. Express the infinite sum as a fraction of integers.

Assessing What You've Learned

- What is the difference between a sequence and a series?
- What is the difference between a geometric series and an arithmetic series?
- If you are given an explicit formula for a series, can you compute a partial sum?
- If you are given a recursive formula for a series, can you compute a partial sum?
- If you are given an explicit formula for an infinite series, can you determine whether it will be possible to determine the sum? When it is possible, do you know how to compute the sum of an infinite series?
- Can you use appropriate technologies, such as the graphing calculator and computer spreadsheet software, to model series?

JOURNAL Look back at the journal entries you have written about Chapter 10. Then, to help you organize your thoughts, here are some additional questions to write about.

- What do you think were the main ideas of the chapter?
- Were there any exercises that you particularly struggled with? If so, are they clear now? If some ideas are still foggy, what plans do you have to clarify them?

Organize your notes on each type of series discussed in this chapter. For each type make sure that you have information about how to determine partial sums and, where possible, infinite sums of the series. Practice writing a few test questions that explore series. You may use sequences that are arithmetic, geometric, or perhaps neither. You may want to include problems that use sigma notation. Be sure to include detailed solution methods and answers.

CHAPTER

11 Probability

American artist Carmen Lomas Garza (b 1948) created this color etching entitled *Lotería–Primera Tabla* (1972), which shows one card game in the traditional Mexican game *lotería*. As in bingo, a caller randomly selects one image that may appear on the game cards, and players try to cover an entire row, column, diagonal, or all four corners of their game cards. For many games, including *lotería,* the chances of winning can be calculated using probability.

OBJECTIVES

In this chapter you will

- learn about randomness and the definition of probability
- count numbers of possibilities for determining probabilities
- determine expected values of random variables
- see how numbers of combinations relate to binomial probabilities

LESSON

11.1

Randomness and Probability

Vowels worth nothing, consonants worth . . . $5000

PAT SAJAK

It isn't fair," complains Noah. "My car insurance rates are much higher than yours." Rita replies, "Well, Noah, that's because insurance companies know the chances are good that I'm a better driver than you, so it will cost them less to insure me." Insurance companies can't know for sure what kind of driving record you will have, but they know the driving records of people in your age group. They use this information to determine your chance of an accident and therefore your insurance rates. This is just one example of how probability theory and the concept of randomness affect your life.

CareerConnection Actuaries use mathematics to solve financial problems. They often work for insurance, consulting, and investment companies, where they might use probability, statistics, and risk theory to develop outcomes such as employee benefit plans, welfare program costs, and funding levels needed by insurance companies to pay expected claims.

Probability theory originated in the 17th century as a means of determining the fairness of games, and it is still used by casinos to make sure that gamblers lose more money than they win. Probability is also important in the study of sociological and natural phenomena.

HistoryConnection Many games are based on random outcomes, or chance. Paintings and excavated material from Egyptian tombs show that dice-like games using *astragali* (ankle bones) were established by the time of the First Dynasty, around 3500 B.C.E. Later, in the Ptolemaic Dynasty (300 to 30 B.C.E.), games with six-sided dice became common in Egypt. The ancient Greeks made icosahedral (20-sided) and other polyhedral dice. The ancient Romans were such enthusiastic dice players that laws were passed forbidding gambling except during certain seasons.

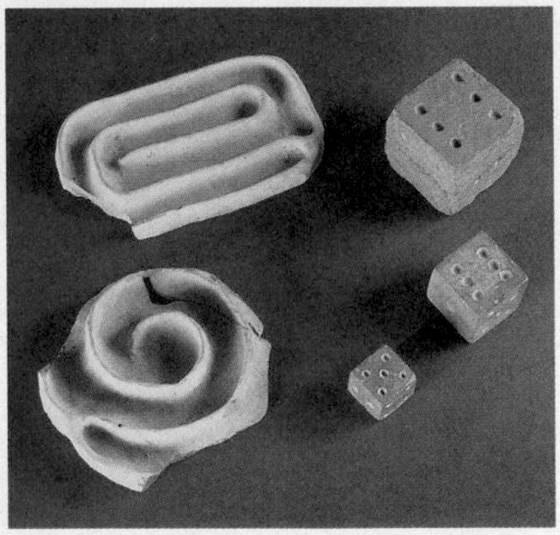

The game of dice, or *tesserae*, was popular during the Roman Empire (ca 1st century B.C.E.–5th century C.E.). The mosaic on the left depicts three men playing *tesserae*. The photo on the right shows ancient Gallo-Roman counters and dice dating from the second half of the 1st century B.C.E.

At the heart of probability theory is **randomness.** Rolling a die, flipping a coin, drawing a card, and spinning a game board spinner are examples of random processes. In a random process, no single event is predictable, even though the long-range pattern of many events often is.

The next Investigation will help you understand predictability of random outcomes.

Investigation
Coin Flip

You will need

- a coin

Step 1 Without flipping a coin, record a random arrangement of H's and T's, as if you were flipping a coin ten times. Call this Sequence A.

Step 2 Now flip a coin ten times, and record the results on a second line. Call this Sequence B.

Step 3 How is Sequence A different from the result of your coin flips? Make at least two observations.

Step 4 Find the longest run of consecutive H's or T's that occurs in Sequences A and B. Then find the 2nd-longest run. Record these lengths for each person in the class as tally marks in a table.

Longest string	Sequence A	Sequence B	2nd-longest string	Sequence A	Sequence B
1			1		
2			2		
3			3		
4			4		
5 or more			5 or more		

Step 5 Count the number of H's in each set. Record the results of the entire class in a table.

Number of H's	Sequence A	Sequence B
0		
1		
2		
3		
4		
5		
6		
7		
8		
9		
10		

Step 6 If you were asked to write a new random sequence of H's and T's, how would it differ from Sequence A?

A random process is often used to generate *random numbers.* Over the long run, each number is equally likely to occur, and there is no pattern in any sequence of numbers generated. [▶ See **Calculator Note 1H** to find out how to generate random numbers with your calculator. ◀]

EXAMPLE A

Use a calculator's random number generator to find the probability of rolling a sum of 6 with a pair of dice.

▶ Solution

As you study this solution, follow along on your own calculator. Your results will be slightly different. [▶ See **Calculator Note 11A.** ◀]

To find the probability of the event "the sum is 6," also written "P(sum is 6)," simulate a large number of rolls of a pair of dice. First, create a list of 300 random integers from 1 to 6 to simulate 300 tosses of the first die. Store this in list L1. Store a second list of 300 outcomes in list L2. Add the two lists to get a random list of 300 sums of two dice. Store this in list L3.

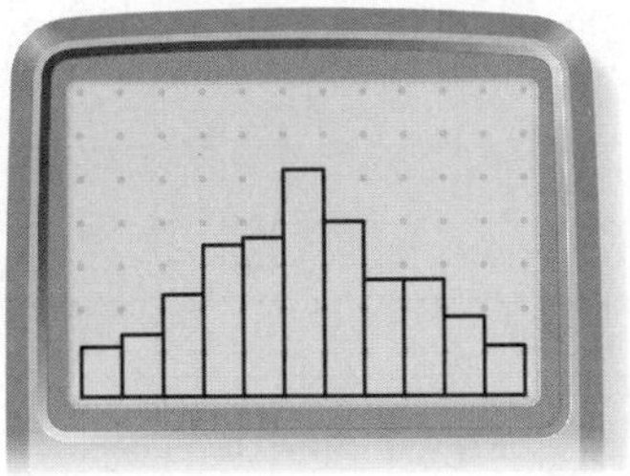

[2, 13, 1, 0, 70, 10]

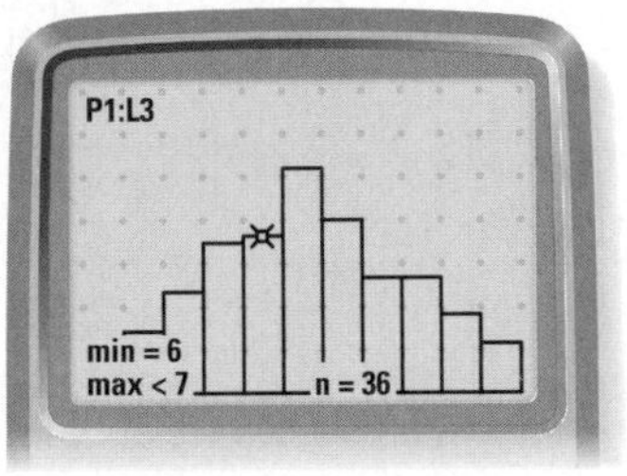

[2, 13, 1, 0, 70, 10]

Create a histogram of the 300 entries in the sum list. The calculator screens here show the number of each of the sums from 2 to 12. Tracing shows that the bin height of the "6" bin is 36. So, out of 300 simulated rolls, $P(\text{sum is } 6) = \frac{36}{300} = .12$.

Repeating the entire process 5 more times gives slightly different results: $\frac{249}{1800} \approx .1383$.

In Example A, rather than actually rolling a pair of dice 1,800 times, we performed a **simulation,** representing the random process electronically. Sometimes statisticians use dice, coins, or spinners to simulate other situations.

You might recall that the probability of an event like "the sum of two dice is 6" must be a number between 0 and 1. The probability of a sure thing, or an event that is certain to happen, is 1. The probability of an impossible event is 0. In Example A you found that P (sum is 6) is approximately .14, or 14%.

Probabilities that are based on trials and observations like this are called **experimental probabilities.** Often a pattern does not become clear until you observe a large number of trials. Find your own results for 300 or 1,800 simulations of a sum of two dice. How do they compare with Example A?

Events are made up of outcomes, for instance, when a red die and a green die are rolled and their sum is found, two possible events are a sum of 3 and a sum of 5. The event "sum is 3" consists of two possible outcomes. One of these outcomes is 1 on the red die and 2 on the green die. The other outcome is 2 on the red die and 1 on the green die.

The event "sum is 5" consists of four possible outcomes. One outcome is 1 on the red die and 4 on the green die. The others are 2 red, 3 green; 3 red, 2 green; and 4 red, 1 green.

If you were to flip three coins—a dime, a nickel, and a quarter—one possible event is that exactly two coins would land heads. The three outcomes that make this up would be

the dime and nickel land heads
the nickel and the quarter land heads
the dime and the quarter land heads

Sometimes it is possible to determine the **theoretical probability** of an event by counting the number of ways a desired event can happen and comparing this number to the total number of equally likely possible outcomes. Outcomes that are equally likely have the same chance of occurring. For example, on a fair die it is equally likely that a 1, 2, 3, 4, 5, or 6 will be rolled.

Experimental Probability

If $P(E)$ represents the probability of an event, then

$$P(E) = \frac{\textit{the number of occurences of an event}}{\textit{the total number of trials}}$$

Theoretical Probability

If $P(E)$ represents the probability of an event, then

$$P(E) = \frac{\textit{the number of different ways the event can occur}}{\textit{the total number of equally likely outcomes possible}}$$

How can you calculate the theoretical probability of rolling a sum of 6 with two dice? At first you might think that there are 11 possible sums of the two dice (from 2 through 12), so $P(\text{sum is } 6) = \frac{1}{11}$. But the 11 sums are not equally likely. If one die is green and other is red, for example, you can get a sum of 5 four ways.

Green	Red
1	4
2	3
3	2
4	1

But you get a sum of 12 only if you roll a 6 on both dice. So a sum of 5 is more likely than a sum of 12.

So what equally like outcomes can we use in this situation to find the theoretical probability?

EXAMPLE B | Find the theoretical probability of rolling a sum of 6 with a pair of dice.

▶ Solution

The possible equally likely outcomes, or sums, when two dice are rolled are represented by the 36 grid points in the diagram on the following page. The point in the upper left corner represents a roll of 1 on the first die and 6 on the second die for a total of 7.

The five possible outcomes with a sum of 6 are labeled A through E in the diagram. Point D, for example, represents an outcome of 4 on the green die and 2 on the white die. What outcome does Point A represent?

The theoretical probability is the number of ways the event can occur, divided by the number of equally likely events possible. So, $P(\text{sum is } 6) = \frac{5}{36} \approx .1389$, or 13.89%.

Before moving on, compare the experimental and theoretical results for this event. Do you think the experimental probability of an event can vary? How about its theoretical probability?

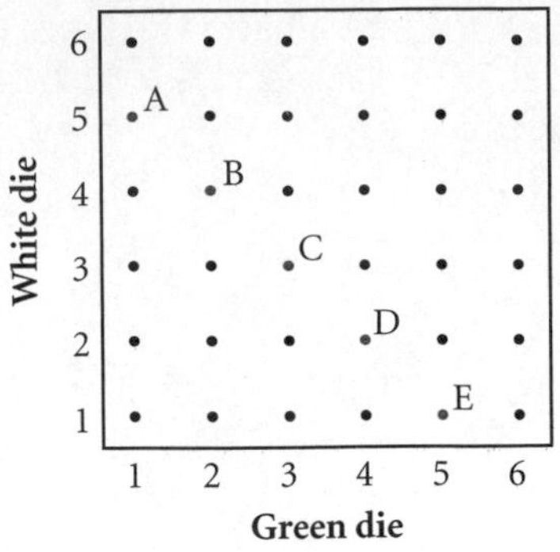

When dice are rolled, the outcomes are integers. What if outcomes could be other kinds of numbers? In those cases, often you can use an area model.

EXAMPLE C What is the probability that any two randomly selected numbers between 0 and 6 have a sum less than or equal to 5?

▸ *Solution*

Because the two values are no longer limited to integers, counting would be impossible. The outcomes are represented by all points within a 6-by-6 square.

In the diagram, Point A represents the outcome $1.47 + 2.8 = 4.27$, and Point B is $4.7 + 3.11 = 7.81$. The points in the triangular shaded region are all those with a sum less than or equal to 5. They satisfy the inequality $n_1 + n_2 \leq 5$, where n_1 is the first number and n_2 is the second number. The area of this triangle is $(0.5)(5)(5) = 12.5$. The area of all possible outcomes is $(6)(6) = 36$. The probability is therefore $\frac{12.5}{36} \approx .347$, or 34.7%.

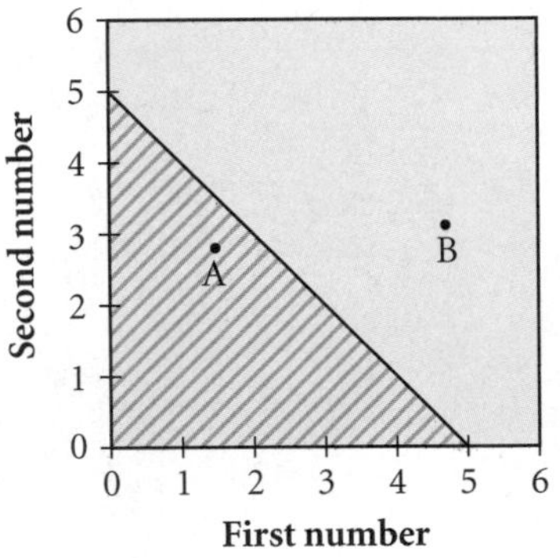

A probability obtained by finding a ratio of lengths or areas is called a *geometric probability.*

Theoretical probabilities can help you predict trends in data. Experimental probabilities can help you estimate a trend if you have enough cases, but obtaining enough data to calculate a long-run trend is not always feasible. In the rest of this chapter, you'll explore different ways to calculate numbers of outcomes to find theoretical probabilities.

EXERCISES

Practice Your Skills

1. Nina has observed that her algebra teacher does not coordinate his sock color with anything else that he wears. Guessing that the color is a random selection, she records these data during three weeks of observation:

black, white, black, white, black, white, black, brown, white, brown, white, white, white, black, black

a. What is the probability that the teacher will wear black socks the next day?

b. What is the probability that he will wear white socks the next day?

c. What is the probability that he will wear brown socks the next day?

2. This table shows numbers of students in several categories at Lee Edsel College. Find the probabilities described below. Express each answer to the nearest .001.

	Male	Female	Total
Freshman	263	249	512
Sophomore	235	242	477
Junior or senior	228	207	435
Total	726	698	1424

a. What is the probability that a randomly chosen student is female?

b. What is the probability that a randomly chosen student is a sophomore?

c. What is the probability that a randomly chosen junior or senior is male?

d. What is the probability that a randomly chosen male is a freshman?

3. The graph of the shaded area shows all possible pairs (x, y). Use the graph and basic area formulas to answer each question. Express each answer to the nearest 0.001.

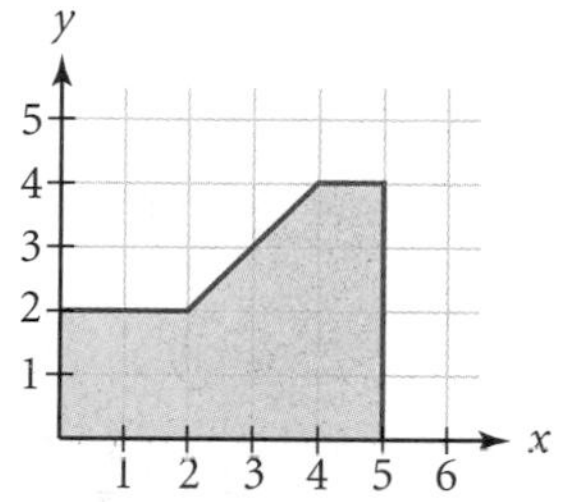

a. What is the probability that x is between 0 and 2?

b. What is the probability that y is between 0 and 2?

c. What is the probability that x is greater than 3?

d. What is the probability that y is greater than 3?

e. What is the probability that $x + y$ is less than 2?

4. Find each probability.

a. Each day your instructor randomly calls on 5 students in your class of 30. What is the probability that you will be called on today?

b. If 2.5% of the items produced by a particular machine are known to be defective, then what is the probability that a randomly selected item will not be defective?

c. What is the probability that the sum of two tossed dice will not be 6?

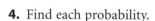

Reason and Apply

5. To prepare necklace-making kits, three camp counselors pull beads out of a box one bead at a time. They discuss the probability that the next bead pulled out of the box will be red. Describe each probability as theoretical or experimental.

a. Alfie says that $P(\text{red}) = \frac{1}{2}$ because 15 of the last 30 beads he pulled were red.

b. Blake says that $P(\text{red}) = \frac{1}{2}$ because the box label says that 1,000 of the 2,000 beads are red.

c. Chris says $P(\text{red}) = \frac{1}{3}$ because 200 of the 600 beads the group has pulled out so far have been red.

6. Rank 6i–iii according to the best method for producing a random integer between 0 and 9, inclusive. Support your reasoning with complete statements.
   i. The number of heads when nine pennies are dropped.
   ii. The length, to the nearest inch, of a standard 9-in. pencil belonging to the next person you meet who has a pencil.
   iii. The last digit of the page number on your left after an open book is spun and dropped from 4 ft.

7. Suppose you are playing a board game for which you need to roll a 6 on a die before you can start playing.
   a. Predict the average number of turns a player should expect to wait before starting to play.
   b. Describe a simulation, using random numbers, that you could use to model this problem.
   c. Do the simulation ten times, and record the number of rolls you need to start playing in each game. (For example, the sequence of rolls 4, 3, 3, 1, 6 means you start playing on the 5th roll.)
   d. Find the average number of rolls needed to start during these ten games.
   e. Combine your results from 7d with those of three other classmates, and approximate the average number of turns a player should expect to wait.

8. Simulate rolling a fair die 100 times with your calculator's random number generator. [▶ See **Calculator Note 11A**. ◀] Display the results in a histogram to see the number of 1's, 2's, 3's, and so on. Do the simulation 12 times.
   a. Make a table storing the results of each simulation for your 1,200 rolls. Find the experimental probability of rolling a 3 after each trial.

Trial number	1's	2's	3's	4's	5's	6's	Ratio of 3's	Cumulative ratio of 3's
1								$\frac{?}{100} = \underline{?}$
2								$\frac{?}{200} = \underline{?}$
...								$\frac{?}{300} = \underline{?}$

   b. What do you think will be the long-run experimental probability?
   c. Make a graph of the cumulative ratio of 3's versus the number of tosses. Plot the points (cumulative number of tosses, cumulative ratio of 3's). Then plot four more points as you extend the domain of the graph to 2,400, 3,600, and 4,800 trials by adding the data from three classmates. Would it make any difference if you considered 5's instead of 3's? Explain.
   d. What is $P(3)$ for this experiment?
   e. What do you think should be the theoretical probability $P(3)$? Explain.

9. Consider rolling a green die and a white die. The roll (1, 5) is different from (5, 1).

a. How many different outcomes are possible for this two-dice experiment?

b. How many different outcomes are possible in which there is a 4 on the green die? Draw a diagram to show the location of these points. What is the probability of this event?

c. How many different outcomes are possible in which there is a 2 or 3 on the white die? What is the probability of this event?

d. How many different outcomes are possible in which there is an even number on the green die and a 2 on the white die? What is the probability of this event?

10. For a two-dice roll, find the number of equally likely outcomes of each event described. Then write the probability of each event.

a. The dice sum to 9.

b. The dice sum to 6.

c. The dice have a difference of 1.

d. The sum of the dice is 6, and their difference is 2.

e. The sum of the dice is at most 5.

11. Consider this diagram.

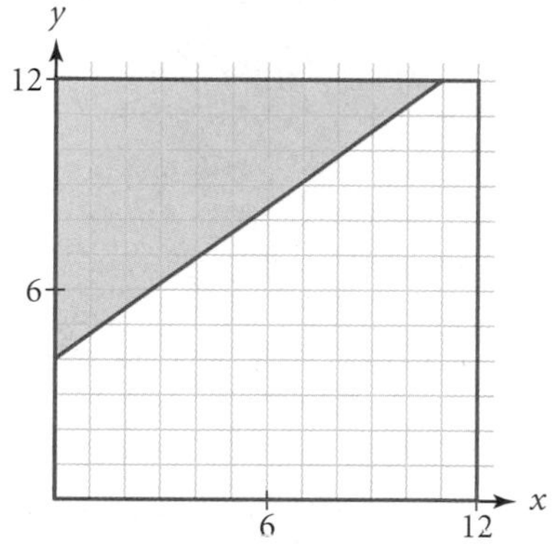

a. What is the total area of the square?

b. What is the area of the shaded region?

c. Over the long run what ratio of random points will be in the shaded area if both the horizontal and vertical coordinates are randomly chosen numbers between 0 and 12, inclusive?

d. What is the probability that a randomly chosen point will land in the shaded area?

e. What is the probability that any given random point will not land in the shaded area?

f. What is the probability that any given random point will land on another specific point? On a specific line?

12. Suppose x is a random number between 0 and 8, and y is a random number between 0 and 8. (The variables x and y are not necessarily integers.)

a. Write a symbolic statement describing the event that the sum of the two numbers is at most 6.

b. Draw a two-dimensional picture of all possible outcomes, and shade the region described in 12a.

c. Determine the probability of the event described in 12a.

13. Use the histogram pictured here for 13a–d.

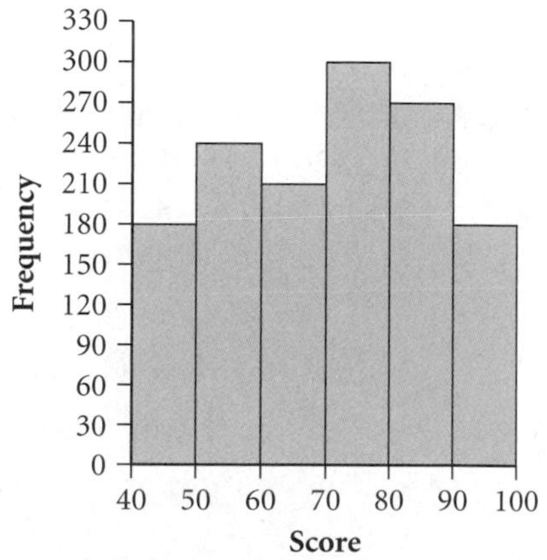

a. Approximate the frequency of the group scoring from 80 to 90.

b. Approximate the sum of all the frequencies.

c. Find P(a score between 80 and 90).

d. Find P(a score that is not between 80 and 90).

14. A 6-in. cube painted on the outside is cut into 27 smaller, congruent cubes. Find the probability that one of the smaller cubes, picked at random, will have the specified number of painted faces.

a. exactly one

b. exactly two

c. exactly three

d. no painted face

15. Where in atoms do electrons live? The graph here shows the probability that a hydrogen atom is at various distances from the nucleus, measured in picometers. A picometer (pm) is 1×10^{-12} meter. Use the graph to answer these questions.

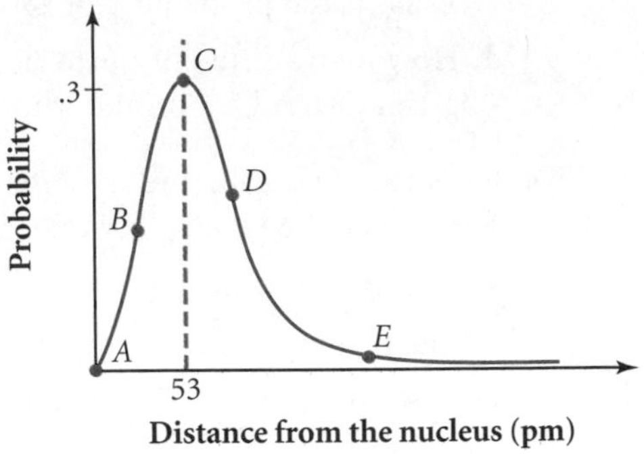

a. The electron is at which distance from nucleus (Points *A–E*) with the highest probability?

b. The electron is at which point with zero probability?

c. As the distance from the nucleus increases, describe what happens to the probability of an electron being that distance from the nucleus.

ScienceConnection Early scientists thought that electrons orbit the nucleus of an atom much like the planets orbit the Sun. Danish physicist Niels Bohr (1885–1962) thought that the electron of a hydrogen atom always orbited at 53 pm from the nucleus. In the mid-1920s, Austrian physicist Erwin Schrödinger (1887–1961) first proposed using a probability model to locate the electron's position. Instead of leading to a predefined orbit, Schrödinger's work led to an electron cloud model. Scientists are concerned with electron location because knowledge of the positions of electrons lays the foundation for understanding chemical changes and their likelihood at the atomic level.

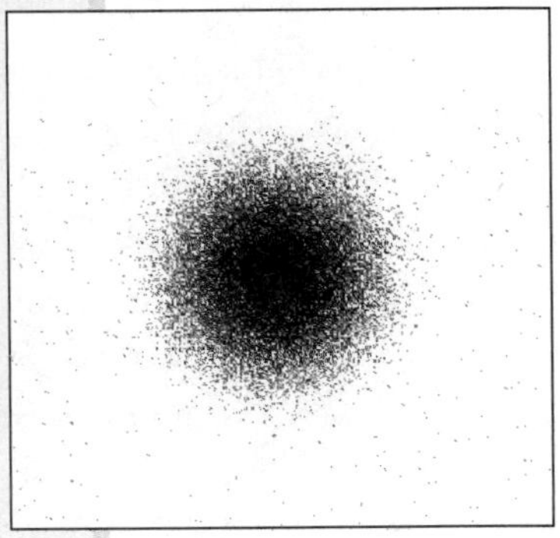

This electron cloud model shows possible locations of an electron in a hydrogen molecule. The density of points in a particular region indicates the probability that an electron is located in that area.

Review

16. (Algebra) Expand $(x - y)^4$.

17. (Lesson 5.7) Write $\log a - \log b + 2 \log c$ as a single logarithmic expression.

18. (Lesson 5.8) Solve $\log 2 + \log x = 4$.

19. (Lesson 7.5) Consider this system of inequalities:

$$\begin{cases} 3x + y \leq 15 \\ x + 6y \geq -2 \\ -5x + 4y \leq 26 \end{cases}$$

a. Graph the triangle defined by this system.

b. Give the coordinates of the vertices of the triangle in 19a.

c. Find the area of the triangle.

EXPLORATION

Geometric Probability

French naturalist George Louis Leclerc, Comte de Buffon (1707–1788), posed one of the first geometric probability problems. If a coin is tossed randomly onto a floor of congruent tiles, what is the probability that the coin will land entirely within a tile, not touching any edge? The answer depends on the size of the coin and the size and shape of the tiles. You'll explore this problem through experimentation, then analyze your results.

Activity

Coin Toss

You will need

- a millimeter ruler
- a penny
- a nickel
- a dime
- a quarter
- grid paper in different sizes: 20 mm, 25 mm, 30 mm, and 40 mm with 5-mm borders

Work with a group, dividing the tasks among yourselves.

Step 1 Each member of your group should choose a coin and a grid. Measure the diameter of your coin, and record your answers in millimeters. Toss your coin 100 times, and count the number of times the coin lands entirely within a square, not touching any lines. Record the number of successes and the experimental probability of success in a table like the one shown. Collect data from members of your group to complete the table.

	Coin diameter (mm)	20-mm grid	30-mm grid	40-mm grid	40-mm grid with 5-mm borders
Penny					
Nickel					
Dime					
Quarter					

Step 2 Where must the center of a coin fall in order to produce a successful outcome? What is the area of this region for each combination of coin and grid paper? What is the area of each square? How do you account for the border in the 4th type of grid paper?

Step 3 Use your results from Step 2 to calculate the theoretical probability of success for each combination of coin and grid paper. How do these theoretical probabilities compare to your experimental probabilities? If they are significantly different, explain why.

Questions

1. Design a grid, different from those you used in the activity, that has a probability of success of .1 for a coin of your choice.
2. Determine a formula that will calculate the theoretical probability of success given a coin with diameter d and a grid of squares with side length a and line thickness t.
3. What would be the theoretical probability of success if you tossed a coin with diameter 10 mm onto a grid paper tiled with equilateral triangles with side length 40 mm, as shown?

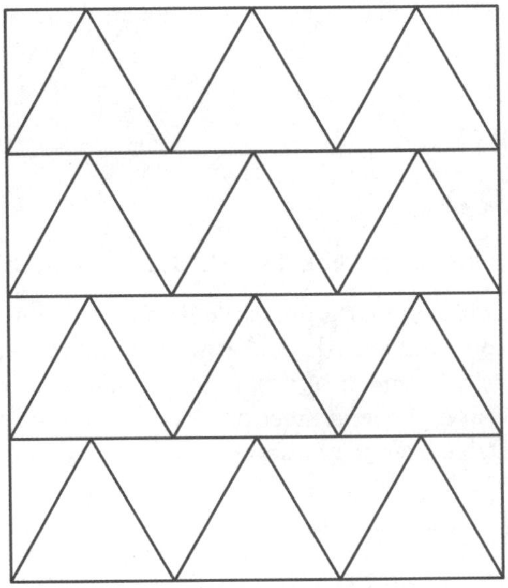

LESSON

11.2 Counting Outcomes and Tree Diagrams

In Lesson 11.1 you determined some theoretical probabilities. In some cases it can be difficult to count the number of possible or desired events. You can make this easier by using **tree diagrams** to organize information and count repeated outcomes.

EXAMPLE A

A national advertisement says that every puffed-barley cereal box contains a toy and that the toys are distributed equally. Talya wants to collect a complete set of the different toys from cereal boxes.

a. If there are two different toys, what is the probability that she will find both in her first two boxes?

b. If there are three different toys, what is the probability that she will have them all after buying her first three boxes?

▶ Solution

Draw tree diagrams to organize the possible outcomes.

a. In this tree diagram the first branching represents the possibilities for the first box, and the second branching represents the possibilities for the second box. Thus, the four paths from left to right represent all possibilities for two boxes and two toys. Path 2 and Path 3 contain both toys. If the advertisement is accurate about equal distribution of toys, then the paths are equally likely. So the probability of getting both toys is $\frac{2}{4} = .5$.

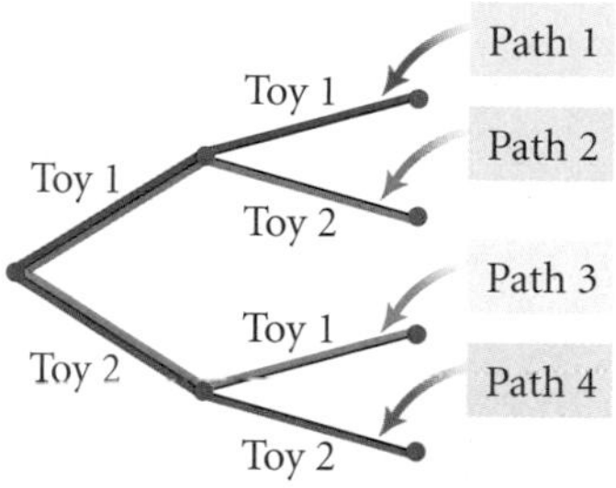

b. This tree diagram shows all the toy possibilities for three boxes. There are 27 possible paths. You can determine this quickly by counting the number of branches on the far right. Six of the 27 paths contain all three toys, as shown. As the paths are equally likely, the probability of having all three toys is $\frac{6}{27}$, or approximately .222.

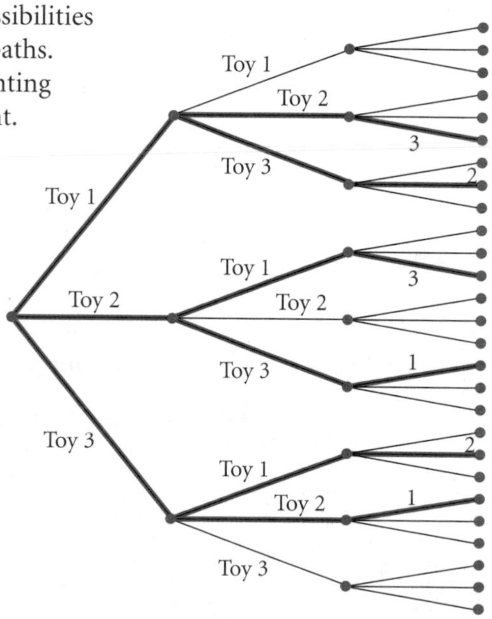

Simple tree diagrams can clearly represent the total number of different outcomes and can allow you to identify those paths representing the desired outcome. Each single branch of the tree represents a **simple event.** A path, or sequence of simple events, is a **compound event.** Tree diagrams can be helpful for organizing complicated situations.

Investigation
The Multiplication Rule

Step 1 On your paper redraw the tree diagram for Example A, part a. This time, include the probability of each simple event. Then, find the probability of each path.

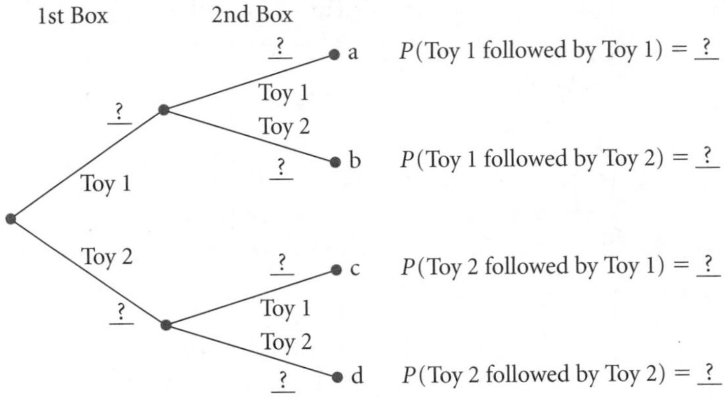

Step 2 Redraw the tree diagram for Example A, part b. Include the probability of each event on each branch, and also write the probabilities of each path. What is the sum of the probabilities of all possible paths? What is the sum of the probabilities of the highlighted paths?

Step 3 Suppose the national advertisement in Example A listed four different toys distributed equally in a huge supply of boxes. Draw only as much of a tree diagram as you need to answer these questions.

a. What would be P(Toy #2) in Talya's first box? Talya's second box? third box? P(any particular toy in any particular box)?

b. In these situations does the toy she finds in one box influence the probability of a particular toy in the next box?

c. One outcome that includes all four toys is Toy #3 followed by Toy #2 followed by Toy #4 followed by Toy #1. What is the probability of this outcome? How many different equally likely outcomes are there?

Step 4 Write a statement explaining how the probabilities on a path's branches can be used to find the probability of the path.

Step 5 What is P(obtaining the complete set in the first 4 boxes)?

In some cases, such as that for four toys and four boxes as described in Step 3 of the Investigation, a tree diagram with equally likely branches is a lot to draw. In some cases, as in Example B, a tree with paths of different probabilities may be practical.

EXAMPLE B

Professor Roark teaches three sections of algebra, and every class has 40 students. His first class has 24 sophomores, his second class has 16 sophomores, and his third class has 20 sophomores. If Professor Roark randomly chooses one student from each class to participate in a competition, what is the probability that he will select 3 sophomores?

▶ Solution

You could consider drawing a tree with 40 branches representing the students in the first class. This would split into 40 branches for the second class, and each of these paths would split into 40 more branches for the third class. This would be a tree with 64,000 paths!

Instead, you can draw two branches for each stage, one representing a choice of a sophomore and one representing a choice of a non-sophomore. This tree would picture all eight possible events. However, the events are not equally likely. For the first class, the probability of choosing a sophomore is $\frac{24}{40} = .6$, and the probability of choosing a non-sophomore is $1 - .6 = .4$. Calculate the probabilities for the second and third classes, and represent them on a tree diagram, as shown.

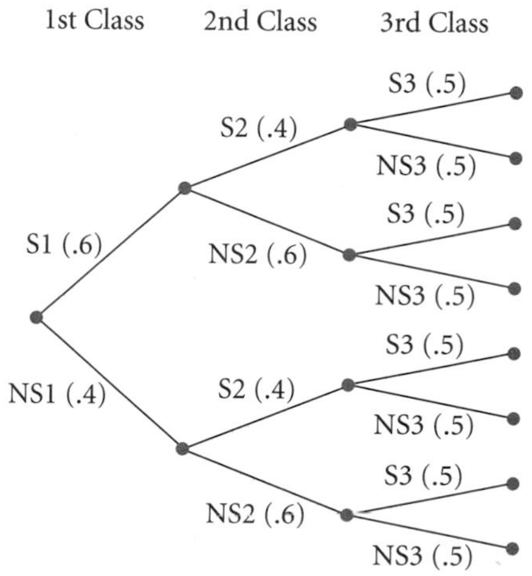

The uppermost path represents the choice of a sophomore from each class. In the Investigation, you learned that the probability of a path can be found by multiplying the probabilities of its branches. So the probability of choosing three sophomores is $(.6) \cdot (.4) \cdot (.5) = .12$.

In Example B the probability of choosing a sophomore in the second class is the same regardless of whether a sophomore was chosen in the first class. These events are called **independent.** Events are independent when the occurrence of one has no influence on the occurrence of the other.

The Probability of a Path
(The Multiplication Rule for Independent Events)

If n_1, n_2, n_3, and so on represent events along a path, the probability that this sequence of events will occur can be found by multiplying the probabilities of the events.

$$P(n_1 \text{ and } n_2 \text{ and } n_3 \text{ and} \ldots) = P(n_1) \cdot P(n_2) \cdot P(n_3) \cdot \ldots$$

Are there events that aren't independent? If so, how do you calculate their theoretical probabilities? Professor Roark's situation provides another example.

EXAMPLE C

Consider the situation from Example B again.

a. What is the probability that only one sophomore will be on the team?

b. Suppose you are a sophomore in Professor Roark's second class, and the competition rules say that only one sophomore can be on the three-person team. What is the probability that you will be selected?

▶ Solution

a. The three highlighted paths represent the different outcomes that include a single sophomore. The first path has probability $(.6) \cdot (.6) \cdot (.5) = .18$; the second path has probability $(.4) \cdot (.4) \cdot (.5) = .08$; and the last path has probability $(.4) \cdot (.6) \cdot (.5) = .12$. The probability that one of these paths will occur is $.18 + .08 + .12 = .34$. So, 34% of the 64,000 total paths contain exactly one sophomore.

b. If only one sophomore is allowed, then the probability that you will be selected depends on what happened in the first class. If a sophomore was selected earlier, then you cannot be chosen. So $P(\text{you}) = 0$. You have a chance only if a sophomore was not selected earlier. In that case $P(\text{you}) = \frac{1}{20} = .05$ because there are 20 students in your class. There is a .4 probability that a non-sophomore will be chosen in the first class, and then a .05 probability that you will be chosen in the second class. So the probability of your being chosen for the competition is $.4(.05) = .02$. You might suggest that Professor Roark use a fairer method of selecting the team!

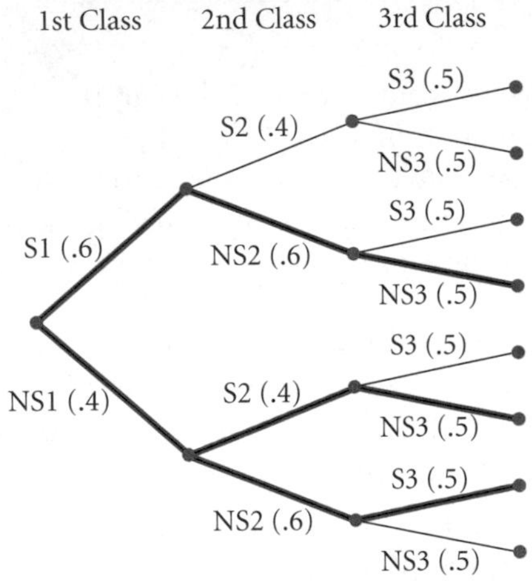

When the probability of an event depends on the occurrence of another, the events are **dependent.** Independent and dependent events can be described using **conditional probability.** When Events A and B are dependent, the probability of A occurring given that B has occurred is different from the probability of A by itself. The probability of A given B is denoted with a vertical line:

$P(A \mid B)$

In Examples B and C the probability of a sophomore in Class 2 given a sophomore in Class 1 would be written $P(S2 \mid S1)$. The fact that events $S1$ and $S2$ were dependent in Example C means that $P(S2 \mid S1) \neq P(S2)$. In Example B, however, $S1$ and $S2$ were independent, so $P(S2 \mid S1) = P(S2)$.

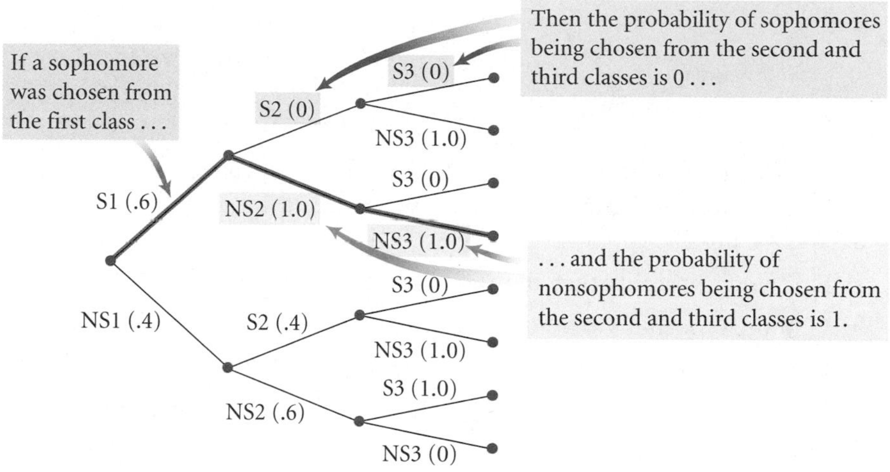

The Multiplication Rule (Again)

If n_1, n_2, n_3, and so on represent events along a path, the probability that this sequence of events will occur can be found by multiplying the probabilities of the events.

$$P(n_1 \text{ and } n_2 \text{ and } n_3 \text{ and } n_4 \ldots) = P(n_1) \cdot P(n_2 \mid n_1) \cdot P(n_3 \mid n_1 \text{ and } n_2) \cdot P(n_4 \mid n_1 \text{ and } n_2 \text{ and } n_3) \ldots$$

In the case of independent events, this statement is the same as the earlier statement of the multiplication rule because $P(n_2 \mid n_1) = P(n_2)$, $P(n_3 \mid n_2) = P(n_3)$, and so on. In later lessons you'll see how to calculate the theoretical probabilities of other sorts of events.

EXERCISES

Practice Your Skills

1. Create a tree diagram showing the different outcomes if the Student Union cafeteria has three main entrees, two vegetable choices, and two dessert choices.

2. Find the probabilities of each path, a–d, in this tree diagram. What is the sum of the values of a, b, c, and d?

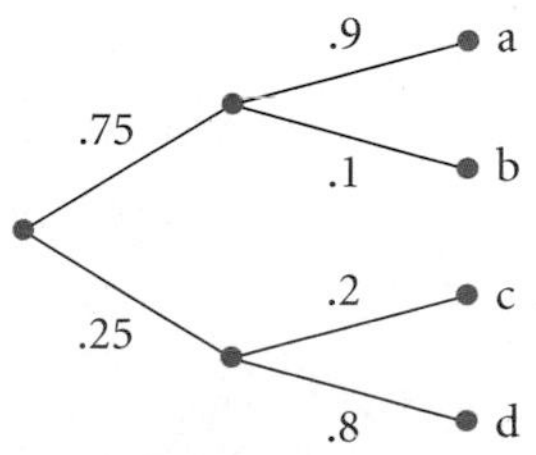

3. Find the probabilities of each path, a–g, in the tree diagram below.

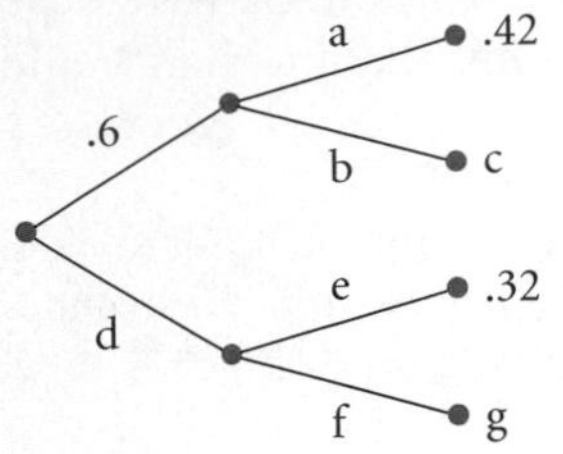

4. Three friends are auditioning for different parts in a play. Each student has a 50% chance of success. Use this tree diagram to answer 4a and 4b.

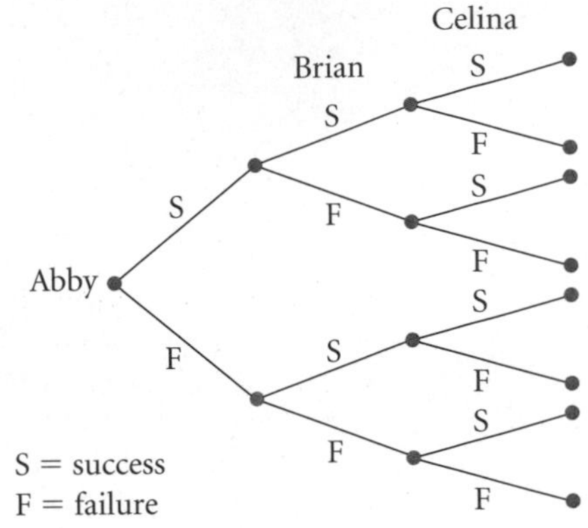

a. Find the probability that all three students will be successful.

b. Find the probability that exactly two students will be successful.

c. If you know that exactly two students have been successful but do not know which pair, what is the probability that Celina was successful?

Reason and Apply

5. Explain the branch probabilities listed in this tree diagram, which models the outcomes of selecting two different students from a class of 7 juniors and 14 sophomores.

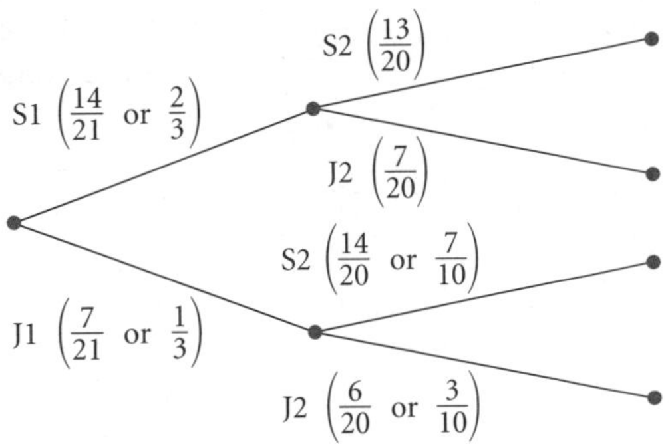

6. Use the diagram from Exercise 5 to answer 6a–c.

a. Use the multiplication rule to find the probability of each path.

b. Are the paths equally likely? Explain.

c. What is the sum of the four answers in 6a?

7. A recipe calls for four ingredients: flour, baking powder, shortening, and milk (F, B, S, M). But there are no directions for the order in which they should be combined. Chris has never followed a recipe like this before and has no idea which order is best, so he chooses the order at random.

a. How many different possible orders are there?

b. What is the probability that milk should be first?

c. What is the probability that flour is first and shortening is second?

d. What is the probability that the order is FBSM?

e. What is the probability that the order isn't FBSM?

f. What is the probability that flour and milk are next to each other?

8. Draw a tree diagram that pictures all possible equally likely outcomes if a coin is flipped as specified.

a. two times b. three times c. four times

9. How many different equally likely outcomes are possible if a coin is flipped as specified?

a. two times b. three times c. four times

d. five times e. ten times f. n times

10. You are totally unprepared for a true-or-false quiz, so you decide to guess randomly at the answers without reading the problems. There are four questions. Find the probabilities described in 10a–e.

a. P(none correct) b. P(exactly one correct)

c. P(exactly two correct) d. P(exactly three correct)

e. P(all four correct) f. What should be the sum of the five probabilities in 10a–e?

g. If a passing grade means that you have answered at least three questions correctly, what is the probability that you have passed the quiz?

11. **APPLICATION** The ratios of phones manufactured at each of three sites, M1, M2, and M3, are 0.2, 0.35, and 0.45 respectively. The diagram also shows some of the ratios of defective (D) and good (G) phones manufactured at each site. The top branch indicates that 0.2 of the phones are manufactured at site M1. The ratio of these phones that are defective is 0.05. Therefore, 0.95 of these phones are good. The probability that a randomly selected phone is both from site M1 and defective is $(.20)(.05) = .01$.

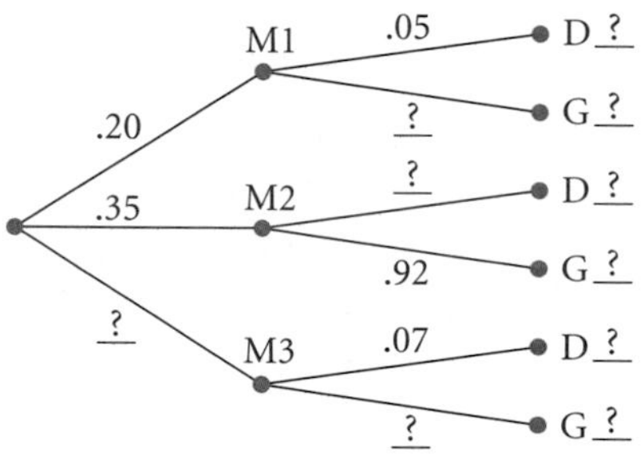

a. Copy the diagram, and fill in the missing probabilities.

b. Find P(a phone from site M2 is defective).

c. Find P(a randomly chosen phone is defective).

d. Find P(a phone is manufactured at site M2 if you already know it is defective).

12. The Detroit Pistons are one point behind the Chicago Bulls, and time has run out in the game. However, the Pistons have a player at the free-throw line, and he has two shots to attempt. He makes 83% of the free-throw shots he attempts. Assume the shots are independent events, so each one has the same probability. Make a tree diagram, and use it to find each probability.

a. Find P(he misses both shots).

b. Find P(he makes at least one of the shots).

c. Find P(he makes both shots).

d. Find P(the Pistons win the game without playing overtime).

13. What is the probability that there are exactly two girls in a family with four children? Assume that girls and boys are equally likely.

14. This table gives numbers of students in several categories.

Number of students	Male	Female	Total
Freshmen	263	249	512
Sophomores	243	234	477
Juniors or Seniors	220	215	435
Total	726	698	1424

Are the events "Freshman" and "Female" dependent or independent? Explain your reasoning.

15. **APPLICATION** In 1963 the U.S. Postal Service introduced the ZIP code to help process mail more efficiently.

a. A ZIP code contains five digits, 0–9. How many possible ZIP codes are there?

b. In 1983 the U.S. Postal Service introduced ZIP + 4. The extra four digits at the end of the ZIP code help pinpoint the destination of a parcel with greater accuracy and efficiency. How many ZIP + 4 codes are there?

c. The Canadian postal service uses a mailing code containing six characters of the form

letter, digit, letter, digit, letter, digit

How many possible Canadian postal codes are there if no restrictions are on the letters and digits?

d. In Canadian postal codes the letters D, F, I, O, Q, U are never used, and the letters W and Z are not used as the first characters. How many postal codes are there now?

16. Braille is code read by the blind. Each Braille character consists of a cell containing six positions that can have either a raised dot or no raised dot. How many different Braille characters are possible?

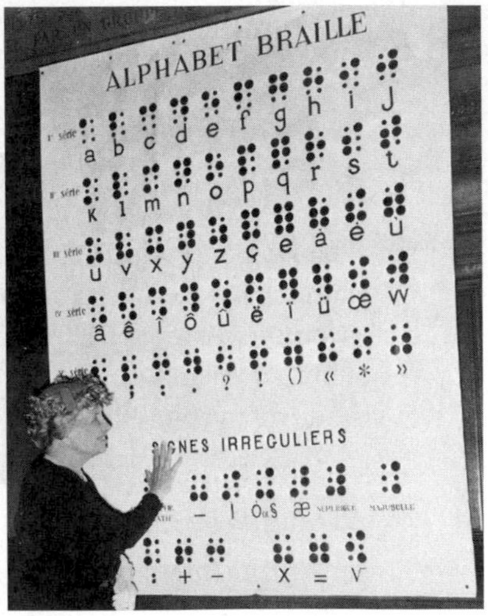

HistoryConnection At the age of 12, French teacher Louis Braille (1809–1852) invented a code that enabled blind people to read and write. He got the idea from a former soldier who used a code consisting of 12 raised dots that enabled soldiers to share information on the battle field without speaking. Braille modified the number of dots to 6 and added symbols for math and music. In 1829 he published the first book in Braille.

American author and lecturer Helen Keller (1880–1968) was both blind and deaf. This photo shows her with a Braille chart.

Review

17. (Lesson 6.5) Write each expression in the form $a + bi$.

a. $(2 + 4i) + (5 + 2i)$

b. $(2 + 4i) - (5 + 2i)$

c. $(2 + 4i)(5 + 2i)$

d. $\dfrac{2 + 4i}{5 + 2i}$

18. (Lesson 11.1) A sample of 230 students is categorized as shown.

	Male	Female
Junior	60	50
Senior	70	50

a. What is the probability that a junior is female?

b. What is the probability that a student is a senior?

19. (Lesson 11.1) What is the probability that a random point selected within the rectangle shown is in the orange region? The blue region?

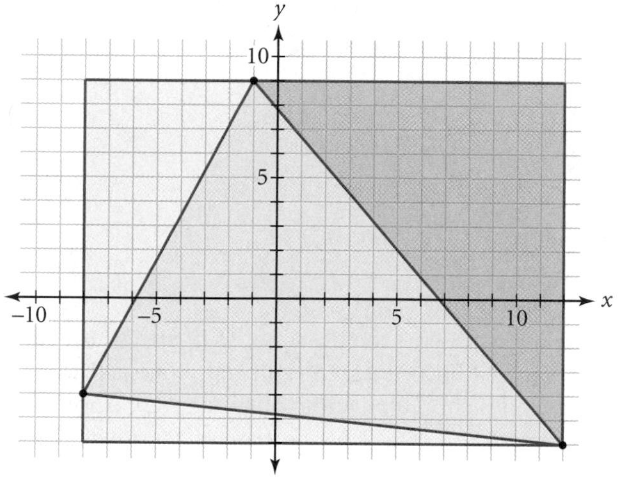

20. (Lesson 9.3) The side of the largest square in the diagram is 4. Each new square has side length equal to half that of the previous one. If the pattern continues forever, what is the long-run length of the spiral made by the diagonals?

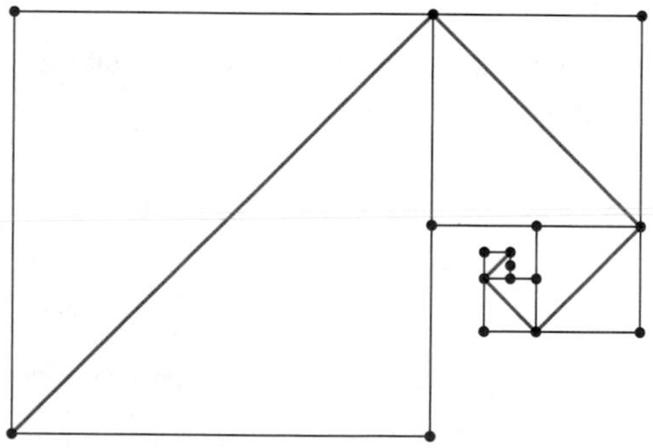

LESSON

11.3 Mutually Exclusive Events and Venn Diagrams

Of course there is no formula for success except perhaps an unconditional acceptance of life and what it brings.

ARTHUR RUBINSTEIN

Two outcomes or events that cannot both occur are *mutually exclusive.* You worked with theoretical probabilities and mutually exclusive events when you added the probabilities of different paths.

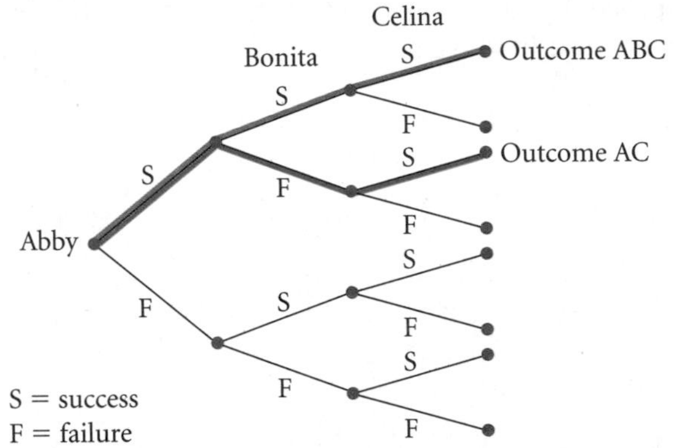

For example, this tree diagram represents the possibilities of successful auditions by Abbey, Bonita, and Celina. One path from left to right represents the outcome that Abbey and Celina are successful but Bonita is not (outcome AC). Another outcome is success by all three (outcome ABC). These two outcomes cannot both take place, so they are mutually exclusive.

Suppose that there is a .5 probability of success for each of Abbey, Bonita, and Celina. Then, the probability of any single path in the tree is $(.5)(.5)(.5) = .125$. So the probability that either *AC* or *ABC* occurs is the sum of the probabilities on two particular paths, $.125 + .125 = .25$.

The tool for breaking down non-mutually exclusive events into mutually exclusive events is the **Venn diagram** which consists of overlapping circles.

EXAMPLE A

Melissa has been keeping a record of probabilities of events involving

i. her guitar string breaking during orchestra rehearsal

ii. a pop quiz in math

iii. her college soccer team losing

Although the three events are not mutually exclusive, they can be broken into eight mutually exclusive events. These events and their probabilities are shown in the Venn diagram.

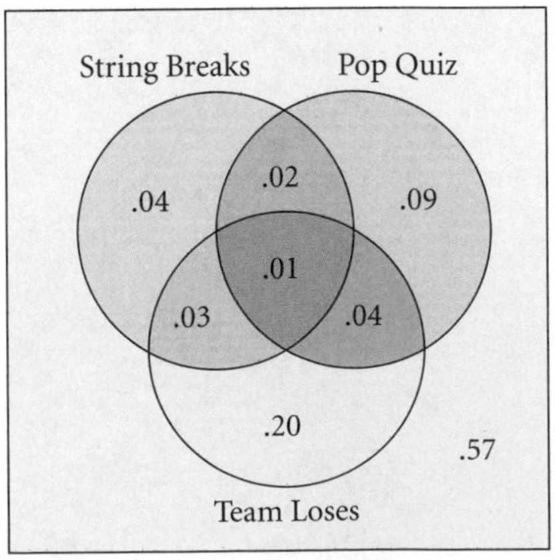

a. What is the meaning of the region labeled .01?

b. What is the meaning of the region labeled .03?

c. What is the probability, $P(Q)$, of a pop quiz today?

d. Find the probability of a pretty good day, $P(\text{not } Q \text{ and not } L \text{ and not } B)$. This means no quiz, no loss, and no breaking string.

▶ Solution

a. The area labeled .01 represents the probability of a really bad day. In this intersection of all three circles, Melissa's string breaks, a pop quiz is given, and her team loses.

b. The area labeled .03 represents the probability that her string will break and her team will lose but no pop math quiz will be given.

c. The probability of a pop quiz being given can be found by adding the four areas that are part of the pop quiz circle: $.02 + .09 + .01 + .04 = .16$.

d. The probability of a pretty good day, $P(\text{not } Q \text{ and not } L \text{ and not } B)$, is pictured by the region outside the circles and is .57.

In general, the probability that one or the other mutually exclusive event will occur is the sum of the probabilities of the individual events.

The Addition Rule for Mutually Exclusive Events

If n_1, n_2, n_3, and so on represent mutually exclusive events, the probability that any of this collection of mutually exclusive events will occur is the sum of the probabilities of the individual events.

$$P(n_1 \text{ or } n_2 \text{ or } n_3 \text{ or} \ldots) = P(n_1) + P(n_2) + P(n_3) + \ldots$$

But what if you don't know all the probabilities that Melinda knew? In the Investigation you'll discover one way to figure out probabilities of mutually exclusive events when you know the probabilities of non-mutually exclusive events.

Investigation
The Addition Rule

A random survey was conducted of 100 students at Sun Coast College. Seventy are enrolled in mathematics, 50 in chemistry, 30 in both subjects, and 10 in neither.

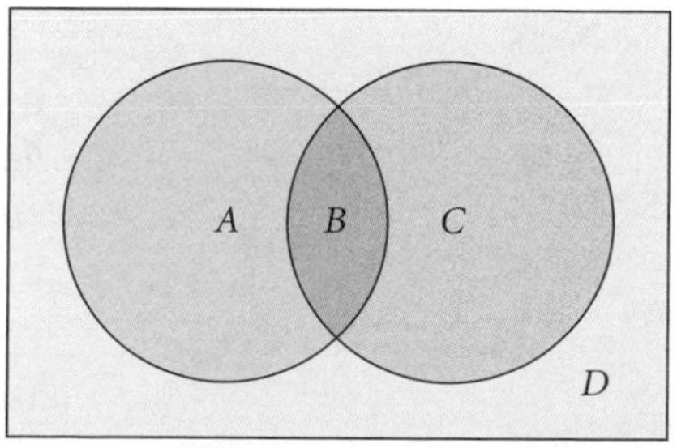

Step 1 "A student takes mathematics" and "a student takes chemistry" are two events. Are these events mutually exclusive? Explain.

Step 2 Complete a Venn diagram depicting enrollments in mathematics and chemistry courses.

Step 3 Convert the numbers of students in your Venn diagram to probabilities.

Step 4 Explain why the probability that a randomly chosen student is in mathematics or science, $P(M \text{ or } C)$, does not equal $P(M) + P(C)$.

Step 5 Create a formula for calculating $P(M \text{ or } C)$ that includes the expressions $P(M)$, $P(C)$, and $P(M \text{ and } C)$.

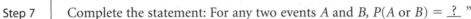

Step 6 Suppose two dice are tossed. Draw a Venn diagram to represent the events.

$A =$ "sum is 7"

$B =$ "both dice > 2"

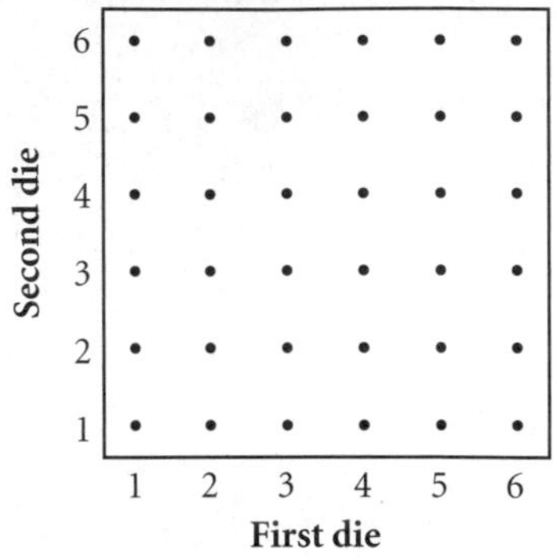

Find the probabilities in parts a–e by counting dots:

a. $P(A)$

b. $P(B)$

c. $P(A \text{ and } B)$

d. $P(A \text{ or } B)$

e. $P(\text{not } A \text{ and not } B)$

f. Find $P(A \text{ or } B)$ by using a rule or formula similar to your response in Step 5.

Step 7 Complete the statement: For any two events A and B, $P(A \text{ or } B) = \underline{?}$ "

A more general form of the addition rule allows you to find the probability of an "or" statement even when two events are not mutually exclusive.

The General Addition Rule

If n_1 and n_2 represent Event 1 and Event 2, then the probability that at least one of the events will occur can be found by adding the probabilities of the events and subtracting the probability that both will occur.

$$P(n_1 \text{ or } n_2) = P(n_1) + P(n_2) - P(n_1 \text{ and } n_2)$$

You might wonder if independent events and mutually exclusive events are the same, because in both cases they seem separate. Independent events don't affect the probabilities of each other. Mutually exclusive events affect each other dramatically: If one occurs, the probability of the other is 0.

But there is a connection between independent events and mutually exclusive events. In calculating the probabilities of non-mutually exclusive events, you use the probability that they both will occur. In the case of independent events, you know this probability.

EXAMPLE B The probability that a rolled die comes up 3 or 6 is $\frac{1}{3}$. What's the probability that a die will come up 3 or 6 on the first and/or second roll?

▸ Solution

If F represents getting a 3 or a 6 on the first roll and S represents getting a 3 or a 6 on the second roll, then $P(F \text{ or } S) = P(F) + P(S) - P(F \text{ and } S)$.

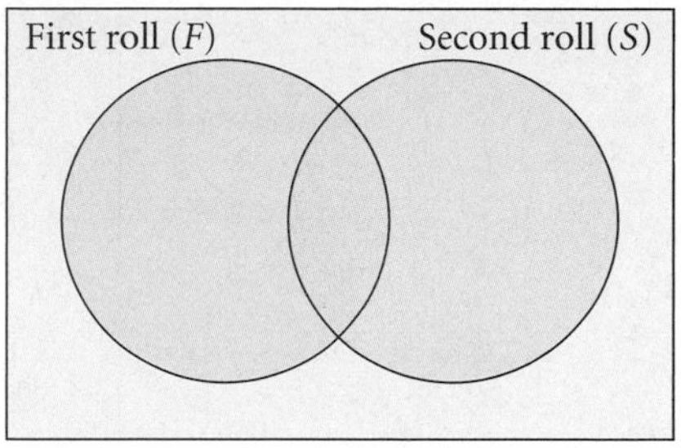

This value is $\frac{1}{3} + \frac{1}{3} - P(F \text{ and } S)$. What's the value of $P(F \text{ and } S)$? Because F and S are independent, $P(F \text{ and } S) = P(F)\,P(S)$, so $P(F \text{ and } S) = \left(\frac{1}{3}\right)\left(\frac{1}{3}\right) = \frac{1}{9}$. So $P(F \text{ or } S) = \frac{1}{3} + \frac{1}{3} - \frac{1}{9} = \frac{5}{9}$. The probability of getting a 3 or 6 on the first and/or second roll is $\frac{5}{9}$.

When two events are mutually exclusive and make up all possible outcomes, they are referred to as **complements.** The complement of "a 1 or a 3 on the first roll" is "not a 1 or a 3 on the first roll," an outcome that is represented by the regions outside the F circle. Because $P(F)$ is $\frac{1}{3}$, the probability of the complement, $P(\text{not } F)$, is $1 - \frac{1}{3} = \frac{2}{3}$.

EXAMPLE C

Every student in the college music program is backstage, and no other students are present. Use O to represent the event that a student is in the orchestra, C to represent the event that a student in the choir, and B to represent the event that a student is in the band. A reporter who approaches a student backstage at random knows some probabilities:

i. $P(B \text{ or } C) = .8$

ii. $P(\text{not } O) = .6$

iii. $P(C \text{ and not } O \text{ and not } B) = .1$

iv. O and C are independent.

v. O and B are mutually exclusive.

a. Turn each of these statements into a statement about percents in regular language.

b. Create a Venn diagram of probabilities describing this situation.

▸ Solution

a. Convert each probability statement into a percentage statement.

i. $P(B \text{ or } C) = .8$ means that 80% of the students are in the band or in the choir.

ii. $P(\text{not } O) = .6$ means that 60% of the students are not in the orchestra

iii. $P(C \text{ and not } O \text{ and not } B) = .1$ means that 10% of the students are in the choir only.

iv. "O and C are independent" means that the percentage of students in the choir is the same as the percentage of orchestra students in the choir. Being in the orchestra does not make a student any more or any less likely to be in the choir, and vice versa.

v. "O and B are mutually exclusive" means that there are no students in both orchestra and band.

b. Because every student backstage is in the music program, $P(\text{not } C \text{ and not } O \text{ and not } B) = 0$. So the region Z in the Venn diagram has a probability of 0.

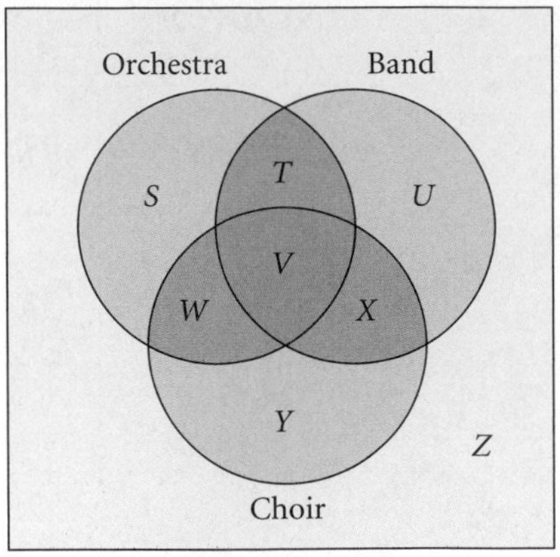

Because O and B are mutually exclusive, $T = 0$ and $V = 0$.

$P(C \text{ and not } O \text{ and not } B) = .1$ says that $Y = .1$.

$P(B \text{ or } C) = .8$ says that $U + W + X + Y = .8$.

$P(\text{not } O) = .6$ says that $U + X + Y = .6$.

The difference in the last two statements indicates that

$(U + W + X + Y) - (U + X + Y) = .8 - .6$, so $W = .2$.

$P(\text{not } O) = .6$ means $P(O) = .4$; therefore $S + W = S + .2 = .4$, so $S = .2$.

Because O and C are independent,

$$P(O) \cdot P(C) = P(O \text{ and } C)$$

$$(S + W)(W + X + Y) = W$$

$$.4(.2 + .1 + X) = .2$$

$$.3 + X = \frac{.2}{.4} = .5$$

$$X = .2$$

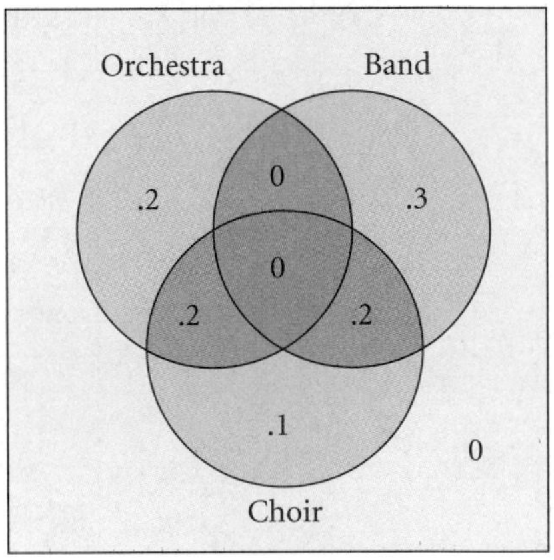

Finally, returning to

$U + X + Y = .6$

$U + .2 + .1 = .6$, so

$U = .3$

Take the time to check that the sum of the probabilities in all the areas is 1. Solving this kind of puzzle gives you a working knowledge of probabilities and the properties of *and*, *or*, *independent*, and *mutually exclusive.*

EXERCISES

Practice Your Skills

Exercises 1–4 refer to this diagram, which gives probabilities related to the two events "sophomore" and "in college algebra."

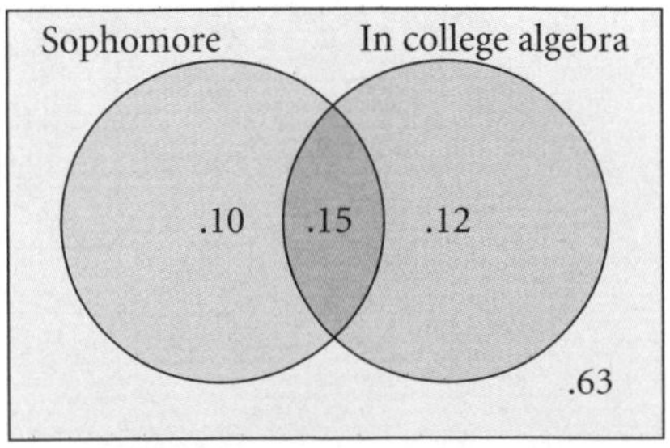

1. Identify in words each of the four regions.

2. Find the probability of each event indicated. Let S represent a sophomore and A represent a student in college algebra.

a. $P(S)$ **b.** $P(A \text{ and not } S)$

c. $P(S \mid A)$ **d.** $P(S \text{ or } A)$

3. Suppose the diagram refers to a campus with 500 students. Change the probabilities of each student into frequencies (number of students).

4. Are the two events, sophomore and in advanced algebra, independent? Show mathematically that you are correct.

5. Events A and B are pictured in this Venn diagram.

a. Are the two events mutually exclusive? Explain.

b. Are the two events independent? Explain.

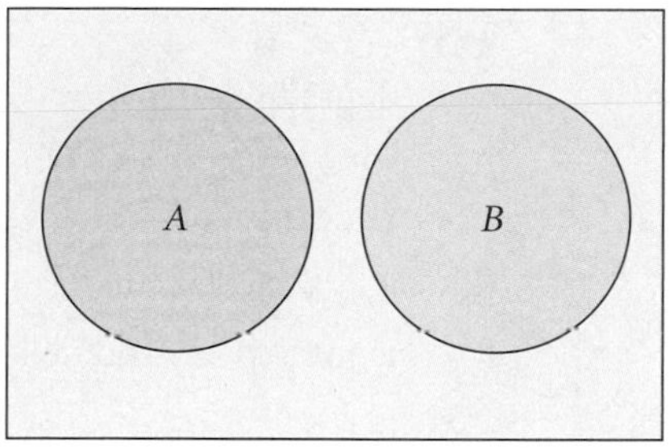

Reason and Apply

6. Of the 4,200 students at El Camino Community College, 1,260 study computer applications, and 2,730 study math. Twenty percent of the math students take computer applications.

a. Create a Venn diagram of this situation.

b. What percentage of the students take both computer applications and math?

c. How many students take neither computer applications nor math?

7. Two events, A and B, have probabilities $P(A) = .2$, $P(B) = .4$, $P(A \mid B) = .2$.

a. Create a Venn diagram of this situation.

b. Find the values of each probability indicated.

i. $P(A \text{ and } B)$

ii. $P(\text{not } B)$

iii. $P(\text{not } (A \text{ or } B))$

8. Kendra needs help on her math homework and decides to call one of her three friends, Amber, Bob, and Carol. Kendra knows that Amber is on the phone 30% of the time, Bob is on the phone 20% of the time, and Carol is on the phone 25% of the time.

a. If each one's phone usage is independent, make a Venn diagram of the situation.

b. What is the probability that all three of her friends will be on the phone when she calls?

c. What is the probability that none of her friends will be on the phone when she calls?

9. If $P(A) = .4$ and $P(B) = .5$, what is the range of values possible for $P(A \text{ and } B)$? What range of values is possible for $P(A \text{ or } B)$? Use Venn diagrams to help explain how this range is possible.

10. Assume that the diagram in Exercise 1 refers to a college with 800 students. Twenty sophomores have moved from geometry to college algebra; draw the diagram showing this.

11. At right are two color wheels. Figure A represents the mixing of light, and Figure B represents the mixing of pigments.

According to Figure A, what color is produced when equal amounts of these colors of light are mixed?

a. Red and green **b.** Blue and green

c. Red, green, and blue

According to Figure B, what color is produced when equal amounts of these pigments are mixed?

d. Magenta and cyan **e.** Yellow and cyan

f. Magenta, cyan, and yellow

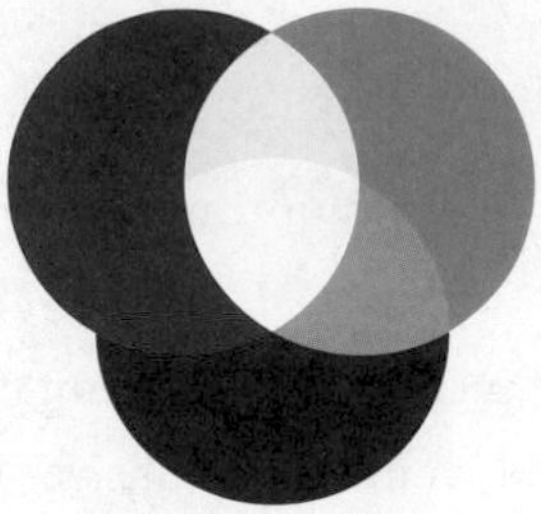

Figure A

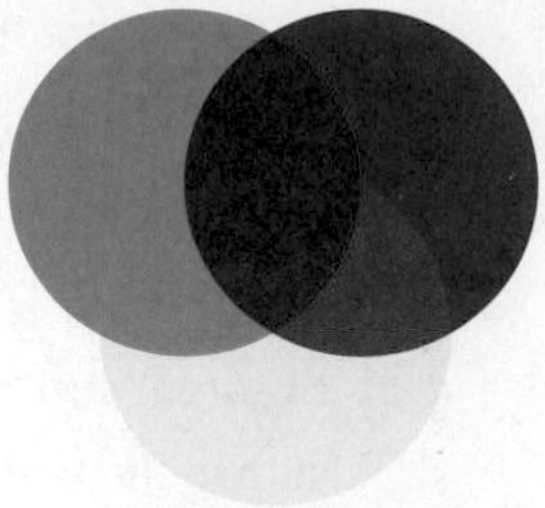

Figure B

ScienceConnection The three primary colors of light are red, green, and blue, and each has its own range of wavelengths. When these waves reach our eyes, we see the color associated with the reflected wave. These colors are also used to project images on TV screens, on computer monitors, and in lighting performances on stage. In the mixing of colors involving paint, ink, or dyes, the primary colors are cyan (greenish-blue), magenta (purplish-red), and yellow. Mixing other pigment colors cannot duplicate these three colors. Pigment color mixing is important in the textile industry and the art and design fields.

Review

12. (Lesson 10.2) The registered voters represented in this table have been interviewed and rated. Assume that this sample is representative of the voting public. Find each probability.

	Liberal	Conservative
Age under 30	210	145
Age 30–45	235	220
Age over 45	280	410

a. P(a randomly chosen voter will be over age 45 and liberal)

b. P(a randomly chosen voter will be conservative)

c. P(a randomly chosen voter will be conservative if under 30)

d. P(a randomly chosen voter will be under 30 if conservative)

13. (Lesson 1.4) The most recent test scores in a chemistry class were 74, 71, 87, 89, 73, 82, 55, 78, 80, 83, and 72. What was the average (mean) score?

14. (Lesson 11.2) If an unfair coin has $P(H) = \frac{2}{5}$ and $P(T) = \frac{3}{5}$, then what is the probability that six flips come up with H, T, T, T, H, T in exactly this order?

15. (Algebra) Rewrite each expression in the form $a\sqrt{b}$ such that b contains no factors that are perfect squares.

a. $\sqrt{18}$ **b.** $\sqrt{54}$ **c.** $\sqrt{60x^3y^5}$

16. (Lesson 3.6) Port Charles and Turner Lake are 860 km apart. At 6:15 A.M. Patrick and Ben start driving from Port Charles to Turner Lake at an average speed of 80 km/h. At 9:30 A.M. Trina and Louis decide to meet them. They leave Turner Lake and drive at 95 km/h toward Port Charles. When and where do the two parties meet?

LESSON 11.4

Random Variables and Expected Value

The real measure of your wealth is how much you'd be worth if you lost all your money.

ANONYMOUS

Imagine that you are sitting near the rapids on the bank of a rushing river. Salmon are attempting to swim upstream. They must jump out of the water to pass the rapids. While you are sitting on the bank, you observe 100 salmon jump; of those, 35 succeed. Having no other information, you can estimate that the probability of success is 35% for each jump.

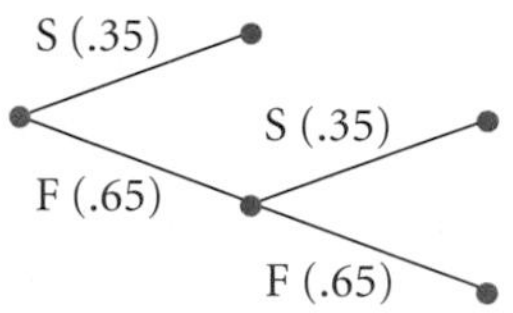

What is the probability that a salmon will succeed on its second attempt? This probability requires that two conditions be met: that the salmon fails on the first jump and that it succeeds on the second. In the diagram you see that this probability is $(.65)(.35) = .2275$, or about 23%. To determine the probability that the salmon succeeds on the first or second jump, sum the probabilities of the two mutually exclusive events: succeeding on the first jump and succeeding on the second jump. The sum is $.35 + .2275 = .5775$, or about 58%.

Probabilities of success such as these are often used to predict the number of independent trials required before the first success (or failure) is achieved. The salmon situation gave experimental probabilities. In the Investigation you'll explore theoretical probabilities associated with dice.

Investigation

"Dieing" for a Four

Each person will need a single die [▶ or see **Calculator Note 11A** to simulate rolling a die ◀]. Imagine that you're about to play a board game in which you must roll a 4 on your die before taking your first turn.

Step 1 Record the number of rolls it takes until you get a 4. Repeat this a total of ten times. Combine your group's results, and find the mean of all values. Then find the mean of all the group results in the class.

Step 2 Based on this experiment, how many rolls would you expect to make on average before a 4 comes up?

Step 3 To calculate the result more theoretically, imagine a "perfect" sequence of rolls with the results 1, 2, 3, 4, 5, 6, 1, 2, 3, 4, 5, 6, etc. On average, how many rolls are needed after each 4 and before the next one?

Step 4 Another theoretical approach uses the fact that the probability of success is $\frac{1}{6}$. Calculate the probability of rolling the first 4 on the first roll, the first 4 on the second roll, the first 4 on the third roll, and the first 4 on the fourth roll. (A tree diagram might help you do the calculations.)

Step 5 Find a formula for the probability of rolling the first 4 on the *n*th roll.

Step 6 Create a calculator list with the numbers 1 through 100. [▶ See **Calculator Note 11B** for a quick way to enter sequences into lists. ◀] Use your formula from Step 5 to make a second list of the probabilities of rolling the first 4 on the first roll, the second roll, the third roll, and so on, up to the 100th roll. Create a third list that is the product of these two lists. Calculate the sum of this third list. [▶ See **Calculator Note 1J.** ◀] What is the meaning of this sum?

Step 7 How close is the sum you found in Step 6 to your estimates in Steps 2 and 3?

The average value you found in Steps 2, 3, and 6 is called the **expected value.** It is also known as the *long-run value* or *mean value.*

But of what is it the expected value? A numerical quantity, the value of which depends on the outcome of a chance experiment, is called a **random variable.** In the Investigation, the random variable gave the number of rolls before getting a 4 on a die. You found the expected value of that random variable. The number of jumps a salmon makes before succeeding is another example of a random variable. In fact, these are both called **discrete random variables** because their values are integers. They're also called **geometric random variables** because we can use partial sums of a geometric series to compute their probabilities.

The next example shows discrete random variables that are not geometric.

EXAMPLE A

When two fair dice are rolled, the sum of the results varies.

a. What is the random variable, what are its values, and what probabilities are associated with those values?

b. What is the expected value of this random variable?

▶ Solution

a. The random variable, usually called x, has as its values all possible sums of the two dice, $2 \leq x_i \leq 12$.

x_i	2	3	4	5	6	7	8	9	10	11	12
probability $P(x)$	$\frac{1}{36}$	$\frac{2}{36}$	$\frac{3}{36}$	$\frac{4}{36}$	$\frac{5}{36}$	$\frac{6}{36}$	$\frac{5}{36}$	$\frac{4}{36}$	$\frac{3}{36}$	$\frac{2}{36}$	$\frac{1}{36}$

b. The expected value of x is the theoretical average you'd expect to have after many rolls of the dice. Your intuition may tell you that the expected value is 7. One way of finding this weighted average is to imagine 36 "perfect" rolls, so every possible outcome occurs exactly once. The mean of the values is

$$\frac{2 + 3 + 3 + 4 + 4 + 4 + \cdots + 11 + 11 + 12}{36} = 7$$

If you distribute the denominator over the terms in the numerator and group like outcomes, you get an equivalent expression that uses the probabilities:

$$\frac{1}{36}\cdot 2+\frac{2}{36}\cdot 3+\frac{3}{36}\cdot 4+\frac{4}{36}\cdot 5+\frac{5}{36}\cdot 6+$$

$$\frac{6}{36}\cdot 7+\frac{5}{36}\cdot 8+\frac{4}{36}\cdot 9+\frac{3}{36}\cdot 10+\frac{2}{36}\cdot 11+\frac{1}{36}\cdot 12=7$$

Note that each term in this expression is equivalent to the product of a value x_i and the corresponding probability $P(x_i)$ in the table above.

Expected Value

The expected value of a random variable, x, is an average value found by multiplying the value of each event by its probability and then summing all of the products. If $P(x_i)$ is the probability of an event with a value of x_i, then the expected value, $E(x)$ is $\sum x_i P(x_i)$

Even if a random variable is discrete, its expected value may not be an integer.

EXAMPLE B

When Nate goes to visit his grandfather, his grandfather always gives him a piece of advice and a bill from his wallet. Grandpa will close his eyes and take out one bill and give it to Nate. On this visit he sent Nate to get his wallet. Nate peeked inside and saw eight bills: two 1-dollar bills, three 5-dollar bills, two 10-dollar bills, and one 20-dollar bill. What is the expected value of his gift?

▶ ***Solution***

The random variable x takes on 4 possible values, and each has a known probability.

Outcomes, x_i	\$1	\$5	\$10	\$20	
Probability $P(x_i)$	.25	.375	.25	.125	**Sum**
Product $x_i \cdot P(x_i)$	0.25	1.875	2.5	2.5	7.125

The expected value is \$7.125.

The other approach, using division, gives the same result:

$$\frac{1+1+5+5+5+10+10+20}{8}=\frac{57}{8}=7.125$$

Nate doesn't actually expect Grandpa to pull out a \$7.125 bill. But if Grandpa repeated the action over and over again with the same bills, he would average \$7.125. The expected value applies to a single trial, but it's based on an average over many imagined trials.

In Example B, suppose Nate always had the same choice of bills, and he had to pay his grandfather \$7 for the privilege of receiving a bill. Over the long run, he'd make \$0.125 per trial. On the other hand, if Nate had to pay \$8 each time, his grandfather would make \$0.875 on average per trial. This is the principle behind how raffles and casinos make money. Gamblers win at times but lose on average.

ConsumerConnection The casino gambling phrase "the house always wins" comes from playing against the "house edge": Casinos earn money because the chances are higher that in total, people bet more money than they win. The expected value is the money people bet per game in relation to the house edge, which is calculated by multiplying the average bet by the number of bets per hour times the house edge.

You can use either the mean or the probability approach to find the expected value if the random variable takes on only finitely many values. But there's no theoretical limit to the number of times you might roll a die before getting a 4. If the random variable has infinitely many values, you must use the probability approach to find the expected value.

Exercises

Practice Your Skills

1. Which of these numbers comes from a discrete random variable? For those that don't, explain why not.

a. The number of children that will be born to members of your class

b. The length of your pencil

c. The number on the jersey of the football player who scores the next touchdown

d. The number of pieces of mail in your mailbox today

2. Which of these numbers comes from a geometric random variable? For those that don't, explain why not.

a. The number of phone calls a telemarketer makes until she makes a sale

b. The number of cats in the home of a cat owner

c. The number of minutes until the radio plays your favorite song

d. The number of songs played until the radio plays your favorite song

3. You have learned that 8% of the students in your school are left-handed. Suppose you stop students at random and ask whether they are left-handed.

a. What is the probability that you will first find a left-handed person on your third try?

b. What is the probability that you will find a left-handed person within three tries?

4. You are taking a multiple-choice test for fun. Each question has five choices (a–e). You roll a six-sided die and mark the answer according to the number on the die, leaving the answer blank if the number on the die is 6. Each question is scored one point for a right answer, minus one-quarter of a point for a wrong answer, and no points for a question left blank.

a. What is the expected value for each question?

b. What is the expected value for a 30-question test?

Reason and Apply

5. Sly asks Les to play a game with him. They will each roll a die. If the sum is greater than 7, Les scores five points. If the sum is less than 8, Sly scores four points.

a. Play the game ten times. [▶ See **Calculator Note 11A** if you have no dice. ◀] Record the final score.

b. What is the experimental probability that Les will win?

c. Draw a tree diagram of this situation showing the theoretical probabilities.

d. If you consider this game from Les's point of view, his winning value is $+5$, and his losing value is -4. What is the expected value of the game from his point of view?

e. Suggest a different distribution of points that would favor neither player.

6. Suppose each box of a certain kind of cereal contains one letter from the word CHAMPION. The letters have been equally distributed in the boxes. You win a prize when you send in all eight letters.

a. Predict the number of boxes you would expect to buy to get all eight letters.

b. Describe a method of modeling this problem using the random number generator in your calculator.

c. Use your method to simulate winning the prize. Do this five times. Record your results.

d. What is the average number of boxes you will have to buy to win the prize?

e. Combine your results with those of several classmates. What seems to be the overall average number of boxes needed to win the prize?

7. In a local concert hall 16% of seats are in the A section, 24% are in the B section, 32% are in the C section, and 28% are in the D section. Section A seats sell for \$35, Section B for \$30, Section C for \$25, and Section D for \$15. You see a ticket stuck high in a tree.

a. What is the expected value of the ticket?

b. The markings look like either A or C section. If this is true, what is the probability that it is C?

c. If the ticket is from the A or C section, then what is the expected value of the ticket?

8. The tree diagram shows a game played by two players.

a. Find a value of x that gives approximately the same expected value for both players.

b. Design a game that could be described with this tree diagram. (You may use coins, dice, spinners, or some other device.) Explain the rules of the game and the scoring of points.

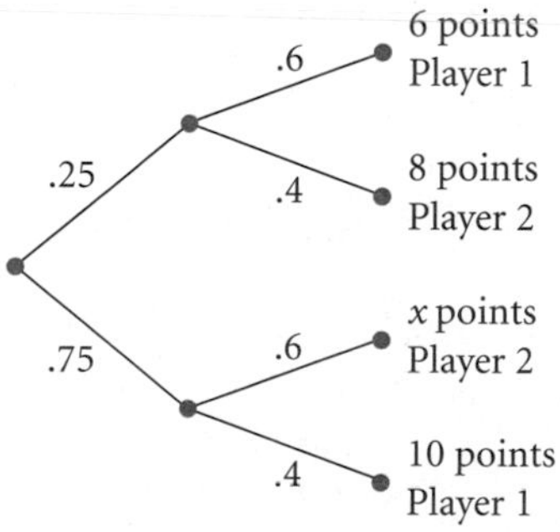

9. A group of friends has a huge bag of red and blue candies with approximately the same number of each color. Each of the friends picks out candies one at a time until he or she gets a red candy. The bag is then passed on to another friend.

a. Devise and describe a simulation of this problem. Use your simulation to approximate the long-run average number of candies each person pulls out in one turn.

b. What is the long-run average number of blue candies pulled out by each person?

10. A quality-control engineer randomly selects five radios from an assembly line. Experience has shown that the probabilities of finding 0, 1, 2, 3, 4, or 5 defective radios is as shown in the table.

Number of defective radios x	Probability $P(x)$	$x \cdot P(x)$
0	.420	
1	.312	
2	.173	
3	.064	
4	.031	
5	.000	

a. What is the probability the engineer will find at least one defective radio in a random sample of five?

b. Complete the entries for the column headed $x \cdot P(x)$.

c. Find the sum of the entries for the column headed $x \cdot P(x)$.

d. What is the real-world meaning of your answer in 10c?

11. What is the probability that a 6 will not appear until the eighth roll of a fair die?

12. **APPLICATION** Bonny and Sally are playing in a tennis tournament. On average, Bonny makes 80% of her shots, and Sally makes 75% of hers. Bonny is serving first. Below are tree diagrams depicting possible sequences of events for one and two volleys of the ball. (A volley is a single sequence in which Bonny hitts the ball and Sally returns it.) Once a ball is missed, that branch ends.

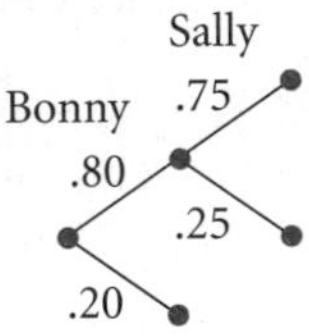

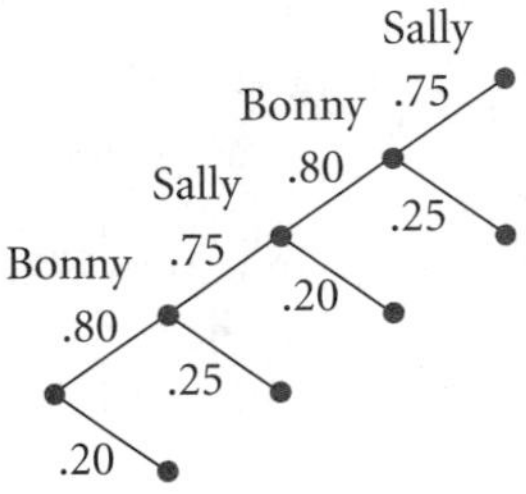

a. What is the probability that Bonny will win the point after just one volley?

b. What is the probability that Bonny will win the point in exactly two volleys?

c. Make a tree diagram to model three volleys. What is the probability that Bonny will win the point in exactly three volleys?

d. What is the probability that Bonny will win the point in at most three volleys?

e. What kind of sequence do the answers to 12a–c form? Explain.

f. What is the probability that Bonny will win the point in at most six volleys?

g. In the long run, what is the probability that Bonny will win the point?

TechnologyConnection In computer simulations of sporting events, software designers enter all the data and statistics they can find so they can model the game as accurately as possible. In tennis, for instance, individual statistics on serves, backhand shots, forehand shots, and positions on the court must be taken into account. When setting up fantasy teams or matches, programmers base their formulas on actual probabilities, and even the best players or teams will lose some of the time.

Review

13. (Lesson 11.2) Find $P(E_1 \text{ or } E_2)$ if E_1 and E_2 are mutually exclusive and complementary.

14. (Lesson 11.3) Two kinds of flu spread through a university one winter. The probability that a student gets both varieties is .18. The probability that a student gets neither variety is .42. What is the probability that a student gets exactly one of the flu varieties?

15. (Lesson 11.3) This table gives counts of different types of paperclips in Maricela's paperclip holder.

	Small	Large
Metal and oval	47	23
Plastic and oval	25	10
Metal and triangular	18	6
Plastic and triangular	10	5

a. Create a Venn diagram of the probabilities of picking each kind of clip if one is selected at random.

b. Create a tree diagram of the outcomes and their probabilities. (Use size on the first branch, shape on the second, and material on the third.)

16. (Lesson 5.8) Solve $\left(\frac{1}{2}\right)^n \leq 10^{-5}$.

17. (Lesson 11.2) Suppose the probability of winning a single game of chance is .9. How many wins in a row would it take before the likelihood of such a string of wins is less than 1%?

PROJECT

COIN TOSS GAME

Imagine a game in which you must toss a penny onto a piece of paper with a grid of 1-in. squares. You win if your penny lands completely within a square, not touching any of the lines. How likely are you to win? What if the squares measure *a* inches on each side? What if the coin has a radius of *r* inches? What if the lines are *t* inches thick? Design a coin toss game board such that the probability of success is .1.

LESSON

11.5 Permutations and Probability

The numerator and denominator of a theoretical probability are numbers of possibilities. Sometimes those possibilities follow regular patterns that allow you to "count" them.

Investigation
Order and Arrange

You will need

- five different objects

Step 1 Suppose you are working on a jigsaw puzzle. You get to a section of the puzzle where there are five spaces to be filled. Unfortunately, it looks like they are all nearly the same shape and color. You grab a handful of the remaining pieces and start trying them. How many different ways could the remaining pieces fit into the remaining spaces? If you try at random to find the right arrangement, how many arrangements are there to try?

Investigate this problem using different objects to represent the puzzle pieces. Let n represent the number of objects and r the number of spaces or slots you need to fill. For example, $n = 3$ and $r = 2$ represents the number of different ways three pieces can fill two spaces.

In this case Object A followed by Object B is different from Object B followed by Object A. Copy and complete the table below for different values of n and r.

Number of spaces, r		Number of items, n				
		$n = 1$	$n = 2$	$n = 3$	$n = 4$	$n = 5$
	$r = 1$					
	$r = 2$			6		
	$r = 3$					
	$r = 4$					
	$r = 5$					

Step 2 Describe any patterns you found in either the rows or the columns of the table.

You might have seen lots of patterns. One could be described as a product of numbers filling slots. For example, for $n = 4$ and $r = 3$ you would make three slots because $r = 3$, then start filling them in with four choices and decrease by one choice for each slot as you go: $\underline{4}$ $\underline{3}$ $\underline{2}$. The product of those numbers is 24, the number in the third row, fourth column of the table you completed.

Why do you multiply the numbers in the slots? The problem might remind you of a familiar situation. How many different outfits—consisting of a sweater, pants, and shoes—could you wear if you were to select from four different sweaters, six different pairs of pants, and two pairs of shoes?

You can visualize a tree diagram with four choices of sweaters. For each of those sweaters, you can select six different pairs of pants. Each of those sweater-and-pants outfits can be matched with two pairs of shoes. (Actually drawing all of the paths would be difficult and messy.) Each different outfit is represented by a path of three segments representing a sweater *and* a pair of pants *and* a pair of shoes. The paths represent all the possible outcomes, or the different ways in which the entire sequence of choices can be made.

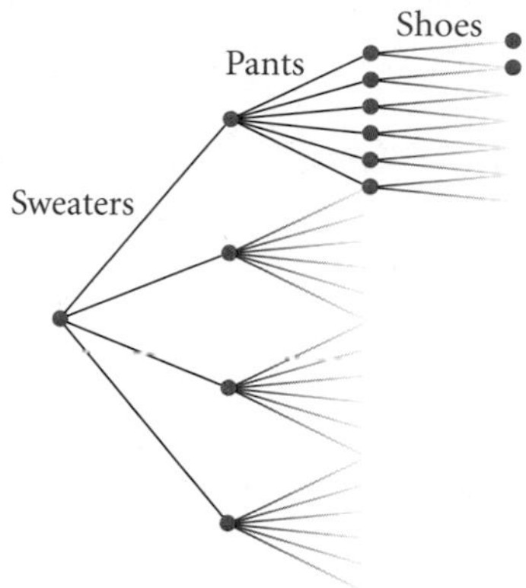

How many outfits are there? From the tree you're imagining, you can see that the total number of outfits with four choices, then six choices, and then two choices can be found by multiplying $4 \cdot 6 \cdot 2$. You are actually using the **counting principle.**

The Counting Principle

Suppose there are n_1 ways to make a choice, and for each of these there are n_2 ways to make a second choice, and for each of these there are n_3 ways to make a third choice, and so on. The product $n_1 \cdot n_2 \cdot n_3 \cdot \ldots$ is the number of possible outcomes.

The counting principle provides a quick method for counting outcomes by using multiplication. Rather than memorizing the formula, you can look for patterns and sketch or visualize a representative tree diagram.

EXAMPLE A

Suppose a set of license plates consists of any three letters of the alphabet followed by any three digits.

a. How many different license plates are possible?

b. What is the probability that a license plate has no repeated letters or numbers?

▶ *Solution*

To better understand the problem, fill in some slots.

a. There are 26 possible letters for each of the first three slots and 10 possible digits for each of the last three slots. Remember that letters and digits can repeat. Using the counting principle, you can multiply the number of possibilities.

$$\underline{26} \cdot \underline{26} \cdot \underline{26} \cdot \underline{10} \cdot \underline{10} \cdot \underline{10} = 17576000$$

There are 17,576,000 possible license plates.

b. Once the first letter is chosen, there are only 25 ways of choosing the second letter to avoid repetition. This pattern continues for the third letter and for the digits. Filling slots gives the product $\underline{26} \cdot \underline{25} \cdot \underline{24} \cdot \underline{10} \cdot \underline{9} \cdot \underline{8} = 11{,}232{,}000$. The probability that a license plate will be one of these arrangements is this number divided by the total number of possible outcomes.

$$\frac{11232000}{17576000} \approx .639$$

This means about 63.9% of license plates would not have repeated letters or numbers.

When the objects cannot be used more than once, the number of possibilities decreases at each step. These are called *arrangements without replacement;* in other words, once an item is chosen, that same item cannot be used again in the same arrangement. An arrangement of some or all of the objects of a set, without replacement, is called a **permutation.** The notation ${}_nP_r$ is read, "the number of permutations of n things chosen r at a time." As in the Investigation and part b of Example A, you can calculate ${}_nP_r$ by multiplying $n(n-1)(n-2)(n-3)\ldots(n-r+1)$. [► See **Calculator Note 11C** for computing ${}_nP_r$ on a calculator. ◄]

EXAMPLE B

Seven flute players are performing in an ensemble.

a. The names of all seven players are listed in the program in random order. What is the probability that the names are in alphabetical order?

b. After the performance the players are backstage. There is a bench with room for only four to sit. How many possible arrangements are there?

c. What is the probability that the four players are sitting in alphabetical order?

► Solution

a. There are seven choices for the first name on the list, six choices remaining for the second name, five for the third name, and so on. ${}_7P_7 = \underline{7} \cdot \underline{6} \cdot \underline{5} \cdot \underline{4} \cdot \underline{3} \cdot \underline{2} \cdot \underline{1} = 5040$. Only one of these arrangements is in alphabetical order. The probability is $1/5040 \approx 0.0002$.

b. There are only four slots to fill. There are seven choices for the first chair, six choices remaining for the second chair, five for the third chair, and four for the fourth chair.

$${}_7P_4 = 7 \cdot 6 \cdot 5 \cdot 4 = 840$$

c. With each arrangement of four there is only one correct order.

$${}_4P_4 = 24$$

There are 24 ways to arrange 4 players, so the probability is $\frac{1}{24} \approx 0.04167$.

Notice that the answer to part c does not depend on the answer to part b.

When $r = n$, you can see that ${}_nP_r$ equals the product of integers from n all the way down to 1. A product like this is called a **factorial** and is written with an exclamation point. For example, 7 factorial, or 7!, is $7 \cdot 6 \cdot 5 \cdot 4 \cdot 3 \cdot 2 \cdot 1$, or 5040. [► See **Calculator Note 11D** for finding factorials on a calculator. ◄]

Look again at Example B. You can write the solution to part a as

$${}_7P_7 = 7! = 5040$$

For part b, you need only the product of integers from 7 down to 4. You can write this as $\frac{7!}{3!}$.

$$\frac{7!}{(7-4)!} = \frac{7!}{3!} = \frac{7 \cdot 6 \cdot 5 \cdot 4 \cdot \cancel{3} \cdot \cancel{2} \cdot \cancel{1}}{\cancel{3} \cdot \cancel{2} \cdot \cancel{1}} = 7 \cdot 6 \cdot 5 \cdot 4 = 840$$

You can use this idea to write a formula for any number of permutations ${}_nP_r$. As shown above, you can write the number as $\frac{n!}{(n-r)!}$.

To avoid division by 0 when $r = n$, 0! is defined to equal 1. So, when r and n are equal,

$${}_nP_r = \frac{n!}{1} = n!$$

Permutations

A *permutation* is an arrangement of some or all of the objects of a set, without replacement.

The *number of permutations* of n objects chosen r at a time ($r \leq n$) is

$${}_nP_r = n(n-1)(n-2)\ldots(n-r+1) = \frac{n!}{(n-r)!}$$

Verify that the formula in the box gives you the same values you found in the Investigation.

Often, the challenge is to decide how to apply the counting strategy in a particular problem. In this lesson you used tree diagrams, the counting principle, your calculator, and perhaps other ways to count permutations. These are good tools for helping you understand each problem before you use any formulas.

EXERCISES

Practice Your Skills

1. Screamers Ice Cream Parlor has ten flavors of ice cream. They offer a triple-scoop cone for $2. Which of these situations are permutations? If any are not, then tell why not.

a. The different cones if all three scoops are different flavors, and vanilla, lemon, then mint is different from a cone with vanilla, mint, then lemon

b. The different cones if all three scoops are different flavors, and vanilla, lemon, then mint is the same as a cone with vanilla, mint, then lemon

c. The different cones if you can repeat a flavor two or three times, and vanilla, lemon, then lemon is different from a cone with lemon, vanilla, then lemon

d. The different cones if you can repeat a flavor two or three times, and vanilla, lemon, then lemon is the same as a cone with lemon, vanilla, then lemon

2. Evaluate the factorial expressions. (Some answers will be in terms of n.)

a. $\frac{12!}{11!}$ **b.** $\frac{7!}{6!}$ **c.** $\frac{(n+1)!}{n!}$ **d.** $\frac{n!}{(n-1)!}$

e. $\frac{120!}{118!}$ **f.** $\frac{n!}{(n-2)!}$ **g.** Find n if $\frac{(n+1)!}{n!} = 15$.

3. Evaluate the numbers of permutations. (Some answers will be in terms of n.)

a. ${}_7P_3$ **b.** ${}_7P_6$ **c.** ${}_{n+2}P_n$ **d.** ${}_nP_{n-2}$

4. Consider making a four-digit ID number using the digits 3, 5, 8, and 0.

a. How many can be formed using each digit once?

b. How many can be formed using each digit once and not using 0 first?

c. How many can be formed if repetition is allowed, and any digit can be first?

d. How many can be formed if repetition is allowed, but 0 is not used first?

ConsumerConnection Access to personal bank, telephone, or e-mail accounts often requires a password, usually one with a minimum of four digits and/or letters. Most businesses recommend that consumers change their passwords frequently and choose a password with more than six characters, in combinations of digits and upper- and lowercase letters, to increase the security of their accounts. In addition, password programs are designed to shut down automatically after a certain number of consecutive incorrect passwords have been entered.

Reason and Apply

5. For what value(s) of n and r does ${}_nP_r = 720$? (Is there more than one answer?)

6. How many factors are in the expression $n(n-1)(n-2)(n-3)\ldots(n-r+1)$?

7. An eight-volume set of reference books is kept on a shelf. The books are used frequently and put back in random order.

a. In how many ways can the eight different books be arranged on the shelf?

b. How many ways can the books be arranged so that Volume 5 will be the rightmost book?

c. Use the answers from 1a and 1b to find the probability that Volume 5 will be the rightmost book if the books are arranged at random.

d. Explain how to compute the probability in 7c by another method.

e. If the books are arranged randomly, what is the probability that the last book on the right is an even number? Explain how you determined this probability.

f. How many ways can the books be arranged so that they are in the correct order with volume numbers increasing from left to right?

g. How many ways can the books be arranged so that they are out of order?

h. What is the probability that the books happen to be in the correct order?

TechnologyConnection Computers are programmed to play games such as chess by considering all possible moves. By working with a tree diagram of possible moves, a computer can choose from millions of board positions in a matter of seconds. A person, on the other hand, usually plays the game by remembering and visualizing previous winning moves. It can take many minutes for a human to decide which move to make.

8. A computer is programmed to list all permutations of N items. Figure out how long it will take for the computer to list all of the permutations for the values of N listed. Use an appropriate time unit for each answer (minutes, hours, days, or years).

N	Number of permutations of N things	Time
5	120	0.00012 sec
10	3,628,800	3.6288 sec
12		
13		
15		
20		

9. You have purchased 4 tickets to a charity raffle. Only 50 tickets were sold. Three tickets will be drawn, and 3 prizes will be awarded.

a. What is the probability that you will win the first prize (and no other prize)?

b. What is the probability that you will win both first and second prizes, but not third prize?

c. What is the probability that you will win the second or third prize?

d. If the prizes are gift certificates for \$25, \$10, and \$5, in that order, what is the expected value of your winnings?

10. Each new student is required to attend an orientation session. How many ways can three new students be assigned to five available orientation sessions?

Review

11. (Lesson 11.3) In a physical education class of 50 students, 28 are sophomores, and 30 are athletes.

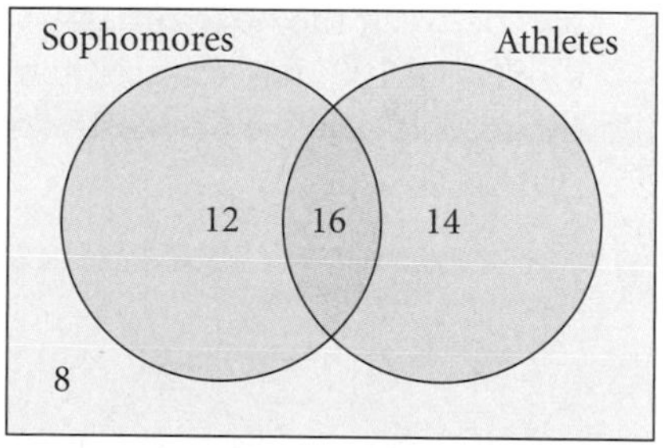

a. What is the probability that a randomly selected member of the class is an athlete?

b. If you already know that a randomly selected member is an athlete, what is the probability that the student is a sophomore?

12. (Lesson 11.4) A six-sided die is rolled. If the number showing is even, you lose a point for each dot showing. If the number showing is odd, you win a point for each dot showing.

a. Find $E(x)$ for one roll if x represents the number of points you win.

b. Find the expected winnings for ten rolls.

13. (Lesson 11.2) Assume that boys and girls are equally likely to be born.

a. If there are three consecutive births, what is the probability that two girls and one boy will be born in that order?

b. What is the probability that two girls and one boy are born in any order?

c. Given that the first two babies are girls, what is the probability that the third will be a boy?

14. For an art project Jesse is planning to make a set of origami nesting boxes (boxes that fit into each other). Each box requires five squares that will fold into an open box with no lid. The largest box will measure 4 in. on each side. The side length of each successive box is 95% of that of the previous box. Jesse wants the smallest box to be no less than 0.5 in. on a side.

a. How many boxes can Jesse make in one set?

b. What is the total amount of paper needed for one set of open boxes?

15. (Lesson 6.2)

Write the equation of this parabola in

a. polynomial form

b. vertex form

c. factored form

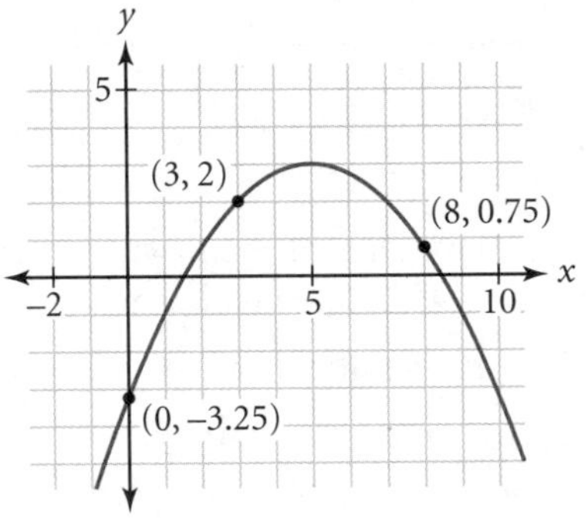

IMPROVING YOUR REASONING SKILLS

Beating the Odds

Tracy and Trish have two boxes. In one box are 50 blue marbles; in the other box are 50 red marbles. Tracy will blindfold Trish and place the two boxes in front of her. Trish will pick one marble from one of the boxes. If Trish picks a red marble, she wins. If she picks a blue marble, Tracy wins. Before being blindfolded Trish requests that she be allowed to distribute the marbles between the boxes in any way she likes. Tracy thinks about the request and says, "Sure, as long as all 100 marbles are there, what difference could it make?" How should Trish distribute the marbles to give herself the greatest chance of winning? What would be the probability that she would win?

LESSON

11.6

Combinations and Probability

Math is like love—a simple idea but it can get complicated.

R. DRABEK

If three coins are flipped, the tree diagram and the counting principle both indicate that there are $\underline{2} \cdot \underline{2} \cdot \underline{2} = 8$ equally likely outcomes: 2 choices, then 2 choices, then 2 choices. But if you are not concerned about the order in which the heads and tails occur, then Paths 2, 3, and 5 can be described as "2 heads and 1 tail," and Paths 4, 6, and 7 can be described as "1 head and 2 tails." So if you're not concerned about order, there are only 4 outcomes, which are not equally likely:

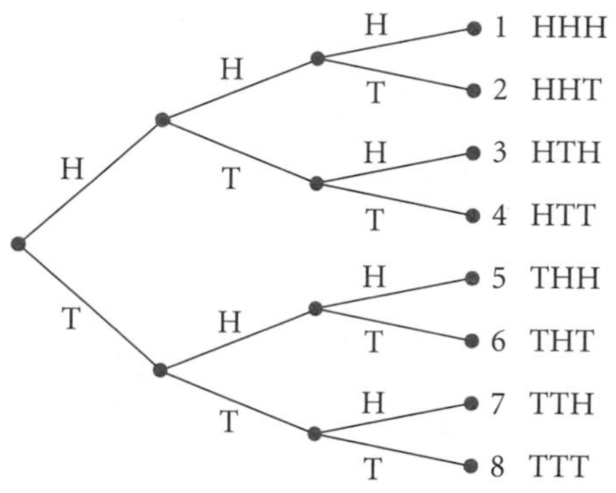

3 heads (one path)

2 heads and 1 tail (three paths)

1 head and 2 tails (three paths)

3 tails (one path)

In this lesson you will learn about counting outcomes and calculating theoretical probabilities when order doesn't matter. There are fewer possibilities when order doesn't matter than there are when order is important.

EXAMPLE A

At the first meeting of the International Club, the members get acquainted by introducing themselves and shaking hands. Each member shakes hands exactly once with every other person in the room. How many handshakes are there in each of the situations listed below?

a. Three people are in the room.

b. Four people are in the room.

c. Five people are in the room.

d. Fifteen people are in the room.

▶ Solution

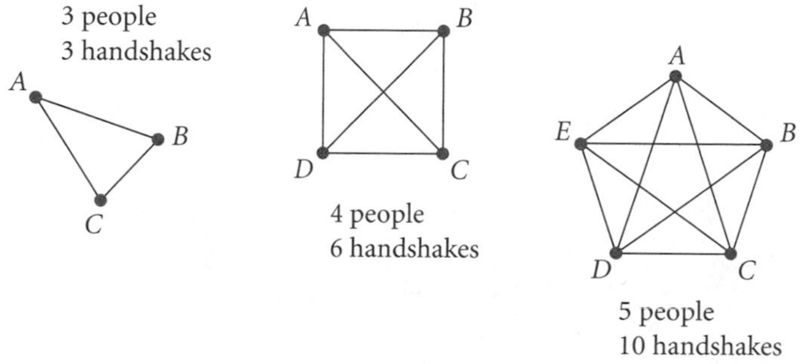

The points (vertices) pictured can represent the 3, 4, or 5 people in a room, and the lines (**edges**) can represent the handshakes. The diagrams show that there are

3 handshakes among 3 people, 6 handshakes among 4 people, and 10 handshakes among 5 people. You can find the number of handshakes by counting edges, but as you add more people to the group, it will become more difficult to draw and count. So look for patterns to determine the number of handshakes among 15 people.

The four edges at each vertex in the 5-person handshake solution might suggest that there are $4 \cdot 5$ edges. However, if you use this method to count edges, then an edge like DB will be counted at Vertex D and again at Vertex B. Because DB is the same as BD, you are counting twice as many edges as the actual total. The number of possibilities in which order doesn't matter is only half the number in which order does matter. Therefore, if 15 people are in the room, there are $\frac{14 \cdot 15}{2} = 105$ handshakes.

You can think of each handshake as a pairing of two of the people in the room, or two of the vertices. When you count collections of objects without regard to order, you are counting *combinations.*

The number of combinations of five people taken two at a time is symbolized by ${}_5C_2$. (Sometimes this notation is read as "five choose two.") Although there are ${}_5P_2 = 20$ permutations of five vertices taken two at a time, you have only half as many combinations:

$$ {}_5C_2 = \frac{{}_5P_2}{2} = \frac{20}{2} = 10 $$

Similarly, ${}_{15}C_2 = \frac{{}_{15}P_2}{2} = \frac{15 \cdot 14}{2} = 15 \cdot 7 = 105$ handshakes.

EXAMPLE B

Ann, Ben, Chang, and Dena are members of the International Club, and they have volunteered to be on a committee that will arrange a reception for exchange students. Usually there are only three students on the committee. How many different three-member committees could be formed with these four students?

▶ Solution

Note that order isn't important in these committees. ABD and BDA are the same committee and shouldn't be counted more than once. The number of different committee combinations will be fewer than the ${}_4P_3 = 24$ permutations listed below.

ABC	**ABD**	**ACD**	**BCD**
ACB	ADB	ADC	BDC
BAC	BAD	CAD	CBD
BCA	BDA	CDA	CDB
CAB	DAB	DAC	DBC
CBA	DBA	DCA	DCB

The four committees in the top row can represent all of the $3! = 6$ arrangements listed in each column. Therefore, the number of permutations, ${}_4P_3$, is six times the number of combinations. That is, ${}_4C_3 = \frac{{}_4P_3}{3!}$. You can evaluate ${}_4C_3$ using the factorial definition of ${}_nP_r$.

$$ {}_4C_3 = \frac{{}_4P_3}{3!} = \frac{4!}{3!(4-3)!} = \frac{4!}{3!1!} = \frac{4 \cdot \cancel{3} \cdot \cancel{2} \cdot \cancel{1}}{\cancel{3} \cdot \cancel{2} \cdot \cancel{1}} = 4 $$

Combinations

A **combination** is a grouping of objects from a set without regard to order. The *number of combinations* of n objects taken r at a time ($r \leq n$) is

$$_nC_r = \frac{_nP_r}{r!} = \frac{n!}{r!(n-r)!}$$

Rather than simply memorizing the formula, try to understand how combination numbers like these connect to the number of permutations and to a tree diagram. You may want to draw one or more representations of each problem you investigate.

You can count combinations to calculate theoretical probabilities.

EXAMPLE C

Suppose a coin is tossed ten times.

a. What is the probability that it will land heads exactly five times?

b. What is the probability that it will land heads exactly five times, including on the third toss?

► Solution

a. The tree diagram of this problem has ten stages (one for each flip) and splits into two possibilities (heads or tails) at each point on the path. It's not reasonable to draw the entire tree diagram, but by the counting principle there are $2 \cdot 2 \cdot 2 \cdot 2 \cdot 2 \cdot 2 \cdot 2 \cdot 2 \cdot 2 \cdot 2 = 2^{10} = 1024$ possibilities. To find the numerator of the probability, you must determine how many of the 1,024 separate paths contain five heads. Because order is not important, you can find the number of paths that fit this description by counting combinations. There are $_{10}C_5$ or $\frac{10!}{5!5!} = 252$ ways of choosing five of the ten branches to contain H's. [► See Calculator Note 11E to find combination numbers on a calculator. ◄] That is, 252 of the 1,024 paths will contain five heads. Therefore, the probability that you will get exactly five heads and five tails is $\frac{252}{1024} \approx .246$, or 24.6%, (a little less than 1 in 4).

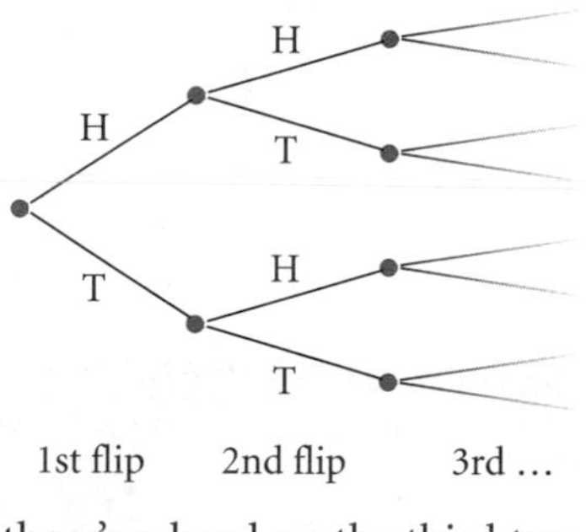

b. If there's a head on the third toss, then the other four heads must be chosen from the other nine tosses. There are $_9C_4 = 126$ ways of making that choice. Therefore, the probability of this event is $\frac{126}{1024} \approx .123$.

In the Investigation you'll count combinations to discover mathematical reasons why playing the lottery is a losing proposition.

Investigation
Winning the Lottery

You will do Step 1 of this Investigation with the whole class. Then you will work with your group to analyze the results.

Consider a state lottery called Lotto 47. Twice a week players select six different numbers between 1 and 47, inclusive. The state lottery commission also selects six numbers from 1 through 47. Selection order doesn't matter, but a player needs to match all six numbers to win the Lotto.

Step 1 Follow these directions with your class to simulate playing Lotto 47.

a. For five minutes write down as many different sets of six numbers as you can. Write only integers between 1 and 47, inclusive.

b. After five minutes of writing, everyone stands up.

c. Your teacher will generate a random integer, 1–47. Cross out all of your sets that do not contain that number. If you have no sets remaining, sit down.

d. Your teacher will generate a second number in this range. (If it's the same number as before, it will be skipped.) Again, cross out any remaining sets that do not contain this number. If you have crossed off all your sets, sit down.

e. Your teacher will continue generating different random numbers until no one is left standing or six numbers have been generated.

Work together with your group to answer the questions below.

Step 2 What is the probability that any one set of six numbers wins?

Step 3 At $1 for each set of six numbers, how much did each group member invest during the first five minutes? What was the total group investment?

Step 4 Estimate the total amount invested by the entire class in the five minutes. Explain how you determined this estimate.

Step 5 Estimate the probability that someone in your class wins. Explain how you determined this estimate.

Step 6 Estimate the probability that someone in your college would win if everyone in the college participated in this activity. Explain how you determined this estimate.

Step 7 If each possible set of six numbers were written on a 1-in. chip, and if all the chips were laid end to end, how long would the line of chips be? Convert your answer to an appropriate unit.

Step 8 Write a paragraph comparing Lotto 47 with some other event the probability of which is approximately the same.

EXERCISES

Practice Your Skills

1. Evaluate each factorial expression without your calculator.

a. $\frac{10!}{3!7!}$ b. $\frac{7!}{4!3!}$ c. $\frac{15!}{13!2!}$ d. $\frac{7!}{7!0!}$

2. Evaluate each expression.

a. ${}_{10}C_7$ **b.** ${}_{7}C_3$ **c.** ${}_{15}C_2$ **d.** ${}_{7}C_0$

3. Consider each expression of the form ${}_{n}P_r$ and ${}_{n}C_r$.

a. What is the relationship between ${}_{7}P_2$ and ${}_{7}C_2$?

b. What is the relationship between ${}_{7}P_3$ and ${}_{7}C_3$?

c. What is the relationship between ${}_{7}P_4$ and ${}_{7}C_4$?

d. What is the relationship between ${}_{7}P_7$ and ${}_{7}C_7$?

e. Describe how you can find ${}_{n}C_r$ if you know ${}_{n}P_r$.

4. Which is larger, ${}_{18}C_2$ or ${}_{18}C_{16}$? Explain.

Reason and Apply

5. For what value(s) of n and r does ${}_{n}C_r = 35$?

6. Find a number r, $r \neq 4$, such that ${}_{10}C_r = {}_{10}C_4$. Explain why this makes sense.

7. Suppose you are to answer any four of seven essay questions on a history test, and your teacher doesn't care in which order you answer them.

a. How many different question combinations are possible?

b. What is the probability that you include Essay Question 5 if you randomly select your combination?

8. Write a short letter to Pika Lock Company and explain why their "combination locks" should be called "permutation locks." Be sure to tell them how a true "combination lock" would work.

9. Find the following sums.

a. ${}_{2}C_0 + {}_{2}C_1 + {}_{2}C_2$

b. ${}_{3}C_0 + {}_{3}C_1 + {}_{3}C_2 + {}_{3}C_3$

c. ${}_{4}C_0 + {}_{4}C_1 + {}_{4}C_2 + {}_{4}C_3 + {}_{4}C_4$

d. Make a conjecture and test it by finding the sum for all possible combinations of five things.

10. Consider the Lotto 47 played in the Winning the Lottery Investigation.

a. If it takes someone ten seconds to fill out a Lotto 47 ticket, how long would it take them to fill out all possible tickets?

b. If someone fills out 1,000 tickets with different numbers, what is their probability of winning the Lotto 47?

11. Draw a circle, and space points equally on its circumference. Draw chords to connect all pairs of points.

a. If you place four points, how many chords are there?

b. If you place five points, how many chords are there?

c. If you place nine points, how many chords are there?

d. If you place n points, how many chords are there?

12. In most state and local courts, 12 jurors and two alternates are chosen from a pool of 30 prospective jurors. If a juror becomes unable to serve, then the first alternate will replace them. The second alternate will be called upon if another juror is dismissed.

a. In how many ways can 12 jurors and a first and second alternate be chosen from 30 people?

b. In federal court cases, 12 jurors and 4 alternates are usually selected from a pool of 64 prospective jurors. In how many ways can this be done?

Review

13. (Algebra) Expand each expression.

a. $(x + y)^2$ **b.** $(x + y)^3$ **c.** $(x + y)^4$

14. (Lesson 11.5) How many speeds does a bicycle have if it has three sprockets in front and five sprockets in the rear?

15. (Lesson 11.2) Use the tree diagram to find each probability and explain its meaning.

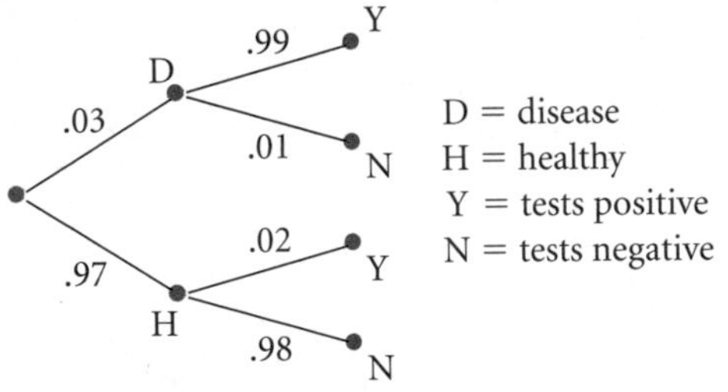

a. $P(H \text{ and } P)$ **b.** $P(Y \mid H)$ **c.** $P(P)$ **d.** $P(H \mid Y)$

16. (Lesson 8.3) While on a birding field trip, Angelo spies what he is sure is a rare spotted owl at the top of a tall tree. Lying on his stomach in the grass, he measures the angle of elevation of his line of sight as 32°. He then crawls 8.6 m closer to the tree. The angle of his line of sight is now 42°. He is afraid to move any closer for fear of disturbing the owl. How close did he get to the tree? How close did he get to the owl?

LESSON

11.7

The Binomial Theorem and Pascal's Triangle

Mathematics is the science of patterns.

LYNN ARTHUR STEEN

Probability is an area of mathematics that is rich with patterns. Many random processes involve patterns in which there are two possible choices, such as flipping coins. In this lesson you will learn about using the binomial theorem and Pascal's triangle to find probabilities in those cases.

Pascal's triangle is shown here. It contains many different patterns that have been studied for centuries. The triangle begins with a 1, then two 1's beneath it. Each successive row is filled with numbers formed by adding the two numbers above it. For example, each of the 4's is the sum of a 1 and 3 on the previous row. Every row begins and ends with a 1.

$$\begin{array}{ccccccccccc} & & & & & 1 & & & & & \\ & & & & 1 & & 1 & & & & \\ & & & 1 & & 2 & & 1 & & & \\ & & 1 & & 3 & & 3 & & 1 & & \\ & 1 & & 4 & & 6 & & 4 & & 1 & \\ 1 & & 5 & & 10 & & 10 & & 5 & & 1 \end{array}$$

In Lesson 11.6 you studied numbers of combinations. Notice that combination numbers occur in the rows of Pascal's triangle. For example, the numbers 1, 5, 10, 10, 5, and 1 in the sixth row are the values of ${}_5C_r$:

$${}_5C_0 = 1, {}_5C_1 = 5, {}_5C_2 = 10, {}_5C_3 = 10, {}_5C_4 = 5, {}_5C_5 = 1$$

Is this the case in all rows? If so, why? In the Investigation you'll explore these questions.

Investigation

Pascal's Triangle and Combination Numbers

A group of five students regularly eats lunch together, but each day only three of them can show up.

Step 1 How many groups of these three students could there be? Express your answer in the form ${}_nC_r$ and as a numeral.

Step 2 If Leora is definitely at the table, how many other students are at the table? How many students are there from which to choose? Find the number of combinations of students possible in this instance. Express your answer in the form ${}_nC_r$ and as a numeral.

Step 3 How many combinations are there that don't include Leora? Consider how many students there are from which to select, and how many are to be chosen. Express your answer in the form ${}_nC_r$ and as a numeral.

Step 4 | Repeat Steps 1–3 for groups of four of the five students.

Step 5 | What patterns do you notice in your answers to Steps 1–3 for groups of three students and four students? Write a general rule that expresses ${}_nC_r$ as a sum of other combination numbers.

Step 6 | How does this rule relate to Pascal's triangle?

Pascal's triangle is often used for expanding binomials. Binomials like $(H + T)$ or $(1 + r)$ or $(x + y)$ appear frequently in mathematics. An **expansion of a binomial** is a rewriting of a binomial raised to a power. For example, the expansion of $(H + T)^3$ is $1H^3 + 3H^2T + 3HT^2 + 1T^3$. Note that the coefficients of this expansion are the numbers in the fourth row of Pascal's triangle.

Why are the numbers in Pascal's triangle equal to the coefficients of a binomial expansion? Remember that the numbers in Pascal's triangle are evaluations of ${}_nC_r$. So the same question could also be worded, "Why are the coefficients of a binomial expansion equal to values of ${}_nC_r$?"

EXAMPLE A

Relate binomial coefficients in the expansion $(H + T)^3$ to combination numbers.

▶ Solution

$(H + T)^3$ can be written as $(H + T)(H + T)(H + T)$. To expand this, multiply the right side in steps. First, multiply the first pair $(H + T)$ by the second pair $(H + T)$.

$$(H + T)\,(H + T) = (HH + HT + TH + TT)$$

Then, multiply each term of that result with the H and the T in the third pair.

$$(HH + HT + TH + TT)(H + T) =$$
$$HHH + HHT + HTH + HTT + THH + THT + TTH + TTT$$

Notice that if H represents flipping a head, and T represents flipping a tail, this result shows you all the possible outcomes of flipping a coin three times. This expression can also be rewritten as

$$1H^3 + 3H^2T + 3HT^2 + 1T^3$$

This process shows you that there is one way of getting three heads, three ways of getting two heads and a tail, and so on. These combinations could also be expressed as ${}_3C_3$, ${}_3C_2$, and so on.

Here are binomial expansions of the first few powers of $(H + T)$. Notice that the coefficients are the same as the rows of Pascal's triangle. Think about what these expansions tell you about the results of flipping a coin 0, 1, 2, and 3 times.

$(H + T)^0 =$	$\mathbf{1}$	1st row
$(H + T)^1 =$	$\mathbf{1}H + \mathbf{1}T$	2nd row
$(H + T)^2 =$	$\mathbf{1}H^2 + \mathbf{2}HT + \mathbf{1}T^2$	3rd row
$(H + T)^3 =$	$\mathbf{1}H^3 + \mathbf{3}H^2T + \mathbf{3}HT^2 + \mathbf{1}T^3$	4th row

You can use Pascal's triangle, or the values of ${}_nC_r$, to expand binomials without multiplying all the terms together.

The Binomial Theorem

If a binomial $(p + q)$ is raised to a whole-number power, n, the coefficients of the expansion are the combination numbers ${}_nC_n$ to ${}_nC_0$, as shown:

$$(p + q)^n = {}_nC_0 p^n q^0 + {}_nC_1 p^{n-1} q^1 + {}_nC_2 p^{n-2} q^2 + \cdots + {}_nC_{n-2} p^2 q^{n-2} + {}_nC_{n-1} p^1 q^{n-1} + {}_nC_n p^0 q^n$$

or

$$(p + q)^n = \sum_{j=0}^{n} {}_nC_j \cdot p^{n-j} q^j$$

Binomial expansions can be used to represent results of random processes with two possible outcomes. In Example A you saw that if a coin is flipped three times, there is 1 possible outcome of 3 heads, 3 outcomes of 2 heads and 1 tail, 3 outcomes of 1 head and 2 tails, and 1 outcome of 3 tails.

When you flip a fair coin, the outcomes H and T are equally likely.

Example B shows how a binomial expansion can be used to find probabilities of outcomes that are not equally likely.

EXAMPLE B

A hatching yellow-bellied sapsucker has a .58 probability of surviving to adulthood. If a nest has six eggs, what are the probabilities that 0, 1, 2, 3, 4, 5, and 6 birds will survive?

► Solution

Represent the event of survival or success by S and the event of nonsurvival by N. The outcome that 4 birds of the 6 survive and the other 2 do not survive is given by S^4N^2. The number of combinations in which 4 birds survive is given by ${}_6C_4$. The expression ${}_6C_4 S^4 N^2$ occurs in the expansion of $(S + N)^6$.

$$(S + N)^6 = {}_6C_0 \cdot S^6 \cdot N^0 + {}_6C_1 \cdot S^5 \cdot N^1 + {}_6C_2 \cdot S^4 \cdot N^2 + {}_6C_3 \cdot S^3 \cdot N^3 + {}_6C_4 \cdot S^2 \cdot N^4 + {}_6C_5 \cdot S^1 \cdot N^5 + {}_6C_6 \cdot S^0 \cdot N^6$$

In fact, each term of the binomial expansion gives a number of ways that a certain number of birds will survive and the others will not. You can substitute .58 for S and .42 for N to find the probability of each outcome.

$$= 1(.58)^6(.42)^0 + 6(.58)^5(.42)^1 + 15(.58)^4(.42)^2 + 20(.58)^3(.42)^3 + 15(.58)^2(.42)^4 + 6(.58)^1(.42)^5 + 1(.58)^0(.42)^6$$
$$\approx .038 + .165 + .299 + .289 + 1.57 + .045 + .005$$

You might put these numbers into a table to organize the information.

Number of birds that survive *x*	6	5	4	3	2	1	0
Probability *P(x)*	.038	.165	.299	.289	.157	.045	.005

So it is most likely that 4 birds survive and very unlikely that 0 will survive.

In a situation like Example B sometimes you want to know the probability that at most 4 birds survive or at least 4 birds survive. Can you determine how to calculate the values in this expanded table? You will calculate the missing table values in Exercise 3 of the exercise set.

Probability of survival	6 birds	5 birds	4 birds	3 birds	2 birds	1 bird	0 birds
Exactly	.038	.165		.289	.157	.045	.005
At most	1	.962			.207	.050	.005
At least	.038				.949	.995	.998

You can think of the 6 birds as a **sample** taken from a **population** of many birds. You know the probability of success (survival) in the population, and you want to calculate various probabilities of success in the sample. You saw in Example B that a survival rate of 4 birds was most likely.

What if the situation were reversed? Suppose you don't know that the probability of success in the population is .58, but you do know that 4 out of 6 birds survive to adulthood. This would indicate a population success probability of $\frac{4}{6}$, or .67; but you saw in the example that it's also possible with a population success probability of .58. What other population success probabilities would have the highest probability of 4 survivors out of 6?

This kind of problem often presents itself when voters are polled. A pollster questions a sample of voters, not the whole population. Statisticians then must interpret the data from that sample to make predictions about the population.

EXAMPLE C

A random sample of 32 voters is taken from a large population. In the sample 24 voters favor passing a proposal. What can you predict about the probability that a random member of the whole population favors the proposal?

▶ Solution

The ratio of supporters in the sample is $\frac{24}{32} = 0.75$. This could happen if the probability that a randomly chosen member of the population supports the proposal is also .75. But it could also happen for other probabilities of success in the population.

You can represent the unknown probability of support in the population with p and the probability of nonsupport with q. Because p and q are complements, $q = 1 - p$. The probability that exactly n people in a sample of 32 will be supporters is ${}_{32}C_n p^n q^{32-n}$.

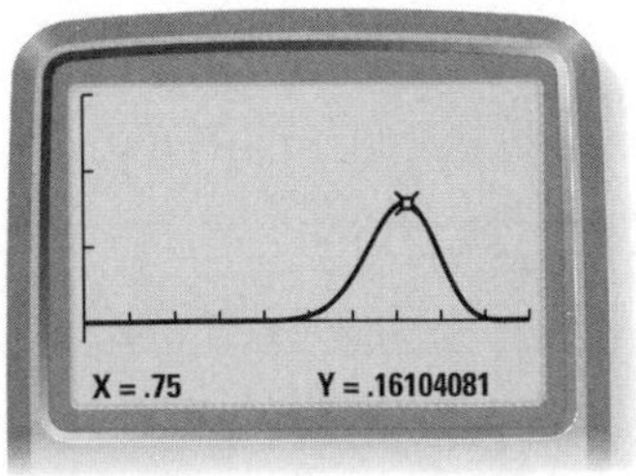

If $n = 24$, as in the sample taken, then $p = .75$ gives the highest probability. But what other values of p give a higher probability of 24 supporters than of 23 or 25 supporters? That is, what values of p will satisfy the two inequalities

$${}_{32}C_{24}\,p^{24}q^{8} > {}_{32}C_{24}\,p^{23}q^{9}$$

$${}_{32}C_{24}\,p^{24}q^{8} > {}_{32}C_{24}\,p^{25}q^{7}?$$

The first inequality can be solved by algebra.

${}_{32}C_{n}p^{24}q^{8} > {}_{32}C_{n}p^{23}q^{9}$	Original inequality.
${}_{32}C_{n}p > {}_{32}C_{n}q$	Divide both sides by p^{23} and by q^{8}.
$\frac{32!}{24!8!}p > \frac{32!}{23!9!}q$	Definition of combination numbers.
$\frac{p}{24!8!} > \frac{q}{23!9!}$	Divide both sides by 32!.
$9p > 24q$	Multiply both sides by 24! and by 9!.
$9p > 24(1 - p)$	Replace q with $1 - p$.
$33p > 24$	Add $24p$ to both sides.
$p > \frac{24}{33}$	Divide both sides by 33.

Solving the other inequality in the same way gives

$$p < \frac{25}{33}$$

You can then say that

$$\frac{24}{33} < p < \frac{25}{33}, \text{ or } .7273 < p < .7676$$

Although $\frac{25}{33}$ is closer to .7676 than to .7675, knowing that $p < \frac{25}{33}$ does not guarantee that it's less than .7675, only that it's less than .7676.

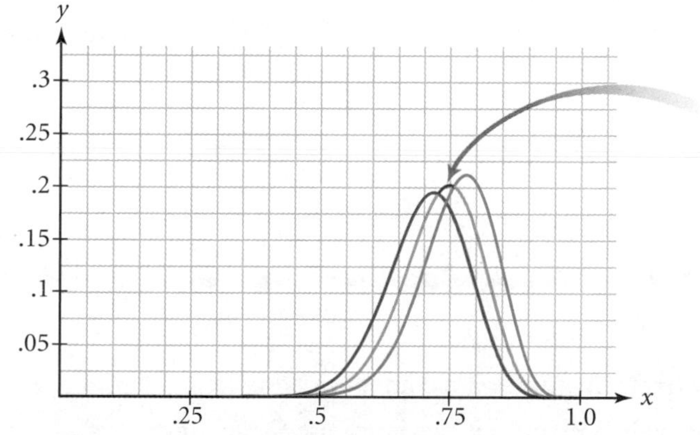

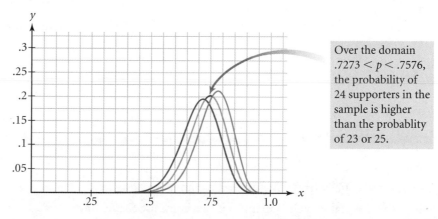

Over the domain $.7273 < p < .7576$, the probability of 24 supporters in the sample is higher than the probablity of 23 or 25.

So a random member of the population has a probability between about 73% and 76% of favoring the proposal. The proposal is likely to pass.

In Example C you found an interval for a population probability. In statistics classes you learn about a similar, more commonly used type of interval called a *confidence interval.* This interval is also based on probability.

EXERCISES

Practice Your Skills

1. Given the expression $(x + y)^{47}$, find the requested term.

a. 1st term **b.** 11th term **c.** 21st term **d.** 31st term

2. Assume that several events are independent and that the probability of each is .25.

a. What is the probability that an event will fail?

b. What is the probability that two events will succeed in two trials?

c. What is the probability that n events will succeed in n trials?

d. What is the probability that there will be some combination of two successes and three failures in five trials?

3. Return to the expanded table after the Investigation in this lesson.

a. Fill in the missing probabilities in the "exactly" row.

b. Find the two missing probabilities in the "at most" row. To find the probability that *at most* two birds survive, find the sum of the probabilities of zero, one, and two birds surviving.

c. To find the *at least* probabilities, you will need to add the probabilities from the right in the "exactly" row. Find the three missing percentages in the "at least" row.

d. Why don't the "at most" and "at least" values for each number of birds add up to 1.0? Make a statement about birds that incorporates the 20.3% entry located in the "at least" row.

4. Suppose that the probability of a success is .62. What is the probability that there would be 35 successes in 50 trials?

5. Solve for p: ${}_{32}C_{24}p^{24}q^{8} > {}_{32}C_{24}p^{25}q^{7}$

Reason and Apply

6. Answer each probability question.

a. List the equally likely outcomes if a coin is tossed twice.

b. List the equally likely outcomes if two coins are tossed once.

c. Draw a tree diagram that pictures the answers to 6a and 6b.

d. Describe the connection between the combination numbers ${}_2C_0 = 1$, ${}_2C_1 = 2$, and ${}_2C_2 = 1$ and the results of your answers to 6a–c.

e. Give a real-world meaning of the equation $(H + T)^2 = 1 \cdot H^2 + 2 \cdot HT + 1 \cdot T^2$.

7. Expand each binomial using the binomial theorem.

a. $(x + y)^4$

b. $(p + q)^5$

c. $(2x + 3)^3$

d. $(3x - 4)^4$

8. **APPLICATION** A survey of 50 people shows that only 10 are in support of a new traffic circle on 8th Street.

a. Which term of $(p + q)^{50}$ would correspond to the results of this sample?

b. What inequality does p satisfy if the probability that 10 of the 50 are supporters is greater than the probability that 11 are supporters?

c. What inequality does p satisfy if the probability that 10 of the 50 are supporters is greater than the probability that 9 are supporters?

d. Solve these inequalities to find an interval for p.

9. Dr. Miller is using a method of treatment that is 97% effective.

a. What is the probability that there will be no failure in 30 treatments?

b. What is the probability that there will be fewer than 3 failures in 30 treatments?

c. Let x represent the number of failures in 30 treatments. Enter an equation in y_1 that will provide a table of values representing the probability $P(x)$ for any value of x.

d. Use the equation and table from 9c to find the probability that there will be fewer than 3 failures in 30 treatments.

10. A university medical research team has developed a new test that is 88% effective at detecting a disease in its early stages. What is the probability that there will be more than 20 false readings in 100 applications of the test? [► See **Calculator Note 11F.** ◄]

11. Suppose the probability is .12 that a penny chosen at random was minted before 1975. [► See **Calculator Note 11F.** ◄]

a. What is the probability that you will find 25 or more such coins in a roll of 100 pennies?

b. What is the probability that you will find 25 or more such coins in two rolls of 100 pennies each?

c. What is the probability that you will find 25 or more such coins in three rolls of 100 pennies each?

12. Research has shown that a blue-footed booby has a 47% chance of surviving from egg to adulthood. Suppose you have found a nest of four eggs.

a. What is the probability that all four will survive to adulthood?

b. What is the probability that none of the four will survive to adulthood?

c. How many birds would you expect to survive?

13. A coin is tossed five times, and it comes up heads four out of five times. In your opinion is this event a rare occurrence? Defend your position.

14. Data collected over the last ten years show that in a particular town it will rain sometime during 30% of the days in the spring. (Assume that that the chance of rain is independent from one day to the next).

a. How likely is it that there will be a week with exactly five rainy days?

b. How likely is it that there will be a week with exactly six rainy days?

c. How likely is it that there will be a week with exactly seven rainy days?

d. How likely is it that there will be a week with at least five rainy days?

15. Consider the function $y = f(x) = \left(1 + \frac{1}{x}\right)^x$. Note that this expression is a binomial raised to a power.

a. Fill in this table using the binomial theorem or Pascal's triangle. Verify using your calculator.

x	1	2	3	4
y				

b. Using your calculator, find $f(10)$, $f(100)$ and $f(1000)$, $f(10000)$.

c. Describe what happens to the values of $f(x)$ as larger and larger values of x are used.

16. Professor Gutierrez has 25 students and randomly sends 4 or 5 to the board each day to solve a homework problem. You can calculate that there are ${}_{25}C_4 = 12650$ ways that 4 students could be selected and ${}_{25}C_5 = 53130$ ways that 5 students could be selected. Suppose the class grows to 26 students. Without using a calculator, determine how many ways 5 students can be selected now. Explain your solution method.

Review

17. (Lesson 11.4) The probability that a CD is error free is .995.

a. In a box of five CDs, what is the probability that none are error free? That one is error free? Two? Three? Four? Five?

b. What is the expected number of error-free CDs?

18. (Lesson 11.1) Suppose that 350 points are randomly selected within the rectangle, and 156 of them fall within the closed curve. What is an estimate of the area within the curve?

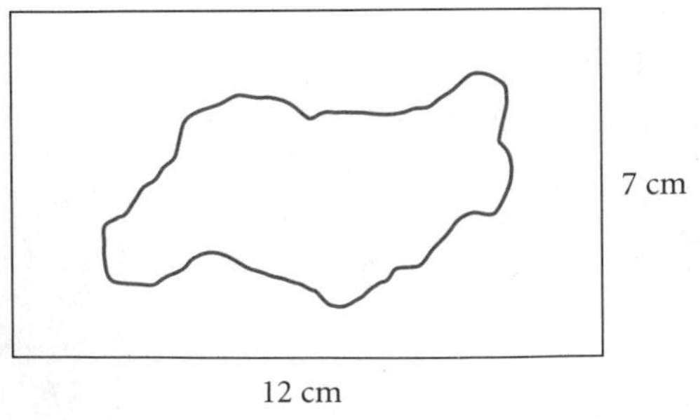

EXPLORATION
with Microsoft® Excel

Dice Experiment

Download the Chapter 11 Excel Exploration from the website at www.keycollege.com/online or from your Student CD.

You will simulate the tossing of two dice a large number of times. Then you will explore the experimental probability of getting any particular sum on the dice.

On the spreadsheet you will find complete instructions and questions to consider as you explore.

Activity
Study Pascal's Triangle

Excel skills required

- Data fill

You will generate many rows of Pascal's Triangle.

You can find instructions for the required Excel skills in the Excel Notes at www.keycollege.com/online or on your Student CD.

Launch Excel.

Step 1 Set up the labels as shown.

	A	B	C
1	Row #	Cumulative Sum	Sum

Step 2 In Cell D2 enter the number 1.

Step 3 In Cell D3 enter the formula `=$D$2`, and in Cell E3 enter the formula `=D2+E2`.

Step 4 You need to sum the numbers in each row and show the results in Column C. The number of items in each row will vary, but there will be fewer than 50 items. Excel assumes an empty cell is equal to 0. In Cell C2 enter the formula `=SUM(D2:BA2)`.

Step 5 In Cell B2 enter the number 1. That is the sum of all the numbers in Column C up to and including that row.

Step 6 Data-fill Cell C2 down to Cell C3.

Step 7 In Cell B3 enter the formula `=SUM($C$2:C3)`.

Step 8 Data-fill Cells D3 and E3 down to Row 28.

Step 9 Highlight Cells E4 to E28, and use that to data-fill across to Column AD.

Step 10	In Cell A2 enter the number 0, and in Cell A3 enter the formula =A2+1. This column will track the row number, starting off with the first row as Row 0.
Step 11	Using Cells A3, B3, and C3, data-fill down to Row 28 in the spreadsheet.

Questions

1. What pattern emerges in Column C? Express this pattern symbolically.
2. What is the relationship between Columns B and C? Express this relationship symbolically.
3. The array of numbers in Column D and the columns to the right of it is called *Pascal's triangle.* What entry appears in the 10th row, 3rd column?
4. Using Pascal's triangle, find the term involving x^8 in the expansion of $(x + 2)^{11}$.
5. Expand $(x + 3)^6$ using Pascal's triangle.
6. A fair coin will be tossed ten times. What is the probability of getting heads exactly five times? What is the probability of getting heads five or six times?
7. Another coin will be tossed ten times. It is an unfair coin, weighted to favor heads with the probability, p, of .6. What is the probability of getting heads exactly five times? Hint: We need ${}_{10}C_5(0.6)^5(1 - 0.6)^5$. What is the probability of getting heads five or six times?

Extension

8. Describe the pattern you see in Column E. (Hint: Consider taking finite differences.)
9. It is known that if a number in Column B is prime, then the product of that number and the number to the right of it in Column C is a perfect number. This product is called a *Mersenne perfect number,* and the prime in Column B is called a *Mersenne prime.* A prime number has exactly two factors: the number itself and one. A perfect number equals the sum of all the proper factors. For instance, 6 has factors of 1, 2, and 3; and $1 + 2 + 3 = 6$.
   a. Find the first four Mersenne prime numbers, and verify that when they are multiplied by the number to the right, the result is a perfect number.
   b. If you randomly pick two rows from the first 25 rows in this table, what is the probability that the entries in column B of both rows will be a Mersenne prime?
   c. Do some Web research to learn about the Great Internet Mersenne Prime Search and the importance of Mersenne primes. These numbers get very large very quickly. The 42nd Mersenne prime was discovered in February 2005 and has 7,816,230 digits.

CHAPTER 11 REVIEW

In this chapter you were introduced to the concept of **randomness,** and you learned how to generate random numbers on your calculator. Random numbers should all have an equal chance of occurring and in the long run should occur equally frequently. You can use random number procedures to simulate situations and determine the **experimental probability** of an event. You can determine **theoretical probability** by comparing the number of successful **outcomes** to the total number of possible outcomes. You represented situations involving probability with **Venn diagrams** and learned the meaning of events that are **dependent, independent,** and **mutually exclusive.** You used **tree diagrams** to help you count possibilities, and you learned that sometimes probability situations can be represented geometrically. Using what you learned about theoretical probability, you were able to calculate **expected value,** which is calculated by multiplying the value of each event by its probability and then summing all the products.

To help find theoretical probabilities, you were introduced to some formal counting techniques. The **counting principle** states that when there are n_1 ways to make the first choice, n_2 ways to make the second choice, n_3 ways to make the third choice, and so on, the product $n_1 \cdot n_2 \cdot n_3 \cdot \ldots$ represents the total number of different ways in which the entire sequence of choices can be made. These arrangements of choices, in which the order is important, are called **permutations.** The notation ${}_nP_r$ indicates the number of ways of choosing r things out of n possible choices. If the order is unimportant, then arrangements are called **combinations,** and the symbol ${}_nC_r$ represents the number of combinations of r things from a set of n choices. Your calculator can automatically calculate permutation numbers and combination numbers, but before you use the calculator, be sure that you understand the situation and can visualize the possibilities. Combination numbers also appear in **Pascal's triangle** and as coefficients in **binomial expansions** that can be used to help calculate probabilities when there are two possible outcomes.

EXERCISES

1. Name two different ways to generate random numbers from 0 to 10.

2. Answer each geometric probability problem.

a. What is the probability that a randomly plotted point will land in the shaded region pictured here?

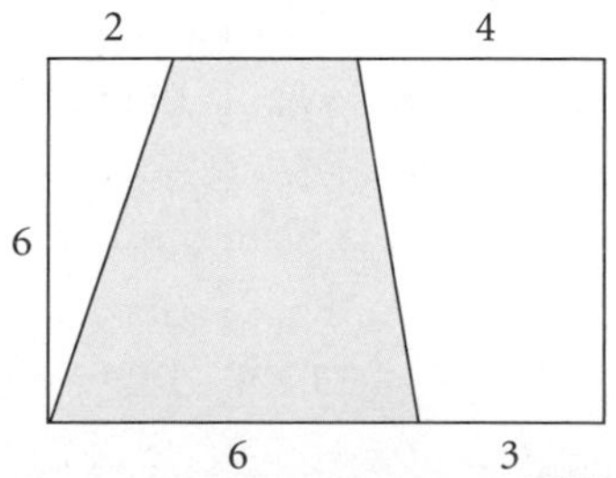

b. One thousand points are randomly plotted in the rectangular region here. Suppose that 374 of the points land in the shaded portion of the region. What is an approximation of the area of the shaded portion?

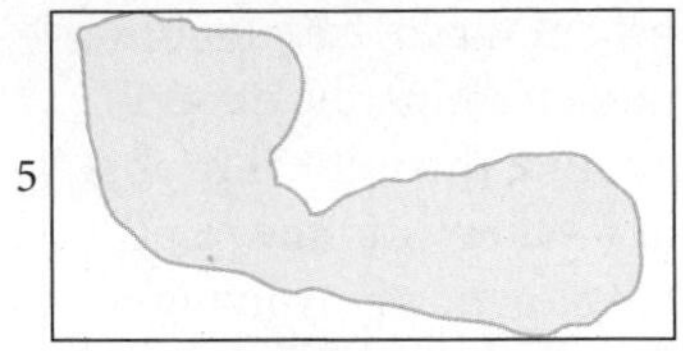

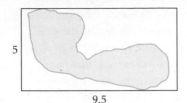

3. Suppose you roll two octahedral (eight-sided) dice.

a. Draw a diagram that shows all possible outcomes of this experiment.

b. Indicate on your diagram all the possible outcomes for which the sum of the dice is less than 6.

c. What is the probability that the sum is less than 6?

d. What is the probability that the sum is more than 6?

4. A true-or-false test has five questions.

a. Draw a tree diagram representing all of the possible results. (Assume all five questions are answered.)

b. How many possible ways are there of getting three true and two false answers?

c. How could you count combinations or permutations to answer 4b?

d. Suppose you knew that the answers to the first two questions on the test were true. What is the probability that there will be three true and two false answers on the test?

5. The local outlet of Frankfurter Franchise sells three types of hot dogs: plain, with chili, and veggie. The owners know that 47% of their sales are chili dogs, 36% are plain, and the rest are veggie. They also offer three types of buns: plain, rye, and multigrain. Sixty-two percent of their sales are plain buns, 27% are multigrain, and the rest are rye.

a. Make a tree diagram showing this information.

b. What is the probability that the next customer will order a chili dog on rye?

c. What is the probability that the next customer will *not* order a veggie dog on a plain bun?

d. What is the probability that the next customer will order either a plain hot dog on a plain bun or a chili dog on a multigrain bun?

6. A survey was taken regarding students' preferences for whipped cream or ice cream to be served with chocolate cake. The results, tabulated by age, are reported below.

	Under 18	18–19	20–25	Over 25	Total
Ice cream	18	37	85	114	
Whipped cream	5	18	37	58	
Total					

a. Complete the table.

b. What is the probability that a randomly chosen 18- or 19-year-old will prefer ice cream?

c. What is the probability that a randomly chosen student between 20 and 25 will prefer whipped cream?

d. What is the probability that someone who prefers ice cream is under 18?

e. What is the probability that a randomly chosen student will prefer whipped cream?

7. Rita is practicing darts. On this particular dart board she can score 20 points for a bull's-eye and 10 points, 5 points, or 1 point for the other regions. Although Rita doesn't know exactly where her five darts will land, she has been a fairly consistent dart player over the years. She figures that she hits the bull's-eye 30% of the time, the 10-point circle 40% of the time, the 5-point circle 20% of the time, and the 1-point circle 5% of the time. What is the expected value of her score if she throws a dart 10 times?

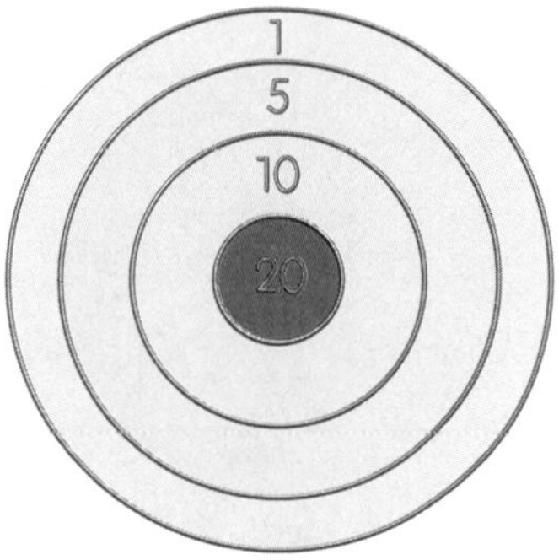

8. Misty polls residents of her neighborhood about the types of pets they have: cat, dog, other, or none. She determines these statistics.

Ownership of cats and dogs is mutually exclusive.

32% of homes have dogs.

54% of homes have a dog or a cat.

16% of homes have only a cat.

42% of homes have no pets.

22% of homes have pets that are not cats or dogs.

Draw a Venn diagram of this data. Include the probability of each outcome.

9. Elliott has time to take exactly 20 more pizza orders before closing. He has enough pepperoni for 16 more pizzas. On a typical night 65% of orders are for pepperoni pizzas. What is the probability that Elliott will run short of pepperoni?

10. Find the term specified for each binomial expansion.

a. The 1st term of $\left(1 + \frac{x}{12}\right)^{99}$

b. The last term of $\left(1 + \frac{x}{12}\right)^{99}$

c. The 10th term of $(H + T)^{21}$

Assessing What You've Learned

- Can you explain the difference between experimental and theoretical probability? What is the difference?
- What are mutually exclusive events? If A and B are mutually exclusive events, how do you find $P(A \text{ or } B)$?
- What are independent events? If A and B are independent events, what do you know about $P(A \text{ and } B)$?
- How can you use Venn diagrams to solve probability problems? How can Venn diagrams help you determine mutually eclusive events?
- What is a random variable? How do you calculate the expected value of a random variable?
- Explain the counting principle.
- What is the difference between a combination and a permutation? How do you calculate a number of permutations? How do you calculate a number of combinations?
- What is the binomial theorem? Can you use the binomial theorem to expand binomials? Can you use the binomial theorem to solve problems with two possible outcomes? How is the binomial theorem related to Pascal's triangle?

JOURNAL Look back at the journal entries you have written about Chapter 11. Then, to help you organize your thoughts, here are some additional questions to write about.

- What do you think were the main ideas of the chapter?
- Were there any exercises that you particularly struggled with? If so, are they clear now? If some ideas are still foggy, what plans do you have to clarify them?

Selected Answers

This section contains answers for the odd-numbered problems in each set of Exercises. When a problem has many possible answers, you are given only one sample answer or a hint on how to begin.

CHAPTER 1 • CHAPTER 1 CHAPTER 1 • CHAPTER

LESSON 1.1

1a. Any whole number answer will work.

1b. For any number that goes into the blank, the solution will always be seven times the number. So, for example, if the missing number is 5, the solution (how much candy the grandfather found) will be $5 \times 7 = 35$.

1c. Yes, any number could still be put into the blank. The solution will be eight times the number. So, for example, if the missing number is 5, the solution (how much candy the grandmother found) will be $5 \times 8 = 40$.

3a. $-\frac{2}{5}$ **3b.** $\frac{3}{2}$

5. The line has slope $\frac{12}{16}$ and goes through (0, 0), so the slope from (0, 0) to any point (x, y) on the line will be $\frac{y}{x} = \frac{12}{16}$.

5a. $a = 12$ **5b.** $b = 7.5$

7a. 1 **7b.** 1.6 **7c.** -2.25 **7d.** 0.625

9a. 32° **9b.** 40° **9c.** 100° **9d.** $-25°$

9e. The slope is $\frac{9}{5}$. This means that if the Celsius temperature increases by one degree, the equivalent Fahrenheit temperature increases by $\frac{9}{5}$ of a degree.

11a. $x^2 + 4x + 7x + 28$

	x	4
x	x^2	$4x$
7	$7x$	28

11b. $x^2 + 5x + 5x + 25$

	x	5
x	x^2	$5x$
5	$5x$	25

11c. $xy + 2y + 6x + 12$

	x	2
y	xy	$2y$
6	$6x$	12

11d. $x + 3x - x - 3$

	x	-1
x	x^2	$-1x$
3	$3x$	-3

LESSON 1.2

1a. Subtract 12 from both sides.

1b. Divide both sides by 5.

1c. Add 18 to both sides.

1d. Multiply both sides by -3.

3a. $c = 27$ **3b.** $c = 5.8$ **3c.** $c = 9$

5a. $x = 72$ **5b.** $x = 24$ **5c.** $x = 36$

7a. $-12L - 40S = -540$

7b. $12L + 75S = 855$

7c. $35S = 315$

7d. $S = 9$. The small beads cost 9¢ each.

7e. $L = 15$. The large beads cost 15¢ each.

7f. $J = 264$. Jill will pay $2.64 for her beads.

9a. Solve Equation 1 for a. Substitute the result, $5b - 42$, for a in Equation 2 to get $b + 5 = 7[(5b - 42) - 5]$.

9b. $b = \frac{167}{17}$ **9c.** $a = \frac{121}{17}$

9d. A has $\frac{121}{17}$ denarii and B has $\frac{167}{17}$ denarii.

11a. $3x + 21$ **11b.** $-12 + 2n$ **11c.** $4x - x^2$

13a. $3a^2 + 3ab$ **13b.** $-12 + 2n$ **13c.** $5x - 36$

15a. 14 **15b.** 13

LESSON 1.3

1a. ≈ 4.3 sec **1b.** 762 cm **1c.** 480 mi

1d. ≈ 6.5 hours **1e.** 292.5 mi **1f.** 377.4 mi

3. 150 mi/hr

5a. 54 in.^{2} **5b.** 1.44m^3 **5c.** 1.20 ft **5d.** 24 cm

7. There are 76 possible seating arrangements.

9. 10.

Position	1 (left)	2	3	4 (right)
Name	Rocky	Sadie	Winks	Pascal
Age	8	5	13	10
Sleeps	Floor	Box	Chair	Sofa
Toy	Stuffed toy	Catnip ball	Silk rose	Rubber mouse

Sadie plays with the catnip-filled ball.

11a. $x^2 + 1x + 5x + 5$ **11b.** $x^2 + 3x + 3x + 9$

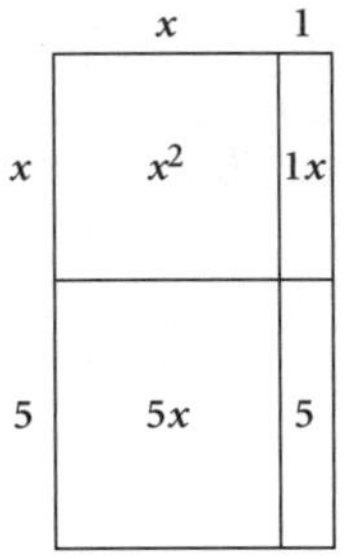

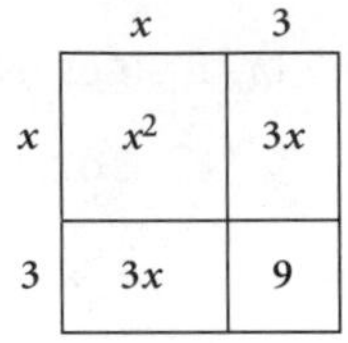

11c. $x^2 + 3x - 3x - 9$

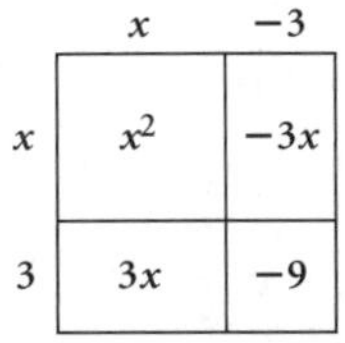

LESSON 1.4

1a. mean: 29.2 min; median: 28 min; mode: 26 min

1b. mean: 17.35 cm; median: 18.0 cm; mode: 17.4 cm

1c. mean: \$2.34; median: \$2.50; mode: \$1.50, \$2.50, and \$3.00

1d. mean: 2; median: 2; mode: 1 and 3

3. minimum: 1.25 days; first quartile: 2.5 days; median: 3.25 days; third quartile: 4 days; maximum: 4.75 days

5. D

7. Answers will vary. The bottom box plot is longer because it reflects the longer range of Oscar's scores. There is no left whisker for the top box plot because the first quarter of Connie's scores is made up of a single repeated score. The median isn't in the center of the top box plot because the second quarter of Connie's scores is spread over a longer interval than the third quarter of her scores. There is a bigger range for the first quarter of Oscar's scores than for the fourth quarter, so the left whisker is longer than the right one in the bottom box plot.

9a. Connie: *range* = 4; *IQR* = 3. Oscar: *range* = 24; *IQR* = 18.

9b. *range* = 47; *IQR* = 14.

11a. sample answer: {62, 63, 64, 65, 70, 70, · 70}

11b. sample answer: {60, 60, 64, 65, 66, 67, 68} Note that there are many other possible correct answers.

13a. The female mean is 10.46 miles; the male mean is 8.76 miles.

13b. The female median is 6 miles; the male median is 7 miles.

13c. In both cases, the mean is higher than the median.

15a. Chemical: 2.29816, 2.29869, 2.29915, 2.30074, 2.30182; Atmospheric: 2.30956, 2.309935, 2.31010, 2.31026, 2.31163

15b.

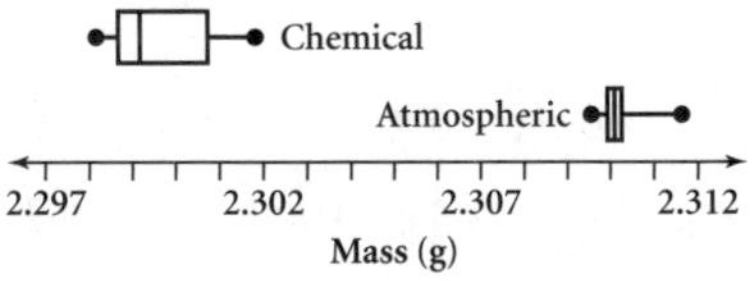

15c. Sample answer: Both graphs are skewed right. There is a significant difference in the range and *IQR* for each set of data. The mass of the nitrogen produced from the atmosphere is heavier than the mass of the nitrogen produced from chemical compounds.

LESSON 1.5

1a. 47.0 **1b.** −6, 8, 1, −3 **1c.** 6.1

3a. 9, 10, 14, 17, 21 **3b.** range = 12; *IQR* = 7

3c. centimeters

5. possible answer: {71, 79, 80, 84, 89, 91, 94} Note that there are many other possible correct answers.

7. 20.8 and 22.1. These are the same outliers found by the interquartile range.

9a.

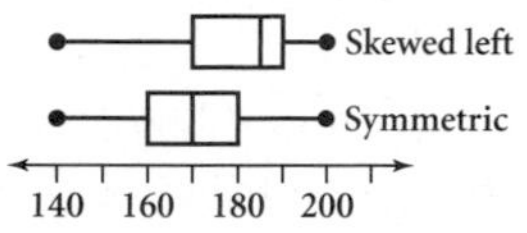

9b. The skewed data set will have a greater standard deviation because the data to the left (below the median) will be spread farther from the mean.

9c. possible answer: symmetric: {140, 160, 165, 170, 175, 180, 200}; skewed left: {140, 170, 180, 185, 187, 190, 200}

9d. Symmetric: 18.5; skewed left: 19.4. These support the answer to 9b.

11a. The 8:00 class appears to have pulse rates most alike because that class had the smallest standard deviation.

11b. The 7:00 class appears to have the fastest pulse rates because that class has both the highest mean and the greatest standard deviation.

13a. median = 75 bagels; $IQR = 19$ bagels

13b. $x = 80.89$ bagels; $s = 24.6$ bagels

13c.

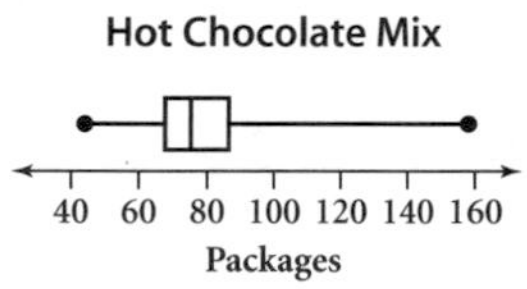

outliers: 147, 158

13d.

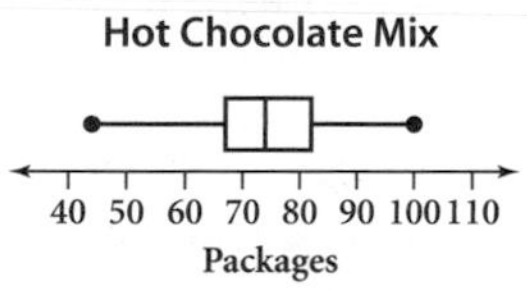

13e. *median* = 74 bagels; $IQR = 15$ bagels; $x = 74.7$ bagels; $x = 12.4$ bagels

13f. The mean and standard deviation are calculated from all data values, so outliers affect these statistics significantly. The median and *IQR,* in contrast, are defined by position and not greatly affected by outliers.

15a. mean: \$80.52; median: \$75.00; mode: \$71.00, \$74.00, \$76.00, \$102.50

15b.

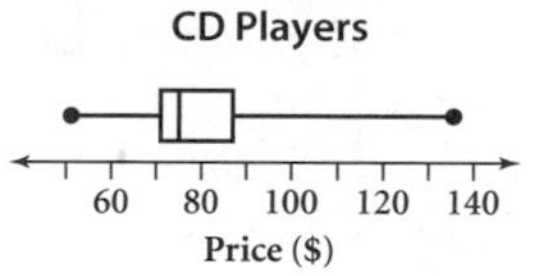

The box plot is skewed right.

15c. $IQR = \$16$; outliers: \$112.50 and \$135.50

15d. The median will be less affected because the relative positions of the middle numbers will be changed less than the sum of the numbers.

15e.

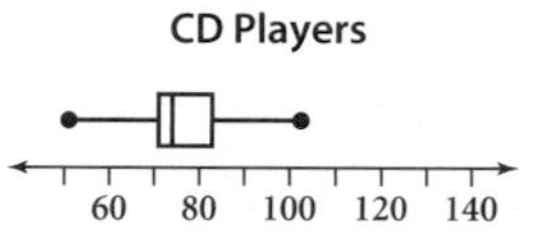

The median of the new data set is \$74.00 and, as can be seen from the box plot, is relatively unchanged.

17a. offensive team: $\bar{x} = 263.8$ lb, *median* = 255 lb; defensive team: $\bar{x} = 245.7$ lb, *median* = 250 lb

17b.

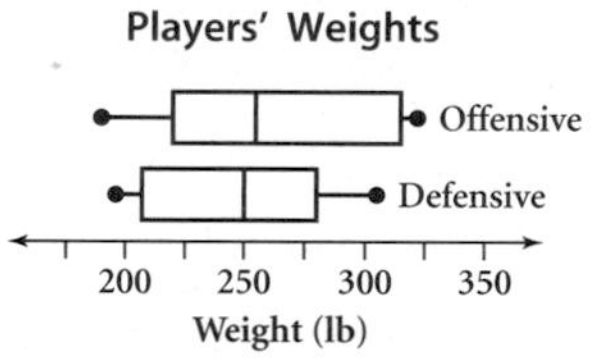

Sample answers: The lightest players are on the offensive team. The offensive team has greater mean and median weights. The spread of the weights of the offensive team is greater than that of the defensive team. Over a quarter of the offensive players weigh more than any of the defensive players.

CHAPTER 1 REVIEW

1a. 168 miles

1b. 9.5 gallons

1c. Car B can go 102 miles farther than Car A.

1d. For Car A the slope is $\frac{120}{5} = 24$, which means the car can drive 24 miles per gallon of gasoline. For Car B the slope is $\frac{180}{5} = 36$, which means the car can drive 36 miles per gallon.

3a. $x^2 + 3x + 4x + 12$

	x	4
x	x^2	$4x$
3	$3x$	12

3b. $2x^2 + 6x$

	x	3
$2x$	$2x^2$	$6x$

3c. $x^2 + 2x + 6x + 12$

	x	2
x	x^2	$2x$
6	$6x$	12

3d. $2x^2 - 8x - 1x + 4$

	x	-4
$2x$	$2x^2$	$-8x$
-1	$-1x$	$+4$

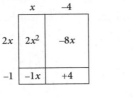

5. Let n represent the unknown number.

5a. $2n + 6$ **5b.** $5(n - 3)$

5c. $2n + 6 = 5(n - 3); n = 7$

7. Amy is 21 years old.

9a. Let w represent the mass in grams of a white block, and let r represent the mass in grams of a red block.

9b. $4w + r = 2w + 2r + 40; 5w + 2r = w + 5r$

9c. $w = 60; r = 80$

9d. The mass of a white block is 60 grams and the mass of a red block is 80 grams.

11a. $4x^{-1}$ or $\frac{4}{x}$ **11b.** $\frac{1}{2}x^{-1}$ or $\frac{1}{2x}$ **11c.** x^{15}

13a. $5 \text{ gal} \cdot \frac{4 \text{ qt}}{1 \text{ gal}} \cdot \frac{4 \text{ c}}{1 \text{ qt}} \cdot \frac{8 \text{ oz}}{1 \text{ c}} = 640 \text{ oz}$

13b. $1 \text{ mi} \cdot \frac{5280 \text{ ft}}{1 \text{ mi}} \cdot \frac{12 \text{ in.}}{1 \text{ ft}} \cdot \frac{2.54 \text{ cm}}{1 \text{ in.}} \cdot \frac{1 \text{ m}}{100 \text{ cm}} =$ 1609.344 m

15a. The sum of all the deviations from the mean must be 0. The sum of this set is 13, so there must be an error.

15b. The correct value is 20.

15c. i. 747, 707, 669, 676, 767, 783, 789, 838; standard deviation = 59.12; median = 757; IQR = 94.5

15c. ii. 850, 810, 772, 779, 870, 886, 892, 941; standard deviation = 59.12; median = 860; IQR = 94.5

17a. mean = 38.1; median = 35

17b. mean = 44.6; median = 43

Academy Awards Winners

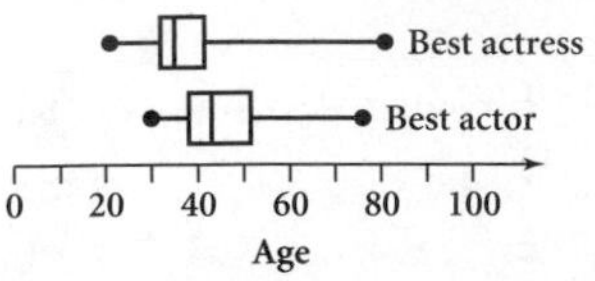

17d. The Best Actress data set should have the greater standard deviation because it has more spread.
standard deviation actress: 12.8
standard deviation actor: 10.2

19. Plot B, because the data have more spread

21. One strategy is to make Set A skewed and Set B symmetric. Possible answer: Set A: {1, 2, 3, 4, 5, 6, 47}; $s = 16.5$; $IQR = 4$. Set B: {1, 5, 7, 9, 11, 13, 17}; $s = 5.2$; $IQR = 8$.

23. Antarctica (59°F) is an outlier for the high temperatures. There are no outliers for the low temperatures.

CHAPTER 2 • CHAPTER 2 CHAPTER 2 • CHAPTER

LESSON 2.1

1a. 20, 26, 32, 38 **1b.** 47, 44, 41, 38

1c. 32, 48, 72, 108 **1d.** −18, −13.7, −9.4, −5.1

3. $u_1 = 40$ and $u_n = u_{n-1} - 3.45$ where $n \geq 2$
$u_5 = 26.2$; $u_9 = 12.4$

5a. $u_1 = 2$ and $u_n = u_{n-1} + 4$ where $n \geq 2$
$u_{15} = 58$

5b. $u_1 = 10$ and $u_n = u_{n-1} - 5$ where $n \geq 2$
$u_{12} = -45$

5c. $u_1 = 0.4$ and $u_n = 0.1 \cdot u_{n-1}$ where $n \geq 2$
$u_{10} = 0.0000000004$

5d. $u_1 = -2$ and $u_n = u_{n-1} - 6$ where $n \geq 2$
$u_{30} = -176$

5e. $u_1 = 1.56$ and $u_n = u_{n-1} + 3.29$ where $n \geq 2$
$u_{14} = 44.33$

5f. $u_1 = -6.24$ and $u_n = u_{n-1} + 2.21$ where $n \geq 2$
$u_{20} = 35.75$

7. $u_1 = 4$ and $u_n = u_{n-1} + 6$ where $n \geq$
2 $u_4 = 22$; $u_5 = 28$; $u_{12} = 70$; $u_{32} = 190$

9a. 399 km

9b. 10 hours after the first car starts, or 8 hours after the second car starts

11a. \$60 **11b.** \$33.75

11c. during the ninth week

13a. Both students earned a B.

13b and c. Students who think this is unfair may refer to the median, or point out that 54% is an outlier.

LESSON 2.2

1a. 1.5 **1b.** 0.4 **1c.** 1.03 **1d.** 0.92

3a. $u_1 = 100$ and $u_n = 1.5u_{n-1}$ where $n \geq 2$
$u_{10} = 3844.3$

3b. $u_1 = 73.4375$ and $u_n = 0.4u_{n-1}$ where $n \geq 2$
$u_{10} = 0.019$

3c. $u_1 = 80$ and $u_n = 1.03u_{n-1}$ where $n \geq 2$
$u_{10} = 104.38$

3d. $u_1 = 208$ and $u_n = 0.92u_{n-1}$ where $n \geq 2$
$u_{10} = 98.21$

5a. $(1 + 0.07)u_{n-1}$ or $1.07u_{n-1}$

5b. $(1 - 0.18)A$ or $0.82A$

5c. $(1 + 0.08125)x$ or $1.08125x$

5d. $(2 - 0.85)u_{n-1}$ or $1.15u_{n-1}$

7a. 0.8 **7b.** 11 in. **7c.** 21 bounces; 31 bounces

9a. Number of new hires for next five years: 2; 3; 3 or 4; 4; and 5.

9b. about 30 employees

11a.

Generations back	0	1	2	3	4	5
Ancestors within the generation	1	2	4	8	16	32

11b. Start with 1 and recursively multiply by 2; $u_0 = 1$ and $u_n = 2u_{n-1}$ where $n \geq 1$

11c. u_{30}; 30 generations ago, Jill had 1 billion living ancestors.

11d. 750 years ago

11e. The population of the planet at that time was less than 1 billion. Jill must have some common ancestors.

13a. $\frac{6.5}{12} \approx 0.542\%$ **13b.** \$502.71

13c. \$533.48 **13d.** \$584.80

15a. 3; possible answer: $\frac{162}{18} = 9, 3^2 = 9$

15b. 2, 6, 54, 486, 1458, 13122

15c. 118,098

17a. $u_1 = 180$ and $u_n = u_{n-1} - 7$ where $n \geq 2$

17b. $u_{10} = 117$ **17c.** $u_{27} = -2$

19a. $u_0 = a$, $u_n + 1 = 1 - .25u_n$

19b. u_3 is how much candy is left after the brother eats. u_3 is $\frac{27}{64}$ of the original amount.

19c. 27 pieces **19d.** 64 pieces

LESSON 2.3

1a. 31.2, 45.64, 59.358; shifted geometric, increasing

1b. 776, 753.2, 731.54; shifted geometric, decreasing

1c. 45, 40.5, 36.45; geometric, decreasing

1d. 40, 40, 40; shifted geometric or arithmetic, neither increasing nor decreasing

3a. 320 **3b.** 320 **3c.** 0 **3d.** 40

5a. The first day, 300 g of chlorine were added. Each day, 15% disappears, and 30 more grams are added.

5b. It levels off at 200 g.

7a. The account balance will continue to decrease (slowly at first, but faster after a while). It does not level off.

7b. A monthly withdrawal of \$68 would hold the account steady at \$24,000.

9. $u_0 = 20$ and $u_n = (1 - 0.25)u_{n-1}$ where $n \geq 1$; 11 days ($u_{11} \approx 0.84$ mg)

11a. Sample answer: After 9 hours there are only 8 mg, after 18 hours there are 4 mg, after 27 hours there are still 2 mg left.

11b.

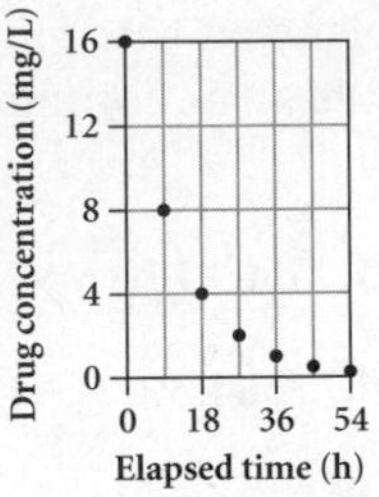

11c. 8 mg

13a. $u_2 = -96$, $u_5 = 240$

13b. $u_2 = 2$, $u_5 = 1024$

15. 23 times

LESSON 2.4

1a. 0 to 9 for n and 0 to 16 for u_n

1b. 0 to 19 for n and 0 to 400 for u_n

1c. 0 to 29 for n and -178 to 25 for u_n

1d. 0 to 69 for n and 0 to 3037 for u_n

3a. geometric, nonlinear, decreasing

3b. arithmetic, linear, decreasing

3c. arithmetic, linear, increasing

3d. geometric, nonlinear, increasing

5a. iii. The graph of an arithmetic sequence is linear.

5b. ii. The graph of a nonshifted geometric sequence will increase indefinitely or have a limit of zero.

5c. i. The graph has a nonzero limit, so it must represent a shifted geometric sequence.

7. The graph of an arithmetic sequence is always linear. The graph increases when the common difference is positive and decreases when the common difference is negative. The steepness of the graph relates to the common difference.

9a.

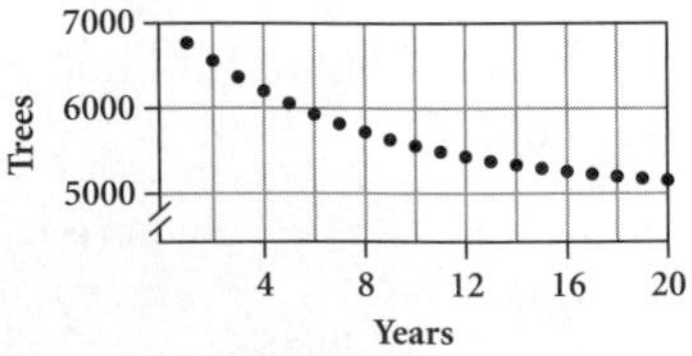

9b. The graph appears to have a long-run value of 5000 trees, which agrees with the long-run value found in Exercise 8 in Lesson 2.3.

11. Possible answer: $u_{50} = 40$ and $u_n = u_{n-1} + 4$ where $n \geq 51$

13a. 547.5, 620.6, 675.5, 716.6, 747.5

13b. $\dfrac{547.5 - 210}{0.75} = 450$; subtract 210 and divide the difference by 0.75.

13c. $u_0 = 747.5$ and $u_n = \dfrac{u_{n-1} - 210}{0.75}$ where $n \geq 1$

15a. The mean is higher because the data seem skewed to the right. The third quartile is further from the median than the first quartile and the highest reported salary is further from the median than the lowest.

15b. The median US household income is less than the income of DVD recorder buyers. People with more disposable income buy higher priced electronic goods.

LESSON 2.5

1a. investment, because a deposit is added

1b. \$450 **1c.** \$50 **1d.** 3.9%

1e. annual (once a year)

3a. \$130.67 **3b.** \$157.33

3c. \$184.00 **3d.** \$210.67

5. \$588.09

7a. \$1877.14 **7b.** \$1912.18 **7c.** \$1915.43

7d. The more frequently the interest is compounded, the more quickly the balance will grow. That is, after 10 years, an investment compounded daily will be greater than an investment compounded monthly, which is greater than an investment compounded annually.

9a. \$123.98

9b. for $u_0 = 5000$ and

$u_n = \left(1 + \dfrac{0.085}{12}\right)u_{n-1} + 123.98$ where $n \geq 1$

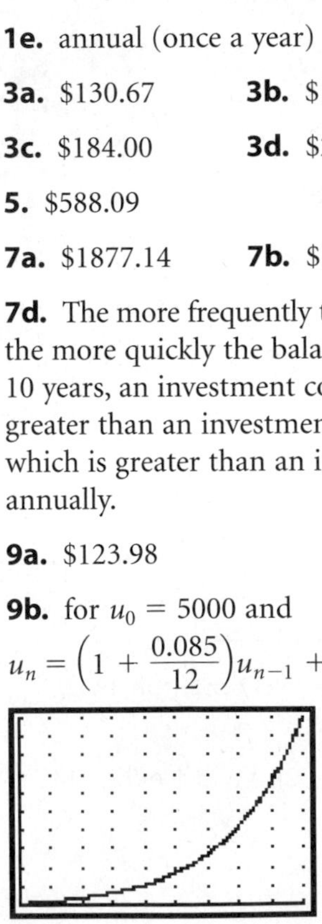

[0, 540, 60, 0, 1000000, 10000]

11a. \$447.30

11b. $u_0 = 60000$ and

$u_n = \left(1 + \frac{.076}{12}\right)u_{n-1} - 447.30$ where $n \geq 1$

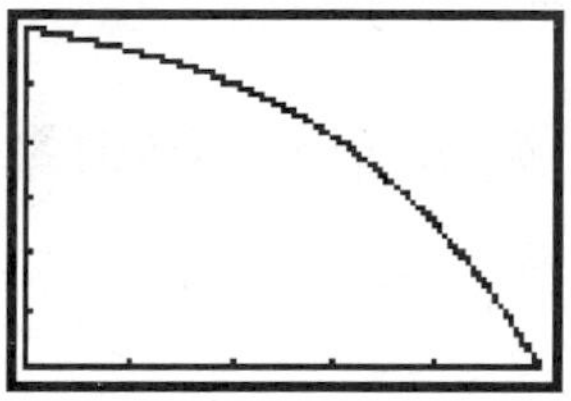

13a. 0

13b. It grows larger and larger.

13c. 4

CHAPTER 2 REVIEW

1a. geometric

1b. $u_1 = 256$ and $u_n = 0.75u_{n-1}$ where $n \geq 2$

1c. $u_8 \approx 34.2$ **1d.** $u_{10} \approx 19.2$ **1e.** $u_{17} \approx 2.57$

3a. $-3, -1.5, 0, 1.5, 3$; 0 to 6 for n and -4 to 4 for u_n

3b. 3, 7, 19, 49, 145; 2 to 6 for n and 0 to 150 for u_n

5a. $u_0 = 14.7$ and $u_n = (1 - 0.20)u_{n-1}$ where $n \geq 1$

5b.

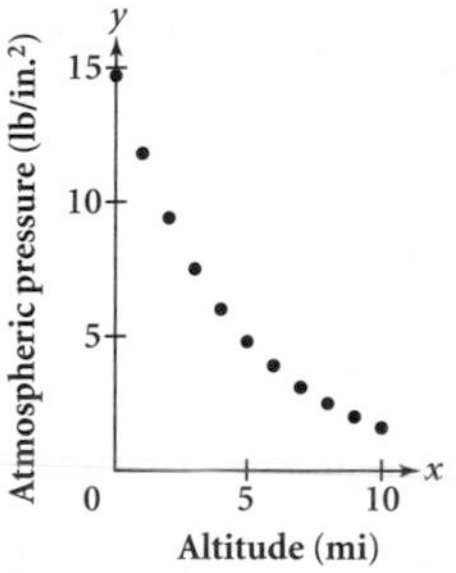

5c. ≈ 3.1 psi **5d.** 11 mi

7a. \$657.03 **7b.** \$4083.21

9a. iv. The recursive formula is that of an increasing arithmetic sequence, so the graph must be increasing and linear.

9b. iii. The recursive formula is that of a growing geometric sequence, so the graph must be increasing and curved.

9c. i. The recursive formula is that of a decaying geometric sequence, so the graph must be decreasing and curved.

9d. ii. The recursive formula is that of a decreasing arithmetic sequence, so the graph must be decreasing and linear.

LESSON 3.1

1a.

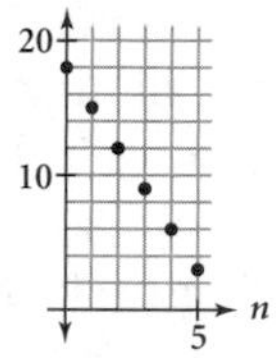

1b. -3; The common difference is the same as the slope.

1c. 18; The y-intercept is the u_0-term of the sequence.

1d. $y = 18 - 3x$

3. $y = 7 + 3x$

5a. 1.7 **5b.** 1 **5c.** -4.5 **5d.** 0

7a. 190 mi **7b.** $y = 82 + 54x$

7c.

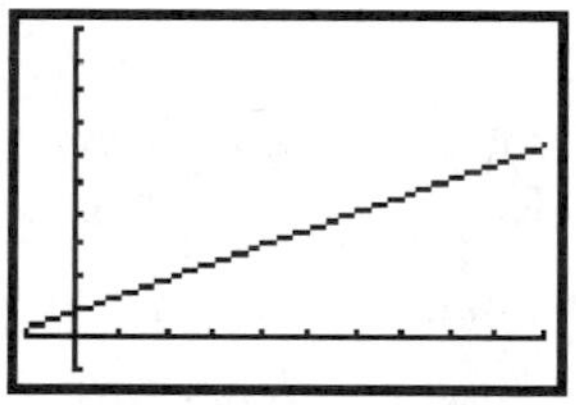

$[-1, 10, 1, -100, 1000, 100]$

7d. The relationship is linear. So, if only distances on the hour are considered, it is an arithmetic sequence. Otherwise, the relationship is continuous rather than discrete.

9a. $u_0 = -2$ **9b.** 5 **9c.** 50

9d. Because you need to add 50 d's to the original height of u_0.

9e. $u_n = u_0 + nd$

11a. possible answers: (0, 7) and (5, 27). The x values should have a difference of 5.

11b. Regardless of the points chosen in 11a, the slope should be 4.

11c. 7, 11, 15, 19, 23, 27

11d. $y = 7 + 4x$. The y-intercept will differ if the original points have different x-values.

13. At the end of one year the salaries are the same. During the third year the second job will have paid more in total.

15. median price: \$93.49; mean: \$96.80; standard deviation: \$7.55. Knowing the median and mean prices informs a shopper of what the mid-price and average price are. The standard deviation indicates whether there is price variation and it is worth shopping around for a good deal—or if all prices are pretty much the same. The median price is the better price to use as a frame of reference. The mean may be low or high if one store is having a sale or is unusually expensive (is an outlier).

LESSON 3.2

1a. $\frac{3}{2} = 1.5$ **1b.** $-\frac{2}{3} \approx -0.67$ **1c.** -55

3a. $y = 14.3$ **3b.** $x = 6.5625$

3c. $a = -24$ **3d.** $b = -0.25$

5a. The equations have the same constant, -2. The lines share the same y-intercept. (The lines are also perpendicular.)

5b. The equations have the same x-coefficient. -1.5. The lines are parallel.

7. -0.5 m/s, or 0.5 m/s toward the motion sensor

9a. \$31,709; \$27,109 **9b.** \$1150 per year

9c. $45000 = 1150n + 27109$; round up to $n = 16$; in her 17th year

9d. It would be lower. Usually percentage raises are better in the long run. Here \$1150 is more than 3% of Anita's salary for the first year.

11a. Possible answer: Using the 2nd and 7th data points, the slope is \$7.62 per ticket. Each additional ticket sold brings in about \$7.62 more in revenue. This value is somewhere between the two different ticket prices offered.

11b. Answers will vary. Generally use points that are not too close together.

13a. 71.7 beats/min

13b. 6.47 beats/min. The majority of the data falls within 6.47 beats/min of the mean.

15a. $a > 0, b < 0$ **15b.** $a < 0, b > 0$

15c. $a > 0, b = 0$ **15d.** $a = 0, b < 0$

LESSON 3.3

1a. possible answer: $y = \frac{-5}{3} + \frac{2}{3}x$

1b. possible answer: $y = \frac{11}{5} - \frac{1}{5}x$

3a. $f(11) = 31$ **3b.** $t = -41.5$ **3c.** $x = 1.5$

5a.

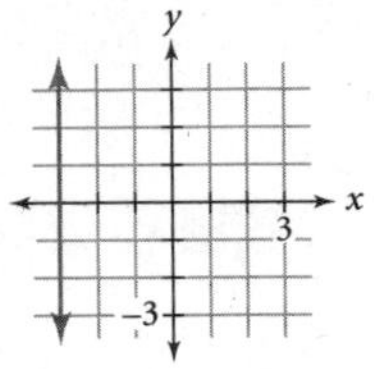

$[-10, 10, 1, -10, 10, 1]$

possible answers: $(-3, 0)$, $(-3, 2)$

5b. undefined **5c.** $x = 3$

5d. Vertical lines have no y-intercepts (unless the vertical line is the y-axis, or $x = 0$; they have undefined slope; all points have the same x-coordinate.

7. possible answers:

7a. y-intercept is about 1.7; $y = 1.7 + 0.58x$

7b. y-intercept is about 7.5; $y = 7.5 - 0.75x$

7c. y-intercept is about 8.6; $y = 8.6 - 0.94x$

9a. possible answer: $[145, 200, 5, 40, 52, 1]$

9b. possible answer: $y = 0.26x + 1.33$

9c. On average, a student's forearm length increases by 0.26 cm for each additional 1 cm of height.

9d. The y-intercept is meaningless because a height of 0 cm will not predict a forearm length of 1.33 cm. The domain should be specified.

9e. 187.2 cm; 42.4 cm

11a. $3x^6y^3$ **11b.** ab^7 **11c.** $\frac{3y^5}{2x^7}$

13a. 16

13b. Add 19.5 and any three numbers greater than 19.5.

LESSON 3.4

1a. $(1.8, -11.6)$ **1b.** $(3.7, 31.9)$

1c. No solution

3. $y = 0.4x + 4.6$

5a. $(4.125, -10.625)$ **5b.** $(-3.16, 8.27)$

5c. $(4.67, 0)$

5d. parallel lines; no intersecting point.

7a. No. At $x = 25$, the cost line is above the income line.

7b. Yes. The profit is approximately \$120.

7c. About 120 dolls. Look for the point where the cost and income lines intersect.

9a. Four Line: $y = 44 + 4([x] + 1)$; Small Business Plan: $y = 6([x] + 1)$.

9b.

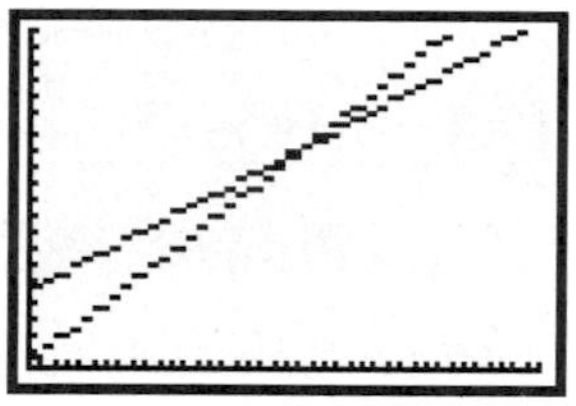

[0, 40, 1, 0, 200, 10]

9c. For calls of over 22 minutes use 4 Line. For calls under 22 minutes use 6 Saver.

11a.

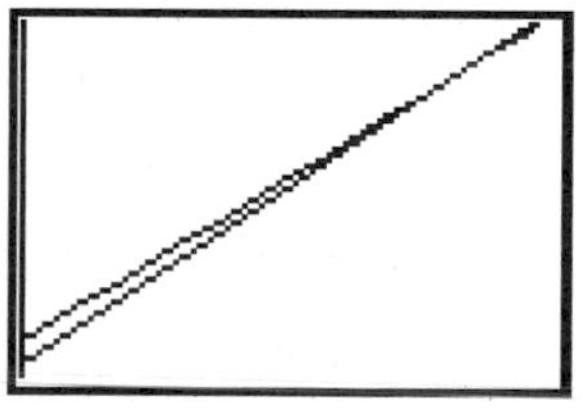

[0, 60, 10, 50, 250, 50]

11b. male: approximately 199.81 cm; female: approximately 199.09 cm

11c. between 44 cm and 45 cm.

13a. 343 ppm **13b.** 382 ppm

13c. approximately the year 2129

LESSON 3.5

1a. $w = 11 + r$ **1b.** $h = \dfrac{18 - 2p}{3} = 6 - \left(\dfrac{2}{3}\right)p$

1c. $r = w - 11$ **1d.** $p = \dfrac{18 - 3h}{2} = 9 - \left(\dfrac{3}{2}\right)h$

3a. $5x - 2y = 12$; passes through the point of intersection of the original pair

3b. $-4y = 8$; passes through the point of intersection of the original pair and is horizontal

5a. $\left(-\dfrac{97}{182}, +\dfrac{19}{7}\right) \approx (-0.5330, 2.7143)$

5b. $\left(8, -\dfrac{5}{2}\right) = (8, -2.5)$

5c. $\left(\dfrac{186}{59}, -\dfrac{4}{59}\right) \approx (3.1525, -0.0678)$

5d. $n = 26, s = -71$ **5e.** $d = -13, f = -34$

5f. $\left(\dfrac{44}{7}, -\dfrac{95}{14}\right) = (6.2857, -6.7857)$

5g. no solution **5h.** no solution

7a. $\left(\dfrac{8}{11}, \dfrac{53}{22}\right) \approx (0.7273, 2.4091)$

7b. $(-\sqrt{2}, -2) \approx (-1.4142, -2)$ and $(\sqrt{2}, -2) \approx (-1.4142, -2)$

7c. no solution

9a. cost for first camera: $y = 47 + 11.5x$; cost for second camera: $y = 59 + 4.95x$

9b. the \$47 camera; the \$59 camera; approximately 2 yr

9c. You could graph each cost function and zoom in for the intersection point; you could use substitution to find the intersection point; or you could look at a table to find the break-even point where the cost for both cameras is the same.

11. $u_{31} = v_{31} = 21$

13a. $x + y = 40$ and $6x = 9y$; $y = 6$. The fulcrum should be positioned 6 in. from the 9 lb weight.

13b. $8J + 5A = 8(150)$ and $8(A) + 5.6(J) = 8(150)$. Solving the system leads to the fact that Justine weighs 100 lb, while Alden weighs 80 lb.

15a.

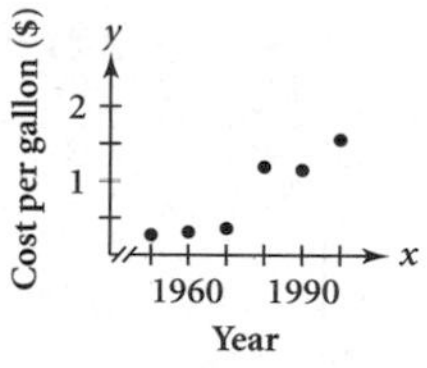

15b. $y = \dfrac{213}{8000}x - 51.78$ or $y = 0.027x - 51.78$

15c. If the same trend continues, the cost of gasoline in 2010 will be \$1.75. Answers will vary.

17a. i. 768, −1024 **ii.** 52, 61 **iii.** 32.75, 34.5

17b. i. geometric **ii.** other **iii.** arithmetic

17c. i. $u_1 = 243$ and $u_n = \left(-\frac{4}{3}\right)u_{n-1}$ where $n \geq 2$

17c. iii. $u_1 = 24$ and $u_n = u_{n-1} + 1.75$ where $n \geq 2$

CHAPTER 3 REVIEW

1. $\frac{-975}{19}$

3a. approximately (19.9, 740.0)

3b. approximately (177.0, 740.0)

5a. Poor fit; there are too many points above the line.

5b. reasonably good; the points are well-distributed above and below the line, and not clumped. The line follows the downward trend of the data.

5c. poor fit; equal number of points above and below the line, but they are clumped to the left and to the right, respectively. Or the line does not follow the trend of the data.

7a. (1, 0)

7b. same line; they intersect at every point

7c. no intersection; lines are parallel

9a. The ratio 0.378 represents the slope; that is, for each pound on Earth, you would weigh 0.378 pound on Mercury.

9b. $m = 0.378(160) = 60.48$. The student's weight on Mercury would be about 60.48 lb.

9c. Moon: **D;** $Y3 = 0.166x$; Mercury: **C;** $Y1 = 0.378x$; Earth: **B;** $Y2 = x$; Jupiter: **A;** $Y4 = 2.364x$

11a. geometric; curved; 4, 12, 36, 108, 324

11b. shifted geometric; curved; 20, 47, 101, 209, 425

13a. possible answer: $u_{2005} = 6486915022$ and $u_n = (1 + 0.015)u_{n-1}$ where $n \geq 2006$. The sequence is geometric.

13b. 6,988,249,788 people

13c. On January 1, 2035, the population will be just above 10 billion. So the population will first exceed 10 billion late in 2034.

13d. Answers will vary. An increasing geometric sequence has no limit. But the model will not work for the distant future because there is a physical limit to how many people will fit on Earth.

15. 0 hours, 25 mg; 12 hours, 37.5 mg; 24 hours, 42.75 mg; 36 hours, 46.375mg; 48 hours, 48.175 mg. 50 mg in the long run.

17a. $u_0 = 6$ and $u_n = u_{n-1} + 7$ where $n \geq 1$

17b. $y = 6 + 7x$

17c. Slope is 7. The slope of the line is the same as the common difference of the sequence.

17d. 223, probably easier to use equation from 17b.

19a. $x = 5.0741$, $median = 4.6$, $mode = 4.5$, $s = 1.734$

19b. Louisiana lies more than 2 standard deviations above the mean.

CHAPTER 4

LESSON 4.1

1a.

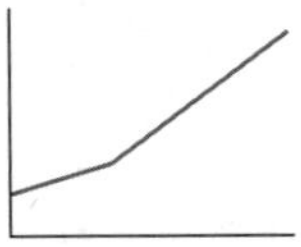

1b.

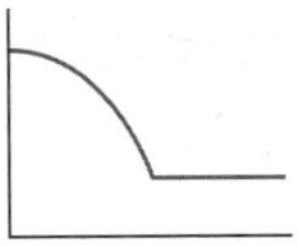

1c.

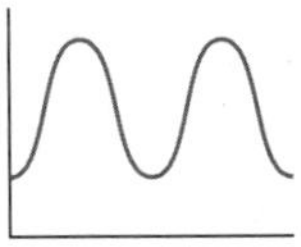

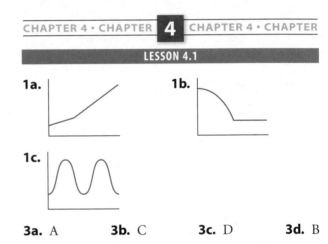

3a. A **3b.** C **3c.** D **3d.** B

5. Sample answer: Zeke, the fish, swam to the bottom of his bowl and stayed there for a while. When Zeke's owner sprinkled fish food into the water, Zeke swam toward the surface to eat. The y-intercept is the fish's depth at the start of the story. The x-intercept represents the fish at the surface of the bowl.

7a. Time in years is the independent variable; the amount of money in dollars is the dependent variable. The graph will be a series of discontinuous segments.

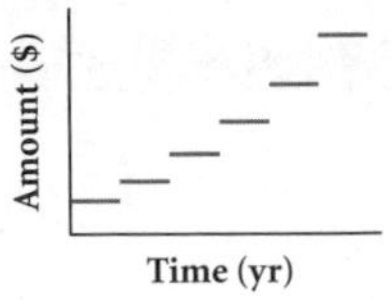

7b. Time in years is the independent variable; the amount of money in dollars is the dependent variable. The graph will be a continuous horizontal segment because your balance increases only at times when interest is compounded.

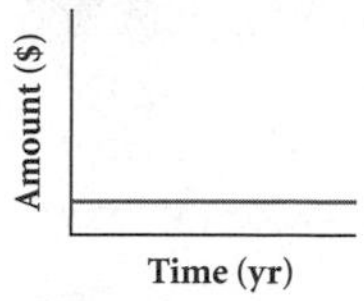

7c. Foot length in inches is the independent variable; shoe size is the dependent variable. The graph will be a series of discontinuous horizontal segments because shoe sizes are discrete.

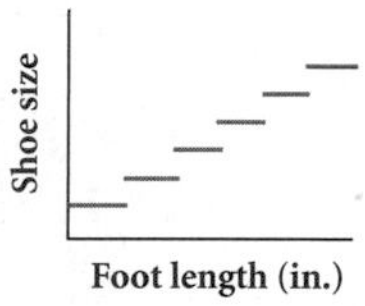

7d. Time in hours is the independent variable; distance in miles is the dependent variable. The graph will be continuous because distance is changing continuously over time.

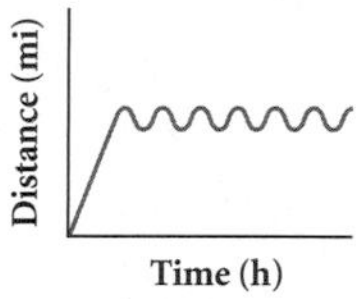

7e. The day of the year is the independent variable; the maximum temperature in degrees Fahrenheit is the dependent variable. The graph will be discrete points because there is just one temperature reading per day.

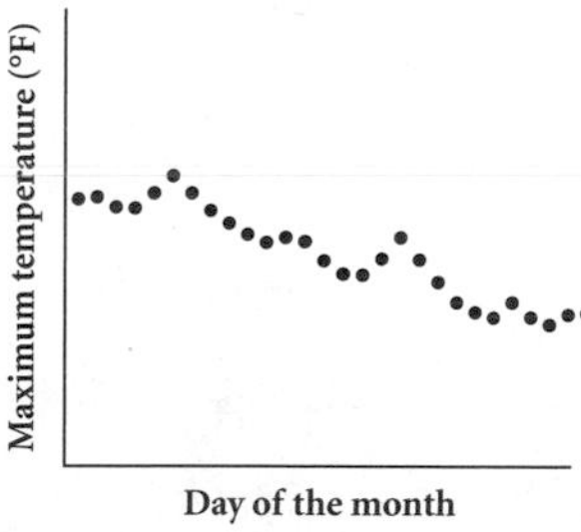

9a. Sample answer: the temperature of a bowl of soup as it cools over a period of time

9b. Sample answer: the speed of a car as it accelerates, then reaches traveling speed over a period of time

9c. Sample answer: the speed of a car as it accelerates, then suddenly brakes over a period of time

11a. Let l represent the length of the rope in meters and let k represent the number of knots; $l = 1.70 - 0.12k$.

11b. Let b represent the bill in dollars and let c represent the number of additional CDs purchased; $b = 7.00 + 9.50(c - 8)$.

13a. \$142,784.22 **13b.** \$44,700.04

13c. \$0; you actually pay off the loan after 19 years 9 months.

13d. Possible answer: By making an extra \$300 payment per month for 20 years, or \$72,000, you save hundreds of thousands of dollars in the long run.

LESSON 4.2

1a. Function; each x-value has only one y-value.

1b. Not a function; there are x-values that are paired with two y-values.

1c. Function; each x-value has only one y-value.

3. B

5a. The price of the calculator is the independent variable; function.

5b. The time the money has been in the bank is the independent variable; function.

5c. The amount of time since your last haircut is the independent variable; function.

5d. The distance you have driven since your last fill-up is the independent variable; function.

7a,c,d.

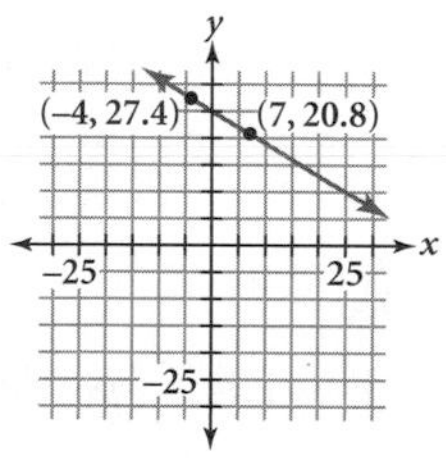

7b. 20.8 **7d.** −4

7e. possible answer: Joe is paid \$25 for babysitting. He decides to walk the children down to the grocery store, and buy them popsicles. Each popsicle costs 60 cents. The input (x) stands for the number of popsicles Joe buys, and the output is Joe's net profit from babysitting.

9a. sample answer:

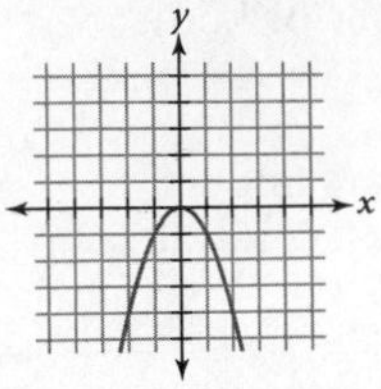

9b. sample answer:

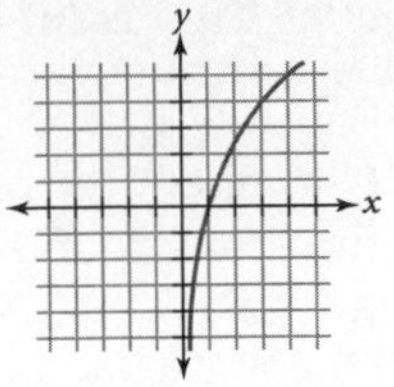

9c. sample answer:

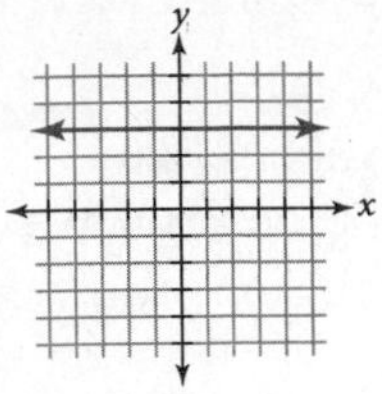

11. The graph is a function; the person can be only at one position at each moment in time, so there is only one y-value for each x-value.

13a. 54 diagonals **13b.** 20 sides

15a.

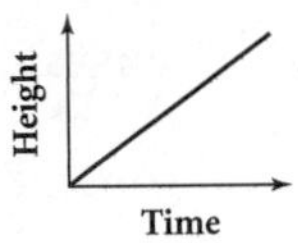

15b.

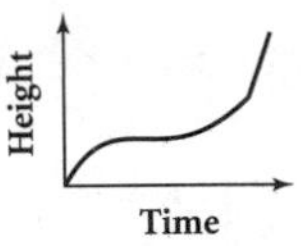

15c.

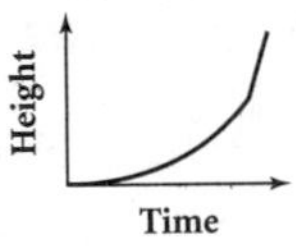

17. (7, 25.5)

LESSON 4.3

1. $y = -3 + \frac{2}{3}(x - 5)$

3a. $-2(x + 3)$ or $-2x - 6$

3b. $-3 + (-2)(x - 2)$ or $-2x + 1$

3c. $5 + (-2)(x + 1)$ or $-2x + 3$

5a. $y = -3 + 4.7x$ **5b.** $y = -2.8(x - 2)$

5c. $y = (x + 1.5) + 4$ or $y = x + 5.5$

7. $y = 47 - 6.3(x - 3)$

9a. $(1533.\overline{3}, 733.\overline{3})$ **9b.** $(x + 533.\overline{3}, y + 233.\overline{3})$

9c. 20 steps

11a. $m(x) = \dfrac{86 + 73 + 76 + 90 + x}{5}$

11b. $m(79) = 80.8$

11c. A score of 95 will give a five-game average of 84 points.

LESSON 4.4

1a. $y = x^2 + 2$ **1b.** $y = x^2 - 6$

1c. $y = (x - 4)^2$ **1d.** $y = (x + 8)^2$

3a. translated down 3 units

3b. translated up 41 units

3c. translated right 2 units

3d. translated left 4 units

5a. $x = 2$ or $x = -2$ **5b.** $x = 4$ or $x = -4$

7a. $y = (x - 5)^2 - 3$ **7b.** $(5, -3)$

7c. $(6, -2), (4, -2), (7, 1), (3, 1)$. If (x, y) are the coordinates of any point on the black parabola, then the coordinates of the corresponding point on the red parabola are $(x + 5, y - 3)$.

7d. Segment b has length 1 unit and segment c has length 4 units.

9a.

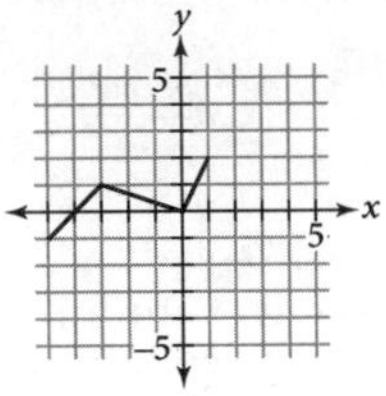

9b.

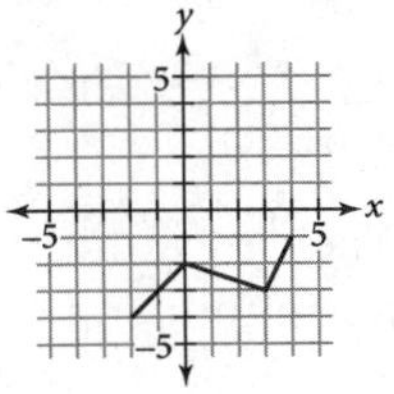

11a.

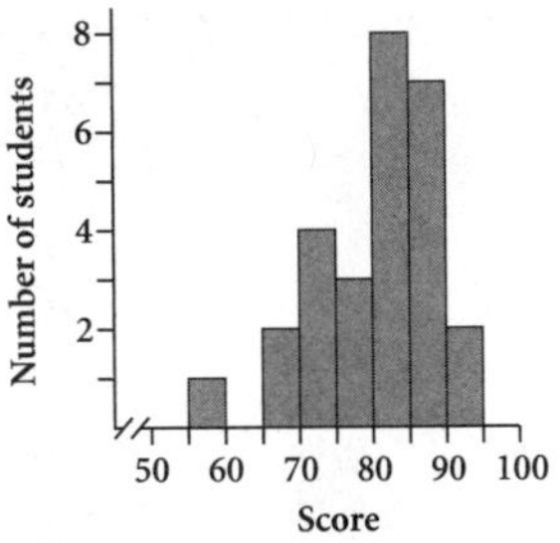

11b.

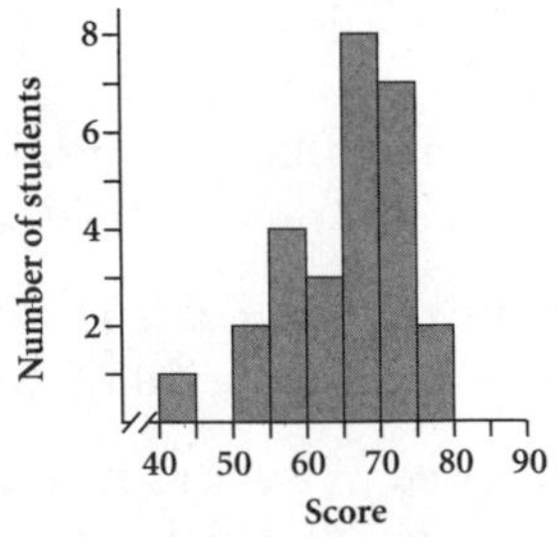

13.

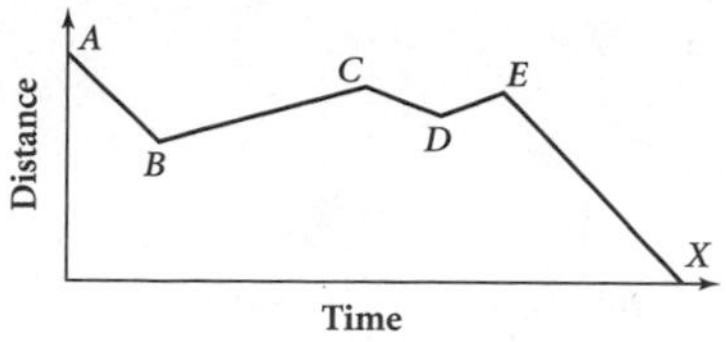

LESSON 4.5

1a. $y = \sqrt{x} + 3$ **1b.** $y = \sqrt{x + 5}$

1c. $y = \sqrt{x + 5} + 2$ **1d.** $y = \sqrt{x - 3} + 1$

1e. $y = \sqrt{x - 1} - 4$

3a. $y = -\sqrt{x}$ **3b.** $y = -\sqrt{x} - 3$

3c. $y = -\sqrt{x + 6} + 5$ **3d.** $y = \sqrt{-x}$

3e. $y = \sqrt{-x + 2} - 3$

5a. possible answers: $(-4, -2)$, $(-3, -1)$, and $(0, 0)$

5b. $y = \sqrt{x + 4} - 2$ **5c.** $y = -\sqrt{x - 2} + 3$

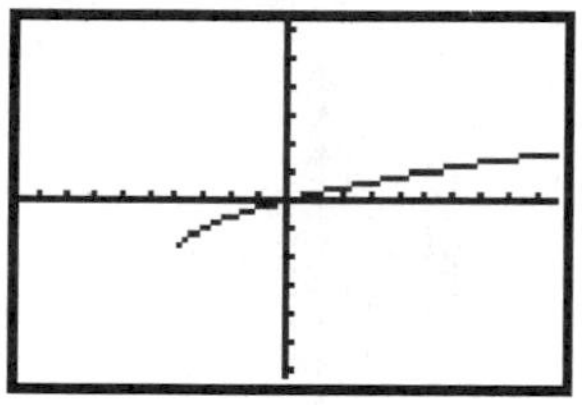

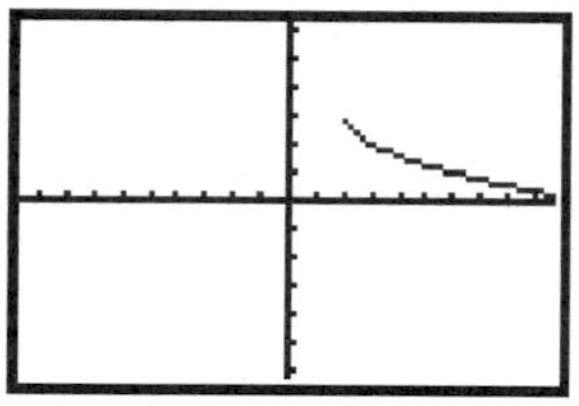

7a. Possible answer: Neither parabola passes the vertical line test.

7b. i. $y = \pm\sqrt{x + 4}$ **7b. ii.** $y = \pm\sqrt{x} + 2$

7c. i. $y^2 = x + 4$ **7c. ii.** $(y - 2)^2 = x$

9. $y + 2 = -((x - 5) + 3)^2 + 4$ or $y = -(x - 2)^2 + 2$

11a. $S = 5.5\sqrt{0.7D}$

11b.

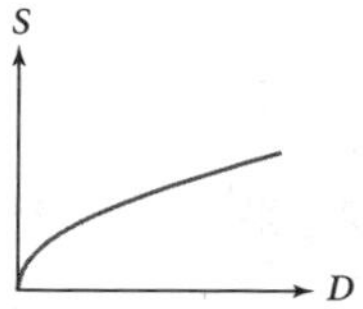

[0, 100, 5, 0, 60, 5]

11c. approximately 36 mi/h

11d. $D = \frac{1}{0.7}\left(\frac{S}{5.5}\right)^2$

11e.

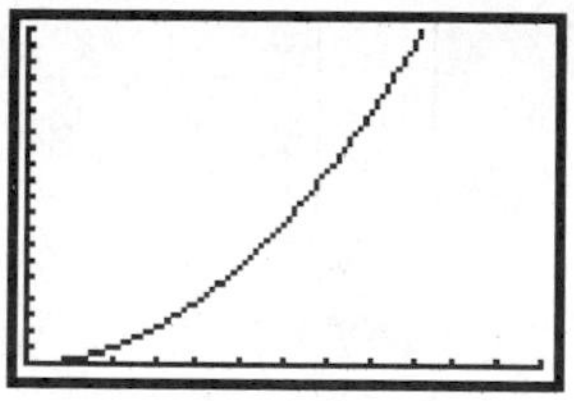

[0, 60, 5, 0, 100, 5]

It is a parabola.

11f. approximately 199.5 ft

13a. $x = 293$ **13b.** no solution

13c. $x = 7$ or $x = -3$ **13d.** $x = -13$

15a. $y = \frac{1}{2}x + 5$

15b. $y = \frac{1}{2}(x - 8) + 5$

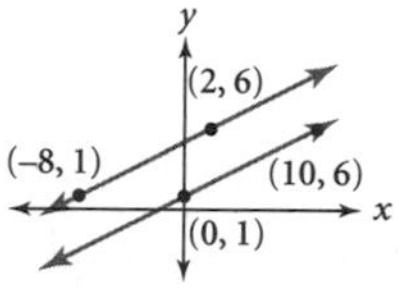

15c. $y = \left(\frac{1}{2}x + 5\right) - 4$ or $y + 4 = \frac{1}{2}x + 5$

15d. Both equations are equivalent to $y = \frac{1}{2}x + 1$.

LESSON 4.6

1a. $y = |x| + 2$ **1b.** $y = |x| - 5$

1c. $y = |x + 4|$ **1d.** $y = |x - 3|$

1e. $y = |x| - 1$ **1f.** $y = |x - 4| + 1$

1g. $y = |x + 5| - 3$ **1h.** $y = 3|x - 6|$

1i. $y = -\left|\frac{x}{4}\right|$ **1j.** $y = (x - 5)^2$

1k. $y = -\frac{1}{2}|x + 4|$ **1l.** $y = -|x + 4| + 3$

1m. $y = -(x + 3)^2 + 5$ **1n.** $y = \pm\sqrt{x - 4} + 4$

1p. $y = -2\left|\frac{x - 3}{3}\right|$

3a. $y = 2(x - 5)^2 - 3$

3b. $y = 2\left|\frac{x + 1}{3}\right| - 5$ **3c.** $y = -2\sqrt{\frac{x - 6}{-3}} - 7$

5a. $x = 1$ and $x = 7$ **5b.** $x = -8$ and $x = 2$

7a. (6, 6)

7b. (2, 5) (8, 5)

7c. (2, 6) (8, 6)

9a. possible answers: $x = 4.7$ or $y = 5$

9b. possible answers: $y = 4\left(\frac{x - 4.7}{1.9}\right)^2 + 5$ or $\left(\frac{y - 5}{4}\right)^2 = \frac{x - 4.7}{-1.9}$

9c. There are at least two parabolas that meet the conditions. One is oriented horizontally and the other is oriented vertically.

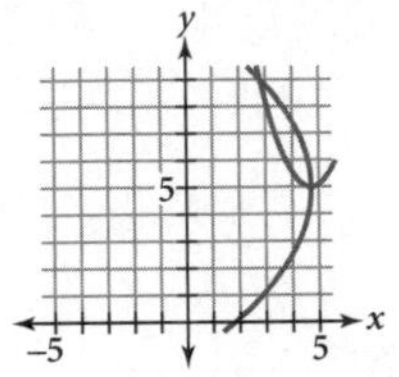

11a.

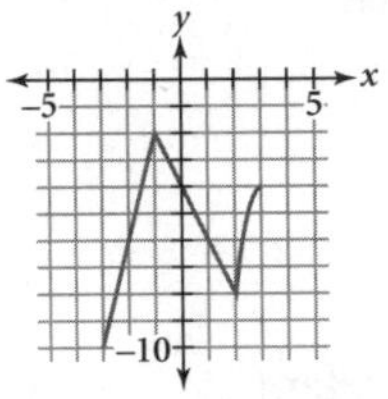

11b.

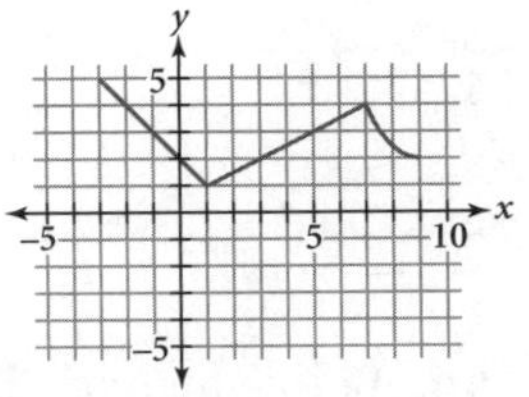

11c.

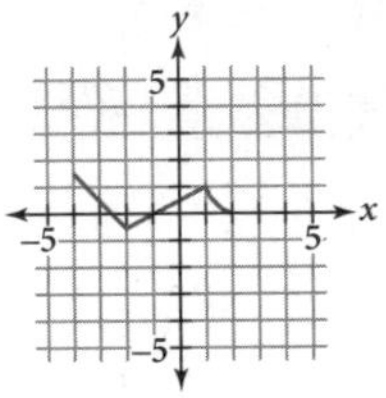

13a. $\bar{x} = 83.75, s = 7.45$

13b. $\bar{x} = 89.75, s = 7.45$

13c. By adding 6 points to each rating, the mean increases by 6, but the standard deviation remains the same.

15a. Possible answer:

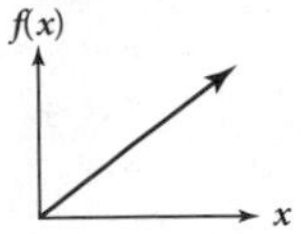

15b. Possible answer:

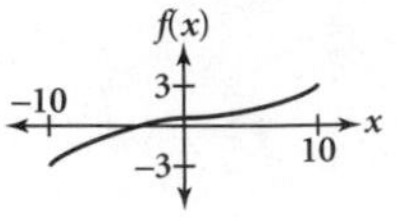

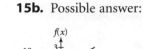

15c. Possible answer:

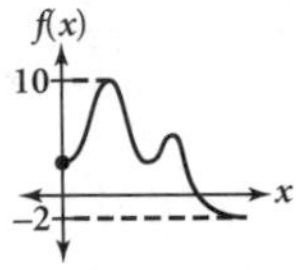

LESSON 4.7

3a.

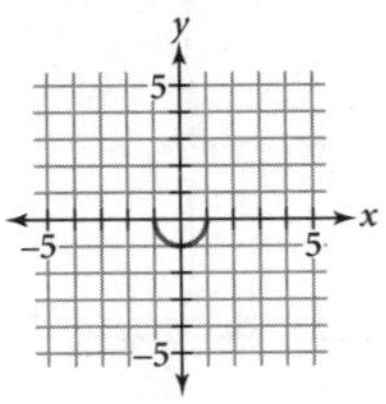

3b.

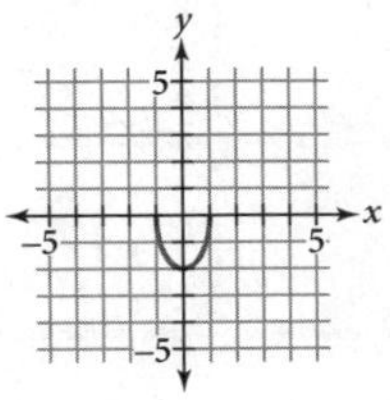

3c.

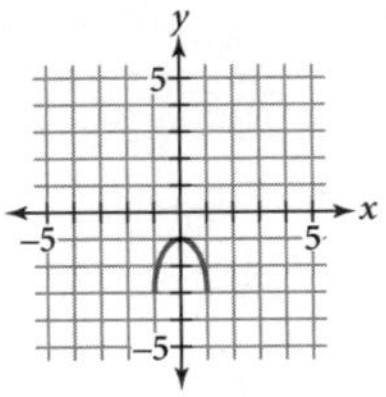

1. *(Lesson 4.7)*

Equation	Transformation (translation, reflection, stretch, shrink)	Orientation (horizontal, vertical)	Direction (up, down, left, right, across x-axis, across y-axis)	Amount
$y = 3 + x^2$	Translation	Vertical	Down	3
$-y = \|x\|$	Reflection	Horizontal	across x-axis	n/a
$y = \sqrt{\frac{x}{4}}$	Stretch	Vertical	n/a	4
$\frac{y}{0.4} = x^2$	Shrink	Horizontal	n/a	0.4
$y = \|x - 2\|$	Translation	Horizontal	Right	2
$y = \sqrt{-x}$	Reflection	Vertical	across y-axis	n/a

5a. $y = \sqrt{1 - x^2} + 2$

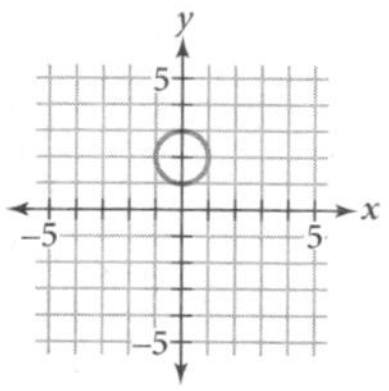

5b. $y = \sqrt{1 - (x + 3)^2}$

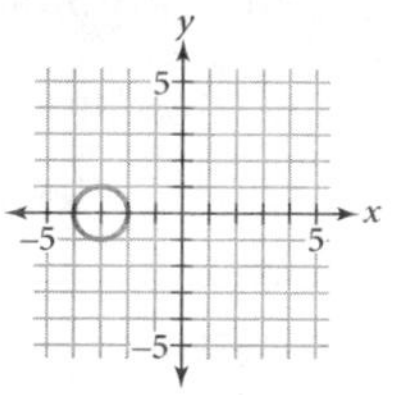

5c. $y = 2\sqrt{1 - x^2}$

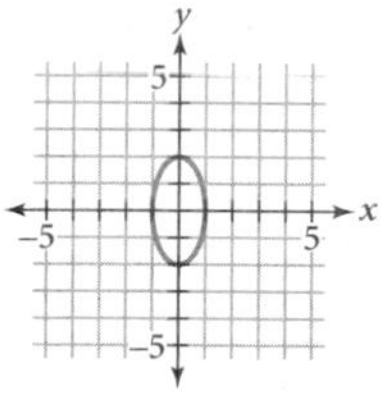

5d. $y = \sqrt{1 - \left(\frac{x}{2}\right)^2}$

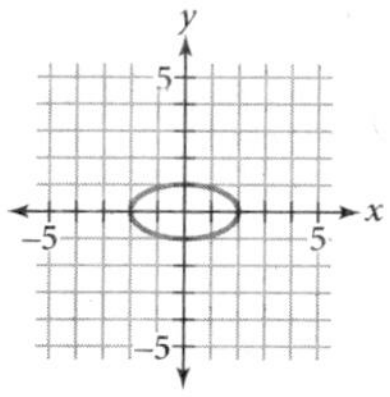

7a. $x^2 + (2y)^2 = 1$

7b. $(2x)^2 + y^2 = 1$

7c. $\left(\frac{x}{2}\right)^2 + (2y)^2 = 1$

9a. (0, 0) and (1, 1)

9b. The box has width 1 and height 1. The width is the change in x-values, and the height is the change in y-values.

9c.

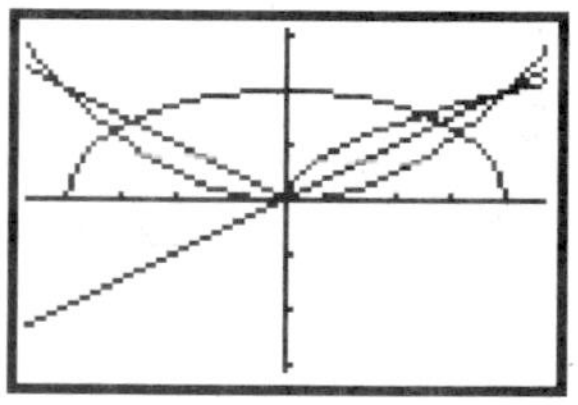

[−0.5, 5, 1, −0.5, 3, 1]

(0, 0) and (4, 2)

9d. The box has width 4 and height 2. The width is the change in x-values, and the height is the change in y-values.

9e.

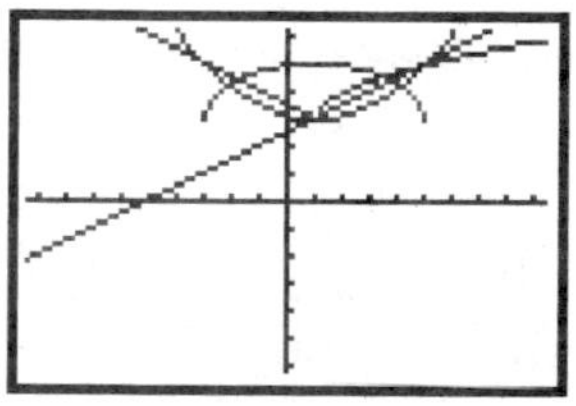

[−0.5, 6, 1, −0.5, 6, 1]

(1, 3) and (5, 5)

9f. The box has width 4 and height 2. The change in x-values is 4, and the change in y-values is 2.

9g. The x-coordinate is the location of the right endpoint, and the y-coordinate is the location of the top of the transformed semicircle.

11. 625, 1562.5, 3906.25

13a. After approximately 14 years she will run out of money.

13b. She would need to withdraw $937.50 each month.

15a. 1,210,000,000 passengers, assuming that all data occurs at midpoints of bins.

15b. 40,333,333 passengers

15c. Five-number summary: 27.5, 32.5, 37.5, 47.5, 82.5; assume that all data occurs at midpoints of bins.

LESSON 4.8

1a. ≈ 1.5 m/s **1b.** ≈ 12 L/min **1c.** ≈ 15 L/min

3a. 2 **3b.** 1 **3c.** 0

5a. 2 **5b.** 6

5c.

x	$g(x)$	$f(x)$	$f(g(x))$	$g(f(x))$
−2	4	6	−2	−2
1	2	Not defined	1	Not defined
2	Not defined	1	Not defined	2
4	Not defined	−2	Not defined	4
5	5	5	5	5
6	−2	Not defined	6	Not defined

7a.

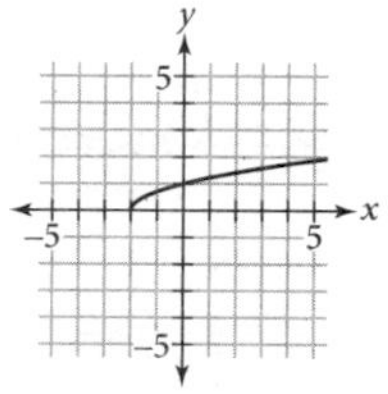

7b.

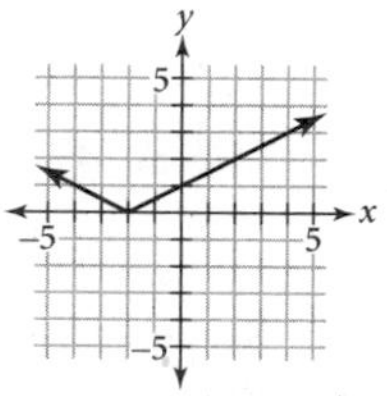

7c.

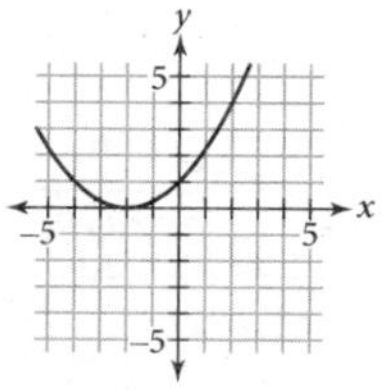

9a. 4 **9b.** 3 **9c.** 3.0625 **9d.** 4

9e. $f(g(x)) = -(x-2)^2 + 2(x-2) + 3 = -x^2 + 6x - 5$

9f. $g(f(x)) = ((-x^2 + 2x + 3) - 2)^2 = x^4 - 4x^3 + 2x^2 + 4x + 1$

11a. Jen: $5.49 - 0.50 = 4.99$; $(1 - 0.10)(4.99) = \$4.49$

Priya: $(1 - 0.10)(5.49) = 4.94$; $4.94 - 0.50 = \$4.44$

11b. $C(x) = x - 0.50$

11c. $D(x) = (1 - 0.10)x$ or $0.90x$

11d. $C(D(x)) = 0.90x - 0.50$

11e. The 10% discount was taken first, so it was Priya's server.

11f. There is no price because $C(D(x)) = D(C(x))$, or $0.90x - 0.50 = 0.90(x - 0.50)$ has no solution.

13a. $\left(\frac{x}{3}\right)^2 + \left(\frac{y}{3}\right)^2 = 1$ or $x^2 + y^2 = 9$

13b.

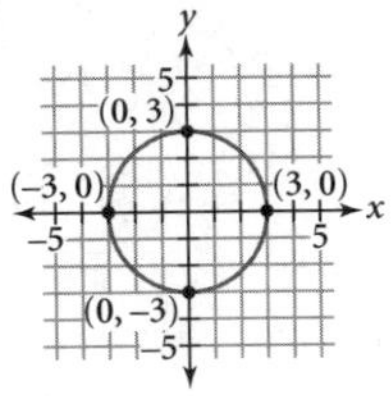

15a. $f'(x) = (x - 3)^2 + 5$

15b. (3, 5) **15c.** (5, 9)

CHAPTER 4 REVIEW

1. Sample answer: For a time there are no pops. Then the popping rate slowly increases. When the popping reaches a furious intensity, it seems to level out. Then the number of pops per second drops quickly until the last pop is heard.

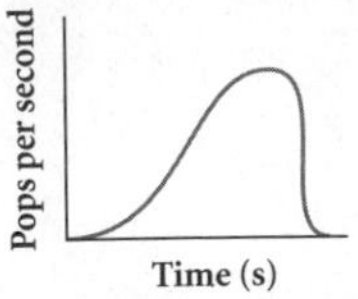

3a.

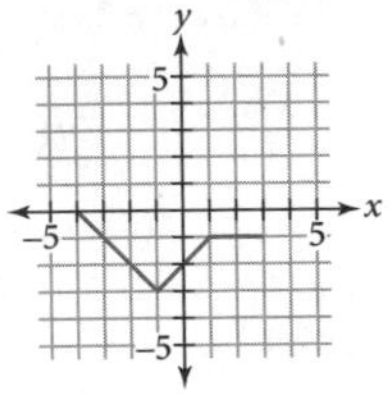

3b.

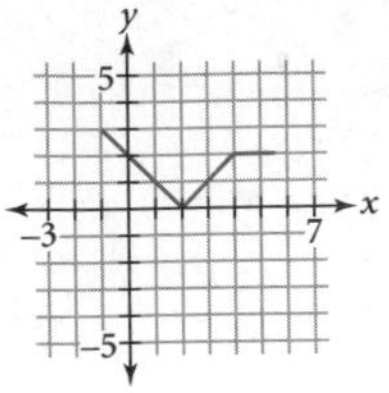

3c.

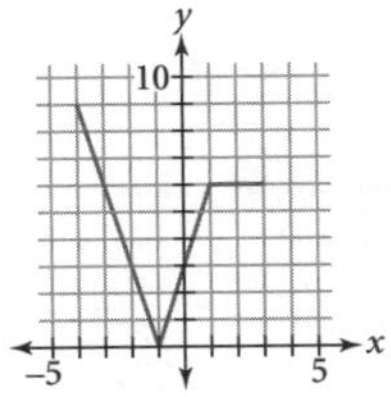

3d.

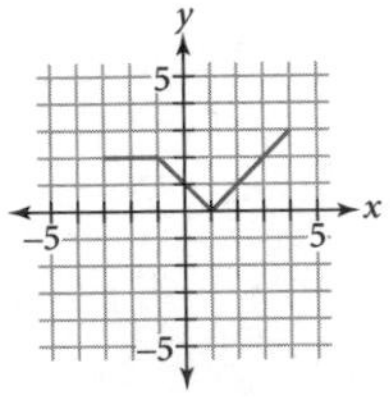

5a.

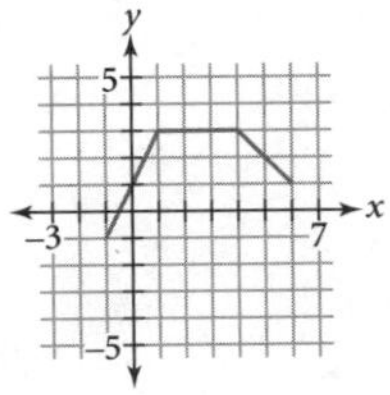

5b.

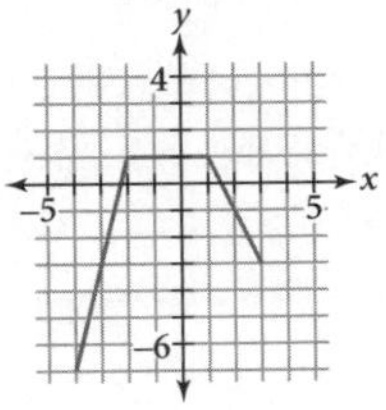

5c.

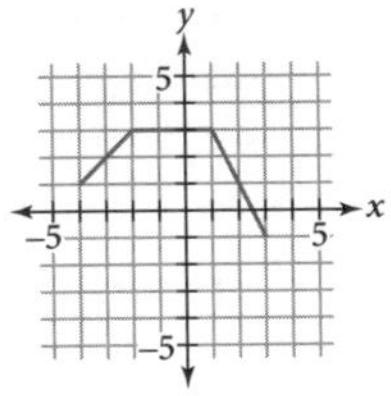

5d.

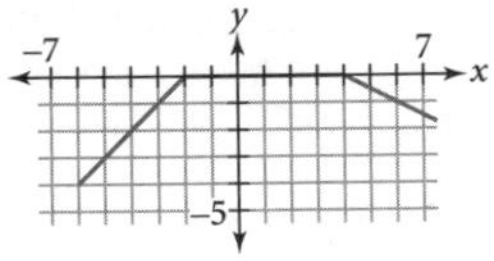

5e.

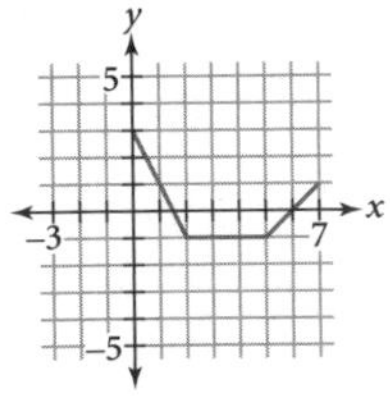

5f.

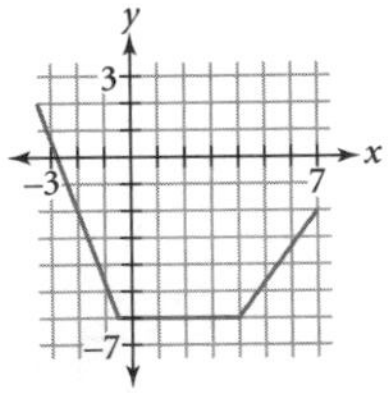

9a. *(Chapter 4 Review)*

Fare (\$), *x*	1.00	1.10	1.20	1.30	1.40	1.50	1.60	1.70	1.80
Number of passengers	18000	17000	16000	15000	14000	13000	12000	11000	10000
Revenue (\$), *y*	18000	18700	19200	19500	19600	19500	19200	18700	18000

7a. $y = \frac{2}{3}x - 2$ **7b.** ; $y = \pm\sqrt{x + 3} - 1$

7c. $y = \pm\sqrt{-(x - 2)^2 + 1}$ or $y = \pm\sqrt{1 - (x - 2)^2}$

9b.

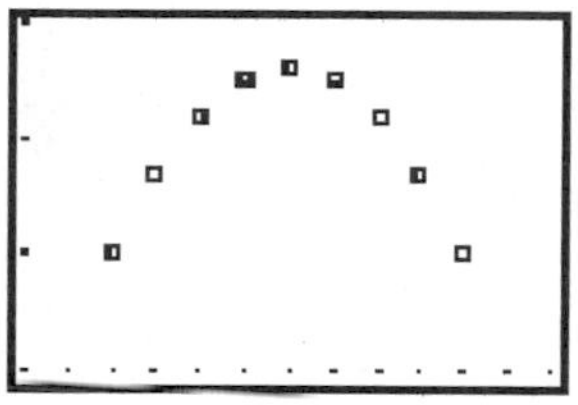

[0.8, 2, 0.1, 17000, 20000, 1000]

9c. (1.40, 19600). By charging \$1.40 per ride, the company maximizes the revenue, \$19,600.

9d. $y = -10000(x - 1.4)^2 + 19600$

9d. i. \$16,000 **9d. ii.** \$0 or \$2.80

CHAPTER 5 • CHAPTER 5 CHAPTER 5 • CHAPTER

LESSON 5.1

1a. $f(5) \approx .52738$ **1b.** $g(14) \approx 19{,}528.32$

1c. $h(24) \approx 22.9242$ **1d.** $j(37) \approx 3332.20$

3a. $f(0) = 125, f(1) = 75, f(2) = 45$; $u_0 = 125$ and $u_n = 0.6u_{n-1}$ where $n \geq 1$

3b. $f(0) = 3, f(1) = 6, f(2) = 12$; $u_0 = 3$ and $u_n = 2u_{n-1}$ where $n \geq 1$

5a. $u_0 = 1.151$ and $u_n = (1 + 0.015)u_{n-1}$ where $n \geq 1$

5c. Let x represent the number of years since 1991, and let y represent the population in billions. $y = 1.151(1 + 0.015)^8$

5d. $y = 1.151(1 + 0.015)^{10} \approx 1.336$; the equation gives a population that is greater than the actual population; sample answer: the growth rate of the Chinese population has slowed since 1991.

7a.–7d.

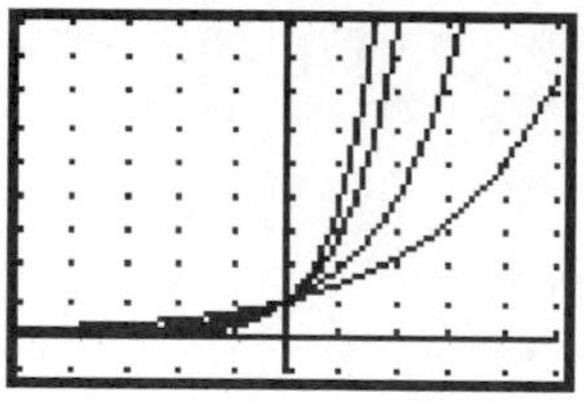

[−5, 5, 1, −1, 9, 1]

7e. As the base increases, the graph becomes steeper. The curves all intersect the y-axis at (0, 1).

7f. The graph of $y = 6^x$ should be the steepest of all of these. It will contain the points (0, 1) and (1, 6).

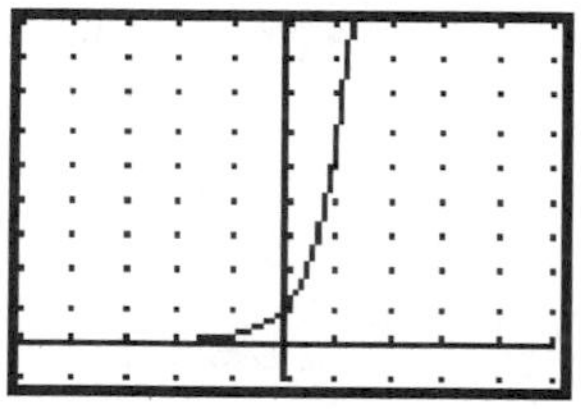

[−5, 5, 1, −1, 9, 1]

9a.–9d.

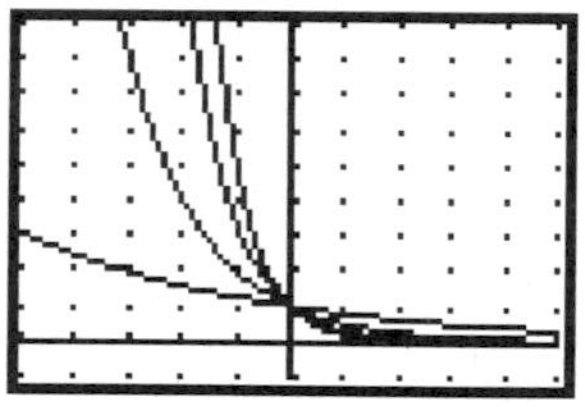

[−5, 5, 1, −1, 9, 1]

9e. As the base increases, the graph flattens out. The curves all intersect the y-axis at (0, 1).

9f. The graph of $y = 0.1^x$ should be the steepest of all of these. It will contain the points (0, 1) and (−1, 10).

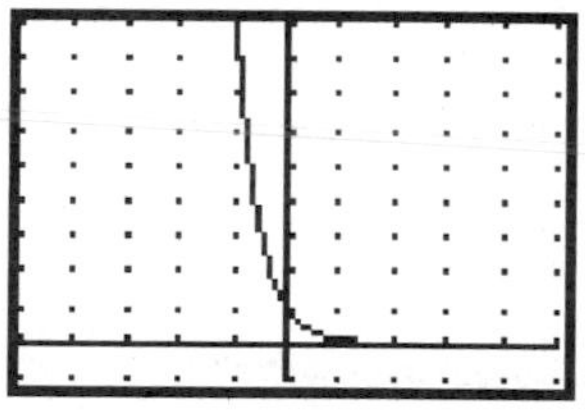

[−5, 5, 1, −1, 9, 1]

5b. *(Lesson 5.1)*

Year	1991	1992	1993	1994	1995	1996	1997	1998	1999	2000
Population (in billions)	1.151	1.168	1.186	1.204	1.222	1.240	1.259	1.277	1.297	1.316

11a. $\frac{27}{30} = 0.9$

11b. $f(x) = 30(0.9)^x$

11c.

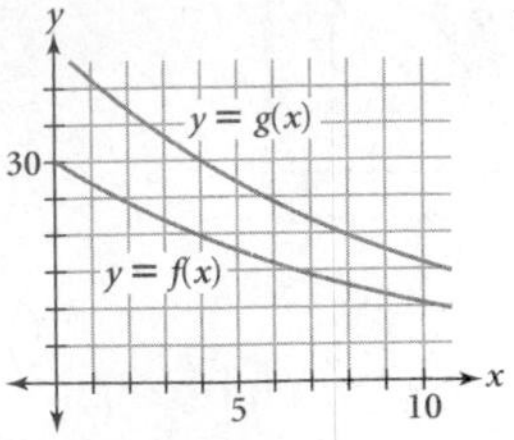

11d. $g(4) = 30$

11e. possible answer: $g(x) = 30(0.9)^{x-4}$

13a. Let x represent time in s, and let y represent distance in m.

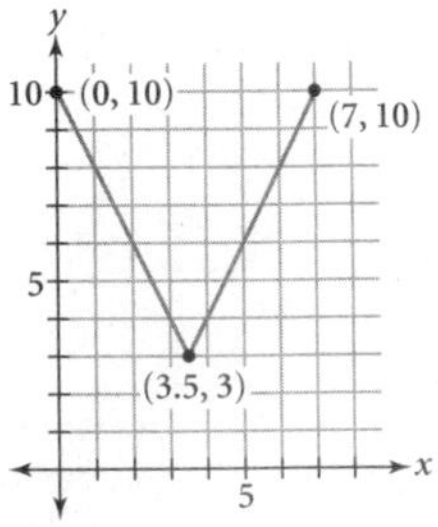

13b. domain: $0 \leq x \leq 7$; range: $3 \leq y \leq 10$

13c. $y = 2|x - 3.5| + 3$

15a. Follow directions in the spreadsheet.

15b. Follow directions in the spreadsheet.

15c. Multiply by 1.5.

15d. Use 1000 as the first value and multiply by 1.3 instead of 1.5.

15e. Up to 37 the number in column C is larger than the number in column B. Eventually column B becomes larger because it has a larger growth rate.

15f. $83,276,496,637

LESSON 5.2

1a. $\frac{1}{125}$

1b. -36

1c. $-\frac{1}{81}$

1d. $\frac{1}{144}$

1e. $\frac{16}{9}$

1f. $\frac{7}{2}$

3a. False. You must have the same base for the multiplication property of exponents.

3b. False. You must raise to the power before multiplying.

3c. False. You must raise to the power before dividing.

3d. True

5a. x^{12}

5b. $8x^{12}$

5c. $10x^7$

5d. $12x^5$

5e. $8x^{12}$

5f. $\frac{1}{25}x^{-12}$

7a. 49, 79.7023, 129.6418, 210.8723, 343

7b. 30.7023; 49.9395; 81.2305; 3132.1277. The sequence is not arithmetic because there is not a common difference.

7c. 1.627; 1.627; 1.627; 1.627. The ratio of consecutive terms is always the same, so the difference is growing exponentially.

7d. Possible answers: Noninteger powers produce noninteger values. Decimal powers form a geometric sequence.

9a. $y - 4 = x^3$ or $y = x^3 + 4$

9b. $y = (x + 2)^3$

9c. $4y = x^3$ or $y = \frac{1}{4}x^3$

9d. $8(y + 2) = x^3$ or $y + 2 = \left(\frac{1}{2}x\right)^3$ or $y = \frac{1}{8}x^3 - 2$

11a. approximately 0.9476

11b. 137.2

11c. 137.2 rads

11d. $y = 137.2(.9476)^x$

11e. 27.30 rads

13a. $x = 7$

13b. $x = -4$

13c. $x = 4$

13d. $x = 4.61$

15a.

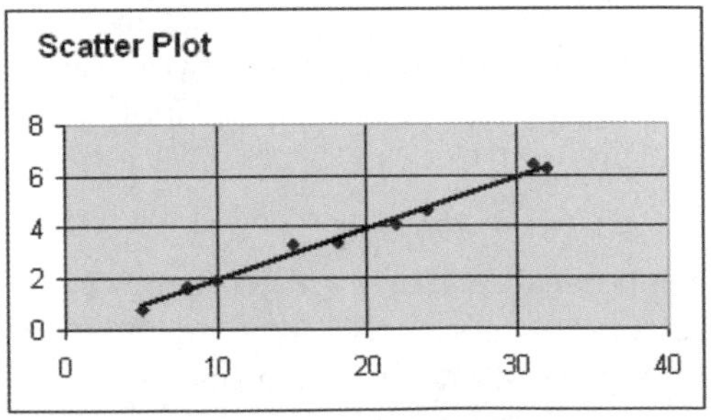

15b. $y = 0.1982x - 0.0292$ is one possibility.

LESSON 5.3

1. a, e, j; b, d, g; c, i; f, h

3a. $a^{1/6}$

3b. $b^{4/5}$ or $b^{8/10}$ or $b^{0.8}$

3c. $c^{-1/2}$ or $c^{-0.5}$

3d. $d^{7/5}$ or $d^{1.4}$

5a. 3.27 **5b.** 784 **5c.** 0.16

5d. 0.50 **5e.** 1.07 **5f.** No solution

7a.

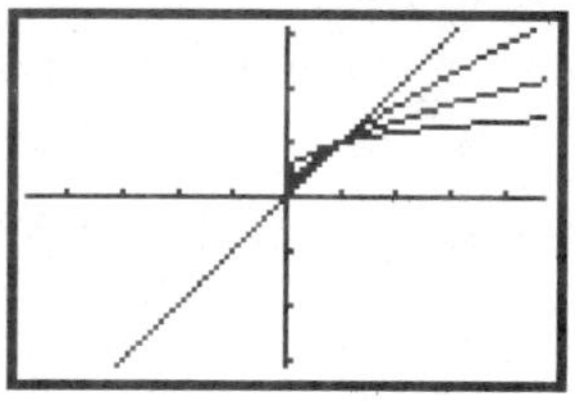

[−4.7, 4.7, 1, −3.1, 3.1, 1]

7b. $y = x^{5/4}$ should be steeper and curve upward.

9a. exponential **9b.** neither

9c. exponential **9d.** power

11. 490 W/cm^2

13a. approximately 0.723 AU

13b. approximately 29.471 yr

13c. Mercury (0.387, 0.2407); Venus (0.7232, 0.615); Earth (1.00, 1.00); Mars (1.523, 1.8795); Jupiter (5.201, 11.861); Saturn (9.542, 29.475); Uranus (19.181, 84.008); Neptune (30.086, 165.02)

15. $y = 98(1.45)^x$

x	y
0	98
2	206.045
3	298.76525
5	628.1539
8	1915.0058

17a.

1	3			
2	12	9		
3	27	15	6	
4	48	21	6	0
5	75	27	6	0

17b.

1	2				
2	16	14			
3	54	38	24		
4	128	74	36	12	
5	250	122	48	12	0
6	432	182	60	12	0

17c. The nth difference quotient is a constant, or the $(n + 1)$st difference quotient is 0.

19a. $27x^9$ **19b.** $16x^9$

19c. $1/5\ x^{-1}$ **19d.** $108x^8$

19e. $18x^2y^4$ (Not possible to write as ax^n)

19f. $2x^2 - x$ (Not possible to write as ax^n)

21a. $u_1 = 20,\ u_{n+1} = 1.2u_n$

21b. 86 rat sightings

21c. $y = 20(1.2)^x$ with x as the number of years that have passed since 20 rats were seen and y as the number of rat sightings.

LESSON 5.4

1a. $x = 50^{1/5} \approx 2.187$ **1b.** $x = 3.1^3 = 29.791$

1c. no real solution

3a. $9x^4$ **3b.** $8x^6$ **3c.** $216x^{-18}$

5a. She must replace y with $y - 7$ and y_1 with $y_1 - 7$; $y - 7 = (y_1 - 7) \cdot b^{x-x_1}$.

5b. a should be approximately 193. b should be approximately 0.510.

7a.

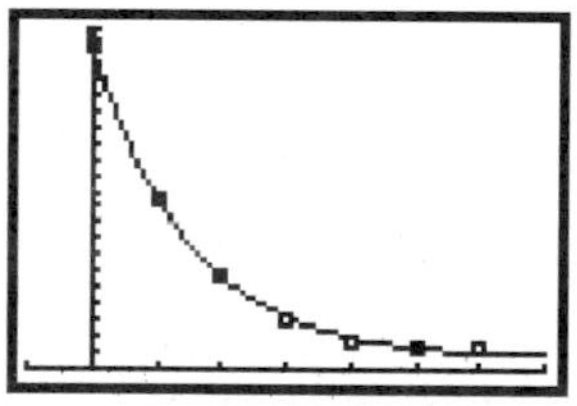

[0, 6, 1, 0, 5, 1]

7b. Sample answer: $y = .2x$

7c. Sample answer: $y = .71x$

7d. Sample answer: $y = 0.37x^{1.5}$. The graph of the data and the equation appear to be a good fit.

7e. d seems to be the best fit. Answers will vary about whether the exponential model is a better fit.

7f. approximately 1,229,200 km

7g. approximately 545.390 days

9a. approximately 19.58 cm

9b. approximately 23.75 m

11a. approximately 0.9534

11b. approximately 6.6 g

11c. $y = 6.6(0.9534)^x$

11d. approximately 0.6 g

11e. about 14.5 yr

LESSON 5.5

1. $(-3, -2), (-1, 0,) (2, 2), (6, 4)$

3. Graph c is the inverse because the x- and y-coordinates have been switched from the original graph so that the graphs are symmetric across line $y = x$.

5a. $f(7) = 4$ and $g(4) = 7$.

5b. They might be inverse functions.

5c. $f(1) = -2$ and $g(-2) = 5$

5d. They might *not* be inverse functions.

5e. $f(x)$ for $x \geq 3$ and $g(x)$ for $x \geq -4$ are inverse functions.

7a.

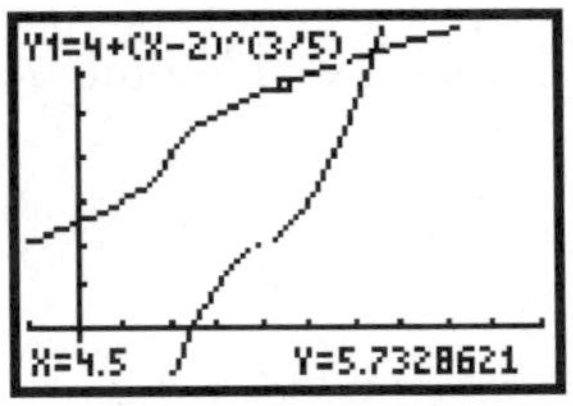

$[-1, 10, 1, -1, 7, 1]$

7b. The inverse function from 6b should be the same as the function drawn by your calculator.

7c. The inverse function you find algebraically should be the same as the one drawn by your calculator.

9a. i. $f^{-1}(x) = \dfrac{x + 140}{6.34}$

9a. ii. $f(f^{-1}(15.75)) = 15.75$

9a. iii. $f^{-1}(f(15.75)) = 15.75$

9a. iv. $f(f^{-1}(x)) = f^{-1}(f(x)) = x$

9b. i. $f^{-1}(x) = \dfrac{x - 32}{1.8}$

9b. ii. $f(f^{-1}(15.75)) = 15.75$

9b. iii. $f^{-1}(f(15.75)) = 15.75$

9b. iv. $f(f^{-1}(x)) = f^{-1}(f(x)) = x$

11a. The equation for converting Celsius degrees into Fahrenheit degrees is $F = 1.8C + 32$.

11b. Solve the equation from 11a for C; $C = 100 - y$. Substitute this expression for C into the equation for F, and solve for y. $F = 1.8(100 - y) + 32$; $F = 180 - 1.8y + 32$; $F = 212 - 1.8y$; $y = \dfrac{F - 212}{-1.8}$.

13a. $c(x) = 7.18 + 3.98x$, where c is the cost and x is the number of thousand gallons

13b. \$39.02

13c. $g(x) = \dfrac{x - 7.18}{3.98}$, where g is the number of thousands of gallons and x is the cost

13d. 12; 12,000 gallons

13e. $g(c(x)) = g(7.18 + 3.98x) = \dfrac{7.18 + 3.98x - 7.18}{3.98} = \dfrac{3.98x}{3.98} = x$

$c(g(x)) = c\left(\dfrac{x - 7.18}{3.98}\right) = 7.18 + 3.98\left(\dfrac{x - 7.18}{3.98}\right)$
$= 7.18 + x - 7.18 = x$

13f. about \$6

13g. Answers will vary, but volume should equal $231 \cdot 1500$, or 346,500 in.^{3} or approximately 200 ft^3.

15. $f(x) = 5.6(1.5)^x$

17. $y = 3(x - 3)^2 + 2$ and $x = \left[\dfrac{1}{3}(y - 2)\right]^2 + 3$

LESSON 5.6

1a. $10^x = 1000$ **1b.** $5^x = 625$ **1c.** $7^x = \sqrt{7}$

1d. $8^x = 2$ **1e.** $5^x = \dfrac{1}{25}$ **1f.** $6^x = 1$

3a. $x = \log_{10} 0.001$; $x = -3$

3b. $x = \log_5 100$; $x \approx 2.8614$

3c. $x = \log_{35} 8$; $x \approx 0.5849$

3d. $x = \log_{0.4} 5$; $x \approx -1.7565$

3e. $x = \log_{0.8} 0.03$; $x \approx 15.7144$

3f. $x = \log_{17} 0.5$; $x \approx -0.2447$

5a. false; $x = \log_6 12$ **5b.** false; $2^x = 5$

5c. false; $x = \dfrac{\log 5.5}{\log 3}$ **5d.** false; $x = \log_3 7$

7a. 1980

7b. approximately 13%

7c. approximately 5.6 yr

9a. $y = 88.7(1.0077)^x$

9b. 23 or 24 clicks

11a. $x = 345$

11b. $x = 7^{1/2.4} \approx 2.25$

13a. $y + 1 = x - 3$

13b. $y + 4 = (x + 5)^2$

13c. $y - 2 = |x + 6|$

13d. $y - 7 = \sqrt{x - 2}$

LESSON 5.7

1a. $g^h \cdot g^k$; product property of exponents

1b. log *st;* log of a product property

1c. f^{w-v}; quotient property of exponents

1d. log *h* − log *k;* log of a quotient property

1e. j^{st}; power of a power property of exponents

1f. *g* log *b;* log of a power property

1g. $k^{m/h}$; definition of rational exponents

1h. $\log_u t$; change-of-base property

1i. w^{t+s}; product property of exponents

1j. $\frac{1}{p^h}$; definition of negative exponents

3a. $x = 3.38$

3b. $x = 11.49$

3c. $x = 11.17$

3d. $x = 42.74$

5a. $y = 14.7(0.8022078)^x$

5b. $y = 8.47$ psi

5c. $x = 6.57$ mi

7a. 96.5% remains after 1 min

7b. $y = 100(0.965)^x$

7c. 19.456 min

7d. In one day the carbon-11 is virtually gone, so you could never date an archaeological find.

9a. $y = 5 + 3x$

9b. $y = 2(3^x)$

11a. False. If everyone got a grade of 86% or better, one would have to have gotten a much higher grade to be in the 86th percentile.

11b. False. Consider the data set {5, 6, 9, 10, 11}. The mean is 8.2; the median is 9.

11c. False. Consider the data set {0, 2, 28}. The range is 28; the difference between the mean, 10, and the maximum, 28, is 18.

11d. true

11e. true

LESSON 5.8

1a.
$$\begin{aligned}\log(10^{n+p}) &= \log[(10^n)(10^p)]\\ (n+p)\log 10 &= \log 10^n + \log 10^p\\ (n+p)\log 10 &= n\log 10 + p\log 10\\ (n+p)\log 10 &= (n+p)\log 10\end{aligned}$$

1b.
$$\begin{aligned}\log\left(\frac{10^d}{10^e}\right) &= \log(10^{d-e})\\ \log 10^d - \log 10^e &= \log 10^{d-e}\\ d\log 10 - e\log 10 &= (d-e)\log 10\\ (d-e)\log 10 &= (d-e)\log 10\end{aligned}$$

3. $t = \frac{\log 3}{\log 1.005625} \approx 195.9$; about 195.9 mo or about 16 yr, 4 mo

5a. $f(20) \approx 133.28$; 133 games are sold on the 20th day.

5b. $f(80) \approx 7969.17$; 7969 games are sold on the 80th day.

5c. $x = 72.09$; on day 72, 6000 games were sold.

5d. $12000/(1 + 499(1.09)^{-x}) = 6000$; $2 = 1 + 499(1.09)^{-x}$; $1 = 499(1.09)^{-x}$; $0.002 = (1.09)^{-x}$; $\log(0.002) = \log(1.09)^{-x}$; $\log 0.002 = -x\log 1.09$; $x = \frac{\log 0.002}{\log 1.09} \approx 72.11$

5e.

[0, 500, 100, 0, 15000, 1000]

Sample answer: The number of games sold starts out slowly, then speeds up. Sales decline as everyone who wants the game has purchased one.

7a.

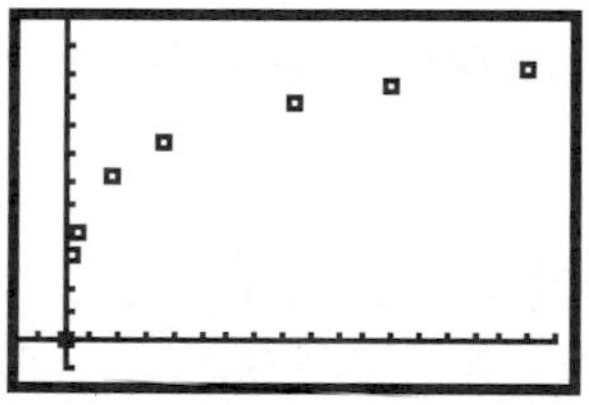

[−16, 180, 10, −5, 60, 5]

7b. Sample Answer: $y = 6 + 20 \log x$

7c. Yes.

9a. $16(1-0.15)+1=3.65$ gal of chlorine after 1 day. $14.6(1-0.15)+1=3.3525$ gal after 2 days, $13.41(1-0.15)+1=3.099625$ gal after 3 days.

9b. $u_0=16,\ u_n=u_{n-1}(1-0.15)+1$

9d. The long-run value is the value of x for which when $x=x(1-0.15)+1$ or $x=0.85x+1$, $0.15x=1$, $x=6.67$ qt. To fit the data, shift the basic exponential decay equation up 6.67 to get $y-6.67=ab^x$. The starting value is 16 and the common ratio is 0.85. The final equation is $y-6.67=16(0.85)^x$.

11a. $y=4.5(1.414)^x$ **11b.** $y=\dfrac{\ln x-\ln 4.5}{\ln 1.414}$

13. Center $(-5, 8)$. Vertices $(-5, 6)$ $(-5, 10)$, $(-4.75, 8)$, $(-5.25, 8)$

CHAPTER 5 REVIEW

1a. $\frac{1}{16}$ **1b.** $-\frac{1}{3}$ **1c.** 125

1d. 7 **1e.** $\frac{1}{4}$ **1f.** $\frac{27}{64}$

1g. -1 **1h.** 12 **1i.** 0.6

3a. $x=\dfrac{\log 28}{\log 4.7}\approx 2.153$

3b. $x=\pm\sqrt{\dfrac{\log 2209}{\log 4.7}}\approx \pm 2.231$

3c. $x=2.9^{1/1.25}=2.9^{0.8}$

3d. $\log_{3.1}x=47;\ 3.1^{47}=x;\ x\approx 1.242\cdot 10^{23}$

3e. $x=14.429^{5/12}$

3f. $x=\dfrac{\log 18}{\log 1.065}\approx 45.897$

3g. $x=10^{3.771}\approx 5902$

3h. $x=47^{5/3}\approx 612$

5. $x=\dfrac{16\log\left(\frac{8}{45}\right)}{\log 0.5}\approx 39.9$; about 39.9 h

7. $y=5\left(\frac{32}{5}\right)^{(x-1)/6}$ or $y=3.669\,(1.363)^x$

9a.

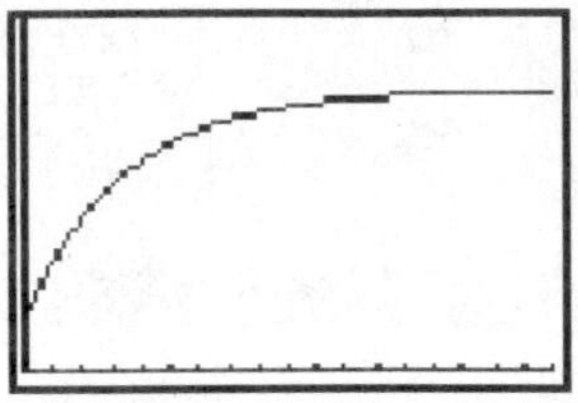

[0, 18, 1, 0, 125, 0]

9b. domain: $x\geq 0$; range: $20\leq y\leq 100$

9c. Vertically stretch by a factor of 80; reflect across the x-axis; vertical shift of 100.

9d. 55% of the average adult size

9e. about 4 years old

11a. about 37 short sessions

11b. about 47.52 wpm

11c. Sample answer: It takes much longer to improve your typing speed as you reach higher levels. 60 wpm is a good typing speed, and very few people type more than 90 wpm, so $0\leq x\leq 90$ is a reasonable domain.

CHAPTER 6 • CHAPTER **6** CHAPTER 6 • CHAPTER

LESSON 6.1

1a. 3 **1b.** 2 **1c.** 7 **1d.** 5

3a. no; {2.2, 2.6, 1.8, −0.2, −3.4}

3b. no; {0.006, 0.007, 0.008, 0.009}

3c. no; {150, 150, 150}

9c. *(Lesson 5.8)*

0	1	2	3	4	5	6	7	8	9	10
16	14.6	13.41	12.40	11.54	10.81	10.19	9.66	9.21	8.83	8.50
11	12	13	14	15	16	17	18	19	20	
8.23	7.99	7.80	7.63	7.48	7.36	7.26	7.17	7.09	7.03	

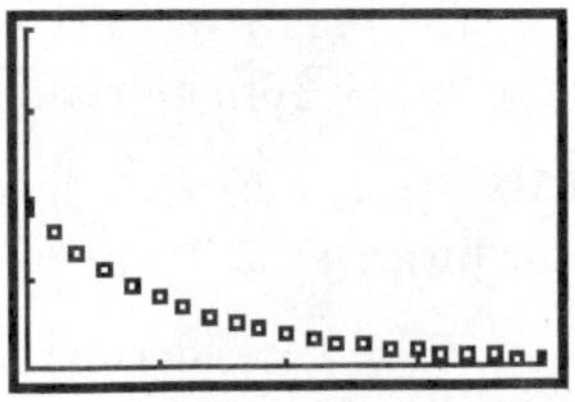

[−2, 22, 1, 5, 18, 1]

5a.

X	−1	0	1	2	3	4
Y	2	0	3	11	24	42

5b. $y = 2.5x^2 + 0.5x$

7a. Degree 2.

7b. The second finite difference is constant. The degree is 2.

7c. Four would give you a good idea. Four data points give you three first difference points which give two second difference points. You cannot begin to see that the second difference is constant until you see two consecutive points. A fifth point is really needed just to make sure the third difference was not 0 for just one value.

7d. $s = 0.5n^2 + 0.5n$, 78.

7e. The s values tell how many pennies are needed to make a triangle with n rows.

9a.

Layers x	1	2	3	4	5	6
Blocks y	1	5	14	30	55	91

9b. $b = \frac{1}{3}h^3 + \frac{1}{2}h^2 + \frac{1}{6}h$

9c. 204 blocks **9d.** 12 layers

11. $D_1 = \{6, 10, 14, 18, 22, 24\}$; $D_2 = \{4, 4, 4, 4, 4\}$. The second differences are constant, so a quadratic function expresses the relationship. Let x represent the energy level, and let y represent the maximum number of electrons. $y = 2x^2$.

13a. $x = 2.5$ **13b.** $x = 3$ or $x = -1$

13c. $x \approx 1.7227$ (Graphing the two sides of the equation and looking for intersections is one way to solve this equation).

15. $x^3 + 9x^2 + 26x + 24$

LESSON 6.2

1a. factored form and vertex form

1b. none of these forms

1c. factored form **1d.** general form

3a. −1 and 2 **3b.** −3 and 2 **3c.** 2 and 5

5a. $y = x^2 - x - 2$ **5b.** $y = 0.5x^2 + 0.5x - 3$

5c. $y = -2x^2 + 14x - 20$

7a. $y = a(x + 2.4)(x - 0.8)$

7b. $y = -1.8(x + 2.4)(x - 0.8)$

7c. $x = -0.8$, $y = 4.608$

7d. $y = -1.8(x + 0.8)^2 + 4.608$

9a. $y = -0.5x^2 - hx - 0.5h^2 + 4$

9b. $y = ax^2 - 8ax + 16a$

9c. $y = ax^2 - 2ahx + ah^2 + k$

9d. $y = -0.5x^2 - (0.5r + 2)x - 2r$

9e. $y = ax^2 - 2ax - 8a$

9f. $y = ax^2 - a(r + s)x + ars$

11a.

Length (m)	35	30	25	20	15
Area (m²)	175	300	375	400	375

11b. $y = x(40 - x)$ or $-x^2 + 40x$

11c. A width of 20 m (and length of 20 m) maximizes the area at 400 m^2.

11d. 0 m and 40 m

13a. $$\begin{cases} u_0 = -8 & \\ u_n = -\frac{1}{4}u_{n-1} & \text{for } n \geq 1. \end{cases}$$

The long run value is 0.

13b. $$\begin{cases} u_0 = 8 & \\ u_n = \frac{1}{2}u_{n-1} + 1 & \text{for } n \geq 1. \end{cases}$$

The long run value is 2.

15a. 0 **15b.** −15 **15c.** −6

15d. $-6\frac{3}{8}$ **15e.** $-23\frac{1}{3}$

LESSON 6.3

1a. $\left(x + \frac{5}{2}\right)^2$ **1b.** $(2x - 3)^2$ or $4\left(x + \frac{3}{2}\right)^2$

1c. $(x - y)^2$

3a. $y = (x + 10)^2 - 6$

3b. $y = (x - 3.5)^2 + 3.75$

3c. $y = 6(x - 2)^2 + 123$

3d. $y = 5(x + 0.8)^2 - 3.2$

5. $(-4, 12)$

7a. Let x represent time in seconds, and let y represent height in meters; $y = -4.9(x - 1.1)(x - 4.7)$ or $y = -4.9x^2 + 28.42x - 25.333$.

7b. 28.42 m/s **7c.** 25.333 m

9a. $y = x(332 - 20x)$ **9b.** \$8.30; \$1377.80

11a. $y = -4.9t^2 + 100t + 25$

11b. 25 m; 100 m/s **11c.** 10.2 s; 535 m

13a. $2x^2 - 2x - 12$ **13b.** $x^3 + 2x^2 + x + 2$

15a. $(x + 5)^2 + \left(\frac{y - 7}{3}\right)^2 = 1$; ellipse

15b.

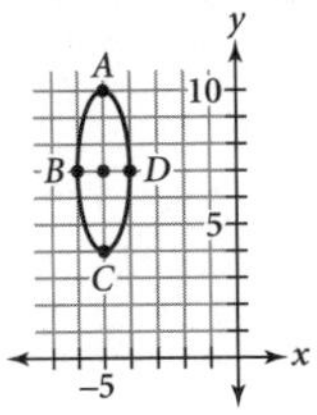

$A(-5, 10)$, $B(-6, 7)$, $C(-5, 4)$, $D(-4, 7)$; center: $(-5, 7)$

LESSON 6.4

1a. $x = 7.3$ or $x = -2.7$

1b. $x = -0.95$ or $x = -7.95$

1c. $x = 2$ or $x = -\frac{1}{2}$

3a. -0.102 **3b.** -5.898

3c. -0.243 **3d.** 8.243

5a. $y = (x - 1)(x - 5)$ **5b.** $y = (x + 2)(x - 9)$

5c. $y = 5(x + 1)(x + 1.4)$

7a. $y = a(x - 3)(x + 3)$ for $a \neq 0$

7b. $y = a(x - 4)(x + 0.4)$ or $a(x - 4)(5x + 2)$ for $a \neq 0$

7c. $y = a(x - r_1)(x - r_2)$ for $a \neq 0$

9. The function can be any quadratic function for which $b^2 - 4ac$ is negative. Sample answer: $y = x^2 + x + 1$.

11a. $y = -4x^2 - 6.8x + 49.2$

11b. 49.2 L **11c.** 2.76 min

13a. $x^2 + 14x + 49 = (x + 7)^2$

13b. $x^2 - 10x + 25 = (x - 5)^2$

13c. $x^2 + 3x + \frac{9}{4} = \left(x + \frac{3}{2}\right)^2$

13d. $2x^2 + 8x + 8 = 2(x^2 + 4x + 4) = 2(x + 2)^2$

15a. $y = 2x^2 - x - 15$

15b. $y = -2x^2 + 4x + 2$

LESSON 6.5

1a. $8 + 4i$ **1b.** 7

1c. $4 - 2i$ **1d.** $-2.56 - 0.61i$

3a. $5 + i$ **3b.** $-1 - 2i$

3c. $2 - 3i$ **3d.** $-2.35 + 2.71i$

5a.

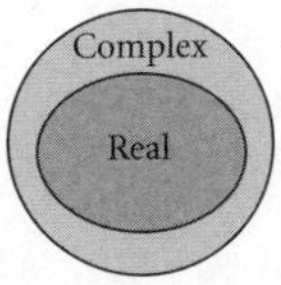

5b.

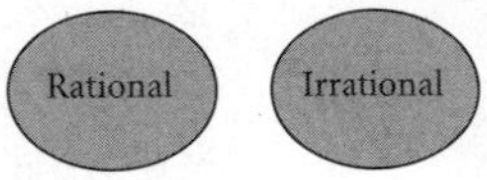

5c.

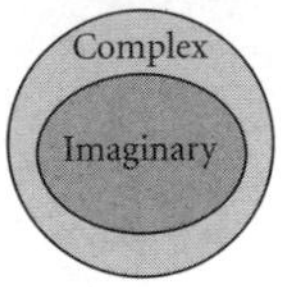

5d.

5e.

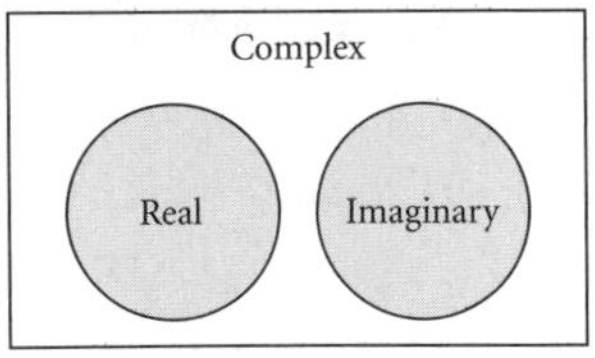

7a. $-i$ **7b.** 1 **7c.** i **7d.** -1

9. $0.2 + 1.6i$

11a. $y = x^2 - 2x - 15$ **11b.** $y = x^2 + 7x + 12.25$

11c. $y = x^2 + 25$ **11d.** $y = x^2 - 4x + 5$

13a. $x = (5 + \sqrt{34})i = 10.83i$ or $x = (5 - \sqrt{34})i = -0.83i$

13b. $x = 2i$ or $x = i$

13c. The coefficients of the quadratic equations are nonreal.

15a. 0, 0, 0, 0, 0, 0; remains constant at 0

15b. $0, i, -1+i, -i, -1+i, -i$; alternates between $-1+i$ and $-i$

15c. $0, 1-i, 1-3i, -7-7i, 1+97i, -9407+193i$; no recognizable pattern in these six terms.

15d. $0, 0.2+0.2i, 0.2+0.28i, 0.1616+0.312i, 0.12877056+0.3008384i, 0.1260781142+0.2774782585i$; approaches $0.142120634+0.2794237653i$

17a. Move right 3 units, stretch vertically by a factor of 2, shift up 5.

17b.

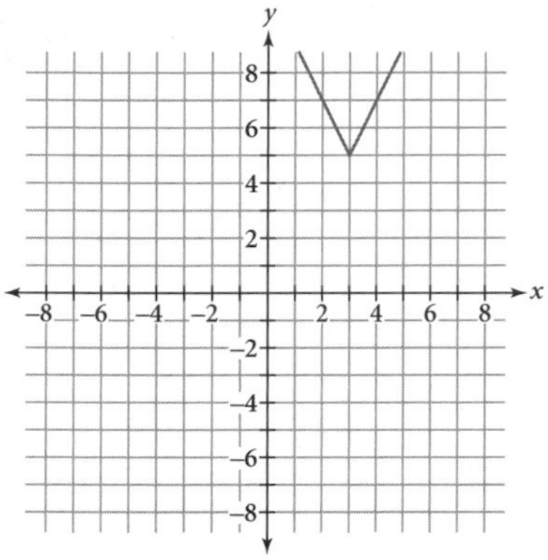

LESSON 6.6

1a. x-intercepts: $-1.5, -6$; y-intercept: -2.25

1b. x-intercept: 4; y-intercept: 48

1c. x-intercepts: $3, -2, -5$; y-intercept: 60

1d. x-intercepts: $-3, 3$; y-intercept: -135

3a. $y = x^2 - 10x + 24$ **3b.** $y = x^2 - 6x + 9$

3c. $y = x^3 - 64x$ **3d.** $y = 3x^3 + 15x^2 - 12x - 60$

5a. approximately 2.94 units; approximately 420 cubic units

5b. 5 and approximately 1.28

5c. The graph exists, but these x- and y-values make no physical sense for this context. If $x \geq 8$, there will be no box left after you take out two 8-unit square corners from the 16-unit width.

5d. The graph exists, but these x- and y-values make no physical sense for this context.

7a. sample answer:

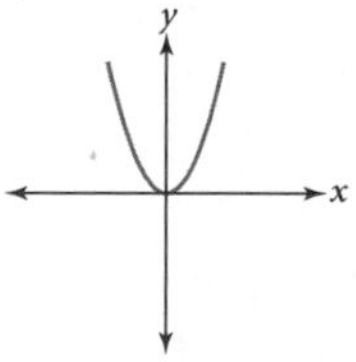

7b. sample answer:

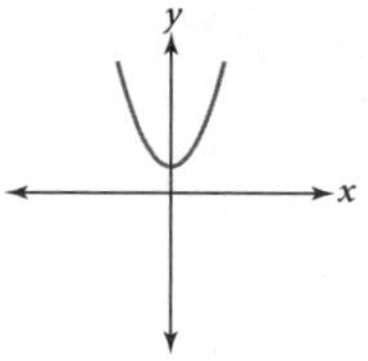

7c. not possible

7d. sample answer:

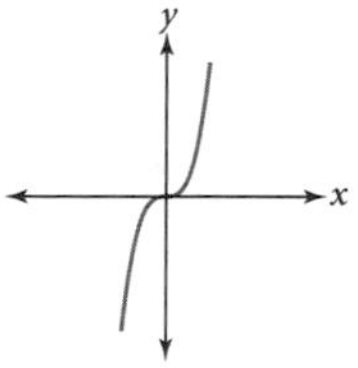

7e. sample answer:

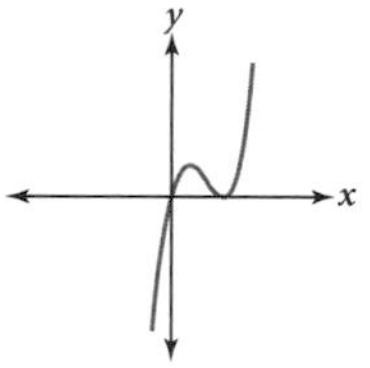

7f. not possible

9. No. These points are collinear.

11a. $x = \pm\sqrt{50.4} \approx \pm 7.1$

11b. $x = \pm\sqrt{13} \approx \pm 3.6$

11c. $x = 2.4 \pm \sqrt{40.2} \approx 2.4 \pm 6.3$; $x \approx 8.7$ or $x \approx -3.9$

11d. $x = -4$

13. $f(x) = -2.5x + 1$

The differences are all -1, so the function is linear. Find an equation for the line through any two of the points.

LESSON 6.7

1a. $x = -5$, $x = 3$, and $x = 7$

1b. $x = -6$, $x = -3$, $x = 2$, and $x = 6$

1c. $x = -5$ and $x = 2$

1d. $x = -5$, $x = -3$, $x = 1$, $x = 4$, and $x = 6$

3a. 3 **3b.** 4 **3c.** 2 **3d.** 5

5a. $y = a(x-4)$ where $a \neq 0$

5b. $y = a(x-4)^2$ where $a \neq 0$

5c. $y = a(x-4)^3$ where $a \neq 0$; or $y = a(x-4)(x-r_1)(x-r_2)$ where $a \neq 0$, and r_1 and r_2 are complex conjugates

[−9.4, 9.4, 1, −2000, 2500, 500]

7a. 4 **7b.** 5

7c. $y = -x(x+5)^2(x+1)(x-4)$

9. The leading coefficient is equal to the y-intercept divided by the product of the roots if the degree of the function is even, or the y-intercept divided by -1 times the product of the roots if the degree of the function is odd.

11a. i. $y = (x+5)^2(x+2)(x-1)$

11a. ii. $y = -(x+5)^2(x+2)(x-1)$

11a. iii. $y = (x+5)^2(x+2)(x-1)^2$

11a. iv. $y = -(x+5)(x+2)^3(x-1)$

11b. i. $x = -5, x = -5, x = -2$, and $x = 1$

11b. ii. $x = -5, x = -5, x = -2$, and $x = 1$

11b. iii. $x = -5, x = -5, x = -2, x = 1$, and $x = 1$

11b. iv. $x = -5, x = -2, x = -2, x = -2$, and $x = 1$

13. A polynomial function of degree n will have at most n x-intercepts and $n - 1$ extreme values.

15. $3 - 5\sqrt{2}$; $0 = a(x^2 - 6x - 41)$ where $a \neq 0$

LESSON 6.8

1a. $3x^2 + 7x + 3$ **1b.** $6x^3 - 4x^2$ **1c.** $8x^2 - 2$

3a. 12 **3b.** 2 **3c.** 7 **3d.** -4

5a. $\pm 15, \pm 5, \pm 3, \pm 1, \pm\frac{15}{2}, \pm\frac{5}{2}, \pm\frac{3}{2}, \pm\frac{1}{2}$

5b. $-5, \frac{1}{2}, 3$

7a. $2(3i)^3 - (3i)^2 + 18(3i) - 9 = -54i + 9 + 54i - 9 = 0$

7b. $x = -3i$ and $x = \frac{1}{2}$

9. $y = (x-3)(x+5)(2x-1)$ or $y = 2(x-3)(x+5)\left(x + \frac{1}{2}\right)$

11a. $f(x) = .5745x^{2/3}$ **11b.** $f^{-1}(x) = \left(\frac{x}{.57450}\right)^{3/2}$

11c. 26 in. **11d.** 144 ft

13a. $y = x^2 - 4x - 8$, $y = (x - 2 - 2\sqrt{3})(x - 2 + 2\sqrt{3})$; vertex: $(2, -12)$; y-intercept: -8; x-intercepts: $2 - 2\sqrt{3}, 2 + 2\sqrt{3}$

13b. $y = 3x^2 + 6x - 24$, $y = 3(x-2)(x+4)$; vertex: $(-1, -27)$; y-intercept: -24; x-intercepts: $2, -4$

13c. $y = -\frac{1}{2}x^2 + 5x + 12$, $y = -\frac{1}{2}(x-12)(x+2)$; vertex: $\left(5, \frac{49}{2}\right)$; y-intercept: 12; x-intercepts: $12, -2$

13d. $y = 2x^2 - 12x + 21$, $y = \left(x - \frac{6 - i\sqrt{6}}{2}\right)\left(x - \frac{6 + i\sqrt{6}}{2}\right)$; vertex: $(3, 3)$; y-intercept: 21; x-intercept: none

CHAPTER 6 REVIEW

1a. $2(x-2)(x-3)$

1b. $(2x+1)(x+3)$ or $2(x+0.5)(x+3)$

1c. $x(x-12)(x+2)$

3. 1; 4; 10; $\frac{1}{6}n^3 - \frac{1}{2}n^3 + \frac{1}{3}n$

5a.

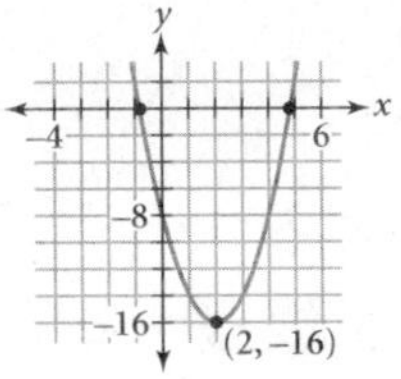

zeros: $x = -0.83$ and $x = 4.83$

5b.

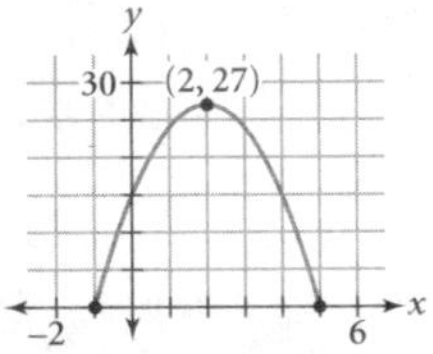

zeros: $x = -1$ and $x = 5$

5c.

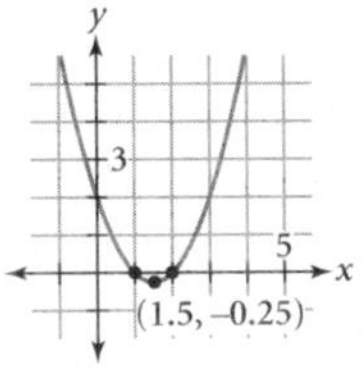

zeros: $x = 1$ and $x = 2$

5d.

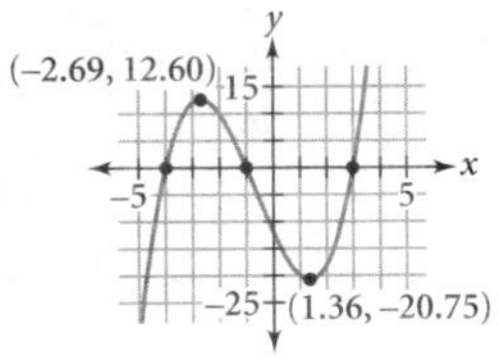

zeros: $x = -4$, $x = -1$, and $x = 3$

5e.

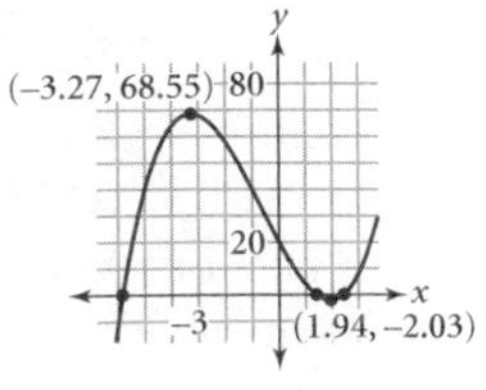

zeros: $x = -5.84$, $x = 1.41$, and $x = 2.43$

5f.

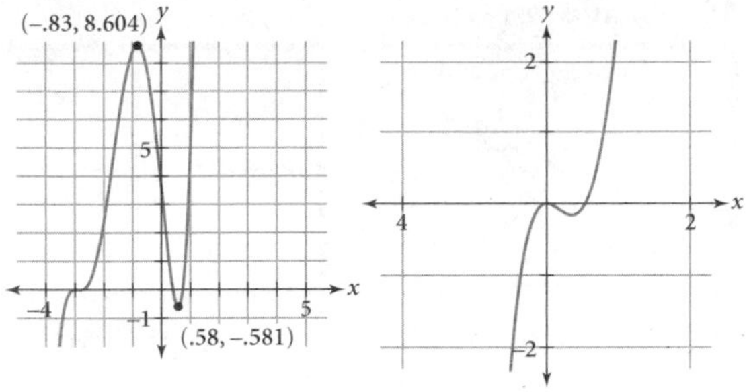

zeros: $x = -3$, $x = 2.73$
$x = 0.4$, and $x = 0.73$

7. 18 in. by 18 in. by 36 in.

9a. $y = 0.5x^2 + 0.5x + 1$

9b. 16 pieces; 56 pieces

11a. $\pm 1, \pm 3, \pm 13, \pm 39, \pm\frac{1}{3}, \pm\frac{13}{3}$

11b. $x = -\frac{1}{3}, x = 3, x = 2 + 3i$, or $x = 2 - 3i$

13. $2x^2 + 4x + 3$

CHAPTER 7 • CHAPTER **7** CHAPTER 7 • CHAPTER

LESSON 7.1

1a.

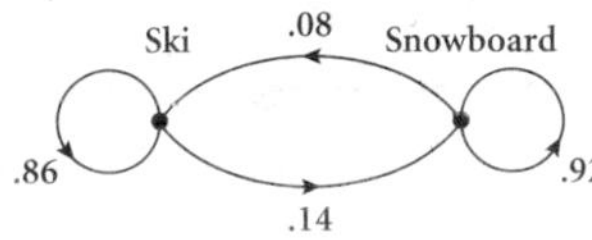

1b. $\begin{bmatrix} .86 & .14 \\ .08 & .92 \end{bmatrix}$

3. $\begin{bmatrix} .60 & .40 \\ .53 & .47 \end{bmatrix}$

5a. 20 women and 25 men

5b. 18 men batted right-handed

5c. 13 women batted right-handed

7a.

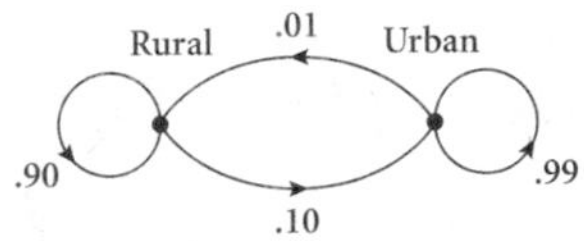

7b. $\begin{bmatrix} .99 & .01 \\ .10 & .90 \end{bmatrix}$

7c. $[16.74 \quad 8.26]$; $[17.3986 \quad 7.6014]$

9a.

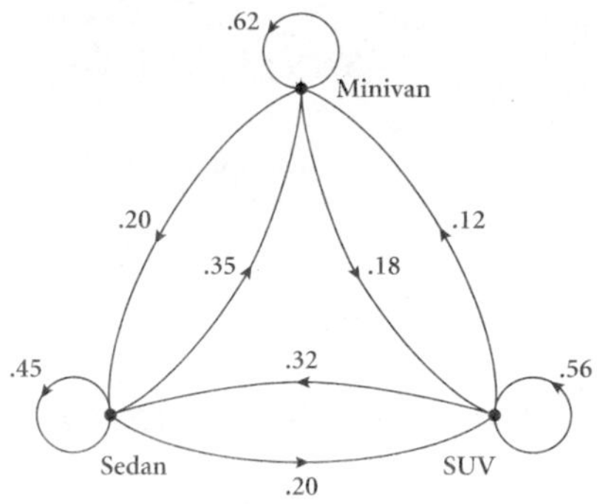

9b. $\begin{bmatrix} .62 & .20 & .18 \\ .35 & .45 & .20 \\ .12 & .32 & .56 \end{bmatrix}$

9c. The sum of each row is 1; percentages should sum to 100%.

11a. 5×5

11b. $m_{32} = 1$; there is one round-trip flight between City C and City B.

11c. Sample answer: City A has the most flights. From the graph, more paths have A as an endpoint than any other city. From the matrix, the sum of row 1 (or column 1) is greater than the sum of any other row (or column).

11d.

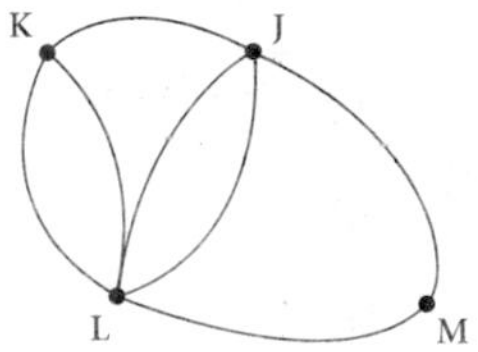

13. $y = 4 - \frac{2}{3}x$

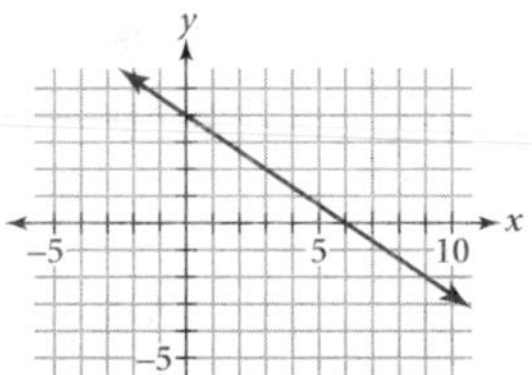

LESSON 7.2

1. $[196.84 \quad 43.15]$; 197 students will prefer dark chocolate, 43 will prefer white chocolate

3a. $\begin{bmatrix} 7 & 3 & 0 \\ -19 & -7 & 8 \\ 5 & 2 & -1 \end{bmatrix}$

3b. $\begin{bmatrix} -2 & 5 \\ 8 & 7 \end{bmatrix}$ **3c.** $[13 \quad 29]$

3d. not possible because the inside dimensions do not match

3e. $\begin{bmatrix} 4 & -1 \\ 4 & -2 \end{bmatrix}$

3f. not possible because the dimensions aren't the same

5a.

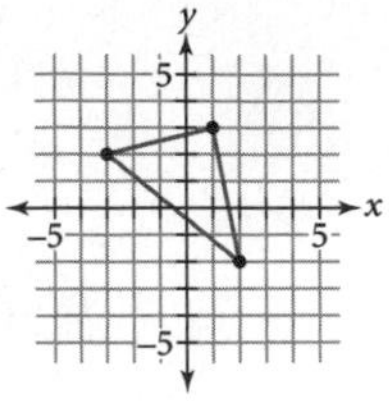

5b. $\begin{bmatrix} 3 & -1 & -2 \\ 2 & 3 & -2 \end{bmatrix}$

5c.

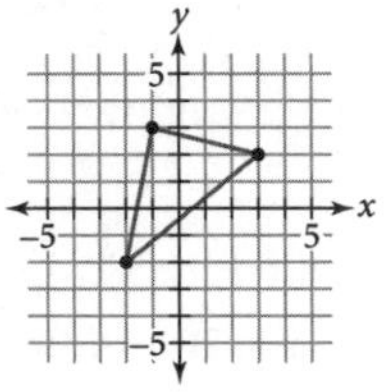

5d. The original triangle is reflected across the y-axis.

7a.

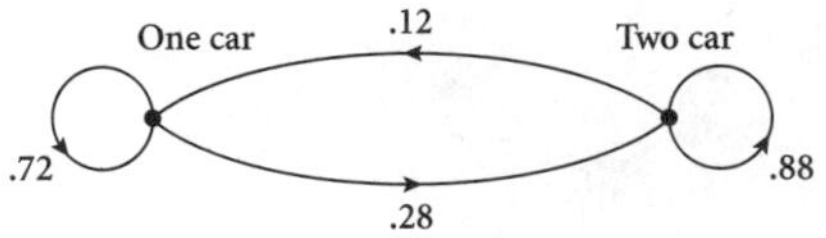

7b. $[4800 \quad 4200]$ **7c.** $\begin{bmatrix} .72 & .28 \\ .12 & .88 \end{bmatrix}$

7d. $[4800 \quad 4200]\begin{bmatrix} .72 & .28 \\ .12 & .88 \end{bmatrix} = [3960 \quad 5040]$

7e. $[3456 \quad 5544]$

9a. $a = 3, b = 2.8$ **9b.** $a = 7, b = 4$

11a. $[.95 \quad .05]\begin{bmatrix} .98 & .02 \\ .45 & .55 \end{bmatrix} = [\textit{recessive dominant}]$

11b. 95.35%

13. The probability that the spider is in room 1 after four room changes is .375; the long-run probabilities for rooms 1, 2, and 3 are $[.\overline{3}\ .\overline{3}\ .\overline{3}\,]$.

15a. i. consistent and independent

15a. ii. inconsistent

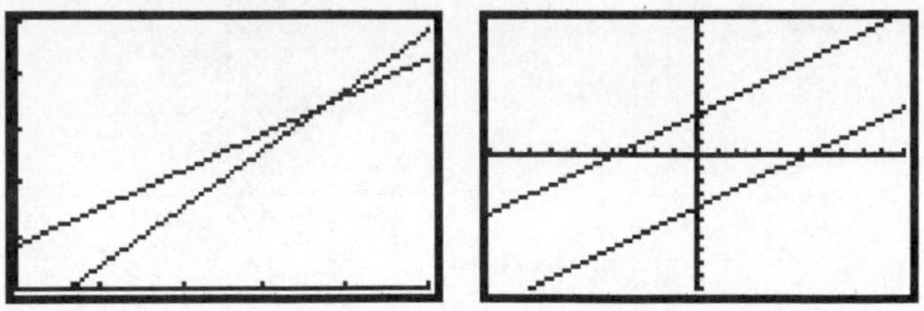

[0, 50, 10, 0, 50, 10] [−10, 10, 1, −10, 10, 1]

15a. iii. inconsistent

15a. iv. consistent and independent

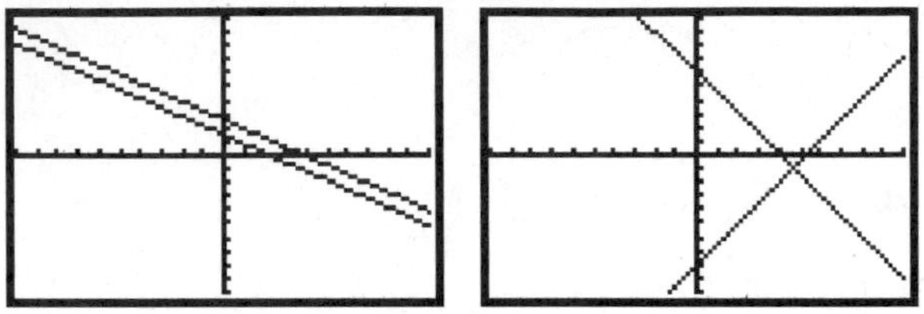

[−10, 10, 1, −10, 10, 1] [−10, 10, 1, −10, 10, 1]

15a. v. inconsistent

15a. vi. consistent and dependent

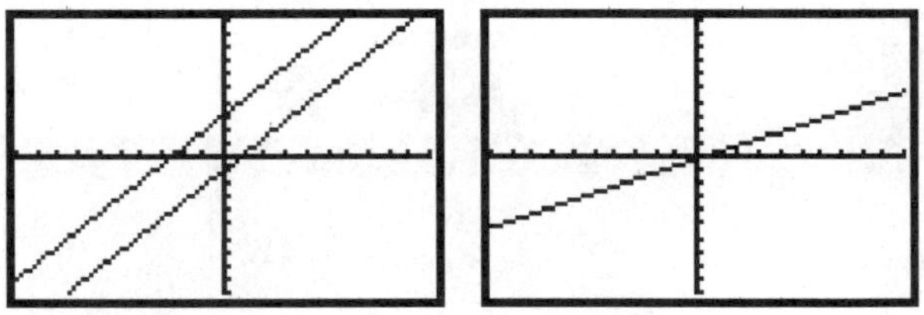

[−10, 10, 1, −10, 10, 1] [−10, 10, 1, −10, 10, 1]

15a. vii. consistent and dependent

15a. viii. consistent and independent

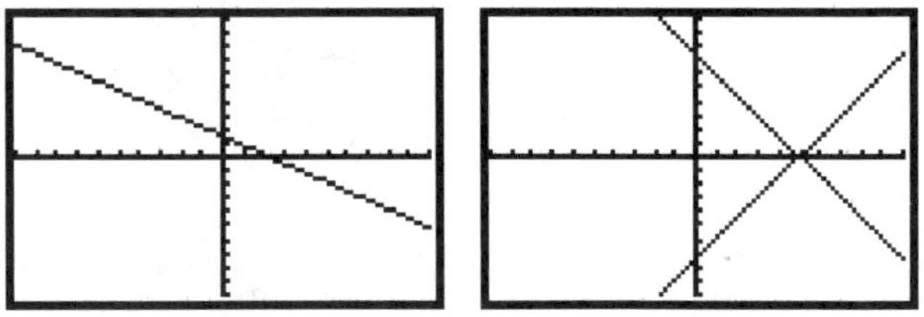

[−10, 10, 1, −10, 10, 1] [−10, 10, 1, −10, 10, 1]

15b. Sample answer: The graphs of inconsistent linear systems are parallel lines.

15c. Sample answer: When you try to solve an inconsistent system, you reach a false statement, such as $-4 = 3$.

15d. Sample answer: You can recognize an inconsistent linear system without graphing it because the lines have the same slope, but a different y-intercept.

15e. Sample answer: The graphs of consistent and dependent linear systems are the same line.

15f. Sample answer: When you try to solve a consistent and dependent system, you get a true statement, such as 0 = 0.

15g. Sample answer: The lines in a consistent and dependent linear system have the same slope and the same intercept, or the equations are multiples of each other.

17. $x = 2, y = \frac{1}{2}, z = -3$

LESSON 7.3

1a. $\begin{cases} 2x + 5y = 8 \\ 4x - y = 6 \end{cases}$ **1b.** $\begin{cases} x - y + 2z = 3 \\ x + 2y - 3z = 1 \\ 2x + y - z = 2 \end{cases}$

3a. $\left[\begin{array}{ccc|c} 1 & -1 & 2 & 3 \\ 0 & 3 & -5 & -2 \\ 2 & 1 & -1 & 2 \end{array}\right]$

3b. $\left[\begin{array}{ccc|c} 1 & -1 & 2 & 3 \\ 1 & 2 & -3 & 1 \\ 0 & 3 & -5 & -4 \end{array}\right]$

5a. $\left[\begin{array}{ccc|c} 1 & 0 & 0 & -31 \\ 0 & 1 & 0 & 24 \\ 0 & 0 & 1 & -4 \end{array}\right]$ **5b.** $\left[\begin{array}{ccc|c} 1 & 0 & 0 & -1 \\ 0 & 1 & 0 & 1 \\ 0 & 0 & 1 & 0 \end{array}\right]$

5c. cannot be reduced to row-echelon form (dependent system)

5d. cannot be reduced to row-echelon form (inconsistent system)

7. 38°, 62°, 80°

9. 3 full-page ads, 7 half-page ads, and 12 business-card size ads

11a. first plan: \$14,600; second plan: \$13,100

11b. $y = 12500 + 0.6x$

11c. $y = 6800 + 1.8x$

11d. more than 4750 tickets

11e. Sample answer: The company should choose the first plan if they expect to sell fewer than 4750 tickets and the second if they expect to sell more than 4750 tickets.

13. $\overline{AB}$: $y = 6 + \frac{2}{3}(x - 4)$ or $y = 4 + \frac{2}{3}(x - 1)$;
$\overline{BC}$: $y = 4 - \frac{2}{3}(x - 7)$ or $y = 6 + \frac{2}{3}(x - 4)$;
$\overline{CD}$: $y = 1 + 3(x - 6)$ or $y = 4 + 3(x - 7)$;
$\overline{DE}$: $y = 1$; $\overline{AE}$: $y = 4 - 3(x - 1)$ or $y = 1 - 3(x - 2)$

LESSON 7.4

1a. $\begin{bmatrix} 3 & 4 \\ 2 & -5 \end{bmatrix}\begin{bmatrix} x \\ y \end{bmatrix} = \begin{bmatrix} 11 \\ -8 \end{bmatrix}$

1b. $\begin{bmatrix} 1 & 2 & 1 \\ 3 & -4 & 5 \\ -2 & -8 & -3 \end{bmatrix}\begin{bmatrix} x \\ y \\ z \end{bmatrix} = \begin{bmatrix} 0 \\ -11 \\ 1 \end{bmatrix}$

1c. $\begin{bmatrix} 5.2 & 3.6 \\ -5.2 & 2 \end{bmatrix}\begin{bmatrix} x \\ y \end{bmatrix} = \begin{bmatrix} 7 \\ 8.2 \end{bmatrix}$

1d. $\begin{bmatrix} \frac{1}{4} & \frac{-2}{5} \\ \frac{3}{8} & \frac{2}{5} \end{bmatrix}\begin{bmatrix} x \\ y \end{bmatrix} = \begin{bmatrix} 3 \\ 2 \end{bmatrix}$

3a. $\begin{bmatrix} 1a + 5c & 1b + 5d \\ 6a + 2c & 6b + 2d \end{bmatrix} = \begin{bmatrix} -7 & 33 \\ 14 & -26 \end{bmatrix}$;
$\begin{bmatrix} a & b \\ c & d \end{bmatrix} = \begin{bmatrix} 3 & -7 \\ -2 & 8 \end{bmatrix}$

3b. $\begin{bmatrix} 1a + 5c & 1b + 5d \\ 6a + 2c & 6b + 2d \end{bmatrix} = \begin{bmatrix} 1 & 0 \\ 0 & 1 \end{bmatrix}$;
$\begin{bmatrix} a & b \\ c & d \end{bmatrix} = \begin{bmatrix} -\frac{1}{14} & \frac{5}{28} \\ \frac{3}{14} & -\frac{1}{28} \end{bmatrix}$

5a. $\begin{bmatrix} 4 & -3 \\ -5 & 4 \end{bmatrix}$

5b. $\begin{bmatrix} -\frac{3}{29} & \frac{4}{29} \\ \frac{5}{58} & \frac{3}{58} \end{bmatrix}$

5c. $\begin{bmatrix} \frac{7}{5} & -\frac{3}{5} \\ -2 & 1 \end{bmatrix}$ or $\begin{bmatrix} 1.4 & -0.6 \\ -2 & 1 \end{bmatrix}$

5d. Inverse does not exist.

7a. Let l represent the length in cm, and let w represent the width in cm. $2l + 2w = 44, l = 2 + 2w$; $w = \frac{20}{3}, l = \frac{46}{3}$

7b. Let l represent the length of a leg in cm, and let b represent the length of the base in cm. $2l + b = 40,\ b = l - 2; l = 14, b = 12$

7c. Let f represent the Fahrenheit reading in °F, and let c represent the Celsius reading in °C. $f = 3c - 0.4, f = 1.8c + 32; c = 27, f = 80.6$

9. \$2300 at 6% and \$2700 at 7.5%

13. An error message means the system is either dependent or inconsistent. In this case, the lines are the same, so the system is dependent.

15a. $\begin{bmatrix} 4 & -3 \\ -5 & 4 \end{bmatrix}$

15b. $\begin{bmatrix} -0.5555 & 1.4444 & 0.1111 \\ 0.5 & -1 & 0 \\ -1.6666 & 2.3333 & 0.3333 \end{bmatrix}$ or

$\begin{bmatrix} -\frac{5}{9} & \frac{13}{9} & \frac{1}{9} \\ \frac{1}{2} & -1 & 0 \\ -\frac{7}{6} & \frac{7}{3} & \frac{1}{3} \end{bmatrix}$

17a. possible answer: $y = 2x + 5$

17b. possible answer: $y = -\frac{1}{3}x - 2$

17b. possible answer: $2x + 5y = 20$

17d. possible answer: $x - 2y = -8$

19. 2

LESSON 7.5

1a. $y < \frac{10 - 2x}{-5}$ or $y < -2 + 0.4x$

1b. $y < \frac{6 - 2x}{-12}$ or $y < -\frac{1}{2} + \frac{1}{6}x$

3a. $y < 2 - 0.5x$ **3b.** $y \geq 3 + 1.5x$

3c. $y > 1 - 0.75x$ **3d.** $y \leq 1.5 + 0.5x$

5. 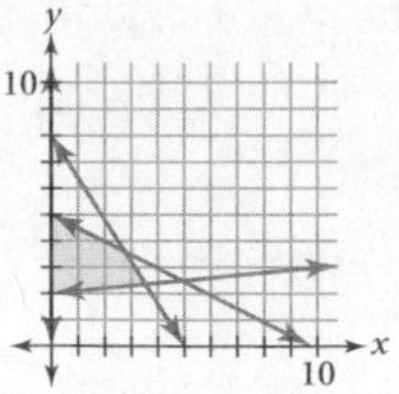

vertices: (0, 2), (0, 5), (2.752, 3.596), (3.529, 2.353)

7. 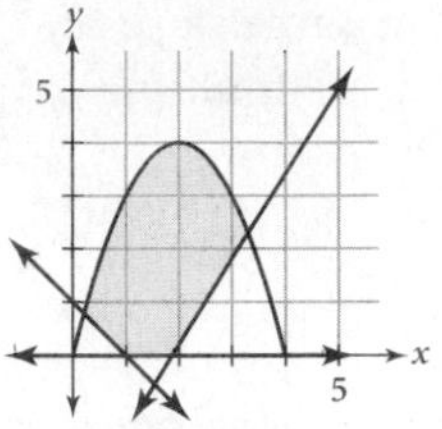

vertices: (1, 0), (1.875, 0), (3.307, 2.291), (0.209, 0.791))

9a. Let x represent length in inches, and let y represent width in inches.

$$\begin{cases} xy \geq 200 \\ xy \leq 300 \\ x + y \geq 33 \\ x + y \leq 40 \end{cases}$$

9b. 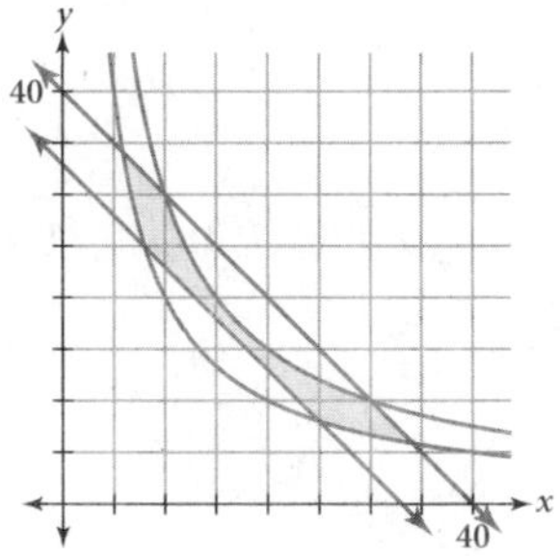

9c. i. no **9c. ii.** yes **9c. iii.** no

11a. $x > 25600$ km **11b.** 110.73 lb

11c. In theory no, because as the denominator grows larger, the value of the fraction approaches zero but never gets to zero.

13. $a = 100, b \approx 0.7$

15a. 2 or 3 spores

15b. about 1,868,302 spores

15c. $x = \frac{\log \frac{y}{2.68}}{\log 3.84}$

15d. after 14 hr 40 min

LESSON 7.6

1.

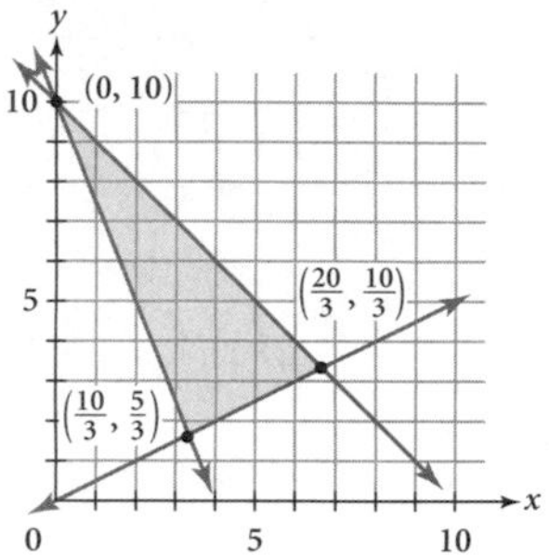

3.

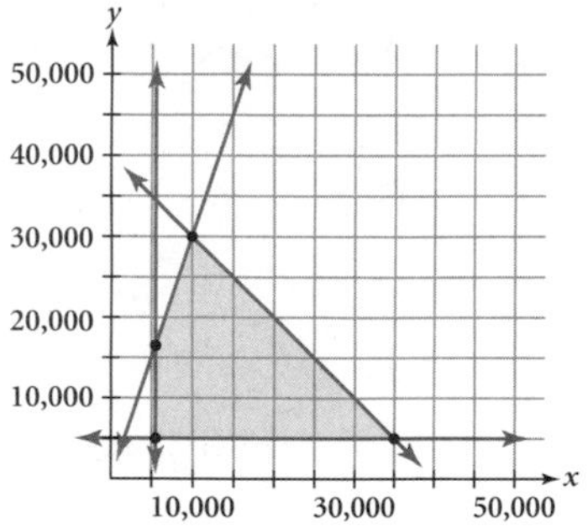

vertices: (5500, 5000), (5500, 16500), (10000, 30000), (35000, 5000); (35000, 5000); maximum: 3300

5a. possible answer: $y \geq 7, y \leq \frac{3}{2}x + 2,$
$y \leq \frac{-3}{5}x + 13$

5b. $x \geq 0, y \geq 7, y \geq (x - 3) + 6, y \leq \frac{-3}{5}x + 13$

5c. $x \geq 0, y \geq 0, x \leq 11, y \leq (x - 3) + 6, y \leq 7$

7. 5 radio minutes and 10 newspaper ads to reach a maximum of 155,000 people. This requires the assumption that people who listen to the radio are independent of people who read the newspaper.

9. 600 barrels of low-sulfur and 600 barrels of high-sulfur oil for a minimum total cost of $19,920

11a. $\left(x = -\frac{7}{11}, y = \frac{169}{11}\right)$

11b. $x = -3.5, y = 74, z = 31$

13. $\begin{cases} x \geq 2 \\ y \leq 5 \\ x + y \geq 3 \\ 2x - y \leq 9 \end{cases}$

15. $y = -\left(\frac{x}{2}\right)^2 - \frac{3}{2}$ or $y = -\frac{1}{4}x^2 - \frac{3}{2}$

CHAPTER 7 REVIEW

1a. Impossible because the dimensions are not the same.

1b. $\begin{bmatrix} -4 & 7 \\ 1 & 2 \end{bmatrix}$

1c. $\begin{bmatrix} -12 & 4 & 8 \\ 8 & 12 & -8 \end{bmatrix}$

1d. $\begin{bmatrix} -3 & 1 & 2 \\ -11 & 11 & 6 \end{bmatrix}$

1e. Impossible because the inside dimensions do not match.

1f. $\begin{bmatrix} -7 & -5 & 6 \end{bmatrix}$

3a. $x = \frac{233}{62}, y = \frac{81}{31}$

3b. $x = 1.22, y = 6.9, z = 3.4$

5a. consistent and independent

5b. consistent and dependent

5c. inconsistent

5d. inconsistent

8a. $\begin{bmatrix} .92 & .08 & 0 \\ .12 & .82 & .06 \\ 0 & .15 & .85 \end{bmatrix}$

9b. Lupton: 774, Claytor: 718, Snowdon: 652

10. about 4.4 yr

11a. $a < 0; p < 0; d > 0$

11b. $a > 0; p > 0; d = 0$

11c. $a > 0; p = 0; d < 0$

13. 20 students in the morning class, 18 students in the afternoon class, and 24 students in the evening class

15a. $x = 245$

15b. $x = 20$

15c. $x = -\frac{1}{2}$

15d. $x = \frac{\log\left(\frac{37000}{15}\right)}{\log 9.4} \approx 3.4858$

15e. $x = 21$

15f. $x = \frac{\log 342}{\log 36} \approx 1.6282$

17a. $y = 50(0.72)^{x-4}$ or $y = 25.92(0.72)^{x-6}$

17b. $b = 0.72$; decay

17c. approximately 186

17d. 0

19a. a translation right 5 units and down 2 units

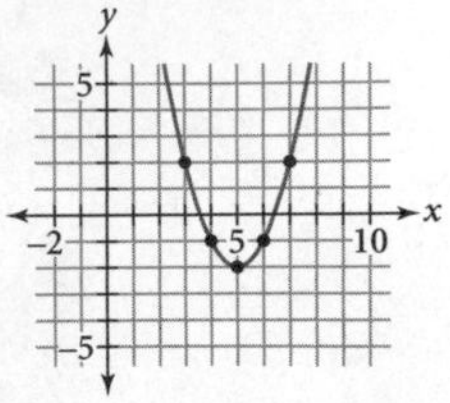

19b. a reflection across the x-axis and a vertical stretch by a factor of 2

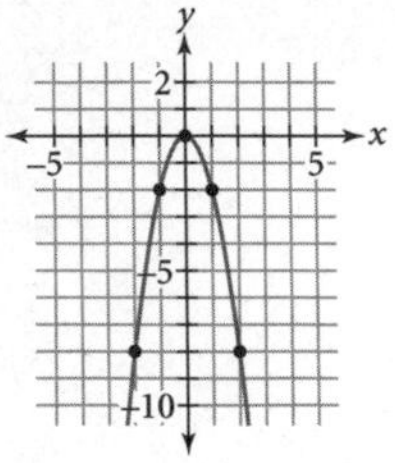

19c. $-1 \cdot [P] = \begin{bmatrix} 2 & 1 & 0 & -1 & -2 \\ -4 & -1 & 0 & -1 & -4 \end{bmatrix}$;
a reflection across the x-axis and a reflection across the y-axis. However, because the graph is symmetric with respect to the y-axis, a reflection over that axis does not change the graph.

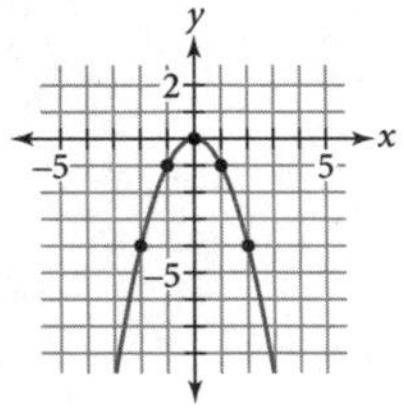

19d. $[P] + \begin{bmatrix} -2 & -2 & -2 & -2 & -2 \\ 3 & 3 & 3 & 3 & 3 \end{bmatrix} =$
$\begin{bmatrix} -4 & -3 & -2 & -1 & 0 \\ 7 & 4 & 3 & 4 & 7 \end{bmatrix}$

CHAPTER 8 • CHAPTER 8 CHAPTER 8 • CHAPTER

LESSON 8.1

1a.

t	x	y
−2	−7	−3
−1	−4	−1
0	−1	1
1	2	3
2	5	5

1b.

t	x	y
−2	−1	4
−1	0	1
0	1	0
1	2	1
2	3	4

1c.

t	x	y
−2	4	1
−1	1	2
0	0	3
1	1	4
2	4	5

1d.

t	x	y
−2	−3	0
−1	−2	1.73
0	−1	2
1	0	1.73
2	1	0

3a.

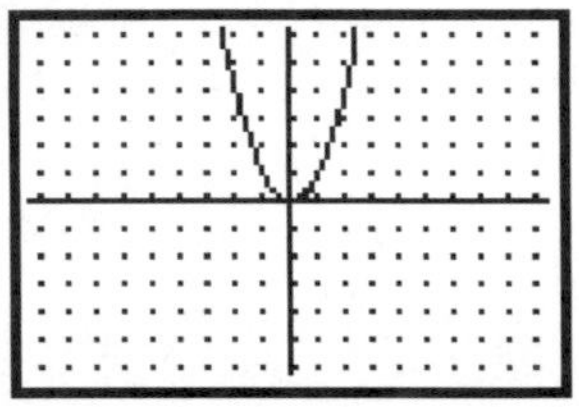

[−9.4, 9.4, 1, −6.2, 6.2, 1]

3b. The graph is translated right 2 units.

3c. The graph is translated down 3 units.

3d. The graph is translated right 5 units and up 2 units.

3e. The graph is translated horizontally a units and vertically b units.

5a. 15 s **5b.** 30 yd **5c.** −2 yd/s

5d. Sample answer: 65 is his starting position, −2 is his velocity, and 50 is his position relative to the sideline.

5e. The graph simulation will produce the graphs pictured in the problem. A good window is [0, 100, 10, 0, 65, 0] with $0 \le t \le 15$.

5f.

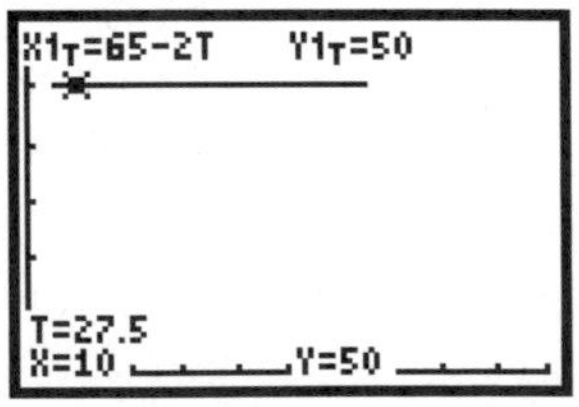

[0, 100, 10, 1, 65, 0]

$0 \le t \le 30$; 27.5 s

5g. $65 - 2t = 10$; 27.5 s

7a. The graph is reflected across the x-axis.

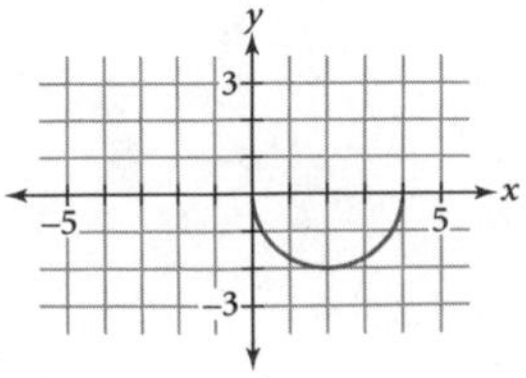

7b. The graph is reflected across the y-axis.

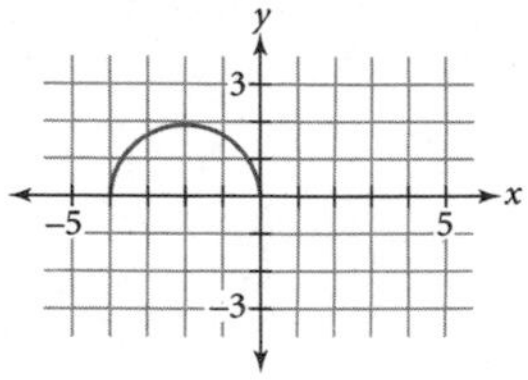

9a. possible answer: $x = 0.4t$ and $y = 1$

9b. possible answer: $[0, 50, 5, 0, 3, 1]$; $0 \leq t \leq 125$

9c. possible answer: $x = 1.8(t - 100)$, $y = 2$

9d. The tortoise will win.

9e. The tortoise takes 125 s and the hare takes approximately 28 s, but because he starts 100 s later, he finishes at 128 s.

11a. The Los Angeles to Honolulu plane flies at 450 mi/h and leaves 2 h later than the Honolulu to Los Angeles plane. The Honolulu to Los Angeles plane flies at 525 mi/h.

11b. 3.5 h; 675 mi west of Los Angeles

11c. $450(t - 2) = 2500 - 525t$; 3.5 h, 675 mi west of Los Angeles

13. $(7, -3)$

15a. $2.5n^2 - 5.5n - 3$ **15b.** 887

LESSON 8.2

1a. $t = x - 1$

1b. $t = \dfrac{x + 1}{3}$

1c. $t = \pm\sqrt{x}$

1d. $t = x + 1$

3a. $y = \dfrac{x + 7}{2}$

3b. $y = \pm\sqrt{x} + 1$

3c. $y = \dfrac{2x - 4}{3}$

3d. $y = 2(x + 2)^2$

5.

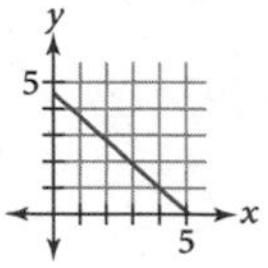

7. $-2.5 \leq t \leq 2.5$

9a. $x = 20 + 2t$ and $y = 5 + t$

9b.

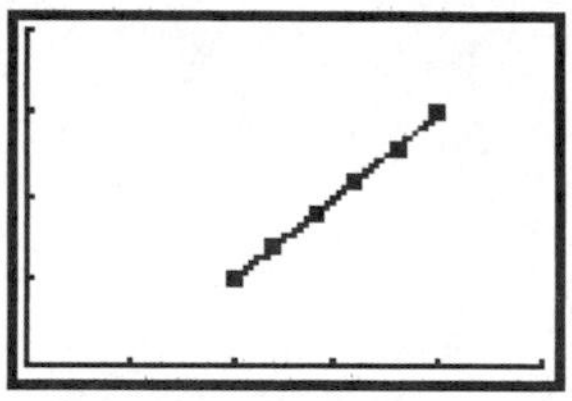

$[0, 50, 10, 0, 20, 5]$; $0 \leq t \leq 10$

The points lie on the line.

9c. $y = \frac{1}{2}x - 5$

9d. The slope of the line in 9c is the ratio of the y-slope over the x-slope in the parametric equations.

11a. possible answer: $x = 1$, $y = 1.5t$

11b. $x = 1.1$, $y = 12 - 2.5t$

11c. possible answer: $[0, 2, 1, 0, 12, 1]$; $0 \leq t \leq 3$

11d.

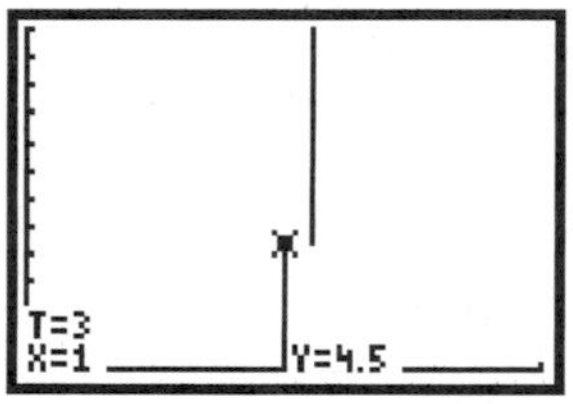

$[0, 2, 1, 0, 12, 1]$; $0 \leq t \leq 3$; they meet after hiking 3 s, when both are 4.5 ft north of the trailhead.

11e. $1.5t = 12 - 2.5t$; $t = 3$; substitute $t = 3$ into either y-equation to get $y = 4.5$.

13. $x = t^2$, $y = t$

15. $y = \left(\frac{2}{3}(x - 5) - 2\right) + 3$ or $y = \frac{2}{3}x - \frac{7}{3}$

LESSON 8.3

1a. 55.5° **1b.** 0° **1c.** 27° **1d.** 37.2°

3a.

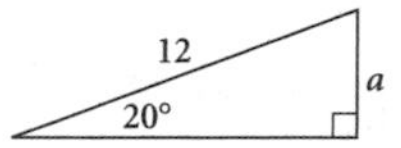

$a = 4.1$

3b.

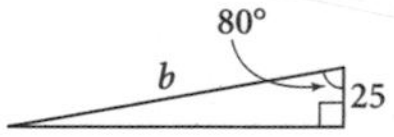

$b = 144.0$

3c.

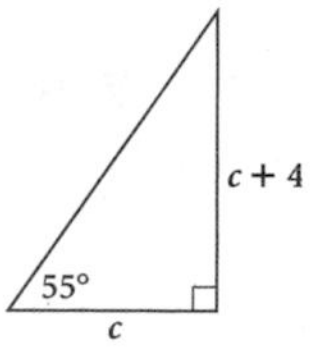

$c = 9.3$, $c + 4 = 13.3$

3d.

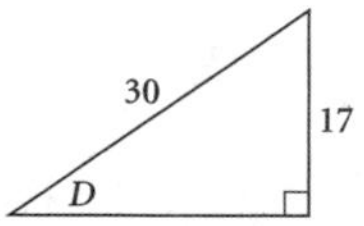

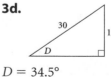

$D = 34.5°$

5.

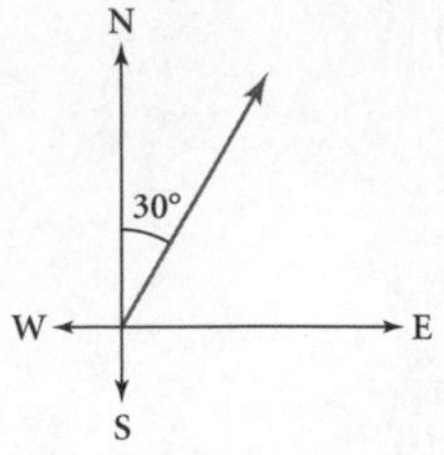

5a. 60°

5b.

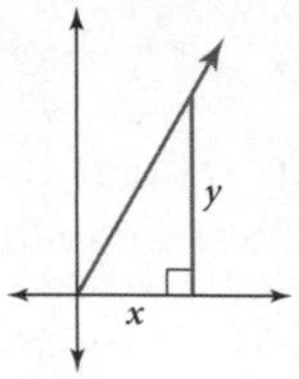

5c. 180 mi east, 311.8 mi north

9a. possible answer:

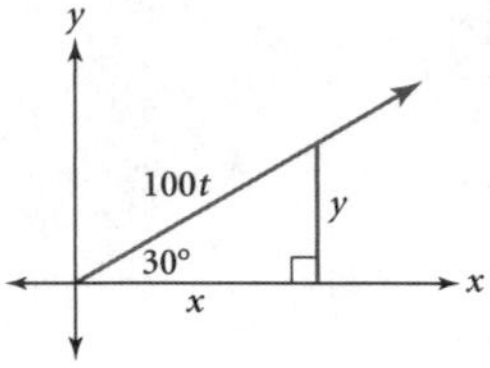

$x = 100t \cos 30°$ and $y = 100t \sin 30°$

9b. $0 \le t \le 5$

9c. The 100 represents the speed of the plane, t represents time in hours, 30° is the angle the plane is making with the x-axis, x is the horizontal position at any time, and y is the vertical position at any time.

11a.

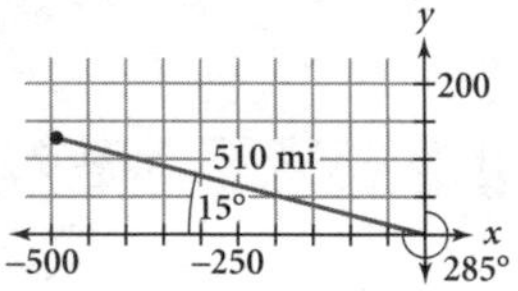

11b. 23.2 h

11c. 493 mi west, 132 mi north

13a. $y = \frac{3}{4}x + \frac{1}{2}$

13b. $y = \frac{3x + 2}{4}$ or $y = \frac{3}{4}x + \frac{1}{2}$.

15. $(x - 2.6)^2 + (y + 4.5)^2 = 12.96$

LESSON 8.4

1. $x = 10t \cos 30°$; $y = 10t \sin 30°$

3a.

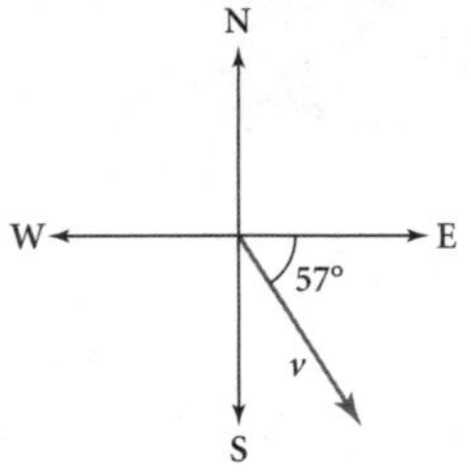

3b.

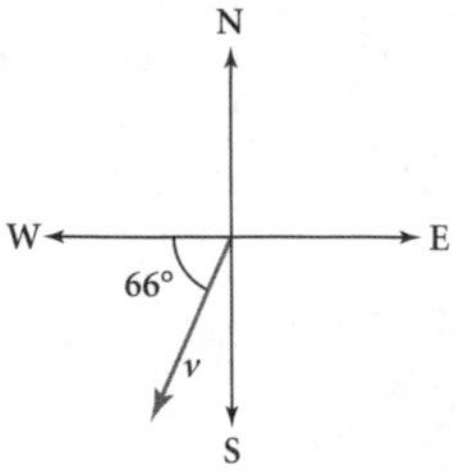

3c.

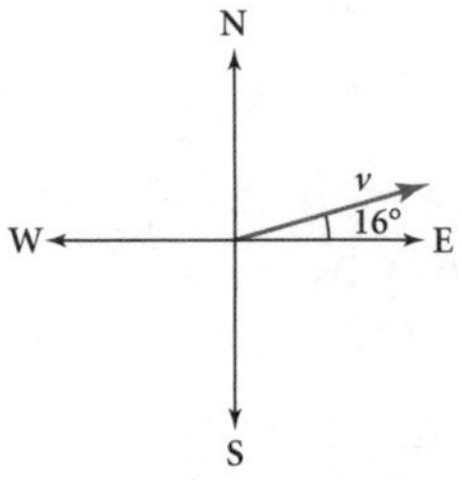

3d.

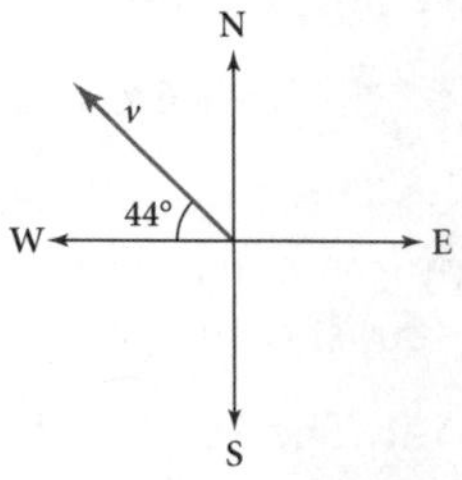

5a. $(-0.3, 0.5)$

5b. $x = -0.3 + 4t$

5c. $y = 0.5 - 7t$

5d.

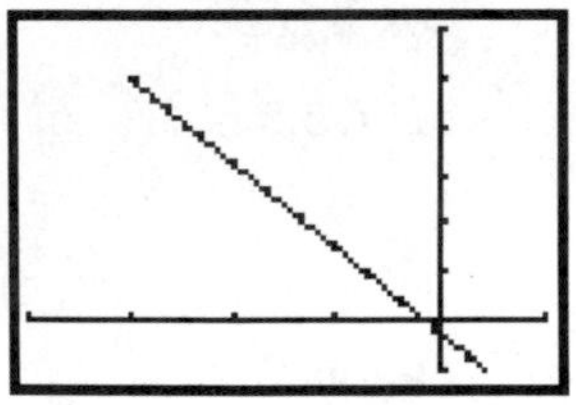

$[-0.4, 0.1, 0.1, -0.1, 0.6, 0.1]$; $0 \le t \le 0.1$

5e. At 0.075 h, the boat lands 0.025 km south of the dock.

5f. 0.605 km

7a. $y = -5t$

7b. $x = st$

7c. $s = 10$ mi/h

7d. 4.47 mi

7e. 0.4 h

7f. 11.18 mi/h

7g. 63.4°

9a. $y = -20t \sin 45°$

9b. $x = 20t \cos 45°$

9c. Both the plane's motion and the wind contribute to the actual path of the plane, so you add the x-contributions and add the y-contributions to form the final equations.

9d. possible answer: $[-1000, 0, 100, -100, 0, 10]$; $0 \le t \le 5$

9e. 4.24. It takes the plane 4.24 h to fly 1000 mi west.

9f. 60 mi

11a.

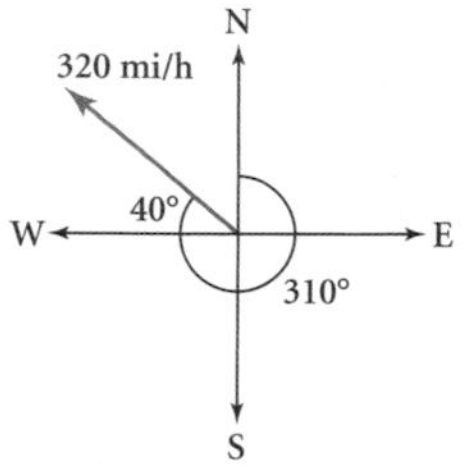

11b. $x = -320t \cos 40°$, $y = 320t \sin 40°$

11c.

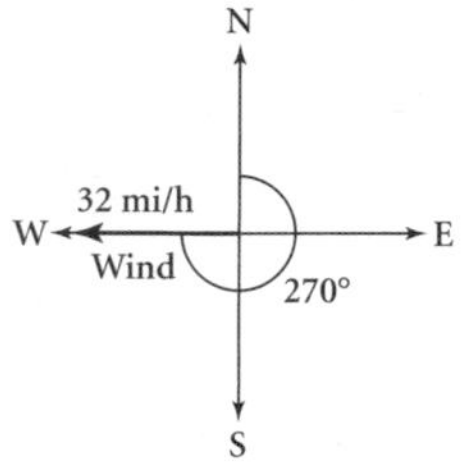

11d. $x = -32t$, $y = 0$

11e. $x = -320t \cos 40° + -32t$, $y = 320t \sin 40°$

11f. 1385.7 mi west and 1028.5 mi north

13a. x-component: $50 \cos 40° \approx 38.3$; y-component: $50 \sin 40° \approx 32.1$

13b. x-component: $90 \cos 140° \approx -68.9$; y-component: $90 \sin 140° \approx 57.9$

13c. x-component: -30.6; y-component: 90.0

13d. 95.1 g **13e.** 109° **13f.** 95.1 g at 289°

15a. two real, rational roots

15b. two real, irrational roots

15c. no real roots

15d. one real, rational root

LESSON 8.5

1a. the moon; centimeters and seconds

1b. right 400 cm and up 700 cm

1c. up-left **1d.** 50 cm/s

3a. $x = 2t$, $y = -4.9t^2 + 12$

3b. $-4.9t^2 + 12 = 0$

3c. 1.56 s, 3.13 m from the cliff

3d. possible answer: [0, 3.13, 1, 0, 12, 1]

5a. possible answer: [0, 5, 1, 0, 3.5, 1], $0 \leq t \leq 1.5$

5b. Sample answer: This projectile was launched from 2 m above the Earth with a velocity of 6 m/s at an inclination of 52°.

7a. $x = 83t \cos 0°$, $y = -4.9t^2 + 83t \sin 0° + 1.2$

7b. No; it will hit the ground 28.93 m before reaching the target.

7c. Answer must be between 2.44° and 3.43°.

7d. at least 217 m/s

9a. $x = 122t\cos 38°$, $y = -16t^2 + 122t\sin 38°$

9b. 451 ft

9c. 378 ft

11. 46 ft from the end of the cannon

13a. $x = 2.3t + 4$, $y = 3.8t + 3$

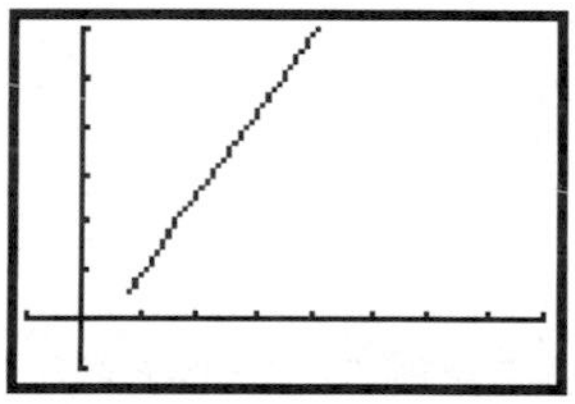

[−5, 40, 5, −5, 30, 5]

13b. 4.44 m/s on a bearing of 31°

14a. 20 ft 1 in. **14b.** 32 m **14c.** 930.9 cm^2

15. $x = 2 \cos t - 3$, $y = 5 \sin t + 4$;
$\left(\frac{x + 3}{2}\right)^2 + \left(\frac{y - 4}{5}\right)^2 = 1$

LESSON 8.6

1. 9.7 cm

3. $X \approx 50.2°$ and $Z \approx 92.8°$

5a. $B = 25.5°$; $BC \approx 6.4$ cm; $AB \approx 8.35$ cm

5b. $J \approx 38.8°$; $L \approx 33.3°$; $KJ \approx 4.76$ cm

7. approximately 27 million mi or approximately 153 million mi

9. 2.5 km

11a. 41° **11b.** 70° **11c.** 0°

13a.

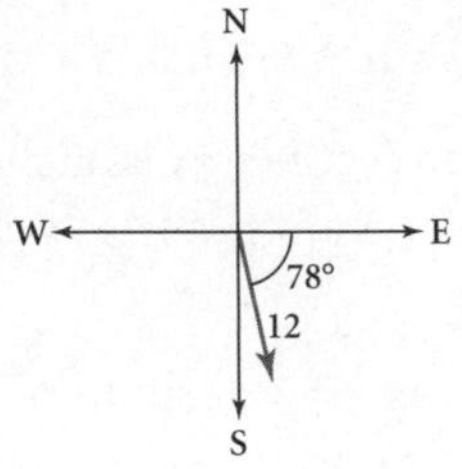

x-component: $12 \cos 78° \approx 2.5$;
y-component: $-12 \sin 78° \approx -11.7$

13b.

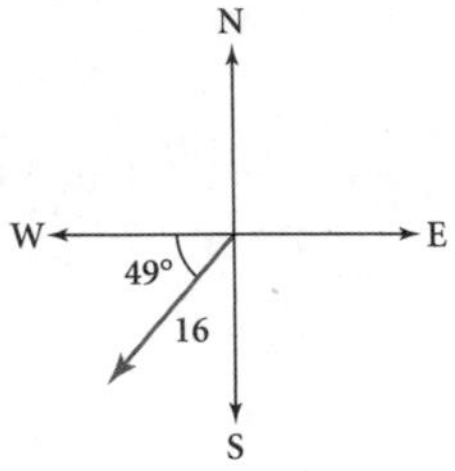

x-component: $-16 \cos 49° \approx -10.5$;
y-component: $-16 \sin 49° \approx -12.1$

15a. \$26,376.31

15b. 20 years 11 months

LESSON 8.7

1. approximately 6.1 km

3a. $A = \cos^{-1}\left(\frac{16 - 25 - 36}{-2(5)(6)}\right) \approx 41.4°$

3b. $b = 8$

5. 1659.8 mi

7. From point A, the underground chamber is at a 22° angle from the ground between A and B. From point B, the chamber is at a 120° angle from the ground. If the truck goes 1.5 km farther in the same direction, the chamber will be approximately 2.6 km directly beneath the truck.

9. 148°–149°

11. 10.3 nautical mi

CHAPTER 8 REVIEW

1a. east 160 m and south 240 m

1b. $x = 20t$, $y = -30t$

3a.

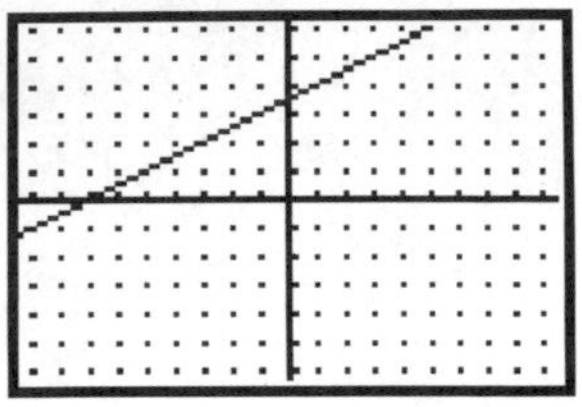

$[-9.4, 9.4, 1, -6.2, 6.2, 1]$

$y = \frac{x + 7}{2}$. The graph is the same.

3b.

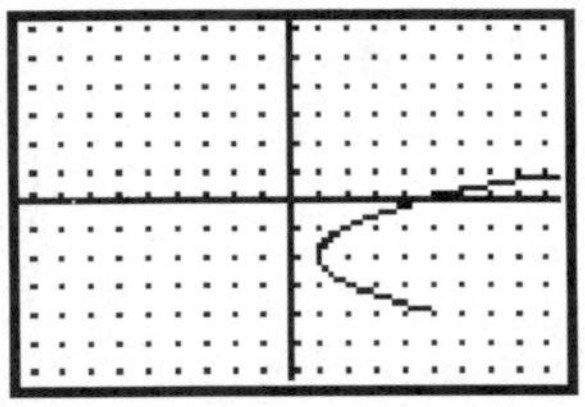

$[-9.4, 9.4, 1, -6.2, 6.2, 1]$

$y = \pm\sqrt{x - 1} - 2$. The graph is the same except for the restrictions on t.

3c.

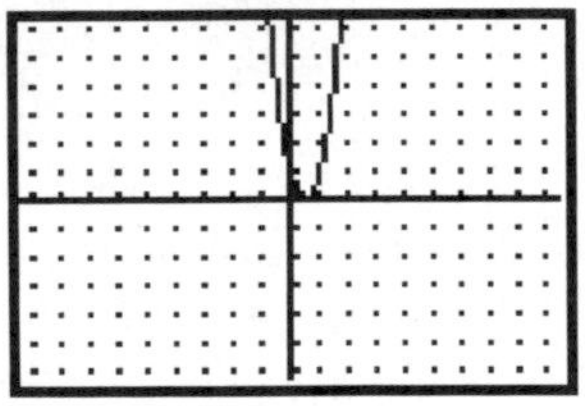

$[-9.4, 9.4, 1, -6.2, 6.2, 1]$

$y = (2x - 1)^2$. The graph is the same. The values of t are restricted, but endpoints are not visible within the calculator screen given.

3d.

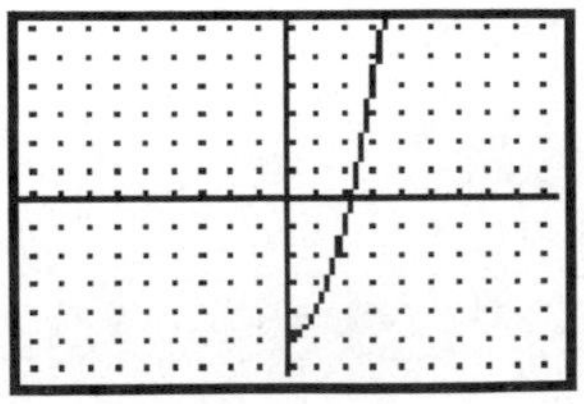

$[-9.4, 9.4, 1, -6.2, 6.2, 1]$

$y = x^2 - 5$. The graph is the same, except the parametric equations will not allow for negative values for x.

5a. $A \approx 43°$ **5b.** $B \approx 28°$ **5c.** $c \approx 23.0$

5d. $d \approx 12.9$ **5e.** $e \approx 21.4$ **5f.** $f \approx 17.1$

7. 7.2 m

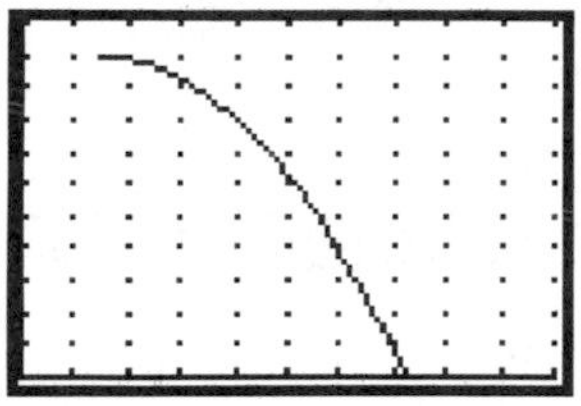

[0, 10, 1, 0, 11, 1]

9. She should fly at a bearing of 107.77° if the wind averages 25 mi/h. If the wind were 30 mi/h continuously, she could miss her destination by as much as 7 mi.

CHAPTER 9 • CHAPTER **9** CHAPTER 9 • CHAPTER

LESSON 9.1

1a. 10 units **1b.** $\sqrt{74}$ units

1c. $\sqrt{85}$ units **1d.** $\sqrt{81 + 4d^2}$ units

3. $x = -1 \pm \sqrt{2160}$ or $x = -1 \pm 12\sqrt{5}$

5. approximately 25.34 units

7. $\frac{6 - \sqrt{3}}{2} < y < \frac{6 + \sqrt{3}}{2}$, or approximately between the points (2.5, 2.134) and (2.5, 3.866)

9a. $y = \sqrt{10^2 + x^2} + \sqrt{(20 - x)^2 + 13^2}$

9b. domain: $0 \le x \le 20$; range: $30 < y < 36$

9c. When the wire is fastened approximately 8.696 m from the 10 m pole, the minimum length is approximately 30.48 m.

11a. $d = \sqrt{(5 - x)^2 + (0.5x^2 + 4)^2}$

11b. approximately 6.02 units; approximately (0.92, 1.42)

13a. $(x + 3)^2 = 14$ **13b.** $(y - 2)^2 = 3$

13c. $(x + 3)^2 + (y - 2)^2 = 17$

15. domain: all real numbers; range: $y \ge -2$

17. 9.56 cm

LESSON 9.2

1a. center: (0, 0); radius: 2

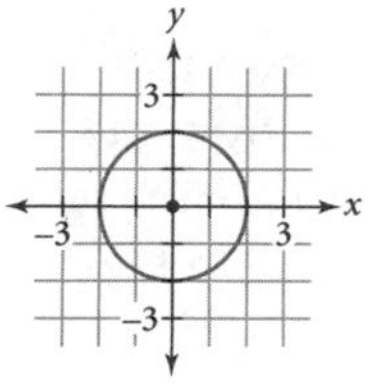

$y = \pm\sqrt{4 - x^2}$

1b. center: (3, 0); radius: 1

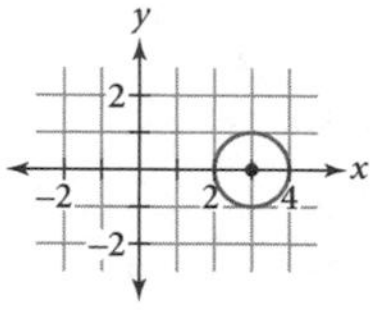

$y = \pm\sqrt{1 - (x - 3)^2}$

1c. center: (1, 2); radius: 3

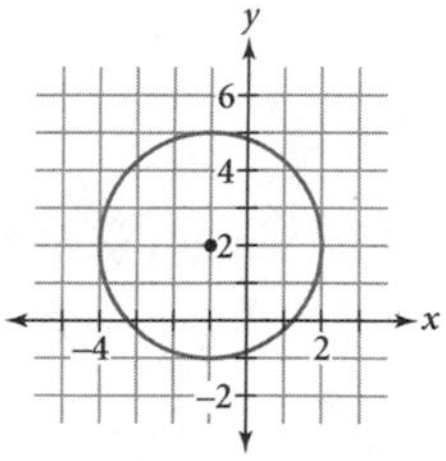

$y = \pm\sqrt{9 - (x + 1)^2} + 2$

1d. center: (0, 1.5); radius: 0.5

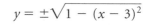

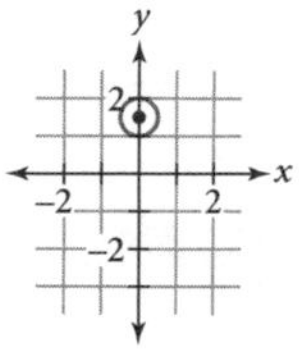

$y = \pm\sqrt{.025 - x^2} + 1.5$

3a. $(x - 3)^2 + y^2 = 25$

3b. $(x + 1)^2 + (y - 2)^2 = 9$

3c. $(x - 2.5)^2 + (y - 0.75)^2 = 16$

3d. $(x - 2.5)^2 + (y - 1.25)^2 = 0.25$

5. An ellipse with center (h, k), horizontal scale factor of a and vertical scale factor of b would have equations: $x = a\cos t + h$ and $y = b\sin t + k$.

5a.

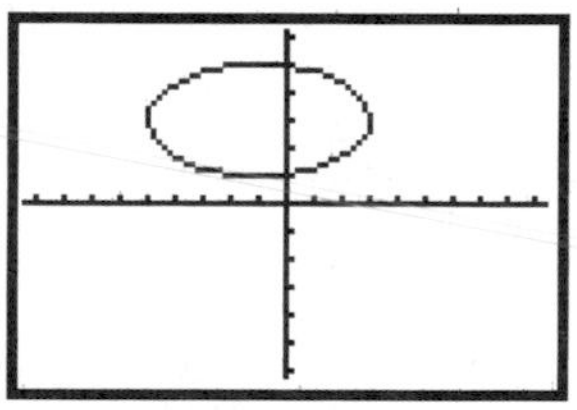

5b.

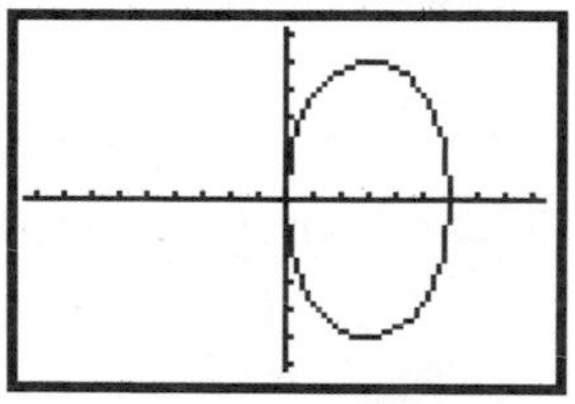

7a. $(\sqrt{27}, 0)(-\sqrt{27}, 0)$

7b. $(3, \sqrt{21})(3, -\sqrt{21})$

7c. $(-1 + \sqrt{7}, 2)(-1 - \sqrt{7}, 2)$

7d. $(3 + \sqrt{27}, -1), (3 - \sqrt{27}, -1)$

9a. 1.0 m **9b.** 1.6 m

11. Let x represent distance in miles measured parallel to the major axis and measured from the intersection of the major and minor axes. Let y represent distance in miles measured parallel to the minor axis and measured from the intersection of the axes.

$$\left(\frac{x}{237{,}177}\right)^2 + \left(\frac{y}{236{,}667}\right)^2 = 1$$

13. $y = -(x + 3)^2 + 2$

15. $y = 2x^2 - 24x + 117$

LESSON 9.3

1a. (1, 0.5) **1b.** $y = 8$ **1c.** (9, 2)

3a. (0, 6); $y = 4$ **3b.** (−1.75, 0); $x = -2.25$

3c. (−3, 0); $y = 1$ **3d.** (3.875, 0); $x = 4.125$

5. $y^2 = -12x$

7a. $y = \frac{1}{8}x^2 + 1$

7b. The graph is a parabola with vertex (0, 1), focus (0, 3), and directrix $y = -1$.

9a.

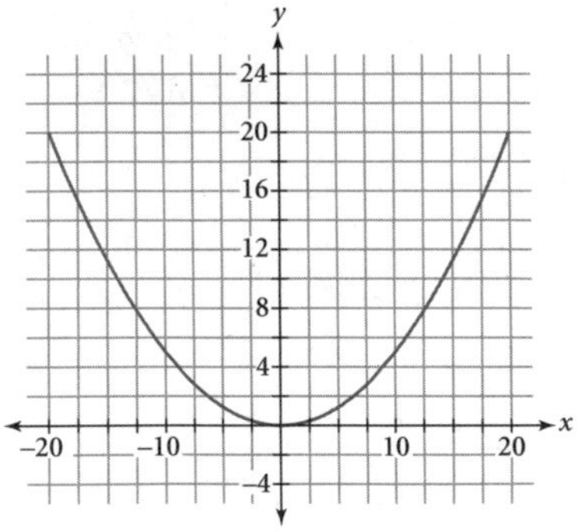

9b. (−20, 20) and (20, 20)

9c. The grill should be located 5 cm from the center of the dish.

11. $-2.4x^2 + 21.12x - 44.164 = 0$

13. $\left(\frac{x-2}{16}\right)^2 + \left(\frac{y-1}{\sqrt{192}}\right)^2 = 1$

$$\left(\frac{x-2}{16}\right)^2 + \left(\frac{y-1}{8\sqrt{3}}\right)^2 = 1$$

15. $x = -1, y = 4, z = -6$

LESSON 9.4

1a. vertices: (−2, 0) and (2, 0); asymptotes: $y = \pm 2x$

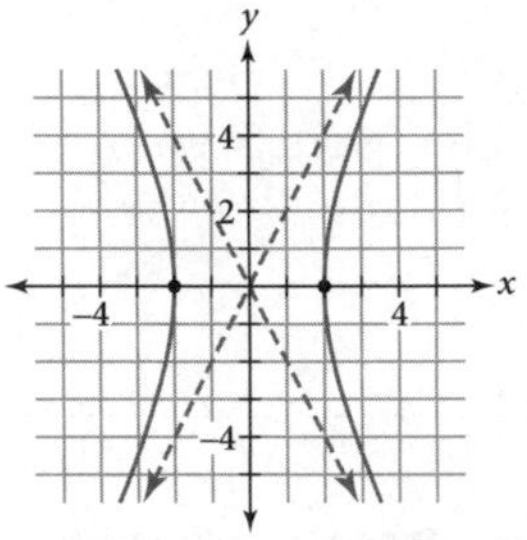

1b. vertices: (2, −1) and (2, −3); asymptotes: $y = \frac{1}{3}x - \frac{8}{3}$ and $y = -\frac{1}{3}x - \frac{4}{3}$

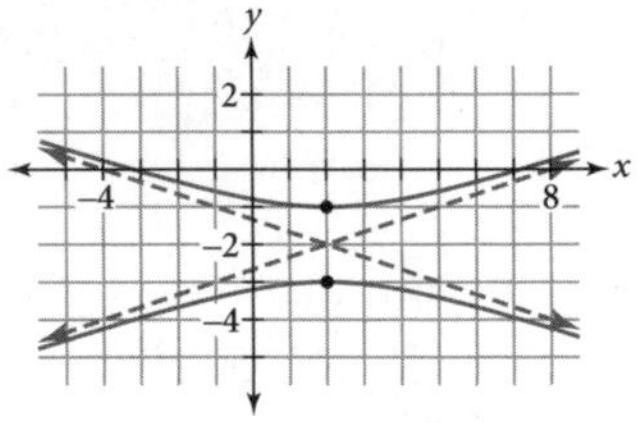

1c. vertices: (1, 1) and (7, 1); asymptotes: $y = x - 3$ and $y = -x + 5$

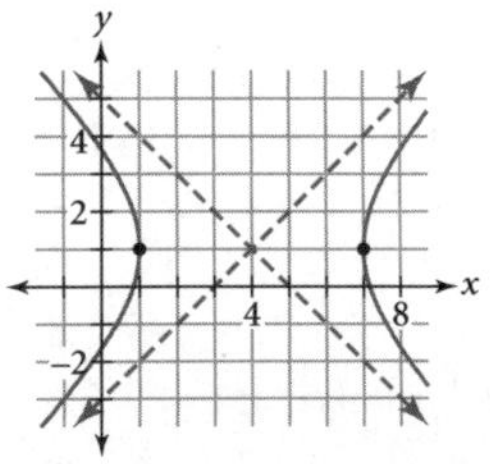

1d. vertices: (−2, 1) and (−2, −3); asymptotes: $y = \frac{2}{3}x + \frac{1}{3}$ and $y = -\frac{2}{3}x - \frac{7}{3}$

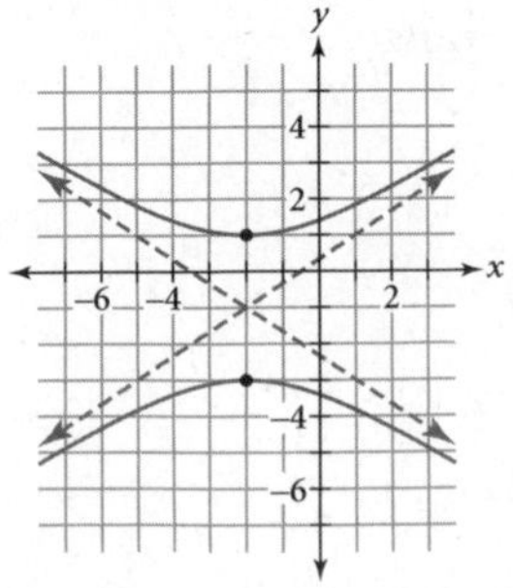

3a. $\left(\frac{x}{2}\right)^2 - \left(\frac{y}{1}\right)^2 = 1$

3b. $\left(\frac{y + 3}{2}\right)^2 - \left(\frac{x - 3}{2}\right)^2 = 1$

3c. $\left(\frac{x + 2}{3}\right)^2 - \left(\frac{y - 1}{4}\right)^2 = 1$

3d. $\left(\frac{y - 1}{4}\right)^2 - \left(\frac{x + 2}{3}\right)^2 = 1$

5a. The lengths of the sides of the asymptote rectangle are $2a$ and $2b$. Using the Pythagorean Theorem, the length of the diagonal is $\sqrt{(2a)^2 + (2b)^2} = \sqrt{4a^2 + 4b^2} = \sqrt{4(a^2 + b^2)} = 2\sqrt{a^2 + b^2}$. Half this distance is $\sqrt{a^2 + b^2}$.

5b. $\left(2, -2 + \frac{\sqrt{10}}{2}\right)$ and $\left(2, -2 - \frac{\sqrt{10}}{2}\right)$

7. $\left(\frac{y + 2.49}{3.95}\right)^2 - \left(\frac{x + 2.35}{2.63}\right)^2 = 1$

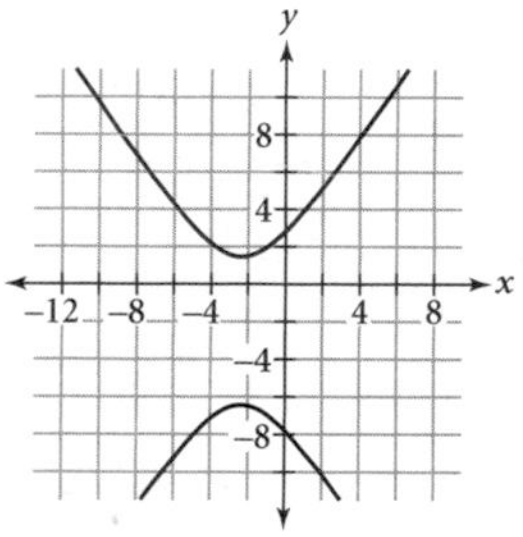

9.

x-value	5	10	20	40
Distance	0.83	0.36	0.17	0.08

11a.

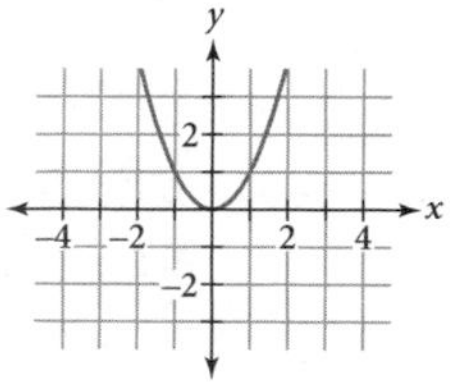

11b.

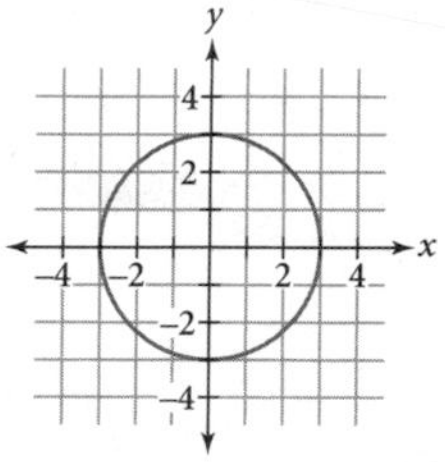

11c.

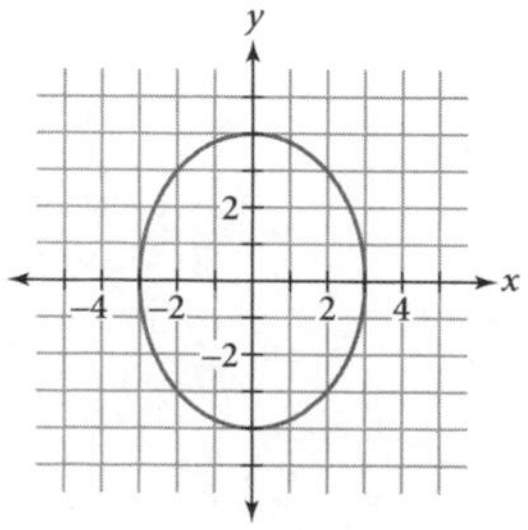

11d.

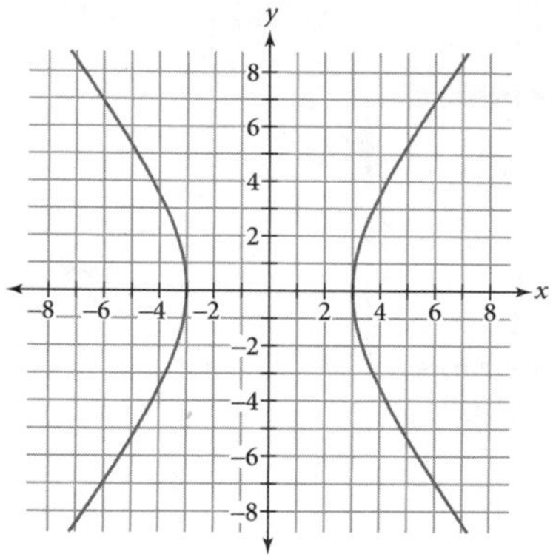

13. $\left(\frac{x - 11900000}{57900000}\right)^2 + \left(\frac{y}{56700000}\right)^2 = 1$

15a. vertex: $(-1, -2)$; focus: $(-1, -2.25)$; directrix: $y = -1.75$

15b. vertex: $(1.5, 2.75)$; focus: $(1.75, 2.75)$; directrix: $y = 1.25$

17. $n^2 + 3n - 24$

LESSON 9.5

1a. $(1/2, 1/4)$ and $(-2, 4)$

1b. $(3/4, 73/16)$ or $(0.75, 4.5625)$

1c. $(1.177, -1.240)$, $(-1.177, -1.240)$, $(2.767, -5.160)$, $(-2.767, -5.160)$

1d. $(\sqrt{7}, 1)$, $(-\sqrt{7}, 1)$, $(\sqrt{7}, -1)$, $(-\sqrt{7}, -1)$

5. $x^2 + y^2 = 11.52$

7a. $(-2, 5 + 2\sqrt{5})$ and $(-2, 5 - 2\sqrt{5})$ or approximately $(-2, 0.53)$ and $(-2, 9.47)$

7b. $\left(1 + \frac{\sqrt{3}}{2}, 2\right)$ and $\left(1 - \frac{\sqrt{3}}{2}, 2\right)$ or approximately $(0.13, -2)$ and $(1.87, -2)$

9. $(x - 1)(x - 2)(x - i\sqrt{2})(x + i\sqrt{2})$

LESSON 9.6

1a. $f(x) = \frac{1}{x} + 2$

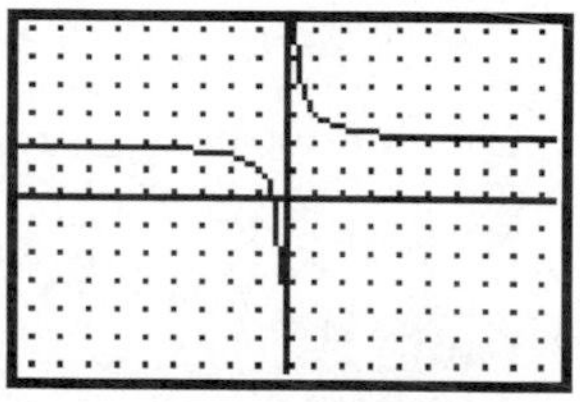

$[-9.4, 9.4, 1, -6.2, 6.2, 1]$

1b. $f(x) = \frac{1}{x-3}$

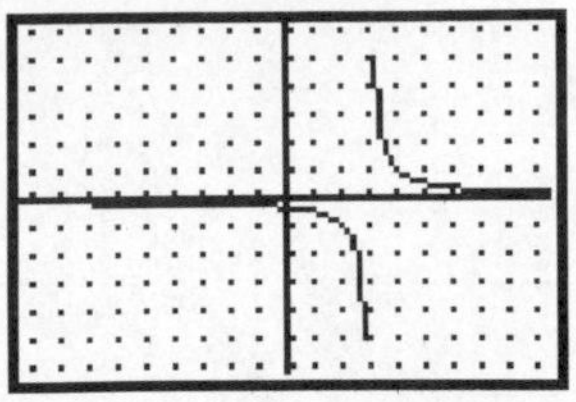

$[-9.4, 9.4, 1, -6.2, 6.2, 1]$

1c. $f(x) = \frac{1}{x+4} - 1$

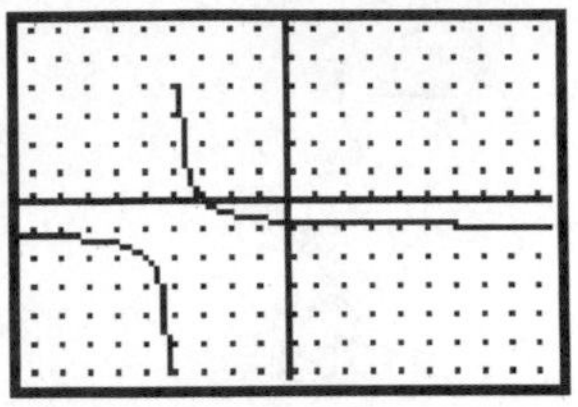

$[-9.4, 9.4, 1, -6.2, 6.2, 1]$

1d. $f(x) = 2\left(\frac{1}{x}\right)$ or $f(x) = \frac{2}{x}$

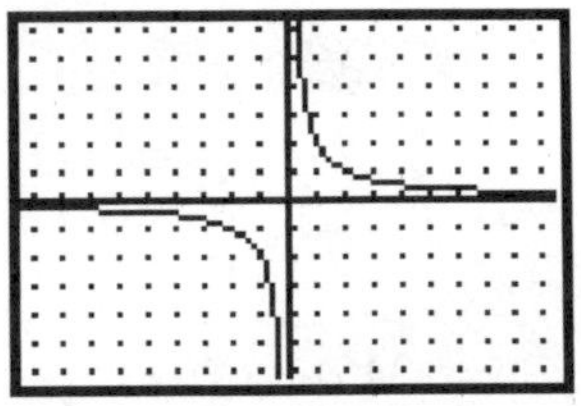

$[-9.4, 9.4, 1, -6.2, 6.2, 1]$

1e. $f(x) = 3\left(\frac{1}{x}\right) + 1$ or $f(x) = \frac{3}{x} + 1$

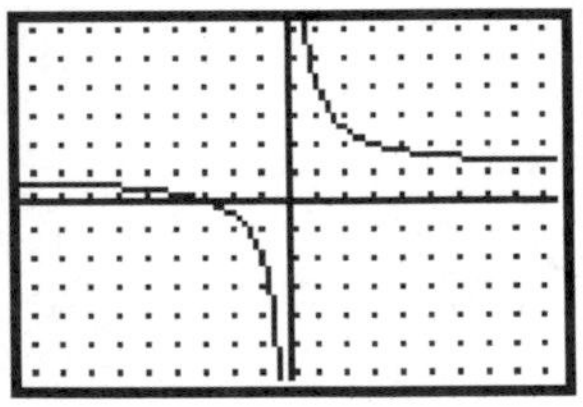

$[-9.4, 9.4, 1, -6.2, 6.2, 1]$

3a. $x = -4$ **3b.** $x = \frac{113}{18}$ (or $x = 6.27$)

3c. $x = 2.6$ **3d.** $x = -8.5$

5. 12 games

7a. 20.9 mL **7b.** $f(x) = \frac{20.9 + x}{55 + x}$

7c. 39.72 mL **7d.** The graph approaches $y = 1$.

9a. i. $y = 2 + \frac{-3}{x-5}$ **9a. ii.** $y = 3 + \frac{2}{x+3}$

9b. i. For $y = 2 + \frac{-3}{x-5}$, stretch vertically by a factor of -3, and translate right 5 units and up 2 units.

9b. ii. For $y = 3 + \frac{2}{x+3}$, stretch vertically by a factor of 2, and translate left 3 units and up 3 units.

9c. i.

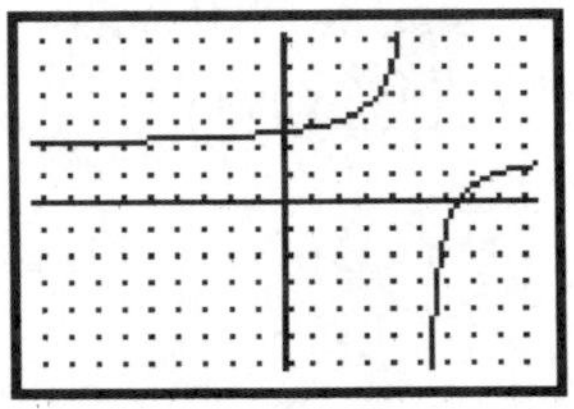

$[-9.4, 9.4, 1, -6.2, 6.2, 1]$

9c. ii.

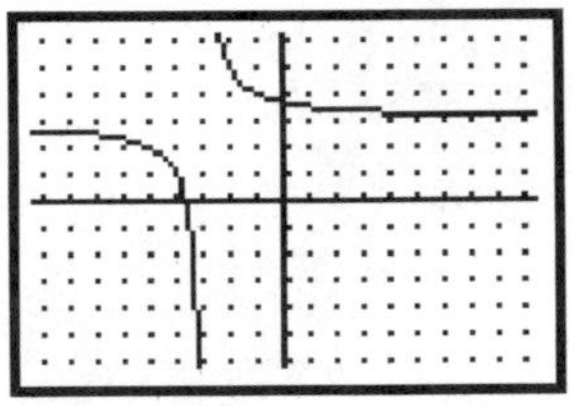

$[-9.4, 9.4, 1, -6.2, 6.2, 1]$

11a. $(x-5)(x-2)$ **11b.** $x(x-3)(x+3)$

13. Either the wire is fastened about 1.47 m from the base of the 2 m rod on the side opposite the 5 m rod, or it is fastened between the two rods 9.8 m from the 2 m rod.

15. $\left(\sqrt{5}, \frac{1}{\sqrt{5}}\right), \left(-\sqrt{5}, -\frac{1}{\sqrt{5}}\right)$

LESSON 9.7

1a. $\frac{(x+3)(x+4)}{(x+2)(x-2)}$ **1b.** $\frac{x(x-7)(x+2)}{(x+1)(x+1)}$

3a. $\frac{7x-7}{x-2}$ **3b.** $\frac{-7x+12}{2x-1}$

5a. $y = \frac{x+2}{x+2}$ **5b.** $y = \frac{-2(x-3)}{x-3}$

5c. $y = \frac{(x+2)(x+1)}{x+1}$

7a.

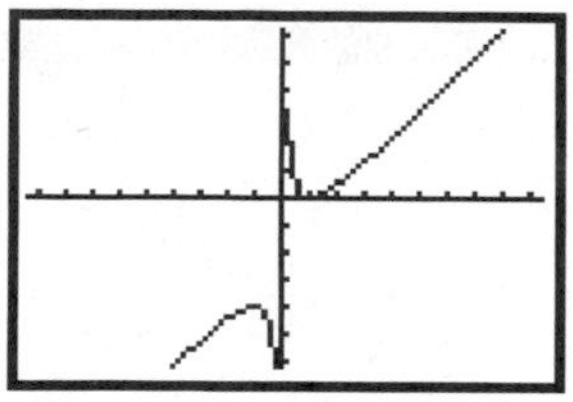

$[-9.4, 9.4, 1, -6.2, 6.2, 1]$

vertical asymptote $x = 0$,
slant asymptote $y = x - 2$

7b.

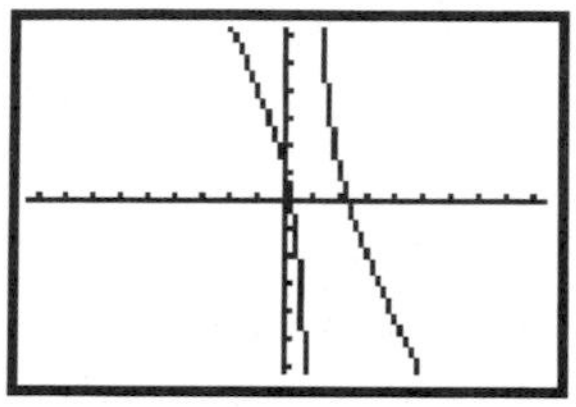

$[-9.4, 9.4, 1, -6.2, 6.2, 1]$

vertical asymptote $x = 1$,
slant asymptote $y = -2x + 3$

7c.

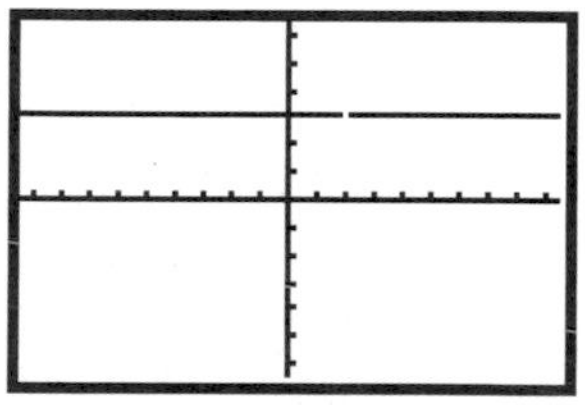

$[-9.4, 9.4, 1, -6.2, 6.2, 1]$

hole at $x = 2$

7d. For 7a: $y = \frac{x^2 - 2x + 1}{x}$. The denominator is 0 when $x = 0$, so the vertical asymptote is $x = 0$. For 7b: $y = \frac{-2x^2 + 5x - 1}{x - 1}$. The denominator is 0 when $x = 1$, so the vertical asymptote is $x = 1$. For 7c: $y = \frac{3x - 6}{x - 2}$. Both the numerator and denominator are 0 when $x = 2$. This causes a hole in the graph.

9. $y = \frac{-(x + 2)(x - 6)}{3(x - 2)}$

11a. $x = 3 \pm \sqrt{2}$

11b. no real solutions; $x = \frac{3 \pm i\sqrt{7}}{2}$

13a. 95

13b.

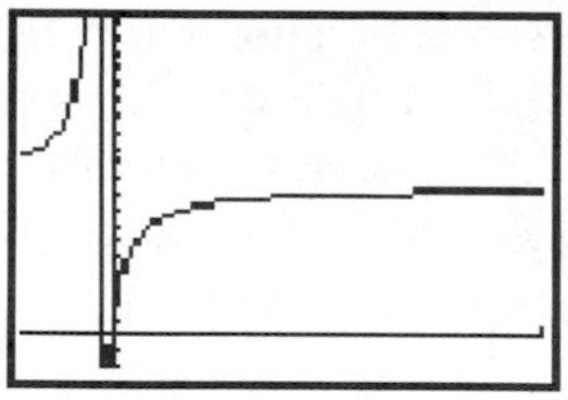

$[-10, 50, 50, -20, 200, 10]$

13c. There is a vertical asymptote, $x = -1.6$, and a horizontal asymptote, $y = 96$.

13d. The vertical asymptote is meaningless because there cannot be a negative number of moose. The horizontal asymptote means that as the density of the moose increases, more will be attacked, up to a maximum of 96 during 100 days. This is the maximum number that the wolves can eat.

15a. $83\frac{1}{3}$ g

15b. approximately 17% almonds and 43% peanuts

15c. 50 g; approximately 27.3% almonds, 27.3% cashews, and 45.4% peanuts

LESSON 9.8

1a. $\frac{x(x + 2)}{(x - 2)(x + 2)} = \frac{x}{x - 2}$

1b. $\frac{(x - 1)(x - 4)}{(x + 1)(x - 1)} = \frac{x - 4}{x + 1}$

1c. $\frac{3x(x - 2)}{(x - 4)(x - 2)} = \frac{3x}{x - 4}$

1d. $\frac{(x + 5)(x - 2)}{(x + 5)(x - 5)} = \frac{x - 2}{x - 5}$

3a. $\frac{(2x - 3)(x + 1)}{(x + 3)(x - 2)(x - 3)}$

3b. $\frac{x^2 + 4x + 6}{(x + 2)(x + 3)(x - 2)}$

3c. $\frac{2x^2 - x + 9}{(x - 3)(x + 2)(x + 3)}$

3d. $\frac{2x^2 - 5x + 6}{(x + 1)(x - 2)(x - 1)}$

5a. $\frac{2(x - 2)}{x + 1}$

5b. 1

7a. There is a zero at $x = 3$, because that value causes the numerator to be 0. The vertical asymptotes are $x = 2$ and $x = -2$, because these values make the denominator 0. The horizontal asymptote is $x = 0$, because this is the value that y approaches when $|x|$ is large.

7b.

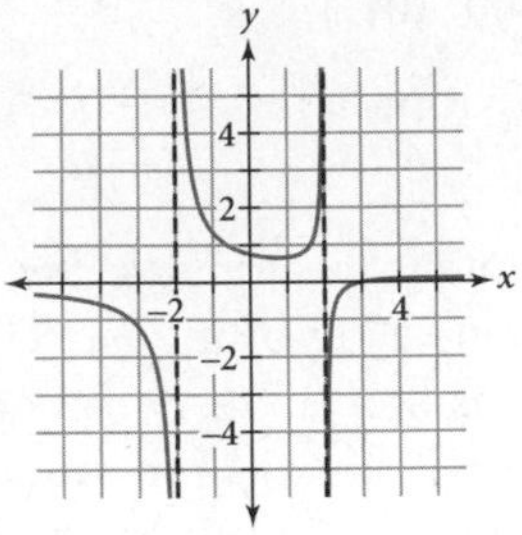

$[-5, 5, 1, -5, 5, 1]$

9.

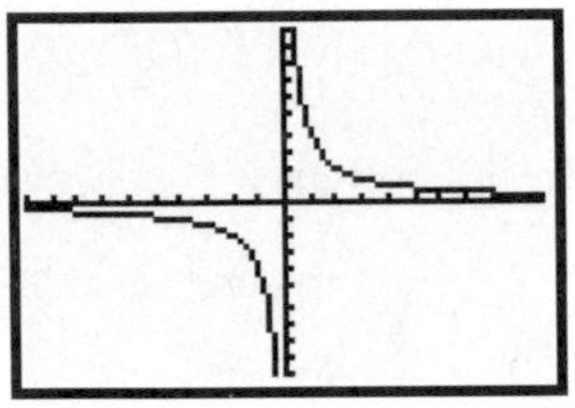

$[-10, 10, 1, -10, 10, 1]$

11a. $x = 3, y = 1$

11b. translation right 3 units and up 1 unit

11c. -2

11d. $y = 1 - \frac{2}{x - 3}$ or $y = \frac{x - 5}{x - 3}$

11e. x-intercept: 5; y-intercept: $\frac{5}{3}$

13a and c.

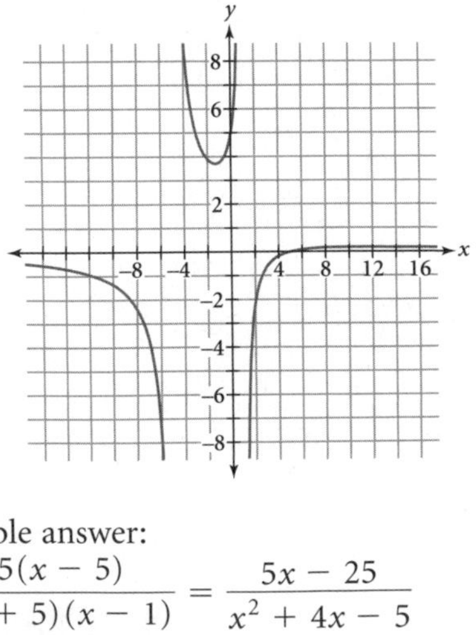

13b. possible answer:

$$h(x) = \frac{5(x - 5)}{(x + 5)(x - 1)} = \frac{5x - 25}{x^2 + 4x - 5}$$

CHAPTER 9 REVIEW

1a.

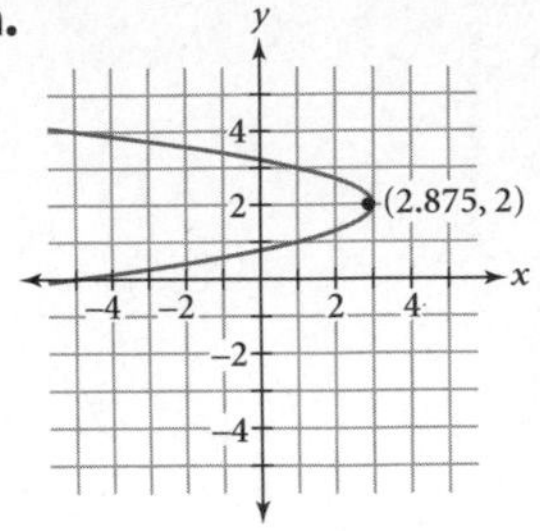

1b.

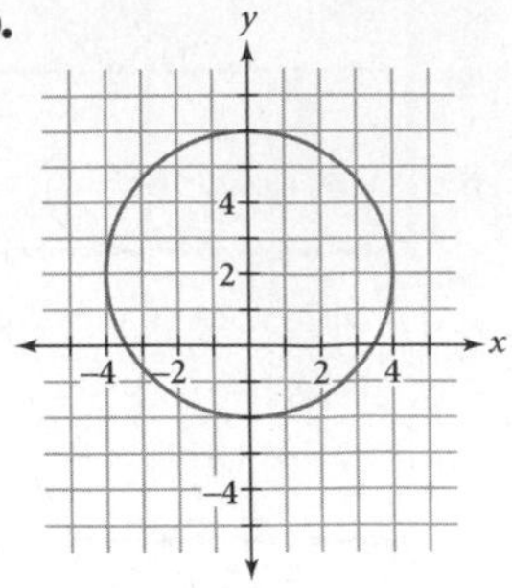

1c.

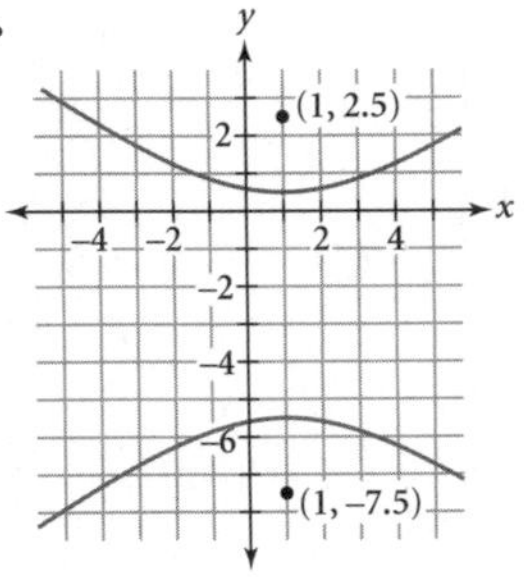

1d.

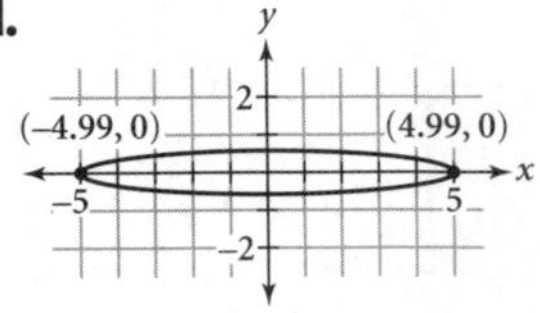

3a. $y = \pm 0.5x$

3b. $x^2 - 4y^2 - 4 = 0$

3c. $d = 0.5x - \sqrt{\frac{x^2}{4} - 1}$

3d.

x	2	10	20	100
distance	1	0.101	0.050	0.010

5. $(\sqrt{12}, 1), (\sqrt{12}, -1), (-\sqrt{12}, 1), (-\sqrt{12}, -1)$

7a. $y = 1 + \frac{1}{x + 2}$ or $y = \frac{x + 3}{x + 2}$

7b. $y = -4 + \frac{1}{x}$ or $y = \frac{-4x + 1}{x}$

9. Multiply the numerator and denominator by the factor $(x + 3)$.

$$y = \frac{(2x - 14)(x + 3)}{(x - 5)(x + 3)}$$

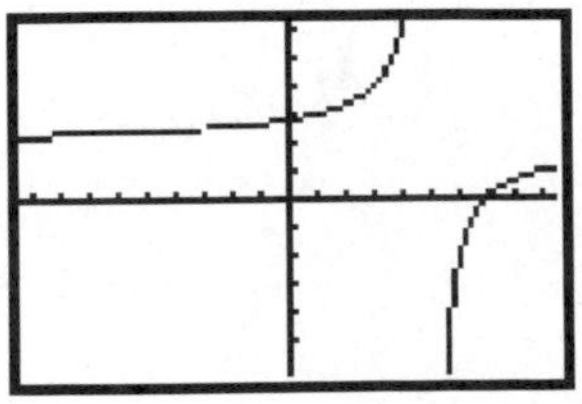

$[-9.4, 9.4, 1\ -6.2, 6.2, 1]$

11a. $\dfrac{3x^2 + 8x + 3}{(x - 2)(x + 1)(x + 2)}$

11b. $\dfrac{3x}{x + 1}$

11c. $\dfrac{(x + 1)^2(x - 1)}{x(x - 2)}$

13a. $y = 2|x|$

13b. $y = 2|x - 4|$

13c. $y = 2|x - 4| - 3$

13.

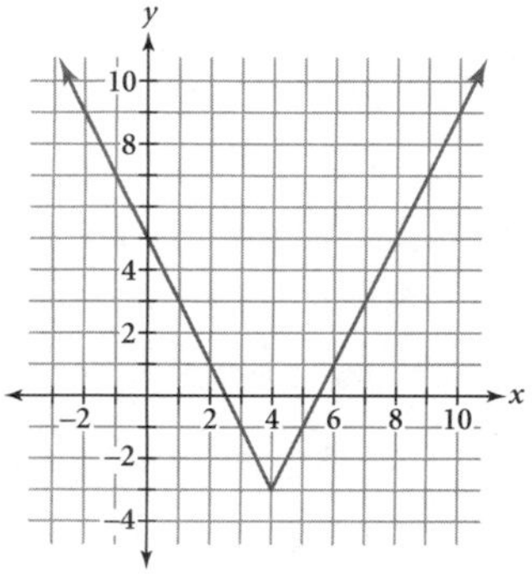

15a. Not possible. The number of columns in [A] must match the number of rows in [B].

15b. Not possible. To add matrices, they must have the same dimensions.

15c. $\begin{bmatrix} -3 & 1 \\ 1 & -5 \end{bmatrix}$

15d. $\begin{bmatrix} -2 & 6 \\ -10 & 0 \\ -7 & 3 \end{bmatrix}$

15e. $\begin{bmatrix} 5 & -3 \\ -1 & 9 \end{bmatrix}$

17a. 7.5 yd/s

17b. 27.5°

17c. $x = 7.5t \cos 27.5°$, $y = 7.5t \sin 27.5°$

17d. $x = 100 - 7.5t \cos 27.5°$, $y = 7.5t \sin 27.5°$

17e. midfield (50, 26), after 7.5 s

19. $y = 0.00115x^2 + 4$

21a. $y = -973 + 0.502x$

21b. Around 35 or 36 mpg depending on how many decimal places are retained from the regression equation's slope in your model.

21c. Answers will vary. Eventually you will not be able to make the car more efficient, or a change in technology will make the cars more efficient.

23a. $-4 + 9i$

23b. $-5 - 5i$

23c. $3 - 4i$

23d. $\frac{1}{5} - \frac{7}{5}i$

LESSON 10.1

1. 12, 27, 42, 57, 72; $u_1 = -3$, $d = 15$

3a. $3 + 4 + 5 + 6$; 18

3b. $-2 + 1 + 6 = 5$

5. $S_{75} = -5700$

7a. $u_{46} = 229$

7b. $u_n = 5n - 1$ or $u_1 = 4$ and $u_n = u_{n-1} + 5$ where $n \geq 2$

7c. $S_{46} = 5359$

9a. 3, 6, 9, 12, 15, 18, 21, 24, 27, 30

9b. $u_1 = 3$ and $u_n = u_{n-1} + 3$ where $n \geq 2$

9c. 3384 cans

9d. 13 rows with 15 cans left over

11. $S_x = x^2 + 64x$

13a. 21 trapezoids

13b. 21 rows

13c. 945 toothpicks

13d. The numbers of toothpicks in each row form a sequence, while the total numbers of toothpicks used form a series.

15a. i. increases without bound

15a. ii. 0

15b. i. 2

15b. ii. $\frac{1}{2}$

15c. i. The sum will be infinitely large.

15c. ii. The sum will approach 4.

17a. 81, 27, 9, 3, 1, $\frac{1}{3}$

17b. $u_1 = 81$ and $u_n = \frac{1}{3}u_{n-1}$ where $n \geq 2$

LESSON 10.2

1a. $0.4 + 0.04 + 0.004 + \ldots$

1b. $u_1 = 0.4$, $r = 0.1$

1c. $S = \frac{4}{9}$

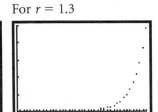

3a. $0.123 + 0.000123 + 0.000000123 + \ldots$

3b. $u_1 = 0.123, r = 0.001$

3c. $S = \frac{123}{999} = \frac{41}{333}$

5. $u_1 = 32768$

7a. 96, 24, 6, 1.5, 0.375, 0.09375, 0.0234375, 0.005859375, 0.00146484375, 0.0003662109375

7b. $S_{10} \approx 128.000$

7c.

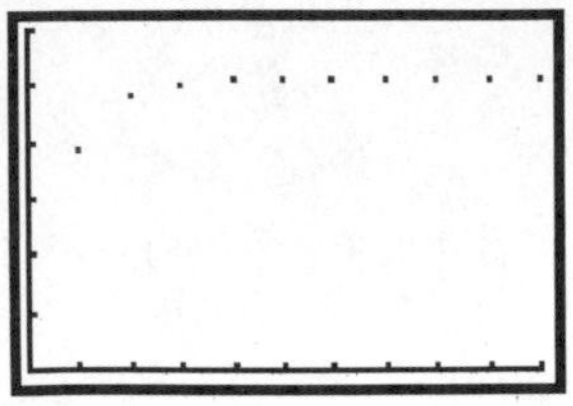

7d. $S = 128$

9a. \$25,000,000

9b. \$62,500,000

9c. 2.5

9d. 44.4%

11a. $\sqrt{2}$ in.

11b. 0.125 in.^{2}

11c. approximately 109.25 in.

11d. 128 in.^{2}

13. 88 gal

15a. \$56, 625

15b. 43 wk

LESSON 10.3

1a. $u_1 = 12, r = 0.4, n = 8$

1b. $u_1 = 75, r = 1.2, n = 15$

1c. $u_1 = 40, r = 0.8, n = 20$

1d. $u_1 = 60, r = 2.5, n = 6$

3a. $S_5 = 92.224$

3b. $S_{15} \approx 99.952$

3c. $S_{25} \approx 99.999$

5a. $S_{10} = 3069$

5b. $n = 22$

5c. $u_1 = 2.8$

5d. $r = 0.95$

7a. $S_{10} = 15.984375$

7b. $S_{20} \approx 15.99998474$

7c. $S_{30} \approx 15.99999999$

7d. They continue to increase, but by a smaller amount each time.

9. The second choice is more profitable by approximately \$4.5 quadrillion.

11a. neither

11b. $S_8 \approx 2.717857$

13a. i. 128

13a. ii. over 9×10^{18}

13a. iii. 255

13a. iv. over 1.8×10^{19}

13b. $\sum_{n=1}^{64} 2^{n-1}$

15. \$637.95

17. Yes. The long-run height is only 24 in.

CHAPTER 10 REVIEW

1a. $u_{128} = 511$

1b. $u_{40} = 159$

1c. $u_{20} = 79$

1d. $S_{20} = 820$

3a. 144; 1728; 20,736; 429,981,696

3b. $u_1 = 12$ and $u_n = 12\ u_{n-1}$ where $n \geq 2$

3c. $u_n = 12^n$

3d. approximately 1.1×10^{14}

5a. approximately 56.49 ft

5b. 60 ft

7a. $S_{10} \approx 12.957$; $S_{40} \approx 13.333$

7b. $S_{10} \approx 170.478$; $S_{40} \approx 481571.531$

7c. $S_{10} \approx 40$; $S_{40} \approx 160$

7d. For $r = 0.7$

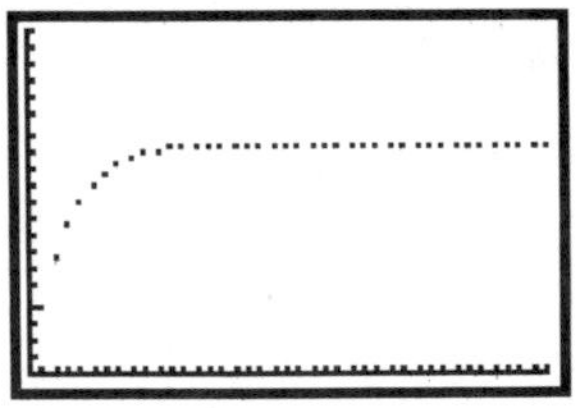

[0, 40, 1, 0, 20, 1]

For $r = 1.3$

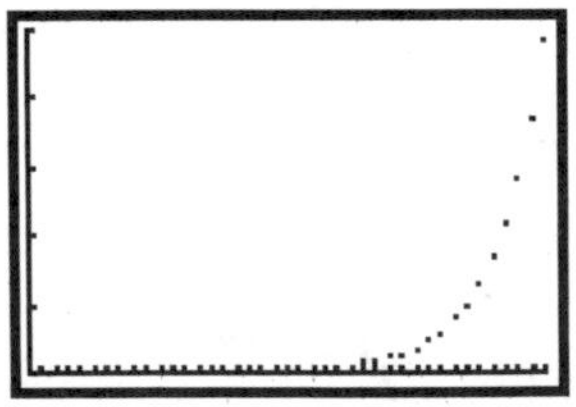

[0, 40, 1, 0, 500000, 100000]

For $r = 1$

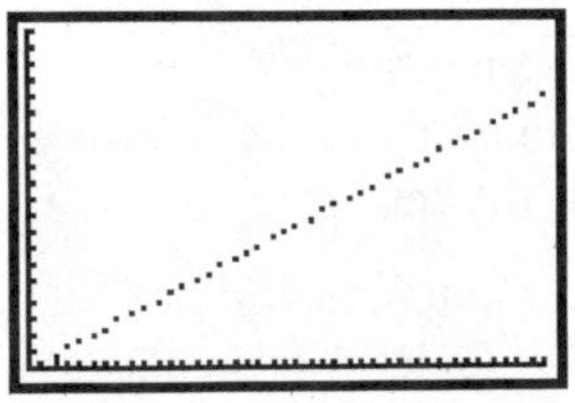

[0, 40, 1, 0, 200, 10]

7e. 0.7

LESSON 11.1

1a. $\frac{6}{15} = .4$ **1b.** $\frac{7}{15} \approx .467$ **1c.** $\frac{2}{15} \approx .133$

3a. $\frac{4}{14} \approx .286$ **3b.** $\frac{10}{14} \approx .714$ **3c.** $\frac{7.5}{14} \approx .536$

3d. $\frac{1.5}{14} \approx .107$ **3e.** $\frac{12}{14} \approx .143$

5a. experimental **5b.** theoretical

5c. experimental

7a. Answers will vary.

7b. Possible answer: Use the random integer command on the calculator to simulate rolling a die.

7c. Answers will vary.

7d. Answers will vary. Sum your answers from 7c and divide the answer by 10.

7e. Answers will vary. Long-run averages should tend toward 6 turns in order to roll a 6.

9a. 36

9b. 6; $\frac{1}{6} \approx .167$

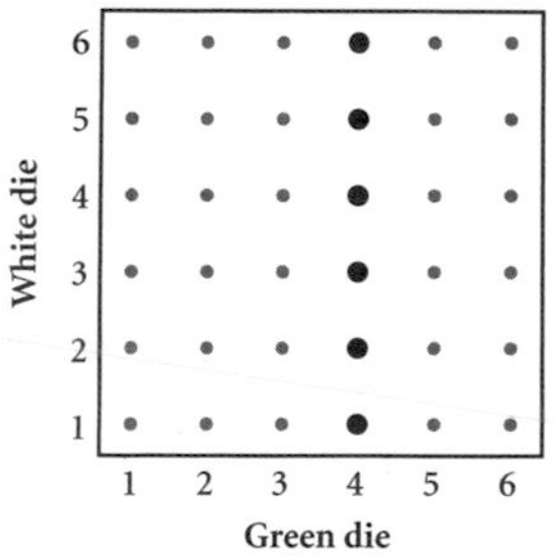

9c. 12; $\frac{12}{36} \approx .333$ **9d.** 3; $\frac{3}{36} \approx .083$

11a. 144 units^2 **11b.** 44 units^2

11c. $\frac{44}{144}$ **11d.** $\frac{44}{144} \approx .306$

11e. $\frac{100}{144} \approx .694$ **11f.** 0; 0

13a. 270 **13b.** 1380

13c. $\frac{270}{1380} \approx .196$ **13d.** $\frac{1110}{1380} \approx .804$

15a. *C* **15b.** *A*, at the nucleus

15c. The probability starts at zero at the nucleus, increases and peaks at a distance of 53 pm, and then decreases quickly, then more slowly, but never reaches zero.

17. $\log\left(\frac{ac^2}{b}\right)$

19a.

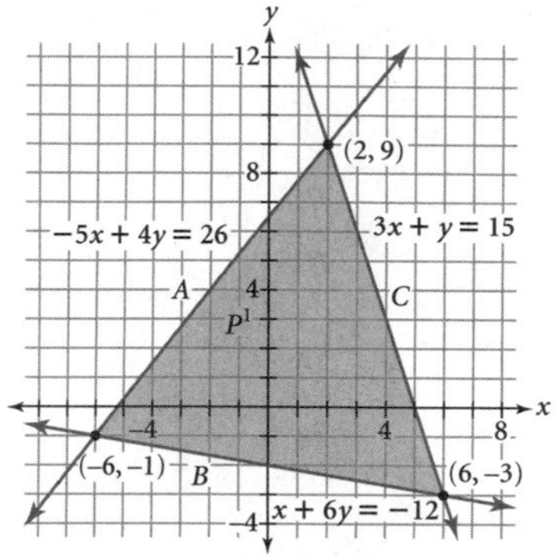

19b. (2, 9), (−6, −1), (6, −3)

19c. 68 units^2

LESSON 11.2

1.

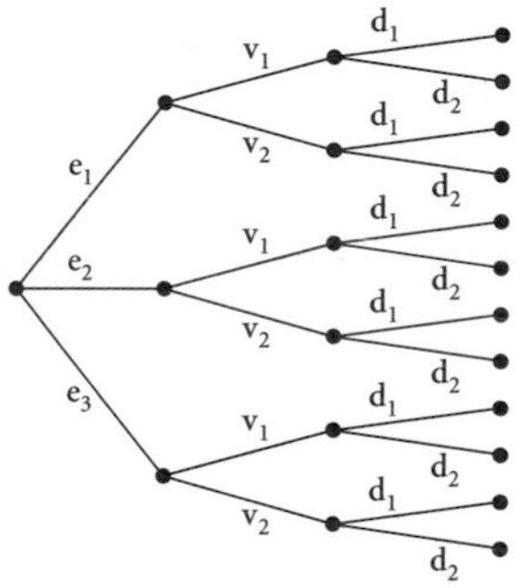

3a. $P(\text{a}) = .7$; $P(\text{b}) = .3$; $P(\text{c}) = .18$; $P(\text{d}) = .4$; $P(\text{e}) = .8$; $P(\text{f}) = .2$; $P(\text{g}) = .08$

5. Once the first student is chosen, the class total is reduced by 1 and either the junior or sophomore portion is reduced by 1.

7a. 24 **7b.** .25 **7c.** $\frac{2}{24} \approx .083$

7d. $\frac{1}{24} \approx .042$ **7e.** $\frac{23}{24} \approx .958$ **7f.** $\frac{12}{24} = .5$

9a. 4 **9b.** 8 **9c.** 16

9d. 32 **9e.** 1024 **9f.** 2^n

11a.

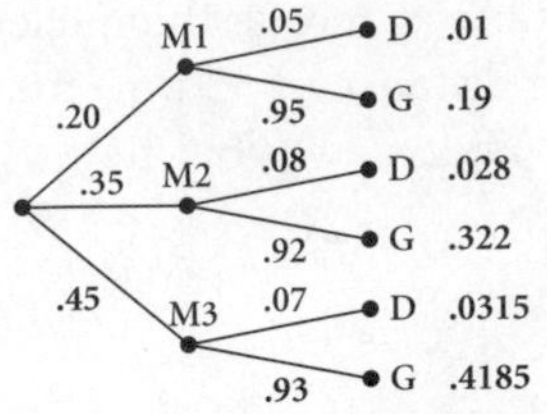

11b. .08 **11c.** .0695 **11d.** .4029

13. $\frac{6}{16} = .375$

15a. 100,000 **15b.** 1,000,000,000

15c. 17,576,000 **15d.** 7,200,000

17a. $7 + 6i$ **17b.** $-3 + 2i$

17c. $2 + 24i$ **17d.** $\frac{18}{29} + \frac{16}{29}i$

19. $P(\text{orange}) = .152$; $P(\text{blue}) = .49$

LESSON 11.3

1. 10% of the students are sophomores and not in advanced algebra. 15% of the students are sophomores in advanced algebra. 12% of the students are in advanced algebra but are not sophomores. 63% of the students are neither sophomores nor in advanced algebra.

3.

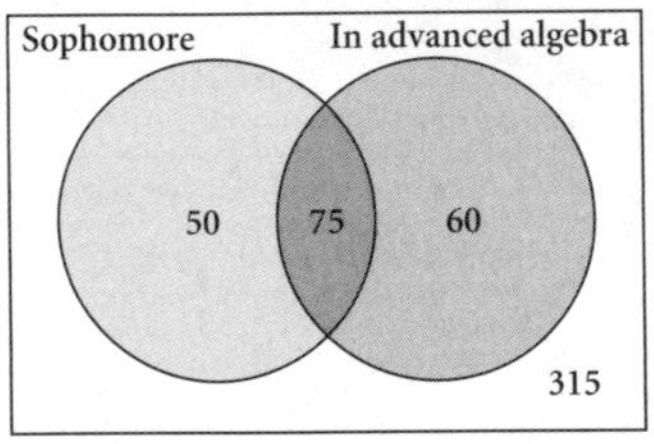

5a. Yes, because they do not overlap.

5b. No. $P(A \text{ and } B) = 0$. This would be the same as $P(A) \cdot P(B)$ if they were independent.

7a.

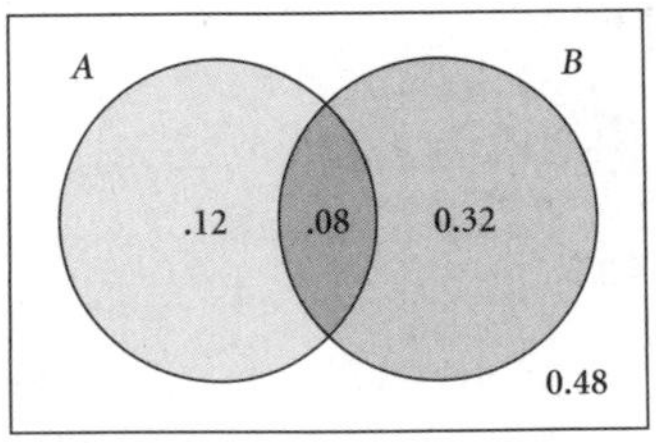

7b. i. .08 **7b. ii.** .60 **7b. iii.** .48

9. $0 \le P(A \text{ and } B) \le .4$, $.5 \le P(A \text{ or } B) \le .9$

11a. yellow **11b.** cyan **11c.** white

11d. blue **11e.** green **11f.** black

13. approximately 77

15a. $3\sqrt{2}$ **15b.** $3\sqrt{6}$ **15c.** $2xy^2\sqrt{15xy}$

LESSON 11.4

1a. Yes; the number of children will be an integer and it is based on a random process.

1b. No; the length may be a noninteger.

1c. No; the jersey number is a random number, but this is not the result of a chance experiment.

1d. Yes; there will be an integer number of pieces of mail, and it is based on random processes of who sends mail when.

3a. approximately .068 **3b.** approximately .221

5. Theoretically, Sly should get about 23 points and Les should get 21.

5a. Answers will vary.

5b. Answers will vary. Theoretically, it should be close to .47.

5c.

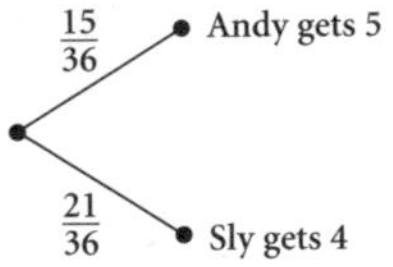

5d. -0.25

5e. Answers will vary. One possible answer is 5 points if Sly wins and 7 points if Les wins.

7a. \$25 **7b.** .67 **7c.** \$28.33

9a. Answers will vary. In the long run, the average should tend toward 2 candies.

9b. The long-run average for number of blue candies should be 1.

11. approximately .0465

13. 1

15a.

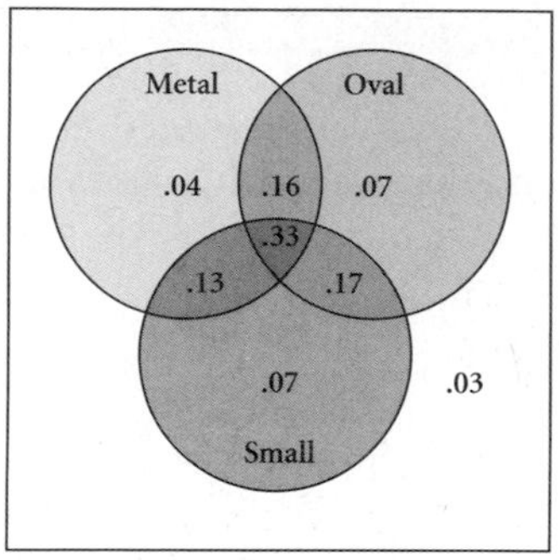

15b.

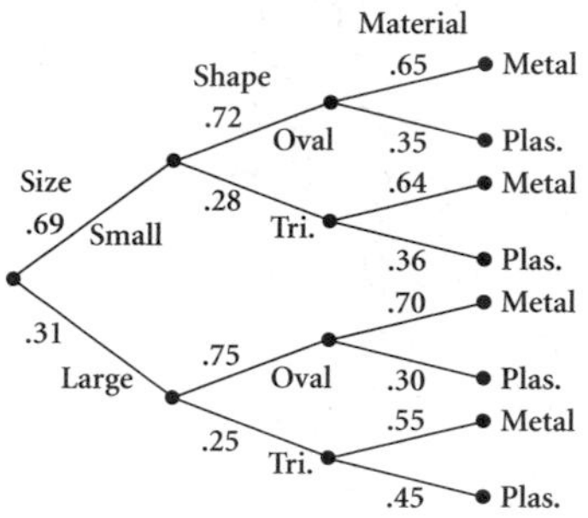

17. 44

LESSON 11.5

1a. Yes. Order matters in permutations.

1b. No. Because the order is not the same, the arrangements should be counted separately if they are permutations.

1c. No. Repetition is not allowed in permutations.

1d. No. Repetition is not allowed in permutations.

3a. 210 **3b.** 5040 **3c.** $\frac{(n+2)!}{2}$ **3d.** $\frac{n!}{2}$

5. $n = r = 6$, or $n = 6$ and $r = 5$

7a. 40,320 **7b.** 5040 **7c.** .125

7d. Sample answer: There are eight possible positions for Volume 5, all equally likely.

So $P(5 \text{ in rightmost slot}) = \frac{1}{8} = .125$.

7e. .5; Sample answer: There are four books that can be arranged in the rightmost position. Therefore, the number of ways the books can be arranged is $7! \cdot 4 = 20{,}160$.

7f. 1 **7g.** 40,319 **7h.** $\frac{1}{40320} \approx .00025$

9a. $\left(\frac{4}{50}\right)\left(\frac{46}{49}\right)\left(\frac{45}{48}\right) \approx .070$

9b. $\left(\frac{4}{50}\right)\left(\frac{3}{49}\right)\left(\frac{46}{48}\right) \approx .005$

9c. $1 - \left(\frac{46}{50} \cdot \frac{45}{49} \cdot \frac{44}{48}\right) - \left(\frac{4}{50} \cdot \frac{46}{49} \cdot \frac{45}{48}\right) \approx .155$

9d. \$3.20

11a. $\frac{30}{50} = .6$ **11b.** $\frac{16}{30} \approx .533$

13a. $\frac{1}{8} = .125$ **13b.** $\frac{3}{8} = .375$ **13c.** $\frac{1}{2} = .5$

15a. $y = -0.25x^2 + 2.5x - 3.25$

15b. $y = -0.25(x - 5)^2 + 3$

15c. $y = -0.25(x - 5 + 2\sqrt{3})(x - 5 - 2\sqrt{3})$

LESSON 11.6

1a. 120 **1b.** 35 **1c.** 105 **1d.** 1

3a. $\frac{{}_7P_2}{2} = {}_7C_2$ **3b.** $\frac{{}_7P_3}{3!} = {}_7C_3$ **3c.** $\frac{{}_7P_4}{4!} = {}_7C_4$

3d. $\frac{{}_7P_7}{7!} = {}_7C_7$ **3e.** $\frac{{}_nP_r}{r!} = {}_nC_r$

5. $n = 7$ and $r = 3$, or $n = 7$ and $r = 4$

7a. 35 **7b.** $\frac{20}{35} = .571$

9a. 4 **9b.** 8 **9c.** 16

9d. The sum of all possible combinations of n things is 2^n. $2^5 = 32$.

11a. 6 **11b.** 10 **11c.** 36

11d. ${}_nC_2 = \frac{n!}{2(n-2)!}$

13a. $x^2 + 2xy + y^2$

13b. $x^3 + 3x^2y + 3xy^2 + y^3$

13c. $x^4 + 4x^3y + 6x^2y^2 + 4xy^3 + y^4$

15a. .0194 is the probability that someone has the disease and tests positive.

15b. .02 is the probability that a healthy person tests positive.

15c. .0491 is the probability that a person tests positive.

15d. .395 is the probability that a person who tests positive is healthy.

LESSON 11.7

1a. x^{47}

1b. $5{,}178{,}066{,}751x^{37}y^{10}$

1c. ${}_{47}C_{20}x^{27}y^{20}$

1d. ${}_{47}C_{30}x^{17}y^{30}$

3a. .299

3b. .797, .498

3c. .203, .502, .791

3d. Both the "at most" and "at least" numbers include the case of "exactly." For example, if "exactly" 5 birds (.165) is subtracted from "at least" 5 birds (.203), the result (.038) is the same as 1 − .962 ("at most" 5 birds).

5. $p < \frac{25}{33}$

7a. $x^4 + 4x^3y + 6x^2y^2 + 4xy^3 + y^4$

7b. $p^5 + 5p^4q + 10p^3q^2 + 10p^2q^3 + 5pq^4 + q^5$

7c. $8x^3 + 36x^2 + 54x + 27$

7d. $81x^4 - 432x^3 + 864x^2 - 768x + 256$

9a. .401

9b. .940

9c. $Y1 = {}_{30}C_x(97)^{30-x}(.03)^x$

9d. .940

11a. .000257

11b. .446

11c. .983

13. Answers will vary. This event will happen in 15.6% of trials.

15a.

x	1	2	3	4
y	2	2.25	≈ 2.370	≈ 2.441

15b. $f(10) \approx 2.5937, f(100) \approx 2.7048,$ $f(1000) \approx 2.7169, f(10000) \approx 2.7181$

15c. There is a long-run value of about 2.718.

17a.

x (error-free)	P(x)
0	Approximately 0
1	Approximately 0
2	Approximately 0
3	.0002
4	.0245
5	.9752

17b. 4.9746

CHAPTER 11 REVIEW

1. Answers will vary. You might number 10 chips or slips of paper and select one. You might look at the random-number table and select the first digit of each number. You could alter the program Generate to :Int 10Rand + 1.

3a–b.

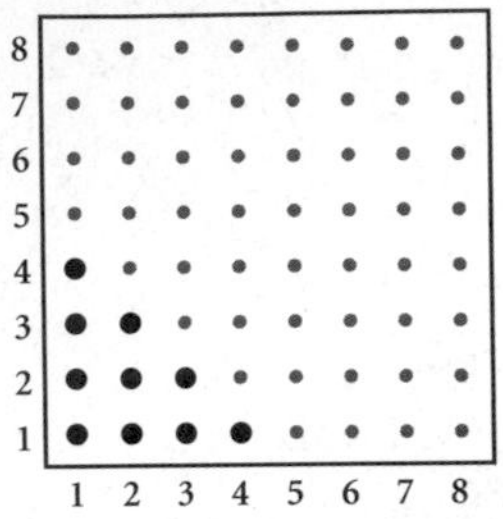

3c. $\frac{10}{64} \approx .156$

3d. $\frac{49}{64} \approx .766$

5a.

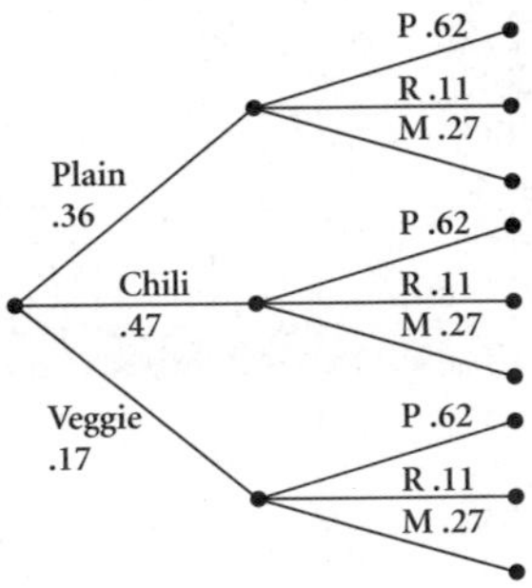

5b. .0517

5c. .8946

5d. .3501

7. 110.5

9. .044

Glossary

A

ambiguous case A situation in which more than one possible solution exists.

amplitude Half the difference of the maximum and minimum values of a periodic function.

angular speed The amount of rotation, or angle traveled, per unit of time.

antilog The inverse function of a logarithm.

arithmetic mean See **mean.**

arithmetic sequence A sequence in which each term after the starting term is equal to the sum of the previous term and a common difference.

arithmetic series A sum of terms of an arithmetic sequence.

asymptote A line that a graph approaches, but does not reach, as *x*- or *y*-values increase in the positive or negative direction.

augmented matrix A matrix that represents a system of equations. The entries include a column for the coefficients of each variable and a final column for the constant terms.

B

base The base of an exponential expression, b^x, is b. The base of a logarithmic expression, $\log_b x$, is b.

bearing An angle measured clockwise from north.

binomial A polynomial with two terms.

binomial expansion The rewriting of an expression of the form $(p + q)^n$ using the binomial theorem.

Binomial Theorem For any binomial $(p + q)$ and any positive integer n, the binomial expansion is $(p + q)^n = {}_nC_n p^n q^0 + {}_nC_{(n-1)} p^{n-1} q^1 + {}_nC_{(n-2)} p^{n-2} q^2 \cdots + {}_nC_0 p^0 q^n$.

bisection method A method of finding an *x*-intercept of a function by calculating successive midpoints of segments with endpoints above and below the zero.

Boolean algebra A system of logic that combines algebraic expressions with "and" (multiplication), "or" (addition), and "not" (negative) and produces results that are "true" (1) or "false" (0).

box plot A one-variable data display that shows the five-number summary of a data set.

box-and-whisker plot See **box plot.**

C

center (of a circle) See **circle.**

center (of an ellipse) The point midway between the foci of an ellipse.

center (of a hyperbola) The point midway between the vertices of a hyperbola.

circle A locus of points in a plane that are located a constant distance, called the radius, from a fixed point, called the center.

coefficient of determination (R^2) A measure of how well a given curve fits a set of nonlinear data.

combination An arrangement of choices in which the order is unimportant.

common base property of equality For all real values of a, m, and n, if $a^n = a^m$, then $n = m$.

common difference The constant difference between consecutive terms in an arithmetic sequence.

common logarithm A logarithm with base 10, written $\log x$, which is shorthand for $\log_{10} x$.

common ratio The constant ratio between consecutive terms in a geometric sequence.

complements Two events that are mutually exclusive and make up all possible outcomes.

completing the square A method of converting a quadratic equation from general form to vertex form.

complex conjugate A number whose product with a complex number produces a nonzero real number. The complex conjugate of $a + bi$ is $a - bi$.

Glossary

complex number A number with a real part and an imaginary part. A complex number can be written in the form $a + bi$, where a and b are real numbers and i is the imaginary unit, $\sqrt{-1}$.

complex plane A coordinate plane used for graphing complex numbers, where the horizontal axis is the real axis and the vertical axis is the imaginary axis.

composition of functions The process of using the output of one function as the input of another function. The composition of f and g is written $f(g(x))$.

compound event A sequence of simple events.

compound interest Interest charged or received based on the sum of the original principal and accrued interest.

conditional probability The probability of a particular dependent event, given the outcome of the event on which it depends.

conic section Any curve that can be formed by the intersection of a plane and an infinite double cone. Circles, ellipses, parabolas, and hyperbolas are conic sections.

conjugate pair A pair of complex numbers whose product is a nonzero real number. The complex numbers $a + bi$ and $a - bi$ form a conjugate pair.

consistent (system) A system of equations that has at least one solution.

constraint A limitation in a linear programming problem, represented by an inequality.

continuous random variable A quantitative variable that can take on any value in an interval of real numbers.

convergent series A series in which the terms of the sequence approach a long-run value, and the partial sums of the series approach a long-run value as the number of terms increases.

correlation A linear relationship between two variables.

correlation coefficient (*r*) A value between -1 and 1 that measures the strength and direction of a linear relationship between two variables.

cosecant The reciprocal of the sine ratio. If A is an acute angle in a right triangle, then the cosecant of angle A is the ratio of the length of the hypotenuse to the length of the opposite leg, or $\csc A = \frac{hyp}{opp}$. See **trigonometric function.**

cosine If A is an acute angle in a right triangle, then the cosine of angle A is the ratio of the length of the adjacent leg to the length of the hypotenuse, or $\cos A = \frac{adj}{hyp}$. See **trigonometric function.**

cotangent The reciprocal of the tangent ratio. If A is an acute angle in a right triangle, then the cotangent of angle A is the ratio of the length of the adjacent leg to the length of the opposite leg, or $\cot A = \frac{adj}{opp}$. See **trigonometric function.**

coterminal Describes angles in standard position that share the same terminal side.

counting principle When there are n_1 ways to make a first choice, n_2 ways to make a second choice, n_3 ways to make a third choice, and so on, the product $n_1 \cdot n_2 \cdot n_3 \cdot \cdots$ represents the total number of different ways in which the entire sequence of choices can be made.

cubic function A polynomial function of degree 3.

cycloid The path traced by a fixed point on a circle as the circle rolls along a straight line.

D

decay A geometric sequence with base between 0 and 1. If a_n decays then the sequence values decrease as n increases.

degree In a one-variable polynomial, the power of the term that has the greatest exponent. In a multivariable polynomial, the greatest sum of the powers in a single term.

dependent (events) Events are dependent when the probability of occurrence of one event depends on the occurrence of the other.

dependent (system) A system with infinitely many solutions.

dependent variable A variable whose values depend on the values of another variable.

determinant The difference of the products of the entries along the diagonals of a square matrix.

For any 2×2 matrix $\begin{bmatrix} a & b \\ c & d \end{bmatrix}$, the determinant is $ad - bc$.

deviation For a one-variable data set, the difference between a data value and some standard value, usually the mean.

dilation A transformation that stretches or shrinks a function or graph both horizontally and vertically by the same scale factor.

dimensions (of a matrix) The number of rows and columns in a matrix. A matrix with m rows and n columns has dimensions $m \times n$.

directrix See **parabola.**

discontinuity A jump, break, or hole in the graph of a function.

discrete graph A graph made of distinct, nonconnected points.

discrete random variable A random variable that can take on only distinct (not continuous) values.

distance formula The distance, d, between points (x_1, y_1) and (x_2, y_2), is given by the formula $d = \sqrt{(x_2 - x_1)^2 + (y_2 - y_1)^2}$.

domain The set of input values for a relation.

double root A value r is a double root of an equation $f(x) = 0$ if $(x - r)^2$ is a factor of $f(x)$.

doubling time The time needed for an amount of a substance to double.

E

e A transcendental number related to continuous growth, with a value of approximately 2.718.

eccentricity A measure of how elongated an ellipse is.

edges Line segments connecting vertices.

elimination A method for solving a system of equations that involves adding or subtracting multiples of the equations to eliminate a variable.

ellipse A shape produced by stretching or shrinking a circle horizontally or vertically. The shape can be described as a locus of points in a plane for which the sum of the distances to two fixed points, called the foci, is constant.

ellipsoid A three-dimensional shape formed by rotating an ellipse about one of its axes.

end behavior The behavior of a function $y = f(x)$ for x-values that are large in absolute value.

entry Each number in a matrix. The entry identified as a_{ij} is in row i and column j.

even function A function that has the y-axis as a line of symmetry. For all values of x in the domain of an even function, $f(-x) = f(x)$.

event A specified set of outcomes.

expanded form The form of a repeated multiplication expression in which every occurrence of each factor is shown. For example, $4^3 \cdot 5^2 = 4 \cdot 4 \cdot 4 \cdot 5 \cdot 5$.

expansion An expression that is rewritten as a single polynomial.

expected value An average value found by multiplying the value of each possible outcome by its probability, then summing all the products.

experimental probability A probability calculated based on trials and observations, given by the ratio of the number of occurrences of an event to the total number of trials.

explicit formula A formula that gives a direct relationship between two discrete quantities. A formula for a sequence that defines the nth term in relation to n, rather than the previous term(s).

exponent The exponent of an exponential expression, b^x, is x. The exponent tells how many times the base, b, is a factor.

exponential function A function with a variable in the exponent, typically used to model growth or decay. The general form of an exponential function is $y = ab^x$, where the coefficient, a, is the y-intercept and the base, b, is the ratio.

extraneous solution An invalid solution to an equation. Extraneous solutions are sometimes found when both sides of an equation are raised to a power.

extrapolation Estimating a value that is outside the range of all other values given in a data set.

extreme values Maximums and minimums.

F

Factor Theorem If $P(r) = 0$, then r is a zero and $(x - r)$ is a factor of the polynomial function $y = P(x)$. This theorem is used to confirm that a number is a zero of a function.

factored form The form $y = a(x - r_1)(x - r_2) \cdot \cdot \cdot (x - r_n)$ of a polynomial function, where $a \neq 0$. The values $r_1, r_2, \ldots, r^n$ are the zeros of the function, and a is the vertical scale factor.

factorial For any integer n greater than 1, n factorial, written $n!$, is the product of the consecutive integers from n decreasing to 1.

fair Describes a coin that is equally likely to land heads or tails. Can also apply to dice and other objects.

family of functions A group of functions with the same parent function.

feasible region The set of points that is the solution to a system of inequalities.

Fibonacci sequence The sequence of numbers 1, 1, 2, 3, 5, 8, . . . , each of which is the sum of the two previous terms.

finite A limited quantity.

finite differences method A method of finding the degree of a polynomial that will model a set of data, by analyzing differences between data values corresponding to equally spaced values of the independent variable.

first quartile **(Q_1)** The median of the values less than the median of a data set.

five-number summary The minimum, first quartile, median, third quartile, and maximum of a one-variable data set.

focus (plural **foci**) A fixed point or points used to define a conic section. See **ellipse, hyperbola,** and **parabola.**

fractal The geometric result of infinitely many applications of a recursive procedure or calculation.

fractional exponents See **rational exponents.**

frequency (of a data set) The number of times a value appears in a data set, or the number of values that fall in a particular interval.

frequency (of a sinusoid) The number of cycles of a periodic function that can be completed in one unit of time.

function A relation for which every value of the independent variable has at most one value of the dependent variable.

function family A group of functions that share important characteristics. Classifying a function according to its function family is often useful in working with the function. Examples of function families are *linear*, *quadratic*, and *exponential.*

function notation A notation that emphasizes the dependent relationship between the variables used in a function. The notation $y = f(x)$ indicates that values of the dependent variable, y, are explicitly defined in terms of the independent variable, x, by the function f.

G

general form (of a polynomial) The form of a polynomial in which the terms are ordered such that the degrees of the terms decrease from left to right.

general form (of a quadratic function) The form $y = ax^2 + bx + c$, where $a \neq 0$.

general term The nth term, u_n, of a sequence.

geometric probability A probability that is found by calculating a ratio of geometric characteristics, such as lengths or areas.

geometric random variable A random variable that represents the number of trials needed to get the first success in a series of independent trials.

geometric sequence A sequence in which each term is equal to the product of the previous term and a common ratio.

geometric series A sum of terms of a geometric sequence.

golden ratio The ratio of two numbers (larger to smaller) whose ratio to each other equals the ratio of their sum to the larger number. Or, the positive number whose square equals the sum of itself and 1. The number $\frac{1+\sqrt{5}}{2}$, or approximately 1.618, often represented with the lowercase Greek letter phi, ϕ.

golden rectangle A rectangle in which the ratio of the length to the width is the golden ratio.

greatest integer function The function $f(x) = [x]$ that returns the largest integer that is less than or equal to a real number, x.

growth A geometric sequence with base between 0 and 1. If a_n grows then the sequence values decrease as n increases.

H

half-life The time needed for an amount of a substance to decrease by one-half.

histogram A one-variable data display that uses bins to show the distribution of values in a data set.

hole A missing point in the graph of a relation.

hyperbola A locus of points in a plane for which the difference of the distances to two fixed points, called the foci, is constant.

hyperboloid A three-dimensional shape formed by rotating a hyperbola about the line through its foci or about the perpendicular bisector of the segment connecting the foci.

I

identity An equation that is true for all values of the variables for which the expressions are defined.

identity matrix The square matrix, symbolized by $[I]$, that does not alter the entries of a square matrix $[A]$ under multiplication. Matrix $[I]$ must have the same dimensions as matrix $[A]$, and it has entries of 1's along the main diagonal (from top left to bottom right) and 0's in all other entries.

image A graph of a function or point(s) that is the result of a transformation of an original function or point(s).

imaginary axis See **complex plane.**

imaginary number A number that is the square root of a negative number. An imaginary number can be written in the form bi, where b is a real number ($b \neq 0$) and i is the imaginary unit, $\sqrt{-1}$.

imaginary unit The imaginary unit, i, is defined by $i^2 = -1$ or $i = \sqrt{-1}$.

inconsistent (system) A system of equations that has no solution.

independent (events) Events are independent when the occurrence of one has no influence on the occurrence of the other.

independent (system) A system of equations that has exactly one solution.

independent variable A variable whose values are not based on the values of another variable.

inequality A statement that one quantity is less than, less than or equal to, greater than, greater than or equal to, or not equal to another quantity.

inference The use of results from a sample to draw conclusions about a population.

infinite A quantity that is unending, or without bound.

infinite geometric series A sum of infinitely many terms of a geometric sequence.

infinite sum The sum of an infinite number of terms of a series.

inflection point A point where a curve changes between curving downward and curving upward or vice versa.

inside dimensions (of a matrix) For two matrices $[A]$ and $[B]$, multiplied $[A] \cdot [B]$, the number of columns in $[A]$ and the number of rows in $[B]$. Multiplication of matrices is only possible when the inside dimensions are the same.

intercept form The form $y = a + bx$ of a linear equation, where a is the y-intercept and b is the slope.

interpolation Estimating a value that is within the range of all other values given in a data set.

interquartile range (*IQR*) A measure of spread for a one-variable data set that is the difference between the third quartile and the first quartile.

inverse The relationship that reverses the independent and dependent variables of a relation.

inverse matrix The matrix, symbolized by $[A]^{-1}$, that produces an identity matrix when multiplied by $[A]$.

inverse variation A relation in which the product of the independent and dependent variables is constant. An inverse variation relationship can be written in the form $xy = k$, or $y = \frac{k}{x}$.

L

Law of Cosines For any triangle with angles A, B, and C, and sides of lengths a, b, and c (a is opposite $\angle A$, b is opposite $\angle B$, and c is opposite $\angle C$), these equalities are true: $a^2 = b^2 + c^2 - 2bc \cos A$, $b^2 = a^2 + c^2 - 2ac \cos B$, and $c^2 = a^2 + b^2 - 2ab \cos C$.

Law of Sines For any triangle with angles A, B, and C, and sides of lengths a, b, and c (a is opposite $\angle A$, b is opposite $\angle B$, and c is opposite $\angle C$), these equalities are true: $\frac{\sin A}{a} = \frac{\sin B}{b} = \frac{\sin C}{c}$.

least squares line A line of fit for which the sum of the squares of the residuals is as small as possible.

limit A long-run value that a sequence or function approaches. The quantity associated with the point of stability in dynamic systems.

line of fit A line used to model a set of two-variable data.

line of symmetry A line that divides a figure or graph into mirror-image halves.

linear In the shape of a line or represented by a line, or an algebraic expression or equation of degree 1.

linear equation An equation characterized by a constant rate of change. The graph of a linear equation in two variables is a straight line.

linear programming A method of modeling and solving a problem involving constraints that are represented by linear inequalities.

local maximum A value of a function or graph that is greater than other nearby values.

local minimum A value of a function or graph that is less than other nearby values.

locus A set of points that fit a given condition.

logarithm A value of a logarithmic function, abbreviated log. For $a > 0$ and $b > 0$, $\log_b a = x$ means that $a = b^x$.

logarithm change-of-base property For $a > 0$ and $b > 0$, $\log_a x$ can be rewritten as $\frac{\log_b x}{\log_b a}$.

logarithmic function The logarithmic function $y = \log_b x$ is the inverse of $y = b^x$, where $b > 0$ and $b \neq 1$.

logistic function A function used to model a population that grows and eventually levels off at the maximum capacity supported by the environment. A logistic function has a variable growth rate that changes based on the size of the population.

M

major axis The longer dimension of an ellipse. Or the line segment with endpoints on the ellipse that has this dimension.

matrix A rectangular array of numbers or expressions, enclosed in brackets.

matrix addition The process of adding two or more matrices. To add matrices, you add corresponding entries.

matrix multiplication The process of multiplying two matrices. The entry c_{ij} in the matrix $[C]$ that is the product of two matrices, $[A]$ and $[B]$, is the sum of the products of corresponding entries in row i of matrix $[A]$ and column j of matrix $[B]$.

maximum The greatest value in a data set or the greatest value of a function or graph.

mean ($\bar{x}$ or μ) A measure of central tendency for a one-variable data set, found by dividing the sum of all values by the number of values. For a probability distribution, the mean is the sum of each value of x times its probability, and it represents the x-coordinate of the centroid or balance point of the region.

measure of central tendency A single number used to summarize a one-variable data set, commonly the mean, median, or mode.

median A measure of central tendency for a one-variable data set that is the middle value, or the mean of the two middle values, when the values are listed in order. For a probability distribution, the median is the number d such that the line $x = d$ divides the area into two parts of equal area.

minimum The least value in a data set or the least value of a function or graph.

minor axis The shorter dimension of an ellipse. Or the line segment with endpoints on the ellipse that has this dimension.

mode A measure of central tendency for a one-variable data set that is the value(s) that occur most often. For a probability distribution, the mode is the value(s) of x at which the graph reaches its maximum value.

model A mathematical representation (sequence, expression, equation, or graph,) that closely fits a set of data.

monomial A polynomial with one term.

multiplicative identity The number 1 is the multiplicative identity because any number multiplied by 1 remains unchanged.

multiplicative inverse Two numbers are multiplicative inverses, or reciprocals, if they multiply to 1.

mutually exclusive (events) Two outcomes or events are mutually exclusive when they cannot both occur simultaneously.

N

natural logarithm A logarithm with base e, written $\ln x$, which is shorthand for $\log_e x$.

negative exponents For $a > 0$, and all real values of n, the expression a^{-n} is equivalent to $\frac{1}{a^n}$ and $\left(\frac{a}{b}\right)^{-n} = \left(\frac{b}{a}\right)^n$.

nonrigid transformation A transformation that produces an image that is not congruent to the original figure. Stretches, shrinks, and dilations are nonrigid transformations (unless the scale factor is 1 or -1).

number of combinations The number of arrangements possible in a combination.

number of permutations The number of arrangements possible in a permutation.

O

oblique (triangle) A triangle that does not contain a right angle.

odd function A function that is symmetric about the origin. For all values of x in the domain of an odd function, $f(-x) = -f(x)$.

one-to-one function A function whose inverse is also a function.

outcome A possible result of one trial of an experiment.

outlier A value that stands apart from the bulk of the data.

outside dimensions (of a matrix) For two matrices $[A]$ and $[B]$, multiplied $[A] \cdot [B]$, the number of columns in $[A]$ and the number of rows in $[B]$. The outside dimensions of $[A]$ and $[B]$ are the dimensions of the product $[A] \cdot [B]$.

P

parabola A locus of points in a plane that are equidistant from a fixed point, called the focus, and a fixed line, called the directrix.

paraboloid A three-dimensional shape formed by rotating a parabola about its line of symmetry.

parameter (in parametric equations) See **parametric equations.**

parameter (statistical) A number, such as the mean or standard deviation, that describes an entire population.

parametric equations A pair of equations used to separately describe the x- and y-coordinates of a point as functions of a third variable, called the parameter.

parent function The most basic form of a function. A parent function can be transformed to create a family of functions.

partial sum A sum of a finite number of terms of a series.

Pascal's triangle A triangular arrangement of numbers containing the coefficients of binomial expansions. The first and last numbers in each row are 1's, and each other number is the sum of the two numbers above it.

percentile rank The percentage of values in a data set that are below a given value.

perfect square A number that is equal to the square of an integer, or a polynomial that is equal to the square of another polynomial.

period The time it takes for one complete cycle of a cyclical motion to take place. Also, the minimum amount of change of the independent variable needed for a pattern in a periodic function to repeat.

periodic function A function whose graph repeats at regular intervals.

permutation An arrangement of choices in which the order is important.

phase shift The horizontal translation of a periodic graph.

polar coordinates A method of representing points in a plane with ordered pairs in the form (r, θ), where r is the distance of the point from the origin and θ is the angle of rotation of the point from the positive x-axis.

polynomial A sum of terms containing a variable raised to different powers, often written in the form $a_nx^n + a_{n-1}x^{n-1} + \cdots + a_1x^1 + a_0$, where x is a variable, the exponents are nonnegative integers, and the coefficients are real numbers.

polynomial function A function in which a polynomial expression is set equal to a second variable, such as y or $f(x)$.

population A complete set of people or things being studied.

power function A function that has a variable as the base. The general form of a power function is $y = ax^n$, where a and n are constants.

power of a power property For $a > 0$, and all real values of m and n, $(a^m)^n$ is equivalent to a^{mn}.

power of a product property For $a > 0$, $b > 0$, and all real values of m, $(ab)^m$ is equivalent to a^mb^m.

power of a quotient property For $a > 0$, $b > 0$, and all real values of n, $\left(\frac{a}{b}\right)^n$ is equivalent to $\frac{a^n}{b^n}$.

power property of equality For all real values of a, b, and n, if $a = b$, then $a^n = b^n$.

power property of logarithms For $a > 0$, $x > 0$, and $n > 0$, $\log_a x^n$ can be rewritten $n \log_a x$.

principal The initial monetary balance of a loan, debt, or account.

principal value The one solution to an inverse trigonometric function that is within the range for which the function is defined.

probability distribution A continuous curve that shows the values and the approximate frequencies of the values of a continuous random variable for an infinite set of measurements.

product property of exponents For $a > 0$ and $b > 0$, and all real values of m and n, the product $a^m \cdot a^n$ is equivalent to a^{m+n}.

product property of logarithms For $a > 0$, $x > 0$, and $y > 0$, $\log_a xy$ is equivalent to $\log_a x + \log_a y$.

projectile motion The motion of an object that rises or falls under the influence of gravity.

Q

quadratic curves The graph of a two-variable equation of degree 2. Circles, parabolas, ellipses, and hyperbolas are quadratic curves.

quadratic formula If a quadratic equation is written in the form $ax^2 + bx + c = 0$, the solutions of the equation are given by the quadratic formula, $x = \frac{-b \pm \sqrt{b^2 - 4ac}}{2a}$.

quadratic function A polynomial function of degree 2. Quadratic functions are in the family with parent function $y = x^2$.

quotient property of exponents For $a > 0$ and $b > 0$, and all real values of m and n, the quotient $\frac{a^m}{a^n}$ is equivalent to a^{m-n}.

quotient property of logarithms For $a > 0$, $x > 0$, and $y > 0$, the expression $\log_a \left(\frac{x}{y}\right)$ can be rewritten as $\log_a x - \log_a y$.

R

radian An angle measure in which one full rotation is 2π radians. One radian is the measure of an arc, or the measure of the central angle that intercepts that arc, such that the arc's length is the same as the circle's radius.

radical A square root symbol.

radius See **circle.**

raised to the power A term used to connect the base and the exponent in an exponential expression. For example, in the expression b^x, the base, b, is raised to the power x.

random number A number that is as likely to occur as any other number within a given set.

random process A process in which no individual outcome is predictable.

random sample A sample in which not only is each person (or thing) equally likely, but all groups of persons (or things) are also equally likely.

random variable A variable that takes on numerical values governed by a chance experiment.

randomness The unpredictability of individual outcomes in random processes.

range (of a data set) A measure of spread for a one-variable data set that is the difference between the maximum and the minimum.

range (of a relation) The set of output values of a relation.

rational Describes a number or an expression that can be expressed as a fraction or ratio.

rational exponent An exponent that can be written as a fraction. The expression $a^{m/n}$ can be rewritten as $\left(\sqrt[n]{a}\right)^m$ or $\sqrt[n]{a^m}$, for $a < 0$.

rational function A function that can be written as a quotient, $f(x) = \frac{p(x)}{q(x)}$, where $p(x)$ and $q(x)$ are polynomial expressions and $q(x)$ is of degree 1 or higher.

Rational Root Theorem If the polynomial equation $P(x) = 0$ has rational roots, they are of the form $\frac{p}{q}$, where p is a factor of the constant term and q is a factor of the leading coefficient.

real axis See **complex plane.**

recursion Applying a procedure repeatedly, starting with a number or geometric figure, to produce a sequence of numbers or figures. Each term or stage builds on the previous term or stage.

recursive formula A starting value and a recursive rule for generating a sequence.

recursive rule Defines the nth term of a sequence in relation to the previous term(s).

reduced row-echelon form A matrix form in which each row is reduced to a 1 along the diagonal, and a solution, and the rest of the matrix entries are 0's.

reference angle The acute angle between the terminal side of an angle in standard position and the x-axis.

reference triangle A right triangle that is drawn connecting the terminal side of an angle in standard position to the x-axis. A reference triangle can be used to determine the trigonometric ratios of an angle.

reflection A transformation that flips a graph across a line, creating a mirror image.

reflective property of an ellipse Ellipses have reflective symmetry about both the major and minor axis.

regression analysis The process of finding a model with which to make predictions about one variable based on values of another variable.

relation Any relationship between two variables.

residual For a two-variable data set, the difference between the y-value of a data point and the y-value predicted by the equation of fit.

rigid transformation A transformation that produces an image that is congruent to the original figure. Translations, reflections, and rotations are rigid transformations.

roots The solutions of an equation in the form $f(x) = 0$.

row reduction method A method that transforms an augmented matrix into a solution matrix in reduced row-echelon form.

S

sample A part of a population selected to represent the entire population. Sampling is the process of selecting and studying a sample from a population in order to make conjectures about the whole population.

scalar A real number, as opposed to a matrix or vector.

scalar multiplication The process of multiplying a matrix by a scalar. To multiply a scalar by a matrix, you multiply the scalar by each value in the matrix.

scale factor A number that determines the amount by which a graph is stretched or shrunk, either horizontally or vertically.

secant The reciprocal of the cosine ratio. If A is an acute angle in a right triangle, the secant of angle A is the ratio of the length of the hypotenuse to the length of the adjacent leg, or $\sec A = \frac{hyp}{adj}$. See **trigonometric function.**

sequence An ordered list of numbers.

series A sum of terms of a sequence.

shape (of a data set) Describes how the data are distributed relative to the position of a measure of central tendency.

shifted geometric sequence A geometric sequence that includes an added term in the recursive rule.

shrink A transformation that compresses a graph either horizontally or vertically.

simple event An event consisting of just one outcome. A simple event can be represented by a single branch of a tree diagram.

simple random sample See **random sample.**

simulation A procedure that uses a chance model to imitate a real situation.

sine If A is an acute angle in a right triangle, then the sine of angle A is the ratio of the length of the opposite leg to the length of the hypotenuse, or $\sin A = \frac{opp}{hyp}$. See **trigonometric function.**

sine wave A graph of a sinusoidal function. See **sinusoid.**

sinusoid A function or graph for which $y = \sin x$ or $y = \cos x$ is the parent function.

skewed (data) Data that are spread out more on one side of the center than on the other side.

slant asymptote An asymptote that is neither horizontal nor vertical.

slope The steepness of a line or the rate of change of a linear relationship. If (x_1, y_1) and (x_2, y_2) are two points on a line, then the slope of the line is $A = \frac{y_2 - y_1}{x_2 - x_1}$, where $x_2 \neq x_1$.

spread The variability in numerical data.

square root function The function that undoes squaring, giving only the positive square root (that is, the positive number that, when multiplied by itself, gives the input). The square root function is written $y = \sqrt{x}$.

standard deviation (s) A measure of spread for a one-variable data set that uses squaring to eliminate the effect of the different signs of the individual deviations. It is the square root of the variance, or $s = \sqrt{\frac{\sum_{i=1}^{n}(x_i - \bar{x})^2}{n - 1}}$.

standard form (of a conic section) The form of an equation for a conic section that shows the transformations of the parent equation.

standard form (of a linear equation) The form $ax + by = c$ of a linear equation.

standard normal distribution A normal distribution with mean 0 and standard deviation 1.

standard position An angle positioned with one side on the positive x-axis.

statistic A numerical measure of a data set or sample.

statistics A collection of numerical measures, or the mathematical study of data collection and analysis.

stem-and-leaf-plot A one-variable data display in which the left digit(s) of the data values, called the stems, are listed in a column on the left side of the plot, while the remaining digits, called the leaves, are listed in order to the right of the corresponding stem.

step function A function whose graph consists of a series of horizontal lines.

stretch A transformation that expands a graph either horizontally or vertically.

substitution A method of solving a system of equations that involves solving one of the equations for one variable and substituting the resulting expression into the other equation.

substitution property The property that says if $a = b$ then a can replace b in another equation.

symmetric (data) Data that are balanced, or nearly so, about the center.

synthetic division An abbreviated form of dividing a polynomial by a linear factor.

system of equations A set of two or more equations with the same variables that are solved or studied simultaneously.

T

tangent If A is an acute angle in a right triangle, then the tangent of angle A is the ratio of the length of the opposite leg to the length of the adjacent leg, or $\tan A = \frac{opp}{adj}$. See **trigonometric function.**

term (algebraic) An algebraic expression that represents only multiplication and division between variables and constants.

term (of a sequence) Each number in a sequence.

terminal side The side of an angle in standard position that is not on the positive x-axis.

theoretical probability A probability calculated by analyzing a situation, rather than by performing an experiment, given by the ratio of the number of different ways an event can occur to the total number of equally likely outcomes possible.

third quartile (Q_3) The median of the values greater than the median of a data set.

transcendental number An irrational number that, when represented as a decimal, has infinitely many digits with no pattern, such as π or e, and is not the solution of a polynomial equation with integer coefficients.

transformation A change in the size or position of a figure or graph.

transition diagram A diagram that shows how something changes from one time to the next.

transition matrix A matrix whose entries are transition probabilities.

translation A transformation that slides a figure or graph to a new position.

tree diagram A diagram whose branches show the possible outcomes of an event, and sometimes probabilities.

trigonometric function A periodic function that uses one of the trigonometric ratios to assign values to angles with any measure.

trigonometric ratios The ratios of lengths of sides in a right triangle. The three primary trigonometric ratios are sine, cosine, and tangent.

trigonometry The study of the relationships between angles and lengths in triangles and other figures.

trinomial A polynomial with three terms.

U

unit circle A circle with radius of one unit. The equation of a unit circle with center (0, 0) is $x^2 + y^2 = 1$.

unit hyperbola The parent equation for a hyperbola, $x^2 - y^2 = 1$ or $y^2 - x^2 = 1$.

V

variance (s^2) A measure of spread for a one-variable data set that uses squaring to eliminate the effect of the different signs of the individual deviations. It is the sum of the squares of the deviations divided by one less than the number of values, or $s^2 = \frac{\sum_{i=1}^{n}(x_i - \bar{x})^2}{n - 1}$.

vector A quantity with both magnitude and direction.

velocity A measure of speed and direction. Velocity can be either positive or negative.

Venn diagram A diagram of overlapping circles that shows the relationships among members of different sets.

vertex (of a conic section) The point or points where a conic section intersects the axis of symmetry that contains the focus or foci.

vertex (of a feasible region) A corner of a feasible region in a linear programming problem.

vertex form The form $y = a(x - h)^2 + k$ of a quadratic function, where $a \neq 0$. The point (h, k) is the vertex of the parabola, and a is the vertical scale factor.

vertices The plural of vertex.

vertical line test A method of using a graph to determine whether a relation is a function. A function has only one output (y-) value for a given input (x-) value. If a completely vertical line can join two points on a graph, that means there are two y-values corresponding to one x-value, and the graph does not represent a function.

Z

zero exponent For all values of a except 0, $a^0 = 1$.

zero-product property If the product of two or more factors equals zero, then at least one of the factors must equal zero. A property used to find the zeros of a function without graphing.

zeros (of a function) The values of the independent variable (x-values) that make the corresponding values of the function ($f(x)$-values) equal to zero. Real zeros correspond to x-intercepts of the graph of a function. See **roots.**

Photo Credits

Abbreviations: top (**t**), middle (**m**), bottom (**b**), left (**l**), right (**r**)

Chapter 1
1: Tony Cenicola/The New York Times; **5:** Michael Nicholson/CORBIS; **9:** Bodleian Library, Oxford, U.K. Copyright (MS Hunt. 214 Title Page); **16:** CORBIS; **17:** Scala/Art Resource, NY; **27:** CORBIS; **30:** CORBIS; **31:** (*t*): AP-Wide World Photos; **31** (*b*): Ken Karp Photography; **32:** Ken Karp Photography; **38:** Bohemian Nomad Picturemakers: **48:** Reuters NewMedia Inc./CORBIS; **50:** Getty Images

Chapter 2
53: Residual Light by Anthony Discenza. Photos courtesy of the artist; **54** (*l*): Sandro Vannini/CORBIS; **54** (*r*): Tom & Dee Ann McCarthy/CORBIS; **56:** Owaki-Kulla/CORBIS; **60:** Archive Photos/PictureQuest; **61:** Gary D. Landsman/CORBIS; **71** (*l*): Duomo/CORBIS; **71** (*r*): Tom & Pat Leeson/DRK Photo; **73:** CORBIS; **75:** RF; **79:** NOAA; **84:** Paul Skelcher/Rainbow

Chapter 3
103: Zaha Hadid Architects; **108:** Alan Schein Photography/CORBIS; **117:** Andre Jenny/Focus Group/PictureQuest; **121:** David Muench/CORBIS; **132:** AP Wide World; **143:** Jose Fuste Raga/CORBIS

Chapter 4
148: Benjamin Edwards; **161:** Robert Holmes/CORBIS; **168:** © 2002 Eun-Ha Paek, stills from L'Faux Episode 7 on MilkyElephant.com

Chapter 5
215: Tree Mountain—A Living Time Capsule—11,000 Trees, 11,000 People, 400 Years 1992–1996, Ylojarvi, Finland, (420 meters long X 270 meters wide X 28 meters high) © Agnes Denes. Photo courtesy of Agnes Denes; **216:** Roger Ressmeyer/CORBIS; **229:** Paul Almasy/CORBIS; **236:** Geoff Tompkinson/Photo Researchers, Inc.; **238:** © 2002 Eames Office (www.eamesoffice.com); **247:** Alice Arnold/Retna Ltd.; **248:** Getty Images; **273:** Cheryl Fenton; **255** (*t*): Tom Bean/CORBIS; **255** (*b*): Getty Images; **257:** Bettmann/CORBIS; **261:** Reuters NewMedia Inc./CORBIS; **266:** Bettmann/CORBIS

Chapter 6
277: Things Fall Apart: 2001, mixed media installation with vehicle; variable dimensions/San Francisco Museum of Modern Art, Accessions Committee Fund purchase © Sarah Sze; **286:** Mike Souther-Eye Ubiquitous/CORBIS; **289:** Ken Karp Photography; **295:** Ken Karp Photography; **299:** Vadim Makarov; **313(t):** Mehau Kulyk/Photo Researchers, Inc.; **313** (*m*): Hank Morgan/Photo Researchers, Inc.

Chapter 7
341: Fragile by Amy Stacey Curtis. Photos of Fragile © 2000 by Amy Stacey Curtis/www.amystaceycurtis.com; **346:** CORBIS; **366:** AP-Wide World Photos; **374:** Lester Lefkowitz/CORBIS; **375:** © The Nobel Foundation; **383:** NASA; **398:** Lowell Georgia/CORBIS

Chapter 8
401: Corey Rich/Coreyography LLC; **409:** David Lawrence/CORBIS; **438:** Ken Karp Photography; **441:** Ken Karp Photography

Chapter 9
466: Gravity in Four Directions, 2001. Leaves, pills, photocopies, acrylic, resin on wood, 72 X 72 inches. Image courtesy of James Cohan Gallery, New York.; **471:** RF; **474** (*tl*): Reunion des Musees Nationaux/Art Resource, NY; © L&M Services B. V. Amsterdam 20030605; (*tr*): Archivo Iconografico, S. A./CORBIS; (*bl*): Lee Snider/CORBIS; (*br*): St. Louis Science Center; **477** (*t*): The Museum of Modern Art, Licensed by SCALA/Art Resource, NY; (*b*): Ken Karp Photography; **480:** Ken Karp Photography; **492:** Lee Foster/Bruce Coleman Inc.; **535:** Julian Calder/CORBIS

Chapter 10
539: Courtesy of the artist and Lehmann Maupin Gallery, New York; **540:** Binod Joshi/AP-Wide World Photos; **545:** Bettmann/CORBIS; **549:** RF

Chapter 11
565: Smithsonian American Art Museum, Washington, D.C./Art Resource, NY; **566** (*l*): Roger Wood/CORBIS; (*r*): National Museum of India, New Delhi/The Bridgeman Art Library; **584:** Bettmann/CORBIS; **601:** Chris Rogers-Index Stock Imagery/PictureQuest

Index

A

absolute-value function,
  transformations of, 184–188
acceleration, 546
addition
  of complex numbers, 309
  inequalities and, 378
  of matrices, 349, 355
  of rational expressions,
    524–525
  row operations of matrices, 361
addition property of equality, 134
addition rule for mutually exclusive
  events, 587–588
adjacency matrix, 347
agriculture and horticulture, 75, 83,
  109, 220, 292, 390
Alcuin of York, 6
allometry, 242
ambiguous case, 449
analytic geometry, 5
angle(s), Greek language for, 161
annual percentage rate (APR), 94
aphelion, 535
Apollonius of Perga, 475
appreciation and depreciation, 64,
  169, 217–218
archaeology and anthropology, 69,
  132, 178, 255, 262
architecture, 54, 56, 116, 143, 305,
  474, 483
area
  of a triangle, 452
  probability and, 570
area model, 570
arithmetic sequences, 57–58
  as basic sequence, 60
  common difference of, 57, 64,
    105, 278
  defined, 57
  explicit formulas for, 104–107
  graphs of, 77–78, 105
  slope and, 105
arithmetic series, 540–544
  formula for partial sum of,
    542–544
art, 53, 168, 217, 286, 305, 341, 382,
  430, 466, 474, 535, 539, 565
astronomy, 174, 235, 241, 450, 451,
  482–483, 484, 501, 535
asymptotes
  of hyperbolas, 496, 498
  of rational functions, 510,
    517–519
  slant, 518
augmented matrix, 360, 364
Austria, 574

B

base
  of exponents, 223
  of logarithms, 252, 253, 269
bearing, 418–422
Bernoulli, Jacob, 169
best-fit line. *See* line of fit
bias, sample standard deviation as
  correction for, 34, 35
binomial expansion, 714–718
binomials, defined, 278
binomial theorem, 615
biology, human, 27, 39, 100, 125,
  274, 275, 284, 358, 437, 583, 606
biology, nonhuman, 4, 71, 242, 266,
  270, 275, 292, 384, 389, 522, 593,
  615–616, 619
bisection method, 332
Bohr, Niels, 574
Boolean algebra, 204
Boole, George, 204
botany, 242
box-and-whisker plots, 26–27, 29,
  36–37
box plots, 26–27, 29, 36–37
Boyle, Robert, 235
Braille, Louis, 584
business, 68, 91, 108, 131, 149–150,
  154, 213, 268, 292, 299, 346,
  346–347, 348, 351–352, 365–366,
  366, 385–386, 389–390, 397, 553,
  624, 625

C

Canada, 398, 584
carbon dating, 69, 255, 262
Cartesian graphs. *See* coordinate
  graphs
Cellarius, Andreas, 474
Celsius, Anders, 249
Celsius and Fahrenheit conversion, 7
center of circle, 426, 475
central tendency, measures of,
  24–27, 36
chemistry, 10–11, 18, 21, 30, 83, 110,
  190, 267, 345–346, 512, 514
China, 219, 280, 374
Chu Shih-chieh, 541
circle(s)
  center of, 426, 475
  as conic section, 474–475
  defined, 426, 475
  parametric equations for,
    426–430
  radius of, 426, 475
  standard form of equation, 476
  transformations and, 192–194
  unit. *See* unit circle
coefficient matrix, 372
color wheels, 592
combinations and probability,
  607–610
common base, 252
common base property of equality,
  224
common difference, 57, 64, 105, 278
common knowledge, 17
common logarithm, 252
common ratio, 60, 64, 65
commutative property, 203
compass, 422
complements, 589–590
completing the square, 294–298
complex numbers, 307–310
  complex conjugates, 307, 310
  conjugate pairs, 307, 317,
    322–323
  defined, 308
  graphing, 310
  imaginary unit (i), 307
  modeling using, 309
  operations with, 309–310
complex plane, 310, 312
composition of functions, 200–203
  of inverse and its function,
    246–247, 253
  multiplication of functions
    distinguished from, 203
compound event, 578
compound interest, 66
computers and Internet, 23, 204,
  306, 332, 598, 604, 605

Index

conditional probability, 580–581
confidence interval, 618
congruent figures, 171
conic sections, 474–475
  construction of, 477, 502, 504
  systems of, solving, 503–506
  transformations of, 192–194, 488
conjugate pairs
  defined, 307
  zeros as, 317, 322–323
consistent systems of equations, 367
constraints, 379, 381
consumer awareness, 23, 40–41, 46, 64, 86, 101, 124, 127–128, 131, 137, 176, 191, 207, 249, 275, 336, 346–347, 357, 358, 373–374, 390, 430, 515, 551, 604
consumer price index (CPI), 101
continuous graphs, 150, 151, 155
contour maps, 151
convergent series, 547–550, 559–561
cooking, 492, 582
coordinate graphs
  complex plane, 310
  history of, 5
  problem solving with, 4–5
cosine (cos), 419 *See also* trigonometry
cosines, law of, 453–456
counting principle, 601–602
Cramer's Rule for 2-by-2 systems, 140–141
cryptography, 376, 377
cubic functions
  defined, 315
  factored form of, 315
  graphs of, 315, 320, 321–322

D

dance, 366, 401
data
  defined, 24
  experimental, finite differences method and, 280, 281
  extrapolation, 123
  five-number summary of, 26
  interpolation, 123
  maximum values in, 26
  measures of center of, 24–27, 36
  minimum values in, 26
  outliers, 25, 36–37, 39
  range of, 26
  samples, 616
  shape of, 26–27
  skewed, 27
  spreads of, 26–27, 31–37
  symmetric, 26–27
  See also graphs; statistics; variables
data tables, 334–335
decay. *See* growth and decay
degree of polynomials
  defined, 278
  finite differences method of finding, 279–280
  shape of graph and finding, 322
Denmark, 574
dependent events, 580–581
dependent systems of equations, 359, 367
dependent variable, 114
  of functions, 155
  graph interpretation and, 150, 151
  of inverse functions, 243, 245
Descartes, René, 5
design and engineering, 171, 315–316, 424–425, 457, 508, 529, 536
deviation, 33–35
diagrams, problem solving with, 2–3
diet and nutrition, 382–383, 386–387, 388, 514
dilations, 351
dimensions of matrix, 343, 354
directrix, 486–487
discontinuities, 158
discrete graphs, 77
  continuous graphs compared to, 150, 151
  of functions, 155
  of sequences, 77–80
discrete random variables, 594
distance
  absolute value function and, 184
  formula for, 467–470
distance calculations, 108, 182, 184, 244–245, 376, 420–422, 431–434, 435–436, 444–449, 450, 451, 458, 464, 471–472, 536, 592
distributive property of exponentiation over division, 224
distributive property of exponentiation over multiplication, 224
division
  of complex numbers, 310
  inequalities and, 378
  of polynomials, 328–330
  of rational expressions, 526–527
  row operations of matrices, 361
  synthetic, 330
division property of exponents, 224
dominant traits, 357
doubling time, 218–219
dynamic systems, 73

E

e, 269–270
eccentricity, 480
economic multiplier, 551
economics, 58, 101, 241, 254, 365–366, 375, 551
edges, probability and, 607–608
education, 51, 58, 100, 146, 275, 378, 397, 398, 571, 584, 591, 605, 620
Egypt, ancient, 305, 504, 566
electrons, 285
element (entry) of matrix, 343
elimination
  matrices and, 360–362
  in systems of equations, 134–136, 360–362
ellipse(s), 477–480
  as conic section, 474–475
  construction of, 477, 502, 504
  defined, 477
  eccentricity of, 480
  foci of, 477
  graphing, 478–479
  major axis, 478
  minor axis, 478
  properties of, 484, 531–532
  reflection property of, 484
  standard form of equation of, 477
  transformations of circles and, 192–194, 477
elliptical orbits, 196
end behavior, 320, 510
energy. *See* resource consumption, exploration, and conservation
England, 6
entertainment, 373–374, 557, 597, 605, 625
entry (element) of matrix, 343
environment, 73, 264–265, 300, 522, 540

Index

equality, properties of, 134, 224
equations
exponential. *See* exponential equations
linear. *See* linear equations
parametric. *See* parametric equations
quadratic. *See* quadratic equations
roots of. *See* roots of an equation
systems of. *See* systems of equations
unit circle, 192
valid conditions for relationship of, 107
See also equations, solving; functions
equations, solving
exponential equations, with logarithms, 251–253, 264–265
exponent properties and, 225
logarithmic properties and, 257–260
power equations, 231, 239–240
quadratic. *See* quadratic equations, solving
systems of. *See* systems of equations, solving
undoing order of operations, 231
Erdos numbers, 4
Erdos, Paul, 4
events
complements, 589–590
compound, 578
defined, 568–569
dependent, 580–581
independent, 579–580, 581, 588–589
mutually exclusive, 586–590
simple, 578
Excel program, 42–44, 94–98, 140–141, 208–210, 271–272, 334–336, 392, 531–532, 621–622
expanded form of exponents, 223
expected value, 593–596
experimental probability, 568–569, 570
experiment design, consistent results from, 31–32
explicit formulas, 104–107
exponential equations, solving, with logarithms, 251–253
exponential functions, 216–219
doubling time, 218–219
general form of, 218
growth and decay modeled with, 216–219, 264–265
as inverse of logarithms, 253
solving, 217–218
exponents
base of, 223
expanded form, 223
notation for, 223
positive bases defined for properties of, 231
properties of, 223–225, 231
rational. *See* fractional (rational) exponents
expressions
Boolean, 204
rational, operations with, 524–528
See also equations; polynomials; terms
extrapolation, 123
extreme values, 320

F

factored form of equations, 314–317, 327–328
factored form of quadratic functions, 288–289
conversion to general form, 288–290
vertex found with, 294–295
factored form of rational functions, 519
factorials, 602–603
factor theorem, 328
Fahrenheit and Celsius conversion, 7
fair-trade certification, 390
family of functions, 171
feasible region, 379, 380–381, 385–388
fiber-optic technology, 399
Fibonacci, Leonardo, 14
film, 49–50, 238
finite differences method, 279–280
finite number of terms, 541
first quartile, 26
five-number summary, 26
foci
of ellipses, 477
of hyperbolas, 494, 497, 498, 500
focus, of parabola, 486–487
formulas
explicit, 104
recursive, 56–57, 104
fractals
complex plane and, 312, 313
Mandelbrot set, 312, 313
self-similarity of, 552
Sierpiński triangle, 59, 552
fractional (rational) exponents, 229–231
defined, 230
as power function, 231–232
as roots (numerator of 1), 230
transformations and, 232–233
France, 286, 584
functions
composition of. *See* composition of functions
defined, 155
end behavior of, 320, 510
exponential. *See* exponential functions
family of, 171
greatest integer, 158
inverse. *See* inverse functions
linear. *See* linear functions
logarithms. *See* logarithmic functions
logistic, 90–93
notation for, 156–158
parent, 171
polynomial, defined, 278
power. *See* power functions
projectile motion, 294
quadratic. *See* quadratic functions
rational. *See* rational functions
reflections of. *See* reflections
square root. *See* square root functions
step, 158
stretching and shrinking of. *See* stretches and shrinks
translation of. *See* translations
vertical line test to determine, 156
zeros of. *See* zeros of a function
See also equations
fundraising, 40, 162, 299–300, 363–364

G

Galileo Galilei, 187, 280
gambling, 566, 596, 610, 612
garbage, 540
gardening. *See* agriculture and horticulture

Gauss, Carl Friedrich, 374, 543
Gaussian elimination, 374
Gauss-Jordan elimination, 374
genealogy, 68
general addition rule, 588
general form of a polynomial, 278
general form of quadratic equations. *See* quadratic equations
general form of quadratic functions, 286, 289
  converting to vertex form, 294–297
  factored form converting to, 288–290
general quadratic equation, 302–304
general term, 56
genetics, 357
geology, 457
geometric probability, 570, 575–576
geometric random variables, 594
geometric sequences, 59–60
  as basic sequence, 60
  common ratio of, 60, 64, 65
  defined, 60
  graphs of, 77–80
  shifted, 73–74
  zero term as choice in, 64
geometric series
  convergent infinite, 547–550, 559–561
  partial sums of, 554–556
geometry
  analytic, 5
  language for, 161
  See also polygon(s)
geosynchronous orbits, 196
Germany, 374, 543
golden ratio, 305
golden rectangle, 305
government, 254, 612
GPS (Global Positioning System), 503
graphing
  of absolute-value functions, 184
  of arithmetic sequences, 77–78, 105
  of circles, 426–428
  of complex numbers, 310
  of composition of functions, 200–201
  of cubic functions, 315, 320, 321–322
  of ellipses, 478–479
  of geometric sequences, 77–80
  of hyperbolas, 496–497
  of linear equations, 367
  of linear equations in three dimensions, 367
  of line of fit, 120
  of parabolas, 170–174
  of parametric equations, 402–406, 411–414
  of partial sum of a sequence, 549
  of polynomial functions, 316–317, 320, 321–323, 327
  of quadratic functions, 287
  of rational functions, 509–511, 516–519
  of sequences, 77–80, 105
  of square root functions, 177–178
  of step functions, 158
  of systems of conic sections, 504–505
  of systems of equations. *See* systems of equations, solving
  of systems of inequalities, 378–381
  of transformations. *See* transformations
graphs
  analyzing, 80
  box plot, 26–27, 29, 36–37
  complex plane and, 310
  continuous, 150, 151, 155
  discontinuities, 158
  discrete. *See* discrete graphs
  holes, 516, 517
  interpretation of, 149–151
  linear, 77
  problem solving with, 4–5
  transformations of. *See* transformations
  vertex-edge, 348
  vertical line test of, 156
graph sketch, 150
graph theory, 348
gravity, 179
greatest integer function, 158
Greece, ancient, 161, 451, 475, 566
group theory, 348
growth and decay
  definitions of, 65
  doubling time, 218–219
  exponential functions modeling, 216–219, 264–265
  half-life, 216, 218–219
  logistic functions and, 90–93
  recursion modeling, 64–66, 79–80
guess-and-check, 22

H

half-life, 216, 218–219
Hawaii, 24, 120–121
al-Haytham, Abú Ali Al-Hasan ibn, 504
Hoffenberg, Marvin, 375
holes, 516, 517
Hollings, C. S., 522
horizontal asymptotes, 510, 517, 518
Human Genome Project, 20
Hungary, 61
hyperbola(s), 494–499
  asymptotes of, 496, 498
  as conic section, 474–475
  definition of, 494
  equation of, 496, 498
  foci of, 494, 497, 498, 500
  graphing of, 496–497
  unit, 495
  vertices of, 494
hyperboloid, 474

I

i, 307
identities, trigonometric, 459–461
identity matrix, 368–369
image, 165
imaginary axis, 310
imaginary unit (*i*), 307
income, 21, 85, 86, 106–107, 110, 116, 159, 557
inconsistent systems of equations, 359, 367
independent events, 579–580, 581, 588–589
independent variable, 114
  of functions, 155
  graph interpretation and, 150, 151
  of inverse functions, 243, 245
India, 261, 419, 557
inequalities, 378
  operations with, 378
  See also systems of inequalities
infinite series, 547, 550
inside dimensions, 354
insurance, 126
intercept form
  defined, 112
  line of fit using, 120–121

intercept form *(continued)*
subsitution to solve systems in, 129, 133
interest, 66, 69, 75, 100, 144–145
interpolation, 123
interquartile range (IQR), 36–37
inverse functions, 243–246
exponents and logarithms as, 253
one-to-one function, 245–246
trigonometric, 421–422
inverse matrix, 368–372, 392–394
inverse relations, 245
inverse variation, 509 *See also* rational functions
Investigation into the Laws of Thought, An (Boole), 204
investments, 87, 88–89, 94–97, 198, 222, 239, 253, 374, 378
IQR (interquartile range), 36–37
Iran, 280
Iraq, 9, 419, 504
Italy, 14, 54, 143, 280

J

Jordan, Wilhelm, 374
journal keeping, 52

K

al-Kashi, Jamshid Masud, 280
Kepler, Johannes, 235
Khwarizmi, Muhammad ibn Musa, 9, 419
Kitab al-jabr w'al-muqâbalah (Khwarizmi), 9, 419
knots (speed), 473

L

language
"algebra" as term, 9
braille code, 584
Greek prefixes in, 161
Latin prefixes in, 171
Navajo code-talkers, 376
"sine" as term, 419
translation and, 164
Latin language, 171
law enforcement, 470
law of cosines, 453–456
law of sines, 447–449, 456
least common denominator, 524
Leclerc, George Louis, 575
Leonardo da Vinci, 110
Leontief, Wassily, 375
Liber Abaci (*Book of Calcuclations*) (Fibonacci), 14
light (optics), 69, 270, 285, 451, 480, 486, 492–493, 592
light year, 451
limits, 71–74
defined, 73
See also logistic functions
line(s)
parallel, 165, 180, 367
See also line of fit
linear equations
arithmetic sequences and, 104–107
formula, 104
intercept form. *See* intercept form
point-slope form, 121–122
standard form, 168
systems of. *See* systems of equations
in three variables, 367
linear functions, translation of, 163–165, 180
linear graphs, 77
linear programming, 385–388
line of fit
defined, 119
estimation of, 119–123
intercept form and, 120–121
point-slope form and, 121–122
line of symmetry, 170–171
Li Shun-Fêng, 280
Liu Hui, 374
local maximum, 320
local minimum, 320
locus, 469
logarithmic functions, 251–253
change-of-base property, 253
common base, 252
common logarithm, 252
definition of, 252
exponents and, 251–253
as inverse of exponents, 253
natural, 269–270
positive numbers required for, 264
properties of, 253, 257–260
solving exponential equations with, 251–253
logistic functions, 90–93
long-run value. *See* expected value

M

major axis, 478
Malthus, Thomas, 92
Mandelbrot, Benoit, 312, 313
Mandelbrot set, 312, 313
manufacturing, 38, 62, 131, 385–386, 397, 522, 583, 598
maps, 151, 448
mass, 144
matrices
addition of, 349, 355
adjacency, 347
augmented, 360, 364
coefficient, 372
cryptography and, 377
defined, 342
dimension of, 343, 354
entry (element) of, 343
identity, 368–369
inverse, 368–372, 392–394
multiplication of, 351–355
reduced row-echelon form, 360, 364
representation of information with, 343
row operations in, 361
row reduction method, 360–364, 374
scalar multiplication, 351, 355
solution, 360
systems of equations solved with, 360–364, 370–372
transformations of, 350–351
transition, 342–344
maximum
of data sets, 26
local, 320
of quadratic function, 294
mean, 25 *See also* standard deviation
mean value. *See* expected value
measures of central tendency, 24–27, 36
median
box plots and, 26, 36
defined, 25
in five-number summary, 26
medicine and health, 72–73, 75–76, 101, 145, 274, 437, 493, 514, 531, 619
Menaechmus, 475
Mersenne perfect number, 622
Mersenne prime, 622

metallurgy, 534
meteorology and climatology, 39–40, 51, 79, 84, 100, 138, 242, 248–249, 545, 620
minimum
of data sets, 26
local, 320
of quadratic function, 294
minor axis, 478
mode, 25
model, defined, 79
monomials, defined, 278
multiplication
of complex numbers, 309
geometric sequences and, 60
inequalities and, 378
of matrices, 351–355
of rational expressions, 526–527
row operations of matrices, 361
scalar, of matrices, 351, 355
multiplication property of equality, 134
multiplication property of exponents, 224
multiplication rule for independent events, 579–581
multiplicative identity, 368
multiplicative inverse, 368
music, 60, 170, 247, 255, 261, 286
mutually exclusive events, 586–590

N

Napier, John, 257
Napier's bones, 257
Native Americans, 376
natural logarithm function, 269–270
nautical mile, 473
Navajo code-talkers, 376
navigation, 422, 431–434, 439, 451, 458
negative numbers, square root of. *See* complex numbers
Netherlands, 502
Newton, Isaac, 179
nonrigid transformations, 185
notation and symbols
explicit formulas, 105
exponents, 223
factorials, 602
function, 156–158
i, 307
infinity, 550
inverse of one-to-one function, 245
linear equations, 105
logarithm, 29, 251, 252
mean, 25
orders of magnitude, 238
recursive formulas, 56, 80
sigma, 25, 541
slope, 112
standard deviation, 34
terms, 56, 64, 80
numbers
chart summarizing types of, 308
complex. *See* complex numbers
prime, 622
rounding of, 422, 456
transcendental, 269
triangular, 283

O

oblique triangle(s), 445
obtuse triangle(s), 446
one-to-one functions, 245–246
inverse of. *See* inverse functions
Optics (al-Haytham), 504
order of operations, undoing, to solve equations, 231
orders of magnitude, 238
Oughtred, William, 257
outcomes, 568–569
outliers, 25, 36–37, 39
outside dimensions, 354

P

parabola(s), 486–490
as conic section, 474–475
defined, 170, 486
directrix of, 486–487
equation of, 489
focus of, 486–487
graphing of, 170–174
line of symmetry of, 170–171
transformations of, 488
translations of, 171–174, 208–210
vertex of, 171, 173
See also quadratic equations
paraboloid, 486
parallel lines
inconsistent system of equations and, 367
translations mapping, 165, 180
parameter
in Excel, 43
in parametric equations, 402, 405, 462
parametric equations
for circles, 426–430
conversion to nonparametric equations, 411–414
defined, 402
graphing, 402–406, 411–414
for projectile motion, 438–441
translation of, 405–406
trigonometry and, 418–421, 426–430
parent functions, 171
partial sums
of arithmetic series, 540–544
of geometric series, 554–556
Pascal's triangle, 613–615, 621–622
Pasteur, Louis, 266
patterns, 54
problem solving with, 45
See also recursion
perfect square, 295
perihelion, 535
periodicity, 459–461
permutations, 600–603
Péter, Rózsa, 61
pets, 22–23, 299, 389
physics, 115, 130, 137–138, 144, 161, 179, 228, 235, 239–240, 261–262, 285, 374, 383, 444, 451, 515, 546, 574
planes, intersection of, 367
Playing with Infinity (Péter), 61
point(s), locus of, 469
point-slope form, 121–122
Poland, 59, 312
polar orbits, 196
polls and surveys, 592, 616–617, 619, 625
polygon(s)
diagonals of, 161, 279–280
sum of interior angles of, 161
as term, 161
polynomials
binomials, 278
degree of. *See* degree of polynomials
division of, 328–330
end behavior, 320
factored form of, 314–317, 327–328
general form of, defined, 278

polynomials *(continued)*
graphs of, 316–317, 320, 321–323, 327
higher-degree, 320–323, 325, 327–330
local minimums and maximums (extreme values), 320
monomials, 278
trinomials, 278
turns of, 340
zeros, finding, 327–330
See also quadratic functions
population, human, 69, 84, 91–92, 128, 145, 219, 242, 346, 400
population, nonhuman, 76, 90, 92–93, 237, 300, 562
population samples, 616
power equations, solving, 231, 239–240
power functions
general form of, 231
rational function as, 231–232
solving, 231
transformations of, 232
power of a power property of exponents, 224
power property of equality, 224
power property of logarithms, 258, 260
prime numbers, 622
probability
area model, 570
arrangements without replacement, 602
binomial expansion and, 614–618
combinations and, 607–610
conditional probability, 580–581
counting principle, 601–602
events. *See* events
expected value, 593–596
experimental probability, 568–569, 570
geometric probability, 570, 575–576
origins of, 566
outcomes, 568–569
Pascal's triangle and, 613–615, 621–622
of a path, 579–581, 586
permutations and, 500–503
randomness and, 566–568
simulations, 568
theoretical probability, 569–570
tree diagrams and, 577–581, 586
Venn diagrams and, 586–590
See also statistics
Problems for the Quickening of the Mind (Alcuin of York), 6
problem solving
algebra solution steps, 12
coordinate graphs and, 4–5
diagrams and, 2–3
group effort and, 4
organizing information, 16–20
strategies for, 4, 45
symbolic representation, 9–13
product property of logarithms, 258, 260
projectile motion
function of, 294
parametric equations and, 438–441
in the plane, 459–461
properties
addition property of equality, 134
common base property of equality, 224
commutative, 203
distributive property of exponentiation over division, 224
distributive property of exponentiation over multiplication, 224
division property of exponents, 224
of the ellipse, 484, 531–532
of exponents, 223–225, 231
logarithm change-of-base property, 253
of logarithms, 253, 257–260
multiplication property of equality, 134
multiplication property of exponents, 224
multiplicative identity, 368
multiplicative inverse, 368
power of a power property of exponents, 224
power property of equality, 224
reciprocal property of exponents, 224
reflection property of an ellipse, 484
substitution, 133
zero exponents, 224
zero-product, 287–288
Pythagorean Theorem, 192, 468
See also trigonometry

Q

quadratic equations, general form, quadratic formula and, 302–304
quadratic equations, solving
completing the square, 294–298
data table, 334–335
graphing, 287
quadratic formula, 302–304
quadratic formula, 301–304
quadratic functions, 171
factored form. *See* factored form of quadratic functions
form of, choosing, 289–290
general form of. *See* general form of quadratic functions
stretches and shrinks of, 186–187
translations of, 170–174, 208–210, 286
vertex form of. *See* vertex form of quadratic functions
zeros of. *See* zeros of a function
quartiles, 26
Quételet, Adolphe, 92
quotient property of logarithms, 258, 260

R

radio, 189, 255
radioactivity, 73, 216, 216–217, 228, 242, 262, 502
radius, 426, 475
randomness, 566–568
random numbers, 568
random samples, 616
random variables, 594
range (of data), 26
rational exponents. *See* fractional (rational) exponents
rational expressions, operations with, 524–528
rational functions, 508–512
asymptotes of, 510, 517–519
defined, 509
factored form of, 519
graphing of, 509–511, 516–519
holes of, 516–517
as power function, 231–232
transformations of, 510–512
rational root theorem, 329
Rayleigh, Lord (John William Strutt), 30

real axis, 310
recessive traits, 357
reciprocal property of exponents, 224
recreation, 117, 153, 175, 422, 443
recursion
defined, 54
formula for. *See* recursive formula
growth and decay modeled with, 64–66, 79–80
loans and investments modeled with, 86–87
notation for, 56, 80
partial sum of a series, 541–542, 554
recursive definition, 54–56
rule for, 56
sequences. *See* sequences
recursive definition, 54–56
recursive formula, 56–57, 104
Recursive Functions in Computer Theory (Péter), 61
recursive rule, 56
reduced row-echelon form, 360, 364
reflection property of an ellipse, 484
reflections
defined, 178
as rigid transformation, 185
of square root family, 177–179
summary of, 194
refraction, Snell's law of, 451
regression, exponential, 264–265
relation
defined, 155
inverse of, 245
replacement, arrangements without, 602
resource consumption, exploration, and conservation, 36–37, 114, 138, 249–250, 390, 396, 398, 403, 457, 492, 537
right triangle. *See* Pythagorean theorem; trigonometry
rigid transformations, 185
Robert of Chester, 419
Rome, ancient, 17, 171, 566
roots of an equation
defined, 288
factored form of polynomials and, 316–317
quadratic formula to find, 301–304
See also zeros of a function
rounding
accuracy reduced by, 456
of trigonometric ratios, 422
row reduction method, 360–364, 374
rule of 72, 222

S

samples, 616 *See also* data; probability; statistics
satellites, 196, 486, 503
scalar multiplication, 351, 355
scalars, 351
Schooten, Frans van, 502
Schrödinger, Erwin, 574
seismology, 506
self-similarity, 552
sequences
arithmetic. *See* arithmetic sequences
defined, 56
geometric. *See* geometric sequences
graphs of, 77–80, 105, 549
shifted, 73–74
summation of terms in. *See* series
valid conditions for relationship of, 107
See also terms
series
arithmetic, 504–544
convergent, 547–550, 559–561
defined, 540
finite number of terms, 541
geometric. *See* geometric series
infinite, 547, 550
partial sum of, 540–544
See also sequences
shape of data sets, 26–27
shifted sequences, 73–74
shipping, mailing, and packaging, 315–316, 336, 390, 584
Sierpiński triangle, 59, 552
Sierpiński , Waclaw, 59
sigma notation, 25, 541
similar figures, 185
simple event, 578
simulations, 568
sine (sin), 419 *See also* trigonometry
sines, law of, 447–449, 456
Sive de Organica Conicarum Sectionum in Plano Decriptione, Tractatus (Schooten), 502
skewed data set, 27
slant asymptotes, 518
slide rule, 257
slope, 111–113
arithmetic sequences and, 105
of asymptotes of a hyperbola, 498
choice of points to determine, 112–113
formula for, 111
point-slope form, 121–122
Snell's law of refraction, 451
solution matrix, 360
sound, 267, 268, 409–410, 483–484, 494
Soviet Union, 9, 196
Spain, 419
speed applications. *See* velocity and speed calculations
sports, 11–12, 42, 48, 71, 122, 128–129, 131, 168–169, 205, 236, 303, 342, 344, 345, 399, 407, 413–414, 416, 431–432, 437, 440–441, 443, 458, 464, 537, 551, 558, 563, 583, 598
spread
measures of, 31–37
shape of data and, 26–27
square root functions
graphing of, 177–178
reflections and, 177–179
stretches and shrinks of, 186
standard deviation, 34–35
standard form of linear equation, 168
statistics
central tendency, measures of, 24–27, 36
defined, 24
deviation, 33–35
outliers, 25, 36–37, 39
predictions with. *See* probability
samples, 616
spread, measures of, 26–27, 31–37
See also data
step functions, 158
stretches and shrinks, 188
absolute-value function, 184–188
circles and, 192–194
dilation, 351
as nonrigid transformation, 185
summary of, 194
sub-problems, problem solving with, 45
substitution
composition of functions and, 201

substitution *(continued)*
property of, 133
in systems of equations, 128–129, 133–134
subtraction
of complex numbers, 309
of rational expressions, 525–526
sum
of data values, 25
partial sum of series, 540–544
supply and demand, 366
Sweden, 249
Switzerland, 269
symbolic algebra, 204
symbols. *See* notation and symbols
symmetric data set, 26–27
symmetry, line of, 170–171
synthetic division, 330
systems of equations, 127
consistent, 367
dependent, 359, 367
inconsistent, 359, 367
number of solutions of, 359, 364, 367
in three dimensions, 367
systems of equations, solving
conic sections, 503–506
Cramer's Rule, 140–141
elimination and, 134–136
graphing, 127–128
with inverse matrices, 368–372, 392–394
with matrices, 360–364, 374
nonlinear, 503–506
no solutions, 129–130
number of equations required for, 372
substitution and, 128–129, 133–134
systems of inequalities
constraints in, 379, 381
feasible region in, 379, 380–381, 385–388
graphing solutions for, 378–381
linear programming and, 385–388
operations with, 378

T

tangent (tan), 419
inverse of, 421–422
See also trigonometry
technology, 257, 332
telecommunications, 69, 196, 348, 399, 458, 471, 486, 501, 503
terms
general, 56
notation for, 56, 64, 80
of polynomials, 278
of recursive sequence, 56, 64
starting, choice of, 64
See also series
tesserae, 566
testing and assessment, 63, 176, 583, 624
theorems
binomial, 615
factor, 328
Pythagorean, 192, 468
rational root, 329
theoretical probability, 569–570
third quartile, 26
time
as independent variable, 114
as parametric variable, 462
zero term in problems with, 64
Tower of Pisa, 143
trajectory, 444
transcendental numbers, 269
transformations
defined, 170
of fractional exponents, 232–233
matrices and, 350–351
nonrigid, 185
of parabolas, 488
of rational functions, 510–512
reflections. *See* reflections
rigid, 185
stretches and shrinks. *See* stretches and shrinks
summary of, 194
translations. *See* translations
transition diagrams, 342–344
transition matrices, 342–344
translations, 163
defined, 164
image, 165
of linear functions, 163–165, 180
of parametric equations, 405–406
of quadratic functions, 170–174, 208–210, 286
as rigid transformation, 185
summary of, 194
transportation, 22, 24–26, 199, 213, 256, 347, 383, 402–405, 409, 417, 424–425, 432–434, 435, 436, 484, 494
Treatise on Conic Sections (Apollonius of Perga), 475
tree diagrams, 577–581, 586
triangle(s)
area of, 452
law of sines, 447–449
oblique, 445
obtuse, 446
right. *See* Pythagorean theorem; trigonometry
triangular numbers, 283
triangulation, 458
trigonometric functions
identities, 459–461
inverses of, 421–422
trigonometric ratios, 418–422
trigonometry
defined, 418
law of cosines, 453–456
law of sines, 447–449, 456
parametric equations and, 418–421, 426–430
projectile motion and, 438–441
ratios of, 418–422
rounding of values, 422, 456
setting a course with, 431–434
trinomials, defined, 278
turns, 340

U

Umi (king of Hawaii), 24
unit circle
defined, 192
equation of, 192
transformation of, 192
unit conversion
Fahrenheit/Celsius temperatures, 7
graphical models for, 7, 8
unit hyperbola, 495
units of measure
astronomical, 235, 451
nautical mile, 473
of standard deviation, 35
Universal Product Codes (UPCs), 358

V

variability. *See* spread
variables
dependent. *See* dependent variable
discrete random, 594
independent. *See* independent variable

number of, and number of equations in system, 372
random, 594
variance, 34
vectors, 434
velocity, defined, 404
velocity and speed calculations, 16–17, 109–110, 111, 112–113, 139, 153, 182, 294–295, 299, 402–405, 407, 408–409, 473, 546
Venn diagrams, 311, 586–590
Venn, John, 311
Verhulst, Pierre François, 92
vertex
of inequalities, 379
of a parabola, 171, 173
vertex-edge graph, 348
vertex form of quadratic functions, 286–287
completing the square to find, 294–298
equation, 286–287, 289
factored form and, 288–290, 295
quadratic formula and, 301–302
vertical asymptotes, 510, 517–518
vertical line test, 155
vertices of a hyperbola, 494
volcanoes, 120–121
volume
of a cube, 314
cubic polynomials and, 320

W

water resources, 83, 249–250, 264–265
weight, 144

X

x-intercepts, finding, bisection method, 332

Y

y-intercepts, of linear equation, 105

Z

zero, exponents of, 224
zero-product property, 287–288
zero slope, 112
zeros of a function
complex conjugates, 317, 322–323
defined, 287
factored form of polynomial and, 315–317
factor theorem to confirm, 328
of higher-degree polynomials, 327–330
quadratic formula to find, 301–304
rational root theorem to find, 329
zero-product property to find, 287–288
See also roots of an equation
zero term, 64